Lecture Notes in Artificial Intelligence   11432

Subseries of Lecture Notes in Computer Science

More information about this series at http://www.springer.com/series/1244

Ngoc Thanh Nguyen · Ford Lumban Gaol ·
Tzung-Pei Hong · Bogdan Trawiński (Eds.)

Intelligent Information and Database Systems

11th Asian Conference, ACIIDS 2019
Yogyakarta, Indonesia, April 8–11, 2019
Proceedings, Part II

 Springer

Editors
Ngoc Thanh Nguyen (iD)
Ton Duc Thang University
Ho Chi Minh City, Vietnam

Wrocław University of Science
and Technology
Wrocław, Poland

Tzung-Pei Hong
National University of Kaohsiung
Kaohsiung, Taiwan

Ford Lumban Gaol (iD)
Bina Nusantara University
Jakarta, Indonesia

Bogdan Trawiński (iD)
Wrocław University of Science
and Technology
Wrocław, Poland

ISSN 0302-9743   ISSN 1611-3349 (electronic)
Lecture Notes in Artificial Intelligence
ISBN 978-3-030-14801-0   ISBN 978-3-030-14802-7 (eBook)
https://doi.org/10.1007/978-3-030-14802-7

Library of Congress Control Number: 2019932916

LNCS Sublibrary: SL7 – Artificial Intelligence

This Springer imprint is published by the registered company Springer Nature Switzerland AG
The registered company address is: Gewerbestrasse 11, 6330 Cham, Switzerland

Preface

ACIIDS 2019 was the 11th event in a series of international scientific conferences on research and applications in the field of intelligent information and database systems. The aim of ACIIDS 2019 was to provide an international forum of research workers with scientific background on the technology of intelligent information and database systems and its various applications. The ACIIDS 2019 conference was co-organized by BINUS University (Indonesia) and Wrocław University of Science and Technology (Poland) in co-operation with the IEEE SMC Technical Committee on Computational Collective Intelligence, European Research Center for Information Systems (ERCIS), University of Newcastle (Australia), Yeungnam University (South Korea), Leiden University (The Netherlands), Universiti Teknologi Malaysia (Malaysia), Quang Binh University (Vietnam), Ton Duc Thang University (Vietnam), and Vietnam National University, Hanoi (Vietnam). It took place in Yogyakarta in Indonesia during April 8–11, 2019.

The ACIIDS conference series is already well established. The first two events, ACIIDS 2009 and ACIIDS 2010, took place in Dong Hoi City and Hue City in Vietnam, respectively. The third event, ACIIDS 2011, took place in Daegu (South Korea), followed by the fourth event, ACIIDS 2012, in Kaohsiung (Taiwan). The fifth event, ACIIDS 2013, was held in Kuala Lumpur in Malaysia while the sixth event, ACIIDS 2014, was held in Bangkok in Thailand. The seventh event, ACIIDS 2015, took place in Bali (Indonesia), followed by the eighth event, ACIIDS 2016, in Da Nang (Vietnam). The ninth event, ACIIDS 2017, was organized in Kanazawa (Japan). The 10th jubilee conference, ACIIDS 2018, was held in Dong Hoi City (Vietnam).

For this edition of the conference, we received more than 300 papers from 38 countries all over the world. Each paper was peer-reviewed by at least two members of the international Program Committee and the international reviewer board. Only 124 papers with the highest quality were selected for an oral presentation and publication in these two volumes of the ACIIDS 2019 proceedings.

Papers included in the proceedings cover the following topics: knowledge engineering and Semantic Web; text processing and information retrieval; machine learning and data mining; decision support and control systems; computer vision techniques; databases and intelligent information systems, collective intelligence for service innovation, technology management, e-learning and fuzzy intelligent systems, data structures modelling for knowledge representation, advanced data mining techniques and applications; intelligent information systems; intelligent methods and artificial intelligence for biomedical decision support systems, intelligent and contextual systems, intelligent systems and algorithms in information sciences, intelligent supply chains and e-commerce, sensor networks and internet of things, analysis of image, video, movements and brain intelligence in life sciences, computer vision and intelligent systems.

The accepted and presented papers focus on new trends and challenges facing the intelligent information and database systems community. The presenters showed in what way research work could stimulate novel and innovative applications. We hope you will find these results useful and inspiring for your future research work.

We would like to extend our heartfelt thanks to Jarosław Gowin, Deputy Prime Minister of the Republic of Poland and Minister of Science and Higher Education, for his support and honorary patronage of the conference.

We would like to express our sincere thanks to the CEO of Bina Nusantara Group, Ir Bernard Gunawan, honorary chairs, Prof. Harjanto Prabowo (Rector of BINUS University, Indonesia), and Prof. Cezary Madryas (Rector of Wrocław University of Science and Technology, Poland), for their support.

Our special thanks go to the program chairs, special session chairs, organizing chairs, publicity chairs, liaison chairs, and local Organizing Committee for their work for the conference. We sincerely thank all the members of the international Program Committee for their valuable efforts in the review process, which helped us to guarantee the highest quality of the selected papers for the conference. We cordially thank the organizers and chairs of special sessions, who contributed to the success of the conference.

We would like to express our thanks to the keynote speakers: Prof. Stephan Chalup from the University of Newcastle, Australia, Prof. Hamido Fujita from Iwate Prefectural University, Japan, Prof. Tokuro Matsuo from Advanced Institute of Industrial Technology, Japan, Prof. Michał Woźniak from Wrocław University of Science and Technology, Poland, and Dr. Alfred Budiman, CEO of Samsung R&D Indonesia (SRIN), for their world-class plenary speeches.

We cordially thank our main sponsors, BINUS University (Indonesia), Wrocław University of Science and Technology (Poland), IEEE SMC Technical Committee on Computational Collective Intelligence, European Research Center for Information Systems (ERCIS), University of Newcastle (Australia), Yeungnam University (South Korea), Leiden University (The Netherlands), Universiti Teknologi Malaysia (Malaysia), Quang Binh University (Vietnam), Ton Duc Thang University (Vietnam), and Vietnam National University, Hanoi (Vietnam). Our special thanks are due also to Springer for publishing the proceedings and sponsoring awards, and to all the other sponsors for their kind support.

We wish to thank the members of the Organizing Committee for their excellent work and the members of the local Organizing Committee for their considerable effort. We cordially thank all the authors, for their valuable contributions, and the other participants of this conference. The conference would not have been possible without their support. Thanks are also due to many experts who contributed to making the event a success.

April 2019

Ngoc Thanh Nguyen
Ford Lumban Gaol
Tzung-Pei Hong
Bogdan Trawiński

Organization

Honorary Chairs

Harjanto Prabowo	Rector of BINUS University, Indonesia
Cezary Madryas	Rector of Wrocław University of Science and Technology, Poland

General Chairs

Ngoc Thanh Nguyen	Wrocław University of Science and Technology, Poland
Ford Lumban Gaol	BINUS University, Indonesia

Program Chairs

Spits Warnars Harco Leslie Hendric	BINUS University, Indonesia
Tzung-Pei Hong	National University of Kaohsiung, Taiwan
Edward Szczerbicki	University of Newcastle, Australia
Bogdan Trawiński	Wrocław University of Science and Technology, Poland

Steering Committee

Ngoc Thanh Nguyen (Chair)	Wrocław University of Science and Technology, Poland
Longbing Cao	University of Science and Technology Sydney, Australia
Suphamit Chittayasothorn	King Mongkut's Institute of Technology Ladkrabang, Thailand
Ford Lumban Gaol	Bina Nusantara University, Indonesia
Tu Bao Ho	Japan Advanced Institute of Science and Technology, Japan
Tzung-Pei Hong	National University of Kaohsiung, Taiwan
Dosam Hwang	Yeungnam University, South Korea
Bela Stantic	Griffith University, Australia
Geun-Sik Jo	Inha University, South Korea
Hoai An Le-Thi	University of Lorraine, France
Zygmunt Mazur	Wrocław University of Science and Technology, Poland
Toyoaki Nishida	Kyoto University, Japan

Leszek Rutkowski   Częstochowa University of Technology, Poland
Ali Selamat   Universiti Teknologi Malaysia, Malyasia

Special Session Chairs

Maciej Huk   Wrocław University of Science and Technology,
            Poland
Yaya Heriyadi   BINUS University, Indonesia

Liaison Chairs

Suphamit Chittayasothorn King Mongkut's Institute of Technology
            Ladkrabang, Thailand
Quang-Thuy Ha   VNU University of Engineering and Technology,
            Vietnam
Mong-Fong Horng   National Kaohsiung University of Applied
            Sciences, Taiwan
Dosam Hwang   Yeungnam University, South Korea
Le Minh Nguyen   Japan Advanced Institute of Science
            and Technology, Japan
Ali Selamat   Universiti Teknologi Malaysia, Malyasia

Organizing Chairs

Harisno   BINUS University, Indonesia
Adrianna Kozierkiewicz   Wrocław University of Science and Technology,
            Poland

Publicity Chairs

Agung Trisetyarso   BINUS University, Indonesia
Marek Kopel   Wrocław University of Science and Technology,
            Poland
Marek Krótkiewicz   Wrocław University of Science and Technology,
            Poland

Publication Chairs

Marcin Maleszka   Wrocław University of Science and Technology,
            Poland
Andrzej Siemiński   Wrocław University of Science and Technology,
            Poland

Webmaster

Marek Kopel	Wrocław University of Science and Technology, Poland

Keynote Speakers

Alfred Budiman	CEO of Samsung R&D Indonesia (SRIN), Indonesia
Stephan Chalup	University of Newcastle, Australia
Hamido Fujita	Iwate Prefectural University, Japan
Tokuro Matsuo	Advanced Institute of Industrial Technology, Japan
Michał Woźniak	Wrocław University of Science and Technology, Poland

Special Session Organizers

1. *Special Session on Intelligent Systems and Algorithms in Information Sciences (ISAIS 2019)*

Martin Kotyrba	University of Ostrava, Czech Republic
Eva Volná	University of Ostrava, Czech Republic
Ivan Zelinka	VŠB – Technical University of Ostrava, Czech Republic
Pavel Petr	University of Pardubice, Czech Republic

2. *Special Session on Intelligent and Contextual Systems (ICxS 2019)*

Maciej Huk	Wrocław University of Science and Technology, Poland
Keun Ho Ryu	Chungbuk National University, South Korea
Goutam Chakraborty	Iwate Prefectural University, Japan

3. *Intelligent Methods and Artificial Intelligence for Biomedical Decision Support Systems (IMAIBDSS 2019)*

Jan Kubicek	VŠB – Technical University of Ostrava, Czech Republic
Marek Penhaker	VŠB – Technical University of Ostrava, Czech Republic
Ondrej Krejcar	University of Hradec Kralove, Czech Republic
Kamil Kuca	University of Hradec Kralove, Czech Republic
Ali Selamat	Universiti Teknologi Malaysia, Malaysia

4. *Special Session on Computer Vision and Intelligent Systems (CVIS 2019)*

Van-Dung Hoang	Quang Binh University, Vietnam
Wahyono	Universitas Gadjah Mada, Indonesia
Kang-Hyun Jo	University of Ulsan, South Korea
Hyun-Deok Kang	Ulsan National Institute of Science and Technology, South Korea
Thi-Lan Le	Hanoi University of Science and Technology, Vietnam

5. *Special Session on Advanced Data Mining Techniques and Applications (ADMTA 2019)*

Chun-Hao Chen	Tamkang University, Taiwan
Bay Vo	Ho Chi Minh City University of Technology, Vietnam
Tzung-Pei Hong	National University of Kaohsiung, Taiwan

6. *Special Session on Intelligent Supply Chains and e-Commerce (ISCEC 2019)*

Arkadiusz Kawa	Poznań University of Economics and Business, Poland
Bartłomiej Pierański	Poznań University of Economics and Business, Poland

7. *Special Session on Intelligent Information Systems (IIS 2019)*

Urszula Boryczka	University of Silesia, Poland
Andrzej Najgebauer	Military University of Technology, Poland
Dariusz Pierzchała	Military University of Technology, Poland

8. *Special Session on Collective Intelligence for Service Innovation, Technology Management, E-learning and Fuzzy Intelligent Systems (CISTEF 2019)*

Pen-Choug Sun	Aletheia University, Taiwan
Gong-Yih Hsieh	Aletheia University, Taiwan

9. *Analysis of Image, Video, Movements and Brain Intelligence in Life Sciences (IVMBI 2019)*

Andrzej Przybyszewski	Polish-Japanese Academy of Information Technology, Poland
Jerzy Nowacki	Polish-Japanese Academy of Information Technology, Poland
Konrad Wojciechowski	Polish-Japanese Academy of Information Technology, Poland

Marek Kulbacki   Polish-Japanese Academy of Information
                        Technology, Poland
Jakub Segen   Polish-Japanese Academy of Information
                        Technology, Poland

10. *Data Structures Modelling for Knowledge Representation (DSMKR 2019)*

Marek Krótkiewicz   Wrocław University of Technology, Poland
Piotr Zabawa   Cracow University of Technology, Poland

International Program Committee

Ajith Abraham   Machine Intelligence Research Labs, USA
Muhammad Abulaish   South Asian University, India
Andrew Adamatzky   University of the West of England, UK
Waseem Ahmad   Waiariki Institute of Technology, New Zealand
Bashar Al-Shboul   University of Jordan, Jordan
Lionel Amodeo   University of Technology of Troyes, France
Toni Anwar   Universiti Teknologi Malaysia, Malaysia
Taha Arbaoui   University of Technology of Troyes, France
Ahmad Taher Azar   Benha University, Egypt
Thomas Bäck   Leiden University, The Netherlands
Amelia Badica   University of Craiova, Romania
Costin Badica   University of Craiova, Romania
Kambiz Badie   ICT Research Institute, Iran
Zbigniew Banaszak   Warsaw University of Technology, Poland
Dariusz Barbucha   Gdynia Maritime University, Poland
Ramazan Bayindir   Gazi University, Turkey
Juri Belikov   Tallinn University of Technology, Estonia
Maumita Bhattacharya   Charles Sturt University, Australia
Leon Bobrowski   Białystok University of Technology, Poland
Bülent Bolat   Yildiz Technical University, Turkey
Veera Boonjing   King Mongkut's Institute of Technology
                        Ladkrabang, Thailand

Mariusz Boryczka   University of Silesia, Poland
Urszula Boryczka   University of Silesia, Poland
Zouhaier Brahmia   University of Sfax, Tunisia
Stephane Bressan   National University of Singapore, Singapore
Peter Brida   University of Zilina, Slovakia
Andrej Brodnik   University of Ljubljana, Slovenia
Grażyna Brzykcy   Poznań University of Technology, Poland
Robert Burduk   Wrocław University of Science and Technology,
                        Poland
Aleksander Byrski   AGH University of Science and Technology,
                        Poland
David Camacho   Universidad Autonoma de Madrid, Spain

Tru Cao	Ho Chi Minh City University of Technology, Vietnam
Frantisek Capkovic	Institute of Informatics, Slovak Academy of Sciences, Slovakia
Oscar Castillo	Tijuana Institute of Technology, Mexico
Dariusz Ceglarek	Poznań High School of Banking, Poland
Zenon Chaczko	University of Technology, Sydney, Australia
Goutam Chakraborty	Iwate Prefectural University, Japan
Somchai Chatvichienchai	University of Nagasaki, Japan
Chun-Hao Chen	Tamkang University, Taiwan
Rung-Ching Chen	Chaoyang University of Technology, Taiwan
Shyi-Ming Chen	National Taiwan University of Science and Technology, Taiwan
Suphamit Chittayasothorn	King Mongkut's Institute of Technology Ladkrabang, Thailand
Sung-Bae Cho	Yonsei University, South Korea
Kazimierz Choroś	Wrocław University of Science and Technology, Poland
Kun-Ta Chuang	National Cheng Kung University, Taiwan
Dorian Cojocaru	University of Craiova, Romania
Jose Alfredo Ferreira Costa	Federal University of Rio Grande do Norte (UFRN), Brazil
Ireneusz Czarnowski	Gdynia Maritime University, Poland
Theophile Dagba	University of Abomey-Calavi, Benin
Quang A Dang	Vietnam Academy of Science and Technology, Vietnam
Tien V. Do	Budapest University of Technology and Economics, Hungary
Grzegorz Dobrowolski	AGH University of Science and Technology, Poland
Habiba Drias	University of Science and Technology Houari Boumediene, Algeria
Maciej Drwal	Wrocław University of Science and Technology, Poland
Ewa Dudek-Dyduch	AGH University of Science and Technology, Poland
El-Sayed M. El-Alfy	King Fahd University of Petroleum and Minerals, Saudi Arabia
Keiichi Endo	Ehime University, Japan
Sebastian Ernst	AGH University of Science and Technology, Poland
Nadia Essoussi	University of Carthage, Tunisia
Rim Faiz	University of Carthage, Tunisia
Simon Fong	University of Macau, SAR China
Dariusz Frejlichowski	West Pomeranian University of Technology, Szczecin, Poland

Takuya Fujihashi	Ehime University, Japan
Hamido Fujita	Iwate Prefectural University, Japan
Mohamed Gaber	Birmingham City University, UK
Ford Lumban Gaol	BINUS University, Indonesia
Dariusz Gąsior	Wrocław University of Science and Technology, Poland
Janusz Getta	University of Wollongong, Australia
Daniela Gifu	University Alexandru Ioan Cuza of Iasi, Romania
Dejan Gjorgjevikj	Ss. Cyril and Methodius University in Skopje, Republic of Macedonia
Barbara Gładysz	Wrocław University of Science and Technology, Poland
Daniela Godoy	ISISTAN Research Institute, Argentina
Ong Sing Goh	Universiti Teknikal Malaysia Melaka, Malaysia
Antonio Gonzalez-Pardo	Universidad Autonoma de Madrid, Spain
Yuichi Goto	Saitama University, Japan
Manuel Graña	University of Basque Country, Spain
Janis Grundspenkis	Riga Technical University, Latvia
Jeonghwan Gwak	Seoul National University, South Korea
Quang-Thuy Ha	VNU University of Engineering and Technology, Vietnam
Dawit Haile	Addis Ababa University, Ethiopia
Pei-Yi Hao	National Kaohsiung University of Applied Sciences, Taiwan
Marcin Hernes	Wrocław University of Economics, Poland
Francisco Herrera	University of Granada, Spain
Koichi Hirata	Kyushu Institute of Technology, Japan
Bogumiła Hnatkowska	Wrocław University of Science and Technology, Poland
Huu Hanh Hoang	Hue University, Vietnam
Quang Hoang	Hue University of Sciences, Vietnam
Van-Dung Hoang	Quang Binh University, Vietnam
Jaakko Hollmen	Aalto University, Finland
Tzung-Pei Hong	National University of Kaohsiung, Taiwan
Mong-Fong Horng	National Kaohsiung University of Applied Sciences, Taiwan
Jen-Wei Huang	National Cheng Kung University, Taiwan
Yung-Fa Huang	Chaoyang University of Technology, Taiwan
Maciej Huk	Wrocław University of Science and Technology, Poland
Zbigniew Huzar	Wrocław University of Science and Technology, Poland
Dosam Hwang	Yeungnam University, South Korea
Roliana Ibrahim	Universiti Teknologi Malaysia, Malaysia

Dmitry Ignatov	National Research University Higher School of Economics, Russia
Lazaros Iliadis	Democritus University of Thrace, Greece
Hazra Imran	University of British Columbia, Canada
Agnieszka Indyka-Piasecka	Wrocław University of Science and Technology, Poland
Mirjana Ivanovic	University of Novi Sad, Serbia
Sanjay Jain	National University of Singapore, Singapore
Jarosław Jankowski	West Pomeranian University of Technology, Szczecin, Poland
Joanna Jędrzejowicz	University of Gdańsk, Poland
Piotr Jędrzejowicz	Gdynia Maritime University, Poland
Janusz Jeżewski	Institute of Medical Technology and Equipment ITAM, Poland
Geun Sik Jo	Inha University, South Korea
Kang-Hyun Jo	University of Ulsan, South Korea
Jason Jung	Chung-Ang University, South Korea
Przemysław Juszczuk	University of Economics in Katowice, Poland
Janusz Kacprzyk	Systems Research Institute, Polish Academy of Sciences, Poland
Tomasz Kajdanowicz	Wrocław University of Science and Technology, Poland
Nadjet Kamel	University Ferhat Abbes Setif1, Algeria
Hyun-Deok Kang	Ulsan National Institute of Science and Technology, South Korea
Mehmet Karaata	Kuwait University, Kuwait
Nikola Kasabov	Auckland University of Technology, New Zealand
Arkadiusz Kawa	Poznań University of Economics and Business, Poland
Rafal Kern	Wrocław University of Science and Technology, Poland
Zaheer Khan	University of the West of England, UK
Manish Khare	Dhirubhai Ambani Institute of Information and Communication Technology, India
Chonggun Kim	Yeungnam University, South Korea
Marek Kisiel-Dorohinicki	AGH University of Science and Technology, Poland
Attila Kiss	Eotvos Lorand University, Hungary
Jerzy Klamka	Silesian University of Technology, Poland
Goran Klepac	Raiffeisen Bank, Croatia
Shinya Kobayashi	Ehime University, Japan
Marek Kopel	Wrocław University of Science and Technology, Poland
Raymondus Kosala	BINUS University, Indonesia
Leszek Koszałka	Wrocław University of Science and Technology, Poland

Leszek Kotulski	AGH University of Science and Technology, Poland
Martin Kotyrba	University of Ostrava, Czech Republic
Jan Kozak	University of Economics in Katowice, Poland
Adrianna Kozierkiewicz	Wrocław University of Science and Technology, Poland
Bartosz Krawczyk	Virginia Commonwealth University, USA
Ondrej Krejcar	University of Hradec Kralove, Czech Republic
Dalia Kriksciuniene	Vilnius University, Lithuania
Dariusz Król	Wrocław University of Science and Technology, Poland
Marek Krótkiewicz	Wrocław University of Science and Technology, Poland
Marzena Kryszkiewicz	Warsaw University of Technology, Poland
Adam Krzyzak	Concordia University, Canada
Jan Kubicek	VSB - Technical University of Ostrava, Czech Republic
Tetsuji Kuboyama	Gakushuin University, Japan
Elżbieta Kukla	Wrocław University of Science and Technology, Poland
Julita Kulbacka	Wrocław Medical University, Poland
Marek Kulbacki	Polish-Japanese Academy of Information Technology, Poland
Kazuhiro Kuwabara	Ritsumeikan University, Japan
Halina Kwaśnicka	Wrocław University of Science and Technology, Poland
Annabel Latham	Manchester Metropolitan University, UK
Bac Le	University of Science, VNU-HCM, Vietnam
Hoai An Le Thi	University of Lorraine, France
Yue-Shi Lee	Ming Chuan University, Taiwan
Florin Leon	Gheorghe Asachi Technical University of Iasi, Romania
Horst Lichter	RWTH Aachen University, Germany
Kuo-Sui Lin	Aletheia University, Taiwan
Igor Litvinchev	Nuevo Leon State University, Mexico
Rey-Long Liu	Tzu Chi University, Taiwan
Doina Logofatu	Frankfurt University of Applied Sciences, Germany
Edwin Lughofer	Johannes Kepler University Linz, Austria
Lech Madeyski	Wrocław University of Science and Technology, Poland
Bernadetta Maleszka	Wrocław University of Science and Technology, Poland
Marcin Maleszka	Wrocław University of Science and Technology, Poland
Petra Maresova	University of Hradec Kralove, Czech Republic

Urszula Markowska-Kaczmar	Wrocław University of Science and Technology, Poland
Mustafa Mat Deris	Universiti Tun Hussein Onn Malaysia, Malaysia
Takashi Matsuhisa	Karelia Research Centre, Russian Academy of Science, Russia
Tamás Matuszka	Eotvos Lorand University, Hungary
Joao Mendes-Moreira	University of Porto, Portugal
Mercedes Merayo	Universidad Complutense de Madrid, Spain
Jacek Mercik	WSB University in Wrocław, Poland
Radosław Michalski	Wrocław University of Science and Technology, Poland
Peter Mikulecky	University of Hradec Kralove, Czech Republic
Marek Miłosz	Lublin University of Technology, Poland
Kazuo Misue	University of Tsukuba, Japan
Jolanta Mizera-Pietraszko	Opole University, Poland
Leo Mrsic	IN2data Ltd. Data Science Company, Croatia
Agnieszka Mykowiecka	Institute of Computer Science, Polish Academy of Sciences, Poland
Paweł Myszkowski	Wrocław University of Science and Technology, Poland
Saeid Nahavandi	Deakin University, Australia
Kazumi Nakamatsu	University of Hyogo, Japan
Grzegorz J. Nalepa	AGH University of Science and Technology, Poland
Mahyuddin K. M. Nasution	Universitas Sumatera Utara, Indonesia
Richi Nayak	Queensland University of Technology, Australia
Fulufhelo Nelwamondo	Council for Scientific and Industrial Research, South Africa
Huu-Tuan Nguyen	Vietnam Maritime University, Vietnam
Le Minh Nguyen	Japan Advanced Institute of Science and Technology, Japan
Loan T. T. Nguyen	Ton Duc Thang University, Vietnam
Quang-Vu Nguyen	Korea-Vietnam Friendship Information Technology College, Vietnam
Thai-Nghe Nguyen	Cantho University, Vietnam
Van Du Nguyen	Wrocław University of Science and Technology, Poland
Yusuke Nojima	Osaka Prefecture University, Japan
Jerzy Nowacki	Polish-Japanese Academy of Information Technology, Poland
Agnieszka Nowak-Brzezińska	University of Silesia, Poland
Mariusz Nowostawski	Norwegian University of Science and Technology, Norway
Alberto Núñez	Universidad Complutense de Madrid, Spain
Manuel Núñez	Universidad Complutense de Madrid, Spain
Richard Jayadi Oentaryo	Singapore Management University, Singapore

Kouzou Ohara	Aoyama Gakuin University, Japan
Tarkko Oksala	Aalto University, Finland
Shingo Otsuka	Kanagawa Institute of Technology, Japan
Marcin Paprzycki	Systems Research Institute, Polish Academy of Sciences, Poland
Rafael Parpinelli	Santa Catarina State University (UDESC), Brazil
Danilo Pelusi	University of Teramo, Italy
Marek Penhaker	VSB - Technical University of Ostrava, Czech Republic
Hoang Pham	Rutgers University, USA
Maciej Piasecki	Wrocław University of Science and Technology, Poland
Bartłomiej Pierański	Poznań University of Economics and Business, Poland
Dariusz Pierzchała	Military University of Technology, Poland
Marcin Pietranik	Wrocław University of Science and Technology, Poland
Elias Pimenidis	University of the West of England, UK
Jaroslav Pokorný	Charles University in Prague, Czech Republic
Nikolaos Polatidis	University of Brighton, UK
Elvira Popescu	University of Craiova, Romania
Piotr Porwik	University of Silesia, Poland
Radu-Emil Precup	Politehnica University of Timisoara, Romania
Małgorzata Przybyła-Kasperek	University of Silesia, Poland
Andrzej Przybyszewski	Polish-Japanese Academy of Information Technology, Poland
Paulo Quaresma	Universidade de Evora, Portugal
Ngoc Quoc Ly	Ho Chi Minh City University of Science, Vietnam
David Ramsey	Wrocław University of Science and Technology, Poland
Mohammad Rashedur Rahman	North South University, Bangladesh
Ewa Ratajczak-Ropel	Gdynia Maritime University, Poland
Leszek Rutkowski	Częstochowa University of Technology, Poland
Tomasz M. Rutkowski	University of Tokyo, Japan
Henryk Rybiński	Warsaw University of Technology, Poland
Alexander Ryjov	Lomonosov Moscow State University, Russia
Keun Ho Ryu	Chungbuk National University, South Korea
Virgilijus Sakalauskas	Vilnius University, Lithuania
Daniel Sanchez	University of Granada, Spain
Rafał Scherer	Częstochowa University of Technology, Poland
Juergen Schmidhuber	Swiss AI Lab IDSIA, Switzerland
Jakub Segen	Polish-Japanese Academy of Information Technology, Poland
Ali Selamat	Universiti Teknologi Malaysia, Malaysia
S. M. N. Arosha Senanayake	Universiti Brunei Darussalam, Brunei Darussalam

Tegjyot Singh Sethi	University of Louisville, USA
Andrzej Siemiński	Wrocław University of Science and Technology, Poland
Dragan Simic	University of Novi Sad, Serbia
Paweł Sitek	Kielce University of Technology, Poland
Adam Słowik	Koszalin University of Technology, Poland
Vladimir Sobeslav	University of Hradec Kralove, Czech Republic
Kulwadee Somboonviwat	Kasetsart University, Thailand
Kamran Soomro	University of the West of England, UK
Zenon A. Sosnowski	Białystok University of Technology, Poland
Bela Stantic	Griffith University, Australia
Stanimir Stoyanov	University of Plovdiv Paisii Hilendarski, Bulgaria
Ja-Hwung Su	Cheng Shiu University, Taiwan
Libuse Svobodova	University of Hradec Kralove, Czech Republic
Tadeusz Szuba	AGH University of Science and Technology, Poland
Julian Szymański	Gdańsk University of Technology, Poland
Krzysztof Ślot	Łódź University of Technology, Poland
Jerzy Świątek	Wrocław University of Science and Technology, Poland
Andrzej Świerniak	Silesian University of Technology, Poland
Ryszard Tadeusiewicz	AGH University of Science and Technology, Poland
Yasufumi Takama	Tokyo Metropolitan University, Japan
Maryam Tayefeh Mahmoudi	ICT Research Institute, Iran
Zbigniew Telec	Wrocław University of Science and Technology, Poland
Aleksei Tepljakov	Tallinn University of Technology, Estonia
Dilhan Thilakarathne	Vrije Universiteit Amsterdam, The Netherlands
Satoshi Tojo	Japan Advanced Institute of Science and Technology, Japan
Krzysztof Tokarz	Silesian University of Technology, Poland
Bogdan Trawiński	Wrocław University of Science and Technology, Poland
Krzysztof Trojanowski	Cardinal Stefan Wyszyński University in Warsaw, Poland
Ualsher Tukeyev	al-Farabi Kazakh National University, Kazakhstan
Aysegul Ucar	Firat University, Turkey
Olgierd Unold	Wrocław University of Science and Technology, Poland
Joost Vennekens	Katholieke Universiteit Leuven, Belgium
Jorgen Villadsen	Technical University of Denmark, Denmark
Bay Vo	Ho Chi Minh City University of Technology, Vietnam
Eva Volná	University of Ostrava, Czech Republic

Wahyono	Universitas Gadjah Mada, Indonesia
Lipo Wang	Nanyang Technological University, Singapore
Junzo Watada	Waseda University, Japan
Konrad Wojciechowski	Silesian University of Technology, Poland
Krystian Wojtkiewicz	Wrocław University of Science and Technology, Poland
Krzysztof Wróbel	University of Silesia, Poland
Marian Wysocki	Rzeszow University of Technology, Poland
Farouk Yalaoui	University of Technology of Troyes, France
Xin-She Yang	Middlesex University, UK
Tulay Yildirim	Yildiz Technical University, Turkey
Piotr Zabawa	Cracow University of Technology, Poland
Sławomir Zadrożny	Systems Research Institute, Polish Academy of Sciences, Poland
Drago Zagar	University of Osijek, Croatia
Danuta Zakrzewska	Łódź University of Technology, Poland
Constantin-Bala Zamfirescu	Lucian Blaga University of Sibiu, Romania
Katerina Zdravkova	Ss. Cyril and Methodius University in Skopje, Republic of Macedonia
Vesna Zeljkovic	Lincoln University, USA
Aleksander Zgrzywa	Wrocław University of Science and Technology, Poland
Qiang Zhang	Dalian University, China
Beata Zielosko	University of Silesia, Poland
Maciej Zięba	Wrocław University of Science and Technology, Poland
Adam Ziębiński	Silesian University of Technology, Poland
Marta Zorrilla	University of Cantabria, Spain

Program Committees of Special Sessions

Special Session on Intelligent Systems and Algorithms in Information Sciences (ISAIS 2019)

Martin Kotyrba	University of Ostrava, Czech Republic
Eva Volná	University of Ostrava, Czech Republic
Ivan Zelinka	VŠB – Technical University of Ostrava, Czech Republic
Hashim Habiballa	Institute for Research and Applications of Fuzzy Modeling, Czech Republic
Alexej Kolcun	Institute of Geonics, Czech Republic
Roman Senkerik	Tomas Bata University in Zlin, Czech Republic
Zuzana Kominkova Oplatkova	Tomas Bata University in Zlin, Czech Republic
Katerina Kostolanyova	University of Ostrava, Czech Republic
Antonin Jancarik	Charles University in Prague, Czech Republic
Petr Dolezel	University of Pardubice, Czech Republic

Igor Kostal	University of Economics in Bratislava, Slovakia
Eva Kurekova	Slovak University of Technology in Bratislava, Slovakia
Leszek Cedro	Kielce University of Technology, Poland
Dagmar Janacova	Tomas Bata University in Zlin, Czech Republic
Martin Halaj	Slovak University of Technology in Bratislava, Slovakia
Radomil Matousek	Brno University of Technology, Czech Republic
Roman Jasek	Tomas Bata University in Zlin, Czech Republic
Petr Dostal	Brno University of Technology, Czech Republic
Jiri Pospichal	University of Ss. Cyril and Methodius (UCM), Slovakia
Vladimir Bradac	University of Ostrava, Czech Republic
Roman Jasek	Tomas Bata University in Zlin, Czech Republic
Petr Pavel	University of Pardubice, Czech Republic
Jan Capek	University of Pardubice, Czech Republic

Special Session on Intelligent and Contextual Systems (ICxS 2019)

Adriana Albu	Politehnica University Timişoara, Romania
Basabi Chakraborty	Iwate Prefectural University, Japan
Dariusz Frejlichowski	West Pomeranian University of Technology, Poland
Erdenebileg Batbaatar	Chungbuk National University, South Korea
Goutam Chakraborty	Iwate Prefectural University, Japan
Ha Manh Tran	Ho Chi Minh City International University, Vietnam
Hong Vu Nguyen	Ton Duc Thang University, Vietnam
Hideyuki Takahashi	RIEC, Tohoku University, Japan
Jerzy Świątek	Wrocław University of Science and Technology, Poland
Józef Korbicz	University of Zielona Gora, Poland
Keun Ho Ryu	Chungbuk National University, South Korea
Kilho Shin	University of Hyogo, Japan
Maciej Huk	Wrocław University of Science and Technology, Poland
Marcin Fojcik	Western Norway University of Applied Sciences, Norway
Masafumi Matsuhara	Iwate Prefectural University, Japan
Michael Spratling	University of London, UK
Musa Ibrahim	Chungbuk National University, South Korea
Nguyen Khang Pham	Can Tho University, Vietnam
Plamen Angelov	Lancaster University, UK
Qiangfu Zhao	University of Aizu, Japan
Quan Thanh Tho	Ho Chi Minh City University of Technology, Vietnam

Rashmi Dutta Baruah	Lancaster University, UK
Tadahiko Murata	Kansai University, Japan
Takako Hashimoto	Chiba University of Commerce, Japan
Tetsuji Kubojama	Gakushuin University, Japan
Tetsuo Kinoshita	RIEC, Tohoku University, Japan
Thai-Nghe Nguyen	Can Tho University, Vietnam
Tsatsral Amarbayasgalan	Chungbuk National University, South Korea
Zhenni Li	University of Aizu, Japan

Special Session on Intelligent Methods and Artificial Intelligence for Biomedical Decision Support Systems (IMAIBDSS 2019)

Ani Liza Asmawi	International Islamic University, Malaysia
Martin Augustynek	VŠB – Technical University of Ostrava, Czech Republic
Martin Cerny	VŠB – Technical University of Ostrava, Czech Republic
Klara Fiedorova	VŠB – Technical University of Ostrava, Czech Republic
Habibollah Harun	Universiti Teknologi Malaysia, Malaysia
Lim Kok Cheng	Universiti Tenaga Nasional, Malaysia
Roliana Ibrahim	Universiti Teknologi Malaysia, Malaysia
Jafreezal Jaafar	Universiti Teknologi Petronas, Malaysia
Vladimir Kasik	VŠB – Technical University of Ostrava, Czech Republic
Ondrej Krejcar	University of Hradec Kralove, Czech Republic
Jan Kubicek	VŠB – Technical University of Ostrava, Czech Republic
Kamil Kuca	University of Hradec Kralove, Czech Republic
Petra Maresova	University of Hradec Kralove, Czech Republic
David Oczka	VŠB – Technical University of Ostrava, Czech Republic
Sigeru Omatu	Osaka Institute of Technology, Japan
Marek Penhaker	VŠB – Technical University of Ostrava, Czech Republic
Lukas Peter	VŠB – Technical University of Ostrava, Czech Republic
Chawalsak Phetchanchai	Suan Dusit University, Thailand
Antonino Proto	VŠB – Technical University of Ostrava, Czech Republic
Naomie Salim	Universiti Teknologi Malaysia, Malaysia
Ali Selamat	Universiti Teknologi Malaysia, Malaysia
Imam Much Subroto	Universiti Islam Sultan Agung, Indonesia
Lau Sian Lun	Sunway University, Malaysia

Takeru Yokoi   Tokyo Metropolitan International Institute
                                of Technology, Japan
Hazli Mohamed Zabil   Universiti Tenaga Nasional, Malaysia

Special Session on Computer Vision and Intelligent Systems (CVIS 2019)

Kang-Hyun Jo   University of Ulsan, South Korea
Van-Dung Hoang   Quang Binh University, Vietnam
The-Anh Pham   Hong Duc University, Vietnam
Thi-Lan Le   Hanoi University of Science and Technology,
                                Vietnam
Wahyono   Universitas Gadjah Mada, Indonesia
Alireza Ghasempour   University of Applied Science and Technology,
                                Iran
Afiahayati   University Gadjah Mada, Indonesia
Byeongryong Lee   University of Ulsan, South Korea
Cheolgeun Ha   University of Ulsan, South Korea
Chi-Mai Luong   University of Science and Technology of Hanoi,
                                Vietnam
Christos Bouras   University of Patras, Greece
Heejun Kang   University of Ulsan, South Korea
Hyun-Deok Kang   Ulsan National Institute of Science
                                and Technology, South Korea
Moh Edi Wibowo   Universitas Gadjah Mada, Indonesia
Mu-Song Chen   Da-Yeh University, Taiwan
My-Ha Le   HCMC University of Technology and Education,
                                Vietnam
Ngoc-Son Pham   HCMC University of Technology and Education,
                                Vietnam
Nobutaka Shimada   Ritsumeikan University, Japan
Pavel Loskot   Swansea University, UK
Thanh-Hai Tran   Hanoi University of Science and Technology,
                                Vietnam
Thanh-Truc Tran   Danang ICT, Vietnam
Trung-Duy Tran   Posts and Telecommunications Institute
                                of Technology, Vietnam
Van-Huy Pham   Tong Duc Thang University, Vietnam
Van Mien   University of Exeter, UK
Vu-Viet Vu   Hanoi National University, Vietnam
Yoshinori Kuno   Saitama University, Japan
Youngsoo Suh   University of Ulsan, South Korea
Yuansong Qiao   Athlone Institute of Technology, Ireland

Special Session on Advanced Data Mining Techniques and Applications (ADMTA 2019)

Tzung-Pei Hong	National University of Kaohsiung, Taiwan
Tran Minh Quang	Ho Chi Minh City University of Technology, Vietnam
Bac Le	University of Science, VNU-HCM, Vietnam
Bay Vo	Ho Chi Minh City University of Technology, Vietnam
Chun-Hao Chen	Tamkang University, Taiwan
Chun-Wei Lin	Harbin Institute of Technology Shenzhen Graduate School, China
Wen-Yang Lin	National University of Kaohsiung, Taiwan
Yeong-Chyi Lee	Cheng Shiu University, Taiwan
Le Hoang Son	Vietnam National University, Vietnam
Thi Ngoc Chau	Ho Chi Minh City University of Technology, Vietnam
Van Vo	Ho Chi Minh University of Industry, Vietnam
Ja-Hwung Su	Cheng Shiu University, Taiwan
Ming-Tai Wu	University of Nevada, USA
Kawuu W. Lin	National Kaohsiung University of Applied Sciences, Taiwan
Tho Le	Ho Chi Minh City University of Technology, Vietnam
Dang Nguyen	Deakin University, Australia
Hau Le	Thuyloi University, Vietnam
Thien-Hoang Van	Ho Chi Minh City University of Technology, Vietnam
Tho Quan	Ho Chi Minh City University of Technology, Vietnam
Ham Nguyen	University of People's Security Hochiminh City, Vietnam
Thiet Pham	Ho Chi Minh University of Industry, Vietnam
Nguyen Thi Thuy Loan	International University, VNU-HCMC, Vietnam

Special Session on Intelligent Supply Chains and e-Commerce (ISCEC 2019)

Bartłomiej Pierański	Poznań University of Economics and Business, Poland
Justyna Światowiec-Szczepańska	Poznań University of Economics and Business, Poland
Carlos Andres Romano	Polytechnic University of Valencia, Spain
Davor Dujak	University of Osijek, Croatia
Paulina Golińska-Dawson	Poznań University of Technology, Poland
Paweł Pawlewski	Poznań University of Technology, Poland
Jakub Bercik	Slovak University of Agriculture in Nitra, Slovakia
Adam Koliński	Poznań School of Logistics, Poland

Special Session on Intelligent Information Systems (IIS 2019)

Ryszard Antkiewicz	Military University of Technology, Poland
Leon Bobrowski	Białystok University of Technology, Poland
Urszula Boryczka	University of Silesia, Poland
Mariusz Chmielewski	Military University of Technology, Poland
Jan Hodický	University of Defense in Brno, Czech Republic
Rafal Kasprzyk	Military University of Technology, Poland
Jacek Koronacki	Polish Academy of Sciences, Poland
Rafał Ładysz	George Mason University, USA
Andrzej Najgebauer	Military University of Technology, Poland
Ewa Niewiadomska-Szynkiewicz	Warsaw University of Technology, Poland
Dariusz Pierzchała	Military University of Technology, Poland
Jarosław Rulka	Military University of Technology, Poland
Khalid Saeed	Białystok University of Technology, Poland
Zenon A. Sosnowski	Białystok University of Technology, Poland
Zbigniew Tarapata	Military University of Technology, Poland

Special Session on Collective Intelligence for Service Innovation, Technology Management, E-learning and Fuzzy Intelligent Systems (CISTEF 2019)

Albim Y. Cabatingan	University of the Visayas, Philippines
Teh-Yuan Chang	Aletheia University, Taiwan
Chi-Min Chen	Aletheia University, Taiwan
Chih-Chung Chiu	Aletheia University, Taiwan
Wen-Min Chou	Aletheia University, Taiwan
Chao-Fu Hong	Aletheia University, Taiwan
Chia-Lin Hsieh	Aletheia University, Taiwan
Gong-Yih Hsieh	Aletheia University, Taiwan
Chia-Ling Hsu	Tamkang University, Taiwan
Fang-Cheng Hsu	Aletheia University, Taiwan
Chi-Cheng Huang	Aletheia University, Taiwan
Rahat Iqbal	Coventry University, UK
Huan-Ting Lin	The University of Tokyo, Japan
Kuo-Sui Lin	Aletheia University, Taiwan
Min-Huei Lin	Aletheia University, Taiwan
Yuh-Chang Lin	Aletheia University, Taiwan
Shin-Li Lu	Aletheia University, Taiwan
Janet Argot Pontevedra	University of San Carlos, Philippines
Shu-Chin Su	Aletheia University, Taiwan
Pen-Choug Sun	Aletheia University, Taiwan
Ai-Ling Wang	Tamkang University, Taiwan
Leuo-Hong Wang	Aletheia University, Taiwan
Hung-Ming Wu	Aletheia University, Taiwan
Feng-Sueng Yang	Aletheia University, Taiwan

Hsiao-Fang Yang	National Chengchi University, Taiwan
Sadayuki Yoshitomi	Toshiba Corporation, Japan

Special Session on Analysis of Image, Video, Movements and Brain Intelligence in Life Sciences (IVMBI 2019)

Andrei Barborica	Research & Compliance and Engineering, FHC, Inc., USA
Artur Bąk	Polish-Japanese Academy of Information Technology, Poland
Konrad Ciecierski	Warsaw University of Technology, Poland
Leszek Chmielewski	Warsaw University of Life Sciences, Poland
Zenon Chaczko	University Technology of Sydney, Australia
Christopher Chiu	University of Technology Sydney, Australia
Aldona Barbara Drabik	Polish-Japanese Academy of Information Technology, Poland
Marcin Fojcik	Sogn og Fjordane University College, Norway
Adam Gudyś	Silesian University of Technology, Poland
Ryszard Gubrynowicz	Polish-Japanese Academy of Information Technology, Poland
Piotr Habela	Polish-Japanese Academy of Information Technology, Poland
Celina Imielińska	Vesalius Technolodgies LLC, USA
Henryk Josiński	Silesian University of Technology, Poland
Mark Kon	Boston University, USA
Wojciech Knieć	Polish-Japanese Academy of Information Technology, Poland
Ryszard Klempous	Wrocław University of Science and Technology, Poland
Ryszard Kozera	The University of Life Sciences - SGGW, Poland
Julita Kulbacka	Wrocław Medical University, Poland
Marek Kulbacki	Polish-Japanese Academy of Information Technology, Poland
Krzysztof Marasek	Polish-Japanese Academy of Information Technology, Poland
Majaz Moonis	UMass Medical School, USA
Radoslaw Nielek	Polish-Japanese Academy of Information Technology, Poland
Peter Novak	Brigham and Women's Hospital, USA
Wieslaw Nowinski	Cardinal Stefan Wyszynski University, Poland
Jerzy Paweł Nowacki	Polish-Japanese Academy of Information Technology, Poland
Eric Petajan	LiveClips LLC, USA
Andrzej Polański	Silesian University of Technology, Poland
Andrzej Przybyszewski	Polish-Japanese Academy of Information Technology, Poland

Zbigniew Ras	University of North Carolina at Charlotte, USA & PJAIT, Poland
Joanna Rossowska	Polish Academy of Sciences, Institute of Immunology and Experimental Therapy, Poland
Jakub Segen	Gest3D LLC, USA
Aleksander Sieroń	Medical University of Silesia, Poland
Dominik Ślęzak	University of Warsaw, Poland
Michał Staniszewski	Polish-Japanese Academy of Information Technology, Poland
Zbigniew Struzik	RIKEN Brain Science Institute, Japan
Adam Świtoński	Silesian University of Technology, Poland
Agnieszka Szczęsna	Silesian University of Technology, Poland
Kamil Wereszczyński	Polish-Japanese Academy of Information Technology, Poland
Alicja Wieczorkowska	Polish-Japanese Academy of Information Technology, Poland
Konrad Wojciechowski	Polish-Japanese Academy of Information Technology, Poland
Sławomir Wojciechowski	Polish-Japanese Academy of Information Technology, Poland

Special Session on Data Structures Modelling for Knowledge Representation (DSMKR 2019)

Marek Krótkiewicz	Wrocław University of Science and Technology, Poland
Piotr Zabawa	Cracow University of Technology, Poland
Jerzy Tomasik	CNRS - LIMOS University Clermont-Auvergne, France
Helena Dudycz	Wrocław University of Economics, Poland
Robert Sochacki	University of Opole, Poland
Wojciech Hunek	Opole University of Science and Technology, Poland
Sophia Katrenko	University of Amsterdam, The Netherlands
Krystian Wojtkiewicz	Wrocław University of Science and Technology, Poland
Marcin Jodłowiec	Wrocław University of Science and Technology, Poland

Contents – Part II

Intelligent Information Systems

**Intelligent Methods and Artificial Intelligence for Biomedical
Decision Support Systems**

Intelligent and Contextual Systems

Intelligent Systems and Algorithms in Information Sciences

Computer Vision and Intelligent Systems

Contents – Part I

Machine Learning and Data Mining

Decision Support and Control Systems

Databases and Intelligent Information Systems

Collective Intelligence for Service Innovation, Technology Management, E-learning, and Fuzzy Intelligent

Increasing the Quality of Multi-step Consensus

Dai Tho Dang[1] ⓘ, Ngoc Thanh Nguyen[2] ⓘ,
and Dosam Hwang[1](✉) ⓘ

[1] Department of Computer Engineering, Yeungnam University,
Gyeongsan, Republic of Korea
daithodang@ynu.ac.kr, dshwang@yu.ac.kr
[2] Department of Information Systems,
Faculty of Computer Science and Management,
Wrocław University of Science and Technology, Wrocław, Poland
ngoc-thanh.nguyen@pwr.edu.pl

Abstract. Determining the consensus of a collective is becoming a popular problem-solving method in our society. However, given that determining the consensus of large collectives is time-consuming, a multi-step consensus approach is necessary. Thus, one important problem is to determine the number of steps required to obtain a reliable consensus in an acceptable time. Execution time depends on the number of steps; determining the number of steps relies on the quality of the consensus in each step. The overall consensus quality depends on the problem of determining consensus in each step. Therefore, it is important to improve the consensus quality and investigate the quality according to the number of smaller collectives in each step. Herein, we improve the basic algorithm used for the multi-step consensus approach. The experiment result shows that the approach based on the improved algorithm is more efficient than that of the basic algorithm in terms of consensus quality (4.9%). Furthermore, the consensus quality was investigated according to the number of smaller collectives in each step.

Keywords: Consensus · Multi-step consensus · Collective intelligence

1 Introduction

The problem of gathering knowledge from different sources is gaining popularity [1]. For example, to solve a problem, people tend to seek expert opinions or information on the internet and often discover a proper solution. However, it is very common that knowledge from different sources on the same subject is inconsistent [2].

In this work, a set of opinions, comments, and solutions to one problem from a group is considered a set of knowledge states in a collective. Determining consensus depends on the knowledge states of collective members and is very important in many fields, such as computer science, medicine, and economics [3, 4]. However, inconsistency in the knowledge states of the collective make determining the consensus a complex task [2]. Today, collective sizes are continuously increasing due to a rapidly growing number of knowledge sources. Determining the consensus of large collectives is becoming general in many fields, such as bioinformatics and medicine [3], where

N. T. Nguyen et al. (Eds.): ACIIDS 2019, LNAI 11432, pp. 3–14, 2019.
https://doi.org/10.1007/978-3-030-14802-7_1

consensus problems, such as protein structure prediction, gene prediction, and disease marker prediction [5] are enormous. Furthermore, determining the consensus of a large collective is very time-consuming. To address this, one feasible method - the multi-step consensus approach - was developed [6].

The multi-step consensus approach is based on dividing the primary collective into smaller ones and determining their consensuses. Consequently, a new collective is created from the consensus of these smaller collectives. This new collective is considered the primary collective, which is the subject of a new step. It is essential to determine the number of steps in order to obtain a reliable consensus in an acceptable amount of time. However, this problem has not been investigated in the literature.

In order to simplify the notation, we assume that the knowledge state of each member is represented as a binary vector. In this case, the complexity of the algorithm to determine the consensus satisfying the postulate 1-Optimality is $O(n)$. The problem of determining the consensus satisfying the postulate 2-Optimality is an NP-hard [7]. The postulate 2-Optimality will be used to determine the knowledge of a collective.

In the multi-step consensus approach, consensus quality depends on the problem of determining consensus in each step. Therefore, in this work, we improve the basic algorithm used for the approach in order to increase the consensus quality in each step. For this, we first determined the consensus satisfying the postulate 1-Optimality of the primary collective [7]. This is used to determine the consensus satisfying the postulate 2-Optimality of all smaller collectives in each step, instead of randomly choosing a binary vector.

In each step, the consensus quality depends on the number of smaller collectives [8]. A study of the consensus quality according to the number of smaller collectives in each step is crucial, because it will allow us to determine the most reasonable number of smaller collectives necessary to obtain a high-quality consensus. Particularly, we are able to estimate the number of steps to obtain a reliable consensus for an input primary collective in an acceptable amount of time. However, this problem has not been studied. In the present work, the primary collective is divided into a number of smaller collectives. The limit to the number of smaller collectives is from 1 to the primary collective size. For each of the smaller collective, there is a consensus. We will investigate the quality of these obtained consensuses according to the number of smaller collectives.

According to the experimental results, the quality of the consensus determined by the multi-step consensus approach based on the improved algorithm is 4.9% higher than that based on the basic algorithm. Investigation on consensus quality according to the number of smaller collectives in each step is currently limited to experiments. To better understand its application, the results will need to be further analyzed by future work.

The study is constructed as follows: A short overview of previous researches is provided in Sect. 2; basic concepts are described in Sect. 3; the proposed method is presented in Sect. 4; experimental results and their evaluation are explained in Sect. 5; conclusions and future work are discussed in Sect. 6.

2   Related Work

Consensus problems have been solved using three approaches [9]. First, in the axiomatic approach, seven conditions have been defined for consensus choice functions [10]. The consensus choice function known as the Kemeny median is the most widely used in determining the consensus of a collective. In [9], Nguyen presented ten conditions of which, 1-Optimality (Kemeny median) and 2-Optimality are the most important criteria. Second, the constructive approach resolves consensus problems including element structure and the relationship between elements [9, 11]. Finally, in the optimization approach, optimality rules, for instance, Condorcet's optimality, global optimality, etc. are the conditions for defining the consensus choice functions [9].

Postulates 1-Optimality and 2-Optimality play an important role in determining the consensus of a collective. Determining a consensus that satisfies these postulates is very important in many fields [4]. In most cases, determining such consensuses are NP-hard problems [4, 9, 12]. To deal with this issue, some heuristic algorithms have been proposed [4, 9].

Determining the consensus of a large collective requires more execution time. Consequently, equilibration between execution time and consensus quality must be regarded. To balance execution time and quality, the multi-step consensus approach is used for determining the consensus of a large collective [6]. However, this approach has not been widely examined [13]. In [7], the authors examined problems of one- and two-step approaches to determining consensus. Experimental analysis demonstrated that the two-step approach results and one-step approach results were 5% and 1% lower, respectively than that of the optimal solution.

In [8, 15], the authors investigated a two-step consensus approach. In [8], the k-means algorithm is applied for clustering a large collective into smaller collectives. The consensuses of the smaller collectives are then determined. Each consensus is assigned a weight value depending on the number of members in the corresponding smaller collective. The second step in the approach creates a consensus of the larger collective from the consensuses of these smaller collectives. The experimental results showed that, compared to a non-weighted approach, the weighted approach is helpful in reducing the difference between the two-stage and one-stage consensus choice in determining the knowledge of a large collective. In [14], the authors improved the two-step consensus approach by taking into account the problem of susceptibility to consensus. Through experiment analysis, the proposed approach proved useful.

A three-step consensus approach was proposed in [15]. In the first step, the sequence partitioning method is applied to divide a large collective into smaller, equal-sized groups. Subsequently, the k-means algorithm is applied to divide each group into smaller clusters. The consensus of each group is determined based on the consensus of these clusters. Finally, the consensus of the large collective is determined based on the consensus of the groups. Simulation results revealed the effectiveness of the three-step method in terms of running time, as well as final knowledge quality of the large collective. In [13], the authors cover a topic concerning the initial classification of data into subgroups in the context of the multi-level approach. The experimental results show that to achieve a higher quality of consensus, the grouping approach should be based on the lowest value of Fleiss' kappa measure. The experiments considered a two-level approach; consensus problems of more than two steps were not investigated.

In [16], the authors examined one-, two-, three-, four-, and five-step approaches. Execution time analysis demonstrated that adding additional steps accelerates the algorithm. The multi-step approach was also used for ontology integration [17]. The experimental results pointed out that the multi-level integration process is shorter by even 20% in comparison to the standard one-level approach.

3   Basic Notions [2, 6, 9]

Let U denote a finite set of objects representing all potential knowledge states for the same subject. The elements of U can represent, for example, tuples, logic expressions etc. Symbol 2^U denotes the powerset of U that is the set of all subsets of U. Let $\Pi_k(U)$ be a set of all k-element subsets of U for $k \in \aleph$ ($\aleph$ is the set of natural numbers), and let

$$\Pi(U) = \cup_{k \in \aleph} \Pi_k(U).$$

Thus, $\Pi(U)$ is the set of all non-empty finite subsets with repetitions of set U. A set $X \in \Pi(U)$ is called a collective.

Elements of U have two structures: macrostructure and microstructure. The microstructure is understood as the structure of elements of the set U, such as linear orders, n-tree, time interval, etc. The macrostructure is understood as the relationship between elements in a collective. In general, the macrostructure is often defined by distance functions for measuring the difference between elements in a collective.

Definition 1. The macrostructure of set U is a distance function

$$d : U \times U \to [0, 1]$$

which is:

- *Nonnegative*: $\forall x, y \in U : d(x, y) \geq 0$,
- *Reflexive*: $\forall x, y \in U : d(x, y) = 0 \Leftrightarrow x = y$,
- *Symmetrical*: $\forall x, y \in U : d(x, y) = d(y, x)$.

where [0, 1] is the closed interval of real numbers between 0 and 1. Pair (U, d) is called a distance space. The definition of a distance function is independent of the structure of elements of U.

Definition 2. By a consensus choice function in space (U, d), we mean function:

$$C : \Pi(U) \to 2^U$$

By $C(X)$, we denote the representation of collective $X \in \Pi(U)$. The element $c \in C(X)$ is the consensus of collective X, where $C(X)$ is a normal set (i.e., without repetitions).

There are many postulates for consensus determination, such as reliability, una-nimity, simplification, quasi-unanimity, consistency, Condorcet consistency, general consistency, proportion, 1-Optimality, and 2-Optimality. Postulates 1-Optimality and 2-Optimality are the most popular criteria in consensus determination because satis-fying one of these two postulates inherently satisfies many other postulates.

Definition 3. For a given collective, $X \in \Pi(U)$, the consensus choice function C satisfies the postulate of:

- 1-Optimality iff $\left(x \in C(X) \Rightarrow d(x, X) = min_{y \in U}d(y, X)\right)$
- 2-Optimality iff $\left(x \in C(X) \Rightarrow d(x, X) = min_{y \in U}d^2(y, X)\right)$

Postulate 1-Optimality requires the consensus to be as near as possible to members of the collective and can be recognized as the best representative of the collective. Postulate 2-Optimality states that the sum of the squared distances between a consensus and a collective member should be minimal.

4 Proposed Approach

4.1 Multi-step Consensus Approach

Determining the consensus of a large collective is very time-consuming. As such, the multi-step consensus approach can be used to divide the large collective into smaller collectives in a random way. The difference between the number of members of any two smaller collectives is not greater than 1. A consensus-based algorithm is utilized for determining the consensuses of these smaller collectives. Next, these consensuses are treated as elements of a new collective. These tasks are complete when a chosen number of steps is attained. Finally, the consensus $C(X^*)$ of the collective X^* is determined. $C(X^*)$ is also the consensus of collective X_1 - called the multi-step con-sensus of X_1. The procedure of the multi-step consensus approach is shown in Fig. 1.

While X_1 is a large collective containing n members, the k-step for determining the consensus of X_1 is defined as follows:

Step 1:
- Divide collective X_1 into n_1 smaller collectives that satisfy the following:

$$X_1 = X_{11} \cup X_{12} \cup \ldots \cup X_{1n_1}$$

- Determine the consensuses of smaller collectives $X_{11}, X_{12}, \ldots, X_{1n_1}$ that are $x_{11}^*, x_{12}^*, \ldots, x_{1n_1}^*$, respectively.
- Create a new collective, $X_2 = \{x_{11}^*, x_{12}^*, \ldots, x_{1n_1}^*\}$

Step 2:
- Divide collective X_2 into n_2 smaller collectives that satisfy the following:

$$X_2 = X_{21} \cup X_{21} \cup \ldots \cup X_{2n_2}$$

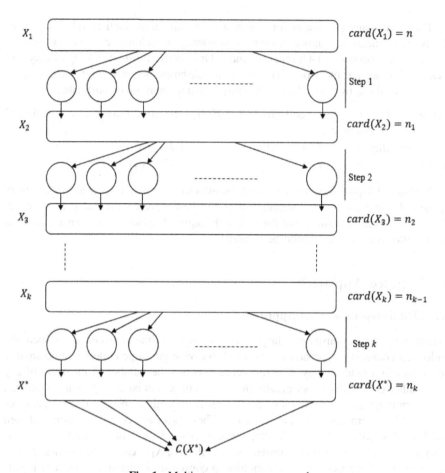

Fig. 1. Multi-step consensus approach

- Determine the consensuses of smaller collectives $X_{21}, X_{22}, \ldots, X_{2n_2}$ that are $x_{21}^*, x_{22}^*, \ldots, x_{2n_2}^*$, respectively.
- Create a new collective, $X_3 = \{x_{21}^*, x_{22}^*, \ldots, x_{2n_2}^*\}$

...

...

Step k:
- Divide collective X_k into n_k smaller collectives that satisfy the following:

$$X_k = X_{k1} \cup X_{k2} \cup \ldots \cup X_{kn_k}$$

- Determine the consensuses of $X_{k1}, X_{k2}, \ldots, X_{kn_k}$ that are $x_{k1}^*, x_{k2}^*, \ldots, x_{kn_k}^*$, respectively.
- Create a new collective, $X^* = \{x_{k1}^*, x_{k2}^*, \ldots, x_{kn_k}^*\}$

Finally, the consensus $C(X^*)$ of collective X^* is determined.

In this work, the collective X_1 contains n expert opinions. Each opinion is represented by a binary vector with length m.

Set U is defined as: $U = \{u_1, u_2, \ldots\}$, where its elements are binary vectors.

For $y, z \in U$, $d(y, z) = \sum_{i=1}^{m} |y^i - z^i|$.

where $y = (y^1, y^2, \ldots, y^m)$, $z = (z^1, z^2, \ldots, z^m)$; $y^i, z^i \in \{0, 1\}$, $i = \overline{1, m}$.

For each collective $X_1 \in \Pi(U)$, $X_1 = \{x_1, x_2, \ldots, x_n\}$, and $x_i = (x_{i1}, x_{i2}, \ldots x_{im})$, $i = \overline{1, m}$.

4.2 Improving the Algorithm

Determining the consensus that satisfies the postulate 2-Optimality is an NP-hard problem. The heuristic algorithm for this problem is presented below [7]:

Algorithm 1. basic algorithm

Input: Collective X

Output: Consensus c of collective X satisfies the postulate 2-Optimality
$$c = (c^1, c^2, \ldots, c^m), c^i \in \{0, 1\}, \ i = \overline{1, m}$$

BEGIN

1. Randomly choosing binary vector $c = (c^1, c^2, \ldots, c^m)$
2. Calculate $tmp = d^2(c, X)$
3. For $i = 1$ to m do
4. $c^i = c^i \oplus 1$
5.   If $d^2(c, X) < tmp$ then
6. $tmp = d^2(c, X)$
7.   Else
8. $c^i = c^i \oplus 1$

END

In Algorithm 1, the first step randomly chooses binary vector c. Because of the time complexity of algorithm [13] for determining the consensus satisfying the postulate 1-Optimality is linear, we determine the consensus satisfies this postulate. After that, this consensus is used to determine the consensus satisfying the postulate 2-Optimality. We improve Algorithm 1 as follows:

Algorithm 2. improved algorithm

    Input: Collective X

                Consensus x_1 of collective X satisfies the postulate 1-Optimality

    Output: Consensus c of collective X satisfies the postulate 2-Optimality

$$c = (c^1, c^2, ..., c^m), c^i \in \{0,1\}, \ i = \overline{1,m}$$

    BEGIN

        1. $c = x_1$

        2.   Calculate $tmp = d^2(c, X)$

        3.   For $i = 1$ to m do

        4. $c^i = c^i \oplus 1$

        5.   If $d^2(c, X) < tmp$ then

        6. $tmp = d^2(c, X)$

        7.   Else

        8. $c^i = c^i \oplus 1$

    END

4.3 Quality of Multi-step Consensus

In previous studies [7, 8, 16, 17], the quality of a collective's consensus is based on the distance from the consensus to the knowledge states of its members. In this work, the consensus quality is based on the distance from the consensus determined by the multi-step consensus approach to the optimal consensus determined by the brute-force algorithm.

$C(X_{opt}^*)$: set of the optimal consensus determined by Brute-force algorithm.

$C(X^*)$: set of the multi-step consensus determined by the multi-step consensus approach.

We define the quality of the multi-step consensus as follows:

$$Q = 1 - \frac{\min d(x, y)}{m}$$

where m is the length of vector, $x \in C(X_{opt}^*)$, $y \in C(X^*)$.

5 Experiments and Evaluation

For the multi-step consensus approach, the collective is divided into smaller collectives at each step. Investigating the consensus quality at each step is the same. In this work, we considered step 1 as a representative.

For a collective, the sufficient requirement for determining a reliable consensus is an odd number of members [18]. Thus, we considered cases wherein collective X_1 is divided into an odd number of smaller collectives. The limit to the odd number of smaller collectives is from 1 to the collective size. For example, if the collective X_1 contains 99 elements, it is divided into 3, 5, ..., 97 small collectives. Note that the difference in the number of members between any two smaller collectives is not greater than 1. For each number of smaller collectives, there is a consensus. Therefore, we determined the consensus according to the numbers of smaller collectives.

In this work, the collective X_1 includes its members' opinions represented as binary vectors of length 20. The test sizes of collective X_1 are 115, 120, and 125 members.

- If collective X_1 contains 115 members, it is divided into 3, 5, ..., 113 smaller collectives;
- If collective X_1 contains 120 members, it is divided into 3, 5, ..., 119 smaller collectives;
- If collective X_1 contains 125 members, it is divided into 3, 5, ..., 123 smaller collectives.

Another aim of this work is to examine the efficiency of the improved algorithm in term of consensus quality for the multi-step consensus approach. The above experiments were performed for the multi-step consensus approach using the basic algorithm and the same using the improved algorithm. Notice that in the multi-step consensus approach using the improved algorithm, the consensus of collective X_1 satisfying the postulate 1-Optimality is used to determine the consensuses of smaller collectives in all steps. We analyzed the test results to evaluate the efficiency of the improved algorithm.

Figures 2, 3, and 4 present the results of conducted experiments. The dotted line represents the quality of the consensus determined by the multi-step consensus based on the improved algorithm, the solid line represents the same based on the basic algorithm.

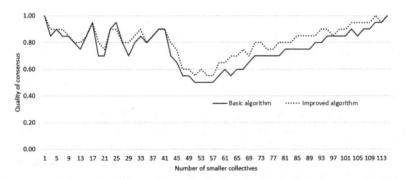

Fig. 2. Consensus quality according to the number of smaller collectives from a collective of 115 members.

We compared the quality of the consensus determined by the multi-step consensus approach using the basic algorithm and the same using the improved algorithm. Before selecting proper tests, sample distributions were analyzed using the Shapiro-Wilk test.

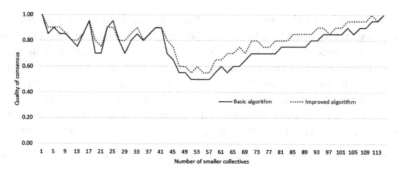

Fig. 3. Consensus quality according to the number of smaller collectives from a collective of 120 members.

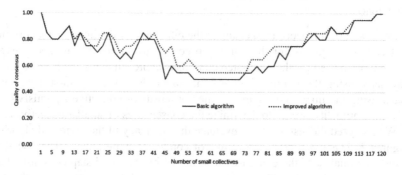

Fig. 4. Consensus quality according to the number of smaller collectives from a collective of 125 members.

The resulting *p-value* of the test was smaller than the significance value ($\alpha = 0.05$); indicating that the samples did not come from a normal distribution.

Using the Kruskal-Wallis test for samples, *p-values* were also smaller than the significance value. Accordingly, the consensus quality determined by the multi-step consensus approach based on the basic algorithm and that based on the improved algorithm was different.

The mean of consensus quality determined by the multi-step consensus approach based on the improved algorithm was found to be 0.798, while the mean of that based on the basic algorithm was 0.749. The mean of consensus quality determined by the multi-step consensus approach using the basic algorithm is inferior to the same using the improved algorithm by 0.049 (equivalent to 4.9%). It means that the improved algorithm is more efficient than the basic algorithm in terms of consensus quality for the multi-step consensus approach.

Consensus quality depends on the number of smaller collectives. For each collective, we divide the qualities of consensuses determined by the multi-step consensus approach using the improved algorithm into quarters. In the quartile range, the percentile scale is divide into four equal parts [19], or quartiles.

- The first quartile extends from zero to the 25th percentile (Q_1).
- The second quartile ranges from Q_1 the 50th percentile (Q_2).
- The third quartile extend from Q_2 to the 75th percentile (Q_3).
- The fourth, or highest, quartile reaches from Q_3 to the 100th percentile (Q_4).

The results show that the consensus quality has a dependence on the number of smaller collectives. Q_1 and Q_2 belong to the second one-third of the number of smaller collectives. Further investigations will be discussed in the future.

6   Conclusions and Future Work

In this paper, we improved the basic algorithm used for the multi-step consensus approach. The consensus quality determined by the multi-step consensus approach based on the improved algorithm is 4.9% higher than which is based on the basic algorithm. Furthermore, the issue related to the quality of the consensus according to the number of smaller collectives is experimentally examined.

Future work is focused on studying consensus quality according to the number of smaller collectives. Additionally, future study will examine other knowledge structures, such as relational structures and interval numbers.

Acknowledgment. This research was supported by Basic Science Research Program through the National Research Foundation of Korea (NRF) funded by the Ministry of Science, ICT & Future Planning (2017R1A2B4009410).

References

1. Dong, Y., et al.: Consensus reaching in social network group decision making: research paradigms and challenges. Knowl. Based Syst. **162**, 3–13 (2018)
2. Nguyen, N.T.: Processing inconsistency of knowledge in determining knowledge of a collective. Cybern. Syst. Int. J. **40**(8), 670–688 (2009)
3. Villaverde, A.F., et al.: A consensus approach for estimating the predictive accuracy of dynamic models in biology. Comput. Methods Programs Biomed. **119**(1), 17–28 (2015)
4. Ali, A., Meilă, M.: Experiments with Kemeny ranking: what works when? Math. Soc. Sci. **64**(1), 28–40 (2012)
5. Yang, B.: Bioinformatics analysis and consensus ranking for biological high throughput data. Ph.D. Dissertation, University of Paris 11 (2015)
6. Maleszka, M., Nguyen, N.T.: Integration computing and collective intelligence. Expert Syst. Appl. **42**(1), 332–340 (2015)
7. Kozierkiewicz-Hetmańska, A.: Comparison of one-level and two-level consensuses satisfying the 2-optimality criterion. In: Nguyen, N.-T., Hoang, K., Jędrzejowicz, P. (eds.) ICCCI 2012. LNCS (LNAI), vol. 7653, pp. 1–10. Springer, Heidelberg (2012). https://doi. org/10.1007/978-3-642-34630-9_1
8. Du Nguyen, V., Nguyen, N.T.: A two-stage consensus-based approach for determining collective knowledge. In: Le Thi, H.A., Nguyen, N.T., Van Do, T. (eds.) Advanced Computational Methods for Knowledge Engineering. AISC, vol. 358, pp. 301–310. Springer, Cham (2015). https://doi.org/10.1007/978-3-319-17996-4_27

9. Nguyen, N.T.: Advanced Methods for Inconsistent Knowledge Management. Springer, London (2008). https://doi.org/10.1007/978-1-84628-889-0
10. Arrow, K.J.: Social Choice and Individual Values. Wiley, New York (1963)
11. Del Moral, M.J., Tapia, J.M., Chiclana, F., Al-Hmouz, A., Herrera-Viedma, E.: An analysis of consensus approaches based on different concepts of coincidence. J. Intell. Fuzzy Syst. **34**(4), 2247–2259 (2018)
12. Amodio, S., D'Ambrosio, A., Siciliano, R.: Accurate algorithms for identifying the median ranking when dealing with weak and partial rankings under the Kemeny axiomatic approach. Eur. J. Oper. Res. **249**(2), 667–676 (2016)
13. Kozierkiewicz-Hetmanska, A., Pietranik, M.: Assessing the quality of a consensus determined using a multi-level approach. In: IEEE International Conference on INnovations in Intelligent SysTems and Applications (INISTA), pp. 131–136. IEEE (2017)
14. Du Nguyen, V., Nguyen, N.T., Hwang, D.: An improvement of the two-stage consensus-based approach for determining the knowledge of a collective. In: Nguyen, N.-T., Manolopoulos, Y., Iliadis, L., Trawiński, B. (eds.) ICCCI 2016. LNCS (LNAI), vol. 9875, pp. 108–118. Springer, Cham (2016). https://doi.org/10.1007/978-3-319-45243-2_10
15. Dang, D.T., Du Nguyen, V., Nguyen, N.T., Hwang, D.: A three-stage consensus-based method for collective knowledge determination. In: Sieminski, A., Kozierkiewicz, A., Nunez, M., Ha, Q.T. (eds.) Modern Approaches for Intelligent Information and Database Systems. SCI, vol. 769, pp. 3–14. Springer, Cham (2018). https://doi.org/10.1007/978-3-319-76081-0_1
16. Kozierkiewicz-Hetmańska, A., Sitarczyk, M.: The efficiency analysis of the multi-level consensus determination method. In: Nguyen, N.T., Papadopoulos, G.A., Jędrzejowicz, P., Trawiński, B., Vossen, G. (eds.) ICCCI 2017. LNCS (LNAI), vol. 10448, pp. 103–112. Springer, Cham (2017). https://doi.org/10.1007/978-3-319-67074-4_11
17. Kozierkiewicz-hetma, A., Marcin, P.: The knowledge increase estimation framework for ontology integration on the concept level. J. Intell. Fuzzy Syst. **32**, 1161–1172 (2017)
18. Kozierkiewicz-Hetmańska, A.: Analysis of susceptibility to the consensus for a few representations of collective knowledge. Int. J. Softw. Eng. Knowl. Eng. **24**(5), 759–775 (2014)
19. William, J.V., Joseph, P.W.: Statistics in Kinesiology, 4th edn. Human Kinetics (2012)

An Independence Measure for Expert Collections Based on Social Media Profiles

Rafał Palak[(✉)] and Ngoc Thanh Nguyen

Faculty of Computer Science and Management,
Wroclaw University of Science and Technology, Wrocław, Poland
{rafal.palak,ngoc-thanh.nguyen}@pwr.edu.pl

Abstract. According to current research, a crowd can outperform experts. Surowiecki in his work, has distinguished decentralization, independence, and diversity as key factors of good crowd performance. Due to lack of mathematical models for modelling these aspects, it is still impossible to prove that they have a big impact on the crowd performance. For solving this problem one of the very important crowd metrics called independence measure should be defined and this is the objective of our paper. Proposed measure allows calculating independence values based on data from social media profiles. The biggest advantage of the measure is the possibility of calculating an independence value for a group of people before it could become a collective for realizing concrete objectives. Currently, known solutions largely simplify the problem by describing independence with a single value. The solution presented in the article assumes that the value of the independence is calculated for a specific topic (the calculated value is part of a vector describing the independence between two experts).

Keywords: Wisdom of crowds · Independence · Collective intelligence · Independence measure

1 Introduction

Today's world is developing at incredible speed. It is becoming much harder to keep up with the developments than it was several years ago. Many companies are aware of this situation and in order to predict the dynamical aspect of markets they have to use new methods. Collective Intelligence delivers methods for this aim. Collective Intelligence is defined [11] as "the capacity of human collectives to engage in intellectual cooperation in order to create, innovate and invent" or "groups of individuals acting collectively in ways that seem intelligent" [18]. In our work, we will call a crowd "wise" when its prediction as a whole is close to the true value. The Collective Intelligence works if estimation errors of individuals are large but unbiased, such that they cancel each other out. Thus, the heterogeneity of numerous decision-makers generates a more accurate aggregate estimates, than estimates of single lay or expert decision-makers. As in [7] authors point out the wisdom of crowd effect as a statistical phenomenon and not a social psychological effect, because it is based on a mathematical aggregation of individual estimates.

© Springer Nature Switzerland AG 2019
N. T. Nguyen et al. (Eds.): ACIIDS 2019, LNAI 11432, pp. 15–25, 2019.
https://doi.org/10.1007/978-3-030-14802-7_2

As research shows, characteristics of a crowd have a huge influence on the results of Collective Intelligence methods [18]. Surowiecki in his work [4] defined rules for wise crowds: *diversity*, *independence*, *decentralization*, and *aggregation*. Collective Intelligence achieves the best efficiency when all four rules are satisfied. To achieve this, it is required to measure each factor. This paper tries to respond to this problem by proposing independence measure. Independence is an important factor for collective intelligence. As authors of work [7] stated: Social influence plays a role in an individual's decision-making process and estimations. Therefore, social influence can also have an impact on a statistical aggregate and result in a collective wisdom of a respective crowd. As social influence among human groups may trigger individuals to revise their estimates, it can have a substantial impact on the statistical wisdom of crowd effects in societies. We try to calculate independence based on social media profiles. Proposed independence measure allows calculating independence in a crowd before it becomes collective (a group of individuals in the literature usually is called a collective if they have a common task). In this article, we will call such crowd "group of people." Solutions known in literature do take into account the fact that people can be less influenced in some topics than in others. Known solutions in literature usually calculate independence based on collective prediction results - often in this moment information about independence is useless. Our research tries to respond to this problem by allowing to calculate independence for a crowd before it will become collective, without knowing the collective predictions results. Proposed measure allows calculating independence for different topics.

In this article, the new independence measure was presented. Measure calculate independence value for each person in collective. Values of independence are calculated based on data gathered from ResearchGate or LinkedIn. The measure has been subjected to in-depth analysis, it has been shown that the obtained results confirm expectations. In this article, some of the properties of the proposed measure have been proven.

The remaining part of this paper is organized as follows: the next section describes the literature review in this field. In Sect. 3, an overview of our method to create intelligence crowd is presented. Section 4 includes the experiment plan and results. The last section is the conclusion.

2 Related Works

Research confirms that a group of normal people could forecast events with more accuracy than one expert [3]. Current results show that crowds can also outperform professional analysts in financial predictions. Investors will earn more based on a recommendation of a crowd than the financial analyst [10]. Collective decision-making is seen as a perfect tool to solve complex problems [5].

To distinguish one crowd from another in literature some measures were proposed. In [16] the author proposes to measure Collective Intelligence similarly as individual IQ. The author proposes IQ tests based on probability functions which extend the present approach to measure individual IQ where only scalar or simple data are used. A similar approach was taken in [1]. Authors proposed measures for group IQ based on a simple average. This approach has one big drawback: such a complex structure as collective described by one scalar loses much of the information. It was noticed by authors in [6] that divided a value of Collective Intelligence into explicit and implicit values. Group cognition, group collaboration, and group cooperation are considered explicit values. Authors point out that there is still a lot of hidden ignored values. Therefore, a value of group wisdom is difficult to accurately estimate. More popular measures were created based on [4] where Surowiecki establishes the rules for a wise crowd:

- The diversity of opinion - each person should have some private information, even if it is just an eccentric interpretation of the known facts [4]
- Independence - people's opinions are not determined by the opinions of those around them [4]
- Decentralization of opinion - people can specialize and draw on local knowledge [4]
- Aggregation - some mechanism exists for turning private judgments into a collective decision [4]

Current studies confirm a huge impact of the above rules in-crowd performance [10, 12, 14, 17]. In our work, we focus only on independence. Independence, as was shown in the introduction, has a huge role in collective intelligence. Despite a great importance in literature, there is not a lot of work that tries to measure independence in collective. Existing work usually does not fulfil the requirements. The lack of mathematical model in Surowiecki's work, causes a situation in which many researchers try to prove their thesis without using any measure.

In [7] authors conduct two experiments first where participants had no information about other participants responses and second where participants had information about other participants responses. Comparison of these two groups shows that social influence is important for collective wisdom. In this work, authors have not used any measure to calculate social influence among collective members. The lack of mathematical model does not allow for an analysis of the results. In [13] authors define a formal model of investors and one of them knew the decisions of the others. The model uses probability to describe the outcomes of situations. Authors define economy which have only two firms A and B, run by managers A and B respectively. Manager A make a decision first in day 1, then on day 2 manager B decide on his or her own knowing the decision of manager A, then on day 3 manager A made his own decision and so on. Authors in their model take into consideration influence managers on each other, but this work is lack of equation that will allow measuring influence directly. Whole work authors define probability base on the fact that manager X is dumb or smart. Dividing people into dumb and smart people is difficult. This simplifies the presented model and it makes it impossible to use in real word applications. In [2] authors calculate participant dependence as the difference between a number of followed people and a number of followers. When a person has more followers, it means that people tend to follow decisions of such person, on other hand when someone follows more people it means that person tends to follow decisions of others. The biggest problem of that approach is the fact that people tend to follow people in some topic and totally ignore others - that was ignored completely in this work. This measure also assumes if a person follows more people than they have followers this situation will always remain true. It is obvious that is a much too simplified situation. In [10] authors calculate independence base on daily return of a recommended stock. This requires collective prediction to make it possible to calculate independence value. Authors used that measure to prove that a more independent crowd tends to make better decisions (crowd has a higher daily return on investments in stocks).

As it was shown, current literature has a different approach to measure independence of a collective. Known measures are too simple to describe such a complex process as an influence. Measures take into account only fraction of the information, therefore do not fully describe influence between people. Above approaches do not recognize the fact that in some topics influence between two people may by higher than in others. A common problem is that often influence is treated as binary: one person can influence others or be influenced, it is not possible to do both. Above approaches also do not allow to have different influence for different topics. These problems are common in literature. Therefore, in this paper, we want to respond to this by presenting our measure. We believe that information about an influence of the collective will be helpful in further research to make more complex analysis of crowds.

3   Proposed Independence Measure

In this work, we propose an independence measure for expert collections. The proposed measure returns a single number that shows how independent one person is from another on a chosen topic. Value of independence could be calculated for a group of people before it becomes collective. The measure will only use information publicly available on social networks. The presented measure will be confronted with current knowledge about interpersonal influence and intuition behind it. We will mathematically prove properties of the proposed measure.

Independence is defined as the capacity to have freedom on the decision-making process regardless of others decisions. It is a number between 0 and 1 (higher means person X more likely affect the decision-making process of another person Y. Therefore, person Y more often makes the same decision as person X). Independence as a number shows in how many situations one person knowing a response of another is willing to accept this response as his or her own. Defined measure uses data gathered from social networks, that data is a vital part of our equation. Most of social networks are not suitable for this task for example: Facebook, Snapchat, Twitter, because they share only a fraction of information about a person. In our case, best social networks suitable for this task are: ResearchGate and LinkedIn.

We analyze all of the available information from the social networks profiles. As a result, we could define the expert profile based on ResearchGate (less obvious elements were explained):

For an expert e the structure of his profile is presented as follows:

$$e = (x_1, x_2, x_3, x_4, x_5, x_6, x_7, x_8, x_9, x_{10})$$

where:

- x_1 - full name of the profile owner,
- x_2 - academic title, e.g. Professor, MSc,
- x_3 - set of skills (the only name of skill is available),
- x_4 - publications list - the structure of publication is below,
- x_5 - affiliation name,
- x_6 - list of followed people with links to their profile,
- x_7 - list of followers with links to their profile,
- x_8 - list of the most common co-authors,
- x_9 - research topics,
- x_{10} - the value of h-index (with self-citations and without),

Representation of a publication p has the following structure:

$$p = (a_1, a_2, a_3, a_4, a_5, a_6, a_7, a_8, a_9, a_{10})$$

- a_1 - title,
- a_2 - the type of publication,
- a_3 - publication date,
- a_4 - publisher,
- a_5 - authors name with a link to the author profile,
- a_6 - the number of reads,
- a_7 - link to full text (optional),
- a_8 - project,
- a_9 - citations,
- a_{10} - references.

The LinkedIn profile of expert e has the following structure:

$$e = (x_1, x_5, x_{11}, x_{12}, x_{13}, x_{14}, x_{15})$$

- x_{11} - contact list - list of contacts with a link to their profile
- x_{12} - current Employer
- x_{13} - list of education, each element is described by:
  - University Name
  - Title
  - Field of study
  - Length of study - information about when studies were started and ended
  - Activities and societies
- x_{14} - list of skills and endorsements, each element is described by:
  - Skill name
  - Number of Endorsements
  - People who endorse the skill
- x_{15} - publication list, the structure of publication was defined below.

Publications in LinkedIn have the following structure:

$$p = (a_1, a_3, a_4, a_5, a_7)$$

Sets of publications from LinkedIn and ResearchGate are the same ($f(x_{15}) = f(x_4)$, where function f extract only title and authors from publications), but in both lists publications have different properties.

As was shown above, social media profiles are the repository of knowledge about people. Profiles rich in information describes people in an accurate way. We use it to define how person x is independent of decisions of person y. In our work, we use only information about publications and followed people (or list of contacts in the case on LinkedIn). This is a new approach to the problem. It is important to emphasize that independence is calculated for a given topic. Therefore, for each topic independence value should be calculated independently, for example: when collective has to solve an economic problem, we take into consideration only economic publications but on other hand when collective has to solve biological problem we take into consideration only biological publications. Known solutions in the literature use an only single value to describe independence (topic was not relevant to the independence). On the rest of the publication, we will call the topic of the collective problem "given topic."

Let C be a collection of experts with data in the social media. By independence measure for C we understand the following function:

$$W : C \times C \to [0, 1].$$

A value $W(x, y)$ represents a degree of independence of expert x from expert y. Below we present some intuitive axioms for this function.

A1. For each $x, y \in C$ there should be:
$W(x, y) > 0$ for x $\neq$ y and $W(x, y) = 0$ iff $x = y$ (this means two experts are totally dependent, that is their independence is equal 0 if and only if they are the same).

A2. For each $x, y \in C$ there should be:
$W(x, y) = 1$ iff experts x and y do not have the same affiliation ($x_5 \neq y_5$), common publication ($x_4 \cap y_4 = \emptyset$) and contacts ($x_{27} \cap y_{27} = \emptyset$ and $x_7 \cap y_7 = \emptyset$).

A3. For each $x, y \in C$ there should be:
$W(x, y) < W(y, x)$ iff experts x, y do not have the same affiliation ($x_5 \neq y_5$), contacts ($x_{27} \cap y_{27} = \emptyset$ and $x_7 \cap y_7 = \emptyset$) and set of publications of expert y: P_y and set of publications of expert x: x_4, are: $|y_4| > |x_4|$ and $x_4 \cap y_4 \neq \emptyset$

Or

experts x, y do not have common publications ($x_4 \cap y_4 = \emptyset$) and set of contacts expert y: $F_y = y_{27} \cup y_7$ and set of contacts of expert x: $F_x = x_{27} \cup x_7$ are: $|F_y| > |F_x|$ and $F_x \cap F_y \neq \emptyset$

A4. For each $x, y \in C$ there should be:
$W(x, y) = W(y, x)$ and $W(x, y) < 1$ and $W(y, x) < 1$ iff set of publications of expert y: y_4 and set of publications of expert x: x_4, are: $|y_4| = |x_4|$ and set of contacts expert y: $F_y = y_{27} \cup y_7$ and set of contacts of expert x: $F_x = x_{27} \cup x_7$ are: $|F_y| = |F_x|$

A5. For each $x, y, z \in C$ there should be:
$W(x, y) < W(x, z)$ iff experts x, z do not have common publications and set of contacts expert y: $F_y = y_{27} \cup y_7$ and set of contacts of expert x: $F_x = x_{27} \cup x_7$ and set of contacts of expert z: $F_z = z_{27} \cup z_7$ are: $|F_x \cap F_y| > |F_x \cap F_z|$

or
experts x, z do not have the same affiliation, contacts and set of publications of expert y: y_4 and set of publications of expert x: x_4 and set of publications of expert z: z_4, are: $|x_4 \cap y_4| > |x_4 \cap z_4|$

To calculate independence, we use the following function:

$$w(x, y) = 1 - \sqrt{\frac{min\left(\left(\sum_{i=r_1}^{r_2}\left(\frac{1}{r_2 - i + 1} \times \frac{|X_i \cap Y_i|}{\max(|X_i|, 1)}\right)\right)^2, 1\right) + \left(\frac{|F_x \cap F_y|}{\max(|F_x|, 1)}\right)^2}{2}} \quad (1)$$

where:

$w(x, y)$ is the number between 0 and 1 that show how person x is independent of person y on a given topic, in other words: how often person x knowing the answer of person y will take this answer as his/her own,

r_1 is the year of the first joint publication of agents x and y on a given topic,

r_2 is the year of last publication of agent x on a given topic,

X_i, Y_i are the sets of publications in the i-th year on a given topic ($X_i \subset x_4$ and $Y_i \subset y_4$)

F_x, F_y set of contacts agents x and y ($F_x = x_{27} \cup x_7$ and $F_y = y_{27} \cup y_7$).

Independence is calculated based on two factors: the number of joint publications on a given topic and the number of mutual contacts. The first factor focuses on cooperation degree. In this case, we distinguish by several types of cooperation:

- master - student - one person is a more experienced researcher as another person ($x_2 \gg y_2$), in this case, a more experienced researcher (x) has a much higher influence on a less experienced (y) researcher than a less experienced researcher (y) on the more experienced researcher (x), in our work we assume a more experienced researcher will have more publications than a less experienced researcher $|x_4| \gg |y_4|$, therefore the influence calculated by our equation will be higher,
- unequal cooperation - one researcher is more experienced than the other ($x_4 > y_4$), this type is similar to master -student relation, one researcher (x) has a higher influence on the other (y), but in this case, differences between influences are smaller, our equation should calculate the influence correctly for the same reasons as for master - student relation,

- mutual cooperation - both sides are equal, we assume they share the same opinions on a given topic in which they have worked together, neither researcher has a higher influence on another researcher $(w(x, y) \cong w(y, x))$, in this case, both researchers should have a similar number of publications $|x_4| \cong |y_4|$. Therefore, an influence value for both should be similar.

The above types show most popular relationships between researchers (other relations are possible, but due to lower popularity it was skipped in this analysis). The equation cannot tell which type of relation two researchers share, but distinguished types of relation is required to show that the equation confirms expectations.

Equation (1) has the following properties:

Property 1: Non-negativity - $w(x, y) \geq 0$ *and* $w(x, y) \leq 1$.

Proof: The proof is trivial therefore will be skipped.

Property 2: Identity of indiscernibles $(x = y) - w(x, y) = 0$.

Proof: The proof is trivial therefore will be skipped.

Property 3: non-symmetry - $w(x, y)! = w(y, x)$.

Proof: Define agents x, y, define set of publications X'', Y'' for agent x and y respectively and set of followers X', Y' for agent x and y respectively, where:

$|X'' \cap Y''| = 0$ and $|X''| > 0$, $|Y''| > 0$
$|X'| = 1$, $|Y'| = 4$, $|X' \cap Y'| = 1$

$$w(x, y) = 1 - \sqrt{\frac{0 + \frac{1^2}{1}}{2}} = 1 - \sqrt{\frac{1}{2}}$$

$$w(y, x) = 1 - \sqrt{\frac{0 + \frac{1^2}{4}}{2}} = 1 - \sqrt{\frac{0 + \frac{1}{16}}{2}} = 1 - \sqrt{\frac{1}{32}}$$

$$1 - \sqrt{\frac{1}{2}} \neq 1 - \sqrt{\frac{1}{32}}$$

Property 4: triangle inequality is not satisfied

Proof: Define agents X, Y, Z define set of publications X'', Y'', Z'' for agent X, Y, Z respectively and set of followers X', Y', Z' for agent X, Y, Z respectively, where:

$|X'' \cap Z''| = 0$, $|X'' \cap Y''| = 0$, $|Y'' \cap Z''| = 0$ and $|X''| > 0$, $|Y''| > 0$, $|Z''| > 0$
$|X'| = 10$, $|Y'| = 20$, $|Z'| = 30$, $|X' \cap Z'| = 0$, $|X' \cap Y'| = 10$, $|Y' \cap Z'| = 10$

$$w(x, z) = 1 - \sqrt{\frac{0 + \frac{0^2}{10}}{2}} = 1 - \sqrt{\frac{0}{2}} = 1$$

$$w(x,y) = 1 - \sqrt{\frac{0 + \frac{10^2}{10}}{2}} = 1 - \sqrt{\frac{1}{2}} \approx 0.3$$

$$w(y,z) = 1 - \sqrt{\frac{0 + \frac{10^2}{20}}{2}} = 1 - \sqrt{\frac{1}{8}} = 0.65$$

therefore equation:

$$w(x,z) \leq w(x,y) + w(y,z)$$

$$1 > 0.3 + 0.65$$

cannot be satisfied.

The second-factor of the equation focuses on how many mutual friends, colleges, observed persons are shared by two researchers. In our equation, we use a "followed people" from ResearchGate profile or connections from LinkedIn. This factor describes a social group of each collective member. This is the result of intuition: two people that share the same job will have more common friends than people that work in other places, which will result in the influence of one person on another will be higher. This factor is needed to detect not obvious connections: the same workplace or the same university. For example, two people working in the same laboratory but working on two different researches and do not have joint publications know each other very well, and one person could depend more on the decision of another person. People that know each other well tend to have more mutual connections than people that don't know each other. Based on this assumption, this part of the equation was created.

4   A Preliminary Analysis

One of the biggest advantages of the proposed Eq. (1) is the fact that it is not required to gather a collective to calculate independence value. This advantage allows building collective with independence value exactly as expected. This approach also makes possible to better understand the phenomena of collective intelligence.

It is vital to show how this mathematical approach could describe a psychological situation. To make this work we made some assumptions:

- A person with more experience in a field will have more publications in this field,
- More experienced researchers tend to work with more people than less experienced researches,
- More experienced experts more likely influence others in their decisions,
- Others influence us more when we belong to the same social group.

Intuition stands behind the above assumptions. It is common knowledge that the most respected researchers have a huge number of publications. Many younger researchers use experience and knowledge of a more experienced researches and based on their expertise follow their research. As an expert researcher can be part of far more projects than other less experienced researchers. More projects in science usually means there are more publications. Therefore, more experienced authors have more publications with different people. More experienced authors become real authorities for less experienced authors, and as many studies show people follow authorities [8, 9]. It is especially true for researchers that begin theirs adventure with research, for example: Ph.D. candidates or young Ph.D. researchers. In their cases, an independence value to more experienced researchers should be significantly smaller than an independence value to less experienced researchers. This is true in the case of our equations, because young unexperienced researchers usually don't have publications or have a small number of them. The second part of equation focuses on social connections, people which we work with, people we have contact with have a huge impact on our personality and our choices. We tend to use our friends' recommendation rather than that of an unknown person [15]. This property is wildly used in recommendation systems.

The equation also has some problems that have to be solved in the future. The biggest problem is that the presented equation considers only researchers that have social network accounts with information about publications. This can be considered as the drawback, but social media sites have become so popular that most people with internet access have at least one social network account.

5 Conclusions

In this paper, an independence measure was presented. We have mathematically proven its properties and shown that return values confirm expectations. Proposed equation present different approach than known approaches, because it approaches the problem in a more complex way. The measure allows for us to calculate independence for a different topic - this was not possible in known measures in literature, besides that it is possible to calculate a value of independence for a group of people before it becomes collective - solutions in a literature often calculate independence based on results of collective prediction. The presented measure allows the following situation: some people influence decisions of others and is influenced by others in his or her own decisions. This situation was omitted in known independence measures. The biggest problem of the equation is the fact that it considers only researchers that have social network accounts with information about publications as potential candidates. It eliminates a large part of the population as possible members of the collective.

Future work focuses on extending measure for other people (not only researchers). We plan to create experiments that prove the usability of the presented measure in real life applicability.

References

1. Fadul, J.A.: Collective learning: applying distributed cognition for collective intelligence. Int. J. Learn. **16**(4) (2009)
2. Hong, H., Du, Q., Wang, G., Fan, W., Xu, D.: Crowd wisdom: the impact of opinion diversity and participant independence on crowd performance (2016)
3. Hong, L., Page, S.E.: Groups of diverse problem solvers can outperform groups of high-ability problem solvers. Proc. Natl. Acad. Sci. U. S. A. **101**(46), 16385–16389 (2004)
4. Surowiecki, J.: The Wisdom of Crowds. Anchor, New York (2005)
5. McHugh, K.A., Yammarino, F.J., Dionne, S.D., Serban, A., Sayama, H., Chatterjee, S.: Collective decision making, leadership, and collective intelligence: tests with agent-based simulations and a field study. Leadersh. Quart. **27**(2), 218–241 (2016)
6. Krause, J., Ruxton, G.D., Krause, S.: Swarm intelligence in animals and humans. Trends Ecol. Evol. **25**(1), 28–34 (2010)
7. Lorenz, J., Rauhut, H., Schweitzer, F., Helbing, D.: How social influence can undermine the wisdom of crowd effect. Proc. Natl. Acad. Sci. **108**(22), 9020–9025 (2011)
8. Milgram, S.: Behavioral study of obedience. J. Abnorm. Soc. Psychol. **67**(4), 371 (1963)
9. Miligram, S.: Obedience to Authority: An Experimental View. Harper & Row, New York (1974)
10. Nofer, M., Hinz, O.: Are crowds on the internet wiser than experts? The case of a stock prediction community. J. Bus. Econ. **84**(3), 303–338 (2014)
11. Lévy, P.: From social computing to reflexive collective intelligence: the IEML research program. Inf. Sci. **180**(1), 71–94 (2010)
12. Palak, R., Nguyen, N.T.: Prediction markets as a vital part of collective intelligence. In: IEEE International Conference on INnovations in Intelligent SysTems and Applications (INISTA), pp. 137–142. IEEE, July 2017
13. Scharfstein, D.S., Stein, J.C.: Herd behavior and investment. Am. Econ. Rev. **80**, 465–479 (1990)
14. Schreier, M., Fuchs, C., Dahl, D.W.: The innovation effect of user design: exploring consumers' innovation perceptions of firms selling products designed by users. J. Market. **76**(5), 18–32 (2012)
15. Sinha, R.R., Swearingen, K.: Comparing recommendations made by online systems and friends. In: DELOS Workshop: Personalisation and Recommender Systems in Digital Libraries (2001)
16. Szuba, T.: A formal definition of the phenomenon of collective intelligence and its IQ measure. Fut. Gener. Comput. Syst. **17**(4), 489–500 (2001)
17. Ho, T.H., Chen, K.Y.: New product blockbusters: the magic and science of prediction markets. Calif. Manag. Rev. **50**(1), 144–158 (2007)
18. Malone, T.W., Bernstein, M.S.: Handbook of Collective Intelligence. MIT Press, Cambridge (2015)

A Pattern Recognition Based FMEA for Safety-Critical SCADA Systems

Kuo-Sui Lin[(✉)]

Department of Information Management,
Aletheia University, Taipei, Taiwan, R.O.C.
au4234@mail.au.edu.tw

Abstract. Failure Mode and Effects Analysis (FMEA) can be used as a structured method to prioritize all possible vulnerable areas (failure modes) for design review of safety-critical supervisory control and data acquisition (SCADA) Systems. However, the traditional RPN based FMEA has some inherent limitations. Thus the main purpose of this study was to propose a new pattern recognition based FMEA to evaluate, prioritize and correct a SCADA system's failure modes. In the new FMEA method, a vague set based Risk Priority Number (RPN) is proposed to measure safety risk status of failure modes and a rule based pattern recognition method is also proposed to prioritize action priorities of failure modes. Finally, a case study was conducted to demonstrate that the proposed new FMEA method is not only capable of addressing its inherent problems but also is effective and efficient to be used as the basis for continuous improvement of a safety-critical SCADA system.

Keywords: FMEA · Pattern recognition · SCADA system · Vague set theory

1 Introduction

An industrial control system (ICS) is a generic term applied to hardware, firmware, communications, and software used to perform vital monitoring and controlling functions of sensitive processes and enable automation of physical systems. The industrial control systems, which include supervisory control and data acquisition (SCADA) systems, distributed control systems (DCS), and other smaller control system configurations such as skid-mounted Programmable Logic Controllers (PLC), are often found in the industrial control sectors (NIST 2015). SCADA systems stand out among other ICSs as systems that monitor and control assets distributed over large geographical areas. SCADA systems are widely used by industries to monitor and control different processes such as telecommunications, power transmission and electrical power grids, oil and gas pipelines, water distribution, chemistry and manufacturing facilities, etc. A SCADA system typically includes monitor and control components such as Master Terminal Units (MTUs), Remote Terminal Units (RTUs), Human Machine Interface (HMI), Programmable Logic Controllers (PLCs), Intelligent Electronic Devices (IEDs), sensors, and actuators [Cherdantseva et al. 2016].

Design review and improvement of a safety-critical SCADA system aims at improving the overall safety risk of a SCADA system. Failure mode and effect analysis

© Springer Nature Switzerland AG 2019
N. T. Nguyen et al. (Eds.): ACIIDS 2019, LNAI 11432, pp. 26–39, 2019.
https://doi.org/10.1007/978-3-030-14802-7_3

(FMEA) is a proactive risk management technique commonly used to identify and eradicate potential failures, problems and errors from a system, design, process or service before they reach customers (Stamatis 2003). The overall risk of a failure mode is determined by calculating the Risk Priority Number (*RPN*) which is the multiplication of the three risk factors: Severity, Occurrence, Detectability (i.e., $RPN = S \times O \times D$). A reduction of any of the above three risk factors will result in a reduction of the RPN. Accordingly, recommended actions can be taken to reduce the number of one or more of the three factors.

However, traditional RPN based FMEA method has some inherent limitations as criticized by many researchers (Bowles and Peláez 1995; Chang et al. 2010; Sankar and Prabhu 2001; Chin et al. 2009). Some of the important disadvantages are restated as follows: (1) The RPN method assumes that the ratings of failure modes are crisp numerical values, but in many real world circumstances much information in the FMEA is vague and difficult to be precisely evaluated; (2) Different combinations of O, S and D may yield exactly the same RPN value, but their hidden risk implications may be totally different. For example, two different events with the (O, S, D) values of values of $(2, 3, 2)$ and $(4, 1, 3)$ respectively, have the same *RPN* value of 12. Yet, the hidden risks of these combinations are not necessarily identical. This may lead to a waste of resources and time. Additionally, in some cases a high-risk event may remain unnoticed; (3) The risk factors O, S, and D are assessed by discrete ordinal scales of measure, but the operation of multiplication is pointless on ordinal scales according to the measurement theory; (4) In the conventional FMEA, three risk factors O, S, D are considered with equal importance. Nonetheless, in the practical risk analysis the relative weightings of FMEA risk factors may be unequal. Thus the main purpose of this study was to propose a new pattern recognition based FMEA to recover above mentioned problems, helping to evaluate, prioritize and correct safety-critical SCADA system's failure modes.

2 Related Works

2.1 Vague Set Theory

Gau and Buehrer (1993) proposed the concept of vague set (VS), where the grade of membership is bounded to a subinterval $[t_A(x), 1 - f_A(x)]$ of [0, 1]. Relevant definition and operations of vague sets introduced in (Gau and Buehrer 1993; Chen and Tan 1994) are briefly reviewed as follows.

Definition: Vague Sets. A vague set A in the universe of discourse X is characterized by a truth membership function, $t_A : X \to [0, 1]$, and a false membership function, $f_A : X \to [0, 1]$, where $t_A(x)$ is a lower bound of the grade of membership of x derived from the "evidence for x", and $f_A(x)$ is a lower bound on the negation of x derived from the "evidence against x", and $0 \leq t_A(x) + f_A(x) \leq 1$. The grade of membership of x in the vague set is bounded to a subinterval $[t_A(x), 1 - f_A(x)]$ of [0, 1]. The interval $[t_A(x), 1 - f_A(x)]$ is called the grade of membership (vague value) of x in A. Let $f_A^*(x) = 1 - f_A(x)$, then the vague value $V_A(x) = [t_A(x), 1 - f_A(x)] = [t_A(x), f_A^*(x)]$. Thus, the unknown part $\pi_A(x)$ of the vague value $V_A(x)$ is defined as: $\pi_A(x) = 1 - f_A(x) - t_A(x) = f_A^*(x) - t_A(x)$.

2.2   Failure Modes and Effects Analysis

FMEA is a team oriented analytic method to identify the functions of a product or a process and the associated potential failure modes, effects, and causes. The aim is to evaluate potential failure modes in order to assess the risk associated with the identified failure modes, and to prioritize the failure modes for identifying and carrying out corrective actions to address the most serious failure modes. In 2017 FMEA Workshop, VDA (Verb and der Automobilindustrie) and AIAG (Automotive Industry Action Group) agreed to harmonize and standardize a common set of FMEA requirements/ expectations. A draft of the AIAG-VDA FMEA Handbook was released in November 2017 and a final version was issued in 2018 (AIAG-VDA 2018), that enable suppliers to have a single co-copyrighted AIAG-VDA FMEA Manual. The new AIAG-VDA FMEA handbook replaces "Fill-in-the-blank" with "Step Analysis" according to the new format. The new version adopts a structured approach and six-step implementation process: Step 1: Scope Definition, Step 2: Structure Analysis, Step 3: Function Analysis, Step 4: Failure Analysis, Step 5: Risk Analysis, Step 6: Optimization. The handbook offers logic on how to prioritize action priorities of failure modes. It recognizes that a higher RPN number may not be point to the correct item for the team to work on next. For example, there is no logic showing how a RPN rating of 90 should be prioritized over a RPN rating of 112. In the handbook, the severity rating is a measure associated with the most serious failure effect for a given failure mode of the function being evaluated. The severity rating shall be used to identify priorities of action plans relative to the scope of an individual FMEA and is determined without regard for occurrence or detection. As shown in Tables 1 and 2, the narrative rules review high severity ranks items first, and make use of occurrence rating and detection rating for classifying action priorities of failure modes.

Table 1. Narrative FMEA rules for action priority (AP)

AP	Justification for action priority - DFMEA
High	High priority due to safety and/or regulatory effects that have a High or Very High occurrence rating
High	High priority due to safety and/or regulatory effects that have a Moderate occurrence rating and High detection rating
High	High priority due to the loss or degradation of an essential or convenience vehicle function that has a Moderate occurrence rating and Moderate detection rating
Medium	Medium priority due to the loss or degradation of an essential or convenience vehicle function that has a Moderate occurrence and Low detection rating
Medium	Medium priority due to perceived quality (appearance, sound, haptics) with a Moderate occurrence and Moderate detection rating
Low	Low priority due to perceived quality (appearance, sound, haptics) with a Moderate occurrence and Low detection rating
Low	Low priority due to no discernible effect

Table 2. FMEA action priority (AP)

AP	Action expectation
High	The team *must* either identify an appropriate action to improve prevention and/or detection controls, or justification on why current controls are adequate
Medium	The team *should* identify appropriate actions to improve prevention and/or detection controls, or justification on why current controls are adequate
Low	The team *could* identify actions to improve prevention or detection controls

3 Proposed Research Method

3.1 Proposed New Pattern Recognition Method

Functional safety risk of a SCADA system is a function of three factors (variables): the likelihood of failure occurrence (O), the effectiveness of failure detection and rectification (D), and the consequences of failure occurrence (S). Thee risk function assign three factors (variables) a real number $R = R(S, O, D)$.

In this study, FMEA Action Priority (AP) created in AIAG-VDA FMEA handbook can be regarded as a narrative rule set to recognize action priorities for failure modes. The narrative rules of AP shown in Table 1 can be used to transform into more formal rules. Suppose there is a data sample A_i to be recognized against a feature space $X = \{x_1, x_2, \ldots, x_n\}$, which is represented by the following formula: $A_i = \{(x_j, r_{ij}) \mid x_j \in X\}$, $j = 1, 2, \ldots, n$. Suppose that there exist p known patterns characterized against a feature space $X = \{x_1, x_2, \ldots, x_n\}$, which is represented by the following formula: $B_p = \{(x_j, r_{pj}) \mid x_j \in X\}$, $p = 1, 2, \ldots, t$. The following steps are proposed to solve fuzzy pattern recognition problems:

Step 1: Soliciting rating vector A$_i$ for the data sample A$_i$

Given a set of evaluation team's rating values against the set of feature X for the data sample A_i can be expressed as A_i, shown as: $A_i = \{(x_1, r_{i1}), (x_2, r_{i2}), \ldots, (x_n, r_{in})\}$. Given a set of predefined rating values against the set of feature X for the data pattern B_i can be expressed as B_p, shown as: $B_p = \{(x_1, r_{p1}), (x_2, r_{p2}), \ldots, (x_n, r_{pn})\}$. The rating value r_{ij} can be expressed by a vague rating value $V_{Ai}(x_j)$. The rating value r_{pj} can be expressed by a vague rating value $V_{BP}(x_j)$. Thus, the rating vectors for the data sample and the rating vector for the data pattern can be expressed by the vague set A_i and B_p, respectively.

$$A_i = [t_A(x_{i1}), 1 - f_A(x_{i1})]/x_{i1} + [t_A(x_{i2}), 1 - f_A(x_{i2})]/x_{i2} + \ldots + [t_A(x_{in}), 1 - f_A(x_{in})]/x_{in}.$$

$$B_p = [t_B(x_{p1}), 1 - f_B(x_{p1})]/x_{p1} + [t_B(x_{p2}), 1 - f_B(x_{p2})]/x_{p2} + \ldots + [t_B(x_{pn}), 1 - f_B(x_{pn})]/x_{pn}$$

Step 2: Transforming numerical scores of the solicited vague rating values

In order to compute similarity measures between vague values of pattern data set B_p and sample data set A_i, the solicited vague values $V_{Ai}(x)$ and $V_{BP}(x)$ must be transformed into comparable numerical scores. In addition to (Lin 2016), the author further proposes a new score function (Eq. 1) (Lin and Chiu 2017; Lin 2018) to transform the vague values into numerical scores.

Employing the proposed new score function, the numerical score of solicited vague value can be transformed, shown as follows:

$$S_L(V_A(x)) = t_A(x)/2 + (1 - f_A(x))/2 = (t_A(x) + f_A^*(x))/2 \tag{1}$$

Thus in this study, the score of the vague value $V_{BP}(x_j)$ can be evaluated by the score function S, shown as follows: $S(V_{BP}(x_j)) = t_{BP}(x_j)/2 + (1 - f_{BP}(x_j))/2 = (t_{BP}(x_j) + f_{BP}^*(x_j))/2$.

Step 3: Computing matching distance measure $d(A_i, B_p)$ between pattern B_p and data sample A_i

Let *A and B* be two vectors on a set of feature variables $X = \{x_1, x_2, \ldots, x_n\}$, *where* $A = <a_1, a_2, \ldots, a_n>$, $j = 1 \ldots n$, $a_j \in [0, 1]$; $\mathbf{B} = <b_1, b_2, \ldots, b_n>$, $j = 1 \ldots n$. $b_j \in [0, 1]$. When the angle θ between the two vectors A and B is not known, the cosine of θ can be calculated as: $\cos \theta = A \bullet B/(\|A\| \|B\|)$, where, $\|A\|$ denotes the length of the vector A. The scalar projection of a vector in a given direction is also known as the component of the vector in the given direction. Thus, the component of A in the B direction (the scalar projection of A onto B) is given by: $comp_B A = \|A\| \cos \theta = A \bullet u_B = A \bullet B/\|B\|$, where, the dot product of vectors A and unit vector u_B is the projection of A onto u_B, i.e., $A \bullet u_B = \| A \| \cos\theta; u_B = B/\|B\|$ is the unit vector which defines the direction of B and the magnitude of u_B is normalized to be length one. Thus, the vector projection of A in the B direction is the unit vector u_B times the scalar projection of A onto B, given as: $proj_B A = u_B \bullet comp_B A = (B/\|B\|) \bullet (A \bullet B/\|B\|)$. Similarly, the vector projection of B in the B direction is the unit vector u_B times the scalar projection of B onto B, given as: $proj_B B = u_B \bullet comp_B B = (B/\|B\|)\bullet (B \bullet B/\|B\|)$. Then, we create a similarity matching function M:

$$M(A/B) \to [0, +\infty], M(A/B) = proj_B A / proj_B B = A \bullet u_B/B \bullet u_B = A \bullet B/B \bullet B.$$

Thus, $M(A/B) = A \bullet B/B \bullet B = \sum_{j=1}^{n} a_j \bullet b_j / \sum_{j=1}^{n} b_j \bullet b_j$. If A and B are identical vectors, the vector product $A \bullet B$ is identical to the vector product $B \bullet B$; therefore, the matching function $M(A/B)$ is equal to 1. If $comp_B A > comp_B B$, the similarity matching function $M(A/B)$ is greater than 1. If $comp_B A < comp_B B$, the similarity matching function $M(A/B)$ is less than 1.

In pattern recognition problems, antecedent portion of rule $R_p (p = 1, \ldots, t)$, comprises a set of antecedent propositions $B_p (p = 1, \ldots, t)$ and a set of observed facts $A_i (i = 1, \ldots, m)$. Then the matching distance $d(A_i, B_p)$ can be calculated as follows:

$$d(A_i, B_p) = \left| M(A_i / B_p) - 1 \right| = \left| \sum_{j=1}^{n} a_{ij} b_{pj} / \sum_{j=1}^{n} b_{pj} \cdot b_{pj} - 1 \right| \qquad (2)$$

The smaller matching distance values $|M(A/B) - 1|$ is, the higher the degree of match between $comp_B A$ and $comp_B B$ is.

Step 4: Compare the matching distance $d(A_i, B_p)$ for $p = 1, 2, \ldots, t$ and select the smallest one, denoted by $d\left(A_i, B_p^*\right)$, from $d(A_i, B_p)$ ($p = 1, 2, \ldots, t$).

According to the recognition principle of minimum matching distance, the process of assigning B_p to A_i is described by

$$B_p^* = \arg \min_{1 \leq p \leq t} d\left(A_i, B_p\right). \qquad (3)$$

A pattern B_p^* can be derived such that $d\left(A_i, B_p^*\right) = min\left\{ d\left(A_i, B_p^*\right) | p = 1, 2, \ldots, t \right\}$. Then the sample data set A_i belongs to the pattern B_p^*.

Step 5: Select the next data sample to proceed until all data samples have been classified.

3.2   New Pattern Recognition Based FMEA for Action Priorities of Failure Methods

Before proceed to the proposed FMEA method, the initial work is to recognize all possible failure modes FM_i, potential cause and frequency occurrence for each failure mode of the evaluated SCADA system. The first stage of the proposed FMEA method is to perform RPN analysis to evaluate safety risk status of the SCADA system. The second stage is to perform pattern recognition for action priorities, which will prompt corrective actions to improve design or process robustness of the evaluated SCADA system.

I. The First Stage: Performing RPN Analysis
During the RPN analysis stage, FMEA team solicited severity ratings, estimated vague occurrence rating values and vague detection rating values of failure modes.

Step 1.1: Soliciting Severity rating score $S(FM_i)$

The severity is related to the seriousness of the effects of a failure mode FM_i. For each failure mode, determine effects and select a severity level for each effect and determine the severity rating S_i. In this study, four linguistic ratings are used for severity levels: S_A, S_B, S_C and S_D.

Step 1.2: Soliciting Occurrence rating score $O(FM_i)$
Step 1.2.1: Soliciting vague Occurrence ratings $O(FM_{ij})$ of the j-th grade for the i-th failure mode FM_i

FMEA is a collective assessment process, and should be done by a team process manner. A FMEA team can employ the new polling method proposed by the author (Lin and Chiu 2017) to solicit vague values $O(FM_{ij})$ of the j-th grade for the i-th failure mode $FM_i(i = 1, 2, \ldots, m)$. The vague membership values $O(FM_{ij})$ of failure modes $FM_i(i = 1, 2, \ldots, m)$ are filled in the vague grade sheet as shown in Table 3, where $O(FM_{ij})$ denotes the solicited vague values of the j-th grade for the i-th failure mode FM_i, and G_j is the j-th linguistic grade for the i-th failure mode.

Table 3. Vague graded evaluation sheet for soliciting vague rating values

Failure Mode	Vague value					
	G_1	G_2	...	G_j	...	G_n
FM_1	$O(FM_{11})$	$O(FM_{12})$	...	$O(FM_{1j})$	...	$O(FM_{1n})$
FM_2	$O(FM_{21})$	$O(FM_{22})$	...	$O(FM_{2j})$	...	$O(FM_{2n})$
$\vdots$	$\vdots$	$\vdots$	$\vdots$	$\vdots$	$\vdots$	$\vdots$
FM_3	$O(FM_{i1})$	$O(FM_{i2})$	...	$O(FM_{ij})$	...	$O(FM_{in})$
$\vdots$	$\vdots$	$\vdots$	$\vdots$	$\vdots$	$\vdots$	$\vdots$
FM_m	$O(FM_{m1})$	$O(FM_{m2})$	...	$O(FM_{mj})$	...	$O(FM_{mn})$

Step 1.2.2: Transforming numerical scores of the solicited vague Occurrence rating values

The solicited vague membership values $O(FM_{ij})$ in the vague grade sheet can be transformed into numerical scores in this step. A new score function (Eq. 1) can be employed to transform the solicited vague values into numerical scores.

Step 1.2.3: Deriving the ranges of linguistic grades into numerical scores

A set of linguistic grades are predefined to represent the rating grades regarding a failure mode portfolio FM, where $FM \in \{FM_1, FM_2, \ldots, FM_i, \ldots, FM_m\}$. Assume that the set of linguistic grade assigned to the failure mode FM_i is G, where $G_j \in \{G_1, G_2, \ldots, G_j, \ldots, G_n\}$ and $0\% \leq m_{1j} \leq E(G_j) \leq m_{2j} \leq 100\%$, then the expected rating values against the j-th linguistic grade G_j are evaluated for each failure mode FM_i as follows:

$$E(G_j) = (1 - \lambda) \times m_{1j} + \lambda \times m_{2j}, \tag{4}$$

where $\lambda \in [0, 1]$ denotes the optimism index determined by the evaluator, and $E(G_j)$ is the expected rating values against the j-th linguistic grade G_j. If $0 \leq \lambda \leq 0.5$, the evaluator is a pessimistic evaluator. If $\lambda = 0.5$, the evaluator is a normal evaluator. If $0.5 \leq \lambda \leq 1.0$, the evaluator is an optimistic evaluator.

Step 1.2.4: Soliciting Occurrence rating score $O(FM_i)$ for each failure mode FM_i

The individual occurrence rating $O(FM_i)$ is evaluated for each failure mode FM_i as follows:

$$O(FM_i) = \sum_{j=1}^{n} [O(FM_{ij}) \times E(G_j)] / \sum_{j=1}^{n} [O(FM_{ij}), \text{ or}$$

$$O(FM_i) = [O(FM_{i1}) \times E(G_1) + O(FM_{i2}) \times E(G_2) + \cdots + O(FM_{ij})$$
$$\times E(G_j) + \cdots + O(FM_{in}) \times E(G_n)] / [O(FM_{i1}) + O(FM_{i2}) + \cdots + O(FM_{in})],$$

where $O(FM_{ij})$ denotes the expected occurrence rating against the j-th linguistic grade for the i-th failure mode FM_i, i denotes the ordinal number of the failure mode, G_j is the j-th linguistic grade, and $E(G_j)$ is the expected occurrence rating against the j-th linguistic grade G_j. In this study, the five linguistic grades used for the evaluation of the system's rating values are: $G = \{G_1, G_2, G_3, G_4, G_5\} = \{Very\,Low(VL), Low(L), Moderate(M), High(H), Very\,High(VH)\}$. Thus, the occurrence rating score $O(FM_i)$ for the failure modes $FM_i(i = 1, 51)$ can be calculated as follows:

$$O(FM_i) = [O(FM_{i1}) \times E(VH) + O(FM_{i2}) \times E(H) + O(FM_{i3}) \times E(M) + O(FM_{i4}) \times E(L) + O(FM_{i5}) \times E(VL)]$$
$$/ [O(FM_{i1}) + O(FM_{i2}) + O(FM_{i3}) + O(FM_{i4}) + O(Q_{i5})].$$

Step 1.3: Soliciting Detection rating score $D(FM_i)$

Repeating the same process as for Step 1.2.1 through Step 1.2.4, the remaining Detection rating scores $D(FM_i)$ for the failure modes $FM_i(i = 1, \ldots, m)$ can be calculated as follows:

$$D(FM_i) = [D(FM_{i1}) \times E(VH) + D(FM_{i2}) \times E(H) + D(FM_{i3}) \times E(M) + D(FM_{i4}) \times E(L) + D(FM_{i5}) \times E(VL)]$$
$$/ [D(O_{i1}) + D(O_{i2}) + D(O_{i3}) + D(Q_{i4}) + D(Q_{i5})].$$

II. The Second Stage: Performing Pattern Recognition for Action Priorities Step 2.1: Recognizing action priority for each failure mode FM_i

The FMEA team can use the new pattern recognition method proposed in Sub-Sect. 3.1 to recognize action priority AP_i for each of the identified failure modes $FM_i(i = 1, \ldots, m)$. Based on the recognized action priority, personnel accountable for exploiting the action plans into corrective actions are assigned.

Step 2.2: Exploiting action plans for each failure mode FM_i

In this step, recommendations of action priorities $AP_i(i = 1, \ldots, m)$ to enhance the performance of failure modes are proposed, which may include preventive actions and corrective actions. By implementing each of the actions associated with AP_i, the RPN_i for each failure mode FM_i can be reduced by lowering any of the three rankings (severity, occurrence, or detection) individually or in combination with one another.

4 Numerical Case Study

4.1 Implementation of the Case Study

Following the procedure of the pattern recognition based FMEA, a numerical case study was conducted to demonstrate the efficiency and effectiveness of the proposed method. In this case study, risk evaluation for the SCADA system was carried out by dividing the whole system into its sub-units. Each sub-unit was further divided up to component level and failure mode of each component was discussed in detail. Total 117 failure modes were identified and evaluated.

As shown in Table 4, to determine the expected rating values $E(G_j)$ in the vague graded evaluation sheet, the five linguistic grades $G_j(j = 1, \ldots, 5)$ used for the evaluation of the system's rating values are: $G = \{G_1, G_2, G_3, G_4, G_5\} = \{Very\,Low(VL),$ $Low(L), Moderate(M), High(H), Very\,High(VH)\}$. In this study, the optimism index λ determined by the evaluator is 0.60 (i.e., $\lambda = 0.60$). Based on Eq. (4), the expected rating values $E(G_j)$ of the assigned linguistic grade G_j for each failure mode can be calculated   as: $E(G_1) = (1 - 0.60) \times 0 + 0.60 \times 2 = 0.12;$ $E(G_2) = (1 - 0.60) \times 0.62 + 0.60 \times 4 = 0.32;$ $E(G_3) = (1 - 0.60) \times 0.4 + 0.60 \times 6 = 0.52;$ $E(G_4) = (1 - 0.60) \times 0.6 + 0.60 \times 8 = 0.72;$ $E(G_5) = (1 - 0.60) \times 0.8 + 0.60 \times 10 = 0.92.$

Using the new polling method proposed by the author (Lin and Chiu 2017; Lin 2018), the FMEA team used a vague graded evaluation sheet to solicit graded vague values for Occurrence of the i-th failure mode $FM_i(i = 1, \ldots, 117)$. The solicited letter-grade vague values for Occurrence of failure mode $FM_i(i = 1, \ldots, 117)$ are summarized in Table 4.

Table 4. Vague graded evaluation sheet and solicited vague graded rating values for occurrence

No of failure mode	Graded vague membership value				
	VL	L	M	H	VH
FM_1	[0, 0]	[0, 0]	[0, 0]	[0.4, 0.5]	[1, 1]
FM_2	[0, 0]	[0, 0]	[0, 0]	[0.4, 0.5]	[0.8, 0.9]
FM_3	[0, 0]	[0, 0]	[0, 0]	[0.4, 0.5]	[0.8, 0.9]
⋮	⋮	⋮	⋮	⋮	⋮
FM_{117}	[0, 0]	[0, 0]	[0, 0]	[0.7, 08]	[1, 1]

It indicated that the solicited letter-grade vague values for Occurrence of the first failure mode FM_1 are: [0.4, 0.5] for High and [1, 1] for Very High. Using Eq. (1), the solicited letter-grade vague membership value for Occurrence of the failure mode FM_{ij} can be transformed into letter-grade numerical score $O(FM_{ij})$. Using Eq. (6), the Occurrence rating scores for the failure modes $FM_i(i = 1, \ldots, 117)$ are illustratively transformed   as: $O(FM_1) = [O(FM_{11}) \times E(VL) + O(FM_{12}) \times E(L) + O(FM_{13}) \times E(M) + O(FM_{14}) \times E(H) + O(FM_{15}) \times E(VH)] / [O(FM_{11}) + O(FM_{12}) + O(FM_{13}) +$

$O(FM_{14}) + O(FM_{15})] = (0.65 * 0.52 + 0.85 * 0.72 + 0.9 * 0.92)/(0.65 + 0.85 + 0.9)$
$= 0.741$. The transformed graded vague membership scores for Occurrence of the failure modes $FM_i(i = 1, \ldots, 117)$ are summarized in Table 5.

Table 5. Vague graded evaluation sheet and transformed vague graded scores for occurrence

Failure mode FM_i	Graded vague membership score					Occurrence rating scores
	VL	L	M	H	VH	
FM_1	0	0	0.65	0.85	0.9	0.741
FM_2	0	0	0.35	0.50	0.85	0.779
FM_3	0	0	0	0.45	0.85	0.851
$\vdots$	$\vdots$	$\vdots$	$\vdots$	$\vdots$	$\vdots$	$\vdots$
FM_{117}	0	0	0.5	0.75	1	0.757

Similarly, the Detection rating scores for the failure modes $FM_i(i = 1, \ldots, 117)$ can be calculated. The Occurrence rating scores and detection rating scores for the failure modes $FM_i(i = 1, \ldots, 117)$ are summarized in Table 6.

Table 6. Transformed occurrence rating scores and detection rating scores for failure modes

No of failure mode	Occurrence rating scores O(FMi)	Detection rating scores D(FMi)
FM_1	0.741	0.932
FM_2	0.779	0.851
FM_3	0.851	0.510
$\vdots$	$\vdots$	$\vdots$
FM_{117}	0.757	0.834

Before proceeding to Action Priority pattern recognition stage, the narrative rules of Action Priority shown in Table 1 can be augmented and used in this study, as given in Table 7. The AP rule #1 in AIAG-VDA FMEA handbook can be developed into two rules, rule #1 and rule #2 because of its "or" logic. Additional rule #8 is added to complete the rules. The narrative rules of Action Priority are then transformed into more formal rules, as shown in Table 8.

Table 7. Narrative FMEA rules for action priority (AP)

AP rules	Action priority	Justification for action priority - DFMEA
R_1	High	High priority due to safety and/or regulatory effects that have a Very High occurrence rating
R_2	High	High priority due to safety and/or regulatory effects that have a High occurrence rating
R_3	High	High priority due to safety and/or regulatory effects that have a Moderate occurrence rating and High detection rating
R_4	High	High priority due to the loss or degradation of an essential function that has a Moderate occurrence rating and Moderate detection rating
R_5	Medium	Medium priority due to the loss or degradation of an essential function that has a Moderate occurrence rating and Low detection rating
R_6	Medium	Medium priority due to the loss or degradation of a convenience function that has a Moderate occurrence rating and Moderate detection rating
R_7	Low	Low priority due to the loss or degradation of a convenience function that has a Low occurrence rating and Moderate detection rating
R_8	Low	Low priority due to the loss or degradation of a convenience function that has a Moderate occurrence rating and Low detection rating
R_9	Low	Low priority due to the loss or degradation of a convenience function that has a Low occurrence rating and Low detection rating
R_{10}	Low	Low priority due to no discernible effect

Table 8. Formal rules of action priority (AP) for failure modes

AP rules	Severity $C_1: S(FM_i)$		Occurrence $C_2: O(FM_i)$		Detection $C_3: D(FM_i)$		Action priority
R_1	S_A		H	0.72	–	–	*High*
R_2	S_A		VH	0.92	–	–	*High*
R_3	S_A		M	0.52	H	0.72	*High*
R_4	S_B		M	0.52	M	0.52	*High*
R_5	S_B		M	0.52	L	0.32	*Medium*
R_6	S_C		M	0.52	M	0.52	*Medium*
R_7	S_C		L	0.52	M	0.32	*Low*
R_8	S_C		M	0.52	L	0.32	*Low*
R_9	S_C		L	0.32	L	0.32	*Low*
R_{10}	S_D		–	–	–	–	*Low*

The formal rules shown in Table 8 can be rewritten as follows:

R_1: IF{(C_1, S_A) And $(C_2, 0.72)$ And $(C_3, -)$}, Then *Action Priority AP* = High,

R_2: IF{(C_1, S_A) And $(C_2, 0.92)$ And $(C_3, -)$}, Then *Action Priority AP* = High,

R_3: IF{(C_1, S_A) And $(C_2, 0.52)$ And $(C_3, 0.72)$}, Then *Action Priority AP* = High,

R_4: IF{(C_1, S_B) And $(C_2, 0.52)$ And $(C_3, 0.52)$}, Then *Action Priority AP* = High,

R_5: IF{(C_1, S_B) And $(C_2, 0.52)$ And $(C_3, 0.32)$}, Then *Action Priority AP* = Medium,

R_6: IF{(C_1, S_C) And $(C_2, 0.52)$ And $(C_3, 0.52)$}, Then *Action Priority AP* = Medium,

R_7: IF{(C_1, S_C) And $(C_2, 0.32)$ And $(C_3, 0.52)$}, Then *Action Priority AP* = Low,

R_8: IF{(C_1, S_C) And $(C_2, 0.52)$ And $(C_3, 0.32)$}, Then *Action Priority AP* = Low,

R_9: IF{(C_1, S_C) And $(C_2, 0.32)$ And $(C_3, 0.32)$}, Then *Action Priority AP* = Low,

R_{10}: IF{(C_1, S_D) And $(C_2, -)$ And $(C_3, -)$}, Then *Action Priority AP* = Low.

In this case study, FM_3 is taken as an illustrative sample data for describing proposed pattern recognition method. "Safety and/or regulatory effects" is the main Severity concern of the FM_3 sample data, which is formally expressed as $S(FM_3) = S_A$, as shown in Table 8. The transformed rating scores $O(FM_3)$ and $D(FM_3)$ for sample data FM_3 are calculated as 0.851 and 0.510, respectively, as shown in Table 6. According to Algorithm 1, rule #1, #2 and #3 are suitable rules to recognize the best suited pattern of the sample data A_3. The vague rating scores for FM_3 in Table 6 are used to represent the sample data A_3. Besides, the rating scores in Table 8 are used to represent the rules B_1, B_2, B_3, shown as: $A_3 = S_A/x_1 + 0.851/x_2 + 0.51/x_3; B_1 = S_A/x_1 + 0.72/x_2; B_2 = S_A/x_1 + 0.92/x_2; B_3 = S_A/x_1 + 0.52/x_2 + 0.72/x_3$.

By applying the proposed matching function (Eq. 2), the matching distance $d(A_3, B_p)$ $(p = 1, 2, 3)$ can be calculated as follows:

$$d(A_3, B_1) = |(0.851^*0.72 + 0.51^*0)/(0.72^*0.72 + 0^*0) - 1| = 0.182,$$

$$d(A_3, B_2) = |(0.851^*0.92 + 0.51^*0)/(0.92^*0.92 + 0^*0) - 1| = 0.075,$$

$$d(A_3, B_3) = |(0.851^*0.52 + 0.51^*0.72)/(0.52^*0.52 + 0.72^*0.72) - 1| = 0.027.$$

The smaller values $d(A_3, B_p)$ is, the higher the degree of matching is. According to the recognition principle of minimum matching distance (Eq. 3), it can be observed that data sample A_5 should be classified to pattern B_3, which states that "Rule 3: High priority due to safety and/or regulatory effects that have a Moderate occurrence rating and High detection rating."

4.2 Findings and Implications of the Case Study

(1) Under vague and uncertain situations, FMEA team members are hesitant in expressing their diverse risk assessments over failure modes. To manage such situations, the novel risk priority approach based on vague set theory was contributed to enhance the assessment capability of FMEA under vague and uncertain environment.

(2) The misconception of action priorities judged from different combinations of O, S and D are also prevented because of a new FMEA action priority is introduced to the new FMEA method.

(3) In view of the ordinal scale multiplication problem, the vague set theory is useful for soliciting numerical scores for SOD risk factors of the failure modes identified by the FMEA team.

(4) The proposed pattern recognition based FMEA also provides relative importance weightings of risk factors O, S, and D to prevent hidden risks of identical combinations.

(5) The results of the case study demonstrated that the proposed pattern recognition based FMEA is useful for evaluating the rating values of failure modes and is also useful for recognizing action priorities for failure modes and further for deploying corrective actions.

5 Conclusions

In order to recover inherent limitations of traditional RPN based FMEA, a new pattern recognition based FMEA is proposed in this study, helping to evaluate, prioritize and correct safety-critical SCADA systems' failure modes. Before proceeding further, a design review was taken to identify failure modes and capture their causes and effects information needed for subsequent FMEA scoring process. In the first stage, a vague set based FMEA provided a quantitative analysis method to quantify the design discussion and review procedure under uncertain and vague environment. In the second stage, a pattern recognition procedure was conducted for recognizing action priorities of failure modes. Finally, a numerical case study was conducted to demonstrate that the proposed method is not only capable of addressing its inherent problems but also is effective and efficient to be used as the basis for continuous improvement of safety-critical SCADA systems.

References

AIAG-VDA: Failure Mode and Effect Analysis (FMEA) Handbook, 1st edn. (2018)

Bowles, J.B., Pelaez, C.E.: Fuzzy logic prioritization of failures in a system failure modes, effects and criticality analysis. Reliab. Eng. Syst. Safety 50(2), 203–213 (1995)

Chang, K.H., Cheng, C.H., Chang, Y.C.: Reprioritization of failures in a silane supply system using an intuitionistic fuzzy set ranking technique. Soft. Comput. 14(3), 285–298 (2010)

Chen, S.M., Tan, J.M.: Handling multicriteria fuzzy decision-making problems based on vague set theory. Fuzzy Sets Syst. 67(2), 163–172 (1994)

Cherdantseva, Y., et al.: A review of cyber security risk assessment methods for SCADA systems. Comput. Secur. 56, 1–27 (2016)

Chin, K.S., Wang, Y.M., Poon, G.K.K., Yang, J.B.: Failure mode and effects analysis by data envelopment analysis. Decis. Support Syst. 48(1), 246–256 (2009)

Gau, W.L., Buehrer, D.J.: Vague sets. IEEE Trans. Syst. Man Cybern. 23, 610–614 (1993)

Lin, K.S., Chiu, C.C.: Multi-criteria group decision-making method using new score function based on vague set theory. In: 2017 International Conference on Fuzzy Theory and Its Applications (iFUZZY 2017), pp. 1–6, Pingtung, Taiwan (2017)

Lin, K.S.: A novel vague set based score function for multi-criteria fuzzy decision making. WSEAS Trans. Math. **8**(6), 1–12 (2016)

Lin, K.S.: Efficient and rational multi-criteria group decision making method based on vague set theory. J. Comput. (2018, Forthcoming)

NIST: Guide to industrial control systems security. Special Publication 800-82 Revision 2. National Institute of Standards and Technology, Gaithersburg, MD. http://dx.doi.org/10.6028/NIST.SP.800-82r2. Accessed 5 Oct 2018

Sankar, N.R., Prabhu, B.S.: Modified approach for prioritization of failures in a system failure mode and effects analysis. Int. J. Qual. Reliab. Manag. **18**(3), 324–335 (2001)

Stamatis, D.H.: Failure Mode and Effect Analysis: FMEA from Theory to Execution, 2nd edn. ASQ Quality Press, New York (2003)

Interactive Genetic Algorithm Joining Recommender System

Po-Kai Wang[1], Chao-Fu Hong[2(✉)], and Min-Huei Lin[2]

[1] Electrical Engineering and Computer Science,
National United University, Miaoli, Taiwan
M0729002@smail.nuu.edu.tw
[2] Department of Information Management, Aletheia University, Taipei, Taiwan
{au4076, au4052}@au.edu.tw

Abstract. The recommender systems tend to face the problem of lacking clues about the new user, as a result, it is difficult for the system to provide users with the right recommendation results. In this study, the Interactive genetic algorithm is adapted to join with recommender system, and this new system is used to solve the problem of the traditional recommender system. In addition, this study proposes a double-layer encoding structure that involves the global and area encoding, which will help the optimization of interactive genetic algorithm (IGA) to solve the user fatigue problem. The case study testified that this recommended framework was able to rapidly filter vast amounts of data and to offer personalized recommendations for each film viewer. For future research, further studies on IGA that integrates analysis of screen genes to achieve a higher quality of recommendation and to further solve IGA user fatigue issues. Finally, this new recommender system was used to test film viewers in Taiwan and Hollywood films released in the past 12 years, and the experimental results show evidence that the new system is useful.

Keywords: Recommender system · Interactive Genetic Algorithms ·
Hybrid recommendation · Genre films

1 Introduction

For most recommender systems, when systems cannot collect sufficient preferred information from the user, it will be difficult to determine and make the appropriate recommendation results. There are two purposes for this research: first, combining Hybrid Recommendation and IGA into a new film recommendation framework to overcome the lack of information collected to understand the user's preference; second, improving the definition of genes and increasing the efficiency of convergence of the IGA algorithm to resolve the user fatigue issues.

In modern society, since the bustling and stressful life that people have, leisure and entertainment activities begin playing an important part in our daily life. According to Vorderer et al. (2004), individuals in modern societies devote remarkable amounts of time to the entertainment experience. Naturally, media, especially film, take the noticeable place in the entertainment industry. However, the world that human beings

© Springer Nature Switzerland AG 2019
N. T. Nguyen et al. (Eds.): ACIIDS 2019, LNAI 11432, pp. 40–48, 2019.
https://doi.org/10.1007/978-3-030-14802-7_4

live is a rapid-growing world, and information overload has become a universal phenomenon and a severe issue as well. Number of genre and film are growing exponentially, which have caused film viewers' difficulties to find the film which they are interested in. Therefore, recommender systems are relied on by human beings more than ever before and many approaches were coined.

Normally, the results that recommender systems generate depends on items' attributes and users' preference data. Nevertheless, existing approaches encounter several issues such as the new item problem, the new user problem, the user fatigue problem and so on. As the reason above, this research is aiming to combine hybrid recommendation approach and IGA algorithm to solve these issues. In this study report, Sect. 2 is the literature review, which reports literature relevant to the present study, Sect. 3 is the description of the model that have IGA joining recommender system, Sect. 4 presents the experimental results of the model, and finally, we conclude with how IGA may solve the user fatigue issue.

2 Literatures Review

In this section, we will review two frequently-used existing recommendation models, namely Similarity Comparison Models and Genetic Algorithm Models. Similarity Comparison Models include content-based recommendation and collaborative Filtering recommendation, while Genetic Algorithm has developed its interactive computation models.

2.1 Similarity Comparison Methods

Content-based Recommendation. The advantages of content-based recommendation approach are easy-to-use, straightforward and intuitive, as well as the easy-to-interpret results. The reason behinds this is that additional data of user preference are not required in this approach. As a result, cold start problem can be resolved easily.

However, according to (Marko and Yoav 1997) and Adomavicius and Tuzhilin (2005), there are three drawbacks in this method. First, limited content analysis. Content-based recommender system relies on the clear attributes of each object and well-defined connection among other objects, which mean the recommendation objects in content-based recommender system must be defined clearly because these attributes are their identities. Furthermore, another problem of limited content analysis is that, if there are two or more items have the exact same attributes, it could confuse the system. In simple words, they are the same things for the system. Second, when the system over-relies on the personal preference data, it could limit the possibilities of users exploring films of different genres, which is known as the overspecialization problem. Last but not least, new user problem, which will occur when the new user begins using the system. Since the system has no clues about the new user, it is difficult for the system to present the right recommendation results. In addition, in content-based recommendation approach, users' rating records are one decisive factor which determines the quality and accuracy of recommendation results. It means when users cannot provide sufficient rating records, it is unlikely for them to get the accurate recommendation results.

Collaborative Filtering Recommendation. Karypis (2001) divided the collaborative filtering recommendation approach into two types based on their object connection. First, user-based recommendation algorithms use Nearest Neighbor Search (NNS) to evaluate the similarities between the target user and other users, then predict the probable results for the target user based on the nearest user(s). Second (Badrul et al. 2001) proposed item-based recommendation solve the problems of scalability, which tend to increase the evaluation complexity since the increased number of users.

However, Adomavicius and Tuzhilin (2005) pointed out that there are three main disadvantages of collaborative filtering recommendation approach. The relatively considerable two problems are new user problem and new item problem, the reasons behind these two issues are similar. The former, like the content-based recommendation approach, is crucial for the systems to collect enough preference information in order to predict what users like; while the latter, as it happens, when the new object is registered into the system will cause new item problems. Since this item has not been rated by any user yet, it is likely that the item will be isolated by the system. One further drawback of the system is the sparsity problem, which occurs when the rating records of all users are insufficient and lead to the difficulties for the system to estimate the matrix.

Summary. According to (Bellogín et al. 2012), who also pointed out that there are eight limitations of single recommendation algorithms: restricted content analysis, new user, overspecialization, portfolio effect, rating data sparsity, grey sheep and new item problem. These issues shall cause the above-mentioned recommender systems hardly to collect new users' preferences or predict their interests for particular genres of film. In order to remedy these drawbacks, a recommender system must be able to extract users' preferences. Therefore, the IGA is recommended in the present study as a good method to address these problems.

2.2   Genetic Algorithm and Interactive Genetic Algorithm

According to (Bäck and Schütz 1996), Genetic Algorithm (GA) is one of the Evolutionary Computation (EC) approaches, and it was first proposed by a biologist named Fraser A. S; later a mature GA approach was proposed by Holland (1975), who was inspired by the DNA coding procedure and the natural selection- proposed by Charles Robert Darwin. Holland applied a similar concept to the computing process in order to optimize the searching approach.

The GA approach has several advantages such as convergence, randomness, and extendibility. Li-Chieh (2005) indicated that the GA approach has a relatively better performance compared with other optimization-solving approaches. However, when it comes to involving the participation of subjective opinions, it could limit the accuracy of the GA approach, as well as defining the best fitness function for the unclear preferences of users or subjective or abstract targets shall be a challengeable task for this approach.

Therefore, Interactive Genetic Algorithm (IGA) - one of the Interactive Evolutionary Computation (IEC) approach - was proposed to replace the fitness function by the interaction with users and create personal preferences database. However, this mechanism is only useful to solve the new user problem and sparsity problem in a recommender system. In this research, we tried to combine IGA with recommender system for formatting new recommender system and to design a new coding method for solving the user fatigue problem in IGA. Detailed interpretations and applications of IGA will be demonstrated in the following section.

3 Research Design

The procedures followed by this research is shown below (Fig. 1). Figure 1 shows that, at the beginning of the search, the system will ask the user *three identifiable questions* in order to distinguish what type of viewer the user belongs.

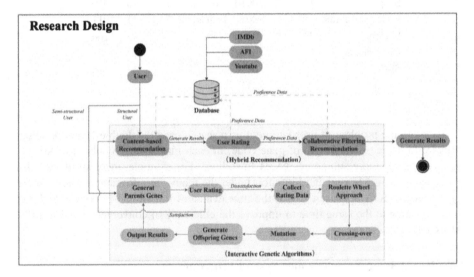

Fig. 1. Research design

As mentioned before, if the user is considered a structural user, then the system will execute the hybrid recommendation approach to recommend films to the user. On the other hand, the IGA approach will be enforced for the semi-structural user and the non-structural user. In the former situation, the user requires to interact with the system for information, such as rating and choosing, in order to aid the system collecting the preference data. Finally, these users will become the structural user after the system collects sufficient information, and subsequently, the hybrid recommendation approach will be initiated by the system.

3.1   Double-Layer Gene Definition

In this research, all the Hollywood film data were collected from IMDb, AFI, YouTube, etc., ranging from year 2005 to 2016. Furthermore, this research not only utilizes the attributes of films, but also the attributes from actors, actresses, directors and writers. We have defined fifteen main genres in this research and arranged them into a clear order from the most panic one to the most delightful one by the psychological effects on the users. The detailed genre information is shown in Table 1 below.

Table 1. Main genres and their codes, decimal encodings (DE) and binary encodings (BE)

Genre	Code	DE	BE	Genre	Code	DE	BE
War	G	0	0000	Epic	O	8	1000
Disaster	F	1	0001	Action	C	9	1001
Suspense	L	2	0010	Adventure	H	10	1010
Police and crime	D	3	0011	Fantasy	I	11	1011
Society	J	4	0100	Romance	N	12	1100
Documentary	P	5	0101	Sports	K	13	1101
Si-Fi	E	6	0110	Musical	B	14	1110
Drama	A	7	0111				

To avoid the problem of user fatigue, this research used the double-layer encoding structure to decrease the length of gene code. In each main genre (Global Encoding), there are one or more sub-genres (Area Encoding). For example, there are three sub-genres called Love, Sexual and Film noir under the main genre Romance. The double-layer structure IGA approach is that the user evaluates the Global Encoding and the Area Encoding at the same time to improve the efficiency of convergence and resolve the user fatigue issues.

3.2   IGA Joining Recommender System Procedure

There are four phases in this recommender system: identifying the user, taking appropriate approach, enforcing the IGA approach and enforcing the Hybrid Recommendation approach. These phases are described below respectively.

Phase 1: Identifying the User. When the new user expects the system recommending the films, the following three identifiable questions must be answered first: 1. What genres do you like? 2. Which film actors do you like? And 3. Which film actresses do you like?

These three questions are presented in the form of multiple choices, and the user is allowed to skip any of these questions if he or she cannot decide to respond. When the user gives up one or more identifiable questions, this user will be identified as a semi-structural user. On the other hand, if the user answers all the questions, he or she will be identified as a structural user.

After the system obtains the user's preference data, it will subsequently convert actors' and actresses' data to genre-type data and evaluate the weight of each genre. These weight and preference data will be utilized in phase 3 and 4.

Phase 2: Taking Appropriate Approach. In this phase, the system will decide which approach need to be executed based on the user's type. The semi-structural users will be required to interact with the system through the IGA approach, while the structural users can skip phase 3 and immediately enter phase 4- enforcing the Hybrid Recommendation Approach- and generate the recommendation results.

Phase 3: Enforcing the IGA Approach. In this phase, the user will have to interact with the system by rating the six films provided (offspring genes). The system utilizes feedback, arranges the wheel percentages of each film genes, executes the cross-over and mutation processes and generates the next generation genes. The procedures of IGA recommendation approach are shown below (Fig. 2).

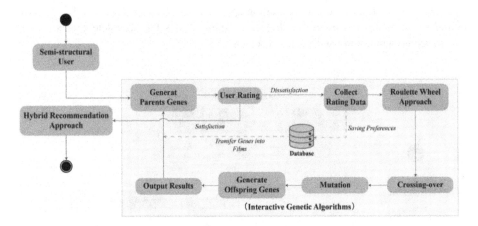

Fig. 2. IGA recommendation approach procedure

Figure 2 shows that, at the end of each cycle, the user can choose whether he or she wants to continue. If the user feels satisfied by one or more of the offspring results or feels tired of rating, he or she can decide to leave the IGA cycle and enter into the Hyper Recommendation Approach procedure.

In this research, the method we used is based on Man's approach while we made some changes to fit our research purpose. Following are eight steps in the IGA cycle used in this research: Step.1: Generate Parents Genes, Step.2: User Rating, Step.3: Collecting Rating Data (Fig. 3), Step.4: Roulette Wheel Approach, Step.5: Cross-over, Step.6: Mutation, and Step.7: Generate Offspring Genes, and Step.8: Output Results.

Phase 4: Enforcing the Hybrid Recommendation Approach. n this phase, all users are considered structural users, which means the problems such as the new user problem or the sparsity problem will unlikely to occur. In this section, we will further explain the concept of the Hybrid Recommendation approach and how it works.

Recommendable Order of Main Genres	Main Genres	Sub-genre(s)	Added up Score of Subgenres		Recommendation Resluts	Priority
3	Drama	Drama	5		P&C - P&C	High
2	War	Military	2		P&C - Spy	
		War	1		P&C - Criminal	
1	Police and Crime (P&C)	P&C	4	⟹	P&C - Gaol	
		Criminal	2		War - Military	
		Spy	3		War - War	
		Gaol	1		Drama - Drama	Low

Fig. 3. The concept of double layer rating structure

Hybrid Recommendation. The Hybrid recommendation approach combines several approaches in order to fix the problems and apply it into different situations.

According to Marko and Yoav (1997) the content-based recommendation approach and collaborative filtering recommendation approach are most common combination of hybrid recommendation approach. Figure 4 below shows the complete procedure of hybrid recommendation used in this research.

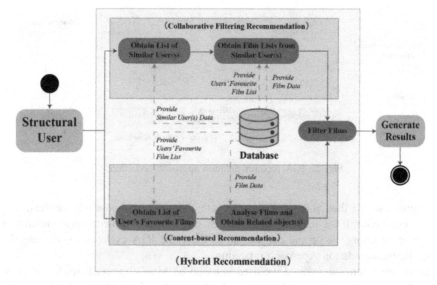

Fig. 4. Hybrid recommendation approach procedure

From the information provided in Fig. 4, it is easy to realize that after users' data are entered into the Hybrid Recommendation approach, the system uses the Collaborative Filtering Recommendation approach and the Content-based Recommendation approach at the same time in order to generate the most suitable results for the user.

4 Experiment Results and Discussion

In order to testify the accuracy and efficiency of IGA Joining Recommender System, the researchers designed a questionnaire to investigate levels of user satisfaction after they used this new recommender system and how they judge the recommendation results. The questionnaire is designed based on Technology Acceptance Model (TAM) (Fred in 1993) to generate valid and reliable results, as shown in Table 2.

Table 2. Questions in the Questionnaire with their factors, question codes (QC), descriptions, and arithmetic means (AM)

Factor	QC	Description	AM
Perceived Usefulness (PU)	A01	Using this film recommender system can aid me to select films more quickly	3.98
	A02	Using this film recommender system can aid me to spend less time in considering the contents of films	3.58
	A03	This film recommender system is quite useful for me in choosing films to watch	3.79
	A04	Using this film recommender system can aid me watching the detailed information of films more quickly	3.60
	A05	Using this film recommender system can add some delights to my daily life when I am choosing the films	3.79
	A06	The results through this film recommender system are suitable for me	3.63
Perceived Ease-of-use (PEOU)	B01	I can operate this film recommender system without having to learn it in advance	3.65
	B02	I can use this film recommender system without having to learn it in advance	3.60
	B03	I do not have to think a lot when I am interacting with this film recommender system	3.56
	B04	This film recommender system is easy to use for me	3.72
	B05*	This film recommender system is useless for me	3.58

*: Question B05 is a reverse coded item, and the result shown above has been reversed and processed

The experimental results in Table 2 show that AM scores in questions A01, A03 and A05 are higher than 3.75, which means that 75% new users are satisfied with the recommendation information. And fortunately, the results of PEOU are higher than 3.5, which indicated that new users do not feel that the system are hard to operate. In addition, from A01, the highest value, it also points to the fact that the double-layer coding can avoid user fatigue problem in IGA Joining Recommender System.

5 Conclusion

The new user problem is the most problematic issue in recommender system. In this study, the authors adopted the IGA to join hybrid recommender system, in which the IGA can clearly find users' characteristics and preferences, and the combination of these well-processed information hybrid recommendation approach can recommend more useful films for new users. Finally, this study used TAM to investigate new users' satisfaction with this new type recommender system. The PU experimental results point out that new users feel that this system is useful to aid them finding preferred films. In addition, the experimental results of PEOU also show evidence that new users feel the system is good to use. In addition, the highest value in question A01 points out that new type of hybrid recommender system can save operating time. Based on these evidences, it is clear that the double-layer encoding structure can solve IGA user fatigue problem. Therefore, double-layer coding IGA joining recommender system is considered an useful new type of hybrid recommender system.

References

Adomavicius, G., Tuzhilin, A.: Toward the next generation of recommender systems: a survey of the state-of-the-art and possible extensions. IEEE Trans. Knowl. Data Eng. **17**(6), 734–749 (2005)

Bellogín, A.: Recommender System Performance Evaluation and Prediction: An Information Retrieval Perspective (2012). https://goo.gl/CACxqn

Bäck, T., Schütz, M.: Intelligent mutation rate control in canonical genetic algorithms. In: Raś, Zbigniew W., Michalewicz, M. (eds.) ISMIS 1996. LNCS, vol. 1079, pp. 158–167. Springer, Heidelberg (1996). https://doi.org/10.1007/3-540-61286-6_141

Badrul, S., George, K., Joseph, K., John, R.: Item-Based Collaborative Filtering Recommendation Algorithms (2001). https://goo.gl/TFkqwD

Marko, B., Yoav, S.: Fab: content-based, collaborative recommendation. Commun. ACM **40**(3), 66–72 (1997)

Fred, D.: User acceptance of information technology: system characteristics, user perceptions, and behavioral impacts. Int. J. Man-Mach. Stud. **38**(3), 475–487 (1993)

Holland, J.H.: Adaptation in Natural and Artificial Systems: An Introductory Analysis with Application to Biology, Control, and Artificial Intelligence. University of Michigan Press, Ann Arbor (1975)

Karypis, G.: Evaluation of item-based top-n recommendation algorithms. In: Proceedings of the Tenth International Conference on Information and Knowledge Management, pp. 247–254. ACM (2001)

Li-Chieh, H.: A study of customer value-orientation and negotiation. Master's thesis, Southern Taiwan University of Science and Technology, Tainan City (2005). https://goo.gl/oFvpKJ

Vorderer, P., Klimmt, C., Ritterfeld, U.: Enjoyment: at the heart of media entertainment. Commun. Theor. **14**(4), 388–408 (2004)

Maturity Level Evaluation of Information Technology Governance in Payment Gateway Service Company Using COBIT

Nilo Legowo$^{(\boxtimes)}$ and Azda Firmansyah$^{(\boxtimes)}$

Information System Management Department,
BINUS Graduate Program – Master of Information Systems Management,
Bina Nusantara University, Jakarta 11480, Indonesia
nlegowo@binus.edu, azda.duplikat@gmail.com

Abstract. As a payment gateway service company or third party service, service is one that must be maintained and enhanced. Data management should also be managed well because the data is an important company asset. The identification of problems and how to solve it should also be improved. IT (Information Technology) Governance is a structure of relationships and processes to direct and control an organization to achieve its goals. Method of research used framework of COBIT (Control Objectives for Information and Related Technology) 4.1 that has the objective to build good IT governance and standards that have been recognized on the international level. This study only focused on one out of four domains contained in the COBIT which is Deliver and Support domain and limit the discussion on DS1 (Define and Manage Service Levels), DS 10 (Manage Problems) and DS 11 (Manage Data) for management awareness and maturity level. The results of management awareness show that the level of performance process for case domain is at the middle level. The Maturity level of DS1, DS10, and DS11 for the current condition (as is) at level 3 which means a condition in a company has had formal procedures and standards in daily activities but never do the monitoring of procedures that have been implemented. Expected maturity level for all (to be) is at level 4. Performance indicators and outcome measures are proposed to achieve IT processes according to expected goals in DS1, DS10, and DS11.

Keywords: IT governance · Deliver and support · COBIT

1 Introduction

Business competition is increasingly competitive. Therefore, to survive in the business competition, an organization should have information system and information technology to support current businesses. Information technology can help the daily operation of companies to become more efficient, information technology is also a major competitive differentiator [7]. The organizations use the information system to develop products, services, and capabilities that provide an advantage in the competitive market.

© Springer Nature Switzerland AG 2019
N. T. Nguyen et al. (Eds.): ACIIDS 2019, LNAI 11432, pp. 49–61, 2019.
https://doi.org/10.1007/978-3-030-14802-7_5

The development of electronics money transaction is increasing every year based on the data published by Bank of Indonesia. The number of transactions in 2007 totaled 586.046 transactions, 2.560.591 transactions in 2008, 17.436.631 transactions in 2009, 26.541.982 transactions in 2010, 41.060.149 transactions in 2011, 100.623.916 transactions in 2012, 137.900.779 transactions in 2013 and 203.369.990 transactions in 2014. Based on these data it appears that the development of electronic money transactions experiencing huge growth that requires businesses in payment gateway increasingly innovative [2].

As a payment gateway company or third party service company, PT SAR becomes a company that connects between Biller with Collecting Agent in online transaction services to the various merchant. This company uses ISO 8583 which is an international standard for financial transactions in exchange of data. This standard is used to facilitate communication between Biller with Collecting Agent. By using this standard of communication, the transaction can still be made between related parties while utilizing the system or different programming languages. The problem that usually occurs is the incompatibility of information systems with business processes and information systems required by the organization [5].

The objectives of this research are; Knowing the condition of the IS/IT governance that is running, Knowing the information system maturity level current for the expected condition and Increase the information system maturity level towards the expected condition (to be). Therefore, there should be a management of information system that has been implemented to ensure the information system is according to business or organization needs or not.

In this study to determine the condition of governance Information Technology in a Payment gateway service company which is one of the private companies in Jakarta. This type of company began to grow in connection with the growing business development of the marketplace startup that made online sales in the type of Business to Customer (B2C). Companies that make online sales transactions need cooperation with third-party companies that can guarantee security and trust in making purchases online. In the case of discussion the company was included in the initial company as a good pioneer in conducting business as a service company in the payment gateway compared to other similar companies.

IT governance using the Cobit framework is widely used in evaluating information technology management and information systems in commerce, service, and telecommunications private companies.

The maturity level of the system needs to be known in the information technology that has been implemented in order to know whether the application of information technology is made in accordance with organizational goals, IT process performance must also be known in relation to the business needs and objectives that can be done using the COBIT framework [8]. COBIT provides best practices for IT governance and control that will help the organization in the optimization of IT investments, ensure the service provided, and provide measurements to check on incompatible things. COBIT is also used to align business objectives with the intended use of IT. COBIT framework is significantly related with overall risk assessment of interconnected COBIT process that can be used to predict the behavior of auditors in terms of finding and providing IT support Audit [11]. IT Governance sustainably needed for development of information technology can go according to the requirements, processes and business goals, because effective IT governance is the most important support of value and success of the organization [12].

2 Identification of Problems

The current problem is unknown maturity level of the implemented system, whereas maturity level of the system is necessary to know where the position of the current system and to make improvements in the future according to business objectives. Business goals and IT goals have not been aligned, but the alignment is necessary to see compatibility between IT objectives with business needs and extent of the contribution of IT to support business processes to improve the quality of information services and realize the vision and mission of the company. IT process also needs to do the identification of information quality, system quality, service quality, satisfaction of users of the system, and the perceived benefits of user-related information systems in IS Governance. Based on identification result will appear process performance IT to business needs and business objectives to improve the quality of information service for internal and external sides.

2.1 IT Governance

One key focus of IT governance is to align IT with business goals with a blend of corporate governance and management of information technology [3]. According to [9] information technology management focuses on achieving internal effectiveness for the support of information technology products and services as well as operational management of information technology at the moment [9].

The scope of IT Governance has a broader, and concentrates on the performance and transformation of information technology to meet today's business needs and the future, both in terms of internal and external business. IT Governance is part of the corporate governance and consists of the leadership, organizational structures, and processes that ensure the information technology companies supporting and expanding the organization's strategies and goals [3].

From the definition above can be concluded that the role of IT Governance to create a business can be aligned with IT, IT investment can support appropriate, and IT can support and expand objectives and the organization's strategies.

ITGI explained that the IT Governance has five main focuses and everything is driven by stakeholder value. Two of them are the result of that namely value delivery and risk management, while the next three are the driving forces namely strategic alignment, resource management and performance measurement [4] (Fig. 1).

COBIT (Control Objectives for Information and Related Technology) was developed by the IT Governance Institute (ITGI). COBIT is an IT governance framework, which applied to the management of IT services, control department, audit functions. It is important for business owners to ensure the accuracy, integrity, availability of data and critical and sensitive information. COBIT is the most appropriate control framework to help organizations to ensure alignment between the use of information

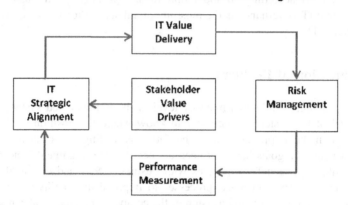

Fig. 1. IT governance focus [4]

technology and business goals [10]. According to Campbell, COBIT is a way to implement IT Governance [1] (Fig. 2).

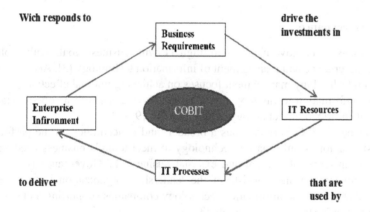

Fig. 2. The basic principles of COBIT [4]

COBIT supports IT governance by providing a framework to ensure [4]:

- IT aligned with business
- IT enabled business and maximize profits
- IT resources are used responsibly
- well managed IT Risk

COBIT provides guidance extensively to obtain the management and control of applications in an enterprise, to illustrate the extent to which the implementation of an information system can compensate for business purposes. COBIT accommodates the translation by providing a process model in four domains: plan and organize (PO), acquire and implement (AI), deliver and support (DS) and monitor and evaluate (ME). The fourth domain has functioning processes for the monitoring of every element of information technology. The basic principle of the COBIT framework is to provide the information necessary to achieve corporate objectives.

3 Methodology

This study describes the implementation of IT governance in Company, This research uses a case study because the authors can get an understanding of an event. The method used in this study is a qualitative method. Qualitative research methods according to Moleong is research that aims to understand the phenomenon of what is experienced by the subject of the study such as behavior, perception, motivation, action, and others in their entirety by means of description in the form of words and language, in an intact of natural context and by utilizing a variety of scientific methods. In this case study, the main data is collected by questionnaires, observations, and documents related to the research in the company [6].

The stage flow of thinking as follows: Questionnaires, observation, document and data collection, Data analysis using the Cobit framework 4.1.

The questionnaire in this research is used as a method in the calculation process to determine the current information system governance. In this study there were 2 (two) types of questionnaires, namely the maturity level questionnaire: Indicators that were asked were 0 (not yet implemented) to 5 (optimized) according to the COBIT maturity model and management awareness questionnaire: namely to raise awareness for company management and identification regarding the entire process related to the quality of internal and external services. Questionnaires are given to employees involved in the system and business processes. to get information about feedback on each domain and the maturity level of the information system.

Observations are made directly on the object of research, in this case study, employees are involved in the system used, in this case the programmer, implementation, system support, DBA (Database Administrator), operational, marketing and finance. The next object is employees involved in business processes, in this case top management, marketing and project management. Related documents that support research are obtained from the company.

Data analysis using Cobit framework 4.1 and the discussion in this study are divided into several topics, namely identification of business goals, identification of IT goals, identification of IT processes, identification of control objectives and maturity levels of information systems implemented.

4 Discussion and Result

This research only uses "Deliver & Support" domain. Domain control used is on DS1 (Define and manage service level), DS10 (Manage Problems) and DS11 (Manage Data) in improving the quality of information services. The number of respondents to the questionnaire of this case study was 40 respondents.

4.1 Management Awareness

Analysis of identification of management awareness of data collection based on completed questionnaires on the detailed control objectives that have been identified in manage service process (DS1), manage problems (DS10) and manage data (DS11). The answer of any questions using multiple choice models, respondents can determine answer by selecting one of the options considered to represent a level of performance to actual conditions of reference as follows (Table 1):

Table 1. Mapping answer of questionnaire and value/level performance DCO (detailed control objective) in DS1, DS10 and DS11 process.

No	Answer	Value performance	Level performance
1	L	1.00	Low
2	M	2.00	Middle
3	H	3.00	High

- H (High) → Good performance
- M (Medium) → Medium performance
- L (Low) → Low performance

Here's the formula for calculating the management awareness:

$$MA = \frac{Nv\ (L, M \text{ or } H)}{Tr} \times 100\%$$

MA   : Result from each management awareness question
Nv   : Number of Value performance
Tr   : Total Responden

Formula for calculating DCO (Detailed Control Objective):

$$DCO = \frac{(MalxNk) + (MamxNk) + (MahxNk)}{100}$$

DCO   : Value Performance of each DCO
Nk   : Identified value performance
Mal   : Result of management awareness for low conditions
Mam   : Result of management awareness for medium conditions
Mah   : Result of management awareness for high conditions

The following results were obtained after doing data analysis (Table 2):

Table 2. Level performance of DCO process DS1 (define and manage service levels)

No	Detailed control objective	Value performance
1	Service level management framework (DS1.1)	2.4
2	Definition of services (DS1.2)	2.3
3	Service level agreements (DS1.3)	2.5
4	Operating level agreements (DS1.4)	2.8
5	Monitoring and reporting of service level achievements (DS1.5)	2.3
6	Review of service level agreements and contracts (DS1.6)	2.4
Average		**2.4**

DCO Level of the manage service level process with a value of performance average for these processes is 2.4. The following radar diagram is based on the analysis result (Fig. 3 and Table 3).

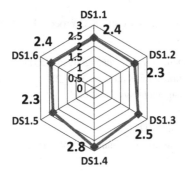

Fig. 3. Radar diagram of DCO DS1 (define and manage service levels)

Table 3. Level performance of DCO (detailed control objective) process DS10 (manage problems)

No	Detailed control objective	Value performance
1	Identification and classification of problems (DS10.1)	2.4
2	Problem tracking and resolution (DS10.2)	2.5
3	Problem closure (DS10.3)	2.5
4	Integration of configuration, incident and problem management (DS10.4)	2.5
Average		**2.5**

DCO Level of the manage problems level process with a value of performance average for these processes is 2.5. The following radar diagram is based on the analysis result (Fig. 4 and Table 4).

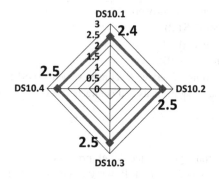

Fig. 4. Radar diagram of DCO DS10 (manage problems)

Table 4. Level performance of DCO process DS11 (manage data)

No	Detailed control objective	Value performance
1	Business requirements for data management (DS11.1)	2.4
2	Storage and retention arrangements (DS11.2)	2.5
3	Media library management system (DS11.3)	2.4
4	Disposal (DS11.4)	2.5
5	Backup and restoration (DS11.5)	2.5
6	Security requirements for data management (DS11.6)	2.5
Average		**2.5**

DCO Level of the manage data level process with a value of performance average for these processes is 2.5. The following radar diagram is based on the analysis result (Fig. 5).

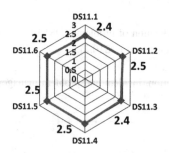

Fig. 5. Radar diagram of DCO DS11 (manage data)

4.2   Maturity Level

Maturity level analysis was obtained from the answers of questionnaires distributed to respondents. These tables and graphs based on respondents answer to provides an overview of the maturity level of service level management (DS1), management problems (DS10) and management data (DS11).

Based on the Table 5, obtained the following information about the attribute of maturity level, namely:

1. Current Maturity level (as is) at DS1 process is at level 3 (**defined**).
2. Expected Maturity level (to be) at DS1 process is at level 4 (**managed**) (Fig. 6).

Table 5. Level of maturity process DS1 (define and manage service levels)

No	Attribute	Maturity level		Maturity level	
		As is	To be	As is	To be
1	AC	2.73	4.45	3	4
2	PSP	2.53	4.23	3	4
3	TA	2.85	4.33	3	4
4	SE	1.95	4.15	2	4
5	RA	2.58	4.20	3	4
6	GSM	3.05	4.43	3	4
Average		**2.61**	**4.29**	**3**	**4**

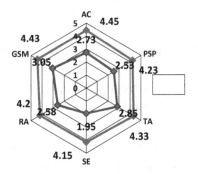

Fig. 6. Representation of maturity level for DS1 process (define and manage service levels) current condition (as is) and expected condition (to be)

Based on the Table 6, obtained the following information about the attribute of maturity level:

1. Current *Maturity level (as is)* at DS10 process is at level 3 (*defined*).
2. Expected *Maturity level (to be)* at DS10 process is at level 4 (*managed*) (Fig. 7).

Table 6. *Level* of *maturity* process DS10 (*manage problems*)

No	Attribute	Maturity level		Maturity level	
		As is	To be	As is	To be
1	AC	3.05	4.50	3	5
2	PSP	2.88	4.55	3	5
3	TA	2.90	4.35	3	4
4	SE	2.25	4.28	2	4
5	RA	2.95	4.38	3	4
6	GSM	3.25	4.38	3	4
Average		**2.88**	**4.40**	**3**	**4**

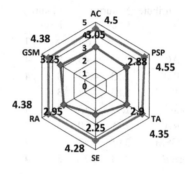

Fig. 7. Representation of maturity level for DS10 process (manage problems) current condition (as is) and expected condition (to be)

Based on the Table 7, obtained the following information about the attribute of maturity level:

1. Current Maturity level (as is) at DS11 process is at level 3 (**defined**).
2. Expected Maturity level (to be) at DS11 process is at level 4 (**managed**) (Fig. 8).

Table 7. Level of maturity process DS11 (manage data)

No	Attribute	Maturity level		Maturity level	
		As is	To be	As is	To be
1	AC	3.03	4.38	3	4
2	PSP	3.03	4.43	3	4
3	TA	3.05	4.28	3	4
4	SE	2.48	4.10	2	4
5	RA	3.05	4.30	3	4
6	GSM	3.08	4.43	3	4
Average		**2.95**	**4.43**	**3**	**4**

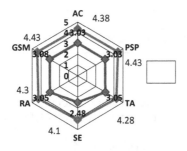

Fig. 8. Representation of maturity level for DS11 process (manage problems) current condition (as is) and expected condition (to be)

The Table 8 shows summary from each result of DCO (Detailed Control Objective)

Table 8. Summary of each DCO

No	DCO	Maturity level		Maturity level		Gap
		As is	To be	As is	To be	
1	DS1	2.61	4.29	3	4	1.68
2	DS10	2.88	4.40	3	4	1.52
3	DS11	2.95	4.32	3	4	1.37
Average		**2.81**	**4.33**	**3**	**4**	**1.52**

4.3   Gap Analysis

Based on questionnaire II result about maturity level, the manage service process for current condition is at level 3 (2.61), and expected condition is at level 4 (4.29), manage problems process for current condition is at level 3 (2.88), and expected condition is at level 4 (4.40), and manage data process for current condition is at level 3 (2.95), and expected condition is at level 4 (4.32). Therefore, improvement is needed to reach the maturity (maturity level) to the expected level (Figs. 9, 10, 11).

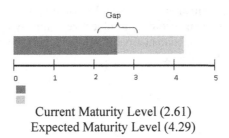

Current Maturity Level (2.61)
Expected Maturity Level (4.29)

Fig. 9. Gap DS1

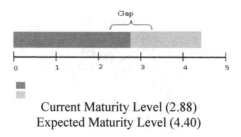

Current Maturity Level (2.88)
Expected Maturity Level (4.40)

Fig. 10. Gap DS10

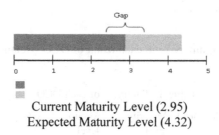

Current Maturity Level (2.95)
Expected Maturity Level (4.32)

Fig. 11. Gap DS11

5 Conclusions

1. The result of management awareness shows that performance level DS1 process is at a medium level, DS10 is at a medium level, and DS11 is at a medium level.
2. Current Maturity level for DS1 (as is) is at 2.61 (level 3), current maturity level for DS10 (as is) is at 2.88 (level 3), and current maturity level for DS11 (as is) is at 2.95 (level 3).
3. Expected maturity level for DS1 (to be) is at 4.29 (level 4), expected maturity level for DS10 (to be) is at 4.40 (level 4) and expected maturity level for DS11 (to be) is at 4.32 (level 4).
4. Current Gap (as is) with expected condition (to be) for maturity level DS1 is at 1.68, current gap (as is) with expected condition (to be) for maturity level DS10 is at 1.52, and current gap (as is) with expected condition (to be) for maturity level DS11 is at 1.37.
5. Implementation of information technology by using COBIT Framework can provide benefits to the system according to objective company business.
6. Suggestion to create any documentation of any work performed and monitoring IT governance to must be enhanced.

References

1. Campbell, P.L.: A COBIT Primer. Sandia National, Philadelphia (2005)
2. Fontana, A.: Innovate We Can! Manajemen Inovasi dan Pencapaian Nilai. Gramedia Widiasarana Indonesia, Jakarta (2009)
3. Grembergen, B., et al.: An exploratory study into IT governance implementations and its impact on business/IT alignment. Inf. Syst. Manag. **26**(2), 123–137 (2009)
4. ITGI: CobiT 4.1. Framework, Control Objectives and Maturity Models. IT Governance Institute, Rolling Meadows, Illinois, SAD (2007)
5. Lucas Jr., H.C., Walton, E.J., Ginzberg, M.J.: Implementing packaged software. MIS Q. **12**(4), 537–549 (1998)
6. Moleong, L.J.: Metodologi Penelitian Kualitatif (Edisi Revisi). PT Remaja Rosdakarya, Bandung (2011)
7. O'Brien, J.A.: Introduction to Information System, 12th edn. Diterjemahkan oleh Dewi Fitriasari, dan Deny Arnos Kwary. Mc Graw Hill, Arizona (2006)
8. Ozkan, S.: A process capability approach to information system effectiveness evaluation. Electron. J. Inf. Syst. Eval. **9**(1), 7–14 (2006)
9. Peterson, R.: Integration Strategies and Tactics for Information Technology Governance dalam Strategies for Information Technology Governance. Idea Group Inc. (2004)
10. Ridley, G., et al.: COBIT and its utilization: a framework from the literature. In: Proceedings of the 37th International Conference on System Sciences, Hawaii (2004)
11. Tuttle, B., Vandervelde, S.D.: An empirical examination of CobiT as an internal control framework for information technology. Int. J. Acc. Inf. Syst. **8**(4), 240–263 (2007)
12. Weill, P., dan Ross, J.W.: IT Governance; How Top Performers Manage IT Decision Rights for Superior Results. Harvard Business School Press, Boston (2004)

Data Structures Modelling for Knowledge Representation

A Framework for Merging Possibilistic Knowledge Bases

Thi Thanh Luu Le[1,2(✉)]

[1] University of Engineering and Technology, Vietnam National University,
Hanoi, Vietnam
lt.thanhluu@gmail.com
[2] University of Finance and Accountancy, Tu Nghia, Quangngai, Vietnam

Abstract. Knowledge base merging is one of active research fields with a large range of applications in Artificial Intelligence. Most of the works in this research field has a lot of restrictions such as in the centralized approach, drowning effect, it is difficult to apply to interactive systems such as multi-agent systems as well as omitting knowledge in the merging process. The purpose of this paper is to focus on the integration of possibilistic knowledge bases in the propositional sense in propositional language. The integration is done through the special possibility distribution of possibilistic knowledge bases. To solve that problem, we introduce a new powerful argumentation framework for merging knowledge bases. In order to model this argument, firstly, the source knowledge bases of each agent are ranked in order of priority and the merging of prior knowledge bases into a priority knowledge base and then infer the final knowledge. An axiomatic model, including a set of rational and intuitive postulates is interested and discussed so that the merging result of knowledge bases needs to be satisfied.

Keywords: Argumentation · Knowledge base merging · Possibilistic logic

1 Introduction

In the past decades, knowledge base merging has become a compelling field in many areas of cooperative information systems. These approaches have been introduced in many papers for merging conflict knowledge bases. It is applied in a large range of areas such as Distributed Database, Multiagent Systems, GroupWare and Distributed Information as Web context. The advantage of knowledge base merging approach is the richness of information so get more from different sources. This knowledge is often contradictory and use priority issues to resolve conflicts, as well as balancing the various factors to achieve the best fusion of inconsistent sources of information is the issue we look forward to.

Uncertainty is almost an attribute of information and knowledge. For this reason, the handling of uncertainty in inference systems has been a long-standing issue of artificial intelligence (AI). When the second expert system was introduced, a proposed setting for the representation of reliability, skepticism and certainty factor was proposed. Then, a series of new proposals have been developed for uncertainty including

© Springer Nature Switzerland AG 2019
N. T. Nguyen et al. (Eds.): ACIIDS 2019, LNAI 11432, pp. 65–76, 2019.
https://doi.org/10.1007/978-3-030-14802-7_6

the theoretical belief function [15], possibilistic theory [16], probability inaccuracy [17], in which the possibilistic theory has emerged at the forefront of AI methods, challenging the maximum right of the representation and the logic installed.

Possibilistic logic is a weighted logic that allows to handle uncertainty (and it also models preferences). In addition, possibilistic logic of inconsistency by taking advantage of the classification for the set of formulas. These are the basic features of standard possibilistic logic. Thus, in the standard possibilistic logic, there are four ways to determine the inference end of a knowledge base in terms of total order [14]: semantic approach based on ranking of representations, the syntactic approach based on Modus Ponens and Weakening, a classical approach based on cuts and rational rejection. They are equivalent and yield the same result as inference. Possibilistic logic is an equivalent framework for determining the ranking of the possible world. In fact, any rank of performance can be represented by a possibilistic knowledge base.

From the promising features of the possibilistic logic evokes us to merging knowledge approaches. Approaches have been published as a two-stage knowledge integration process [12, 13], negotiated game solutions [18], stratified knowledge merging model [9–11, 19, 20], a merging model for weighted knowledge bases [8], a model for knowledge merging by argumentation [21], and knowledge merging by argumentation in possibilistic logic [1, 3, 23].

In this model, we propose a new argumentation framework for merging possibilistic knowledge bases. The contribution of this paper is twofold. First, we introduce a framework for merging possibilistic knowledge-based in which a common argument framework is applied in the knowledge base to achieve meaningful results in comparison with other knowledge base merging techniques for knowledge base merging in possibilistic logic such as [3–7]. Second, an axiomatic model, including rational and intuitive propositions for merging results is introduced and many logical attributes are discussed.

The rest of this paper is organized as follows: We review about possibilistic logic in Sect. 2. Knowledge integration operators for the prioritized belief bases, presented by possibilistic logic is presented in Sect. 3. Section 4 introduces a general argumentation framework and a model to merge knowledge bases. Postulates for knowledge base merging by argumentation and logical properties is introduced and discussed in Sect. 5. Some conclusions and future work are presented in Sect. 6.

2 Possibilistic Logic

The theory of possibility has been studied for several decades. Scenarios of uncertainty may use rules or theoretical possibilities. We consider a propositional language $\mathcal{L}$, based on the limited set of propositional variables denote to the Greek letters $\alpha, \beta, \gamma, \ldots$, the formulas denoted by the letters $\phi, \psi, \xi, \ldots$, and Ω are the finite set of representations of $\mathcal{L}$. The set of models of ϕ notation $[\phi]$, which is a subset of Ω. We denote $\vdash$ the classical syntactical inference and $\models$ the classical semantic inference. The possibility distribution, denoted by π, is a function from Ω to [0, 1]. We also denote $\pi(\omega)$ the degree of compatibility of ω with real world available knowledge. When $\pi(\omega) = 1$, it means that ω is completely consistent with the available knowledge, while $\pi(\omega) = 0$ means that ω is not possible. $\pi(\omega) > \pi(\omega')$ means that ω is more priority than ω'.

Each possibilistic logic formula has the form (ϕ_i, α_i) such that ϕ_i is a propositional formula and $\alpha_i \in [0, 1]$. The weight α_i is the degree of certainty or priority that the propositional formula ϕ_i is true, namely $N(\phi_i) \geq \alpha_i \geq 0$, where N is a necessary measure. When the weight is 1, the possibilistic logic is the classical logic. A possibility distribution that allows two functions from $\mathcal{L}$ to $[0, 1]$ called possibility and necessity measures, denoted by Π and N, and defined by:

$$\Pi(\phi) = \max\{\pi(\omega) : \omega \in \Omega, \omega \vDash \phi\}, \text{and}$$
$$N(\phi) = 1 - \Pi(\neg\phi).$$

Constraints of the form $\prod(\phi) \geq \alpha$ can also be processed in logic, but they correspond to poor information, whereas $N(\phi) \geq \alpha \Leftrightarrow \prod(\neg\phi) \leq 1 - \alpha$ show that $\neg\phi$ somewhat can not and have more information.

A possibilistic formula (ϕ, α) consists of a propositional formula ϕ and a weight $\alpha \in [0, 1]$. A possibilistic knowledge base is a finite set of the formulas $P = \{(\phi_i, \alpha_i) | i = 1, \ldots, n\}$. We denote P^* an aggregation knowledge base for P defined as follows: $P^* = \{\phi_i | (\phi_i, \alpha_i) \in P\}$. Obviously, a possibilistic knowledge base P is consistent if P^* is consistent and vice versa.

By a semantic approach, with a possibilistic knowledge base, there are generally many possibility distributions π on the set of representation Ω such that the necessary measure is determined from this possibility distribution satisfying $N(\phi_i) \geq a_i$ with all formulas ϕ_i. Among these possibility distributions, there is a special distribution, this probability distribution is found by the minimum specification principle, which is the maximum measure of entropy determined as follows:

Definition 1 [5, 23]. $\forall \omega \in \Omega$

$$\pi_P(\omega) = \begin{cases} 1 \text{ if } \forall(\phi_i, \alpha_i) \in P, \omega \vDash \phi_i \\ 1 - \max\left\{\alpha_i : (\phi_i, \alpha_i) \in P \text{ and } \omega \nvDash \phi_i\right\} \text{ otherwise} \end{cases} \quad (1)$$

Example 1. Suppose that

$$P = \{(a, 0.8), (\neg c, 0.7), (b \to a, 0.6), (c, 0.5), (c \to \neg b, 0.4)\}.$$

According to Definition 1, we can determine the possibility distribution for P as follows:

$$\pi_P(a\neg b\neg c) = \pi_P(ab\neg c) = 0.5, \pi_P(abc) = \pi_P(a\neg bc) = 0.3 \text{ and}$$
$$\pi_P(\neg abc) = \pi_P(\neg ab\neg c) = \pi_P(\neg a\neg bc) = \pi_P(\neg a\neg b\neg c) = 0.2$$

Definition 2. Given a possibilistic knowledge base P and $\alpha \in [0, 1]$, the α-cut of P is denoted by $P_{\geq \alpha}$ and defined: $(P_{\geq \alpha} = \{\phi \in P^* | (\phi, \beta) \in P, \beta \geq \alpha\})$. Similarly, a strict α-cut of P is denoted by $P_{> \alpha}$ and defined: $(P_{> \alpha} = \{\phi \in P^* | (\phi, \beta) \in P, \beta > .\alpha\})$.

Definition 3. Possibilistic knowledge base P_1 is equivalent to possibilistic knowledge base P_2, written $P_1 \equiv P_2$ if and only if $\pi_{P_1} = \pi_{P_2}$.

Definition 4. The inconsistency degree of a possibilistic knowledge base P is as follows:

$$Inc(P) = max\{\alpha_i : P_{\geq \alpha_i} \text{ is inconsistent}\} \tag{2}$$

The inconsistency degree of possibilistic knowledge base P is the maximal value α_i where $\alpha_i - cut$ of P is inconsistent. Conventionally, if P is consistent, then $Inc(P) = 0$.

Definition 5. Given a possibilistic knowledge base P and $(\phi, \alpha) \in P, (\phi, \alpha)$ is a subsumption in P if:

$$(P \setminus \{(\phi, \alpha)\})_{\geq \alpha} \vdash \phi \tag{3}$$

Further, (ϕ, α) is a strict subsumption in P if $P_{> \alpha} \vdash \phi$.

The following lemma indicates that tautologies can be removed from possibilistic knowledge bases [2].

Lemma 1. For $(\top, a)$ be a tautological formula in P then P and $P' = P - \{(\top, a)\}$ are equivalent.

Definition 6. For a knowledge base P, the formula ϕ is a reasonable (reliable) consequence of P if

$$P_{> Inc(P)} \vdash \phi \tag{4}$$

Definition 7. For a possibilistic knowledge base P, the formula (ϕ, a) is a possibilistic consequence of P, denoted $P \vdash_\pi (\phi, a)$, if:

- $P_{\geq a}$ is consistent
- $P_{\geq a} \vdash \phi$
- $\forall b > a, P_{\geq b} \nvdash \phi$

With any inconsistent possibilistic knowledge base P, all formulas with certainty degrees smaller than or equal to $Inc(P)$ will be omitted in the merging process.

Example 2. Continuing Example 1, obviously P is equivalent to

$$P' = \{(a, 0.8), (\neg c, 0.7), (b \rightarrow a, 0.6), (c, 0.5)\}.$$

Formula $(c \rightarrow \neg b, 0.4)$ is missed because of $Inc(P) = 0.5$.
We have:

- Reasonable inferences of P: $\neg a, c \rightarrow a, b \rightarrow a, \ldots$
- Possibilistic consequences of: $(c \rightarrow a, 0.7), (b \rightarrow a, 0.6), \ldots$

The two possibilistic knowledge bases P and P' are said to be equivalent, denoted $P \equiv_s P'$, if and only if $\forall a \in (0, 1], P_{\geq a} \equiv P'_{\geq a}$. The two possibilistic knowledge profiles δ_1 and δ_2 are said to be equivalent ($\delta_1 \equiv_s \delta_2$) iff there is a bijection between them that each possibilistic knowledge base of δ_1 is equivalent to its image in δ_2.

3 A Merging Operator in Possibilistic Logic

In this paper, the knowledge bases of the agents are arranged in order of priority. Merging knowledge is organized in rounds, at each round, agents submit their most preferred beliefs, depend on the current state of argumentation process, some arguments may be eliminated and others are added to the set of accepted beliefs.

We consider a merging process for priority knowledge base by argumentation as follows:

1. Each agent arranges its knowledge bases in the order of priority, the more important knowledge is, the higher priority it has. Argumentation process is held in multiple rounds.
2. At each round, the agents give concurrently their knowledge.

   - If all the given knowledge bases are jointly consistent with the temporary results in the previous rounds, they will be given an accept set (temporary output).
   - If the knowledge of some of the joint agents conflicts with the temporary result, the knowledge of these agents is ignored and the remaining knowledge is added to the temporary result set.
   - If an agent proposes a knowledge and other agent can defeat it, this knowledge will be rejected.

3. The argument process will end when no agent has made any arguments. The final temporary result set will be the result of merging process.

We know that each agent corresponds to a knowledge base with a finite set of formulas, i.e., a finite set of arguments, so that the process of argumentation always ends. In addition, agents encourage their knowledge base to prioritize and submit priority arguments as soon as possible.

A possibilistic merging operator, denoted by $\oplus$ is a function from $[0, 1]^n$ to $[0, 1]$. Operator $\oplus$ is used to aggregate the degree of certainty associated with the knowledge provided by different sources. Let $\mathcal{P} = \{P_1, \ldots, P_n\}$ be set of n possibilistic knowledge base (may be jointly inconsistent). The result of knowledge base merging P using $\oplus$, denoted $\mathcal{P}_\oplus$ is defined as follows:

Definition 8 [4]. (Aggregated base) Let $\mathcal{P} = \{P_1, \ldots, P_n\}$ is a set of possibilistic knowledge bases and $\oplus$ a merging operator. The result of merging $\mathcal{P}$ with $\oplus$ is defined by:

$$\mathcal{P}_{\oplus} = \left\{ (H_j, \oplus(x_1, \ldots, x_n)) : j = 1, \ldots, n \right\}$$

where H_j are disjunctions of size J between formulas taken from different $P_i (i = 1, \ldots, n)$ and x_i is either equal to a_i or to 0 based respectively on ϕ_i belongs to H_j or not.

For the two knowledge bases P_1 and P_2 and the integration operator $\oplus$, the semantic combination rule combines the possibility distributions π_{P_1} and π_{P_2} using $\oplus$ where $\pi_{\oplus}(\omega) = \pi_{P_1}(\omega) \oplus \pi_{P_2}(\omega)$. Its syntactical counterpart is the following possibilistic knowledge bases [4]:

$$P_1 \oplus P_2 = \{(\phi_i, 1 - (1 - \alpha_i) \oplus 1): (\phi_i, \alpha_i) \in P_1\} \cup \{(\psi_j, 1 - 1 \oplus (1 - \beta_j)): (\psi_j, \beta_j \in P_2\} \cup \{(\phi_i \vee \psi_j, 1 - (1 - \alpha_i) \oplus (1 - \beta_j)): (\phi_i, \alpha_i) \in P_1 \text{ và } (\psi_j, \beta_j) \in P_2\}$$

$$(5)$$

Two attributes for $\oplus$ are recognized in this definition are as follows [2, 7]:

1. $\oplus(0, \ldots, 0) = 0$
2. If $a_i \geq b_i$ for all $i = 1, \ldots, n$ then $\oplus (a_1, \ldots, a_n) \geq \oplus (b_1, \ldots, b_n)$

The first attribute says that if the formula does not appear in any knowledge base, it does not appear in the merging result. The second attribute is the attribute of complete agreement (called monotonic property) which means that if formula ϕ is more reliable (or preferred to) to formula ψ, then the result of the merging also prefers ϕ to ψ.

Example 3. Let $P_1 = \{(\phi \vee \psi, 0.9), (\neg\phi, 0.8), (\xi, 0.1)\}$
và $P_2 = \{(\neg\psi, 0.7), (\phi, 0.6)\}$. For $\oplus$ be the probabilistic sum defined by $\oplus(a, b) = a + b - ab$. Following Definition 8 and formula (5), we get:

$$\mathcal{P}_{\oplus} = \{(\phi \vee \psi, 0.9), (\neg\phi, 0.8), (\xi, 0.1)\} \cup \{\neg\psi, 0.7), (\phi, 0.6)$$
$$\cup \{(\phi, 0.97), (\phi \vee \psi, 0.96), (\neg\phi \vee \neg\psi, 0.94), (\xi \vee \neg\psi, 0.73), (\xi \vee \phi, 0.64)\}$$

Which is equivalent to

$$\{(\phi \vee \psi, 0.96), (\neg\phi \vee \neg\psi, 0.94), (\neg\phi, 0.8), (\xi \vee \neg\psi, 0.73), (\neg\psi, 0.7), (\xi \vee \phi, 0.64), (\phi, 0.97), (\xi, 0.1)\}$$

Lemma 2, rewritten $\mathcal{P}_{\oplus}$ given in Definition 8, will be useful in the rest of the paper, but first we give the following definition:

Definition 9. (Existence of results) Let P be the knowledge base. The formula (ϕ, a) is a consequence of P, denoted $P \vdash (\phi, a)$ if and only if:

1. $\exists P' \subseteq P$ such that $P' \vdash_{\pi} (\phi, a)$,

2. P' is consistent,
3. $a = \min\{a_i : (\phi_i, a_i) \in P'\}$,
4. P' is a minimal for set inclusion,
5. $\nexists P'' \subseteq P$ satisfying the above conditions with $P'' \vdash_\pi (\phi, b)$ and b > a.

This definition focuses on the foundation of the most preferred formulas.

Example 4. Let $P = \{(\phi \lor \psi, 0.9), (\neg\phi, 0.7), (\xi \lor \psi, 0.6), (\neg\xi, 0.5)\}$.
Then $P \vdash (\phi \lor \psi, 0.9)$, $P \vdash (\neg\phi, 0.7)$ and $P \vdash (\psi, 0.7)$. However, $P \vdash (\neg\psi, 0)$.

4 Knowledge Merging by Argumentation in Possibilistic Logic

In this section, we consider an implementation of general framework above in order to solve the inconsistencies occur when we combine belief bases $(P_1, \ldots, P_n)$. Let us start with the concept of argument [1, 22].

Definition 10. A set of arguments $\mathcal{A}$ is said to be conflict-free if there is no argument A and B in $\mathcal{A}$ where A attacks B.

Definition 11. An argumentation framework is a set of $\langle \mathcal{A}, \mathcal{R}, \succcurlyeq \rangle$ where $\mathcal{A}$ is a finite set of arguments, $\mathcal{R}$ is a binary relation representing the relation among the arguments in $\mathcal{A}$, and $\succcurlyeq$ is a preorder on $\mathcal{A} \times \mathcal{A}$. We also use $\succ$ to represent the strict order.

An argument is a message that the agents want to interact with many other agents. So, when arguments arise, there are states between arguments, that is, support, attack, and rejection.

Definition 12. *(Support)* Given two arguments $A_1 = \langle P_1, a_1 \rangle$, $A_2 = \langle P_2, a_2 \rangle$ and A_2 is a dialogue of A_1, if the knowledge set P_1 contains sentence a_2, then P_2 is called a the dialogue support A_1.

Definition 13. *(Attack)* For two arguments $A_1 = \langle P_1, a_1 \rangle$, $A_2 = \langle P_2, a_2 \rangle$, and A_2 is a dialogue of A_1, if the knowledge set P_1 does not contain sentence a_2, then A_2 is called an attack dialogue A_1.

Definition 14. *(Rejected)* For two arguments $A_1 = \langle P_1, a_1 \rangle$, $A_2 = \langle P_2, a_2 \rangle$, and A_2 is a dialogue of A_1, if sentence a_2 is not equal to the a_1, then the dialogue A_2 is called reject A_1.

Definition 15. *(Dialogue)* If there is any relationship that supports attack or rejection between arguments, then it is called a dialogue.

Definition 16. Each argument is a pair $\langle K, k \rangle$, where k is a formula and K is set of formulas, in which:

1. $K \subseteq K^*$,
2. $K \vdash k$,
3. K is consistent and K is minimal w.r.t. set inclusion.

K is the support and k is the conclusion of this argument. We denote $\mathcal{A}(\mathcal{K})$ the set of all arguments constructed from $\mathcal{K}$.

Definition 17. Let X and Y be two arguments in $\mathcal{X}$.

- Y attacks X if $Y \not\succ X$ and Y $\mathcal{R}$ X.
- If Y $\mathcal{R}$ X but X $\succ$ Y then X can defend itself.
- A set of arguments $\mathcal{A}$ defends X if Y attacks X then there always exists Z $\in \mathcal{A}$ and Z attacks Y.

Assume that P_1 and P_2 are in conflict and if they are equally prioritized, then the merging result neither infer P_1 nor P_2. The question now is "How do two representations of knowledge base have the same priority in the possibilistic logic?". We first need to introduce the concept of a priority degree of a sub-knowledge base.

Definition 18. Let $\mathcal{A}$ be a sub-knowledge base of P. We define its priority degree, denoted $Deg_B(\mathcal{A})$, by:

$$Deg_P(\mathcal{A}) = \min\left\{a : (\phi, a) \in \mathcal{A} \bigcap P\right\}$$

and

$$Deg_P(\mathcal{A}) = 1 \text{ if } \mathcal{A} \bigcap P \text{ is empty}$$

This definition implies that $Deg_P(\mathcal{A})$ is equal to the weight of the lowest priority formulas in $\mathcal{A}$. Now, we define the preference between two knowledge bases as follows:

Definition 19. P_1 is said to be more prioritized than P_2 if for every conflict $\mathcal{E}$ in $P_1 \bigcup P_2$, we have: $Deg_{P_1}(\mathcal{E}) > Deg_{P_2}(\mathcal{E})$

That is, P_1 is more prioritized than P_2 if the least priority (i.e. least weighted) of the formulas in each conflict $\mathcal{E}$ between P_1 and P_2 is P_2.

Let $\mathcal{LP}(\mathcal{E})$ be the set of least prioritized formulas in $\mathcal{E}$.

Two possibilistic knowledge bases P_1 and P_2 are said to have the same priority if for every conflict $\mathcal{E}$ in $P_1 \bigcup P_2$, there exists at least one formula in $\mathcal{LP}(\mathcal{E})$ in P_1, and at least one formula in $\mathcal{LP}(\mathcal{E})$ belongs to P_2.

Example 6. Let P_1 and P_2 be two possibilistic bases defined as follows:

$$P_1 = \{\phi \vee \psi \vee \xi, 0.9, \neg\psi, 0.7, \neg\delta, 0.5\} \text{ and}$$
$$P_2 = \{(\neg\phi, 0.8), (\neg\xi, 0.7), (\xi \vee \delta, 0.5), (\sigma \vee \psi, 0.4)\}$$

There are two conflicts in $P_1 \bigcup P_2$:

$$\mathcal{E}_1 = \{\phi \vee \psi \vee \xi, \neg\phi, \neg\xi, \neg\psi\}, \text{and}$$
$$\mathcal{E}_2 = \{\neg\xi, \xi \vee \delta, \neg\delta\}$$

We have $Deg_{P_1}(\mathcal{E}_1) = Deg_{P_2}(\mathcal{E}_2) = 0.7$ and $Deg_{P_1}(\mathcal{E}_2) = Deg_{P_2}(\mathcal{E}_2) = 0.5$. Then P_1 and P_2 are equal priority.

However, if we have $P_2' = \{(\neg\phi, 0.8), (\neg\xi, 0.6), (\xi \vee \delta, 0.4), (\sigma \vee \psi, 0.4)\}$ then P_1 is more prioritized than P_2'.

From $Deg_{P_1}(\mathcal{E}_1) > Deg_{P_2}(\mathcal{E}_1)$ and $Deg_{P_1}(\mathcal{E}_2) > Deg_{P_2}(\mathcal{E}_2)$.

Lemma 2. Let $\mathcal{P}_\oplus$ be the merging result $\mathcal{P} = \{P_1, \ldots, P_n\}$ with $\oplus$. Then $\mathcal{P}_\oplus$ is equivalent to

$$\{(\phi, \oplus(a_1, \ldots, a_n)) : \phi \in \mathcal{L} \text{ and } P_i \vdash (\phi, a_i)\}$$

Now we define the merging result. This corresponds to the possibility distributions consequences of $\mathcal{P}_\oplus$:

Definition 20. *(Result of merging)* The merging result is:

$$\mathcal{T} = \{(\phi_i, a_i) | \mathcal{P}_\oplus \vdash_\pi (\phi_i, a_i)\}$$

5 Postulates and Logical Properties

In this section we give the postulates and the logical properties and focus only on the set of reasonable results without putting the weights into the inference. That is, a set of logical properties that this argumentation framework needs to satisfy.

CON *(Consistent):* $\oplus(\mathcal{P}_\oplus)$ is consistent.

In possibilistic logic does not require every knowledge base consistently. So, we do not need to require that all knowledge base merging be consistent.

ADI *(Additional Information):*

If $P_1 \cup \ldots \cup P_n$ is consistent then $\oplus(\mathcal{P}_\oplus) \equiv \oplus(P_1 \cup \ldots \cup P_n)$.

The ADI says that in each round of arguments, if all the information from the agents submitted simultaneously does not cause conflicts, they will be added to the results of the merging. Such a request can be satisfied if the merging operator ensures that each agent has at least one knowledge base contributing to the merging result. This defines as follows:

Definition 21. $\oplus$ is called a conjunctive operator if $\oplus(a_1, \ldots, a_n) > 0$ with $\forall a_i > 0$.

EQU *(Equality):*

$$\oplus(\mathcal{P}_\oplus(\{P_1, \ldots, P_n\})) = \oplus(\mathcal{P}_\oplus(\{P_{\pi(1)}, \ldots, P_{\pi(n)}\})), \text{ where } \pi \text{ be a permutation}$$
on $\{1, \ldots, n\}$.

The EQU ensures that all agents are treated fairly in the process of argumentation.

CAU *(Cautiousness):* If P_1 and P_2 are inconsistent and P_1 and P_2 have the same priority, then

$$\oplus(\mathcal{P}_\oplus) \nvdash \oplus(P_1) \text{ and } \oplus(\mathcal{P}_\oplus) \nvdash \oplus(P_2).$$

The idea in the CAU postulate is that when we merge two conflict knowledge bases, the result of merging does not give preference to any knowledge base. This requirement is natural in propositional logic from formulas and there is no way of giving preference between them. Therefore, this cannot be true in the possibilistic logic framework.

REA *(Reaccept):* If $\oplus(\mathcal{P}'_\oplus) \cup \oplus(\mathcal{P}''_\oplus)$ is consistent then $\oplus(\mathcal{P}_\oplus) \vdash \oplus (\mathcal{P}'_\oplus) \cup \oplus(\mathcal{P}''_\oplus)$

If the consequence of the two subgroups is consistent, the result of knowledge integration is the result of this combination.

$\mathbf{P_{Maj}}$ *(Majority):* $\forall P', \exists n, \oplus\left(\left(\mathcal{P} \sqcup P'^n\right)_\oplus\right) \vdash \oplus(P')$

Intuitively, majority is related to the idea of reinforcement, that is if the formulas have the same weight a from two different agents, they will gain a larger weight in the integration result.

$\mathbf{P_{Arb}}$*(Arbitration):* $\forall P', \forall n, \oplus\left(\left(\mathcal{P} \sqcup P'^n\right)_\oplus\right) \equiv \oplus((\mathcal{P} \sqcup P')_\oplus)$

This postulate states that, if we combine n knowledge bases with the same weight, then the integrated result holds only one. That is to say that the postulate arbitration ignores redundancies.

From the propositions satisfying the merging operator, in comparison with the proposed knowledge merging framework by argumentation, we propose the following theorem.

Theorem 1. The Argumentation knowledge base merging solution satisfies **CON, ADI, EQU, REA** and $\mathbf{P_{Arb}}$ attributes, and does not satisfy **CAU** and $\mathbf{P_{Maj}}$.

6   Conclusion

In this paper, a framework for merging possibilistic knowledge bases by argumentation is introduced and discussed. The key idea in this work that we presented an argumentation-based framework for resolving conflicts between knowledge bases in a prioritized case of possibilistic logic framework. The proposed approach is different from the classical way used in the literature to deal with conflicting multiple sources information. The main result of the work presented in this paper is that the argumentation framework base on merging operator defined and is an interesting problem to merge the bases in multi-agent systems. In such a system, each agent has its own base which may conflict with the bases of the other agents. A set of postulates is introduced and logical properties are mentioned and discussed. They assure that the proposed model is sound and complete. However, our approach is still affected by drowning effect. In the future, we will propose a merging approach based on multiple operators, combining consistent and conflict information using different operators and evaluation of computational complexities of knowledge base merging operators.

References

1. Dung, P.M.: On the acceptability of arguments and its fundamental role in nonmonotonic reasoning, logic programming and n-person games. Artif. Intell. **77**(2), 321–357 (1995)
2. Benferhat, S., Dubois, D., Prade, H., Williams, M.-A.: A practical approach to fusing prioritized knowledge bases. In: Barahona, P., Alferes, J.J. (eds.) EPIA 1999. LNCS (LNAI), vol. 1695, pp. 222–236. Springer, Heidelberg (1999). https://doi.org/10.1007/3-540-48159-1_16
3. Amgoud, L., Kaci, S.: An argumentation framework for merging conflicting knowledge bases. Int. J. Approximate Reasoning **45**(2), 321–340 (2007). An argumentation framework for merging conflicting knowledge bases
4. Benferhat, S., Dubois, D., Kaci, S., Prade, H.: Possibilistic merging and distance-based fusion of propositional information. Ann. Math. Artif. Intell. **34**(1–3), 217–252 (2002)
5. Nguyen, T.H.K., Tran, T.H., Nguyen, T.V., Le, T.T.L.: Merging Possibilistic Belief Bases by Argumentation. In: Nguyen, N.T., Tojo, S., Nguyen, L.M., Trawiński, B. (eds.) ACIIDS 2017. LNCS (LNAI), vol. 10191, pp. 24–34. Springer, Cham (2017). https://doi.org/10.1007/978-3-319-54472-4_3
6. Qi, G., Du, J., Liu, W., Bell, D.A.: Merging knowledge bases in possibilistic logic by lexicographic aggregation. In: Grünwald, P., Spirtes, P. (eds.) Proceedings of the Twenty-Sixth Conference on Uncertainty in Artificial Intelligence, UAI 2010, Catalina Island, CA, USA, 8–11 July 2010, pp. 458–465. AUAI Press (2010)
7. Benferhat, S., Kaci, S.: Fusion of possibilistic knowledge bases from a postulate point of view. Int. J. Approximate Reasoning **33**, 255–285 (2003)
8. Gabbay, D., Rodrigues, O.: A numerical approach to the merging of argumentation networks. In: Fisher, M., van der Torre, L., Dastani, M., Governatori, G. (eds.) CLIMA XIII 2012. LNCS, vol. 7486, pp. 195–212. Springer, Heidelberg (2012). https://doi.org/10.1007/978-3-642-32897-8_14
9. Tran, T.H., Nguyen, N.T., Vo, Q.B.: Axiomatic characterization of belief merging by negotiation. Multimed. Tools Appl. 1–27 (2012)
10. Tran, T.H., Vo, Q.B.: An axiomatic model for merging stratified belief bases by negotiation. In: Nguyen, N.-T., Hoang, K., Jędrzejowicz, P. (eds.) ICCCI 2012. LNCS (LNAI), vol. 7653, pp. 174–184. Springer, Heidelberg (2012). https://doi.org/10.1007/978-3-642-34630-9_18
11. Qi, G., Liu, W., Bell, D.A.: Combining multiple prioritized knowledge bases by negotiation. Fuzzy Sets Syst. **158**(23), 2535–2551 (2007)
12. Booth, R.: A negotiation-style framework for non-prioritised revision. In: Proceedings of the 8th Conference on Theoretical Aspects of Rationality and Knowledge, TARK 2001, pp. 137–150. Morgan Kaufmann Publishers Inc. (2001)
13. Booth, R.: Social contraction and belief negotiation. Inf. Fusion **7**, 19–34 (2006)
14. Dubois, D., Prade, H.: Possibilistic logic: an overview. In: Gabbay, D.M., Siekmann, J., Woods, J. (eds.) Computational Logic, Volume 9 of Handbook of the History of Logic, pp. 197–255. Elsevier (2014)

15. Yager, R.R., Liu, L.P. (eds.): Classic Works of the Dempster-Shafer Theory of Belief Functions. Springer, Heidelberg (2008). https://doi.org/10.1007/978-3-540-44792-4
16. Dubois, D.: Belief structures, possibility theory and decomposable measures on finite sets. Comput. AI **5**, 403–416 (1986)
17. Walley, P.: Statistical Reasoning with Imprecise Probabilities. Chapman and Hall, London (1991)
18. Zhang, D.: A logic-based axiomatic model of bargaining. Artif. Intell. **174**, 1307–1322 (2010)
19. Tran, T.H., Vo, Q.B., Kowalczyk, R.: Merging belief bases by negotiation. In: König, A., Dengel, A., Hinkelmann, K., Kise, K., Howlett, R.J., Jain, L.C. (eds.) KES 2011. LNCS (LNAI), vol. 6881, pp. 200–209. Springer, Heidelberg (2011). https://doi.org/10.1007/978-3-642-23851-2_21
20. Tran, T.H., Vo, Q.B., Nguyen, T.H.K.: On the belief merging by negotiation. In: 18th International Conference in Knowledge Based and Intelligent Information and Engineering
21. Tran, T.H., Nguyen, T.H.K., Ha, Q.T., Vu, N.T.: Argumentation framework for merging stratified belief bases. In: Nguyen, N.T., Trawiński, B., Fujita, H., Hong, T.-P. (eds.) ACIIDS 2016. LNCS (LNAI), vol. 9621, pp. 43–53. Springer, Heidelberg (2016). https://doi.org/10.1007/978-3-662-49381-6_5
22. Deng, Y., OuYang, Y.: A belief revision method based on argumentative dialogue model. In: The 11th International Conference on Computer Science & Education, ICCSE 2016, Nagoya University, Japan, 23–25 August (2016)
23. Benferhat, S., Benferhat, J., Hué, J., Lagrue, S., Rossit, J.: Interval-based possibilistic logic. In: Walsh, T. (ed.) Proceedings of 22nd International Joint Conference on Artificial Intelligence (IJCAI 2011), Barcelona, 16–22 July, pp. 750–755 (2011)

Conceptual Modeling of Team Development

Marcin Jodłowiec[1]([✉]) [iD] and Julia Piecuch[2] [iD]

[1] Faculty of Computer Science and Management,
Wrocław University of Science and Technology, Wrocław, Poland
`marcin.jodlowiec@pwr.edu.pl`
[2] Faculty of Management, Computer Science and Finance,
Wrocław University of Economics, Wrocław, Poland
`julia.piecuch@ue.wroc.pl`

Abstract. Team development is crucial aspect of human resource management. This study is dedicated to the literature review and the conceptual modeling of this domain in the context of the creation of information systems, which takes into consideration concepts defined within plurality of the models and theories of team development. In order to achieve that, authors have proposed a conceptual UML-based Metamodel of Team Development, abstracting the most important concepts of team development whilst providing decent level of models' formality and unambiguity. Authors have proposed the mapping procedures to Association-Oriented Metamodel, which preserves precise model semantics and is implementable into actual data structures, in order to verify implementability of undertaken research. To exemplify the approach, example of the team development model for the purposes of training groups has been provided.

Keywords: Conceptual modeling · Organization modeling ·
Group processes · Group development · Association-oriented modeling ·
Model transformation

1 Introduction

Team work and cooperation is an inseparable element of society since beginning of civilization. In order to solve problems effectively and benefit from synergy units connect or are connected in teams. Group forming takes place in business world, in education, in psychotherapy [4] and other areas. To conceptualize phenomena occurring in groups a number of models describing them has been proposed and empirically verified. Phases are particularly important here, in which the group changes and develops. These models organize the information necessary for i.a. evaluation of groups as the entities realizing specific tools which form larger projects.

In spite of scientific maturity and long-term use in research and practice, described in literature group development models are informal [4]. The language

© Springer Nature Switzerland AG 2019
N. T. Nguyen et al. (Eds.): ACIIDS 2019, LNAI 11432, pp. 77–88, 2019.
https://doi.org/10.1007/978-3-030-14802-7_7

used to describe them is a natural language and graphical models with informal syntax. In the era of information society lack of the use of the apparatus with well-formed semantics makes harder to create efficiently tools which would support description and evaluation of team development. The team development should be here understood as the process which reflects interpersonal relationships and behaviours inside the team which takes part in a business process.

The following article is an original proposition of the authors in terms of the structural aspects of team development formalization. The aim of the study is to make an analysis of the group development domain in such a way that helps to precisely create an information system, which will be capable of describing and evaluating groups in action. The conceptualization is carried out in the following layers:

1. The conceptual metamodel for team development expressed in the UML language, which provides concept schema for group process definition. This metamodel abstract from proposed theories with specific phases, group roles, created norms.
2. Building group development models, which conform to proposed metamodel. In this stage we create theory-specific model with chosen phases, types of group roles, team structure, etc.
3. The method for transformation of conceptual group model to physically implementable data structures expressed within Association-Oriented Metamodel (AOM) [13], which provides to create implementable models with high level of unambiguity and high level semantic coherence with the conceptual schema.

The paper's contribution is the exemplification of the process of transformation of domain-specific metamodel into AOM, which shows its usefulness in conceptual modelling by showing its main characteristics: semantic capacity, expressiveness implementability and unambiguity. The rest of the paper is structured as follows: Sect. 2 summarizes information from existing theories of group development. Subsequently, in Sect. 3 the approach to group process on the meta-level is elaborated along with example of building MTD-conforming models. Next section describes the method for transformation conceptual TDM models into AOM. Final section is summary.

2   Team Development – A Literature Review

2.1   The Concept of Group and Its Dynamics

In order to obtain an answer what a group dynamics is, we need to define the notion of a group. In psychology, a group is understood as *"a set of two or more individuals, which perceive themselves as members of the same social category"* [20] or *"share the same goal and activity"* [11]. In other approaches, a set of individuals can be perceived as a group when its cardinality is greater or equals three [15]. In [12] a broader definition of a group is given, where at least 3 individuals: (a) perceive themselves as a group, (b) have independent, individual goals (except shared goals and interdependences resulting from behavior, which affects individuals in group), (c) communicate with each other and have direct "face-to-face" contact. We believe that in the era of global internet communication, the third point is not relevant anymore. Other theory focus on the context of self-perceiving in society, i.e. sense of belonging to group, cooperation and commitment to the group unity [5,10]. Hackman and Katz describe a group as a social system with specified boundaries, individual goals, which compose a group goal and different roles of group members [6]. Older group definition [17] draws attention to the concept of values and norms, which are brought by the group members and stresses their influence on the group.

In the area of information systems modeling, there is many concepts that can be similar to the above, but no explicit definition of a group, however there are the concepts that hold similar semantics. Instead of individuals, in the centre of interests are entities, which *represent* object from real world. They can be *grouped* with each other by the means of different kinds of relationships, e.g. they can be held in collections or aggregated. In this focus we will consider task-oriented and goal-oriented groups. We will refer to them as *teams*. However we are aware, that the concept of group is more general and broader.

The other important concept that needs to be defined is group dynamics. It is a complex process of changing nature and personality of group members, often dramatically. Hence, the level of cooperation also evolves [19]. McGrath notices that in groups with cardinality greater than three exists the dynamics of relations [14]. Important concepts used by practitioners to describe the group are among others: structure, norm, cohesion and group roles. Belbin describes a group norm as consensual and often implicit standard of behavior accepted or not accepted in a given social context [1].

In teams and their development some group roles can be distinguished. In the social context, they create common set of behaviors, features and responsibilities expected from individuals, which hold specified position or type of position in group. This happens due to playing specific roles. Team members settle regular patterns of mutual exchange, which increase predictability and social coordination [1]. For the proper description of group process practitioners use the notion of cohesion, which describes solidarity or unity of team, stemming from the development of strong mutual bonds between members and forces, which unite the group, such as mutual commitment into group goals.

2.2 Models of Team Development

For several dozen years researchers have developed a number of group development models. The basic classification involves their sequentiality. Each of the models can be classified as sequential or non-sequential. Sequential models focus on the order of group behaviour taking time into consideration, whereas non-sequential move the center of gravity to the explanation of relationships between different random factors, which lie at the root of group development. In this section we have described the main assumptions of those models, witch were developed for training groups purposes.

Progressive Models. Progressive models, as the name suggests, focus on the group improvement. In each of the phases group is more mature, as it develops. One of the approaches to group development modeling for training group purposes has been proposed by Shepard and Bennis [2]. They distinguished the following phases of group dynamics: 1. *Dependence phase* – members are focused on relationship with the authority who leads the group, 2. *Submission* – dependence on the leader, 3. *Counterdepencence* – ambivalent attitude, 4. *Resolution* – intense involvement in task paring, groups develop an internal authority system, 5. *Independence phase* – members concern with personal relations - the level of positive behavior lowers, 6. *Enchantment* – group revered and affection, expressed - this is followed by conflict and increase negative acts, 7. *Disenchantment* – anxiety reactions, distrust of others, 8. *Consensual Validation* – understanding and acceptance – finally consensus become easier on vital issues.

Other concept used in training groups and psychotherapy is three-phase Kaplan and Roman's model [9]. 1. *Dependency theme* – highlights central role of leader - at first members exhibited exaggerated feelings of helplessness, 2. *Power theme* – increased tension and hostility - this was followed by critical attitude toward the leader, enthusiasm for the task also is declined, 3. *Intimacy theme* – increased sense of involvement feeling of having "settled in", in the end there was more direct communication.

The most popular theory of group process, which at the beginning has been used mostly to describe therapeutic groups is Tuckman's model [19]. Later, it has been used as the entry point to development of other theories and describing small groups formed in other contexts. In its original version it had 4 phases, but later Tuckman added the fifth one: *Adjourning* [19]: 1. *Forming* – testing which behaviors are accepted and the limits of the task set, 2. *Storming* – forming a structure of the group units and individuality, 3. *Norming* – increasing acceptance and openness within the group, 4. *Performing* – task-oriented perseverance, 5. *Adjourning* – the end of project.

Cyclical Models. Cyclical models represent group dynamics as a cycle breaking the linearity of development. One of such theories is Dunphy's model [7] comprising three phases: 1. *Dependency* – group members look to the leader for direction of their activity, 2. *Fight/Flight* – resist leader's directions and develop their ability to work independently, formation of subgroups, marked increase in group cohesiveness, 3. *Pairing Work* – execution of task and concern for loss of group. Other approach described in [3] distinguishes 5 phases The main difference between this and the former is in two first phases: they are oriented on members' behavior, instead of searching the leader. 1. *Initial complaining* – anxiety by members, 2. *Premature enactment* – internal caution of individuals against threats coming from the group, 3. *Confrontation* – rebellion and complaining of group members, 4. *Internalization* – group collaboration and productive work, 5. *Separation* – terminal phase of the group.

In [18] Slater distinguished 4 phases of group dynamics. His model includes the recurrence between phases. It ignores the end of a process, wheres the last phase in the process description summarizes emotional value that has been added to the group. Phases of Slater's model are as follows: 1. *Inhibition* – limiting hostile behavior during training, 2. *Revolt* – raising consciousness self awareness through rebellion against authority, 3. *Closeness* – increased group cohesiveness, 4. *Paring* – creating a bond between members.

3   Metamodeling of Team Development

In the previous section we have presented a number of approaches which treat group processes in distinct way. Each of them offers different set of phases with different semantics. This section contains a conceptual metamodel of team development (MTD) (Fig. 1). The conceptualization process is based on literature research and own knowledge of authors from practice. We have extracted the following concepts: Process, Phase, Role, Informal Role, Formal Role, Team, Temporal Team, Agent, Individual, Goal, Personal Goal, Group Goal. These interrelated domain concepts have been modeled as the specialization of abstract metaclass Entity.

There is a lot of methods of conceptualization of domain-specific metamodeling languages [21]. Our metamodel has been expressed in the UML 2.5.1 meta-metamodel [16] as a class diagram. Besides the issues abstracting the process phases from the model, MTD covers additional aspects, which have been described below.

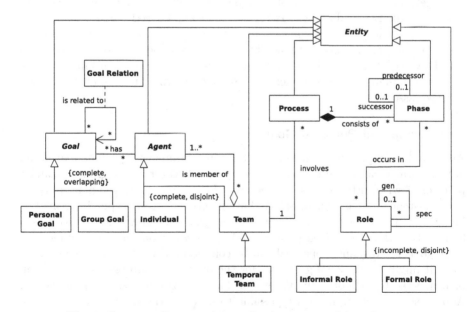

Fig. 1. Conceptualization of team development modeling domain

3.1   Process Structure

The element connecting each of the group development theories is a fact that they comprise a coherent process (`Process`) which consist of phases (`Phase`). The group development is often called a *group process*. The model of process is quite obvious: whilst defining the process structure, the modeler needs to specify its phases including their order. Each of the phases can have maximally one *successor* and at most one `predecessor`. This gives the possibility to define the flow of the process unequivocally, but also both linear-like and recurrent approaches can be modeled.

3.2   Agents: Teams and Individuals

MTD defines abstract category `Agent`, which models an entity having goals, taking actions, fulfilling roles in terms of team development process' phases. However, its most important aspect is the thing that it can belong to group. Two distinct types of agents have been specified: `Individual` and `Team`. An `Individual` specifies an actual type of entity, which takes part in the phases of team development, e.g. an employee, therapy participant, student, training participant. `Team` is the category, which models a group – gathers other agents together. A specific kind of `Team` is `Temporal Team`, which models *ad hoc* groups with temporal limitations.

3.3   Group Roles

Group `Role`s are entities, which describe the state, which can be held by particular `Agent`s in different process of team development phases. The category `Role` has two distinct specializations: `Informal Role` and `Formal Role`. The first one models such roles, which are not assigned to the agents in advance, but are the result of the group processes flow. They can vary depending on adopted theory. The second category focus on formal roles connected with group structure, e.g. project manager, leader, member. Roles can form a taxonomy by defined *gen_spec* association.

3.4   Group and Personal Goals

An important aspect of team work is ability to obtain synergy while realization of tasks, which can be perceived as group goals (`Goal`) from the group dynamics point of view. In other words, irrespective of the group type, the main purpose of its existence is goal realization. We have identified two concepts depicting two kinds of goals: Personal goals of particular group members (`Personal Goal`) and goals of the group itself (`Group Goal`). Moreover, one cannot exclude that these goals are disjoint – one goal can be common for many agents, for some of them it might be personal goal, and for the others a group one. The goals can be connected to each other by the use of `Goal Relation` association metaclass. Adding such category makes it possible to describe the edges flexibly in goal graph.

3.5   Graphical Notation

For the sake of simplicity and understandability by non-IT professionals, a custom a custom graphical model notation for models that conform to MTD has been proposed (Fig. 2). We adopted informal convention to shape the process structure and represent a process as a rectangle, and its phases are represented as ovals contained inside it. To represent the precedence, we used filled directed arrow. The agents are represented either as stickman if they are of individual metaclass or icon depicting group of people if they represent team. Containment is represented as a line with filled dot on the side of containing team. Goals are connected to agents with dashed line and are represented with cloud shapes. Roles are represented with circles, whereas double circle represent formal role and single circle – informal one. `Gen-spec` relationship has analogical notation to UML-like languages. They are connected with the dashed line to the corresponding phases, or to the process rectangle, if they correspond to all the phases of the process.

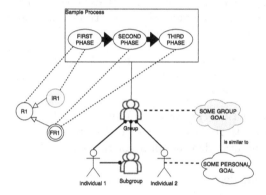

Fig. 2. Sample model conforming to MTD exposing its graphical notation

3.6   Example of Training Groups Model

In this section we present the model of training group that conforms to the proposed metamodel. We chose the revised Tuckman model of group process [19] and prepared conceptual instances for the training groups combined with group role types proposed by Belbin [1]. The results of the model is show in the Fig. 3. Goals are omitted in this model.

4   Transformation to Association-Oriented Model

The following section describes the method for transformation a conceptual model conforming to MTD into AOM data structures. Conceptual to association-oriented model transformation has been designed by the definition of transformation convention $C_{MTD \mapsto AOM}$ considering mapping patterns. The patterns

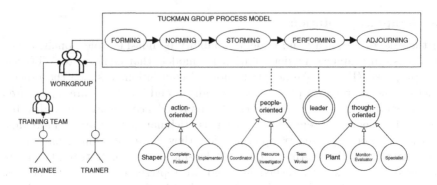

Fig. 3. Example of the model conforming to MTD

transform given constructions expressed within source metamodel (MTD) in semantically equivalent constructions of the other (AOM).

The P mapping pattern is a mapping that allows you to map the structural elements of a source metamodel to the structural elements in the target metamodel, such that:

$$P : E_{M_{src}} \mapsto E_{M_{trg}}, E_M = K_M \cup S_M, \tag{1}$$

where:

K_M – set of all syntactical categories of M metamodel,
S_M – set of all syntactical constructions possible to create over M metamodel,
E_M – set of all structural elements of M metamodel,
M_{scr} – source metamodel,
M_{trg} – target metamodel.

Thus, we can define $C_{M_{src} \mapsto M_{trg}}$ as follows:

$$C_{M_{src} \mapsto M_{trg}} = \{P_{(1)}, \ldots, P_{(n)}\}, n \in \mathbb{N} \tag{2}$$

where each of P_i is a procedure transforming conceptual MTD construction into physical AOM data structure. In this study, we propose the methodology, when group process conceptual design patterns are identified and subsequently mapped into their representation in AOM. We can identify the following MTD design patterns:

SoP (*Structure of the Process*) – this pattern combines the elements classified by `Process` metaclass with its composed elements classified by `Phase` metaclass and their order incuced from *predecessor_successor* metaassociation.
RT (*Role Taxonomy*) – this pattern describes the elements classified by `Role` metaclass taking into consideration the *gen_spec* metaassociation.
SoT (*Structure of the Team*) – this pattern describes the elements classified by `Agent` metaclass, including the aggregated elements in elements classified by `Team` metaclass.

GN (*Goal Network*) – this pattern describes the elements classified by Goal metaclass, including directed relationships between goals described by Goal Relation association metaclass.

4.1 Mapping Procedures

In order to express fixed syntactical constructions of MTD by the use of association-oriented structures, authors proposed procedures of mapping the conceptual design patterns SoP, RT, SoT, GN to corresponding structures in AOM. To provide the notation of AOM expressions the AOM formalism[1] has been used in combination with pseudocode.

The procedures of mapping patterns for each of the design pattens has been shown in the Algorithms 1–4. As the input of each of the procedures is passed a structure denoted as p. This symbol means an instantiation of concrete design pattern. It has to be separated from the model conforming to MTD and comprise the elements which are part of the design pattern. The output of the procedure are the elements of intensional part of AOM data model which is composed of associations, roles and collections. The data model itself implements some meta-classes of the conceptual model as abstract nullary associations. As described in [8], inheriting from nullary associations is used for type substitutability. This helps to preserve the semantics contained in conceptual types of the MTD concepts.

Algorithm 1. Procedure of mapping SoP to AOM metamodel

---

procedure SoPToAOM(p)   ▷ p – SoP pattern instance

    proc ← p.Process.name

    p.Process ↦ $\Diamond proc\,[1] \xleftrightarrow{+proc} [1]\,\square\,proc$

    for each phase ∈ p.Process.''consists of'' **do**

        phName ← phase.name

        phase ↦ $\Diamond phName\,[1] \xleftrightarrow{+phName} [1]\,\square\,phName$

        $CreateRole\left(\Diamond phName, \xrightarrow{successor} \Diamond\,(phase.successor.name)\right)$

        $CreateRole\left(\Diamond phName, \xrightarrow{predecessor} \Diamond\,(phase.predecessor.name)\right)$

    end for

end procedure

---

5 Summary

In this paper we have presented the method for conceptualization of team development. The conceptualization is focused on the creation of the metamodel of

---

[1] See [13] for syntax and semantics description.

Algorithm 2. Procedure of mapping RT to AOM metamodel

---

procedure RTToAOM(p)   ▷ p – RT pattern instance
   for each r ∈ p.Roles **do**   ▷ Taxonomy is represented as a list of roles
       $r \mapsto \Diamond(r.name)\,[1] \overset{+(r.name)}{\longleftrightarrow} [1]\,\Box\,(r.name)$
       if r.gen **is not null then**
          $\Diamond(r.name) \overset{r,f^v}{\longrightarrow} \Diamond(r.gen.name)$
       end if
       for each ph ∈ r.''occurs in'' **do**   ▷ Connect roles to phases
          $CreateRole\Big(\Diamond(ph.name), \overset{occursIn}{\longleftrightarrow} \Diamond(r.name)\Big)$
       end for
       if Informal Role ∈ instanceOf(r) **then**   ▷ Role is informal
          $\Diamond(r.name) \longrightarrow \Diamond InformalRole$
       else if Formal Role ∈ instanceOf(r) **then**   ▷ Role is formal
          $\Diamond(r.name) \longrightarrow \Diamond FormalRole$
       end if
   end for
end procedure

Algorithm 3. Procedure of mapping SoT to AOM metamodel

---

procedure SoToAOM(p)   ▷ p – SoT pattern instance
              ▷ Structure is represented as the top-level agent
    $p.Agent \mapsto \Diamond(p.Agent.name)\,[1] \overset{+(p.Agent.name)}{\longleftrightarrow} [1]\,\Box\,(p.Agent.name)$
   if Team ∈ instanceOf(p.Agent) **then**   ▷ Agent is a team
       for each proc ∈ p.Agent.''involves'' **do**
          $CreateRole\Big(\Diamond(p.Agent.name), \overset{involves}{\longleftrightarrow} \Diamond(proc.name)\Big)$
       end for
       if Temporal Team ∈ instanceOf(p.Agent) **then**
          $\Diamond(p.Agent.name) \longrightarrow \Diamond TemporalTeam$
          $CreateAttribute\,(\Box\,(p.Agent.name)\,, start\_date : datetime)$
          $CreateAttribute\,(\Box\,(p.Agent.name)\,, end\_date : datetime)$
       else
          $\Diamond(p.Agent.name) \longrightarrow \Diamond Team$
       end if
       for each ag ∈ (Team)p.Agent.''consist of'' **do**
          SoToAOM(pattern{ ag })
       end for
   else if Individual ∈ instanceOf(p.Agent) **then** ▷ Agent is an individual
       $\Diamond(p.Agent.name) \longrightarrow \Diamond Individual$
   end if
end procedure

Algorithm 4. Procedure of mapping GN to AOM metamodel

procedure GNToAOM(p)   ▷ p – GN pattern instance
   for each g ∈ p.Goals **do**

$$g \mapsto \Diamond(\text{g.name})\,[1] \xrightarrow{\;+(\text{g.name})\;} [1]\,\square\,(\text{g.name})$$

      if Personal Goal ∈ instanceOf(g) **then**   ▷ Goal is personal
$$\Diamond(g.name) \longrightarrow \Diamond\, PersonalGoal$$
      end if
      if Group Goal ∈ instanceOf(g) **then**   ▷ Goal is group
$$\Diamond(g.name) \longrightarrow \Diamond\, GroupGoal$$
      end if
      for each gr ∈ g.''is related to'' **do**

$$gr \mapsto \Diamond(gr.name)\,[1] \xrightarrow{\;+(gr.name)\;} [1]\,\square\,(gr.name)$$

$$CreateRole\left(\Diamond(gr.name), [*] \xrightarrow{\;+relating\;} [*]\,\Diamond\,(g.name)\right)$$

$$CreateRole\left(\Diamond(gr.name), [*] \xrightarrow{\;+related\;} [*]\,\Diamond\,(rg.Goal)\right)$$

      end for
      for each a ∈ g.has **do**

$$CreateRole\left(\Diamond(a.name), [*] \xrightarrow{\;+has\;} [*]\,\Diamond\,(g.name)\right)$$

      end for
   end for
end procedure

---

group process models. We have also prepared the method of conceptual model transformation to data structures called association-oriented models. Hence, the paper touches the following issues: 1. it constitutes the literature review in terms of theories, models and approaches for group development modeling, 2. it presents novel approach in terms of modeling techniques for group development by moving it into higher level in a formal way, 3. it constitutes a case study of the realization of information system model, beginning from domain analysis, by its conceptualization, to the implementation of transformation procedures into target data model, expressed by the use of novel approach to data modeling: Association-Oriented Metamodel.

As the future work we plan to conceptualize behavioral aspects of group dynamics. In particular, methods for evaluation of measurable team development factors will be elaborated. They will allow to observe such group process measures as e.g. degree of goal fulfillment, evaluation of synergy, level of creativity. For this reason subsequent stages of work will focus on the behavioral modeling of group processes as well as creation of research tools compatible with developed method in order to empirically verify the models in group environments such as corporate, therapeutic, academic and others.

References

1. Belbin, R.M.: Team Roles at Work. Routledge, New York (2012)
2. Bennis, W.G., Shepard, H.A.: A theory of group development. Hum. Relat. **9**(4), 415–437 (1956)
3. Brehm, J.W., Mann, M.: Effect of importance of freedom and attraction to group members on influence produced by group pressure. J. Pers. Soc. Psychol. **31**(5), 816–824 (1975)
4. Chidambaram, L., Bostrom, R.: Group Decis. Negot. **6**(2), 159–187 (1997)
5. Fine, G.A.: Tiny Publics: A Theory of Group Action and Culture. Russell Sage Foundation, New York (2012)
6. Hackman, J.R., Katz, N.: Group behavior and performance. In: Handbook of Social Psychology (2010)
7. Heslin, R., Dunphy, D.: Three dimensions of member satisfaction in small groups. Hum. Relat. **17**(2), 99–112 (1964)
8. Jodłowiec, M.: Complex relationships modeling in association-oriented database metamodel. In: Nguyen, N.T., Hoang, D.H., Hong, T.-P., Pham, H., Trawiński, B. (eds.) ACIIDS 2018. LNCS (LNAI), vol. 10752, pp. 46–56. Springer, Cham (2018). https://doi.org/10.1007/978-3-319-75420-8_5
9. Kaplan, S.R., Roman, M.: Phases of development in an adult therapy group. Int. J. Group Psychother. **13**(1), 10–26 (1963)
10. Kerr, N.L., Tindale, R.S.: Group performance and decision making. Annu. Rev. Psychol. **55**(1), 623–655 (2004)
11. Keyton, J., Stallworth, V.: On the verge of collaboration: identifying group structure and process. In: Group communication in context: studies of bona fide groups, pp. 235–260 (2002)
12. Konieczka, S.: Debating hating: the response and responsibility of a student government. In: Group communication: cases for analysis, appreciation, and application, pp. 167–174 (2010)
13. Krótkiewicz, M.: A novel inheritance mechanism for modeling knowledge representation systems. Comput. Sci. Inf. Syst. **15**(1), 51–78 (2018)
14. McGrath, J.E.: Groups: Interaction and Performance, vol. 14. Prentice-Hall, Englewood Cliffs (1984)
15. Moreland, R.L.: Are dyads really groups? Small Group Res. **41**(2), 251–267 (2010)
16. OMG: OMG Unified Modeling Language (OMG UML). Technical report, Object Management Group, December 2017. http://www.omg.org/spec/UML/2.5.1
17. Sherif, M.: Experiments in group conflict. Sci. Am. **195**(5), 54–59 (1956)
18. Slater, P.: Microcosm: structural, psychological, and religious evolution in groups (1966)
19. Tuckman, B.W., Jensen, M.A.C.: Stages of small-group development revisited. Group Organ. Stud. **2**(4), 419–427 (1977)
20. Turner, J.C.: Towards a cognitive redefinition of the social group. In: Social Identity and Intergroup Relations, pp. 15–40 (1982)
21. Zabawa, P., Hnatkowska, B.: CDMM-F – domain languages framework. In: Świątek, J., Borzemski, L., Wilimowska, Z. (eds.) ISAT 2017. AISC, vol. 656, pp. 263–273. Springer, Cham (2018). https://doi.org/10.1007/978-3-319-67229-8_24

Ontological Information as Part of Continuous Monitoring Software for Production Fault Detection

Marek Krótkiewicz[1]([✉]) [ID], Krystian Wojtkiewicz[1] [ID], Marcin Jodłowiec[1] [ID], Jan Skowroński[1,2], and Maciej Zaręba[1,2]

[1] Faculty of Computer Science and Management,
Wrocław University of Science and Technology, Wrocław, Poland
marek.krotkiewicz@pwr.edu.pl
[2] DSR S.A., Wrocław, Poland

Abstract. The monitoring of manufacturing processes is an important issue in nowadays ERP systems. One of the most important issues is to identify and analyze appropriate data for each of the production units taking part in the process. In the paper authors introduce a new approach towards modelling the relation between production units, signals and factors possible to obtain from the production system. The main idea for the system is based on the ontology of production units. The design of the system using advanced knowledge engineering is elaborated. Since, the implementation of proposed system was one of key assumptions, the relational model is presented that ensures possibility to deploy the system in the future.

Keywords: Manufacturing operation management · OWL · Ontology implementation · Ontology modeling

1 Introduction

Continuous monitoring of the production process consists in collecting and analyzing big volume of data. Nowadays, manufacturing industries are becoming increasingly competitive in use of data in order to achieve and maintain high productivity and quality. At the manufacturing operations management (*MOM*) level of an enterprise, data acquisition is carried out using manufacturing execution systems (*MES*) in which many different approaches can be used in data analysis process. However, there is an International Standard for KPIs (*Key Performance Indicators*) – ISO 22400 [3]. Those KPIs are used for comparing enterprise operations over extended periods of time in order to evaluate production efficiency and its cost. We know how to measure the production efficiency but that's not enough. In order to increase productivity we need to know which elements have influence on it and how strong it is. Furthermore, we have to take control of elements, which affect the manufacturing resources. Therefore, *Production Unit Performance Management Tool* has been proposed and is described

© Springer Nature Switzerland AG 2019
N. T. Nguyen et al. (Eds.): ACIIDS 2019, LNAI 11432, pp. 89–102, 2019.
https://doi.org/10.1007/978-3-030-14802-7_8

in detail in this paper. The main contribution of proposed solution is the ability to describe each production unit taking part in production process with detail. This in effect allows to identify elements that directly influence productivity.

The paper is organize as follows. The second section is a study of related work in terms of OEE importance and use of knowledge base solutions in ERP systems. Next, the system architecture is introduced that is later supplemented by description of OWL ontology used for definition of key system elements. The system description concludes with the implementation of the system data structure in relational model. Last section is a summary.

2   Related Work

It is a quite obvious nowadays that to remain competitive in a global market companies must continuously improve their productivity. This truth works even more among manufacturing companies, where competition is focused on minimizing of production costs and effectiveness of equipment plays a major role to achieve higher productivity [12].

One of the necessary conditions to achieve this goal is using advanced Enterprise Resource Planning (*ERP*) system which is meant to increase the organizational efficiency by enhancing the information processing capability of the enterprise [11]. But improving of effectiveness requires also a reliable and accurate measurement which can bu used as a compass on the road to increasing efficiency. This indicator should let organizations to measure and compare their efficiency in different moments and across different departments but also should help them understand what they should do to improve it and in consequence help to achieve the desired outcomes [2].

2.1   OEE Importance and Reliability

The Total Productive Maintenance (*TPM*) concept, launched by Nakajima ([10]), provided a quantitative metric called Overall Equipment Efficiency Index (*OEE*) which became one of the simplest and popular way to measure of the efficiency or effectiveness in manufacturing companies [9]. OEE is calculated by multiplying the values of following three key production factors: Availability (time), Performance (quantity) and Quality (rejects). The main strength of the OEE index is in highlighting areas of improvement and in ability to perform diagnostic [2,15].

OEE is not only metric, but it also provides a framework to improve the entire process [1]. Automating OEE gives a company the ability to collect and classify data from the shop floor into knowledge base that can help managers understand the root causes of production inefficiency. Integration of OEE into general tools as ERP ones give greater visibility to make more informed decisions on process improvement [2].

Good understanding of OEE concept should help companies during the ERP implementation process [5]. It could help also to design knowledge base which help users understand the current situation and provide useful tool to make reliable forecasts [11].

2.2   Knowledge Base in ERP Solution

In general, a knowledge base is not a static collection of information, but a dynamic resource [18]. This resource itself can have the capacity to learn. A well-organized knowledge base can save enterprise money. For example as a customer relationship management (*CRM*) tool, a knowledge base can provide customers easy access to information that would otherwise require contact with an organization's staff [16].

Enterprise Resource Planning (*ERP*) systems integrate business functions throughout the entire enterprise by facilitating the flow of information across the departments [8]. Usually ERP systems run off a single database and enable firms to better manage their knowledge by control of data in the organization. Thus, the ERP implementation is very knowledge-intensive and success of the project relies heavily upon effective management of knowledge into implementation team [5,11,13].

3   System Architecture

The aim of the developed system is the continuous diagnosis and monitoring of production processes. Operating on specific signal data acquired from production units, the expert configures the diagnostic system to capture process-specific semantics from measured values. The system performs a chain of transformations, and finally enables monitoring capabilities with data visualization. The data flow in system's process is depicted in the Fig. 1.

The diagnosis part consist of a chain of transformations of empirically measured production signal. The system transforms the signal values into classes based on defined semantic criteria, which combined with signal data or stream constitute a signal configuration. Each of the measured signal values is deterministically mapped to a class. Then, a sequence of classes is transformed into the sequence of class transitions. The expert can choose, which class transitions are irrelevant to the process and these are subsequently removed. This is applied to all of the input and output signals and the derived factors (e.g. computed indicators). The next step involves the computation of coincidence between each pair of the class transition sets of analyzed signals. The pairs of input and output signals, which have the coincidence value greater than the threshold given than expert are detected as potentially dependent. If the result is not acceptable, then the signal configuration is verified and refined.

The other module of system is the capability for signal monitoring. This part is responsible for real-time messaging the information about potentially risky alternations in the levels of monitored signals.

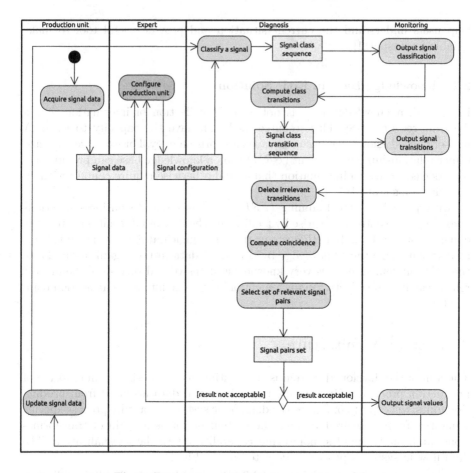

Fig. 1. Diagnosis and monitoring process model

4   Knowledge Representation in OWL Based Ontology

4.1   Representation of Domain Concepts

Well designed ontologies play a significant role in knowledge-based business applications. It is extremely crucial to develop the business domain ontology, which captures knowledge about the domain of interest, before the software application is designed. An ontology describes the concepts in the domain and the relationships between those concepts.

In order to evaluate an ontology of the production units, the OWL language was used, which had become a W3C recommendation in February 2004 [17]. The production units ontology describes a hierarchical structure of the production units, named also *Production Unit Classification*, as well as a hierarchical structure of key performance indicators, simple signals and status reasons of particular state of a production unit. Those four elements are structured as the highest level classes in the taxonomy of the ontology (Fig. 2).

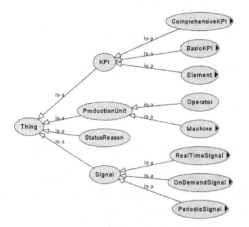

Fig. 2. Main taxonomy of the ontology

ProductionUnit class describes elements of evaluated *Production Unit Classification*. It contains two main types of production units: operators and machines based on ISO 22400 [3]. Machines are divided into 2 kinds, 4 classes, 33 categories and 201 groups at the lowest level. One of the 4 classes are technological machines, which are described by KPIs.

KPI class represents *Key Performance Indicators*. All KPIs used in the ontology are selected from ISO 22400 [4] and describe both a single production unit and a cost of its' efficiency. KPIs are divided into three disjoint classes which describe: *comprehensive KPIs* – the most complex indicators which are defined using *basic KPIs* (e.g. OEE), *basic KPIs* - reveal an aspect of performance for a production unit, derived from *supporting elements*, *supporting elements* - direct measurements derived from the monitored data (*signals*) [6].

Signal class is the conceptual representation of all objects which have influence on a single production unit. This class includes *control signals* (e.g. operator, power, speed), *noise signals* which are not controlled (e.g. temperature, humidity, vibrations), *output signals* which can be observed (e.g. *PQ* - produced quantity, *SQ* - scrap quantity, failure) and can be *supporting elements* as well. In the Fig. 3 the polymorphism in the ontology is shown.

StatusReason class allows to assign specified reason to the signal occurrence. For example a failure can be caused by a crack in the construction element of a machine.

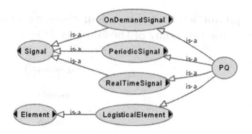

Fig. 3. Polymorphism in signal definition

4.2 Object Properties and Datatype Properties

In the production systems there are many connections and relationships between its' elements, which can be described in OWL using *object properties*. In the ontology several dozen object properties are defined in order to describe main dependencies between production units, KPIs, *signals* and *status reasons* (e.g. *describedBy, describes, affects, dependsOn, equals, isCausedBy, hasComponent, isComponentOf, worksOn*). Values can be stored using *datatype properties* such as *hasValue, timestamp* and *humanFactor*, which tells us how strong of an impact does a human have on the operation of the machine. Figure 4 shows how easily *dependsOn* property can describe direct impact between values of KPIs.

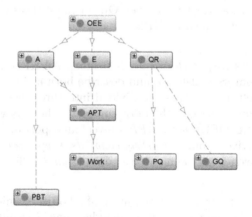

Fig. 4. Direct impact between values of KPIs

5 Data Layer in Relational Model

Implementation of the system presented in Sect. 3, that is dedicated for the use in manufacture environment, has to be efficient. It is supposed to be working close to real time, thus it requires the data structure appropriate to this task.

OWL knowledge bases are great for flexible definition of structures and logical inference, however they are far from relational databases solutions in case of their speed and reliability. Presented system is build in the way that it uses knowledge stored in ontological structures for detection tool as well as provides for monitoring aspect. After thorough research on various available database solutions it has been decided that the data layer of the system will constitute of two modules (Fig. 5). First of them is the relational database used for the storage of ontology defined in Sect. 4. The main categories of information stored there include production units definition, signals and factors. The second module is a file storage for signals and factors.

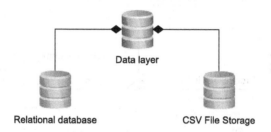

Fig. 5. Data layer architecture of the proposed system

5.1 Main Packages Included in the Relational Schema

The part of the data layer implemented as the relational database is the basis of the system's operation. As part of its structures, 31 tables were distinguished, the in-depth analysis of which goes beyond the scope of this study. Nevertheless, it was decided to present the concept and basic patterns that served to transfer the ontological structures to the relational metamodel. The structure distinguishes 3 main elements, namely *production units*, *signals* and *factors*. The analysis is performed on appropriate data that is defined within *analysis environment*. The description of the relational structure includes its conceptual model, which has been modeled by the use of the Unified Modeling Language [14] due to the complexity of thought constructions used. The final relational database schema stems from the process of transformation of the recurring patterns. The UML class diagram depicting the relationships among identified semantic elements has been presented in the Fig. 6. Two main associations PAS and PAF provide complex relationship that connect *Production unit*, *Analysis Environment* and *Signal*, as well as, *Production unit*, *Analysis Environment* and *Factors* appropriately. Each association should be understood as a set of triples built from objects of classes that are connected to them. The n-ary relationships should not be considered as any kind of binary relationships set, as it was clearly depicted in [7].

Following subsections provide the description of relational database parts (modules) that were designed to store information in regard to identified elements and their relations. Due to complexity of actual schemata only appropriate

tables will be shown on diagrams for each module elaborated. For the simplification sake, all attributes that are not used as key values will be omitted. The
first attribute in each and every table is a primary key, while the foreign keys
names were taken from the source tables names.

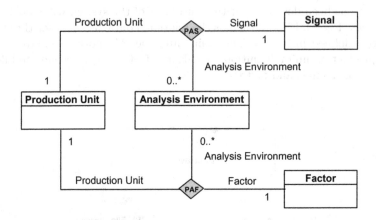

Fig. 6. UML class diagram depicting simplified conceptual database schema

Analysis Environment. The design of this module (Fig. 7) provides definition
of the dimensions, and then the units describing them. For those entities it is
possible to create the dependency structures built on the basis of the predefined binary relationships. The module allows also for specification of values for
each unit of dimension, e.g. for the time unit of *daysofweek* the values will be
sequentially listed, i.e. Monday, Tuesday, Wednesday etc. In most cases time will
be the basic dimension for analysis to be performed and the examples of units
for it are: year, month, day of month, day of week, hour. The construction of
this module has been inspired by the approach towards analysis of data in data
warehouses. Thus it is possible to define dimension of e.g. operators, material,
etc. That allows to define analysis environment with those custom dimension.

Signals. This is one of main modules used for storage of ontological knowledge.
Signals are understood as series of values describing measured or calculated
quantities in previously determined units. Most often they constitute a dynamic
description of the observed phenomenon at specific time points. Sequences of
values are stored in files, while access paths are stored in the database. These
files store data that has been taken from the ERP system, production management system or directly from the production monitoring system. The rules
for storing data in CSV files depends on specific implementation and it won't
be elaborated here. As such signals describe the dynamics of a given production unit both in terms of its work parameters and its effects. The assumption

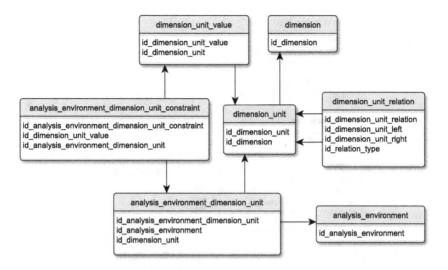

Fig. 7. Analysis environment module database schema part

resulting directly from the ontology design states that the system should provide dynamic and unrestricted definition of signal properties. These properties should be possible to be added in any manner and at any time without the need of database structure modification. The design pattern that has been used for introducing solutions that would satisfy such needs involves definition of certain *type* tables and appropriate *value* tables. The conceptual structure of the module is presented in the Fig. 6 and it presents relationships between conceptual entities elected in the process of ontology modeling.

Production Unit. The main goal of this module is to introduce functionality of production unit definition. It involves defining types of production units along with taxonomic hierarchy of types, possibility to add instances of types, and possibility to assign signal types as well as factor types (Fig. 9). It has to be stated that this system provides detection and monitoring tool as it was defined in Sect. 3 and as such it has no obligation to introduce advanced methods for production unit description, especially in terms of their properties and features. However, it is assumed that each and every production unit instance that appear in this system will be defined and easily identified within the scope of any ERP software being used in the company.

Factor. The tool designed operates on factors that characterize the work of production unit. In this module the types of factors are defined according to taxonomy introduced in the ontology design phase. It is assumed that each factor can be defined as *statistical quantity*, *work factor* or *classifier*. Each type has its own purpose in the system. In this module the definition on how to obtain each

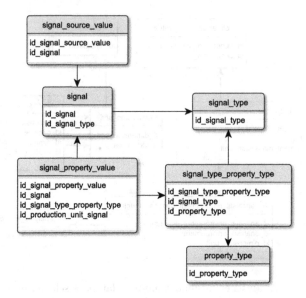

Fig. 8. Signals module database schema part

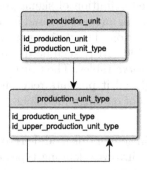

Fig. 9. Production units module database schema part

and every on of them is provided. However, as for relational database schema, this module is represented by only one table, namely *factor_type* (Fig. 8).

5.2 Production Unit – Signal

The purpose of this module is to store information in regard to which signal is assigned to a particular production unit. It is done on two levels. In the higher one the types are connected, i.e. production unit type with the signal type, while on the lower one the actual instances of production units are assigned to available signals (Fig. 10).

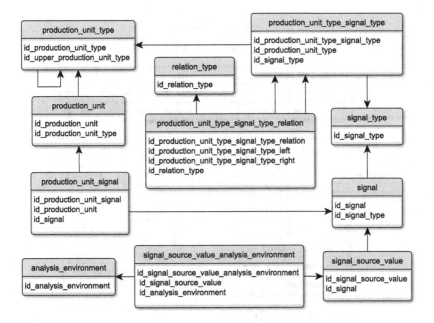

Fig. 10. PAS module database schema part

5.3   Production Unit – Factor

In the last one of the distinguished modules factors are assigned to production units. The functionality is same as with signals. At first the connection is defined at the level of types, so that it could be used for identifying the connections among instances in the system (Fig. 11).

6   Summary

The paper introduces a new approach towards analysis of production processes. The sole idea of the analysis is the identification of process elements that directly influence the productivity. One of the main issues in case of production processes is proper description of their complexity. ERP systems provide synthetic information on inputs and outputs of the process, however they do not concern physical features monitoring of elements taking part in the process, namely production units. This in turn is mainly in the focus of production management systems. Proposed solution aims at feeling the gap between those two types of systems. The idea stems from the use of knowledge engineering techniques that utilizes ontology design for definition of complex relations that occur among production units, signals and factors. The paper introduces ontology that is used for storing definition of production units, signals and factors. Since, the solution aims at implementation in actual production environment the hybrid database solution has been proposed. It is based on file base holding the values of signals and

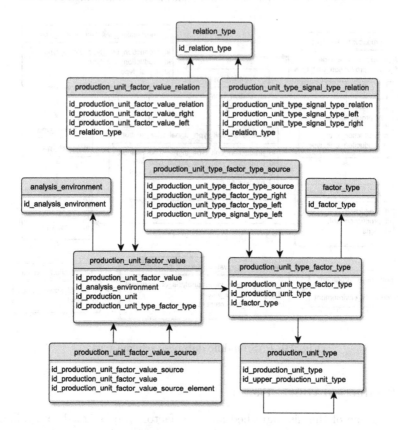

Fig. 11. PAF module database schema part

factors, and a relation database that stores ontology definition and information in regard the implementation. Implementation here is understood as features of production units used in production process as well as definition of signals and factors that are crucial from the analysis point of view.

The solution elaborated in this paper aims at using appropriate structures based on the needs. The future work will focus on performance analysis in computing factors. Another issue is a decision support system that uses the proposed structure as a source of information. One of the drawbacks of the solution is lack of unified APIs that might be used for importing data to the system, that leads to the need of building customized interfaces for each source independently.

Acknowledgements. This work has been created based on the results of the project *Production Unit Performance Management Tool (PUPMT)* co-financed by the European Union under the European Regional Development Fund, based on a contract between DSR S.A. and The National Centre for Research and Development in Poland. Project No. POIR.01.01.01-00-0687/17.

References

1. De Ron, A., Rooda, J.: OEE and equipment effectiveness: an evaluation. Int. J. Prod. Res. **44**(23), 4987–5003 (2006)
2. Iannone, R., Nenni, M.E.: Managing OEE to optimize factory performance. In: Operations Management. InTech (2013)
3. ISO: International Standard ISO 22400-1. Automation Systems and Integration - Key Performance Indicators (KPIs) for Manufacturing Operations Management - Part 1: Overview, Concepts and Terminology. Standard, International Standard Organization (ISO), Geneva (2014)
4. ISO: International Standard ISO 22400-2. Automation Systems and Integration - Key Performance Indicators (KPIs) for Manufacturing Operations Management - Part 2: Definitions and Descriptions. Technical report, International Standard Organization (ISO), Geneva (2014)
5. Jones, M.C.: Tacit knowledge sharing during ERP implementation: a multi-site case study. Inf. Res. Manag. J. (IRMJ) **18**(2), 1–23 (2005)
6. Kang, N., Zhao, C., Li, J., Horst, J.: A hierarchical structure of key performance indicators for operation management and continuous improvement in production systems. Int. J. Prod. Res. **54**(21), 6333–6350 (2016)
7. Krótkiewicz, M.: Association-oriented database model - n-ary associations. Int. J. Softw. Eng. Knowl. Eng. **27**(02), 281–320 (2017). https://doi.org/10.1142/S0218194017500103
8. Markus, M.L., Tanis, C., Van Fenema, P.C.: Enterprise resource planning: multisite ERP implementations. Commun. ACM **43**(4), 42–46 (2000)
9. Muchiri, P., Pintelon, L.: Performance measurement using overall equipment effectiveness (OEE): literature review and practical application discussion. Int. J. Prod. Res. **46**(13), 3517–3535 (2008)
10. Nakajima, S.: Introduction to TPM: total productive maintenance (preventative maintenance series). Hardcover. ISBN 0-91529-923-2 (1988)
11. Newell, S., Huang, J.C., Galliers, R.D., Pan, S.L.: Implementing enterprise resource planning and knowledge management systems in tandem: fostering efficiency and innovation complementarity. Inf. Organ. **13**(1), 25–52 (2003)
12. Ramesh, C., Manickam, C., Prasanna, S.: Lean six sigma approach to improve overall equipment effectiveness performance: a case study in the indian small manufacturing firm. Asian J. Res. Soc. Sci. Humanit. **6**(12), 1063–1072 (2016)
13. Robey, D., Ross, J.W., Boudreau, M.C.: Learning to implement enterprise systems: an exploratory study of the dialectics of change. J. Manage. Inf. Syst. **19**(1), 17–46 (2002)
14. Selic, B., et al.: OMG unified modeling language (version 2.5). Technical report, ODMG, March 2015
15. Sharma, A.K., Shudhanshu, A.B.: Manufacturing performance and evolution of TPM. Int. J. Eng. Sci. Technol. **4**(03), 854–866 (2012)
16. Vandaie, R.: The role of organizational knowledge management in successful ERP implementation projects. Knowl. Based Syst. **21**(8), 920–926 (2008)

17. W3C: OWL Web Ontology Language Reference, Dean, M., Schreiber, G. (eds.) w3c recommendation, 10 February 2004. http://www.w3.org/tr/2004/rec-owl-ref-20040210/. Latest version available at http://www.w3.org/tr/owl-ref/. Technical report, W3C (2004)
18. Krótkiewicz, M., Wojtkiewicz, K., Jodłowiec, M.: Towards semantic knowledge base definition. In: Hunek, W.P., Paszkiel, S. (eds.) BCI 2018. AISC, vol. 720, pp. 218–239. Springer, Cham (2018). https://doi.org/10.1007/978-3-319-75025-5_20

Social Media Influence as an Enabler of a Sustainable Knowledge Management System Inside PT. ABC Organization

Benhard Hutajulu$^{(\boxtimes)}$ and Harisno$^{(\boxtimes)}$

Information System Department, BINUS Graduate Program-Master's
in Information System Management, Bina Nusantara University Jakarta,
West Jakarta, Indonesia
bhutajulu@gmail.com, harisno@binus.edu

Abstract. The goals of this research were to know the influences of social media as enabler for sustainable knowledge management inside organization of PT. ABC. Data were collected from 40 respondents and analyzed using multiple linear regression and Pearson correlation. It can conclude that the six variables—attention, personalization, context, integration, knowledge management lifecycle and outcome—on social media have influence to sustainable knowledge management inside the PT. ABC organization.

Keywords: Social media · Knowledge management ·
Sustainable knowledge management

1 Introduction

Much social media information that can be used as a reference supports a company's knowledge management system [1]. Processed data from social media interactions also becomes usable information and then knowledge [2]. Currently, in the new form, the PT. ABC company does not have a knowledge management system, and it relies on employees' knowledge that supports PT. ABC's use of social media.

The continuously existing direct knowledge management system that is free from management's control is defined as sustainable knowledge management [3]. There are six variables that can be used to achieve sustainable knowledge management [3]: attention, personalization, context, integration, knowledge management lifecycle, and outcome. This study focuses on the correlation between these six social media variables with sustainable knowledge management. The goal of this study is to analyze all six variables on social media to establish if they can be used as enablers for a sustainable knowledge management system.

The anticipated benefits of the results are to establish whether social media usage from an employee's knowledge and all the six variables can be used as supporting factors in the organization when it wants to develop a knowledge management system.

© Springer Nature Switzerland AG 2019
N. T. Nguyen et al. (Eds.): ACIIDS 2019, LNAI 11432, pp. 103–114, 2019.
https://doi.org/10.1007/978-3-030-14802-7_9

2 Literature Review

2.1 Social Media

Creativity published via social media platforms contributes to the knowledge gathered [1]. Design platforms on social media also make computers with an internet connection a communication tool that is available for anyone [4]. The five characteristics based on the activities of social media users—participation, openness, conversation, community, and connectedness [5]—make the technical features attributes on social media, which can benefit in building relations to encourage only communities and deliver knowledge to users.

2.2 Knowledge Transfer

The following are five mechanisms of knowledge transfer: (i) Serial transfers occur when knowledge is transferred from one team to another; (ii) Near transfers happen where a group gains knowledge from another team's tasks to do in similar work; (iii) Far transfers happen when the team develops knowledge from available non-routine tasks, such as typical work from other parts in the organization; (iv) Strategic transfers occur when the team requires collective knowledge to complete rare tasks that are important for the entire organization; (v) Expert transfers happen when the team faces something beyond their knowledge and requires other experts in the organization [6]. Social media offers a promising arena to increase motivation levels of employees and to exchange personal knowledge that ensures the efficiency of interactions [1]. Through these five social media mechanisms, knowledge can transfer to each user, from each user to a group and vice versa, and from one group to another.

2.3 Knowledge Management in Social Media

Knowledge development, use and sharing on social media deals with knowledge evolution, reuse, and participation and support the processes of knowledge management [1]. The ability of social media to allow communication, collaboration, and information exploration, and as a place to learn, carries a significant role in knowledge management. Knowledge management applications benefit information technology too by improving individual performances, improving the organization's business performance, and increasing internal data organization [7].

2.4 Sustainable Knowledge Management System

The translation of an information system's knowledge use as current information and data collection updates refers to the sustainability of a company's knowledge [2]. Sustainable knowledge management reduces the dependence on knowledge championing, monitoring, and reconstruction [3]. The connection between information systems and knowledge management will embody the knowledge cycle on social media as a place for data collection, management and updated. This knowledge cycle will run and be sustained in social media, to make it an enabler of sustainable knowledge management systems achieved with the six variables: attention, personalization, context, integration, knowledge management lifecycle, and outcome.

3 Method

3.1 Framework

The study's first stage highlights the issues raised. The second stage discusses concepts and theories developed to supply solutions to issues raised in the organization. The third stage performs a discussion, interviews, and direct observation to obtain the vision and scope of knowledge management. The fourth stage creates hypotheses. The fifth stage creates a research model. The questionnaire is created and distributed among respondents in the sixth stage. The seventh stage forms the respondent data collection. The eighth stage analyzes data using SPSS software. The ninth stage tests the hypotheses. The tenth stage formulates the conclusions and suggestions for future work (Fig. 1).

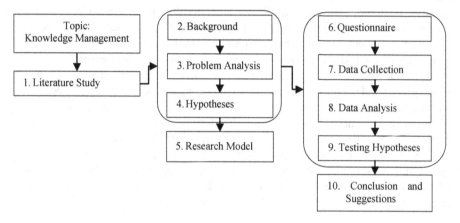

Fig. 1. Framework of the study

3.2 Population and Sample

PT. ABC organization has 40 employees. The researcher distributed the questionnaire to all employees who use social media in their daily work activities. The census method used for saturation means all employees data were sampled.

3.3 Research Model and Variable

The model research in this case study was developed from earlier research on a sustainable knowledge management system [3] as follows:

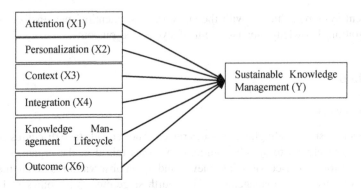

Fig. 2. Construct research model

The research study was conducted to analyse the dependent variable, Y, as sustainable knowledge management with the indicators data, information, knowledge [2]; update, creation, criticism, integration, independent [3] against the independent variable, X, as follows: (i) attention (X1), with indicators: personalization, emotionally evocative, trustworthy and easy to digest [3], concentration on the subject of knowledge, selection of knowledge topics [8]; (ii) personalization (X2), with indicators: experience, knowledge sharing, participation, connectedness and usefulness [1], relevance [9]; (iii) context (X3), with indicators: as discourse, as references [3], conversation, interaction; social appearance and commitment [10], place of creation [7]; (iv) integration (x4), with indicators: communication; individual knowledge becoming group knowledge; professional discussions and time and cost savings [11]; recording activity, organize, meeting arrangement, file transfer, as work reference, delivered responsibility, decision making, functional at work [3]; (v) knowledge management lifecycle (x5), with indicators: acquisition, creation, refinement, transfer [12]; mobilization, diffusion, commoditization [3]; (vi) outcome (x6), with indicators: actions, decisions, knowledge increased, newly formed knowledge [3]; learnings, experiences, strategic alignments, connectivity, value management [13] (Fig. 2).

3.4 Hypotheses

The current social media content is an expectation from its use as a support and enabler of sustainable knowledge management. Therefore, the following developed hypotheses:

Attention (X1), H_1: Social media attention correlates with the sustainable knowledge management system. H_0: Social media attention does not correlate with the sustainable knowledge management system.

Personalization (X2), H_1: Social media personalization correlates with the sustainable knowledge management system. H_0: Social media personalization does not correlate with the sustainable knowledge management system.

Context (X3), H_1: Social media knowledge context correlates with the sustainable knowledge management system. H_0: Social media knowledge context does not correlate with the sustainable knowledge management system.

Integration (X4), H_1: Social media integration into daily activities correlates with the sustainable knowledge management system. H_0: Social media integration into daily activities does not correlate with the sustainable knowledge management system.

Knowledge Management Lifecycle (X5), H_1: Social media knowledge management lifecycle correlate with the sustainable knowledge management system. H_0: Social media knowledge management lifecycle does not correlate with the sustainable knowledge management system.

Outcome (X6), H_1: Social media knowledge outcomes correlate with the sustainable knowledge management system. H_0: Social media knowledge outcomes do not correlate with the sustainable knowledge management system.

3.5   Data Collection Technique

The develops questionnaire in this study use a Likert scale. The bipolar scale method measures either positive or negative responses to the statement. In this study, the author supplies five scale choices as follows: Very Agreed is 5; Agreed is 4; Neutral is 3; Not Agreed is 2; Not Very Agreed is 1.

3.6   Data Process Technique

The data obtained from the process was analyzed with the Statistical Package for the Social Sciences (SPSS) version 25 software, which processes statistical data with efficiency and accuracy to supply the necessary output range and create results.

3.7   Data Analysis Technique

Data obtained from the circulated questionnaires were processed to obtain results and offer conclusions. Multiple linear regression and correlation statistical methods were used for the data analysis to assess any relationship between dependent and independent variables.

4   Results and Discussion

4.1   Survey Results

The sample consisted of 40 PT. ABC employees. The hardcopy questionnaires were distributed directly to the prospective respondents, and all were completed with the following result: participation respondents were 57.6% women and 42.5% men. Social media frequency use was 97.5% on a daily basis and 2.5% rarely. All respondents (100%) agreed that there was knowledge support on social media. Employees accessed knowledge primarily through Instagram (82.5%), Twitter (72.5%); and Facebook (65%).

4.2   Validity Test

The purpose of the validity test to establish that the statements in the questionnaire are valid. The correlation, r, calculated score was obtained from the SPSS software data process. The r table score obtained in this study with 40 respondents has 2 degrees of freedom, whereby an "n" score of 38 has a Cronbach's $\alpha = 5\%$, which results in a t table score of 0.312. The required validity statement is as follows: If the r calculated score is greater than the r table score, the statement is valid. If the r calculated score is less than the r table score, the statement is not valid. From the data processed, all the r calculated scores were greater than the r table scores, meaning all statements were valid.

4.3   Reliability Test

The reliability test used was the Cronbach's α score of internal consistency in the SPSS data process. The results compared with the score should be 0.6. The requirement of a variable reliability test is as follows: If the Cronbach's α score is greater than the coefficient score, the variable is reliable. If the Cronbach's α score is less than the coefficient score, the variable is not reliable. From the data processed, all Cronbach's α scores were greater than 0.6, meaning all variables were reliable.

4.4   Normality Test

The regression analysis used in this study requires all error assumptions to be normally distributed. The Kolmogorov-Smirnov (K-S) test was used to assess this. The processed data output describes the normal distribution with criteria as follows: If the absolute K-S count is less than the K-S table, the data is distributed normally. If the absolute K-S count is greater than the K-S table, the data distribution is not normal. A score of 0.210 was obtained from the K-S table from 40 samples.

From Table 1, all absolute scores from the K-S count were less than the K-S table (0.210), meaning all data was distributed normally.

Table 1. Normality Kolmogorov-Smirnov test

		Attention (X1)	Personalization (X2)	Context (X3)	Integration (X4)	KM Life cycle (X5)	Outcome (X6)	SKM (Y)
Normal parameters[a,b]	Mean	3.925	3.858	3.882	3.837	3.864	3.727	3.837
	Std Dev	0.612	0.547	0.526	0.498	0.539	0.573	0.546
Most extreme differences	Absolute	0.149	0.144	0.096	0.103	0.078	0.063	0.136
Asymp. Sig. (2-tailed)		0.026[c]	0.037[c]	0.200[c,d]	0.200[c,d]	0.200[c,d]	0.200[c,d]	0.059[c]

[a]Test distribution is normal
[b]Calculated from data
[c]Lilliefors significant correction
[d]This is a lower band of the true significance

4.5   Multi-co-linearities Test

Multi-co-linearities assess the connection between independent variables and provides a good model regression when two of the variables correlate. Multi-co-linearities are obtained from the VIF (Varian Inflation Factor) score: If the VIF score is greater than 10, the data has multi-co-linearities. If the VIF score is less than 10, then the data has no multi-co-linearities.

From Table 2, all VIF scores were less than 10, meaning there was no multi-co-linearities in the data.

Table 2. Coefficients[a]

	Attention (X1)	Personalization (X2)	Context (X3)	Integration (X4)	KM life cycle (X5)	Outcome (X6)
VIF	2.492	2.439	3.842	4.494	3.993	3.273

[a]Dependent variable: Sustainable knowledge management system (Y)

4.6   Heteroskedasticity Test

The Heteroskedasticity test establishes whether there is any deviation of an inequality variant from the residual regression model. The method used a Glejser test, with criteria as follows: If the significance score is greater than 0.05, then the data has no Heteroskedasticity. If the significance score is less than 0.05, then the data has Heteroskedasticity.

From Table 3, all variables score greater than 0.05, meaning the data had no Heteroskedasticity.

Table 3. Glejser test[a]

	Attention (X1)	Personalization (X2)	Context (X3)	Integration (X4)	KM lifecycle (X5)	Outcome (X6)	SKM (Y)
Significances	0.627	0.284	0.868	0.632	0.995	0.255	0.204

[a]Dependent variable: ABS_RES_1

4.7   Autocorrelation Test

The Durbin-Watson (DW) score is used to detect autocorrelation. The basic provisions for the autocorrelation occurrence are as follows: If the DW is less than 1 and greater than 3, the data has autocorrelation. If the DW is greater than 1 and less than 3, the data has no autocorrelation.

From Table 4, the Durbin-Watson score of 2.205 is in the range of DW > 1 and < 3, meaning the data has no autocorrelation.

Table 4. Model summary[b].

	R	R Square	Adjusted R Square	Std. Error of the estimate	Durbin-Watson
Model	0.863[a]	0.744	0.698	0.30073	2.205

[a]Predictors: Attention (X1), Personalization (X2), Context (X3), Integration (X4), KM
Lifecycle (X5), Outcome (X6)
[b]Dependent variable: Sustainable knowledge management system (Y)

5   Hypothesis Test

5.1   Pearson Correlation Analysis

The Pearson test analyzes the correlation between the dependent and independent
variables with a classification correlation score as follows [14]: Score 1 for perfect
correlation; Score 0.9–0.99 for a very strong correlation; Score 0.7–0.9 for a strong
correlation; Score 0.5–0.7 for a sufficiently strong correlation; Score 0.3–0.5 for a weak
correlation; Score 0.1–0.3 for a very weak correlation; and Score 0–0.1 for no
correlation.

From Table 5, the correlation scores between the dependent and independent
variables are as follows: Attention (X1) score of 0.690 means there is a strong correlation. A Personalization (X2) score of 0.435 means there is a weak correlation.
A Context (X3) score of 0.530 means there is quite a strong correlation. An Integration
(X4) score of 0.694 means there are strong correlations. A Knowledge Management
Lifecycle (X5) score of 0.717 means there is a strong correlation. An Outcome (X6)
score of 0.802 means there is a strong correlation.

Table 5. Pearson correlation

	Attention (X1)	Personalization (X2)	Context (X3)	Integration (X4)	KM Lifecycle (X5)	Outcome (X6)
Pearson correlation	0.690	0.435	0.530	0.694	0.717	0.802

5.2   F Test Analysis

An F test is used to analyze the correlation between the dependent variable and the
independent variables simultaneously with the hypotheses as follows: H_0 = all independent variables simultaneously influence dependent variables and vice versa.
H_1 = all independent variables simultaneously do not influence dependent variables
and vice versa. The F test criteria are as follows: If the F calculated score is less than the
F table, then H_0 is accepted and H_1 is denied. If the F calculated score is greater than
the F Table, H_0 is denied and H_1 is accepted.

The F Table score has a significance of 0.05 and a degree of freedom (df) with a
regression numerator score of 6 and a residual score of 33, so from the F Table, the

Table 6. ANOVA[a]

	Sum of squares	df	Mean square	F	Sig[b]
Regressions	8.678	6	1.446	15.992	0.000
Residual	2.995	33	0.090		

[a]Dependent variable: Sustainable knowledge management system (Y)
[b]Predictors: Attention (X1), Personalization (X2), Context (X3), Integration (X4), KM Lifecycle (X5), Outcome (X6)

score is 2.39. From Table 6, the F calculated score (15.992) is greater than the F Table score (2.39), so H_0 is denied and H_1 is accepted, meaning all independent variables simultaneously influence all dependent variables and vice versa. A significance score of $0.00 < 0.05$ means attention (X1), personalization (X2), context (X3), integration (X4), knowledge management lifecycle (X5), and outcome (X6) simultaneously influence sustainable knowledge management system as an enabler.

5.3 Multiple Regression Analysis

A strong r (correlation) value between variables indicates a significant influence. From Table 4, the model summary showed that all independent variables influenced the dependent variables, as indicated by the r values. The r value 0,863 means the 86.3% of attention (X1), personalization (X2), context (X3), integration (X4), knowledge management lifecycle (X5), and outcome (X6) influences the sustainable knowledge management system (Y). The coefficient r^2 score of 0,744 means that 74.4% of sustainable knowledge management is caused by the six variables and the remainder (16.6%) is caused by other factors.

From Table 7, non-standardized B values form the multiple regression formula as follows: Y = 0.516 + 0.316 X1 + (−0.182) X2 + (−0.153) X3 + 0.335X4 + 0.161 X5 + 0.394X6.

Table 7. Coefficients[a]

	Attention (X1)	Personalization (X2)	Context (X3)	Integration (X4)	KM life cycle (X5)	Outcome (X6)
Non-standardized B	0.316	−0.182	− 0.153	0.335	0.161	0.394
t	2.541	−1.329	0.851	1.636	0.902	2.597

The conclusion from the above formula shows that the regression between the independent and dependent variables is as follows: Attention (X1) has a positive regression of 0.316, meaning attention influences the sustainable knowledge management system. Personalization (X2) has a negative regression of −0.182, meaning personalization does not affect the sustainable knowledge management system. Context

(X3) has a negative regression of −0.153, meaning context has no effect on the sustainable knowledge management system. Integration (X4) has a positive regression of 0.335, meaning integration influences the sustainable knowledge management system. The knowledge management lifecycle variable (X5) has a positive regression of 0.161, meaning the KM lifecycle influences the sustainable knowledge management system. Outcome (X6) has a positive regression of 0.394, meaning outcome influences the sustainable knowledge management system.

5.4   T-Test Analysis

A t-test analyzes the influence of the dependent variable on each of the independent variables. The hypotheses for the t-test are as follows: H_0 = significant coefficient regression; H_1 = no significant coefficient regression, with criteria: If t is calculated be less than the t table, then H_0 is accepted and H_1 is denied. If the t calculated is greater than the t table, then H_0 is denied and H_1 is accepted. From an n of 40 sampled, with an $\alpha = 0.05$, a degree of freedom (DF) = 2 then 40−2 = 38; therefore, the t table = 1.685. From Table 7, the t value was calculated and concluded as follows: Attention (X1): the t value calculated (2.541) is greater than the t table (1.685), so H_0 is denied and H_1 is accepted, meaning attention significantly affects sustainable knowledge management. Personalization (X2): the t value calculated (−1.329) is less than the t table (1.685), so H_0 is accepted and H_1 is denied, meaning personalization does not significantly affect sustainable knowledge management. Context (X3): the t value calculated (−0.851) is less than the t table (1.685), so H_0 is accepted and H_1 is denied, meaning context does not significantly affect sustainable knowledge management. Integration (X4): the t value calculated (1.636) is less than the t table (1.685), so H_0 is accepted and H_1 is denied, meaning integration does not significantly affect sustainable knowledge management. Knowledge management lifecycle (X5): the t value calculated (0.902) is less than the t table (1.685), so H_0 is accepted and H_1 is denied, meaning the knowledge management lifecycle affects sustainable knowledge management. Outcome (X6): the t value calculated (2.597) is greater than the t table (1.685), so H_0 is denied and H_1 is accepted, meaning outcome significantly affects sustainable knowledge management.

6   Conclusion

Attention to knowledge distributed on social media influences employee follow up, and the correlation with sustainable knowledge management (KM) means it acts as an enabler of sustainable knowledge management systems in PT. ABC.

Social media personalization of knowledge gained has no influence on employee follow up; however, personalization does correlate with sustainable KM, thus making it an enabler of sustainable knowledge management systems in PT. ABC.

Knowledge context in social media has no influence on employee follow up; however, the correlation between knowledge context and sustainable KM makes it an enabler of sustainable knowledge management systems in PT. ABC.

Social media integration into daily activities influences employees to follow up and the correlation with sustainable KM makes it an enabler of sustainable knowledge management systems in PT. ABC.

The knowledge management lifecycle on social media influences employees to follow up and the correlation with sustainable KM makes it an enabler of sustainable knowledge management systems in PT. ABC.

Social media knowledge management lifecycle influences employees to follow up and the correlation with sustainable KM makes it an enabler of sustainable knowledge management systems in PT. ABC.

Social media knowledge outcomes influences employees to follow up and the correlation with sustainable KM make it an enabler of sustainable knowledge management systems in PT. ABC.

Thus, all attention, personalization, context, integration, knowledge management lifecycle, and outcome variables simultaneously influence sustainable knowledge management systems, meaning that knowledge of effective social media use can enable sustainable knowledge management systems in PT. ABC. Most employees use the social media platforms, Instagram, Twitter and Facebook, which thus influences them as enablers of the sustainable knowledge management system for PT. ABC as it supports employees' ability to gain additional knowledge. All six variables should be considered by PT. ABC when building a knowledge management system to make it workable and sustainable in the long term.

Acknowledgments.   Many people contributed to writing this paper, with sincere thanks to Bina Nusantara, the University Graduate Program Rector and Director, all PT. ABC employees, and PT. ABC's R&D Head for allowing the authors to conduct this field research.

References

1. Zheng, Y., Li, L., Zheng, F.: Social media support for knowledge management. In: International Conference on Management and Service Science, pp. 1–4. IEEE, Wuhan (2010)
2. Mǎruşter, L., Niels, R.F., Kristian, P.: Sustainable information systems: a knowledge perspective. J. Syst. Inf. Technol. **10**(3), 218–231 (2008)
3. Lichtenstein, S., Swatman, P.M.: Sustainable knowledge management systems: integration personalisation and contextualisation. In: Proceedings of the 11th European Conference on Information Systems, ECIS, Naples, Italy, pp. 1–8 (2003)
4. David, B.T., Mike, B.: The Executive's Guide to Enterprise Social Media Strategy. Wiley, Hoboken (2011)
5. Mayfield, A.: What is Social Media?. ICrossing, London (2008)
6. Matthews, J.H., Shulman, A.: Questioning knowledge transfer and learning processes across R & D project teams. In: Organizational Learning and Knowledge Management: New Directions, 1–4 June 2001, Richard Ivey School of Business, The University of Western Ontario, Canada (2001)
7. Peter, G.: Strategic Knowledge Management Technology. Idea Group Inc (IGI), Hershey (2005)

8. Hansen, M., Haas, M.: Competing for attention in knowledge markets: electronic document dissemination in a management consulting company. Adm. Sci. Q. **46**(1), 1–28 (2001)
9. Won, K.: Personalization: definition, status, and challenges ahead. J. Object Technol. **1**(1), 29–40 (2002)
10. Lin, X., Featherman, M., Sarker, S.: Information sharing in the context of social media: an application of the theory of reasoned action and social capital theory. In: SIGHCI Proceedings (2013)
11. Gaál, Z., Szabó, L., Obermayer-Kovács, N., Csepregi, A.: Exploring the role of social media in knowledge sharing. Electron. J. Knowl. Manag. **13**, 185–197 (2015)
12. Stenholm, D., Landhal, J., Bergsjo, D.: Knowledge management life cycle: an individual's perspective. In: DS 77: Proceedings of the DESIGN 2014 13th International Design Conference (2014)
13. Massingham, P.R., Massingham, R.K.: Does knowledge management produce practical outcomes? J. Knowl. Manag. **18**(2), 221–254 (2014)
14. Supranto, J.: Statistik. Teori dan Aplikasi. Edisi Keenam. Erlangga, Jakarta (2004)

Advanced Data Mining Techniques and Applications

Content-Based Music Classification by Advanced Features and Progressive Learning

Ja-Hwung Su[1]([⊠]), Chu-Yu Chin[2], Tzung-Pei Hong[3,4], and Jung-Jui Su[5]

[1] Department of Information Management, Cheng Shiu University, Kaohsiung, Taiwan
bb0820@ms22.hinet.net
[2] Telecommunication Laboratories, Chunghwa Telecom Company Ltd., Taoyuan, Taiwan
[3] Department of Computer Science and Information Engineering, National University of Kaohsiung, Kaohsiung, Taiwan
[4] Department of Computer Science and Engineering, National Sun Yat-sen University, Kaohsiung, Taiwan
[5] Department of Computer Science and Information Engineering, Chinese Culture University, Taipei, Taiwan

Abstract. Recently, content-based music information retrieval has been proposed as a support to facilitate music recommendation, music recognition and music retrieval. For music recognition, although it has been investigated by many studies, it is still a challenging issue for how to effectively learn from music contents. Actually, effective music recognition is achieved by considering two factors, namely feature content and learning strategy. Therefore, in this paper, a content-based music classifier named Progressive-Learning- based Music Classifier (PLMC) is proposed to aim at issues of feature content and learning strategy. In terms of feature content, the audio features are upgraded as the advanced features to enhance quality of features. In terms of learning strategy, a progressive learning strategy is proposed by fusing K-Nearest-Neighbor learning and Support Vector Machine learning. Through the proposed progressive learning, the better classification precision can be reached. The experimental results on real music data show the proposed idea performs better than the state-of-the-arts methods in classifying music.

Keywords: Music classification · Advanced features · Progressive learning · K-Nearest-Neighbor · Support Vector Machine

1 Introduction

Nowadays, music has been an important part of our everyday life because the music is easier to access than before by music retrieval techniques. Therefore, a large demand of music retrieval grows explosively. For this demand, a number of online music websites such as Spotify, iTunes and Amazon provide the online listening service that allows the

© Springer Nature Switzerland AG 2019
N. T. Nguyen et al. (Eds.): ACIIDS 2019, LNAI 11432, pp. 117–130, 2019.
https://doi.org/10.1007/978-3-030-14802-7_10

user to retrieve the interested music by concepts. In this service, conceptualizing music is a critical work that needs a large manual cost for annotations. To decrease the conceptualization cost, the automated music recognition is proposed by a set of literatures. However, it is not easy to achieve high-quality music recognition because of semantic gap. To narrow down the semantic gap, music information retrieval plays a mainstay role that discovers valuable knowledge from the musical content. Figure 1 shows the scenario of music information retrieval that supports the real applications such as music recommendation, music retrieval and music recognition. This incurs two further issues: (1) how to approximate the better features to increase sensitivity of acoustic features, and (2) how to learn from the music content to effectively recognize the music. In other words, an effective music recognition system depends on two critical factors, namely feature content and learning strategy. The good feature content can bridge the semantic gap more effectively, while a good learning strategy can mine more valuable patterns. To address the aforementioned points, in this paper, we propose a content-based music classification method named Progressive-Learning-based Music Classifier (PLMC) that understands the music by advanced features and a progressive learning. On the whole, the main contribution of this paper is summarized as follows.

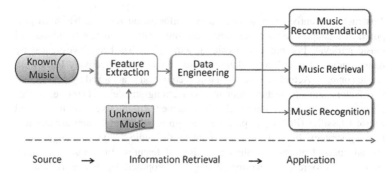

Fig. 1. Music information retrieval.

- For feature content, because the original audio features are not discriminative enough, in this work, they are improved as advanced features to enhance the quality of audio features.
- For learning strategy, because traditional single classifiers are highly limited in problem of semantic gap, in this work, a progressive learning paradigm is proposed to increase the classification precision.

The remaining of this paper is structured as follows. Previous work is briefly reviewed in Sect. 2. In Sect. 3, we present the notion of our proposed method for content-based music classification in detail. Empirical studies are shown in Sect. 4. Finally, conclusions and future work are elaborated in Sect. 5.

2   Related Work

Actually, there have been a number of previous works conducted in the field of music recognition over the past few decades. Basically, it can be further divided into two sets, namely music multi-labeling and music classification. In the followings, they are briefly reviewed to point out the main relevance to our proposed method.

- Music Multi-labeling

This is the recognition type that tags music with multiple concepts, which can also be called music annotation. The most popular application is music emotion annotation. Saari et al. [16] conducted an effective annotator that fuses semantic computing and audio-based modeling as a genre-adaptive computation to achieve music mood annotation. Chen et al. [5] provided a detailed discussion of deficiency of music datasets and then proposed a music emotion dataset named AMG1608. In addition to music emotion annotation, music multi-labeling can actually be viewed as a multi-label classification. Abe [1] improved Support Vector Machine (SVM) as a Fuzzy Support Vector Machine (FSVM) to achieve the multi-label classification. Ahsan et al. [2] proposed a one-vs-all strategy to annotate the music by using classifiers K-Nearest-Neighbor (KNN) and SVM. Moreover, Non-negative Matrix Factorization (NMF) was used to approximate better features to enhance multi-label classifiers such as Multi-Label KNN (ML-KNN) and Max-Margin Multi-Label Classification (M3L). Jao et al. [10] proposed two exemplar-based approaches to tag the music. One is based on the correlations of the musical short-time features, and the other is the sparse linear combinations of the short-time features over the audio exemplars. In [14, 18, 20], the temporal features of music were considered to attack the lack of song-level features. In overall, the past studies [7, 15, 19] investigated recent methods of music annotation methods and made a set of comparative experiments.

- Music Classification

In contrast to above music multi-labeling, this is the original music recognition type that classifies a music piece into a genre or a concept. Basically, this type can further be divided into two sub-types, namely content-based music classification and text-based music classification. For content-based music classification, the main idea is to conceptualize the music by music low-level features. In this type, Bergstra et al. [3] aggregated a number of low-level audio features such as Fast Fourier transform coefficients (FFTCs), Real cepstral coefficients (RCEPS), Mel-frequency cepstral coefficients (MFCCs), Zero-crossing rate (ZCR), Spectral spread, Spectral centroid, Spectral rolloff and Autoregression coefficients (LPC). Then the ADABOOST Freund and Schapire was performed to predict the music genres and artists. Goienetxea et al. [8] proposed an interesting idea that generates new music by similarity distances. In this work, the music data were grouped into several clusters first. For each cluster, the coherence structure was extracted from the template piece based on the statistical model derived from the corpus. Finally, the new melodic information was generated and a new music piece was made thereby. Huang et al. [9] presented a music genre classifier by integrating the Self-Adaptive Harmony Search (SAHS) algorithm and ensemble SVMs. In this method, 5 audio features including intensity, pitch, timbre, tonality and rhythm were extracted first. Next, the features are filtered by SAHS. Finally, the ensemble SVMs were performed to recognize the music genres. Mandel *et al.* [12] performed SVM to classify the unknown music, which transforms the low-level audio features MFCCs into song-level features by Gaussian Mixture Model (GMM). Lee *et al.* [11] also focused on the musical feature representation for music classification. This work presented a feature formalization called Bag Of Words (BOF). Accordingly, the Spectral features were converted into BOF features and thereupon the genre classification was conducted by BOF features and the general classifier. Pálmason *et al.* [13] conducted a case study by a comparative analysis of KNN classifiers. In addition to above content-based music classification, the other type of music classification is text-based music classification. The main idea behind this type [4, 6, 21] is to discover the confident relations between music and lyrics and therefore implement a high-quality music classifier.

To be concluded above, the previous works paid attention on two issues: (1) how to generate near-optimal features and (2) how to conduct an effective classification by integrating state-of-the-arts methods in classification. These can be viewed as an echo of our idea for proposing advanced features and high-performance classification strategy. Although these related works were shown to be very effective, they still lack the integration of solutions of two issues.

3 Proposed Method

3.1 Basic Concept of the Proposed Method

Before presenting the proposed method, the basic idea is shown in this subsection. As mentioned above, the main attention of this paper is focused on feature optimization and learning improvement. For feature optimization, because the low-level audio features are complex, it is not easy to reach high effectiveness and efficiency of music classification. In this paper, the audio features are transformed into song-level features, and then converted into advanced features. The advanced features are shown to be sensitive by the following evaluations. Although the feature optimization is one of our aims, another aim in this paper is to propose a progressive learning strategy instead of improving or combining the learning algorithms. This idea is motivated by our education, which can be viewed as level-wise learning from studying, researching to teaching. Figure 2 depicts the expected scenario that the effectiveness will increase as the learning level increases. In this paper, KNN-based (K-Nearest-Neighbor-based) learning is adopted as the basic learning. Through the basic learning, the unknown music is primarily identified for each concept. That is, different concepts are represented by different degrees for the unknown music. Then, based on the basic learning results, SVM is performed to classify the unknown music into a concept.

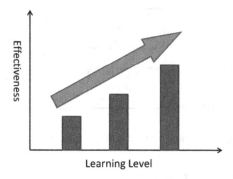

Fig. 2. The expected scenario of the progressive learning strategy.

3.2 Overview of the Proposed Method

As we can recall from above, the goal of this paper is to enhance the music classification by improving the features and learning progressively. To reach this goal, in this paper, we present a content-based music classifier that proposes the advanced features and level-wise learning strategy. Figure 3 shows the framework of the proposed method which is divided into two phases, namely training phase and prediction phase.

- Training phase:
  The goal of this phase is to ensure the performance of prediction, which contains two main operations, namely generation of advanced features and construction of learning models. Through advanced features, the learning can be more effective, and through the level-wise learning, the prediction will be more powerful.
- Prediction phase:
  Based on the learning model, the unknown music can be classified into the expected concept more successfully. Note that, although the adopted classifier is SVM in this paper, it will be a further issue for using other existing powerful classifiers.

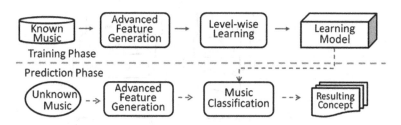

Fig. 3. Framework of the proposed method.

3.3 Training Phase

This is a fundamental but critical phase in this paper. Without successful training, the high quality of music classification is not easy to achieve. Basically, this phase can further be split into two steps, including generation of advanced features and construction of learning models. In the followings, this phase will be described step by step.

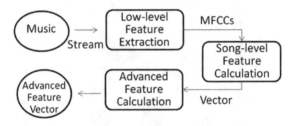

Fig. 4. Workflow of generating advanced features.

A. Generation of Advanced Features

In general, music features can be categorized into three types [7], including low-level, mid-level and song-level features. As shown in Fig. 4, this step contains 3 operations, namely low-level feature extraction, song-level feature calculation and advanced feature calculation. First, the music is transformed into the Timbre low-level feature named Mel-scale Frequency Cepstral Coefficients (MFCCs). Figure 5 shows the structure of MFCCs. In this structure, a music piece is divided into a set of frames and a frame is composed of 26 coefficients in this paper. Next, the frame-level MFCCs are averaged into a song-level vector. Finally, the song-level vector is normalized as an advanced feature vector [17]. In this step, a music piece is finally represented as a vector that contains 26 normalized coefficients. Here, the advanced feature vector is defined in Definition 1.

Definition 1. Given two music training datasets $T1 = \{m_1, m_2, \ldots, m_{|T1|}\}$ and $T2 = \{m_1, m_2, \ldots, m_{|T2|}\}$ containing a set of music pieces, a music piece m_i is defined as: $m_i = \{a_{i,1}, a_{i,2}, \ldots, a_{i,26}\}$, where a indicates the normalized coefficient.

Note that, the general MFCCs might be represented by different numbers of coefficient dimensions in general, and in this paper, 26 coefficients are selected as the adopted features.

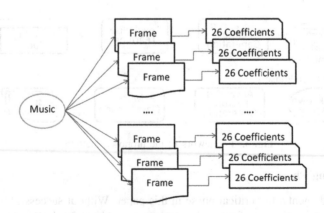

Fig. 5. Low-level feature structure.

B. Construction of Learning Models

After generating the advanced features, the next step of training phase is to perform the progressive learning. In this paper, the progressive learning is a two-stage learning. The first one is KNN-based learning and the second one is SVM-based learning.

- KNN-Based Learning

Basically, this learning can also be regarded as a similarity-driven or distance-based learning. The basic idea of this learning is to identify the unknown music initially by calculating the similarity-driven or distance-based likeness. From K-Nearest-Neighbor viewpoint, the K value is set as the number of concepts and the likeness will be kept as a vector in this paper. Figure 6 shows the procedure of KNN-based learning. In this learning, the centroid of each concept needs to be calculated first, as shown in Lines 1–2. It can be defined in Definition 2.

Input: Two training datasets $T1=\{m_1, m_2, ..., m_{|T1|}\}$ and $T2=\{m_1, m_2, ..., m_{|T2|}\}$, where $T1$ is classified into K conceptual groups and any music piece m_i is represented as $\{a_{i,1}, a_{i,2}, ..., a_{i,26}\}$;
Output: A KNN-based learning dataset *KNND*;
Algorithm: KNN-based learning
1.   **for** each concept c_j in $T1$ **do**
2.   calculate the concept centriod as cen_j by Definition 1;
3.   let *KNND* be the KNN-based learning dataset.
4.   **for** each music m_i in $T2$ **do**
5.   **for** each concept c_j **do**
6.   calculate the likeness between m_i and c_j into the vector $\bm{knnv_{m_i}}$ by Definitions 3-5;
7.   insert $\bm{knnv_{m_i}}$ into *KNND*;
8.   **end for**
9.   **return** *KNND*;

Fig. 6. Algorithm for KNN-based learning.

Definition 2. Following Definition 1, assume a training dataset $T1$ contains K concepts and the jth concept $c_j = \{m_1, m_2, ..., m_x\}$ contains x training music pieces, where the ith music piece is $m_i = \{a_{i,1}, a_{i,2}, ..., a_{i,26}\}$ and a training music piece belongs to one concept only. The centroid of the jth concept c_j is defined as:

$$cen_j = \{a_{cj,1}, a_{cj,2}, ..., a_{cj,d}, ..., a_{cj,26}\},$$

where the dth dimention $a_{cj,d}$ is

$$a_{cj,d} = \frac{\sum_{1 \le i \le x} a_{i,d}}{x}.$$

Next, as shown in Lines 4–8, the likeness between each centroid and each music in T2 is calculated into a vector. To make the likeness more comprehensive, in this paper, three measures are employed as the likeness, which are defined in Definitions 3–5, namely Euclidean likeness, Cosine likeness and Pearson-Correlation-Coefficient, respectively.

Definition 3. Consider a music piece $m_y = a_{y,1}, a_{y,2}, \ldots, a_{y,26}\}$ in $T2$ and a centroid $cen_j = \{a_{cj,1}, a_{cj,2}, \ldots, a_{cj,d}, \ldots, a_{cj,26}\}$, the Euclidean likeness between m_y and cen_j is defined as:

$$Elikes(m_i, cen_j) = \sqrt{\sum_{1 \le d \le 26} (a_{y,d} - a_{cj,d})^2}.$$

Definition 4. Consider a music piece $m_y = \{a_{y,1}, a_{y,2}, \ldots, a_{y,26}\}$ in $T2$ and a centroid $cen_j = \{a_{cj,1}, a_{cj,2}, \ldots, a_{cj,d}, \ldots, a_{cj,26}\}$, the Cosine likeness between m_y and cen_j is defined as:

$$Clikes(m_i, cen_j) = \frac{\sum_{1 \le d \le 26} a_{y,d} * a_{cj,d}}{\sqrt{\sum_{1 \le d \le 26} (a_{y,d})^2} * \sqrt{\sum_{1 \le d \le 26} (a_{cj,d})^2}}.$$

Definition 5. Consider a music piece $m_y = \{a_{y,1}, a_{y,2}, \ldots, a_{y,26}\}$ in $T2$ and a centroid $cen_j = \{a_{cj,1}, a_{cj,2}, \ldots, a_{cj,d}, \ldots, a_{cj,26}\}$, the Pearson-Correlation-Coefficient likeness between m_y and cen_j is defined as:

$$PCClikes(m_i, cen_j) = \frac{\sum_{1 \le d \le 26} (a_{y,d} - \overline{m_i}) * (a_{cj,d} - \overline{cen_j})}{\sqrt{\sum_{1 \le d \le 26} (a_{y,d} - \overline{m_i})^2} * \sqrt{\sum_{1 \le d \le 26} (a_{cj,d} - \overline{cen_j})^2}},$$

where

$$\overline{m_i} = \frac{\sum_{1 \le d \le 26} a_{y,d}}{26},$$

and

$$\overline{cen_j} = \frac{\sum_{1 \le d \le 26} a_{cj,d}}{26}.$$

After calculating the likeness between the unknown music and all centroids of concepts, the KNN-based learning vector of the music mi can be derived as

$$knnv_{m_i} = \{f_{m_i,1}, f_{m_i,2}, \ldots, f_{m_i,26}\} \tag{1}$$

Finally, all the KNN-based learning vectors for the training dataset $T2$ are gathered as the KNN-based learning dataset *KNND*, as shown in Lines 3 and 7.

- SVM-Based Learning

As shown in Fig. 7, the final step of the training phase is to construct a learning model named SVM-based learning model by the KNN-based learning dataset, which can be a support to the prediction phase.

3.4 Prediction Phase

Based on the SVM-based learning model, an unknown music piece can be predicted using the SVM classifier [22]. Because SVM is a well-known classifier widely used by recent researches in the field of machine learning, it will not be described here further.

Fig. 7. SVM-based learning.

4 Experiments

In above sections, the proposed method has been presented in great detail. To reveal the contributed effectiveness, a number of experiments were made and they will be shown in this section.

Table 1. Experimental data.

Data type	#music pieces	Genre
KNN-based Learning Dataset *T*1	450	Piano, saxophone, erhu, flute, harmonica, guitar, symphony, techno, zither, violin, drum, soundtracks, harp, dulcimer, lounge music, rock, classical chorus, rap, disco dance, vocal, country, pop, latin, folk, blues, jazz, new age, metal, r&b and children's music
Data 1	450	
Data 2	600	

4.1 Experimental Settings

The experimental data came from Web and CDs, which contains 30 genres. Table 1 shows the experimental data setting. Overall, the data was further split into three sets, namely *T*1, Data 1 and Data 2, where *T*1 is the KNN-based learning dataset, Data 1 and Data 2 are used for the 2-fold cross-validation. For *T*1, Data 1 and Data 2, each genre contains 15, 15 and 20 music pieces, respectively. That is, there are totally 1500 music pieces used in the experiments. The low-level audio feature is the frame-level MFCC and the final advanced feature vector is composed of 26 dimensions. By referring to Definitions 3–5, the proposed KNN-based learning can be classified into three categories, as shown in Table 2. To conduct the evaluations, *precision* is adopted to measure the experimental methods.

Table 2. Definitions of the KNN-learnings.

Method	Name	Description
Euclidean KNN-based learning	EKNN	KNN-based learning by Euclidean likeness
Consine KNN-based learning	CKNN	KNN-based learning by Consine likeness
Pearson-Correlation-Coefficient KNN-based learning	PCCKNN	KNN-based learning by Pearson-Correlation-Coefficient likeness

4.2   Experimental Results

Because the proposed method was materialized by integrating the advanced features and progressive learning, the evaluation was divided into 3 parts, including (1) evaluations of the advance features, (2) evaluations of the KNN-based learning, and (3) evaluations of the progressive learning.

A. Evaluations of the Advance Features

By referring to above sections, the main composition of the proposed method contains advanced features, KNN-based learning and progressive learning-based classifiers. Hence, the first issue we want to clarify is the impact of the advanced features. Figures 8 and 9 depict the experimental results for this issue. By comparing them, an important point derived is that, whatever the method is, the precisions with advanced features are much better than those with the song-level feature MFCCs. Note that, the KNN-based learning methods were used as music classifiers under $K = 1$. That is, in Figs. 8 and 9, the number of the nearest neighbors for all compared KNN-based learning methods is set as 1.

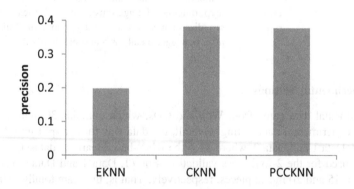

Fig. 8. Precisions of KNN-based learnings with the song-level feature MFCCs.

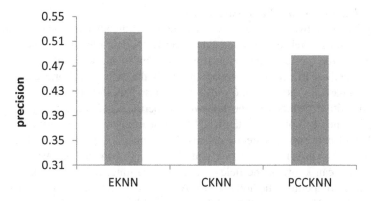

Fig. 9. Precisions of KNN-based learnings with the advanced features.

B. Evaluations of the KNN-Based Learning with Different Likeness Functions

In addition to the advanced features, KNN-based learning can be viewed as another foundation in this work. Therefore, what we want to show in this evaluation is how much the impact of the likeness function is, and this issue can be clarified by Figs. 8 and 9. Figure 8 reveals that, by the original MFCCs, the Euclidean distance is the worst, and the CKNN and PCCKNN perform pretty close. From the advanced feature viewpoint, Fig. 9 reveals that EKNN performs better than the others. In overall, EKNN with the advanced features can achieve the best precision of music classification. Therefore, EKNN with advanced features is selected as the first level learning in the proposed progressive learning.

C. Evaluations of the KNN-Based Learning with Different Likeness Functions

In addition to the advanced features, KNN-based learning can be viewed as another foundation in the proposed method. Therefore, what we want to show in this evaluation is how much the impact of the likeness function is, and this issue can be clarified by Figs. 8 and 9. Figure 8 reveals that, by the original MFCCs, the Euclidean distance is the worst, and the CKNN and PCCKNN perform pretty close. From the advanced feature viewpoint, Fig. 9 reveals that EKNN performs better than the others. In overall, EKNN with the advanced features can achieve the best precision of music classification. Therefore, EKNN with advanced features is selected as the first level learning in the proposed progressive learning.

Table 3. Definitions of the compared methods.

Method	Name
BayesNet [15]	BN
DecisionTree (J48) [15]	DT
K-Nearest-Neighbor [15]	KNN
AdaBoost-DecisionTree (J48)	ADT
Support Vector Machine [7, 12]	SVM
AdaBoost-Support Vector Machine [3]	ASVM
Progressive-Learning-based Music Classifier (proposed)	PLMC

D. Evaluations of the Progressive Learning

From the above two evaluations, we can know that the advanced features are better than original song-level features, and Euclidean KNN-based learning is very sensitive in recognizing the music. In this evaluation, we want to show the effectiveness of the integrating method PLMC via comparisons with the state-of-the-arts classifiers. To have the comparisons more solid, 6 well-known classifiers shown in Table 3 were used as the competitive methods in the following experiments. Note that, by referring to Eq. (1) and Lines 4–8 of Fig. 6, the KNN-based learning vector is derived while K is the number of concepts. Figure 10 reveals the comparison results that deliver some aspects. First, BayesNet is the worst and the proposed method PLMC is the best. Second, as we can know in the field of machine learning, SVM still performs better than the other competitive methods. Third, AdaBoost is very helpful to Decision Tree, but not to SVM. However, without AdaBoost, SVM is till better than Decision Tree with AdaBoost. In overall, the experimental results show that, our proposed idea is more effective than the competitive methods by integrating advanced features and progressive learning.

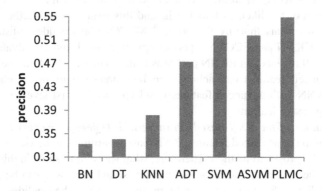

Fig. 10. Precisions of compared methods.

E. Discussion

After showing the experimental results, some analytical discussions are now lifted as follows.

- Figures 8 and 9 show the effectiveness of the advanced features. An important discovery has to be shown here. EKNN is very poor in contrast to the other KNN-based learning methods using original MFCCs, but it is the best using the advanced features. From another viewpoint, CKNN and PCCKNN perform more stably than EKNN. The reason is that, the original features are not normalized so as to make the distance dominated by some features clearly. On the contrary, it does not impact the PCC and Cosine likeness clearly because the likeness idea is not inspired by distance.

- Because the proposed method PLMC is the one integrating advanced features, KNN-based learning and SVM, an issue needs to be clarified for the impacts of the advanced features, KNN-learning and SVM. From Figs. 8 and 9, we can know that the advanced features play an important role for KNN-based learning. Further, Figs. 9 and 10 reveal that, without KNN-based learning, SVM cannot achieve the higher precision. In summary, the integrated components cannot perform well separately.
- In fact, the progressive learning can be interpreted as a hybrid method which accumulates the recognition results level by level. It further incurs some interesting issues. First, how many levels are the best? Second, what is the optimal sorting of predictors? Third, how to accumulate the level-wised results? In this paper, the progressive learning idea is just the beginning. In the future, the further investigation will be conducted to make the issues clearer.

5   Conclusion

Actually, a perfect music classification is not easy to achieve because a single classifier seems not powerful enough. To aim at this issue, in this paper, we propose a music classification method that integrates advanced features and progressive learning. For advanced features, the original audio features are upgraded as advanced features to deal with the problem of semantic gap. For progressive learning, the two-stage learning is proposed, including KNN-based learning and SVM-based learning. It is because the classification is enhanced level by level that the final prediction can reach higher level of quality. The experimental results reveal that our proposed method PLMC can achieve higher quality of music classification than the state-of-the-arts classifiers. Although the proposed method is shown to be more effective than other recent classifiers, there remain some issues to be addressed further. As mentioned above, the optimal settings for the number of involved classifiers, the sorting of involved classifiers and the way to accumulate the level-wise results will be investigated in the future.

Acknowledgement. This research was supported by Ministry of Science and Technology, Taiwan, R.O.C. under grant no. MOST 107-2221-E-230-010.

References

1. Abe, S.: Fuzzy support vector machines for multilabel classification. Pattern Recogn. **48**(6), 2110–2117 (2015)
2. Ahsan, H., Kumar, V., Jawahar, C.V.: Multi-label annotation of music. In: Proceedings of the Eighth International Conference on Advances in Pattern Recognition (ICAPR) (2015)
3. Bergstra, J., Casagrande, N., Erhan, D., Eck, D., Kégl, B.: Aggregate features and AdaBoost for music classification. Mach. Learn. **65**(2–3), 473–484 (2006)
4. Choi, K., Lee, J.H., Downie, J.S.: What is this song about anyway? Automatic classification of subject using user interpretations and lyrics. In: Proceedings of the 14th ACM/IEEE-CS Joint Conference on Digital Libraries (2014)

5. Chen, Y.A., Yang, Y.H., Wang, J.C., Chen, H.: The AMG1608 dataset for music emotion recognition. In: Proceedings of IEEE International Conference on Acoustics, Speech and Signal Processing (ICASSP) (2015)
6. Fang, J., Grunberg, D., Litman, D., Wang, Y.: Discourse analysis of lyric and lyric-based classification of music. In: Proceedings of the 18th International Society for Music Information Retrieval Conference (2017)
7. Fu, Z., Lu, G., Ting, K.M., Zhang, D.: A survey of audio-based music classification and annotation. IEEE Trans. Multimedia 13(2), 303–319 (2011)
8. Goienetxea, I., MartõÂnez-Otzeta, J.M., Sierra, B., Mendialdua, I.: Towards the use of similarity distances to music genre classification: a comparative study. PLoS ONE 13(2), e0191417 (2018)
9. Huang, Y.F., Lin, S.M., Wu, H.Y., Li, Y.S.: Music genre classification based on local feature selection using a self-adaptive harmony search algorithm. Data Knowl. Eng. 92, 60–76 (2014)
10. Jao, P.K., Yang, Y.H.: Music annotation and retrieval using unlabeled exemplars: correlation and sparse codes. IEEE Signal Process. Lett. 22(10), 1771–1775 (2015)
11. Lee, C.H., Lin, H.S., Chen, L.H.: Music classification using the bag of words model of modulation spectral features. In: Proceedings of the 15th International Symposium on Communications and Information Technologies (ISCIT) (2015)
12. Mandel, M.I., Ellis, D.P.W.: Song-level features and support vector machines for music classification. In: Proceedings of the 18th International Society for Music Information Retrieval Conference (2005)
13. Pálmason, H., Jónsson, B.Þ., Amsaleg, L., Schedl, M., Knees, P.: On competitiveness of nearest-neighbor-based music classification: a methodological critique. In: Beecks, C., Borutta, F., Kröger, P., Seidl, T. (eds.) SISAP 2017. LNCS, vol. 10609, pp. 275–283. Springer, Cham (2017). https://doi.org/10.1007/978-3-319-68474-1_19
14. Reed, J., Lee, C.H.: On the importance of modeling temporal information in music tag annotation. In: Proceedings of International Symposium on Acoustics, Speech and Signal Processing (2009)
15. Santos, A.M., Canuto, A.M.P., Neto, A.F.: A comparative analysis of classification methods to multi-label tasks in different application domains. Int. J. Comput. Inf. Syst. Ind. Manag. Appl. 3, 218–227 (2011)
16. Saari, P., Fazekas, G., Eerola, T., Barthet, M., Lartillot, O., Sandler, M.: Genre-adaptive semantic computing and audio-based modelling for music mood annotation. IEEE Trans. Affect. Comput. 7(2), 122–135 (2016)
17. Su, J.H., Hong, T.P., Chen, Y.T.: Fast music retrieval with advanced acoustic features. In: Proceedings of IEEE International Conference on Consumer Electronics (2017)
18. Serrà, J., Müller, M., Grosche, P., Arcos, J.L.: Unsupervised music structure annotation by time series structure features and segment similarity. IEEE Trans. Multimedia 16(5), 1229–1240 (2014)
19. Su, J.H., Tsai, Y.C., Tseng, V.S.: Empirical analysis of multi-labeling algorithms for music emotion annotation. In: Proceedings of ICME 2013 Workshop on Affective Analysis in Multimedia (AAM) (2013)
20. Tian, M., Sandler, M.B.: Towards music structural segmentation across genres: features, structural hypotheses, and annotation principles. ACM Trans. Intell. Syst. Technol. 8(2), 23 (2017)
21. Tsaptsinos, A.: Lyrics-based music genre classification using a hierarchical attention network. In: Proceedings of the 18th International Society for Music Information Retrieval Conference (2017)
22. http://www.csie.ntu.edu.tw/~cjlin/libsvm

Appliance of Social Network Analysis and Data Visualization Techniques in Analysis of Information Propagation

Leo Mrsic$^{(\boxtimes)}$ ⓘ, Srecko Zajec, and Robert Kopal

Algebra University College, Ilica 242, 10000 Zagreb, Croatia
{leo.mrsic,robert.kopal}@algebra.hr,
srecko.zajec@gmail.com

Abstract. This paper explains appliance of social network analysis and data visualization techniques in analysis of information propagation. Context of information (news) propagation through social network is an extremely dynamic and complex area to study. Due to topic actuality and a very small number of works on the similar topic this paper required a comprehensive and systematic approach. Thus, for practical reasons this work is based on the usage of Social Network Analysis (SNA) and visualization of social networking data obtained through Facebook covering 145 + public pages linked to 2.6 million fans. The main hypothesis is based on the premise whether is possible to find any similarities between the real-life social, economic and political entities/processes and online information propagation. The process consists of the development of the underlying model, the retrieval of data, data processing and consequential analysis & visualization which has been elaborated in detail along with the comments related to the methods of application.

Keywords: Advanced visualization · Social networks ·
Information propagation · Information warfare ·
SNA (Social Network Analysis) · Facebook

1 Introduction

The main reason for conducting the research that will be presented on the following pages is the growing popularity of the term "hybrid warfare" (especially in the Republic of Croatia where the research was conducted), as well as the growing popularity of the term that partially covers the same meaning, namely information warfare. This is nothing new, the aforementioned concepts represent known things that with the help of the development of the Internet and social networks have found their place in the mainstream media. The US Defense Ministry defines information warfare as a process that enables the control of information in order to achieve certain goals before a particular opponent and at the same time makes it impossible for them to do the same [1]. Although the Croatian mainstream media outlets recently started to emphasize the importance of hybrid and information warfare, no systematic study on this topic was conducted in the Republic of Croatia to present day. In regards to crucial social, economic and political events/processes, it is very hard to comprehensively map and analyze them in the meaningful way. On the other hand, a large amount of data

N. T. Nguyen et al. (Eds.): ACIIDS 2019, LNAI 11432, pp. 131–143, 2019.
https://doi.org/10.1007/978-3-030-14802-7_11

generated by social networks can be interesting for different types of analyses. What makes such networks so interesting is certainly the fact that they create social interaction data which can be of use to us in order to understand what is happening in our sphere of interest when the already mentioned information is taken into account. In this respect, social interaction data is actually relational data. Than it is important to know who creates and who consumes information [2], all in order to be able to follow the course of it and to study the causal links, or the potential for disseminating information. Regardless of the fact that most of the data shared on social networks is owned by their owners (companies), a certain part of it is publicly available and can be analyzed with various tools. That data can then be used in drawing conclusions that we are interested in, namely finding any similarities between the real-life processes and online information propagation. This paper will be based on this kind of publicly available data, and their consequent processing using Social Network Analysis techniques (SNA).

2 Methods and Analysis Techniques

As we face a large number of entities in the real world and their interconnectivity (which make up the various networks), that interconnection is not easy to understand, so we need to use analytic methods to help us. The so-called network analysis offers a solution to this type of problem, and this research relies on the highest level of such analysis which is called Social Network Analysis (SNA). This is a structural analysis referring to the various links between nodes we find everywhere around us wherever there is some form of social interaction. In what follows, we will address the problem at the level of individual nodes but calculations related to parts and the network as a whole will also be taken into account. To analyze some entities, we use the so-called centrality measures, while the analysis of subgroups shows to which degree the parts of the network are interconnected, which will be shown in the examples. After calculating the metrics related to individual nodes, as well as those related to the network as a whole, a graphical representation of the network based on such calculations can greatly help us perceive more comprehensively the shape and size of the network. It can also help us locate clusters and key locations of linked entities [2]. Since the position and importance of the node in relation to other nodes is one of the main questions of the social network analysis [2], the graphical representation is what makes it easier to point to such relations. This paper is therefore based on the research and identification of the characteristic features of the network that can help us gain new knowledge in order to draw potentially important conclusions that could point us to the right course of action in the information space [14, 15].

3 Information Propagation Model for Digital Age

The flow of information and information warfare as such do not occur solely on social networks. However, the dynamics and speed of change on these platforms make them an ideal subject of study. Social networks within the scope of this paper have been selected as the subject of study for their ability to absorb and disseminate information extremely quickly. Analytical methods will be used to clarify the context in which processes in certain segments of the information space on Facebook are taking place, most specifically, how those processes correlate with real-life events. Looking at the current economic and political relations at the macro level of the European Union, the current issue of Russian influence in the field of media manipulation is extremely popular. This media manipulation is closely related to the economic growth of that superpower and the return of a part of the power lost after the collapse of the Soviet Union. So, we can see a series of initiatives and moves aimed at suppressing the strategy of propagating the Russian version of the "truth". In this respect, the European External Action Service (EEAS) established the Strategic Communications Division (StratCom) to efficiently cope with the disinformation being deployed through the EU media landscape [3]. When the US presidential elections at the end of 2016 are considered, it is possible to outline several concepts that are directly related to misinformation activities. For example, the phrase "fake news" was named as the most popular word in 2017 by the well-known Collins Dictionary. Additionally, the popularization of the so-called trolls and bots has contributed to spreading disinformation during the aforementioned presidential elections [4], especially through social networks. These include the profiles maintained by people (troll) [5] and automated (bot) [6] profiles which have their own ulterior motives pursued most often by representing certain interests. Finally, if we are talking about Croatia which is part of the EU, it is interesting to observe various influence factors ranging from the EU (as whole) to America and Russia. On the micro level, social, economic and political events in Croatia are highly influenced by those (super) powers - often not directly but through different proxy entities. Those proxy entities are also often highly interconnected and are acting in coordinated manner (for example organizing various public demonstrations, public referendums, etc.) Since the potential effects of the mentioned processes can unquestionably lead to different changes within the territory of a country (and beyond), timely observation and proactive control of the information flow can certainly contribute to greater control over such events. The hypothesis in this paper is based on such assumptions. Specifically, on recognizing the patterns in disseminating information on social networks and whether those patterns have any similarity with the real-life events. As the context in which Facebook pages function is primarily of public character, the level of availability of data about them is far greater than the information about personal profiles. In terms of privacy, the various restrictions imposed by the Facebook API, or The Graph API [7], determine which data can be collected or analyzed. That is to say, it is possible to include the information regarding which pages are followed by another page which will also form the basis of this model. Consequently, the basic model will include only a closed network of Facebook fan pages.

In addition to technical limitations, this paper was written with limited budget resources so it postulates that a slightly modified model can work when applied to a network that does not only consist of pages, but also of the personal profiles or groups. In order to understand the way those pages function and communicate information within such a network we can assume that one page follows the other under two conditions: (1.) because the content interest's admins & (2.) because of the desire to receive information, regardless of whether we are interested in it or not. If we assume that such a network is closed, the inflow of information is not possible from the outside. That corresponds to a simplified basic model for conducting the analysis [16–18].

4   Contextualization of the Basic Model and Data Retrieval

Prior to data retrieval and subsequent processing, it was necessary to elaborate the concept by which the pages would be grouped (structured) in accordance with the previously stated assumptions referring to the functioning of the Facebook pages. In this respect, the main principle underlying this process is homophilia [8], i.e. the linkage between nodes (pages) based on their similarity, which ultimately directly affects the structure of the network itself. For this reason, it was extremely important to group pages of similar characteristics, and it should be emphasized that the homophilic principle we applied was based on nominal characteristics, not numerical [8]. It has been mentioned earlier that a Facebook page follows another page for two main reasons, and the first reason is closely related to homophilia - pages follow other pages containing similar themes and interests. Ego networks were grouped according to this principle - for example, grouping together organizations that support abortion ban, formal associations of various interest groups, etc. When mentioning an egocentric network [2] it needs to be stated that we are dealing with a selected Facebook page and belonging outbound links. To ensure maximum impartiality and balance, the entities that will be the subject of the study below will be divided into three groups. Two of these groups are the most important ones - the one that is closer to the left in the political context (designation L) and the other closer to the right (designation D). The third group consists of the so-called non-aligned pages (designation N). This division reflects general social and political developments. The subgroups of these three general groups are made up of different interest organizations - associations, media outlets, prominent public figures, political parties and so on. In this context, it is important to note that the large mainstream media outlets managed by private capital and large political parties are deliberately left out of the analysis and the emphasis is placed on a number of smaller participants that often proclaim themselves as independent and nonprofit and are not necessarily well known to the entire public but have a significant impact if they act coordinated. The reason for this lies precisely in the fact that their activity cannot be concealed in terms of scope and goals, as is the case with large media outlets. The fact that their existence depends on donations by the state or the private sector makes them sufficiently vulnerable to external influences, so their self- declared independence and impartiality are easily called into question. Their interconnectedness is then an interesting subject of analysis (Table 1).

The data was gathered on 146 Facebook pages divided into three basic groups and then grouped into 15 subgroups. The pages had to meet several criteria; they had to be used quite recently and contain outbound links to other sites, they were supposed to use the Croatian language, etc. Once the data is structured, the information space, which also constitutes the main area of research, is defined. To achieve the consistency of our experimental data, scraping time frame was shortened as much as possible (Table 2):

Table 1. Facebook fan pages subgrouping and total number of fans

Subgroup name	Number of Facebook fan pages	Number of fans
N - Events	3	27.539
N - Unions and alliances	3	13.131
Total group N	6	40.670
D - Interest organizations - pro religious	14	159.454
D - Media	10	354.434
D - Independent propaganda	16	415.683
D - Persons	4	174.761
D - Political parties	10	69.930
Total group D	54	1.174.262
L - Interest organizations - anti religious	11	59.579
L - Interest organizations - other - subgroup 01	9	42.688
L - Interest organizations - other - subgroup 02	11	73.807
L - Media	11	117.638
L - Media - subgroup 01	9	116.251
L - Independent propaganda	18	231.958
L - Persons	8	635.068
L - Political parties	9	282.580
Total group L	86	1.559.569

Table 2. Info regarding data retrieval period

Facebook group	Data scraping period
Group N	27.09. - 05.10.2017.
Group L	27.09. - 05.10.2017.
Group L corrections	18.10.2017.
Group D	20.10. - 25.10.2017.
Group D corrections	30.10.2017.

5   Results

In practical terms, the analytical process consisted of data retrieval, its preparation (which usually takes up most of the time), and key analyses and visualizations. Two well-known analysis tools were used - NodeXL (for data scraping) - and Gephi for the network analysis and visualization. In that sense, we were dealing with the static data. The files used ranged from the standard and widely known Excel format (for raw retrieved data) to combinations of various ego networks written in the graphml file format [9]. The data collected after the grouping process was exported in the .gefx file format in which most of the analytical operations were performed [10]. The last two formats are based on the popular .xml standard, enabling us to keep track of a number of other features (for example, the number of fans of particular pages) [11]. It should be also noted that the theory of the scale-free networks and preferential attachment mechanism [8] must be also taken into consideration. In other words, that implies the existence of a large number of nodes with few links, or a smaller number with a lot of links. In the context of this research, the fact that the described networks are extremely vulnerable to attacks against important nodes is particularly interesting (i.e. stopping important entities/pages from information spreading). It should also be emphasized that in the visualization sense the size of an entity is the value of the calculated metric for that same entity (larger circle = greater value). Secondly, a color scheme was used in the process to mark entities and their outbound links to highlight their affiliation to a particular subgroup. This often implies marking different components (clusters) in different colors, but in this particular case, colors are used to emphasize the affiliation of entities and their links to particular subgroups. A necessary part of the network's visualization are the so-called energy layouts which are used in the final stage of visualization [12]. They work in such a way that the network is visually set up so that the nodes which are more closely interconnected are attracted to each other and vice-versa. In our research, the ForceAtlas2 and Yifan Hu [13] algorithms were used. Once the pages are divided into groups and subgroups, it is possible to conduct analyses based on different combinations, and the analyses that were chosen were the one depicting the interconnection between the pages from group D and the analysis based on combining pages from groups L and N. This network analysis was selected because of the immediate social, political and economic influence exhibited by the entities impersonated by the Facebook pages. The following questions arose: how are organizations belonging to either the left or right side of the political spectrum interlinked and which nodes are the most important in terms of centrality metrics; what does the network overview of all entities on the left side of the political spectrum look like, and what could be said of their counterparts on the right side of the same political spectrum; how are the entities, that should, in principle, be independent and unrelated, interlinked (e.g., political parties, media outlets and associations); what is the role of independent propaganda outlets (pages maintained by unknown authors) (Tables 3 and 4).

Table 3. Color legend of the Facebook fan pages and their subgroups - group D fan page network

Subgroup color	Subgroup name
	Not imported/analyzed
	D - Interest organizations - pro religious
	D - Media
	D - Independent propaganda
	D - Persons
	D - Political parties

Visualizations with the numerical designation 1 and 2 relies on the network data from the group D fan pages (moderate grouping coefficient 0,664 which means that this network is mostly robust against random attacks). In this respect Fig. 1 shows the importance of individual pages according to the size of the circles for the Betweenness centrality. The largest orange circle in the middle of the image shows the independent right-wing internet medium which plays the most important role in controlling the flow of information between the individual parts of the network. Behind this portal is a

Table 4. Color legend of the Facebook fan pages and their subgroups - group L and N fan page network

Subgroup color	Subgroup name
	Not imported/analyzed
	N - Events
	N - Unions and alliances
	L - Interest organizations - anti religious
	L - Interest organizations - other - subgroup 01
	L - Interest organizations - other - subgroup 02
	L - Media
	L - Media - subgroup 01
	L - Independent propaganda
	L - Persons
	L - Political parties

non-profit entity, and its role in this regard exemplifies the significant control of information flow between the religious pages (cyan color) and independent propaganda (purple) in the assumed model. Although they emphases their neutrality, this centrality proves the opposite (Table 5).

Table 5. Top 5 Facebook fan pages regarding Betweenness centrality and their subgroup color - group D fan page network

Fan page	Betweenness
Betweenness Page Top 1 (Media)	15433,54
Betweenness Page Top 2 (Media)	12503,47
Betweenness Page Top 3 (Independent propaganda)	12067,06
Betweenness Page Top 4 (Independent propaganda)	10104,72
Betweenness Page Top 5 (Political parties)	6181,50

Fig. 1. Network visualization based on various Facebook fan page subgroups (color) and Betweenness centrality (node size) - group D fan page network. (Color figure online)

138   L. Mrsic et al.

Visualization number 2 refers to PageRank centrality and the pages which draws their influence from other influential pages. This measure of centrality helps us to identify the knots for which we didn't necessarily think they were important. Several larger circles marked with gray color are clearly visible in the picture and data for those pages were not scraped at all. This means that some pages definitely must be re-examined in context of their ability to propagate information (Fig. 2).

Table 6. Top 5 Facebook fan pages regarding PageRank centrality and their subgroup color - group D fan page network

Page	PageRank
PageRank Page Top 1 (Independent propaganda)	0,00295
PageRank Page Top 2 (Interest organizations - pro religious)	0,00253
PageRank Page Top 3 (Not analyzed)	0,00220
PageRank Page Top 4 (Not analyzed)	0,00217
PageRank Page Top 5 (Not analyzed)	0,00209

Fig. 2. Network visualization based on various Facebook fan page subgroups (color) and PageRank centrality (node size) - group D fan page network. (Color figure online)

Figure 3 refers to the analysis of the L and N group combination (grouping coefficient similar to the group D - 0,619), more precisely on the Closeness measure and shows how individual nodes can reach the rest of the network in the least number of steps. Such pages are therefore suitable for rapid dissemination of information. In this very case, primary emphasis is on the pages from the independent propaganda (light green) subgroup, anti-religious organizations (brown), publicly exposed individuals (olive green) and political parties (purple). Interestingly, a politically oriented collective composed of different independent interest organizations (red) has a significant number of pages with a relatively strong value of this metric. There are in total 12 different pages with the highest closeness centrality of 1,0. This heterogeneous group of pages with the highest closeness value suggests that a certain information agenda can be spread rapidly from completely different subgroups (especially if coordinated) (Table 6).

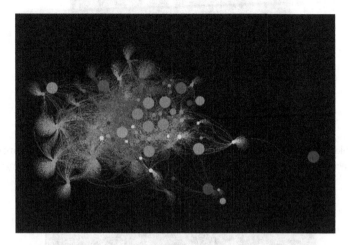

Fig. 3. Network visualization based on various Facebook fan page subgroups (color) and Closeness centrality (node size) - group L and N fan page network. (Color figure online)

Figures 4 (group D) and 5 (groups L and N) refers to component analysis, i.e. stronger connected parts of each network. In regards to group D and Fig. 4 it has been shown that primarily pro religious interest organizations and media with their outgoing connections builds this component which covers over 20% of the network. Their irrefutable connectivity in this regards gives a completely new context to the processes in which information is distributed in this information space.

Looking at the combination of fan pages from the group L and N (Fig. 5), the largest component wasn't the only one which was analyzed. Instead, three largest components were investigated due to a similar share in the whole network (in the total just below 40% of the whole network - 14,94%, 12,36% and 10,94% respectively). In order to get an even clearer insight into the type of analyzed pages that dominates the mentioned components, those components are shown separately and all three together. It is easy to see that the first component is dominated by the green color (independent propaganda), the second largest by red color (subgroup 2 of interest organizations) and the third largest light purple (media). Such direct and/or indirect connectivity gives a significant context to the information propagation in this case as well.

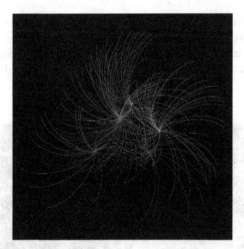

Fig. 4. Display of the biggest component from the group D - pro religious interest organizations (cyan) and media (orange color) are dominantly interconnected (Color figure online)

Fig. 5. Separate and joint display of three biggest components from the combination of groups L and N - independent propaganda (green), interest organizations from the subgroup 2 (red) and media (pink) are dominantly connected (Color figure online)

By comparing the given results between group D and group L + N it is easy to come to the conclusion that the common feature that occurs in both cases is the more pronounced direct and indirect connections between the media and the various interest organizations. Given the fact that the most of the analyzed entities of these types emphasize their independence, according to this model the credibility of such claims can be considered suspicious.

6 Conclusion

Referring to the introductory assumptions and the whole research we can conclude that in the conditions of fast and complex changes there is often a need for creative solutions not visible at first sight. Performing research on the basis of publicly available data can also be an advantage and a disadvantage at the same time. Regardless of everything, through previous chapters specific framework was exhibited which can be upgraded in multiple directions. The same represents a quality ground for further research analysis based on network theory and data visualization techniques. Initial question was set on the premises of interconnectivity between the similar entities in real-life vs. virtual environment and consequentially their influence through information propagation. With the usage of network analysis, we showed that in many cases entities are indeed interlinked on quite a high level. We should also note that certain influential entities proclaimed their neutrality and independence (mostly media outlets). Our results showed the complete opposite - that they are not only highly linked to many other nodes (media to political entities, media to independent propaganda, etc.), but also play a significant role in information distribution. Given the fact that this kind of analysis was done in Croatia for the very first time, results could potentially be more than significant for the interpretation of local and regional real-life social, economic and political processes. Finally, the set of techniques presented through this work enables us to manage our own information flows or neutralize unwanted ones. Simplified, the mentioned techniques enable us to understand those flows as we could handle them in the most appropriate way. This was particularly visible in the above-mentioned analysis of differently grouped Facebook pages, which resulted in interesting findings in terms of their strength and mutual interconnectivity. Gained knowledge can undoubtedly help the variety of interested parties to understand complex social changes, from state bodies (intelligence, police), to scientists who study similar areas of interest [19, 20]. Not only that, but this approach can be deepened and developed in many different directions (for example bow-tie modelling, NLP analysis, etc.)

References

1. Global Information Assurance Certification: Information Warfare. https://www.giac.org/paper/gsec/1870/information-warfare/103284. Accessed 30 Sept 2017
2. Hansen, D., Shneiderman, B., Smith, M.: Analyzing Social Media Networks with NodeXL. Morgan Kaufmann, Burlington (2011). Kindle Edition
3. European External Action Service (EEAS): Strategic Communications Division (StratCom). https://eeas.europa.eu/headquarters/headquarters-homepage_en/100/Strategic%20Communications. Accessed 13 Dec 2017

4. Phruksaphanrat, B.: Preemptive possibilistic linear programming: application to aggregate production planning. In: Proceedings of World Academy of Science, Engineering and Technology, vol. 80, pp. 473–480 (2011)
5. Purnomo, H.D., Wee, H.: Soccer game optimization: an innovative integration of evolutionary algorithm and swarm intelligence algorithm. In: Vasant, P. (ed.) Meta-Heuristics Optimization Algorithms in Engineering, Business, Economics, and Finance, pp. 386–420. Information Science Reference, Hershey (2013). https://doi.org/10.4018/978-1-4666-2086-5.ch013
6. Sadeghi, M., Hosseini, H.M.: Evaluation of fuzzy linear programming application in energy models. Int. J. Energy Optim. Eng. (IJEOE) 2(1), 50–59 (2013). https://doi.org/10.4018/ijeoe.2013010104
7. Vasant, P.: Hybrid LS-SA-PS methods for solving fuzzy non-linear programming problem. Math. Comput. Model. 57(1–2), 180–188 (2013)
8. Vasant, P.: Hybrid simulated annealing and genetic algorithms for industrial production management problems. Int. J. Comput. Methods 7(2), 279–297 (2010)
9. Vasant, P., Barsoum, N.: Hybrid pattern search and simulated annealing for fuzzy production planning problems. Comput. Math Appl. 60(4), 1058–1067 (2010)
10. Vasant, P., Ganesan, T., Elamvazuthi, I., Webb, J.F.: Fuzzy linear programming for the production planning: the case of Textile Firm. Int. Rev. Model. Simul. 4(2), 961–970 (2011)
11. Vasant, P.: Fuzzy decision making of profit function in production planning using S-curve membership function. Comput. Ind. Eng. 51(4), 715–725 (2006)
12. Vo, D.N., Schegner, P.: An improved particle swarm optimization for optimal power flow. In: Vasant, P. (ed.) Meta-Heuristics Optimization Algorithms in Engineering, Business, Economics, and Finance, pp. 1–40. Information Science Reference, Hershey (2013). https://doi.org/10.4018/978-1-4666-2086-5.ch001
13. Xiao, Z., Xia, S., Gong, K., Li, D.: The trapezoidal fuzzy soft set and its application in MCDM. Appl. Math. Model. 36(12), 5844–5855 (2012)
14. Klepac, G.: Data mining models as a tool for churn reduction and custom product development in telecommunication industries. In: Vasant, P. (ed.) Handbook of Research on Novel Soft Computing Intelligent Algorithms: Theory and Practical Applications, pp. 511–537. Information Science Reference, Hershey (2014). https://doi.org/10.4018/978-1-4666-4450-2.ch017
15. Mršić, L.: Widely applicable multi-variate decision support model for market trend analysis and prediction with case study in retail. Vasant, P. (ed.) Handbook of Research on Novel Soft Computing Intelligent Algorithms: Theory and Practical Applications, pp. 989–1018. Information Science Reference, Hershey (2014). https://doi.org/10.4018/978-1-4666-4450-2.ch032
16. Lavoix H.: Developing an Early Warning System for Crises, December 2008
17. Li, L., Goodchild, M.F.: The role of social networks in emergency management: a research agenda. Int. J. Inf. Syst. Crisis Response Manage. (IJISCRAM) 2(4), 48–58 (2010). https://doi.org/10.4018/jiscrm.2010100104
18. Lombardo, R.: Data mining and explorative multivariate data analysis for customer satisfaction study. In: Koyuncugil, A., Ozgulbas, N. (eds.) Surveillance Technologies and Early Warning Systems: Data Mining Applications for Risk Detection, pp. 243–266. Information Science Reference, Hershey (2011). https://doi.org/10.4018/978-1-61692-865-0.ch013

19. Meissen, U., Voisard, A.: Current state and solutions for future challenges in early warning systems and alerting technologies. In: Asimakopoulou, E., Bessis, N. (eds.) Advanced ICTs for Disaster Management and Threat Detection: Collaborative and Distributed Frameworks, pp. 108–130. Information Science Reference, Hershey (2010). https://doi.org/10.4018/978-1-61520-987-3.ch008
20. Miller, G.A.: The magical number seven - plus or minus two: some limits on our capacity for processing information. Psychol. Rev. **63**(2) (1959)

A Temporal Approach
for Air Quality Forecast

Eric Hsueh-Chan Lu[✉] and Chia-Yu Liu

Department of Geomatics, National Cheng Kung University,
No. 1, University Rd., Tainan City 701, Taiwan (R.O.C.)
luhc@mail.ncku.edu.tw, gelosei@gmail.com

Abstract. Recently, air pollution caused by particulate matter that the diameter is less than or equal to 2.5 $\mu g/m^3$ has become an important issue. It is so tiny that it can go through alveolar microvascular and enter our body. PM2.5 makes a significant impact on human health. Therefore, monitoring and forecasting the air quality is an indispensable task for human society. Nowadays, we can easily acquire Air Quality Indices (AQIs) by installing a small-scale air quality sensor or downloading from some freely authorized databases. However, people demand farther PM2.5 information to plan their route. This research aims to forecast PM2.5 value in the future hours. Previous studies indicated that the air quality varies nonlinearly in urban areas and depends on several factors such as temperature, humidity and wind speed. Therefore, we combine air quality data from AirBox and meteorology data to forecast PM2.5 value. Air quality is a continuous data. If monitored air quality is good at the last time stamp, the next monitored air quality has high possibility to be good at the same location. And air quality may have some regular in the history data. We forecast PM2.5 values via the algorithm similar to weighted average method. It can figure out the time intervals with similar weather condition. Finally, the error is calculated to examine the accuracy of our method. In contrast to a famous method, Pearson's Correlation Coefficient, our method preforms well and stable with farther forecast.

Keywords: Air quality forecast · City dynamics · AirBox

1 Introduction

With the Industrial Revolution, advancement of technology, coal combustion and traffic popularization, air pollution is getting more serious. Air quality has serious impact on human health, especially, atmospheric particulate matter. Atmospheric particulate matter is so small that it can enter human body anywhere though blood circulation. It may lead to cardiovascular disease, cancer and even death. Countries formulate relevant specifications to decrease air pollution. Researches in air quality issues are rapidly increasing these years, and it also means that people take air quality issues more seriously. Nowadays, we can easily acquire air quality information by installing an environmental sensor or downloading from some freely authorized databases. There are many kinds of air quality monitors are invented. Rather than the

© Springer Nature Switzerland AG 2019
N. T. Nguyen et al. (Eds.): ACIIDS 2019, LNAI 11432, pp. 144–151, 2019.
https://doi.org/10.1007/978-3-030-14802-7_12

government air quality monitoring station, they are cheaper and portable. Even though we can easily get current air quality, it cannot meet our needs. We need to know farther air quality, and then we can plan our route. For example, if air quality is getting bad in the afternoon, it is not a good idea to jog in the afternoon. This research makes effort on PM2.5 value forecast.

We combine weather forecast data, meteorology data and PM2.5 data for forecast. Some reference data will be chosen to evaluate future PM2.5 value through a weighted-average method. We choose some recent data which are the top-latest observed data and some similar data whose meteorology factors are similar to the target we want to forecast. We evaluate weights between the target and their references by the difference of the target's and references' features. To show the performance of our forecast method, we compare our method with a famous method, Pearson's Correlation Coefficient. Our method performs well and more stable than other method. And forecast result on farther hours converges but not diverges as the forecast hour getting farther.

The remaining of this research is organized as follows. Section 2 reviews related work on air quality analysis and prediction issues. In Sect. 3, we explain the details of air quality forecast and new air quality device allocation recommendation modules. The experimental evaluations are shown in Sect. 4. Finally, the conclusions and future work are mentioned in Sect. 5.

2   Related Work

In this section, we review some important studies related to air quality forecast issues. There are many factors which have connection to air quality. Because of lack of air quality monitoring station, we need other factors which are correlated to air quality. Features are the key that have bearing on accuracy of air quality prediction, so it must be carefully to choose features. According to the reference we collected, air quality prediction issues can be divided to two kinds. One is combining features and air quality information to predict real-time air quality in location without air quality station. The other also combines features and air quality information and forecast future air quality. No matter real-time prediction or future forecast, these references all propose method and mention importance of features choosing. If features are more complete, prediction will be more accurate. Zheng et al. forecast air quality in future 48 h [5]. Air quality may be influenced by any factor, so it is hard to forecast. Zheng et al. combined meteorology and weather forecast information to forecast air quality. They divided 48 h into four periods and give a forecast range to each period. Lu et al. forecast air quality by linear model in future 12 h [4]. They combined meteorology, traffic flow, human mobility, point of interest, road network and city form forecast air quality. Air pollution may come from other city and also may form in own city. Zhu et al. told where air pollution come from by Bayesian Gaussian model and pattern mining [6]. They forecast air quality in future 1 h, and the method perform well. Domańska et al. used 16 kinds of dimensionality reduction to forecast air quality [2]. Although there are many air quality prediction methods, there still lack of overall methods so far. Bouarar et al. build several models to monitor and forecast air quality [1]. They analyzed distribution of chemical pollution, ground emission and atmospheric composition, and forecast. All the references mentioned that importance of features. To predicate air quality, features' information is necessary. There may still be other features which may impact on air quality. In this research, we prepare weather forecast, meteorology and air quality information for air quality forecast.

3 Proposed Method

In this section, we will introduce the temporal features for calculating the weight between two instances and our proposed weighted-average strategy for air quality forecast in the following sub-sections.

3.1 Temporal Features

For forecast PM2.5, we need more features which may impact on PM2.5. Many research papers mentioned that meteorological data have some connection with PM2.5 [1–6]. For example, when relative humidity increases but it has yet to rained, PM2.5 value will increase with relative humidity. However, when it starts to rain, PM2.5 value will usually decrease rapidly. Furthermore, when wind speed is high, the particulate matter in air will be blown away by wind. Nevertheless, when there is landward wind and bring some particulate matter in, the PM2.5 will be bad. We also found the connection in our data. The meteorological data contain temperature, relative humidity, wind speed, wind direction, pressure and daily rainfall. There are introduces of temporal features below:

1. Observed Time of AirBox (OT_A): The feature OT_A is AirBox observed time, including time of day, date, and day of week. AirBox observe temperature, humidity and PM2.5 value about every 5 min.
2. Temperature (T): The feature T is observed temperature. The unit of temperature is degree Celsius.
3. Relative Humidity (RH): The feature RH is observed relative humidity. The unit of relative humidity is percentage.
4. Wind Speed (WS): The feature WS is observed wind speed. The unit of wind speed is m/s.
5. Wind Direction (WD): The feature WD is observed wind direction. The unit of wind direction is bearing angle.
6. Pressure (P): The feature P is observed pressure. The unit of pressure is hPa.
7. Rainfall (RF): The feature RF is observed Daily accumulated rainfall. The unit of pressure is mm.
8. PM2.5 value (PM25): The feature PM25 is observed PM2.5 value. The unit of PM2.5 is µg/m3.

3.2 Weighted-Average Strategy

It is usually happened that when air quality is good in the last period, air quality in next period will be good, too. And we have mentioned in previous section that when weather difference between two data is small, PM2.5 values of them may be similar. The air quality values of different locations and different time are correlated with each other on historical data. We propose to draw a graph to model such temporal correlations over locations. At this part, we pick some reference data to forecast PM2.5 values, including recent data and similar data. We define "recent" as which data is at the last time period before the target data we want to forecast. And we define "similar" as which's weather error with the target weather error is small. The construction of Fig. 1 consists of how we pick recent and similar data. Since the PM2.5 value of the same location is highly correlated to its historical PM2.5 values, we connect each target d^u to the its previous m corresponding data of the same device. That is, the data of the same device but with different time stamps: t_i, t_{i-1}, ..., and t_{i-m+1} are connected. It is represented by the blue lines in Fig. 1. Since the environmental factors can repeat themselves within certain period, it is also possible that the PM2.5 value of a data correlates with that farther away from the future. Our idea is to connect a data in the future layer to the corresponding nodes of n certain past data with the most similar environmental features. The similarity between layers is computed based on the features. See the green line in Fig. 1 as an example.

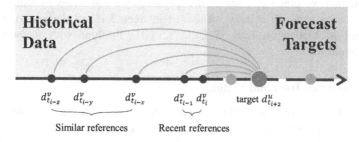

Fig. 1. An example of choosing recent and similar reference data of a station. (Color figure online)

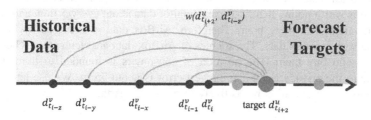

Fig. 2. Weight concept.

Before designing the method, we found that difference degree of meteorology data has some relationship with PM2.5 difference. The forecast method is similar to weighted-average method. Formula (1) is the weight of the inference method between nodes. In Fig. 2, the algebra d^u is the future data which we want to forecast, and the algebra d^v is chosen references of d^u. The algebra $w(d^u, d^v)$ is weight between the target d^u and its reference d^v at time t. The algebra f is features. If there are k features, they will be included to evaluate weight. Before evaluation, all of features will be normalized to 0–1 scale. We use reciprocal of weather difference as weight of edge. After all weights of edges are evaluated, we can forecast PM2.5 value $d^u.p$ of target with Formula (2). We will calculate $d^u.p$ for each forecast hour of each device.

$$w(d^u, d^v) = \frac{1}{\sqrt{\sum_1^k (d^u.f_k - d^v.f_k)^2}} \tag{1}$$

$$d^u.p = \frac{1}{\text{sum}(W)} \sum w(d^u, d^v) \times d^v.p \tag{2}$$

4 Experimental Evaluation

To evaluate the performance of air quality forecast method and new device allocation recommendation we propose, a series of experiment are conducted by using the real monitored data which are already mentioned in Sect. 3.1. All the experiments are implemented in Matlab on Intel i5 CPU 3.3 GHz machine with memory 8 GB Microsoft Window 7.

4.1 Experimental Data and Setting

To evaluate the performance of PM2.5 value forecast, we utilize a series of real data which we already mentioned in Sect. 3.1. AirBox data is provided by Data. Taipei (http://data.taipei/), and weather data is provided by Central Weather Bureau. We use 146 AirBoxes in Taipei, Taiwan (R.O.C), and 21 weather stations in or around. To ignore the error of weather forecast data, we use monitored meteorological data as weather forecast data. We combine these two kinds data depend on observed time and distance. It is worth noting that AirBoxes monitor data about every 5 min, and weather stations monitor data every hour. Therefore, AirBox data is averaged by hours for combining with weather data. The sources of AirBox data are elementary schools and consumers who buy them. The position of consumers is rounded to third decimal places due to the privacy right. The biggest error is about 75 m. We use latitude and longitude to search elevation respectively from google earth.

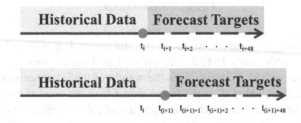

Fig. 3. The result of Naïve Bayes using categorical variables.

We have data from 2017/3 to 2018/5. We divide data on the third day of all period into historical and future data of each device. We forecast PM2.5 from the first hour of the third day of the whole data, and evaluate 48 h at a time. It's worth noting that we only forecast continuous 48 h. If there are some data miss, we will skip to the next complete and continuous 48 h. After a 48 h forecast, it will move to the next 48 h of the next hour to forecast like Fig. 3 shows. When each 48 h of data are forecasted, we will calculate the root-mean-square-error (RMSE) of each hour to evaluate the performance of the forecast method.

4.2 Air Quality Forecast

To forecast PM2.5 values, we have many parameters experiment. First, there are three parameters for weight-average method reference. We have recent r and similar s. First, we experiment the parameter of temporal. Note that this experiment use only r and s to forecast PM2.5, and n and smooth are not involved yet. We study the impact of parameter r and s on forecast PM2.5 values respectively. Figure 4 shows the impact of parameter r on PM2.5 forecast. The x-axis is the volume of r, and the y-axis is RMSE of forecast PM2.5 value. The 6th, 12th, 24th, 36th and 48th hour are displayed respectively in the figure. We can see that as the volume of r gets greater, RMSE of PM2.5 gets smaller on the 12th, 24th, 36th and 48th hour. However, the degree of declining of RMSE is smaller, too. And as the volume of r gets greater, the degree of declining is seen to be convergence. It is worth to note that the 6th hour line is opposite to the other. As the volume of r gets greater, RMSE of the 6th hour gets greater. It is because air quality is a continuous data, and air quality is usually highly similar to its last observed value. It explains why RMSE is smaller when r is set as 1. Besides, as the forecast hour is farther, the RMSE gets bigger. It tells that when the forecast hour is farther, it will be harder to forecast.

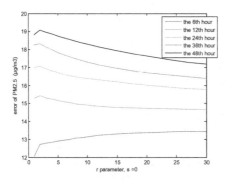

Fig. 4. The impact of parameter r on PM2.5 value forecast.

Figure 5 shows the impact of parameter s on PM2.5 forecast. The convergence degree is more obvious than r's on RMSE of forecast. Although RMSE is bigger when $s = 1$ than it when $r = 1$, RMSE is much better than r when both they are 30. Besides, the declining degree of s is greater than r's, it represents that similar reference data is benefit on PM2.5 forecast. According to Fig. 5, we can see that RMSE lines of forecast hours are close. The value of reference of r gets smaller when forecast farther hour, but s depends. Although s may also pick recent data, it is able to pick data which is far away in historical data. It tells why as forecast hour is farther, RMSE gets bigger when we use only similar reference data.

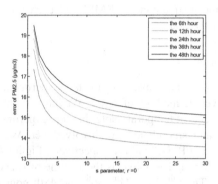

Fig. 5. The impact of parameter s on PM2.5 value forecast.

After these analyses of experiment, we decide to use only one recent reference data to forecast first 6 h, and use 15 recent data references and 15 similar data references to forecast the 7th to 48th hour respectively. We don't use 30 reference data respectively because of the convergence situation and time cost consideration. We don't use the best combination for each forecast hour for preventing from over-fitting. Therefore, we set the same reference volume for the 7th to 48th hour.

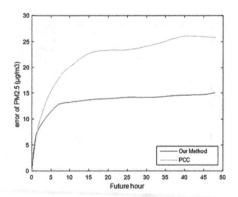

Fig. 6. PM2.5 RMSE of different method.

To display the performance of our method, we compare other famous method. PCC is compared with our method. In terms of PCC method, we compute continuous meteorology RMSE period in the historical data. And find which is similar to continuous meteorology RMSE of the target period. PM2.5 values difference of similar periods will used for PM2.5 forecast. Figure 6 is the performance of two different methods. Our method performs much well than PCC obviously. We can see that our method can converge the RMSE and forecast farther hour stably.

5   Conclusion and Future Work

Recently, air pollution mainly induced by particle matters becomes an important issue. PM2.5 makes a significant impact on human health, especially. We can easily get current PM2.5 values, but we still need farther PM2.5 forecast to plan our route. This research makes effort in PM2.5 value forecast. We found the connection between meteorological data and PM2.5 value. We use recent and similar references. Because air quality is continues data, it needs only one recent reference for first 6 h forecast. To forecast father PM2.5 values, we utilize 15 recent references and 15 similar references. Although the result of experiment shows that RMSE gets smaller as references data gets greater, we also found the obvious convergence situation. For reducing computation load, we don't use greater reference volume. We also compare our method with famous method PCC, and the result of our method performs well than the result of PCC. Our result converges obviously and performs steady of forecast. Although we only forecast future 48 h in this research, our method can forecast farther PM2.5 value. Our method can forecast not only data of Taipei, but also anywhere with meteorology and air quality data.

References

1. Bouarar, I., et al.: Monitoring and forecasting air quality over china: results from the PANDA modeling system. In: IGAC 2016 Science Conference (International Global Atmospheric Chemistry) (2016)
2. Domańska, D., Łukasik, S.: Handling high-dimensional data in air pollution forecasting tasks. Ecol. Inform. **34**, 70–91 (2016)
3. Hsieh, H.P., Lin, S.D., Zheng, Y.: Inferring air quality for station location recommendation based on urban big data. In: Proceedings of the 21th ACM SIGKDD International Conference on Knowledge Discovery and Data Mining, pp. 437–446 (2015)
4. Lu, X., Wang, Y., Huang, L., Yang, W., Shen, Y.: Temporal-spatial aggregated urban air quality inference with heterogeneous big data. In: Yang, Q., Yu, W., Challal, Y. (eds.) WASA 2016. LNCS, vol. 9798, pp. 414–426. Springer, Cham (2016). https://doi.org/10.1007/978-3-319-42836-9_37
5. Zheng, Y., et al.: Forecasting fine-grained air quality based on big data. In: Proceedings of the 21th ACM SIGKDD International Conference on Knowledge Discovery and Data Mining, pp. 2267–2276 (2015)
6. Zhu, J.Y., et al.: pg-causality: identifying spatiotemporal causal pathways for air pollutants with urban big data. IEEE Trans. Big Data (2017)

Quad-Partitioning-Based Robotic Arm Guidance Based on Image Data Processing with Single Inexpensive Camera For Precisely Picking Bean Defects in Coffee Industry

Chen-Ju Kuo[1], Ding-Chau Wang[2], Pin-Xin Lee[2], Tzu-Ting Chen[1],
Gwo-Jiun Horng[3], Tz-Heng Hsu[3], Zhi-Jing Tsai[2], Mao-Yuan Pai[4(✉)],
Gen-Ming Guo[2], Yu-Chuan Lin[1], Min-Hsiung Hung[5], and Chao-Chun Chen[1]

[1] IMIS/CSIE, National Cheng Kung University, Tainan, Taiwan
{P96074090,P96084126,chaochun}@mail.ncku.edu.tw, duke@imrc.ncku.edu.tw
[2] MIS, Southern Taiwan University of Science and Technology, Tainan, Taiwan
{dcwang,9A490021,4A590147,sambuela}@stust.edu.tw
[3] CSIE, Southern Taiwan University of Science and Technology, Tainan, Taiwan
grojium@gmail.com, hsuth@mail.stust.edu.tw
[4] General Research Service Center,
National Pingtung University of Science and Technology, Neipu, Taiwan
mypai@mail.npust.edu.tw
[5] CSIE, Chinese Culture University, Taipei, Taiwan
hmx4@faculty.pccu.edu.tw

Abstract. In this paper, we propose a bean defect picking system with the quad-partitioning-based robotic arm guidance method, aimed at automatically and precisely picking bean defects in coffee industry. We assume the adopted inexpensive devices, including a robotic arm, a camera, and an IoT (Internet of Things) device, have only basic functions. For successfully picking the small size of beans as possible, stably moving the arm head to the target bean is the key technique in this topic. To achieve this goal under hardware limits, we design an iterative robotic arm guidance method to move the arm head close to the target with quad-partitioning relationships in the camera's visual space by using image data processing techniques. The error distance after k iterations of the proposed method is approximately estimated as $\sqrt{(\frac{d_x}{2^{k+1}})^2 + (\frac{d_y}{2^{k+1}})^2}$, where d_x and d_y are the width and the length of the field of view.

Authors thank the "Intelligent Service Software Research Center" from STUST for providing robotic arms used in our experiments and many helps on control of arm devices during development. This work was supported by Ministry of Science and Technology of Taiwan under Grants MOST 107-2221-E-006-017-MY2, 107-2218-E-006-055, 107-2221-E-218-024, and 107-2221-E-034-013. This work was also supported by the "Intelligent Manufacturing Research Center" (iMRC) from The Featured Areas Research Center Program within the framework of the Higher Education Sprout Project by the Ministry of Education in Taiwan.

© Springer Nature Switzerland AG 2019
N. T. Nguyen et al. (Eds.): ACIIDS 2019, LNAI 11432, pp. 152–164, 2019.
https://doi.org/10.1007/978-3-030-14802-7_13

We conduct a case study to validate the proposed method. Testing results show that the proposed system successfully picks bean defects with our proposed robotic arm guidance method.

Keywords: Spatial data analysis · Robotic control · Industrial automation · Iterative adjustment · Fault removal

1 Introduction

In the Industry 4.0 era, adding intelligence to the equipment becomes a main investment nowadays to strengthen their global competitiveness for enterprises [3,4]. Certain small and medium enterprises use intelligent technologies to increase the smart functions for spanning applications of existing products [2]. For example, some robot companies use the image technologies to increase arm automation for fitting various manufacturing or agriculture applications. Such data-driven automation fits robotic devices into different industries by merely replacing software packages, instead of recreating new products. Hence, many enterprises establish data-driven automation technologies to earn value-added profits in the current smart manufacturing trend.

The coffee industry contains many labor-intensive companies in the production chain, covering various bean processing stages, such as planting, harvesting, roasting, and brewing. Among these stages, one key factor in determining the coffee quality is the green bean defect removal before the roasting process. For launching specialty coffee products, most bean companies hire employees to sort beans into different levels. Even, bean companies have further defect removal processes to roasted beans for ensure the bean quality.

Due to the enormous business profit, the automation requirement for removing bean defects are emerged among coffee bean enterprises in past years. Most current solutions to bean sorting are of the batch processing manner, and can be mainly divided into two categories [1,5]: one is the sorting by color, the other is the sorting by density. For sorting-by-color solutions, machines divided beans into many batches and remove batches of beans which contain miscellaneous stuff (e.g., small stones or withered branches) by examining colors with the computer image technology. For sorting-by-density solutions, machines use fans to blow beans in each batch, and beans of less density (usually indicate defects) are blown away. Nevertheless, these batch-mannered solutions still leave certain bean defects inside the selected beans. Since one bean defect may affect 50 fine beans during the roasting process, many experts explicitly exposed that these solutions insufficiently improve the flavor of brewing coffee.

Thus, the bean-level sorting methods are needed to fit requirements from the coffee industry. However, the bean-level sorting may encounter some technical difficulties. In order to dealing with an individual bean, more devices, e.g., robotic arms and sensors, are needed and increase the financial cost. In addition, due to the small size of beans, most economical robotic arms cannot precisely pick the target bean. The price of robotic arms that can precisely manipulate on beans usually exceeds the budget that can be acceptable to small and medium coffee companies. This motivates us to develop a budget bean-picking solution (i.e., adopting inexpensive hardware devices) to the coffee industry.

In this paper, we propose a bean defect picking system with the quad-partitioning-based robotic arm guidance method, for automatically and precisely picking bean defects. The adopted inexpensive devices, including a robotic arm, a camera, and an IoT (Internet of Things) device, in our system provide only basic functions for satisfy business requirement. For successfully picking the small size of beans as possible, how to move the arm head to the target bean is the key technique in this topic. To achieve this goal under hardware limits, we design an iterative robotic arm guidance method to move the arm head close to the target with quad-partitioning relationships in the camera's visual space by using image data processing techniques. Firstly, our proposed method decomposes the current visual space obtained from the camera into four equal quadrants. Secondly, the proposed method moves the head to the intersection of the four quadrants, and identifies the quadrant that the target bean resides. Thirdly, the arm head adjusts its position according to the quad-partitioning relationship between the head and the target bean. This adjustment is called the compensation stage. Fourthly, the head moves down vertically for reducing distance between the head and the target bean. The process continues until the head is sufficiently close to the target bean. In addition, the approximate error distance after certain iterations of the proposed method is also derived in this work. Finally, we develop a prototype of the bean defect picking system for conducting integrated tests. Testing results show that the proposed quad-partitioning-based arm guidance method can successfully move the arm head to the top of target bean, and the measured error distances follow the derived error estimation theorem. This paper can be a useful reference for small and medium enterprises to build bean defects removal systems for the coffee industry.

The rest of this paper is organized as follows. Section 2 describes the architecture of the proposed bean defect picking system. Next, Sect. 3 presents the proposed quad-partitioning-based arm guidance method. Then, we conduct a case study and show the results in Sect. 4. Finally, we conclude the paper in Sect. 5.

2   Architecture of Proposed Bean Defect Picking System

Figure 1 shows the reference architecture of the proposed bean defect picking system, which mainly includes three hardware components: the robotic arm, the camera, and the IoT (Internet of Things) device, aiming at picking coffee bean defects from good ones. The camera is fixed by the head of the robotic arm to provide the head's neighboring spatial status. The robotic arm and the camera all connect to the IoT device, which is the controller of the whole system. For achieving high degree of automation, the design system uses the image data processing techniques from its vision capability to align the head position with our proposed arm guidance method (will be presented in the next section) to the target bean as possible in certain iterations in order to maximize the successfully picking possibility.

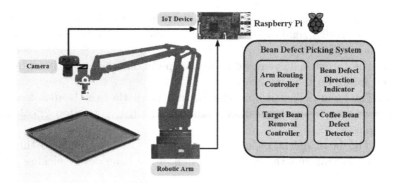

Fig. 1. Architecture of the proposed bean defect picking system.

Recall that the hardware devices used in this work need to be restricted within an acceptable budget limit. Hence, our adopted robotic arm, camera, and IoT device are assumed to provide merely basic functions. For example, the camera does not equip with ranging capacity and has only fixed focus, and the robotic arm may incur less precise movement due to simple mechanical designs. In addition, the IoT device is a single-board computer with limited computation power, e.g., Raspberry Pi 3 in this work. Hence, we need to create the required software modules by recruiting advance data processing techniques in order to achieve the bean defect removal job. In this way, the developed bean-picking solution provides the value-added profit in the smart manufacturing trend.

The proposed system contains four decisive software modules: the *coffee bean defect detector* identifies all bean defects and finds out all their positions, the *bean defect direction indicator* finds out the direction of given target bean to the current visual space, the *arm routing controller* moves the head to the top of the target bean, the *target bean removal controller* picks up the given target bean. The key mechanisms of the bean defect direction indicator and the arm routing controller include: (1) the transformation between pixels and distance and (2) the quad-partitioning-based arm guidance method, which will be described in details in the next section.

The operational flow of picking coffee bean defects are described as follows.

Step 1: The user sends a bean defect picking request to the system, and the coffee bean defect detector is activated to identify the positions of bean defects, which are composed in the bean-defect list and sent to the arm routing controller.

Step 2: The arm routing controller gets a position out from the bean-defect list. Then the arm routing controller moves the arm head to the top of the bean defect by the given position. During arm routing, our proposed arm guidance method utilizes the visual data, obtained from the bean defect direction indicator connecting to the camera, to increase precision of the arm routing.

Step 3: Once the arm head moves to the top of the target bean, the arm routing controller sends a bean-picking command to the target bean removal controller.

Step 4: The target bean removal controller picks out the target bean by using the air bump on the arm head.

Step 5: The bean defect picking system checks whether the bean-defect list is empty or not. If any bean defect exists in the list, goto Step 2; otherwise, goto Step 6.

Step 6: The bean defect picking system moves the head back to the initial position, and informs the user completion of the bean defect picking job.

3   Quad-Partitioning-Based Arm Guidance Method

This section presents the quad-partitioning-based arm guidance method, which enables the robotic arm can automatically remove all identified bean defects.

3.1   Transformation Between Pixels and Distances

Since our bean defect picking system needs to estimate real-world distance with the camera, we discuss the transformation between pixels in the camera's image plane and the associated distance. The transformation between pixels and distances is based on the knowledge of geometrical optics. This mechanism is designed for the bean detect direction indicator module mentioned in the last section. Figure 2 shows the image forming of a rectangle shape in the camera's visual space. The right-hand side of the figure represents the transformation relationship, which indicates that the two triangles corresponding to the field of view and image plane, respectively, are similar in geometry [7]. Hence, the relationship $P : W = F : H$ holds, where F is the focus, W is the width of the rectangle shape (field of view), P is the pixels of the width of the shape in the image plane, and H is the distance between the camera and the rectangle shape in the real world. The relationship can be further rewritten as

$$F \times W = P \times H \tag{1}$$

which is the key formula in the following derivation.

Let w be the width of the tray, p_w be the pixels of the tray width in the image plane, and h_0 be the height of the arm head, where the tray can be completely covered in camera's visual space. Then the focus of the camera, denoted as f, can be estimated by using Eq. (1), and be expressed as the following equation:

$$f = \frac{p_w \times h_0}{w} \tag{2}$$

Assume the camera does not change the focus. Then, the estimated f can be treated as a constant in the following derivations. After obtaining f, the function $pixel2dist(p,h)$, transforming the given pixels p with the arm head in the height h to the real-world distance (denoted as d), can be implemented by using the following equation:

$$d = \frac{p \times h}{f} = \frac{(p \times h) \times w}{p_w \times h_0} \tag{3}$$

Note that when the arm head moves downward vertically, the current height H decreases, as shown in Fig. 3. In this situation, Eq. (1) still holds. Considering the situation that the arm head moves down with distance $\frac{H}{2}$, the maximal distance that can be captured by the camera after movement (denoted as V), i.e., the maximal width in the updated field of view, can be estimated as:

$$\frac{V}{H/2} = \frac{P}{F} = \frac{W}{H} \tag{4}$$

$$\Rightarrow V = \frac{W}{2} \tag{5}$$

That is, the visible distance becomes one half after a vertical movement. Following the derivation, the updated field of view becomes one quarter (i.e., $\frac{1}{2} \times \frac{1}{2} = \frac{1}{4}$) while the arm head moves down with distance $\frac{H}{2}$.

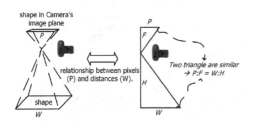

Fig. 2. Illustration of transformation between pixels and distances.

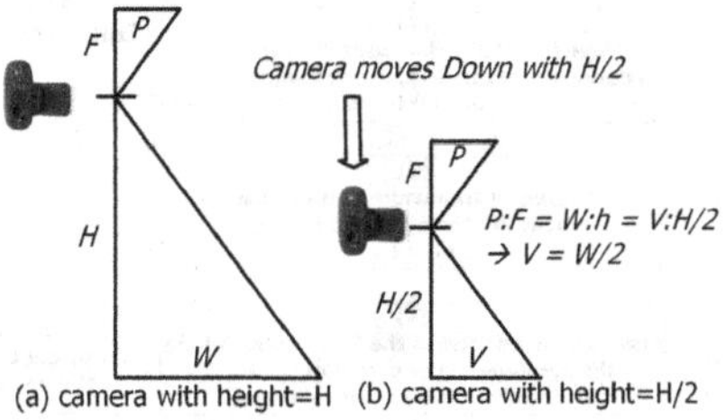

Fig. 3. Illustration of transformation between pixels and distances for a vertical movement.

3.2 Flowchart of the Proposed Method

This method is designed for the arm routing controller module mentioned in the last section. The goal of the quad-partitioning-based arm guidance method is to move the arm head to the top of the given bean defect for precisely aligning it to the target bean, so that the air bump on the arm head can pick the target bean with high success probability. Notice that the arm device equipped with a camera is aware of the neighboring status through assistance of the visual capability. Thus the arm knows the relative spatial relationship to beans after it adjusts its head position. By using this property, one key foundation of our arm guidance method is to let the arm head stays in the south-west of the target bean as possible during adjustment in multiple iterations. With such the key foundation, the process design can be greatly simplified in each iteration.

Figure 4 shows the flowchart of the proposed quad-partitioning-based arm guidance method. The basic idea of the method is to adjust positions of the arm head in multiple iterations, and the iterative process continues until the arm head is very close to the target bean defect. For each iteration, three are two phases to adjust the arm head: the horizontal movement and the vertical movement. For the horizontal movement (i.e., Phase 1), the quadrants with size $\frac{W_x}{2^i} \times \frac{W_y}{2^i}$ are used in the i-th iteration, where W_x and W_y are the width and the length of the field of view to the camera in the initial height. There are two stages in the horizontal movement phase. In the first stage, the arm head moves toward its north-east (NE) from the bottom-left corner to the upper-right corner of the quadrant, i.e., the intersection of four quadrants in the current visual space. Then the camera updates the current visual space by taking an image and detects the relative direction of the target bean. In the second stage, called the compensation stage, the arm head moves to the new adjustment point, which supposes to be closer to the bean than previous positions. The compensation principle is based on the relative direction of the target bean, represented by the quad-partition

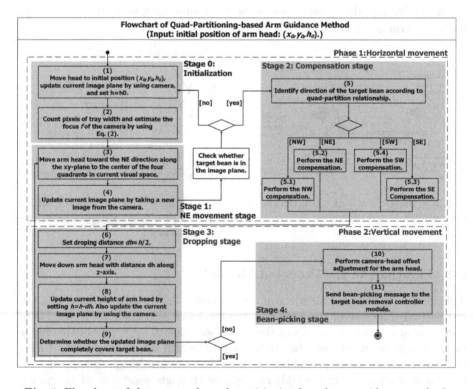

Fig. 4. Flowchart of the proposed quad-partitioning-based arm guidance method.

relationship between the arm head and the target bean, and will be presented later. After the compensation stage, the arm head is always in the south-west (i.e., bottom-left corner) of the target bean.

After the horizontal movement, the arm moves follows the vertical movement (Phase 2), which also contains two stages. In the third stage, called the dropping stage, the arm head moves vertically with distance of the half current height of the arm head. After the dropping stage, the camera again updates the current visual space, and checks whether the arm-head adjustment process continues. If the target bean is completely inside the updated visual space, meaning that the arm head can be aligned further through the visual information, then the method goes to Stage 1 to continue the adjustment process; otherwise, goes to Stage 4. In the forth stage, the arm head performs the camera-head offset adjustment to correct the position difference between the camera and the air bump on the arm head, and then the method informs the target bean removal controller to pick out the bean defect.

Recall that the compensation in Stage 2 is performed based on the direction relationship between the arm head and the target bean. Figure 5 shows direction definitions of four quadrants in the camera's visual space. The horizontal axis in a dashed line gives the east and the west; the vertical axis in a dashed line gives the north and the south. Then, the four quadrants in the quad-partition relationship are named according to their direction to the camera. For example, the NE quadrant stands for the one in the north-east of the camera, i.e., the upper-left one. Other three quadrants, NW (north-west), SW (south-west), SE (south-east), follow such representation. In this way, we can represent the direction of the target bean relative to the arm head.

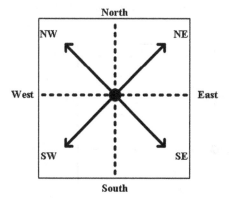

Fig. 5. The quad-partition relationship.

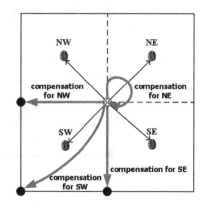

Fig. 6. Illustration of the compensation stage in Fig. 4.

Figure 6 illustrates the compensation principle, mentioned in the flowchart of Fig. 4. Recall that one key foundation of our arm guidance method is to let the arm head stays in the south-west of the target bean as possible. The compensation is needed because the head moves first toward its north-east to the center of four quadrants in Stage 1, and such movement amount may be too far and violate the foundation mentioned above. Hence, the compensation are performed to adjust the head to the south-west of the target bean according to the relative direction of the target bean to the arm head. The compensation stage can be discuss in the following four cases, each of which are drawn in a bold arrow line in the figure:

- North-western (NW) direction: the arm head moves left with the width of a quadrant.
- North-eastern (NE) direction: no compensation is needed to the arm head.
- South-eastern (SE) direction: the arm head moves down with the height of a quadrant.
- South-western (SW) direction: the arm head moves to the bottom left corner of the quadrant that the target bean resides.

An example of the quad-partitioning-based arm guidance method, shown in Fig. 7, demonstrates the arm head movement projecting onto the horizontal plane by using the proposed arm guidance algorithm after three iterations. Since only the horizontal movement behavior can be visualized in this situation, Stages 3 and 4 in the vertical movement are not shown in the figure. In the first iteration, the head moves to the center of the visual space, marked as "Stage 1 ($k = 1$)". The head then finds out the target bean is inside the NW quadrant, and thus it moves left to the point marked as "Stage 2 ($k = 1$)" in the compensation step. In the second iteration, the head moves toward NE to the center of four

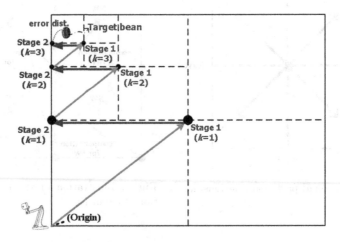

Fig. 7. An illustration of three-round arm head movement projecting onto the horizontal plane by using the proposed arm guidance algorithm.

subquadrants, marked as "Stage 1 ($k = 2$)". Again, the head finds out the target bean is inside the NW quadrant, and then it moves left to the point marked as "Stage 2 ($k = 2$)" for compensation. The arm head continues the movement in the third iteration, and gradually approaches the target bean. After movements, the error distance exists between the head and the target bean after k iterations (e.g., $k = 3$ in this example), and it is estimated in the next subsection.

3.3 Error Estimation of the Proposed Arm Guidance Method

During the arm head approaching the bean defect, the quad-partitioning-based arm guidance method will select a quadrant for further aligning the arm head to the target bean, and other three quadrants are ignored. Such guidance property leads the arm head gradually close to the bean defect in each iteration, until the bean is hard to be classified into a quadrant in the camera's visual space. In practice, this usually happens in cases that the virtual space cannot completely cover the target bean. The following theorem describes the approximate error distance between arm head and the target bean after movement of k iterations.

Theorem 1. *By using the proposed quad-partitioning-based arm guidance method, the error distance ϵ_k between the arm head and the target bean in the k-th iteration, can be estimated by the following equation:*

$$\epsilon_k \approx \sqrt{\left(\frac{d_x}{2 \cdot 2^k}\right)^2 + \left(\frac{d_y}{2 \cdot 2^k}\right)^2} \tag{6}$$

where d_x and d_y are the width and length to the field of view of the camera in the initial status.

Notice that the above theorem only provides the theoretic error distance. the inexpensive arm device may incur error distance in practices due to the simple mechanical design issues, which will be studied in the next Section.

4 Case Study

We developed a prototype of the proposed bean defect picking system with three inexpensive devices, including robotic arm, camera, and IoT device, whose total cost is less than 700 USD (purchased in Sep., 2018) and can be acceptable by most small/medium coffee companies. The arm device is uArm Swift Pro 3-axes robotic arm. The camera is the Logitech C270 HD webcam with maximal 720p resolution. The IoT device is a single-chip computer of Raspberry Pi 3 with 1.2 GHz processor, 1 GB RAM, and 16 GB disk. The software modules mentioned in Sect. 2 are developed with Python, and some image processing functions are achieved with the OpenCV [6] library. The size of coffee tray is $175 \times 125 \, \text{mm}^2$, and the initial height of the arm head is 250 mm. In such setting, the camera's image plane completely covers the bean tray, so that the focus f can be estimated for experiments with Eq. (2) in Sect. 3.1.

4.1   Testing Snapshots

In order to verify the effectiveness of the proposed bean defect picking system with the quad-partitioning-based arm guidance method, we design certain operational scenarios to test the developed system. Figure 8 demonstrates some snapshots during performance of the test scenario in different steps. Due to space limit, we display only the coffee bean defects picking service. The testing snapshots show that the proposed bean defect picking system works successfully.

4.2   Integrated Testing Results

The main performance concern of integrated testing results is the error distance incurred by the proposed system, which indicates the usability of the arm picking bean defects. Thus, the experiment is designed to measure the error distance in different iterations for validating efficacy of the quad-partitioning-based arm guidance method. Figure 9 shows the experimental results, where The horizontal axis is iteration k, and the vertical axis is the error distance in millimeters (mm). Except evaluating our proposed method, the error distances obtained by Theorem 1 and manual measurement are also shown in the figure.

Three observations are obtained from Fig. 9. Firstly, the error distances of our method is quite close to the manual measurement, which indicates hardware of the arm device used in the experiments is under healthy status. In addition, the error distances estimated by Theorem 1 also provide useful references for designed methods for most k's, but are less precise in the first two iterations. Secondly, the error distance decreases with the number of iterations. This shows the proposed arm guidance method indeed gradually lead the arm head to approach the target bean. Thirdly, the different amount of error distances in two

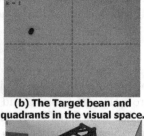

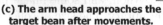

(b) The Target bean and quadrants in the visual space.

(a) The arm head in initial step.   (c) The arm head approaches the target bean after movements.   (d) The arm head picks the target bean with air bump.

Fig. 8. Snapshots of the robotic arm in different steps.

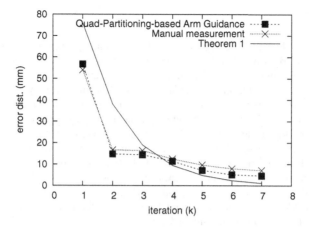

Fig. 9. The error distance in each iteration.

consequent iterations also decreases as iterations in most cases. This means that the proposed arm guidance method is stable and the error distance does not vary irregularly. This property holds until the camera is hard to distinguish the bean in the four quadrants, which usually happens when the camera is very close to the target bean. In our experiments, this happens after $k \geq 5$. From the above observations, the proposed quad-partitioning-based arm guidance method indeed efficiently and effectively leads the arm head very close to the target bean.

5   Conclusions

It has been known that coffee bean defect removal is critical to the quality of brewing coffee. However, there is currently few automated and economic solutions to remove coffee bean defects. In this paper, we proposed a bean defect picking system for automatically picking out bean defects with inexpensive devices. The most difficulty in the system is how a robotic arm precisely picks bean defects. Our proposed quad-partitioning-based arm guidance method provides an iterative process to adjust the head position with the quad-partitioning relationship to align the air bump to the target bean as possible. We implemented a prototype of the proposed system and conducted experiments to demonstrate its bean picking capability. The case studies show that the proposed system successfully and efficiently picks bean defects. This paper can be a useful reference for small and medium enterprises to build bean defects removal systems for the coffee industry.

References

1. Arboleda, E.R., Fajardo, A.C., Medina, R.P.: An image processing technique for coffee black beans identification. In: 2018 IEEE International Conference on Innovative Research and Development (ICIRD), pp. 1–5, May 2018
2. Huang, B., Li, C., Yin, C., Zhao, X.: Cloud manufacturing service platform for small- and medium-sized enterprises. Int. J. Adv. Manuf. Technol. **65**(9), 1261–1272 (2013)
3. Lee, J., Bagheri, B., Kao, H.A.: A cyber-physical systems architecture for industry 4.0-based manufacturing systems. Manuf. Lett. **3**, 18–23 (2015)
4. Lee, J., Kao, H.A., Yang, S.: Service innovation and smart analytics for industry 4.0 and big data environment. Procedia CIRP **16**, 3–8 (2014)
5. Pinto, C., Furukawa, J., Fukai, H., Tamura, S.: Classification of green coffee bean images based on defect types using convolutional neural network (CNN). In: 2017 International Conference on Advanced Informatics, Concepts, Theory, and Applications (ICAICTA), pp. 1–5, August 2017
6. Pulli, K., Baksheev, A., Kornyakov, K., Eruhimov, V.: Real-time computer vision with OpenCV. Commun. ACM **55**(6), 61–69 (2012)
7. Salih, Y., Malik, A.S.: Depth and geometry from a single 2D image using triangulation. In: 2012 IEEE International Conference on Multimedia and Expo Workshops, pp. 511–515, July 2012

On the Analysis of Kelly Criterion and Its Application

Mu-En Wu[1], Wei-Ho Chung[2], and Chia-Jung Lee[3]([✉])

[1] Department of Information and Finance Management,
National Taipei University of Technology, Taipei, Taiwan
mnwu@ntut.edu.tw
[2] Department of Electrical Engineering,
National Tsing Hua University, Hsinchu, Taiwan
whchung@ee.nthu.edu.tw
[3] School of Big Data Management, Soochow University, Taipei, Taiwan
leecj2009@gm.scu.edu.tw

Abstract. We analyze the return of a game for a gambler after bidding T time steps. Consider a gamble with known odds and win rate, the optimal solution is to use Kelly criterion which determines the optimal fraction in each bidding step. In this paper we show that the logarithm of return when bidding optimal fraction is $KL(R||P(b)) - KL(R||P)$, where R is the proportion of winning \losing outcome in T time steps, $P(b)$ is the risk-neutral probability corresponding to odds b, and P is the gambler's individual belief about the win probability of the game. This argument shows that, in a gamble with fixed odds, the KL divergence of the win\lose proportion, say R, and the win rate, say P, determines the portion of the losing amount. On the other hand, the profit is determined by the proportion R and the odds b, irrelevant to win probability P. Any improvement is not obtainable even when the win probability is estimated precisely in advance.

Keywords: Kelly criterion · Optimal fraction · KL-divergence

1 Introduction

Financial trading is an interesting research issue to many experts and scholars. How to develop a good trading strategy to achieve long-term stable profit is the dream of many people [14]. For a single commodity, the most critical skill is the *position sizing* [12, 13], and for multiple commodities is the *portfolio optimization* [7, 11]. Both these two topics are belonging to the area of money management, the most important issues for financial trading or gambling [4]. The basic knowledge comes from the "Kelly criterion," which was provided by Kelly Jr. [1] at Bell Labs in 1956. Kelly put forward the Kelly formula while conducting research on the probabilistic errors in communications. Thorp [3] later applied this formula to the Blackjack poker and then to the global financial markets. William [2], who invented the William indicator in technical analysis, also used the Kelly formula to win several U.S. trading champions. This work was inspired via the follow-up money management approaches, and the methods

© Springer Nature Switzerland AG 2019
N. T. Nguyen et al. (Eds.): ACIIDS 2019, LNAI 11432, pp. 165–172, 2019.
https://doi.org/10.1007/978-3-030-14802-7_14

originating from variants of Ralph Vince's work. It was proved that capital grow fastest under the assumption of bidding the optimal fraction. The calculation of the optimal fraction requires the distribution of outcomes realized. Vince's work are trying to fill the gap between the theoretical bidding of gambles and practical trading in financial markets. However, for most cases such as horses racing, poker games, or financial market, the distributions of outcomes are unpredictable.

Although Kelly criterion is the process of optimization of bidding fraction in the gambling or trading, there are lots of shortcomings shown in the literatures [8–10]. One of the most serious problems is that the gambler's estimated win probability is different from the actual win probability distribution. Actually, the win probability of financial trading strategies changes with different time segments. This leads to a problem that when the actual win probability is unavailable to determine the optimal fraction, we can only make an estimation and calculate the win probability using the Kelly formula. It is straightforward to speculate that if the estimated win probability does not match the actual win rate, the error will be naturally reflected in the accumulated profit and loss of the fund; otherwise, the Kelly formula would not have been the optimal fraction under fixed odds and win rate. Therefore, this paper discusses how the profit and loss of funds are impacted when the estimated win probability deviates from the actual win probability.

Consider a gamble for $R = (R_w, R_l) \in [0, 1]^2$ be the win probability of the distribution of the actual outcome, where R_w is the actual fraction of winning and R_l is the fraction of losing, and $P \in [0, 1]^2$ be the distribution of the gambler's expected win-loss rate. According to our results, the average log return that can be obtained by the gambler is $KL(R||P(b)) - KL(R||P)$, where $P(b)$ is the corresponding fair probability to the odds of the bet, and $KL(\cdot)$ denotes the KL divergence [5]. Notice that the maximum value of this formula occurs at $R = P$; in other words, when the estimated win probability is the same as the actual win rate, and the average log return will reach its maximum value, that is $KL(R||P(b))$. In addition, the formula $KL(R||P(b)) - KL(R||P)$ is similar to the result delivered form the pricing mechanism in the prediction market discussed in [6]. Based on this knowledge, we can conclude that in a bet with fixed odds and an unknown win rate, the upper limit of the return is predetermined, while the error in win probability estimation will only lead to the relative entropy calculated by the upper limit of the return minus the win probability error. We can call this the "cost of estimation error."

The remaining of this paper is shown as follows: Sect. 2 provides the detail the Kelly criterion, which is the main problem we discussed in this paper. In Sect. 3 we show the calculation of our analysis, and we give some simulation in Sect. 4. Finally, we conclude and show some future work in Sect. 5.

2   Preliminaries

Consider a gamble with win probability p and odds b. The odds b means that the gambler bids m dollars before the gamble and will receive $m(1 + b)$ dollars if the realized outcome is winning, and the gambler will lose 1 dollar if the outcome is losing.

In other words, if a gambler bets 100 dollars, he either loses 100 dollars or wins the profit of $100(1+b)$ dollars. An interesting question here is that in such bets, if the gambler can play infinite times and bet his money on hand arbitrarily, how should a gambler bet to maximize his profit growth?

Kelly [1] gave a complete solution to this problem. Assume that a gambler plays T times, during which he wins W times and loses L times ($T = W + L$). Without loss of generality, we may assume that his initial wealth is 1 dollar and that each time he places a bet, he uses a fixed proportion, f, of his wealth. In the paper, the gambler's existing wealth in the t-th step is assumed to be A_t. Hence, we have the following derivations:

If the gambler wins at the t-th step, we have $A_t = A_{t-1}(1 + bf)$. Similarly, if the gambler loses at the t-th step, we have $A_t = A_{t-1}(1 - f)$.

Since the gambler wins W times and loses L times in playing T times, we can conclude that

$$A_T = (1 + bf)^W (1 - f)^L.$$

Note that we want to maximize A_T by determining the bidding fraction f. We can take the log (natural logarithm) of A_T and divide it by T as follows:

$$\frac{1}{T}\log A_T = \frac{W}{T}\log(1 + bf) + \frac{L}{T}\log(1 - f)$$

Recall that the win probability of the gamble is p. This means if we take $T \to \infty$, we have $\frac{W}{T} \to p$ and $\frac{L}{T} \to 1 - p$. Therefore, we can get

$$\lim_{T \to \infty} \frac{1}{T}\log A_T = p\log(1 + bf) + (1 - p)\log(1 - f)$$

Since finding f to maximize A_T also implies maximizing $\lim_{T \to \infty} \frac{1}{T}\log A_T$, we differentiate the right side of the above equation with respect to f and obtain the maximum value of A_T when setting

$$f = \frac{p(1 + b) - 1}{b},$$

which is the famous Kelly formula [1]. Take the coin tossing for example. By setting the "Head" denotes win and "Tail" denotes lose, we consider a gamble with win probability 50% and the odds of 2. According to the above argument, the optimal fixed strategy for the gambler is to bid 25% ($= \frac{50\%(1+1)-1}{2}$) of his wealth each time.

In this paper we consider the problem: How much profit we can make when we always bidding the optimal fraction in a gamble. We show that the profit depends on the distance between the real proportions of winning and losing after T times bidding and the win probability the gambler predicts.

3 Log Return by Bidding Optimal Fraction

Before discussing the question mentioned above, we have to define the fair probability corresponding to a gamble. Usually a gamble accompanies with the win probability and the odds. However, the actually rule for a gamble is only the odds without the win rate. For example, a coin tossing with the odds 2 and the win probability 50% is a common gamble. The win probability 50% comes from there are 50% probability the outcome is "Head" and 50% probability the outcome is "Tail". However, this is not always the true when playing T times. You may toss a coin 10 times with 6 times "Head" and 4 time "Tail" happened. In this case the results of the outcome come from the binomial distribution and we cannot do anything improved on this uncertainty even we know the win probability is about 50%. On the other words, the win-rate 50% in the gamble is just a predicted value for the individual gambler. The unique rule of a gamble remains only the odds.

Now we consider a gamble with fixed odds of b. The odds b implies the reasonable and fair probability of the gamble, satisfying the zero expectation. We can calculate the corresponding fair probability of b, denoted by $P(b)$, where $P(b)$ and b satisfy

$$P(b) \times (1+b) - 1 = 0.$$

In other words, if the bet is fair with the fixed odds of b, the win probability should be $P(b) = \frac{1}{1+b}$, which is unprofitable to the gambler and the dealer both. For notational convince, we denote $P(b)$ as the fair probability distribution of a gamble with the odds. That is,

$$P(b) = \left(\frac{1}{1+b}, \frac{b}{1+b} \right)$$

In this paper, it is assumed that the gamble has been played T times, and the distribution of outcomes are W times of winning and L times of losing. We denote $E_T(f)$ as the return after playing T times by bidding the fraction, f, at each time step. Thus, the calculation of $E_T(f)$ is as follows:

$$E_T(f) = (1+bf)^W \times (1-f)^L$$

According to the previous argument, we may assume the gambler's individual belief for the win probability of the gamble is p, and he will place a bet according to the Kelly formula. Thus, the gambler's optimal fraction to wager is as follows:

$$f^* = \frac{p(1+b) - 1}{b}.$$

Since the gamblers' profit is $E_T(f)$ by bidding f fraction, we can substitute f^* into $E_T(f)$ as follows:

$$E_T(f^*) = \left(1 + b\frac{p(1+b)-1}{b}\right)^W \left(1 - \frac{p(1+b)-1}{b}\right)^L.$$

Using the similar technique to the deduction of Kelly formula, we then take the natural logarithm of $E_T(f^*)$ and then divide the equation by T. We have

$$\frac{1}{T}\log E_T(f^*) = \frac{1}{T}\log\left((p(1+b))^W \left((1-p)\frac{1+b}{b}\right)^L\right)$$

$$= \frac{W}{T}(\log(p) + \log(1+b)) + \frac{L}{T}\left(\log(1-p) + \log\left(\frac{1+b}{b}\right)\right).$$

We add $\frac{W}{T}\log\left(\frac{W}{T}\right)$ and $\frac{L}{T}\log\left(\frac{L}{T}\right)$ in the beginning and then subtract them in the end of the equation to obtain the following:

$$\frac{1}{T}\log E_T(f^*) = \left(\frac{W}{T}\left(\log\left(\frac{W}{T}\right) - \log\left(\frac{1}{1+b}\right)\right) + \frac{L}{T}\left(\log\left(\frac{L}{T}\right) - \log\left(\frac{b}{1+b}\right)\right)\right)$$

$$- \left(\frac{W}{T}\left(\log\left(\frac{W}{T}\right) - \log(p)\right) + \frac{L}{T}\left(\log\left(\frac{L}{T}\right) - \log(1-p)\right)\right)$$

Take $R = \left(\frac{W}{T}, \frac{L}{T}\right)$ as the realized proportion of winning and losing after T times bidding, $P = (p, 1-p)$ as the gambler's estimate of the win probability of the gamble, and $P(b) = \left(\frac{1}{1+b}, \frac{b}{1+b}\right)$ as the fair probability distribution corresponding to the odds b. Then, we simplify the above equation to the following form.

$$\frac{1}{T}\log E_T(f^*) = KL(R\|P(b)) - KL(R\|P),$$

where $KL(\cdot\|\cdot)$ is the relative entropy (also called KL divergence) [5]. The above equation shows the relation between the gambler's return and the belief about the estimation of the win probability after playing T times.

When the gambler's estimate P is close to the actual proportion R of winning and losing with the measurement of KL divergence, the right side of the equation is small and the profit is relatively high. Notice that when P and R are the same, the maximum profit is $KL(R\|P(b))$. We can draw a conclusion with the following theorem.

Theorem: Given that the odds of the bet are b, if a gambler wins W times and loses L times after playing T times and the gambler thinks the win probability is p, the average log return under the optimal fraction will be as follows:

$$KL(R||P(b)) - KL(R||P),$$

where $R = \left(\frac{W}{T}, \frac{L}{T}\right)$ is the actual distribution ratio of win and loss, $P = (p, 1-p)$ is gambler's estimate of the distribution, and $p(b) = \left(\frac{1}{1+b}, \frac{b}{1+b}\right)$ is the probability distribution of a fair gamble.

4   Simulations

Consider a game with win probability 50% and odds 2 while playing 10 times. We consider two cases for simulations. Figure 1(a), (b), (c) are distributions of returns under different bidding fractions when the realized outcomes are winning 0 times \losing 10 times, winning 1 times\losing 9 times, and winning 2 times\losing 8 times, respectively. Note all bidding fractions are non-profitable in these three cases. Figure 2 (a) is the distribution of returns under different bidding fraction when the realized outcomes winning 3 times\losing 7 times. In the cases of Fig. 1(a), (b), (c) and Fig. 2(a) are non-profitable. Figure 2(b), (c) are the distributions of returns under different bidding fraction when the realized outcomes winning 4 times\losing 6 times, and winning 5 times\losing 5 times. Figure 3(a), (b), (c), (d), (e) are the distributions of returns under different bidding fraction when the realized outcome is winning 6 times \losing 4 times, winning 7 times\losing 3 times, winning 8 times\losing 2 times, winning 9 times\losing 1 times, winning 10 times\losing 0 times, respectively. The green horizontal line represents the cost line, based on 1. The blue vertical line is the optimal bidding fraction under this distribution (R). The gray vertical line is the Kelly fraction at which the player believes that the win rate of the game is 50%. Under this experiment, the Kelly fraction is 25% (P).

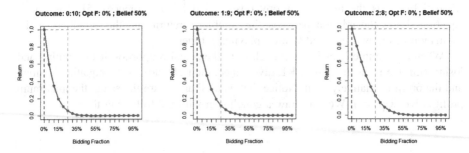

Fig. 1. (a), (b), (c) The distributions of returns under different bidding fractions when the realized outcomes are winning 0 times\losing 10 times, winning 1 times\losing 9 times, and winning 2 times\losing 8 times, respectively. Note all bidding fractions are non-profitable in these three cases.

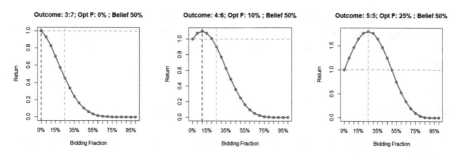

Fig. 2. (a) The distribution of returns under different bidding fraction when the realized outcome is winning 3 times\losing 7 times. There is still no profit in this case. (b), (c) The distribution of returns under different bidding fraction when the realized outcome is winning 4 times\losing 6 times, and winning 5 times\losing 5 times. In these two cases, there are existing profit. For the case of winning 5 times\losing 5 times, the optimal bidding fraction is the same as the bidding fraction for the Kelly fraction at which the player believes the win rate 50%.

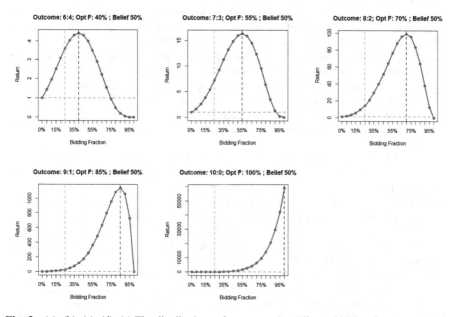

Fig. 3. (a), (b), (c), (d), (e) The distributions of returns under different bidding fraction when the realized outcome is winning 6 times\losing 4 times, winning 7 times\losing 3 times, winning 8 times\losing 2 times, winning 9 times\losing 1 times, winning 10 times\losing 0 times, respectively

5 Conclusions

This paper proves that, when a gambler estimates the win probability distribution and bids the optimal fraction based on his believed distribution, the expected log return can be represented by the KL divergence. For a gamble with fixed odds, the closer the actual win probability and the estimated probability P are, the larger the expected profit is. According to our results, for a fair gamble, if the corresponding probability is the same as the actual win and loss rate, the optimal condition for the gambler is breakeven.

References

1. Kelly, J.L.: A new interpretation of information rate. Bell Labs Tech. J. **35**(4), 917–926 (1956)
2. William, L.R.: How I Made One Million Dollars Last Year Trading Commodities. Windsor Books, Publisher (1979)
3. Thorp, E.: Beat the Dealer: A Winning Strategy for the Game of Twenty-One
4. Vince, R.: The Mathematics of Money Management, Risk Analysis Techniques for Traders, A Wiley Finance Edition. Wiley (1992)
5. Cover, T.M., Thomas, J.A.: Elements of Information Theory, Wiley Series in Telecommunications and Signal Processing (2006)
6. Chou, J.-H., Lu, C.-J., Wu, M.-E.: Making profit in a prediction market. In: Gudmundsson, J., Mestre, J., Viglas, T. (eds.) COCOON 2012. LNCS, vol. 7434, pp. 556–567. Springer, Heidelberg (2012). https://doi.org/10.1007/978-3-642-32241-9_47
7. Chen, C.-H., Chen, Y.-H., Wu, M.-E., Hong, T.-P.: A sophisticated optimization algorithm for obtaining a group trading strategy portfolio and its stop-loss and take-profit points. In: Proceedings of the IEEE International Conference on Systems, Man, and Cybernetics, IEEE SMC 2018, 7–10 October 2018, Miyazaki, Japan (2018)
8. Hsieh, C.H., Barmish, B.R.: On Kelly betting: some limitations. In: Proceedings of Annual Allerton Conference on Communication, Control, and Computing, Monticello, pp. 165–172 (2015)
9. Hsieh, C.H., Barmish, B.R., Gubner, J.A.: Kelly betting can be too conservative. In: Proceedings IEEE 55th Conference Decision Control, December 2016, vol. 476, pp. 3695–3701 (2016)
10. Hsieh, C.H., Barmish, B.R.: On drawdown-modulated feedback in stock trading. In: Proceedings of the IFAC World Congress, Toulouse, France, pp. 975–981 (2017)
11. Chen, C.-H., Chen, Y.-H., Wu, M.-E.: A GGA-based algorithm for group trading strategy portfolio optimization. In: Proceedings of the 4th Multidisciplinary International Social Networks Conference (MISNC 2017), 17–19 July 2017, Bangkok, Thailand (2017)
12. Wu, M.-E., Tsai, H.-H., Tso, R., Weng, C.-Y.: An adaptive Kelly betting strategy for finite repeated games. In: Proceedings of the 9th International Conference on Genetic and Evolutionary Computing (ICGEC 2015), 26–28 August 2015, Yangon, Myanmar (2015)
13. Wu, M.-E., Chung, W.-H.: A novel approach of option portfolio construction using the Kelly criterion. Accepted and to appear in IEEE Access (SCI)
14. Wu, M.-E., Wang, C.-H., Chung, W.-H.: Using trading mechanisms to investigate large futures data and its implications to market trends. Soft. Comput. **21**(11), 2821–2834 (2017). (SCI)

Single Image Super-Resolution with Vision Loss Function

Yi-Zhen Song, Wen-Yen Liu, Ju-Chin Chen$^{(\boxtimes)}$, and Kawuu W. Lin

National Kaohsiung University of Science Technology,
Kaohsiung, Republic of China
Kuas.chenjuchin@gmail.com

Abstract. Super-resolution is the use of low-resolution images to reconstruct corresponding high-resolution images. This technology is used in many places such as medical fields and monitor systems. The traditional method is to interpolate to fill in the information lost when the image is enlarged. The initial use of deep learning is SRCNN, which is divided into three steps, extracting image block features, feature nonlinear mapping and reconstruction. Both PSNR and SSIM have significant progress compared with traditional methods, but there are still some details in detail restoration. defect. SRGAN will generate anti-network applications to SR problems. The method is to improve the image magnification by more than 4 times, which is easy to produce too smooth. In this study, we hope to improve the EnhanceNet by training with different loss functions and different types of images to achieve better reconstruction results.

Keywords: Super-resolution · Deep learning · Generative adversarial network

1 Introduction

With the rapid development of technology, most people are familiar with smart phones or cameras to record their lives. The effects of using a phone or camera are more vivid than using a pen or paper. But most of the time, the images are not as good as expected due to the problems of shaking, ambient light or other degradation factors. Therefore, it is hoped that the image can be restored to a much clear appearance through some image processing technologies, such as deblurring, de-nosing, or super resolution. Among these technologies, super-resolution becomes important in recent years and is useful for the surveillance system. In Taiwan, surveillance systems are established in most regions, and when an accident occurs within the capturing range of the camera, the video can be used to clarify the cause and effect. But many cameras may be set up at different locations or angles, resulting in unsatisfied images. In our study, we would like to enhance image appearances via a deep network and three vision loss function are used to measure the similarity between the reconstruction result from the low-resolution image and the ground-truth high-resolution image. In order to measure the reconstruction performance, different types of images were applied in the experiments.

© Springer Nature Switzerland AG 2019
N. T. Nguyen et al. (Eds.): ACIIDS 2019, LNAI 11432, pp. 173–179, 2019.
https://doi.org/10.1007/978-3-030-14802-7_15

2 Related Work

Super resolution is to use single or multiple low-resolution images to calculate or learn to produce a high-quality super resolution image. This is ill-posed problem, since the reconstructed object is a low-resolution image that lacks high frequency information, the results can be varied with different solutions. we will explore common practices in the field of image super resolution, from the traditional Interpolation method to scale images to deep learning, which is highly concerned by many fields in recent years. Deep learning through computer autonomous learning feature extraction and feature mapping to produce super resolution images. Initially proposed deep convolutional neural networks are used to rebuild the super resolution image which is SRCNN [1] was proposed by Dong et al. in 2014. First, give a low-resolution image X, produce corresponding to the low-resolution image $\tilde{X}$ and record minification n. Then, use Bicubic interpolation to amplify n times to make it the same size as the original high-resolution image. These steps are all preprocessing. After completion, the image was taken as the input low-resolution image Y. The nonlinear mapping of F is achieved through a three-layer convolutional neural network. F(Y) produces an image which is the more similar to the original high-resolution image X the better. FSRCNN [2] was proposed by Dong et al. in 2016, and is the same team as SRCNN. The authors redesigned SRCNN's architecture, which is 40 times faster than SRCNN and has better recovery.

DRCN [3] was proposed by Kim et al. in 2016, and was the first to apply the Recursive Neural Network to image super resolution, and then to alleviate the effect of gradient disappearance with supervised learning recursion. More convolutional layers are used to increase the Receptive Field of the network, and skip connection in the ResNet [4] is combined to deepen the network structure and improve the performance. VDSR [5] was proposed by Kim et al. in 2016. In the paper, the authors mentioned that there was a great similarity between low-resolution images and high-resolution images, which would take a lot of time in training. In fact, we only need to learn the high frequency residuals between high and low resolution images. The authors adopt with the Residual Network similar to the stack filters, and deepen the Network layer to 20 layers. Cascading small filters are used in deep network structure many times to effectively utilize contextual information in large image blocks, and after completing the convolution layer, fill in 0 to the reduced part of image to keep the image size consistent. SRGAN [6] was proposed by Christian et al. in 2017 which is the first network that can reconstruct four times of upscaling images. Although a good PSNR effect is achieved when using traditional CNN to implement a single super resolution, but the magnification rate can only be less than 4. This is because previous MSE was minimized as the target parameter, once the magnification rate exceeds 4 times, the resulting high-resolution image will be too smooth. The authors propose a new perceptual loss function, consisting of content loss and adversarial loss, which enable the network to restore the realistic image texture and details.

3 System Overview

In our system, each image of the training set is divided into three types: one is the original image, one is the image which will be loaded into the network, and the other is magnified image. First, resize the original images to 128×128 and then reduce images by 4 times as input low-resolution images. The image was magnified to 128×128 using Bicubic method, which would be used in the reconstruction procedure. The images generated are computed by a trained convolutional neural network, then the residual image of the input image is obtained, and the training model is applied. The final reconstruction step of the experimental architecture is to combine the output images and the pre-processed Bicubic images to obtain the reconstruction results.

4 Super-Resolution with Vision Loss Minimization

The process of the whole system will be introduced in this section, including the architecture of the network and the training process.

4.1 Super Resolution Network

This study is based on Sajjadi et al. proposed EnhanceNet [7] system, which architecture is a deep convolution neural network, introduced the concept of GAN. The network is divided into generation network G and distinguish network D. Super-resolution image produced by generation network, and then input the image to the discrimination network to calculate the authenticity of the image. If the output value is closer to 1, the higher the authenticity of the image is; otherwise, the calculated value is closer to 0, the image is more inconsistent with the real image.

Network architecture contains five convolution layer [7], four ReLU activation function, ten Residual block and two Upsampling. This network adopts fully convolutional network (FCN), compared to normal CNN networks that connect the full connection layer after convolution layer, then map the feature map generated by the convolution layer to a fixed length feature vector. The FCN network can accept input images of any size. In addition, inspired by the VGG network, each Filter adopts the size of 3×3 to maintain a certain number of parameters while constructing a deeper network. The parameters were initialized by Xarvier, and the Residual block in the middle layer of the network was used to calculate the difference between the input feature map and the output feature map in the concept of ResNet. The feature of higher dimensions is extracted and amplified to the size of high-resolution images by the Nearest Neighbor Upsampling at the end of the network, which effectively reduces the computational complexity.

Residual Block
The concept of the residual module is proposed by ResNet [4]. It is proposed to use skip connection (also known as shortcut) to prevent neural network from being too deep to be trained. That is to make a reference to the input of each layer, learn the residual between input and output. This method is easier to optimize.

4.2   Training Process

The training database uses Microsoft COCO [8] dataset, total of 200,000 images, and the test database adopts dataset Set5 and Set14. The first step of training is to prepare high resolution images of different sizes, through 128×128 and 32×32 image interpolation method as after a merger with residual image and low-resolution images input. Input the low-resolution images to the main network G, the image size is $32 \times 32 \times 3$. First, the input images go through a convolutional layer and the ReLU activation function, the filter kernel size is 3×3, and then enter the layer of residual block. Then, the image is amplified through upsampling to output the residual image, and finally merged with the preprocessed 128×128 image to obtain the super-resolution image. The loss function consists of three parts, Perceptual loss, Texture synthesis loss and Adversarial loss. The super resolution images are thrown into each loss function to update the network parameters.

Perceptual Loss
Input the super resolution images and original high-resolution images into the trained VGG network [9], load the feature part of the network, extract the output results of Max pooling in the second and fifth layers, and calculate the MSE (Mean Squared Error) of both [7].

$$\mathcal{L}_P = \left\| \emptyset_2(I_{est}) - \emptyset_2(I_{HR})_2^2 + \emptyset_5(I_{est}) - \emptyset_5(I_{HR}) \right\|_2^2 \tag{1}$$

where $\emptyset_2$ denote as the second layer of feature map, $\emptyset_5$ denote as the fifth layer of feature map, I_{est} denote as super resolution image, I_H denote as high-resolution image.

Texture Synthesis Loss
Texture synthesis loss and Perceptual loss both use VGG network, but the loss function extracts the first convolution layer and the other first convolution after the first two max pooling layers. The extracted feature map was cut into 16×16 blocks, and Gram Matrix was adopted to perform mean square error calculation. The loss function formula is as follows [7]:

$$\mathcal{L}_T = \left\| G(\emptyset_{1.1}(I_{est})) - G(\emptyset_{1.1}(I_{HR})) \right\|_2^2 + \left\| G(\emptyset_{2.1}(I_{est})) - G(\emptyset_{2.1}(I_{HR})) \right\|_2^2 \\ + \left\| G(\emptyset_{3.1}(I_{est})) - G(\emptyset_{3.1}(I_{HR})) \right\|_2^2 \tag{2}$$

The formula of Gram Matrix is as follows [10]:

$$G_{i,j}^l = \sum_k F_{ik}^l F_{jk}^l \tag{3}$$

In the ℓ layer, the i-th feature map and the j-th feature map are intervolved to calculate the image style. For the feature difference between the original image and the super-resolution image, if the feature map with a large amount of spatial information is calculated directly by the Euclidean distance, the result will have great error. Therefore, Gram Matrix is used to filter its spatial information.

Adversarial Loss

The adversarial loss is the probability produced by inputting both super resolution and high-resolution images to the discrimination network. The Loss function adopts the cross-entropy BCE Loss of binary classification, which formula is as follows:

$$\mathcal{L}_A = -log(D(G(z))) \tag{4}$$

Discrimination network D input for high resolution and super resolution images. The architecture is mostly for the two kinds of combinations, one is convolution layer whose stride is 1 and uses LeakyReLU activation function, the other is convolution layer whose stride is 2 and uses LeakyReLU activation function. The image size of the feature map is reduced by half through the stride is 2, and double the number of filters, making each convolution layer the same computational complexity. The Filter size of each layer is 3 × 3. After five sets of convolution layers and LeakyReLU, the data are rearranged through the Flatten layer. That is, transform the multi-dimensional input into one-dimensional input, then input to the full connection layer, and finally calculate the value of label between 0 and 1 through sigmoid function as classification. If the value of label is closer to 1, it means that it looks more like the real image; otherwise, it looks less like it.

5   Experimental Results

5.1   Environment Setup

This study is performed by the convolution neural network which uses multiple cpus/gpus to speed up calculations. The environment used is the most compatible with deep learning processing that is Ubuntu operating system with Ubuntu16.04 and Pytorch as the development framework for deep learning. Pytorch is a set of open source software proposed in 2017 which developed and used by a team that includes big brands like FaceBook, NVIDIA and Twitter, combined with Torch and Python, the concise syntax, intuitive concept and easy-to-use features make it one of the most popular frameworks today. It is compatible with Windows and Linux operating systems and provides CPU/GPU core operation and pre-training model. There are also many online discussion communities available on the web for researchers or developers to discuss research in this area. Hardware is the CPU of Intel i7-7700 four-core 3.6 Hz, 16 GB DDR4 memory and NVIDIA GTX1080Ti display card containing 3584 CUDA cores.

5.2   Dataset and Evaluation

The database used in this study is MSCOCO training database 2014 [8], with a total of 82,738 images, of which 2,000 were taken for training. The test images used Set5, Set14, and vehicle images. Set5 had five images, and Set14 had 14 images. This experimental evaluation method used PSNR (peak signal noise ratio) to measure the reconstructed image quality, whose unit was db. The higher the PSNR, the more similar it was to the Ground Truth. The evaluation formula is as follows:

$$PSNR = 20 \cdot log_{10}^{\left(\frac{255}{\sqrt{MSE}}\right)} \tag{5}$$

The smaller the MSE, the better, shown as the follows:

$$MSE = \frac{1}{mn}\sum_{i=0}^{m-1}\sum_{j=0}^{n-1}\|I(i,j) - K(i,j)\|^2 \tag{6}$$

where M and n are the length and width of the image, I is the original image, and K is the reconstructed image.

5.3   Experimental Results

it is mentioned that many different loss functions are used for calculation. Many methods are tried in the Adversarial loss section, and the results are slightly different, as shown in the following tables. SR-net contains three different loss functions, PER is perceptual loss, TEX is texture loss and ADV is adversarial loss. In addition, the loss function of ADV was tested by different calculation methods. The training images were

Table 1. PSNR results with different types of loss function

Types of loss function	ADV-loss function	PSNR
PER	–	26.95
PER-TEX	–	27.27
PER-ADV	MSE	27.43
	Cross entropy	27.35
	BCE	27.95
PER-ADV-TEX	MSE	28.10
	Cross entropy	27.38
	BCE	27.03
	L2 loss	27.94

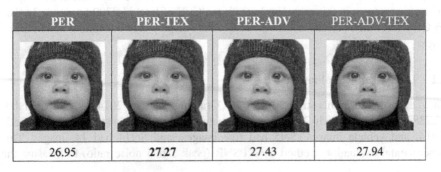

PER	PER-TEX	PER-ADV	PER-ADV-TEX
26.95	27.27	27.43	27.94

Fig. 1. Reconstructed images with different types of loss function

2,000, and the epoch was 200. Table 1 shows the PSNR results and Fig. 1 shows the reconstructed images via the networks trained with different loss function.

6 Conclusion

Along with the vigorous development of deep learning. There are more and more researches on the super resolution in related fields, from the original CNN network to GAN, which has become a topic of discussion recently. In this paper, various parameters and loss functions are used to test different types of images. The effects on the reconstructed images were observed by adjusting various parameters and fine tuning the network architecture. Randomly test Perceptual loss, Texture loss and Adversarial loss, according to the results, if less training images were used, there is no significant difference in the effect to reduce the amount of Residual block to 8 layers. Through experiments, we know that Perceptual loss function takes the output of the full connection layer of VGG network as the new loss function, and the results are not ideal.

References

1. Dong, C., Loy, C.C., He, K., Tang, X.: Learning a deep convolutional network for image super-resolution. In: Fleet, D., Pajdla, T., Schiele, B., Tuytelaars, T. (eds.) ECCV 2014. LNCS, vol. 8692, pp. 184–199. Springer, Cham (2014). https://doi.org/10.1007/978-3-319-10593-2_13
2. Dong, C., Loy, C.C., Tang, X.: Accelerating the super-resolution convolutional neural network. In: Leibe, B., Matas, J., Sebe, N., Welling, M. (eds.) ECCV 2016. LNCS, vol. 9906, pp. 391–407. Springer, Cham (2016). https://doi.org/10.1007/978-3-319-46475-6_25
3. Kim, J., Kwon Lee, J., Mu Lee, K.: Deeply-recursive convolutional network for image super-resolution. In: CVPR, pp. 1637–1645 (2016)
4. He, K., Zhang, X., Ren, S., Sun, J.: Deep residual learning for image recognition. In: CVPR, pp. 770–778 (2016)
5. Kim, J., Kwon Lee, J., Mu Lee, K.: Accurate image super-resolution using very deep convolutional networks. In: CVPR (2016)
6. Ledig, C., et al.: Photo-realistic single image super-resolution using a generative adversarial network. In: CVPR (2016)
7. Sajjadi, M.S.M., Schölkopf, B.: Enhancenet: single image super-resolution through automated texture synthesis. CoRR (2017)
8. Lin, T.-Y., et al.: Microsoft COCO: common objects in context. In: Fleet, D., Pajdla, T., Schiele, B., Tuytelaars, T. (eds.) ECCV 2014. LNCS, vol. 8693, pp. 740–755. Springer, Cham (2014). https://doi.org/10.1007/978-3-319-10602-1_48
9. Simonyan, K., Zisserman, A.: Very deep convolutional networks for large-scale image recognition. In: International Conference on Learning Representations (2015)
10. Gatys, L.A., Ecker, A.S., Bethge, M.: Image style transfer using convolutional neural networks. In: CVPR (2016)

Content-Based Motorcycle Counting
for Traffic Management by Image Recognition

Tzung-Pei Hong[1,2(✉)], Yu-Chiao Yang[1], Ja-Hwung Su[3],
and Chun-Hao Chen[4]

[1] Department of Computer Science and Information Engineering,
National University of Kaohsiung, Kaohsiung, Taiwan
tphong@nuk.edu.tw
[2] Department of Computer Science and Engineering,
National Sun Yat-sen University, Kaohsiung, Taiwan
[3] Department of Information Management,
Cheng Shiu University, Kaohsiung, Taiwan
[4] Department of Computer Science and Information Engineering,
Tamkang University, Taipei, Taiwan

Abstract. Over the past few decades, advanced technologies have increased the number of vehicles, including cars and motorcycles. Because of the large increase of vehicles, the traffic flow becomes more complex and the traffic accidents increase as rapidly. To decrease the number of traffic accidents, a number of studies has been made for how to manage the traffic flow. Especially for motorcycles, in this paper, we propose a method that counts the motorcycles by Convolutional Neural Network (CNN). To reveal the effectiveness of the proposed method, a set of experiments were conducted and the experimental results show the proposed method can bring out a good performance that provides a good support for traffic management systems.

Keywords: Motorcycle counting · Deep learning ·
Convolutional Neural Network · Traffic management · Video surveillance

1 Introduction

Along with the growth of population and the development of industrial technology, the number of vehicles is increasing rapidly every year. These numerous vehicles cause not only heavy air pollution but also many traffic accidents. The number of deaths by traffic accidents reaches 1250000 in the world in 2013 [3]. There are many reasons leading to traffic accidents. Among them, one critical matter is that the lanes on a road are usually not wide enough, such that many motorcycles and cars usually congest on the lanes. For effective traffic control on this issue, it is necessary to know the number of vehicles on the road. We can recognize the vehicles and use the count of vehicles to improve the road planning and traffic management. For example, the government may limit the number of vehicles on a road or widen the road.

Currently, there have been many researches in cars detection and they got good performances. Compared with car detection, motorcycles are more difficult to detect

© Springer Nature Switzerland AG 2019
N. T. Nguyen et al. (Eds.): ACIIDS 2019, LNAI 11432, pp. 180–188, 2019.
https://doi.org/10.1007/978-3-030-14802-7_16

because there are different numbers of people on motorcycles, and their shapes are not as regular as cars. The top 5 countries using motorcycles for regular weekday journeys in the world are Vietnam, Indonessia, Taiwan, India and Pakistan, as shown in Fig. 1. As one can see, Vietnam is the country with the highest motorcycle usage rate. For example in Ho Chi Minh City, which is the capital of Vietnam, there are about 8000000 persons in 2017 [3], while there are about 7000000 motorcycles. In order to deal with the above problems, in this paper, we make attempts to adopt the deep learning to count the motorcycles. Through the CNN, the images in surveillance videos are recognized for the number of motorcycles. The evaluation results show that, the proposed method does achieve a good accuracy for counting motorcycles on real datasets.

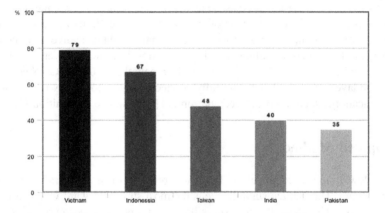

Fig. 1. Top 5 countries for motorcycle usage rates.

The structure of this paper is shown as follows: the related work is briefly reviewed in Sect. 2. The proposed method for counting motorcycles is presented in Sect. 3. The experimental results on real datasets are shown in Sect. 4, and conclusions are made in Sect. 5.

2 Related Work

Actually there have been many researches about vehicle recognition. However, the study for motorcycles counting is few because motorcycles are more difficult to recognize. The shape of various motorcycles and the number of passengers on the motorcycle make it look different. This situation makes machine learning difficult to identify the motorcycles in an image. To identify motorcycle positions in an image, Wen et al. [11] utilized the SVM ensemble approach to deal with the imbalanced datasets. The training process consists of three steps, which are image pre-processing, feature extraction and K partitions of the negative samples. For RGB images, they are scaled the region of interest (ROI) to an RGB image of 32 * 32. The scaled RGB image is transformed into the HSV images. Then, the improved Haar Wavelet algorithm is used to extract features. Finally, the SVM classifier is performed for training.

Mukhtar and Tang [9] use the histogram of oriented gradients (HOG) to extract features. Then classified using the SVM. Silva et al. [10] use Local Binary Pattern to extract features. Then classified using the SVM.

An artificial neural network (ANN) is an important field in machine learning. It is the simulation of the biological brain [7]. It simulates the action of the brain getting environment information, then calculates the problem to be solved, and finally outputting the result of calculation. An ANN has three types of layers, which are the input layer, the hidden layer and the output layer. An ANN has three types of layers: the input layer, the hidden layer and the output layer. There is one layer in the input layer and one layer in the output layer. The hidden layer may have one or more layers. Each layer has one or more nodes (neurons). Except the output layer, every node in one layer is connected to all nodes of the next layer by a line associated with one weight. Except the input layer, every node is used by an activate function to describe the operation of neurons. Activate function converts the input signal to the output signal, which can solve more complex problems [8]. There may be different activate function in different layer. The Convolutional Neural Networks (CNN) is a kind of deep neural network. A CNN consists of convolution layers, pooling layers and fully-connected layers [6]. CNN is employed in many applications, such as speech recognition [11], and image recognition [4].

3 Proposed Method

In fact, a low traffic accident rate is heavily dependent on a good traffic flow management system. The manual surveillance is the most straightforward way that needs a high-priced cost. Therefore, the auto surveillance is proposed to decrease the cost by a set of recent studies. However, a good auto surveillance is not easy to achieve. For this purpose, in this paper, we propose a auto-counting method for motorcycles by using CNN, the basic idea behind this method is to regard the counting as a classification. That is, the numbers of motorcycles are recognized into a set of quantity categories, and the CNN is performed to classify the quantity category for an image. As shown in Fig. 2, the details of the proposed method are described step by step in the following.

- **Step 1:** Collect the images with different numbers of motorcycles for training.
  In this step, we collect the images with different numbers of motorcycles from the surveillance video taped in Ho Chi Minh City [2] and California Institute of Technology from the Web [1]. We extracted about 1500 image frames from the video. From each original image, we cut and get a sub-image of size 500 * 250. Then, we resize each image to the input size of 50 * 25, which is then adopted in the CNN training model. We count the motorcycles in each sub-image and label each sub-image as its motorcycle number manually. Each image is labeled in the range of 0 to 16.
- **Step 2:** Train a CNN classification model for counting motorcycles
  We use the standard CNN model with convolution layers, pooling layers, and fully-connected networks to train the classification model for the number of motorcycles. The CNN model is shown in Fig. 3.

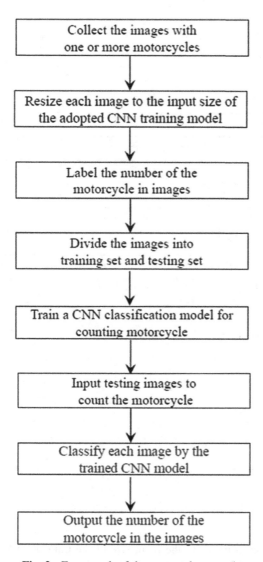

Fig. 2. Framework of the proposed approach.

The model contains 6 convolution layers, 2 max-pooling layers and a fully-connected layer with 3 hidden layers. Each three convolution layers is accompanied with a pooling layer. In the CNN model, the dropout mechanism is used, which is a way usually used to prevent neural networks from over-fitting [10]. Over-fitting means the training result is too close to a particular part of the training dataset and may cause the testing accuracy not as high as the training accuracy. The dropout ratio is set as $d\%$ in our CNN model. That is, there are $d\%$ random neurons dropout. A dropout example is shown in Fig. 4, in which there are 17 final output nodes at the end of the CNN. Each node represents a motorcycle number in an image within 0 to 16.

Layer (type)	Output Shape	Param #
conv2d_1 (Conv2D)	(None, 25, 50, 32)	896
conv2d_2 (Conv2D)	(None, 25, 50, 32)	9248
conv2d_3 (Conv2D)	(None, 25, 50, 32)	9248
conv2d_4 (Conv2D)	(None, 25, 50, 32)	9248
max_pooling2d_1 (MaxPooling2	(None, 12, 25, 32)	0
dropout_1 (Dropout)	(None, 12, 25, 32)	0
conv2d_5 (Conv2D)	(None, 12, 25, 64)	18496
conv2d_6 (Conv2D)	(None, 12, 25, 64)	36928
conv2d_7 (Conv2D)	(None, 12, 25, 64)	36928
conv2d_8 (Conv2D)	(None, 12, 25, 64)	36928
max_pooling2d_2 (MaxPooling2	(None, 6, 12, 64)	0
dropout_2 (Dropout)	(None, 6, 12, 64)	0
flatten_1 (Flatten)	(None, 4608)	0
dense_1 (Dense)	(None, 1024)	4719616
dropout_3 (Dropout)	(None, 1024)	0
dense_2 (Dense)	(None, 512)	524800
dropout_4 (Dropout)	(None, 512)	0
dense_3 (Dense)	(None, 256)	131328
dense_4 (Dense)	(None, 17)	4369

Fig. 3. The adopted CNN model.

■ **Step 3:** Resize each image to the input size of the trained CNN model
Because the size of the image to be judged may be not the same as the input of the trained CNN model, it must be first resized to be fed into the model.

■ **Step 4:** Classify an image with the trained CNN as its motorcycle number
After the image to be judged are resized, the motorcycle number in the image is then directly determined by the trained CNN model. Each class denotes a unique number of motorcycles in an image.

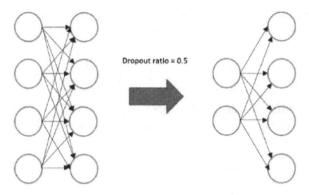

Fig. 4. An example for dropout of CNN.

4   Experiments

4.1   Datasets and Preprocessing

There are two datasets used in the experiments. The first dataset was collected by students at California Institute of Technology from the Web [1]. The dataset contains 826 images of a single motorcycle from the side. The second dataset was generated by a six-minute surveillance video in the Ho Chi Minh City [2], which is one of the city with high motorcycle density in the world. We describe the generation of the second dataset as follows. First, we downloaded a six-minute video from the Youtube. The video is a surveillance video for a certain road in the Ho Chi Minh City, Vietnam and was uploaded in 2013. The vehicles in the video contain motorcycles, buses and bicycles. Second, we extracted about 1500 image frames from the video. Each image is of size 1706 * 959. Third, from each original image, we cut and got a sub-image of size 500 * 250. We obtained 700 motorcycle sub-images, while the other 800 original images were discarded because in these discarded images, some motorcycles cannot be fully contained in the cutting sub-images. An example of the sub-image cutting process is shown in Fig. 5. Finally, 700 images were labeled a number of the motorcycles quantity, where the number is viewed as a category for CNN.

Fig. 5. An example for sub-image cutting.

In the experiments, two evaluation measures are used, namely error-rate and accuracy for motorcycle counting. For error-rate, it is defined as:

$$\text{error rate} = \begin{cases} \frac{|A-P|}{A}, & \text{if } A \neq 0 \\ P, & \text{if } A = 0 \end{cases}, \qquad (1)$$

where P is the predicted number of the motorcycles and A is the actual number of motorcycles. For accuracy, it indicates the classification accuracy.

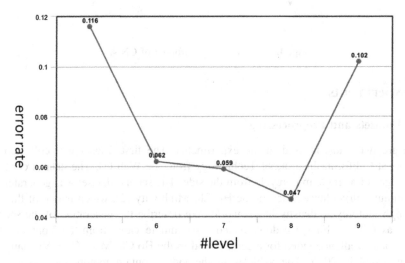

Fig. 6. The counting error rates with different levels of CNN.

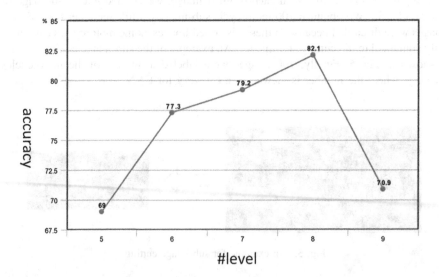

Fig. 7. The experimental accuracies with different levels of CNN.

4.2   Evaluation Results

In the experiments, we compared different numbers of convolutional layers, which includes 2 max-pooling layers and a fully-connected layer with 3 hidden layers and the batch size is set to 250. The comparison of different numbers of convolutional layers is shown in Figs. 6 and 7. Whatever for error-rate or accuracy, the results show that, 8 layers can bring out the best performance in the proposed method. This is because less convolutional layers cannot be learn the details in the image, more convolutional layers may cause gradient vanishing problem [5]. Finally, we conducted data augmentations and compared the accuracies with and without them. Figure 8 shows the execution time with and without data augmentations. From the experimental results, we can obtain that, the execution time increases as the data augmentations are enlarged. However, larger data augmentations will lift the accuracies and low the error rates.

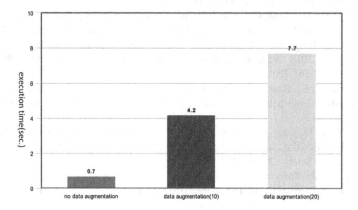

Fig. 8. The execution time with and without data augmentations.

5   Conclusions and Future Work

In this paper, we have presented the deep learning model to recognize motorcycles and count the related quantities in an image. In this approach, the motorcycles extracted from the video frames are counted by CNN. The experimental results show that the proposed approach can really the low error-rate and high accuracy. Through the proposed method, the traffic management will be more smart to decrease the traffic accidents. In the future, we will continuously improve the accuracy by using different segmentation methods and using more data. Currently, our one data set was obtained from a video on the same road during 6 min. In the future, we will establish the other data sets in different environments, such as different roads, different weather (sunny or raining), and different illumination (day or night). Besides, a data set with high diversity may make the trained model more practical and useful.

References

1. http://www.vision.caltech.edu/html-files/archive.html
2. https://www.youtube.com/watch?v=Op1hdgzmhXM
3. https://en.wikipedia.org/w/index.php?title=Induction_loop&oldid=826349762
4. Connie, T., Al-Shabi, M., Cheah, W.P., Goh, M.: Facial expression recognition using a hybrid cnnsift aggregator. In: Proceedings of International Workshop on Multi-disciplinary Trends in Artificial Intelligence (2017). [12]
5. Hu, Y., Huber, A., Anumula, J., Liu, S.: Overcoming the vanishing gradient problem in plain recurrent networks. In: Proceedings of 6th International Conference on Learning Representations (2018). [13]
6. Krizhevsky, A., Sutskever, I., Hinton, G.E.: ImageNet classification with deep convolutional neural networks. In: Proceedings of the 25th International Conference on Neural Information Processing Systems (2012). [16]
7. McCulloch, W.S., Pitts, W.: A logical calculus of the ideas immanent in nervous activity. Bull. Math. Biophys. 5(4), 115–133 (1943). [18]
8. Mhaskar, H.N., Micchelli, C.A.: How to choose an activation function. In: Proceedings of the 6th International Conference on Neural Information Processing Systems (1993). [19]
9. Mukhtar, A., Tang, T.B.: Vision based motorcycle detection using hog features. In: Proceeding of 2015 IEEE International Conference on Signal and Image Processing Applications (2015). [20]
10. Silva, R., Aires, K., Santos, T., Abdala, K., Veras, R., Soares, A.: Automatic detection of motorcyclists without helmet. In: Proceeding of 2013 XXXIX Latin American Computing Conference (2013). [21]
11. Wen, X., Yuan, H., Song, C., Liu, W., Zhao, H.: An algorithm based on SVM ensembles for motorcycle recognition. In: Proceeding of 2007 IEEE International Conference on Vehicular Electronics and Safety (2007). [24]
12. Zhang, Y., et al.: Towards end-to-end speech recognition with deep convolutional neural networks. In: Interspeech, pp. 410–414 (2016). [26]

Intelligent Information Systems

Use of Blockchain in Education: A Systematic Literature Review

Hafiza Yumna$^{(\boxtimes)}$, Muhammad Murad Khan$^{(\boxtimes)}$, Maria Ikram,
and Sabahat Ilyas

Department of Computer Science,
Government College University, Faisalabad, Pakistan
{yumna435,muhammadmurad,mariaikram035}@gcuf.edu.pk

Abstract. Blockchain technology enables the formation of a distributed record of a digital event in decentralized manner where data and related transactions are not under the control of any third party. This technology was early used for value transfer but now it has vast range of applications in different fields such as healthcare, banking, internet of things and many others. In the field of education, it also provides numerous opportunities for decentralized management of records in educational institutions in an interoperability manner. The main objective of this research is to highlight the existing issues related to the educational institutes and to find suitable blockchain features that could resolve them. We have adopted a systematic literature review approach for the identification and the extraction of relevant information from the shortlisted studies. This study describes existing issues in three aspects physical, digital and financial. The results of the analysis shows that the manipulation risk, difficulty in verification and exchanging record between institutions are the major issues faced by the educational institutions. This study, then explores blockchain features including decentralization, traceability and consensus mechanism that can be used to address the issues related to the educational institution. Finally, due to unique and underlying technology, it has still some technical challenges and boundaries along with immutability feature, disclosure of personal privacy and scalability issue also discussed in this study.

Keywords: Blockchain · Education · Systematic review

1 Introduction

The blockchain first showed up in 2008 when Satoshi Nakamoto published "Bitcoin: A Peer-to-Peer Electronic Cash System". Proposed system was based on cryptographic proof instead of reliance, enabling any two parties to execute transactions without the requirement for a trusted third party. This proposal solved the double-spending problem [1] and was the first application of blockchain.

Versatile features of blockchain such as transparent, information exchange in a decentralized manner, smart contracts, speed of transaction etc., can help improve variety of applications. The wide and promising scope of blockchain technology provides better assistance in the education field to students and other entities of

© Springer Nature Switzerland AG 2019
N. T. Nguyen et al. (Eds.): ACIIDS 2019, LNAI 11432, pp. 191–202, 2019.
https://doi.org/10.1007/978-3-030-14802-7_17

educational institutions, i.e., in credentials and certification [6, 7, 17], verified and authenticate record [13], helps recruiters in the recruitment process [19], maintenance of records [13, 19, 23] and ease of accessing these records [18]. Taking everything into consideration, the scope of blockchain technology is not limited to above-mentioned applications but also opens the door for university administration for managing their finance and accounting department by simply putting all the dues and charges of campus in it [6].

This review is about how the blockchain supports the education field and show its worth. This paper highlights some of the existing issues in the education field which can be fixed by blockchain features. But there are still some shortcomings including disclosure of personal privacy [2, 11], protection of private and public keys [18, 23] and scalability issues [2, 7, 10, 12, 13, 24]. We can mold this technology by fixing these shortcoming to have fruitful benefits.

The rest of the paper is structured as follows. Section 2 is about related research and explores the blockchain features. In Sect. 3, research methodology, research questions, and search strategy is discussed. Section 4 discovers the result of research questions. Section 5 presents applications of blockchain in education. Section 6 is about the various threats to validity. Lastly, Sect. 7 concludes the assumptions and constraints of our review and also present the future directions.

2   Literature Review

This section presents some core concepts and theories from existing research related to the blockchain and its several practical implementations in the education field. It also makes a comparison between the existing secondary studies.

2.1   Blockchain

Blockchain is a distributed setup that allows the formulation of distributed digital record of transactions, shared among the nodes of the network instead of being stored on a central server [4]. Current blockchain frameworks are categorized into three types: public or permission less blockchain, private or permission blockchain and consortium blockchain [2, 18].

2.2   Features of Blockchain

Blockchain technology has the following major features.

Decentralized: Blockchain is a decentralized shared public ledger in which all nodes are connected to each other in a mesh network where all the data and decision making is placed and distributed among various nodes [1, 3, 6, 12].

Traceability: Blockchain traceability feature promotes the audibility of an event as it stores information in blocks which are secured by uni-directional cryptographic hash function [19]. Complete chain of blocks is maintained by mining pools, which provide cloud based websites for exploring the blocks [3].

Consensus mechanism: Consensus mechanism refers to the mutual approval of all the nodes associated to the blockchain network [7]. Thus, it does not rely on mediators. Proof-of-work (POW), proof-of-stake (POS), delegated-proof-of-stake (DPOS) are some techniques of consensus mechanism [6].

Immutability: In blockchain, data is stored in ledger form and if there is any modification by external nodes, the hash key values would be changed because these keys are cryptographically linked with previous and preceding blocks and modification in data will interrupt the continuance of the keys [7, 18, 19, 23].

Smart contract: Smart contract is a self-executing computer program, running on a blockchain distributed network [7].

Currency: The blockchain technology has a property of cryptocurrency, which is a type of digital or virtual currency that guarantees the end-to-end transaction making it protected and trustworthy. The formation of this currency is generated by different mining algorithms [19]. Thus, the joined form of blockchain and cryptocurrency can be used in several aspects such as dealings of finance and accounting.

The abovementioned, identified features will be used to classify the primary studies together to conduct this review.

2.3   Use of Blockchain in Education

There is no denying the importance of education for the development and advancement of a country. Therefore, it is always been a struggle to find innovative technologies such as blockchain, to assist in the education field. The most noticeable advantages are seen in the form of data storage management [7, 11–13, 17, 21]. Others benefits are observed in data security [11, 19, 21], system trust [11, 13, 19, 23], Global ubiquitous database [13, 18, 19, 23], formative evaluation [19, 21] and some have got benefit in payments using smart contract [7, 17, 19, 23]. There has been numerous initiatives undertaken by the educational institutions that are using blockchain technology to store the data of their students and faculties. For the very first time, the University of Nicosia used the blockchain technology to manage the record of students, i.e. certificates that they received from MOOC platforms [7, 9, 19]. Massachusetts Institute of Technology (MIT) has developed a learning machine technology based on blockchain technology and they have formed a wallet for their students containing the educational records of a student [7, 19, 23]. Holberton school is also applying blockchain technology to save the educational record of students, i.e. their credential, learning behavior and activities in class [18, 19].

Table 1. Compared secondary studies

	Existing issues in education (aspects)			Blockchain features	Challenges and issues to BC implementation
	Physical	Digital	Financial		
[7]	✓	✓	✗	✓	✓
[8]	✗	✗	✓	✓	✓

2.4 Compared Secondary Studies

Blockchain is the fastest growing domain and provides a great research value from the last few years. However, while looking for secondary studies, we found only two studies that discuss research trends. Firstly, we found a report that was published by European Commission and we identified that it has only focused on physical and digital aspect. Second, a qualitative analysis was found that discussed the perceptions of distributed ledger technology by financial professionals with fiduciary responsibilities at select institutions of higher education. Table 1 enlists the attributes of these secondary studies with respect to the classification of its context.

3 Research Methodology

A Systematic Literature Review is defined as a method of gathering, identifying and interpreting all available research in order to answer a specific research question [5]. We have performed a systematic literature review by following the guidelines provided by Barbara Kitchenham to search for relevant studies. Steps of the guidelines are discussed in the following subsections.

3.1 Need of Conducting SLR

Table 1 presents the attributes of secondary studies in accordance with the classification of this study. By examining the existing secondary studies, it is identified that there are some gaps that have not been discussed yet and need to be filled for further practical implications. For example, Grech et al. [7] has mainly focused on paper certificate and digital certificate whereas, Harpool [8] has purely focused on financial aspect discussing the perceptions of distributed ledger technology by financial professionals. Moreover, both have used the qualitative research method using case studies, interviews, observations and literature reviews. No single study was found that covers all the aspects of existing issues in educational institutions and used a quantitative research method approach exploring the existing primary studies. Hence, we conducted this systematic literature review to achieve our goals and used a quantitative research approach.

3.2 Motivation and Research Question

The first phase of this systematic research is to define the research questions. Hence, following are the focused research questions:

RQ1: What are issues pertaining to area? (**Aim:** to illustrate the issues faced by the current educational domain that can be solved by blockchain.)

RQ2: What are the Blockchain features used to solve the identified issues? (**Aim:** to search out the blockchain features that precisely resolve the issues of the current education system.)

RQ3: What are the unaddressed issues? (**Aim:** to highlight the unresolved issues that could be fixed in future.)

3.3   Search Strategy

We used the tag-based approach to search for related papers in keyword form. On Google Scholar, these keywords i.e. "blockchain", "education", and "review" were searched. All the papers (ignoring the publication year and quality) that were published from start to up to date were collected and downloaded. We explored different papers, reports, and articles published in different journals and conferences.

3.4   Inclusion and Exclusion Criteria

After setting the research questions, all the primary studies were scrutinized to find appropriate data related to our context. For inclusion, 12 primary studies were tagged for data extraction and the remaining were excluded because they were out of our context and we figured out that these studies were specifically on "blockchain" instead of "blockchain in education". Table 2 presents the papers with corresponding ID.

Table 2. Paper ID with corresponding reference

#	Paper title	Year	ID
1	ECBC: a high-performance educational certificate blockchain with efficient query	2017	[11]
2	An introduction to the blockchain and its implications for libraries and medicine	2017	[12]
3	The blockchain and kudos: a distributed system for educational record, reputation, and reward	2016	[13]
4	Disciplina: blockchain for education	2018	[14]
5	The emerging trend of blockchain for validating degree apprenticeship certification in cyber security education	2018	[15]
6	Blockchain for education : lifelong learning passport	2018	[16]
7	Trustless education? A blockchain system for university grades	2017	[17]
8	EduCTX: a blockchain-based higher education credit platform	2018	[18]
9	Exploring blockchain technology and its potential application for education	2018	[19]
10	Higher education in an age of innovation, disruption, and anxiety	2018	[20]
11	Towards blockchain-enabled school information hub	2017	[21]
12	The case for a data bank: an institution to govern healthcare and education	2017	[22]

3.5   Data Extraction

After shortlisting the studies that were to be included in our SLR, the tagged primary studies were then proceeded to data extraction process. In this phase, the relevant data was extracted from the selected primary studies primarily focusing on existing issues in educational institution and to find blockchain features that exceptionally resolve them. Issues and features were covered in 12 primary studies [11–16, 18–22], whereas, unaddressed issues in 10 [2, 7, 10–13, 18, 19, 23, 24].

3.6   Publication Trend

Although research in blockchain was started with the invention of bitcoin in 2008. From then, there has seen a rapid advancement and adoption. However, the very first paper that covered both blockchain and its application in the field of education was found to be published in 2016. In 2017, 5 papers were found, whereas, till March 2018, 6 papers covered area under observation.

4   Results

This section is divided into three subsections. The first section classifies the common issues in the education system. The second subsection identifies the blockchain features that resolve the issues of the current education system and the third subsection high-lights the unsettled issues that could be fixed in future.

4.1   RQ1: What are the Issues Pertaining to the Area?

The objective of this Research Question is to identify the common issues faced by educational institutions. To make this discussion more convenient and clear, these issues are divided into three aspects: physical, digital and financial. The detail is as follows.

Physical Aspect: Physical aspect includes the attributes that are characterized by some physical activity or manner done by people. It contains the issues that are caused by manual activity or physical handling of the educational records.

Manipulation Risk: There is a series of human involvement in the creation of physical records i.e., taking and scoring the exams where the exams are paper-based. Manip-ulation attack happens when these academic records are created by an unauthorized body, i.e., they can do alteration in the marks of a student which can cause the social evil favoritism. The existing mechanism to secure academic record such as degree and transcript are reproducible [21] and hard to differentiate from the original record. Hence, these are more prone to manipulation attack [11].

Difficult to Verify: It is difficult to verify the student's record in manual system [20]. This problem is especially faced by third world countries which do not maintain a centralized record of all the universities. [7, 16]. On the contrary, an applicant face resistance from the institute because it's an additional burden. So, document verifica-tion in those universities, where records are maintained manually, becomes difficult [11, 14, 16, 18, 21].

Demand Human Resource: In traditional education system, there is always need a human resource for students as well as for the institution. For example, in case of a student, he could lose his academic certificate. In this situation, he has to write many applications for the issuance of the certificate and has to go through a costly and time-consuming process. On the other side, the institution needs to verify this application

through various steps e.g. check and match the previous record of a student and also has to maintain a physical record of students for a long period of time [7, 16, 18].

Single Point Failure: The record of a student in the educational institution is centralized and organized by a single entity. Even if some kind of distributed architecture is used within its boundaries, it can still be directly shared by a group of non-trusted parties. Conventional educational records are maintained at a certain place which become a single point of failure. This means, if, because of unforeseen reason, physical records are burnt, the recovery becomes impossible [7, 21].

Digital Aspect: Digital aspect refers to any record or activity that is stored or performed electronically or online. Issues related to digital aspect are given below.

Third Party Approval is Needed: The grading on academic record is produced by a teacher for a student. However, the evidence of the record either digital or physical is generated by a third party [7]. This gives power to the third party to produce a fake academic record [12].

Security Breach: Similar to single failure point of physical records, digital records are accessed from a centralized source which can lead to the changes in the data either intentionally, accidentally or other illegal means by third party. This means if source is compromised, then all digital data can be lost [7, 11, 15].

Difficult to Exchange Record Between Institutions: Record exchanging is very sophisticated and time-consuming process. In some cases, it becomes impossible. For example, if a student wants to migrate from one institution to another, it is difficult to exchange record. The students also face difficulty when they apply for admission as they have to submit all their educational records [18, 21]. No global standard system exists which can offer such services to all the universities [7, 18].

Financial Aspect: Word "finance" refers to the study and management of money. In financial aspect, issues that are related to money and budget are highlighted.

Middleman Commission: Universities conduct several of transactions every month with students, employees, vendors, suppliers and government agencies. There are chances of corruption by higher authorities. It can only be stopped if decentralized auditing is integrated into the financial system of the institutions [21].

Proof of Performance In educational institutions, there is no well-defined structure for monitoring and evaluating the activities of students and teachers [21]. Also, students and teachers feel demotivated when the rewards are not given to them for their hard work. Rewards must be given to them when they show good performance and there must be a proof and record to encourage and motivate them [17, 19, 22] (Table 3).

Table 3. Issues vs features

Issues		Blockchain features				
		Decentalized	Traceability	Consensus mechanism	Smart contract	Currency
Physical	Manipulation risk	✓	✓	✓		
	Difficult to verify	✓				
	Demand human resource	✓			✓	
	Single point failure	✓				
Digital aspect	Third party approval required			✓		
	Security breach		✓			
	Difficult to exchange record between institutions	✓				
Financial	Middleman commission				✓	
	Proof of performance				✓	✓

4.2 RQ2: Blockchain Features Used to Solve the Identified Issue

Blockchain features were discussed in Sect. 2 and this section highlights how these features can help address issues identified in previous section.

Decentralized: Decentralization attributes to a distributed network maintaining redundant records. Decentralization can help reduce:

Manipulation Risk: By making it difficult for an attacker to alter blockchain record maintained by number of nodes. This is difficult as compared to a single node which maintain all records as done in today's centralized system [21].

Difficult to Verify: By providing an open source and distributed platform which contains a multiple copies of transactions and distributes them across all the nodes in the network [11, 18, 21].

Demand Human Resource: By reducing the administrative load in different cases, i.e., in verification or migration case, which minimizes the cost and effort of traditional manual work. For example, if a student is migrating from one institution to another, then the faculty of the desired university can easily verify the record of a student from the blockchain database [18, 21]. It also provides a facility for employers to read and verify a certificate during a hiring process on a single application instead by asking the issuing institution [16].

Single Point Failure: By keeping a copy of data on each node. If any node gets offline, the data will not lost as it is maintained on a redundant network [7].

Difficult to Exchange Record Between Institutions: By helping the institutions to exchange information openly through the open source blockchain in an easy and convenient manner, e.g., in migration case and for further study case [18].

Traceability: Traceability refers to the ability of tracking and reaching out to everything back to its root. Traceability can solve the following issues.

Manipulation Risk: Can be reduced by traceability feature in a way that if someone tries to make illegal transaction or changes in the blockchain, then it can be tracked back by obtaining the block information linked by hash keys from the chronicle blockchain. Thus, any modification or fraudulent activity can be detected immediately against a particular instance [21].

Security Breach: by making a model of proof-of-existence and possession [15]. Blockchain technology creates a digital signature for every transaction which impossible to recreate as compared to electronic signature [7].

Smart Contract: Smart contract is a self-executing computer program under some conditions that is distributed across the blockchain nodes. Smart contract can resolve:

Proof of Performance: By making the real-time payments under the smart contract, hence, payments can be automatically executed via smart contract and real-time rewards could be given to students and teachers on the basis of their performance [13, 19].

Demand Human Resource: By automating human operations into smart contract, such as internal audits, student promotion to new class once the fee is paid etc., [21].

Middleman Commission: By automating the middle man operation into smart contracts, such as, degree verification can be implemented between different blockchain platforms etc., [21]. Systems will charge minimal fee for the contract execution, while making the whole process transparent.

Consensus Mechanism: Consensus Mechanism refers to the mutual approval of all the nodes associated to the blockchain network. Consensus mechanism can facilitate in solving the following issues:

Manipulation Risk: As the data is summed up in blockchain through a different consensus mechanism and so, is not handled by a single entity. There are fewer chances of fraud and mistakes because every new incoming transaction is verified by the other nodes of the network.

Third Party Approval is Needed: Can be solved as the blockchain framework works on consensus mechanism without any intermediary. Instead of using central authority to manage transactions, blockchain allow governance protocol, which work as smart contract.

Currency: Cryptocurrency is a type of digital or virtual currency that uses strong cryptographic techniques and generated by different mining algorithms. It could be used to solve the following issues:

Proof of Performance: By introducing an educational currency, the reward could be given in the form of cryptocurrency through a smart contract to the best-performing students and teachers [17, 19]. This kind of money could be stored in the education wallet and exchangeable with other currencies [13, 17].

4.3   RQ3: What are the Unaddressed Issues?

Although blockchain technology has a great potential to resolve issues in the education field by its tremendous feature, still research is in its infancy. So, it brings challenges and risks while its implementation. This research question highlights some unaddressed issues that could be resolved in the future.

Immutability Feature: In blockchain, once data has been placed on it, it cannot be changed or modified. This immutable feature can affect its useful functioning as it does not allow any change or modification which is often required [19]. Appropriately implementation of blockchain technology significantly improves these criteria, allowing fewer unwanted side effects.

Who Will Give the Approval of the First Network Node? At the beginning, some institution has to be the first network node and that time there will be no existing node to verify it, such an attribute can be seen as a security risk. However, we expect that with the increase in numbers of nodes, such security concerns will be minimized [18].

Disclosure of Personal Privacy: As it is open source and transparent technology, the record or personal information about the students can be accessed or shared without the willingness of students [2, 11, 23].

Scalability Issue: Blockchain has to scale for improving network transactions per second [2, 7, 10, 12, 13, 24], hence, areas such as "side chains" is being explored.

5   Applications of Blockchain in Education

After having the distinguish benefits of blockchain in education, demand for it has become global. Currently, various applications have been running on blockchain and others are in processing stage. Echolink is a global standard blockchain platform that stores verified credentials, skills and work experience in a hashed and unalterable way. All information is entered by the authoritative institutions and thus provides a trustworthiness of such information. **Echolink** recognized a partnership with Microsoft to offer blockchain application cloud service on Azure [25].

Another application **Disciplina** projected by Teach Me Please is a multifunction blockchain platform creates and stores verified personal profiles related to academic and professional career. It helps recruiting services by providing digital CV of student generated during the educational career, accompanied by the authenticity proofs [14]. An application named **Open certificates** developed by Attorneys, assigns block-proof to the educational certificates using Ethereum smart contracts. They have declared their partnership with educational institutions of Singapore [7].

6 Threats to Validity

To summarize the existing evidence related to the use of blockchain in the education field, we tried to gather as many related primary studies as possible for the extraction of knowledge. As the related research was in the exploratory stages, therefore, little peer-reviewed literature was found in this area. As our extraction scenario was based on the perception of the defined research questions, so there might be chances that the reader can identify some attributes that we did not consider and can be helpful in the future. Also, most of the work was on the features and innovative applications; less was on its limitations. Finally, there might be some work done that we could not refer in our paper during the period of publication, as the researchers are continuously focusing to fix the problems in educational institutes through blockchain.

7 Conclusion

In this study, we decided to map all possible relevant primary studies by using a systematic literature approach. By exploring and examining all the features of block-chain, we have presented the suitable solutions to deal education related problems in a precise way. Since this technology is in initial experimental stages, so, it still has to go through an evolutionary process. In future, it is believed that a better review could be written as the world is moving towards innovation and the people are becoming more technology oriented.

References

1. Nakamoto, S.: Bitcoin: A Peer-to-Peer Electronic Cash System, p. 9 (2008). www.Bitcoin.Org
2. Zheng, Z., Xie, S., Dai, H., Chen, X., Wang, H.: An overview of blockchain technology: architecture, consensus, and future trends. In: Proceedings - 2017 IEEE 6th International Congress Big Data, BigData Congress 2017, pp. 557–564, June 2017
3. Cao, S., Cao, Y., Wang, X., Lu, Y.: Association for Information Systems AIS Electronic Library (AISeL) a review of researches on blockchain. Rev. Res. Blockchain, 108–117 (2017)
4. Karafiloski, E., Mishev, A.: Blockchain solutions for big data challenges: a literature review. In: 17th IEEE International Conference on Smart Technol. EUROCON 2017 – Conference Proceedings, pp. 763–768, July 2017
5. Software Engineering Group: Guidelines for Performing Systematic Literature Reviews in Software Engineering (2007)
6. Blockchain for Education & Research Webinar (2016)
7. Grech, A., Camilleri, A.F.: Blockchain in Education (2017)
8. Harpool, R.: Perceptions of Distributed Legder Technology by Financial Professionals with Fiduciary Responsibilites at Select Institutions of Higher Education. ProQuest LLC, Ann Arbor (2017)
9. Sharples, M., et al.: Innovating Pedagogy 2015 (2016)
10. Zhao, J., Fan, S., Yan, J.: Overview of business innovations and research opportunities in blockchain and introduction to the special issue. Int. J. Prod. Econ. 2(1), 28 (2016)

11. Xu, Y., Zhao, S., Kong, L., Zheng, Y., Zhang, S., Li, Q.: ECBC: a high performance educational certificate blockchain with efficient query. In: Hung, D., Kapur, D. (eds.) ICTAC 2017. LNCS, vol. 10580, pp. 288–304. Springer, Cham (2017). https://doi.org/10.1007/978-3-319-67729-3_17
12. Hoy, M.B.: An introduction to the Blockchain and its implications for libraries and medicine. Med. Ref. Serv. Q. **36**(3), 273–279 (2017)
13. Sharples, M., Domingue, J.: The blockchain and kudos: a distributed system for educational record, reputation and reward. In: Verbert, K., Sharples, M., Klobučar, T. (eds.) EC-TEL 2016. LNCS, vol. 9891, pp. 490–496. Springer, Cham (2016). https://doi.org/10.1007/978-3-319-45153-4_48
14. Kuvshinov, K., Nikiforov, I., Mostovoy, J., Mukhutdinov, D.: Disciplina: Blockchain for Education, pp. 1–17 (2018)
15. Bandara, I., Ioras, F., Arraiza, M.P.: The emerging trend of blockchain for validating degree apprenticeship certification in cyber security education, pp. 7677–7683, March 2018
16. Gräther, W., et al.: Blockchain for Education: Lifelong Learning Passport (2018)
17. Rooksby, J.: Trustless education? A blockchain system for university grades. In: New Value Transactions: Understanding and Designing for Distributed Autonomous Organisations, Workshop, DIS 2017, June 2017, p. 4 (2017)
18. Turkanović, M., Hölbl, M., Košič, K., Heričko, M., Kamišalić, A.: EduCTX: a blockchain-based higher education credit platform. IEEE Access **6**, 5112–5127 (2018)
19. Chen, G., Xu, B., Lu, M., Chen, N.-S.: Exploring blockchain technology and its potential applications for education. Smart Learn. Environ. **5**(1), 1 (2018)
20. World Educators: Higher Education in an Age of Innovation, Disruption, and Anxiety, March 2018
21. Bore, N., Karumba, S., Mutahi, J., Darnell, S.S., Wayua, C., Weldemariam, K.: Towards blockchain-enabled school information hub. In: Proceedings of Ninth International Conference and Communication Technologies and Development - ICTD 2017, pp. 1–4 (2017)
22. Raju, S., Rajesh, V., Deogun, J.S.: The case for a data bank: an institution to govern healthcare and education. In: Proceedings of the 10th International Conference on Theory and Practice of Electronic Governance - ICEGOV 2017, pp. 538–539 (2017)
23. Skiba, D.J.: The potential of blockchain in education and health care. Nurs. Educ. Perspect. **38**(4), 220–221 (2017)
24. Lemieux, V.L.: Trusting records: is blockchain technology the answer? Rec. Manag. J. **26**(2), 110–139 (2016)
25. Chen, S.X.: Blockchain Based Professional Networking and Recruiting Platform, pp. 1–14 (2017)

A Comparative Study of Techniques for Avoiding Premature Convergence in Harmony Search Algorithm

Krzysztof Szwarc$^{(\boxtimes)}$ and Urszula Boryczka

Institute of Computer Science, University of Silesia,
ul. Bedzinska 39, 41-200 Sosnowiec, Poland
{krzysztof.szwarc,urszula.boryczka}@us.edu.pl
http://ii.us.edu.pl/

Abstract. The present article summarizes two techniques allowing to avoid premature convergence in Harmony Search algorithm, which was adapted for solving the instances of the Asymmetric Traveling Salesman Problem (ATSP). The efficiency of both approaches was demonstrated on the basis of the results of statistical test and 'test bed' consisting of nineteen instances of ATSP. The conclusion was that the best results were obtained in case of applying mechanisms which enable to reset the components of harmony memory at the moment of reaching stagnation. This process is controlled by parameters which are depended on the problem size.

Keywords: Harmony Search ·
Asymmetric Traveling Salesman Problem ·
Avoiding premature convergence

1 Introduction

From the moment when Harmony Search (HS) algorithm was formulated by Geem in the paper [4], metaheuristic has found numerous applications in solving many optimisation problems (e.g. Dynamic Vehicle Routing Problem with Time Windows [3], Pipe Network Design [5] and 0–1 Knapsack Problem [14]). In the paper [1] an attempt was made to adapt the technique to effective solving the Asymmetric Traveling Salesman Problem (ATSP), which due to its $\mathcal{NP}$-hard nature and huge practical significance (it reflects the characteristics of linear infrastructure present in urbanised areas, enabling to use it for example in the processes occurring in reverse logistics – during optimisation of mobile collection of waste electrical and electronic equipment [9], as well as transport of municipal waste [13]) became an object of interest for many researchers from various fields of science. The authors of the above-mentioned article introduced the mechanism for resetting the elements of harmony memory in order to avoid premature convergence; however, the above-mentioned paper does not demonstrate the relationship between the size of the problem and the recommended rule for its application, thus forming a new research gap.

© Springer Nature Switzerland AG 2019
N. T. Nguyen et al. (Eds.): ACIIDS 2019, LNAI 11432, pp. 203–214, 2019.
https://doi.org/10.1007/978-3-030-14802-7_18

The problem of getting stuck in local optimum concerns both techniques operating on a single solution and algorithms working on many results (population-based methods). Among metaheuristics belonging to the first group, two particularly significant techniques can be distinguished: Tabu Search (TS) proposed by Glover in the paper [6] and Simulated Annealing (SA) described by Kirkpatrick et al. in the paper [7]. TS enabled to exit the local optimum by moving towards the solution described by a less favourable value of objective function, based on the structure referred to as tabu list, whereas SA used the parameter called temperature, applied in order to determine the probability of accepting a less favourable result.

Among the popular techniques used for maintaining a variety of populations, it is worth emphasizing the methods intended for Genetic Algorithm mentioned in the paper [12]: the use of variable probability value of mutation occurrence, creation of random descendant after fulfilling specific conditions, or – best adjusted to HS – use of *Social Disasters Technique*, which assumes that all population members will be replaced by randomly created individuals, with the exception of the member described by the most favourable value of objective function.

The purpose of this article is to determine the recommended method to avoid premature convergence in HS algorithm, adapted for solving the instances of ATSP. Selected for the research was the method which assumes resetting of the population (with the exception of the best individual) as a result of reaching stagnation, as well as the technique based on mechanisms occurring in SA (the ineffectiveness of the approach used in TS was assumed due to the structure of the analysed HS).

The paper consists of the following seven parts: introduction to the discussed subject, description of HS, formulation of ATSP, approximation of HS structure adapted for ATSP, presentation of the methodology of empirical studies, analysis of the obtained results, as well as conclusions and recommendations for further work on the techniques for avoiding premature convergence in HS, which was intended for solving the instances of ATSP.

2   Harmony Search Algorithm

HS is based on the similarity between the process of searching for global optimum by means of algorithmic methods and jazz improvisation. The method assumes the existence of HM structure called harmony memory, which consists of HMS of harmonies containing a given number of pitches (representing the values of decisive variables). Each element belonging to HM constitutes a complete solution of the problem, whose value of objective function is determined on the basis of its components.

In the classic version of algorithm, the initial content of harmony memory is generated in a random manner, which is followed by its sorting, based on the values of objective function of particular HM elements (in such manner that the result in the first position is characterised by the best outcome). After completing the initial stage, iterative development of new solutions begins.

Creation of a new harmony assumes iterative selection of subsequent pitches, in accordance with two parameters – $HMCR$ and PAR. The selection of i pitch occurs with the probability equal to $HMCR$ and it uses the values which were located in the position i in harmonies belonging to HM (otherwise pseudorandom generation of permissible value takes place). In case of selecting values on the basis of $HMCR$ probability, modification of the pitch with PAR probability may occur (the change takes place on the basis of bw parameter, whose value depends on the features describing the instance of the problem).

Development of a new solution enables to compare its value of objective function with the relevant parameter describing the ultimate component of HM (the worst stored solution). If a better result is determined, it will replace the worst result situated in the harmony memory and HM elements will be sorted.

The procedure of developing a new solution is performed for IT iterations, which is followed by returning the best result.

3   Formulation of ATSP

The formulation of Traveling Salesman Problem (TSP) presented in article [11] was adapted for the purpose of this paper. On this basis, we assumed the existence of directed graph $G = (V, A)$, whose edge weights were marked as c_{ij} ($i, j \in \{1, 2, \ldots, n\}$). The problem assumes determination of route – oriented cycle containing n of all cities – with minimum length. The asymmetric variant of TSP analysed in this paper is characterised by the possible occurrence of different weights between the edge connecting vertices i and j and the edge between j and i ($c_{ij} \neq c_{ji}$).

The model assumes the occurrence of one decisive variable (x_{ij}), which represents the existence of an edge connecting i and j nodes in the created solution. It may adapt the following values:

$$x_{ij} = \begin{cases} 1 \text{ if edge } (i, j) \text{ is part of the route constructed,} \\ 0 \text{ otherwise.} \end{cases} \tag{1}$$

The objective function which assumes minimisation of the travel route was formulated in the following way:

$$\sum_{i=1}^{n} \sum_{j=1}^{n} c_{ij} x_{ij} \rightarrow min. \tag{2}$$

The following limiting conditions were added in order to ensure that the salesman will visit each city only once:

$$\sum_{i=1}^{n} x_{ij} = 1, \quad j = 1, \ldots, n,$$

$$\sum_{j=1}^{n} x_{ij} = 1, \quad i = 1, \ldots, n. \tag{3}$$

Additional restrictions (referred to as MTZ) were introduced in order to eliminate the possibility of creating many separate cycles (instead of one cycle):

$$1 \leq u_i \leq n - 1, \quad u_i - u_j + (n-1)x_{ij} \leq n - 2, \quad i,j = 2,\ldots,n. \qquad (4)$$

4   HS Adjusted to ATSP

This paper was based on the approach enabling the adjustment of HS to ATSP, proposed in article [1]. It assumes that the pitch values are integers corresponding to the numbers of the cities which the salesman is supposed to visit (their order of appearance indicates the sequence of travel).

When creating another harmony, the order of occurrence of vertices is examined, which is done by selecting the subsequent pitch value based on the previously generated list of available nodes occurring in the stored solutions directly after the last city, belonging to the currently constructed results. Using the formed structure, the city is selected by means of the popular roulette wheel method (the probability of accepting a particular element depends on the value of objective function of the entire solution - which is represented by the route length – analogously to the approach presented in article [8]), or any available (unvisited) node is drawn (when the created list of vertices is empty). Selection of the city (made from the unvisited nodes) located nearest to the recently visited city in the developed solution was adapted as the pitch modification related to PAR parameter (the method was supplemented by the greedy approach).

In order to avoid premature convergence, the article [1] proposes the possibility of resetting HM elements (in line with the *Social Disasters Technique* approach described in Sect. 1) at the moment of performing a specific number of R iterations from the latest replacement of result in the harmony memory; however, this paper, due to its purpose, was based on a structure deprived of the above-mentioned mechanism. The pseudocode of HS adjusted to ATSP was presented in Algorithm 1.

5   Methodology of Research

The algorithms were implemented in language $C\#$, whereas the research was conducted on laptop Lenovo Y50-70 with the following configuration: Intel Core i7-4720HQ (4 cores, from 2.60 GHz to 3.60 GHz, 6 MB cache), 16 GB RAM (SO-DIMM DDR3, 1600 MHz), HDD 1000 GB SATA 5400 RPM Express Cache 8 GB, Windows 7 Professional N Service Pack 1 64-bit.

Based on article [1], the following parameter values describing specific HS variants were adapted: $HMS = 5$, $HMCR = 0.98$ and $PAR = 0.25$. Nineteen tasks, whose characteristics were presented in Table 1, were selected as the 'test bed'. When analysing the number of iterations after which the algorithm achieved convergence for the selected tasks (presented in the paper [2]), $IT = 1000000$ was determined. Each instance of the problem was solved 30 times, using different seed every time (the seed was identical within different

HS variants, in order to ensure the reliability of formulated conclusions). The assessment of the algorithms performance was made on the basis of the average error determined in the following way: *average error = (average result − optimum)/optimum · 100%.*

Algorithm 1. The Harmony Search for ATSP pseudocode [1]

1: $iterations = 0$
2: $iterationsFromTheLastReplacement = 0$
3: **for** $i = 0; i < HMS; i + +$ **do**
4:     $HM[i]$=stochastically generate feasible solution with repetitions
5: **end for**
6: Sort HM
7: **while** $iterations < IT$ **do**
8:     $H[0]$=first city
9:     **for** $i = 1; i < n; i + +$ **do**   ▷ n - number of cities
10:         Choose random $r \in (0, 1)$
11:         **if** $r < HMCR$ **then**
12:             $list$=generate list containing vertices occurring after $H[i − 1]$ in HM
13:             **if** $list.length > 0$ **then**
14:                 $H[i]$=choose element $\in list$ according to the roulette wheel
15:             **else**
16:                 $H[i]$=choose randomly available city $\notin H$
17:             **end if**
18:             Choose random $k \in (0, 1)$
19:             **if** $k < PAR$ **then**
20:                 $H[i]$=find nearest and available city from $H[i − 1]$
21:             **end if**
22:         **else**
23:             $H[i]$=choose randomly available city $\notin H$
24:         **end if**
25:     **end for**
26:     **if** $f(H)$ is better than $f(HM[HMS − 1])$ **then**
27:         $HM[HMS − 1] = H$
28:         Sort HM
29:         $iterationsFromTheLastReplacement = 0$
30:     **else**
31:         $iterationsFromTheLastReplacement + +$
32:     **end if**
33:     **if** $iterationsFromTheLastReplacement = R$ **then**
34:         **for** $i = 1; i < HMS; i + +$ **do**
35:             $HM[i]$=stochastically generate feasible solution
36:         **end for**
37:         Sort HM
38:         $iterationsFromTheLastReplacement = 0$
39:     **end if**
40:     $iterations + +$
41: **end while**
42: return $HM[0]$

Table 1. Characteristics of 'test bed' based on [10]

No.	Name	Number of vertices	Optimum
1	br17	17	39
2	ftv33	34	1286
3	ftv35	36	1473
4	ftv38	39	1530
5	p43	43	5620
6	ftv44	45	1613
7	ftv47	48	1776
8	ry48p	48	14422
9	ft53	53	6905
10	ftv55	56	1608
11	ftv64	65	1839
12	ft70	70	38673
13	ftv70	71	1950
14	kro124p	100	36230
15	ftv170	171	2755
16	rbg323	323	1326
17	rbg358	358	1163
18	rbg403	403	2465
19	rbg443	443	2720

The following HS configurations were used in the research:

1. HS - algorithm without the mechanism for preventing premature convergence.
2. RHS - HS in which all harmonies belonging to HM are reset after performing R iterations from the last acceptance of the constructed solution, with the exception of the best of them.
3. SHS - HS which enables to accept a worse solution with given probability.

In SHS variant, the probability of replacing solution X, located in the last position in the harmony memory, by result X', is expressed with the following formula:

$$P(X') = \begin{cases} 1 & \text{if } f(X') < f(X), \\ 0 & \text{if } f(X') \geq f(X) \text{ and } t \leq 0, \\ e^{(\frac{f(X)-f(X')}{t})} & \text{otherwise,} \end{cases} \tag{5}$$

where t is the parameter referred to as temperature, which controls the probability of accepting a worse solution (an increase of the above-mentioned probability occurs together with the increase of its value). In this paper, it was assumed that together with the performance of subsequent iterations, the value

t is to be reduced in such manner that it should aim at 0 (thus reducing the exploration force at the subsequent stages of method operation). The adapted goal was achieved by using the following formula for determining temperature in iteration l:

$$t = t_0 - (t_0 \cdot l/IT), \tag{6}$$

where t_0 is the initial temperature. It may be determined in a manner which enables it to adjust automatically to the characteristics of the problem instance, using the following formula:

$$t_0 = \frac{-\overline{\Delta f}}{ln(P_0)}, \tag{7}$$

where P_0 is the probability of moving towards a worse solution in the first iteration, whereas $\overline{\Delta f}$ is the average worsening of the value of objective function, determined on the basis of a specific number of checks of adjacent solutions. For the purpose of this paper, it was assumed that the algorithm will create 20 solutions in order to determine value $\overline{\Delta f}$ (only the results described with a less favourable value of objective function are included in the calculations). In the situation where all routes in the neighbourhood were not worse than the base solution, it was concluded that $t_0 = 100$. The manner of determining the adjacent solution implied that three SHS variants would be distinguished: SHS_{swap}, SHS_{HMS} and SHS_{HMS^2}. The first of them is a universal method used in local search algorithms (for TSP) and it assumes determination of a new route by random swap of two nodes. The second and third variants are adjusted to HS and regard the created harmonies as adjacent solutions. The differences between them arise from the moment of determining the initial temperature - SHS_{HMS} determines it after HMS of iterations, whereas SHS_{HMS^2} - after HMS^2 iterations. Both approaches assume that until the moment of determining t_0, it has the value of 0 (based on HM, the ineffectiveness of determination of t_0 at the beginning of method operation results from the significant effectiveness of Nearest Neighbor heuristic present in HS [2], which implies huge probability to determine all 20 harmonies described with a more favourable value of objective function than the result generated pseudorandomly).

6   Obtained Results

Table 2 presents a summary of average error determined by HS and RHS (with different values of R parameter). On this basis, it was concluded that the introduction of the mechanism for resetting HM elements enables to increase the algorithm efficiency; however, the non-adjustment of R values causes an increase of the average error (a premature removal of relatively good solutions for huge tasks representing ATSP causes the impossibility of constructing good travel routes).

In line with the obtained results, it is recommended to determine $R = 1000$ for instances of ATSP described with the occurrence of maximum 48 nodes, 2,500 iterations for maximum 171 vertices and $500,000$ iterations for problems with 323–443 cities (identical results can be achieved for the largest instances by withdrawing from the mechanism for resetting HM elements). RHS using the above-mentioned recommended values for a specific number of cities was marked in the paper as RHS_{opt}. Taking into consideration only the aggregate average error for the constant value R, it is recommended to reset harmony after the following number of iterations from the last result replacement in harmony memory: 10,000 or 25,000 (the average error at the level of 12.8% was achieved in both cases).

Table 2. Comparison of the average error determined by HS and RHS

Test	Average error												
	HS	RHS (R value)											
		1000	2500	5000	7500	10000	25000	50000	75000	100000	250000	500000	750000
br17	0	0	0	0	0	0	0	0	0	0	0	0	0
ftv33	4.76	**3.63**	3.72	3.74	3.82	4.11	4.5	4.47	4.47	4.53	4.64	4.67	4.7
ftv35	1.46	**1.35**	1.47	1.46	1.43	1.47	1.46	1.44	1.47	1.44	1.48	1.45	1.46
ftv38	2.75	**1.44**	2.07	2.26	2.31	2.42	2.44	2.6	2.81	2.71	2.85	2.88	2.66
p43	**0.05**	**0.05**	0.06	**0.05**	0.06	0.06	**0.05**	**0.05**	**0.05**	**0.05**	**0.05**	**0.05**	**0.05**
ftv44	2.69	**1.76**	2.35	2.48	2.84	2.82	2.68	2.92	3.09	2.99	2.74	2.69	2.69
ftv47	2.97	**1.95**	2.28	2.11	2.22	2.49	2.37	2.52	2.88	2.92	3.09	2.91	3
ry48p	1.83	**0.89**	1.16	1.21	1.27	1.27	1.47	1.3	1.76	1.76	1.84	1.82	1.87
ft53	12.14	10.06	**8.98**	9.4	10.41	9.34	10.21	11.29	11.34	11.13	11.79	12.4	12.14
ftv55	5.06	2.92	**2.39**	2.74	3.74	3.62	3.83	3.98	4.13	4.01	5.13	4.99	5
ftv64	6.53	3.09	**2.75**	3.73	3.66	4.07	4.96	4.91	5.23	5.76	5.68	6.05	6.56
ft70	5.25	4.35	4.36	**4.32**	4.5	4.51	4.48	4.78	4.81	4.85	5.18	5.16	5.2
ftv70	6.57	**4.89**	5.73	5.91	6.02	6.14	6.02	6.53	6.09	6.39	6.96	6.49	6.64
kro124p	11.29	8.88	**8.82**	9.17	9.08	9.08	10.22	10.64	10.92	10.89	11.47	11.38	11.29
ftv170	20.07	19.15	15.71	15.93	**15.68**	16.17	17.5	18.45	18.54	18.26	19.63	19.56	20.07
rbg323	**46.65**	53.84	52.55	51.17	50.7	50.12	47.51	47.51	47.61	47.28	**46.65**	**46.65**	**46.65**
rbg358	66.78	77.01	74.43	71.78	71.59	69.5	68.25	67.71	67.25	**66.44**	67.06	66.78	66.78
rbg403	**26.15**	29.27	29.07	28.27	27.71	27.33	27	26.45	26.45	26.47	26.19	**26.15**	**26.15**
rbg443	**27.46**	30.45	29.99	29.17	28.81	28.7	28.19	27.85	27.69	27.51	27.49	**27.46**	**27.46**
Average	13.18	13.42	13.05	12.89	12.94	**12.8**	**12.8**	12.92	12.98	12.92	13.15	13.13	13.18

The detailed results determined by RHS_{opt} were presented in Table 3. On their basis, occurrence of very large diversification of the quality of routes generated by the examined technique was not detected, which implied the possibility of predicting the results determined by the technique (this characteristic is desired in utilitarian applications due to the frequently one-time only algorithm use, in order to solve a given optimisation problem).

The summary of obtained results for HS and different SHS variants was presented in Table 4. According to the summary, the most favourable type of SHS is SHS_{swap}, whereas the worst type is SHS_{HMS}. Among the examined P_0 values, it is recommended to use the following values: 0.4 for SHS_{swap} and 0.3 for SHS_{HMS} and SHS_{HMS^2}.

Table 3. Detailed results for RHS_{opt}

Test	Objective function value			
	Average	Minimal	Maximal	Sample std. dev.
br17	39	39	39	0
ftv33	1332.73	1286	1388	29.3
ftv35	1492.87	1473	1499	6.8
ftv38	1552.03	1536	1581	10.93
p43	5622.73	5620	5627	1.44
ftv44	1641.43	1613	1728	27.83
ftv47	1810.7	1785	1853	17.52
ry48p	14550.13	14507	14790	61.45
ft53	7525.33	7200	7853	177.12
ftv55	1646.37	1608	1699	28.75
ftv64	1889.57	1850	1963	32.19
ft70	40357.63	39471	40800	267.84
ftv70	2061.73	2012	2129	30.2
kro124p	39426.73	38251	40317	566
ftv170	3187.9	3056	3307	71.64
rbg323	1944.53	1849	1992	33.67
rbg358	1939.6	1866	2002	33.69
rbg403	3109.7	3015	3164	31.85
rbg443	3466.9	3377	3520	30.26

It is worth noting the observation, according to which determination of value t_0 after performing only HMS of iterations prevented generating worse harmonies (than the solution located in the last position in HM) for the tasks described with occurrence of at least one hundred nodes (and for 65 cities in test ftv64), in consequence causing the adaption of $t_0 = 100$ (regardless of the value P_0, the determined results are identical). Particular attention should also be drawn to the fact of the lack of determination, by any SHS variant, of the initial temperature value, which would enable to construct routes not worse than HS for test ft53. Additionally, a decrease in effectiveness of the examined approach to diversification of the space of solutions for the tasks described with at least 323 vertices was reported.

The average error obtained by particular techniques was subject to Wilcoxon Signed-Rank test (the value of 0.05 was assumed as the statistical significance – lower p-values indicate adoption of an alternative hypothesis, according to which $A1$ is better than $A2$). The designated p-values are shown in the Table 5. They do not allow to undermine the hypothesis according to which RHS_{opt} obtained better results than HS and SHS_{swap} ($P_0 = 0.4$), whereas SHS_{swap} ($P_0 = 0.4$) determined better results than HS.

Table 4. Comparison of the average error determined by HS and SHS

Test	Average error												
	HS	SHS_{swap} (P_0)				SHS_{HMS} (P_0)				SHS_{HMS2} (P_0)			
		0.1	0.2	0.3	0.4	0.1	0.2	0.3	0.4	0.1	0.2	0.3	0.4
br17	0	0	0	0	0	0	0	0	0	0	0	0	0
ftv33	4.76	4.76	4.77	4.98	4.55	4.03	3.9	3.77	**3.72**	4.29	3.93	3.96	4.67
ftv35	1.46	1.41	1.36	**1.26**	1.48	1.4	1.33	1.35	1.37	1.55	1.46	1.5	1.4
ftv38	2.75	2.43	2.66	2.44	**2.01**	2.83	3	2.79	2.86	2.52	2.29	2.39	2.27
p43	0.05	0.05	0.05	**0.04**	0.06	0.05	0.05	0.05	0.05	0.05	0.05	0.05	0.05
ftv44	2.69	3.61	3.04	3.67	2.91	3.17	3.38	3.26	3.89	2.94	2.9	**2.44**	3.86
ftv47	2.97	2.41	2.68	3.02	**2.17**	2.76	2.74	2.55	2.56	2.69	2.55	2.46	2.99
ry48p	1.83	1.59	1.94	1.78	**1.58**	2.03	1.98	2.06	1.98	2.06	1.96	1.54	2.04
ft53	**12.14**	12.94	12.63	13.46	12.64	13.23	12.97	13.43	12.99	12.92	13.21	12.61	12.99
ftv55	5.06	4.68	4.52	4.34	4.22	4.26	4.22	3.8	4.35	4.8	3.83	3.82	**3.76**
ftv64	6.53	5.05	4.57	5.2	4.45	4.95	4.95	4.95	4.95	**4.11**	4.14	4.13	4.18
ft70	5.25	**4.87**	5.23	5.11	5.12	5.35	5.28	5.41	5.37	5.23	5.18	5.15	5.13
ftv70	6.57	5.96	6.19	6.07	6.19	6.29	6.28	6.29	6.4	6.28	6.54	6.6	**5.82**
kro124p	11.29	11.23	10.88	10.8	10.41	10.99	10.99	10.99	10.99	11.04	10.78	**10.39**	10.78
ftv170	20.07	18.97	18.97	19.09	20.2	**17.92**	**17.92**	**17.92**	**17.92**	19.54	19.92	19.61	20.11
rbg323	46.65	46.97	**46.3**	46.75	46.91	48.84	48.84	48.84	48.84	48.36	48.07	49.15	49.07
rbg358	66.78	66.09	**65.57**	66.45	66.24	69.1	69.1	69.1	69.1	68.03	68.26	68.6	69.25
rbg403	26.15	26.36	26.47	25.86	**25.68**	27.06	27.06	27.06	27.06	26.16	26.98	27.17	27.33
rbg443	27.46	**27.27**	27.49	27.55	27.49	28.17	28.17	28.17	28.17	27.5	28.06	28.29	28.19
Average	13.18	12.98	12.91	13.05	**12.86**	13.29	13.27	13.25	13.29	13.16	13.16	13.15	13.36

Table 5. Wilcoxon signed-rank test results for the average error

$A1$	$A2$		
	HS	RHS_{opt}	SHS_{swap}
HS	N/A	1	1
RHS_{opt}	**4.2382E-60**	N/A	**4.70407E-44**
SHS_{swap}	**2.81199E-20**	1	N/A

Figure 1 presents the summary of total average error for the examined techniques. On this basis, it was concluded that the most favourable results were determined by RHS_{opt}, whereas the worst results were determined by HS, which was not supplemented with the mechanism for avoiding premature convergence.

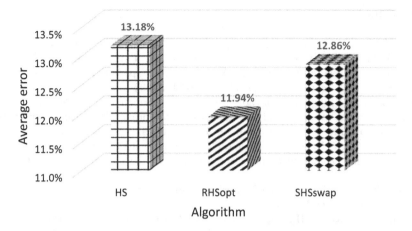

Fig. 1. A summary of average error for different ways to avoid premature convergence in HS

7   Conclusions and Planned Further Work

As a result of the performed work, the efficiency of both examined approaches to avoiding premature convergence in HS was demonstrated, as well as particular attention was drawn to the necessity of appropriate determination of the values of parameters R and P_0. On the basis of analysis of the obtained results, it is recommended to use the mechanism assuming the resetting of all HM elements, apart from the best harmony, at the moment of performing $1,000$ iterations from the last acceptance of the solution for ATSP instances described with the occurrence of maximum 48 nodes, $2,500$ iterations for maximum 171 vertices and $R = 500,000$ for problems with $323 - 443$ cities. For the technique enabling to increase exploration, using the parameter referred to as temperature, it is recommended to determine the value t_0, based on the swap of two cities and to assign the value $P_0 = 0.4$.

On the basis of the performed research, it was concluded that SHS approach is characterised by higher flexibility and stability than RHS – the determination of constant value of parameter P_0 enabled to achieve minor deviations from the value of objective function HS regardless of the problem size, whereas for RHS, premature resetting of HM elements caused a significant increase of the average error for large instances of ATSP. To sum up, for a 'test bed' of unknown size, it is recommended to use SHS (in particular variant SHS_{swap}).

Further work, including the examination of effectiveness of different techniques enabling to avoid premature convergence in HS may concern the analysis of other methods for leaving the local optimum (e.g. determination of the period of harmony occurrence in HM) and methods of reducing temperature value in SHS. It is additionally recommended to check the effectiveness of the above-mentioned approaches in HS, adjusted to solving other optimisation problems.

References

1. Boryczka, U., Szwarc, K.: The adaptation of the harmony search algorithm to the ATSP. In: Nguyen, N.T., Hoang, D.H., Hong, T.-P., Pham, H., Trawiński, B. (eds.) ACIIDS 2018. LNCS (LNAI), vol. 10751, pp. 341–351. Springer, Cham (2018). https://doi.org/10.1007/978-3-319-75417-8_32
2. Boryczka, U., Szwarc, K.: The adaptation of the harmony search algorithm to the ATSP with the evaluation of the influence of the pitch adjustment place on the quality of results. J. Inf. Telecommun. **3**, 2–18 (2019)
3. Chen, S., Chen, R., Gao, J.: A modified harmony search algorithm for solving the dynamic vehicle routing problem with time windows. Sci. Program. **2017**, 13 (2017)
4. Geem, Z.W.: Optimal design of water distribution networks using harmony search. Ph.D. thesis, Korea University (2000)
5. Geem, Z.W., Kim, J.H., Loganathan, G.V.: Harmony search optimization: application to pipe network design. Int. J. Model. Simul. **22**(2), 125–133 (2002)
6. Glover, F.: Future paths for integer programming and links to artificial intelligence. Comput. Oper. Res. **13**(5), 533–549 (1986)
7. Kirkpatrick, S., Gelatt, C.D., Vecchi, M.P.: Optimization by simulated annealing. Science **220**(4598), 671–680 (1983)
8. Komaki, M., Sheikh, S., Teymourian, E.: A hybrid harmony search algorithm to minimize total weighted tardiness in the permutation flow shop. In: 2014 IEEE Symposium on Computational Intelligence in Production and Logistics Systems (CIPLS), Orlando, FL, pp. 1–8 (2014)
9. Nowakowski, P., Szwarc, K., Boryczka, U.: Vehicle route planning in e-waste mobile collection on demand supported by artificial intelligence algorithms. Transp. Res. Part D: Transp. Environ. **63**, 1–22 (2018)
10. Osaba, E., Diaz, F., Onieva, E., Carballedo, R., Perallos, A.: A population meta-heuristic with adaptive crossover probability and multi-crossover mechanism for solving combinatorial optimization problems. Int. J. Artif. Intell. **12**, 1–23 (2014)
11. Öncan, T., Altınel, I.K., Laporte, G.: A comparative analysis of several asymmetric traveling salesman problem formulations. Comput. Oper. Res. **36**(3), 637–654 (2009)
12. Rocha, M., Neves, J.: Preventing premature convergence to local optima in genetic algorithms via random offspring generation. In: Imam, I., Kodratoff, Y., El-Dessouki, A., Ali, M. (eds.) IEA/AIE 1999. LNCS (LNAI), vol. 1611, pp. 127–136. Springer, Heidelberg (1999). https://doi.org/10.1007/978-3-540-48765-4_16
13. Syberfeldt, A., Rogstrom, J., Geertsen, A.: Simulation-based optimization of a real-world travelling salesman problem using an evolutionary algorithm with a repair function. Int. J. Artif. Intell. Expert Syst. (IJAE) **6**(3), 27–39 (2015)
14. Zou, D., Gao, L., Li, S., Wu, J.: Solving 0–1 knapsack problem by a novel global harmony search algorithm. Appl. Soft Comput. **11**(2), 1556–1564 (2011)

Analysis of Different Approaches to Designing the Parallel Harmony Search Algorithm for ATSP

Krzysztof Szwarc[✉] and Urszula Boryczka

Institute of Computer Science, University of Silesia,
ul. Bedzinska 39, 41-200 Sosnowiec, Poland
{krzysztof.szwarc,urszula.boryczka}@us.edu.pl
http://ii.us.edu.pl/

Abstract. This article proposes three approaches to designing the Parallel Harmony Search Algorithm, adjusted to effectively solve the Asymmetric Traveling Salesman Problem. The paper contains a comparative study of the developed models, based on 19 varied instances of the problem, as well as recommendations concerning their appropriate configuration. The quality of developed models was assessed through determination of the percentage difference between the surplus of the values of objective function of solutions and the values describing the sequential algorithm variant. Each of the examined Parallel Harmony Search variants created better results than the sequential Harmony Search algorithm.

Keywords: Parallel Harmony Search · Harmony Search ·
Asymmetric Traveling Salesman Problem

1 Introduction

The Harmony Search (HS) technique is a modern metaheuristic applied successfully in solving many utilitarian problems (it was used in nurse rostering problems [1], as well as in flood protection system management [2]). Its efficiency also caused the creation of different variants of the algorithm, such as Fuzzy Harmony Search [11], which was applied in the process of optimization of controllers [12,13]. The method's popularity and indisputable practical significance of the Asymmetric Traveling Salesman Problem (ATSP; which represents e.g. the process of transport of municipal waste [14]), led to conducting research on adjustment of the above-mentioned algorithm for solving the described combinatorial optimization problem, which resulted in a publication [3] presenting proposals for modifications enabling the algorithm to obtain good results for instances of the ATSP. We assumed that it is possible to improve the method by using parallel calculations (through increasing the ability to find global minima), which contributed to beginning research on Parallel Harmony Search (PHS).

The subject of designing a parallel version of HS was explored by Ceylan et al., who presented a model assuming HS paralleling by sending information

© Springer Nature Switzerland AG 2019
N. T. Nguyen et al. (Eds.): ACIIDS 2019, LNAI 11432, pp. 215–227, 2019.
https://doi.org/10.1007/978-3-030-14802-7_19

both to the adjacent and the specified processor by each algorithm module (corresponding to a particular processor) in the publication [5]. Hong et al. presented results of research concerning PHS in the article [7] about the impact of the frequency of communication on the effectiveness of PHS. It assumed that operations would be performed independently by particular instances of technique, whereas the best determined result would be provided to all HS occurrences after a specific number of iterations.

The purpose of this paper is to design and to define recommended values of parameters as well as to compare the effectiveness of three approaches to paralleling the HS algorithm, adjusted to solving the ATSP (based on the method proposed in the study [3]). The quality of developed models was assessed through determination of the percentage difference between the surplus of the values of objective function of solutions and the values describing the sequential algorithm variant (thereby analyzing the ability of the parallel method to reducing the impact of non-determinism and increase the efficiency of exploitation in some of its variants) and selected techniques, applied in the process of solving ATSP instances.

The article consists of the following chapters: first - including the introduction to the discussed subject, second - describing the classic HS algorithm, third - containing the formulation of the ATSP, fourth - regarding the adjustment of HS to the ATSP, fifth - presenting the characteristics of proposed approaches to paralleling the HS algorithm, sixth - concerning the methodology of research work, seventh - presenting the results of research on the adjustment of parameter values for PHS, eighth - describing the obtained results and ninth - focusing on the conclusions and planned future work.

2   Classic Harmony Search Algorithm

The HS was proposed by Geem in [6]. Its foundation is the process of jazz improvisation, which is compared to search for the global optimum by means of algorithmic methods. The technique assumes the existence of HM structure (referred to as harmony memory), storing HMS harmonies, consisting of a specific number of pitches (representing the values of decisive variables of a particular result). Each HM element shall be regarded as a problem solution.

At the beginning, the content of HM is generated randomly, which is followed by its sorting, based on the value of the objective function of harmony (in such a manner that the result in the first position is the best). Performance of the above-mentioned steps triggers the iterative development of new solutions.

The procedure for creating another solution uses the knowledge collected in HM and is based on the similarity to the process of harmony improvisation in music. Development of a solution consists of iterative selection of another pitch, in accordance with two parameters - $HMCR$ and PAR. The pitch i is selected on the basis of $HMCR$ probability, using the values located in i position in harmonies belonging to HM (otherwise the permissible value is generated randomly). When creating a solution based on HM component, the pitch may be modified with the given PAR probability (the value is changed on the basis of bw parameter, whose value depends on the problem representation).

Development of a new solution results in comparing its value of objective function with the relevant parameter describing HM component located in the last position. If a better result is obtained, it replaces the worst result located in the harmony memory and the HM elements belonging to the structure are sorted again. Generation of a new solution is done for IT iterations, which is followed by returning the best result (located in the first position in the HMS).

3 The Formulation of the ATSP

Öncan et al. proposed the following definition of Traveling Salesman Problem (TSP) in their study [10]: for the directed graph $G = (V, A)$, with arc weights c_{ij} $(i, j \in \{1, 2, \ldots, n\})$, a route (oriented cycle containing all n of cities) is searched, characterized by the minimal length. The possibility of occurrence of imbalances $c_{ij} \neq c_{ji}$ is permissible in ATSP.

The occurrence of edges between i and j vertices in the constructed route is represented by the decisive variable x_{ij} adapting the following values:

$$x_{ij} = \begin{cases} 1 \text{ if edge } (i,j) \text{ belongs to constructed route,} \\ 0 \text{ otherwise.} \end{cases} \tag{1}$$

TSP assumes determination of the shortest route for a commercial agent - the objective function was formulated as:

$$\sum_{i=1}^{n} \sum_{j=1}^{n} c_{ij} x_{ij} \to min. \tag{2}$$

The constraints intended to ensure that the salesman will visit each city only once was formulated as follows:

$$\sum_{i=1}^{n} x_{ij} = 1, \quad j = 1, \ldots, n, \quad \sum_{j=1}^{n} x_{ij} = 1, \quad i = 1, \ldots, n. \tag{3}$$

In order to avoid the possibility of occurrence of solutions representing separate cycles instead of just one, it is necessary to introduce extra restrictions (MTZ) to the model (therefore allowing the proper formulation of the problem):

$$1 \leq u_i \leq n - 1, \quad u_i - u_j + (n - 1)x_{ij} \leq n - 2, \quad i, j = 2, \ldots, n. \tag{4}$$

4 Harmony Search Adjusted to ATSP

This paper is based on the modification of HS, proposed in publication [3]. It assumes that each pitch comprising harmony is represented by integers corresponding to the numbers of particular cities that are to be visited by the salesman. Their sequence of appearance indicates the sequence of travel.

While creating a new harmony, the sequence of vertices is considered, which is done through selection of another pitch value based on the generated list of available nodes appearing in the stored solutions immediately after the last city that belongs to the constructed result. On the basis of the formed structure, the location is selected in accordance with the roulette wheel method (the probability of acceptance of a particular element depends on the value of objective function of a solution, represented by the route length, analogically to the approach presented in article [8]; this way ensuring the proper balance between exploration and exploitation), or any unvisited node is drawn (in case the formed list of vertices is empty; hence increasing non-determinism of the method). The selection (made among the available nodes) of the city situated nearest to the last visited location was adapted as modification of the pitch (related to *PAR* parameter) in the created solution (common knowledge about the problem was applied, by means of introducing the greedy movement).

In order to avoid premature convergence, the possibility of resetting *HM* elements was introduced at the moment of executing a number R of iterations from the last result replacement in the harmony memory. The mechanism assumes preserving the best result and drawing all other solutions, as a result diversifying the structure of created harmonies. The pseudocode of the proposed approach to the creation of HS was presented in Algorithm 1.

5   Parallel Harmony Search

This section consists of three parts describing specific approaches to designing PHS. The first of them presents the model assuming cooperation in terms of transferring the best determined result (in a similar way to the method discussed in the paper [7]), the second shows the approach based on creating common harmony memory on the basis of particular HS instances, whereas the third one outlines the variant without communication.

The example presented in Fig. 1 was used in order to characterize the principle of operation of particular models. The F(X) entry means the value of objective function represented by particular harmonies.

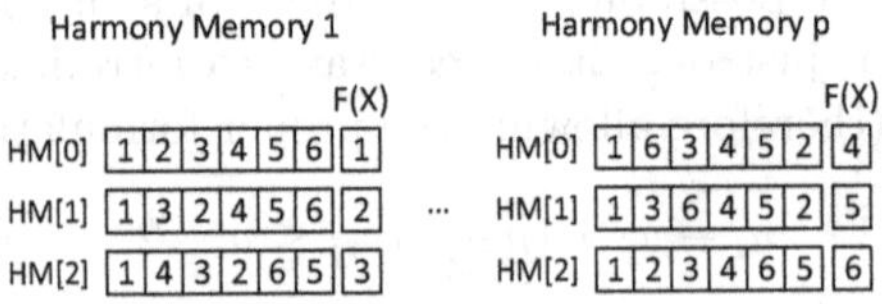

Fig. 1. Content of particular harmony memories

Algorithm 1. The Harmony Search for ATSP pseudocode [3]

1: *iterations* = 0
2: *iterationsFromTheLastReplacement* = 0
3: **for** $i = 0; i < HMS; i + +$ **do**
4:     $HM[i]$=stochastically generate feasible solution
5: **end for**
6: Sort *HM*
7: **while** *iterations* $< IT$ **do**
8:     $H[0]$=first city
9:     **for** $i = 1; i < n; i + +$ **do**   ▷ n - number of cities
10:         Choose random $r \in (0, 1)$
11:         **if** $r < HMCR$ **then**
12:             *list*=generate list containing vertices occurring after $H[i-1]$ in HM
13:             **if** *list.length* > 0 **then**
14:                 $H[i]$=choose element $\in$ *list* according to the roulette wheel
15:             **else**
16:                 $H[i]$=choose randomly available city $\notin H$
17:             **end if**
18:             Choose random $k \in (0, 1)$
19:             **if** $k < PAR$ **then**
20:                 $H[i]$=find nearest and available city from $H[i-1]$
21:             **end if**
22:         **else**
23:             $H[i]$=choose randomly available city $\notin H$
24:         **end if**
25:     **end for**
26:     **if** $f(H)$ is better than $f(HM[HMS-1])$ **then**
27:         $HM[HMS-1] = H$
28:         Sort *HM*
29:         *iterationsFromTheLastReplacement* = 0
30:     **else**
31:         *iterationsFromTheLastReplacement* $+ +$
32:     **end if**
33:     **if** *iterationsFromTheLastReplacement* $= R$ **then**
34:         **for** $i = 1; i < HMS; i + +$ **do**
35:             $HM[i]$=stochastically generate feasible solution
36:         **end for**
37:         Sort *HM*
38:         *iterationsFromTheLastReplacement* = 0
39:     **end if**
40:     *iterations* $+ +$
41: **end while**
42: return $HM[0]$

5.1 PHS with the Best Harmony Migration

The PHS with the best harmony migration (PHS1) assumes creation of p HM structures and execute steps included in the HS algorithm for each of them. After performing IK iterations, the best harmony is copied to all harmony memories in

Algorithm 2. Pseudocode of PHS with the best harmony migration

```
1:  parallelIterations = 0
2:  parfor i = 0; i < p; i + + do
3:      for j = 0; j < HMS; j + + do
4:          HM[i][j]=stochastically generate feasible solution
5:      end for
6:      Sort HM[i]
7:      iftlr[i] = 0                    ▷ iftlr - iterationsFromTheLastReplacement
8:  end parfor
9:  while parallelIterations < IT do
10:     if IK > IT − parallelIterations then
11:         IK = IT − parallelIterations
12:     end if
13:     parfor i = 0; i < p; i + + do
14:         for j = 0; j < IK; j + + do
15:             Execute lines 8-40 from Algorithm 1 for HM[i] and iftlr[i]
16:         end for
17:     end parfor
18:     best=return the best harmony from HM
19:     for i = 0; i < p; i + + do
20:         HM[i][0]=best
21:     end for
22:     parallelIterations+ = IK
23: end while
24: return the best harmony from HM
```

such a way that the result in the first position in each of them is replaced by it. The technique makes it possible to focus on a promising solution, maintaining the diversity of harmony memories, in order to balance exploration and exploitation. The approach was presented in Algorithm 2.

Assuming the occurrence - after IK iterations - of harmony memories presented in Fig. 1, the solution $HM[0]$ will be replaced in each harmony memory with the result located in Harmony Memory 1 (characterized by the best value of objective function). The result of the operation was shown in Fig. 2.

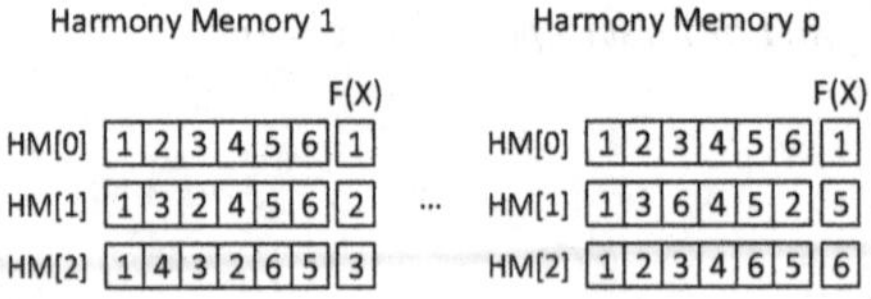

Fig. 2. Content of particular *HM*s for the PHS with the best harmony migration

5.2   PHS with Collective Harmony Memory

The PHS with collective harmony memory (PHS2) is based on the idea present in the PHS1. However, after performing IK iterations, HMS of the best harmonies from the p harmony memory structures are selected in order to create a new HM, which replaces all existing harmony memories. The method assumes intensive exploitation by eliminating the diversity of particular harmony memories. The pseudocode of the described approach has been presented in Algorithm 3.

Three harmony memories (in the analyzed example $HMS = 3$) with the lowest value of objective function were selected for the instance presented in Fig. 1 and a new structure was formed on their basis, replacing the p of harmony memories (the operation result was shown in Fig. 3).

Algorithm 3. Pseudocode of PHS with collective harmony memory

---

```
1:  parallelIterations = 0
2:  parfor i = 0; i < p; i + + do
3:      for j = 0; j < HMS; j + + do
4:          HM[i][j]=stochastically generate feasible solution
5:      end for
6:      Sort HM[i]
7:      iftlr[i] = 0                    ▷ iftlr - iterationsFromTheLastReplacement
8:  end parfor
9:  while parallelIterations < IT do
10:     if IK > IT − parallelIterations then
11:         IK = IT − parallelIterations
12:     end if
13:     parfor i = 0; i < p; i + + do
14:         for j = 0; j < IK; j + + do
15:             Execute lines 8-40 from Algorithm 1 for HM[i] and iftlr[i]
16:         end for
17:     end parfor
18:     solutions = ∅
19:     for i = 0; i < p; i + + do
20:         for j = 0; j < HMS; j + + do
21:             solutions.add(HM[i][j])
22:         end for
23:     end for
24:     Sort solutions
25:     for i = 0; i < HMS; i + + do
26:         for j = 0; j < p; j + + do
27:             HM[j][i] = solutions[i]
28:         end for
29:     end for
30:     parallelIterations+ = IK
31: end while
32: return the best harmony from HM
```

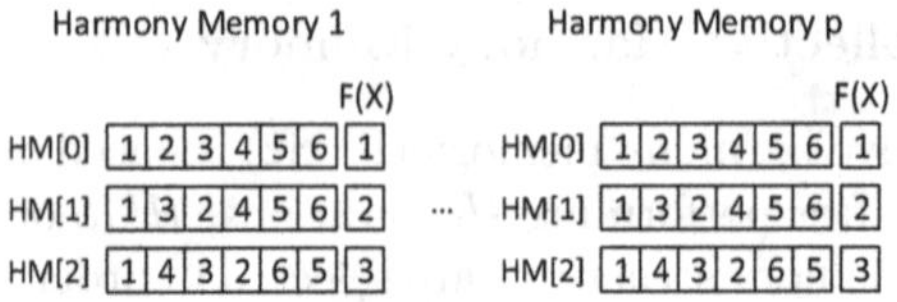

Harmony Memory 1

F(X)

HM[0] 1 2 3 4 5 6 1
HM[1] 1 3 2 4 5 6 2
HM[2] 1 4 3 2 6 5 3

Harmony Memory p

F(X)

HM[0] 1 2 3 4 5 6 1
HM[1] 1 3 2 4 5 6 2
HM[2] 1 4 3 2 6 5 3

...

Fig. 3. Content of particular *HM*s for the PHS with collective *HM*

5.3 PHS Without Communication

The PHS without communication (PHS3) is based on the fact that HS is a non-deterministic algorithm and assumes launching a specific number p of instances of HS algorithm and determination of the result on the basis of the value of objective function of the best harmonies constructed by particular technique occurrences. The pseudocode of the approach was presented in Algorithm 4. Harmony *HM*[0], belonging to Harmony Memory 1 (characterized by the lowest value of objective function), will be returned for the analyzed example (presented in Fig. 1).

Algorithm 4. Pseudocode of PHS without communication

1: **parfor** $i = 0; i < p; i + +$ **do**
2:     $HS[i]$=execute Harmony Search Algorithm
3: **end parfor**
4: return the best harmony from HS

---

6   Methodology of Research

The assessment of effectiveness of particular variants of PHS was made by using 19 instances of the ATSP, the characteristic features of which were presented in [4]. Each test was solved 30 times with the same seed among methods and the quality of particular solutions was assessed on the basis of the average error, expressed with the formula: *average error = (average result − optimum)/optimum · 100%*.

On the basis of paper [3], the values of parameters describing all variants of PHS were determined as follows: $R = 1000$, $HMS = 5$, $HMCR = 0.98$, $PAR = 0.25$. Additionally, by analyzing the number of iterations at which convergence was achieved (presented at work [4]), it was established that $IT = 1000000$. The algorithms were implemented in $C\#$ language, whereas the research was conducted on Lenovo Y50-70 with Intel Core i7-4720HQ (4 cores, from 2.6 GHz to 3.6 GHz, 6 MB cache) and 16 GB RAM (SO-DIMM DDR3, 1600 MHz).

The solutions constructed through sequential HS (with the same parameter values as PHS) were selected as the comparative results for the solutions determined by different variants of PHS. Additionally, obtained results were compared with results taken from subject literature and determined by Nearest Neighbor Algorithm (NNA), Greedy Local Search (GLS), Hill Climbing (HC), Genetic Algorithm (GA) as well as Adaptive Multi-Crossover Population Algorithm (AMCPA).

7   Selection of Parameter Values for PHS

The recommended parameter values for particular models of PHS were determined on the basis of three tasks representing the ATSP (ftv33, p43 and kro124p). For each analyzed parameter value (from the set of $p = \{2, 3, 4\}$ and $IK = \{500, 1000, 1500, 2000, 2500, 3000, 3500, 4000, 4500\}$) 30 solutions of each task were made (within checked value of one parameter, the values of remaining parameters were identical). The same seeds were also applied for various values of tested parameters. The results for the PHS1 were presented in Table 1, for the PHS2 - in Table 2, whereas for the PHS3 - in Table 3. As it was determined on this basis, the application of the following configuration is recommended: $p = 4$, $IK = 3500$ (for the PHS1) and $IK = 4000$ (for the PHS2). It was additionally observed that, regardless of the examined model, the effectiveness of the analyzed method increased together with the increase of p parameter value.

Table 1. Impact of the values of parameters on average error for the PHS1

Test	Average error											
	p			IK								
	2	3	4	500	1000	1500	2000	2500	3000	3500	4000	4500
ftv33	2.93	2.2	1.76	3	2.34	2.2	2.34	2.34	1.92	1.76	1.92	1.92
p43	0.04	0.05	0.04	0.04	0.03	0.03	0.03	0.04	0.03	0.04	0.04	0.04
kro124p	8.41	7.33	6.84	6.76	6.6	6.8	6.94	6.82	7.09	6.84	6.73	7.01
Average	3.79	3.19	**2.88**	3.27	2.99	3.01	3.1	3.07	3.01	**2.88**	2.9	2.99

Table 2. Impact of the values of parameters on average error for the PHS2

Test	Average error											
	p			IK								
	2	3	4	500	1000	1500	2000	2500	3000	3500	4000	4500
ftv33	2.47	2.61	2.2	3.46	2.69	2.61	2.34	2.34	2.38	1.92	2.2	2.47
p43	0.05	0.05	0.04	0.05	0.04	0.05	0.03	0.04	0.03	0.04	0.04	0.04
kro124p	8.29	7.31	6.75	9.96	8.44	8.16	7.95	7.94	7.45	7.5	6.75	6.84
Average	3.6	3.32	**3**	4.49	3.72	3.61	3.44	3.44	3.29	3.16	**3**	3.12

8   Obtained Results

The average error, determined by means of the tested methods, was presented in Table 4. On the basis of the obtained results it was concluded that the PHS3 is characterized by the lowest value, whereas the PHS1 has the highest value (among the PHS variants). After the 1,000,000 iterations, the first of them determined the best results (among the tested models) for 10 tasks, whereas

Table 3. Impact of the values of parameters on average error for the PHS3

Test	Average error		
	p		
	2	3	4
ftv33	2.18	1.72	0.9
p43	0.04	0.03	0.03
kro124p	7.87	7.43	6.85
Average	3.36	3.06	**2.59**

the second - only for 4 tests. For the specified criteria, the PHS2 enabled to create the best solutions for 7 tasks (in particular for tests described by a minimum of 100 vertices, where intensive exploitation proved to be particularly effective). Thanks to paralleling the calculations, regardless of the analyzed variant PHS, the obtained results were better than the results obtained by the sequential algorithm. Analyzed techniques are also characterized by significant efficiency

Table 4. Summary of the average error

Test	Average error								
	NNA [3]	GLS [3]	HC [3]	AMCPA [9]	GA [9]	HS	PHS1	PHS2	PHS3
br17	135.9	7.69	7.69	0.26	1.54	**0**	**0**	**0**	**0**
ftv33	30.87	23.64	23.64	7.77	7.79	3.63	1.76	2.2	**0.9**
ftv35	21.59	21.38	21.38	6.52	6.2	1.35	1.26	1.26	**0.78**
ftv38	16.21	10	10	5.33	6.92	1.44	1.11	0.99	**0.92**
p43	2.63	0.53	0.96	0.15	0.27	0.05	0.04	0.04	**0.03**
ftv44	24.86	24.18	24.18	10.49	8.38	1.76	1.66	1.72	**0.71**
ftv47	33.67	28.89	28.89	7.21	4.86	1.95	1.67	1.57	**1.18**
ry48p	16.19	15.21	13.81	2.79	4.67	0.89	0.67	0.7	**0.61**
ft53	37.78	30.93	30.14	12.64	11.98	10.06	6.89	**6.31**	6.6
ftv55	25.12	23.2	23.2	11.41	14.25	2.92	2.01	2.28	**0.94**
ftv64	43.5	36.54	36.22	13.18	14.86	3.09	**1.1**	1.62	1.42
ft70	11.67	8.15	8.98	4.4	5.45	4.35	**3.78**	3.92	3.86
ftv70	31.85	23.85	23.28	13.06	9.97	4.89	3.99	3.92	**3.13**
kro124p	31.12	26.82	24.77	7.67	10.58	8.88	6.84	**6.75**	6.85
ftv170	42.4	38.73	36.88	46.02	43.28	19.15	**14.13**	16.05	14.44
rbg323	30.77	**12.37**	12.67	43.4	60.11	53.84	50.08	48.97	52.06
rbg358	55.8	**17.02**	20.55	64.05	74.89	77.01	73.08	71.2	72.31
rbg403	43.41	6.98	**4.87**	17.52	20.83	29.27	28.51	28.01	28.43
rbg443	44.19	**7.21**	7.57	25.56	24.29	30.45	29.48	29.43	29.54
Average	35.77	19.12	18.93	15.76	17.43	13.42	12	11.94	**11.83**

Table 5. Detailed summary of obtained results (objective function value)

Test	PHS1				PHS2				PHS3			
	Average	Minimal	Maximal	Sample std. dev.	Average	Minimal	Maximal	Sample std. dev.	Average	Minimal	Maximal	Sample std. dev.
br17	39	39	39	0	39	39	39	0	39	39	39	0
ftv33	1308.63	1286	1339	26.39	1314.27	1286	1339	26.89	1297.57	1286	1339	21.72
ftv35	1491.5	1473	1499	10.85	1491.6	1473	1499	10.11	1484.47	1473	1492	8.01
ftv38	1547	1532	1603	12.8	1545.17	1530	1549	5.55	1544.1	1530	1549	5.68
p43	5622.23	5620	5623	1.14	5622.23	5620	5623	1.1	5621.83	5620	5623	1.09
ftv44	1639.83	1613	1713	27.24	1640.73	1613	1683	23.27	1624.43	1613	1638	7.86
ftv47	1805.6	1776	1835	14.39	1803.8	1780	1835	13.17	1797	1776	1814	9.65
ry48p	14518.57	14507	14707	40.11	14523.57	14507	14707	47.36	14509.9	14459	14556	16.11
ft53	7381	7153	7731	155.17	7341.03	7111	7641	129.20	7360.67	7137	7702	135.82
ftv55	1640.33	1608	1687	22.31	1644.6	1608	1664	11.65	1623.13	1608	1649	17.92
ftv64	1859.2	1839	1963	24.17	1868.7	1846	1909	20.38	1865.1	1848	1905	15.52
ft70	40136.17	39426	40411	212.92	40188.30	39868	40554	179.63	40164.33	39501	40439	186.98
ftv70	2027.87	1967	2096	27.61	2026.40	2001	2062	15.59	2010.97	1959	2063	25.54
kro124p	38708.43	37634	39756	585.38	38676	37756	39843	536.02	38711.9	37924	39638	438.64
ftv170	3144.2	2931	3388	97.13	3197.13	2986	3367	111.67	3152.8	3088	3231	38.03
rbg323	1990.03	1927	2068	38.46	1975.3	1875	2059	41.72	2016.33	1937	2056	25.99
rbg358	2012.97	1924	2073	34.82	1991.07	1904	2089	57.23	2004	1903	2057	38.22
rbg403	3167.87	3108	3217	24.21	3155.5	3089	3203	31.72	3165.83	3105	3200	23.67
rbg443	3521.9	3468	3567	22.48	3520.43	3444	3586	34.65	3523.47	3465	3557	24.27

Objective function value

Table 6. Results of Wilcoxon Signed-Rank Test for the conducted research

O1	O2		
	PHS1	PHS2	PHS3
PHS1	N/A	0.9123	1
PHS2	0.0877	N/A	0.9989
PHS3	**1.74797E−07**	**0.0011**	N/A

in comparison to selected methods described in subject literature. The detailed results obtained by particular PHS models were presented in Table 5.

The obtained percentage surpluses of the values of objective function for particular PHS models were also subjected to Wilcoxon Signed-Rank Test, using R. The following method variants (marked as $O1$ and $O2$) were subjected to the process, whereas the value of 0.05 was assumed as the level of test significance (the obtained p-values described with the lower result were distinguished by using bold and they indicate to accept the alternative hypothesis according to which $O1$ obtained lower results than $O2$). The test results were presented in Table 6. According to them, it is recommended to apply the PHS3.

9   Conclusions and Future Work

On the basis of the results of conducted research, it is recommended to use the PHS3 (with $p = 4$) and avoid the PHS1. Regardless of the applied model, a noticeable increase of the method effectiveness was reported in comparison with its sequential equivalent (the average error was reduced from 13.42 to 11.83%). It is possible to reduce the average error to 11.77% by using the PHS3 (with $p = 4$) for tasks described by a maximum of 48 vertices and the PHS2 (with $p = 4$ and $IK = 4000$) otherwise.

The occurrence of correlations between the p parameter value and the average error was discovered for each of the examined algorithm variants (the quality of determined solutions increases along with the increase of p value), which proves the possibility of increasing the effectiveness of the proposed approaches, by additionally increasing the p value.

Further studies related to the PHS may include development of new communication models and determining the maximum recommended p value. It is also recommended to check the efficiency of researched models for other problems.

References

1. Ayob, M., Hadwan, M., Ahmad Nazri, M.Z., Ahmad, Z.: Enhanced harmony search algorithm for nurse rostering problems. J. Appl. Sci. **13**(6), 846–853 (2013)
2. Bashiri-Atrabi, H., Qaderi, K., Rheinheimer, D., Sharifi, E.: Application of harmony search algorithm to reservoir operation optimization. Water Resour. Manage. **29**(15), 5729–5748 (2015)
3. Boryczka, U., Szwarc, K.: The adaptation of the harmony search algorithm to the ATSP. In: Nguyen, N.T., Hoang, D.H., Hong, T.-P., Pham, H., Trawiński, B. (eds.) ACIIDS 2018. LNCS (LNAI), vol. 10751, pp. 341–351. Springer, Cham (2018). https://doi.org/10.1007/978-3-319-75417-8_32
4. Boryczka, U., Szwarc, K.: The Adaptation of the Harmony Search Algorithm to the ATSP with the evaluation of the influence of the pitch adjustment place on the quality of results. J. Inf. Telecommun. **3**, 2–18 (2019)
5. Ceylan O., Liu, G., Tomsovic, K.: Parallel harmony search based distributed energy resource optimization. In: 2015 18th International Conference on ISAP, pp. 1–6 (2015)
6. Geem, Z.W.: Optimal design of water distribution networks using harmony search. Ph.D. thesis, Korea University (2000)
7. Hong, A., Jung, D., Choi, J., Kim, J.H.: Sensitivity analysis on migration parameters of parallel harmony search. In: Del Ser, J. (ed.) ICHSA 2017. AISC, vol. 514, pp. 3–7. Springer, Singapore (2017). https://doi.org/10.1007/978-981-10-3728-3_1
8. Komaki, M., Sheikh, S., Teymourian, E.: A Hybrid Harmony Search algorithm to minimize total weighted tardiness in the permutation flow shop. In: CIPLS, pp. 1-8 (2014)
9. Osaba, E., Diaz, F., Onieva, E., Carballedo, R., Perallos, A.: A population meta-heuristic with adaptive crossover probability and multi-crossover mechanism for solving combinatorial optimization problems. IJAI **12**, 1–23 (2014)
10. Öncan, T., Altınel, I.K., Laporte, G.: A comparative analysis of several asymmetric traveling salesman problem formulations. Comput. Oper. Res. **36**(3), 637–654 (2009)
11. Peraza, C., Valdez, F., Garcia, M., Melin, P., Castillo, O.: A new fuzzy harmony search algorithm using fuzzy logic for dynamic parameter adaptation. Algorithms **9**(4), 69 (2016)
12. Peraza, C., Valdez, F., Melin, P.: Optimization of intelligent controllers using a Type-1 and interval Type-2 fuzzy harmony search algorithm. Algorithms **10**, 82 (2017)
13. Peraza, C., Valdez, F., Castro, J.R., Castillo, O.: Fuzzy dynamic parameter adaptation in the harmony search algorithm for the optimization of the ball and beam controller. Adv. Oper. Res. **2018**, 16 (2018)
14. Syberfeldt, A., Rogstrom, J., Geertsen, A.: Simulation-based optimization of a real-world travelling salesman problem using an evolutionary algorithm with a repair function. IJAE **6**(3), 27–39 (2015)

Differential Evolution in Agent-Based Computing

Mateusz Godzik[✉], Bartlomiej Grochal, Jakub Piekarz, Mikolaj Sieniawski,
Aleksander Byrski[✉], and Marek Kisiel-Dorohinicki

Department of Computer Science, Faculty of Computer Science,
Electronics and Telecommunications, AGH University of Science and Technology,
Al. Mickiewicza 30, 30-059 Kraków, Poland
{godzik,olekb,doroh}@agh.edu.pl, bartek.grochal@gmail.com,
jpiekarz12@outlook.com, mikolajsieniawskii@gmail.com

Abstract. Evolutionary multi-agent systems (EMAS) turned out to be quite efficient technique for solving complex problems, both benchmark ones (as well-known multi-dimensional functions, e.g. Rastrigin, Schwefel etc) and more practical ones (like Optimal Golomb Ruler or Low Autocorrelation Binary Sequence). However the already classic design of the EMAS (these metaheuristics have been developed for over 15 years) has still many places for improvement. Hybridization is one of such means, and it turns out that incorporating Differential Evolution mechanisms into EMAS (altering the reproduction strategy by making it more social-aware) improves the accuracy of the search. This paper deals with discussion of selected means for hybridization of EMAS with DE, and provides an insight into the efficacy of the novel algorithm compared with classic techniques based on multidimensional benchmark problems.

Keywords: Metaheuristics · Agent-based computing ·
Differential evolution · Hybrid algorithms

1 Introduction

Some of the problems are so difficult that it is necessary to use special algorithms. If the problem cannot be solved in a deterministic way in a reasonable time, it is a good idea to use metaheuristics, also known as last chance methods. One of the advantages of metaheuristics is that they do not need any information about the search space. This is useful in solving combinatorial problems or other complex ones. There is not and there will not be one method that will solve any problems with the same accuracy. It is necessary to look for new metaheuristics (cf. Wolpert and MacReady [24]). However, it is worth considering and not produce the metaheuristics only for the sake of using another inspiration (cf. Sörensen [22]). Krzysztof Cetnarowicz proposed the concept of an Evolutionary Multi-Agent System (EMAS) in 1996 [6].

© Springer Nature Switzerland AG 2019
N. T. Nguyen et al. (Eds.): ACIIDS 2019, LNAI 11432, pp. 228–241, 2019.
https://doi.org/10.1007/978-3-030-14802-7_20

This metaheuristics consists of a set of agents—entities that bear appearances of intelligence and are able to make decisions autonomously. The basic mechanism is the decentralization of evolution and decomposition of the population. The task is divided into parts, and each of them goes to another agent. EMAS have several useful functions. First one is lack of global control—agents co-evolve independently of any superior management. High speed of finding solutions is owing to parallel ontogenesis—agents may die, reproduce, or act at the same time. Many discrete and continuous problems was successfully solved by EMAS. It was thoroughly theoretically analyzed, along with preparing of formal model proving its potential applicability to any possible problem (capability of being an universal optimizer, based on Markov-chain analysis and ergodicity feature) [1].

1996 was also the year of the publication of Differential Evolution (DE) by Storn and Price [23]. This stochastic algorithm is based on a population of solutions. DE optimizes real-valued multidimensional functions without gradients. It is based on cyclical attempts to improve the candidate solutions. It is possible to use DE to optimize discontinuous or time-varying problems.

The hybrid method presented in this paper is based on coupling two metaheuristics, namely EMAS and DE. It's allow the EMAS agents to run DE steps to improve solutions. It should be noted, that DE not use gradients, and thus its synergy with EMAS seems to be even more attractive than e.g. introducing of a certain steepest-descent method that we have already done in the past [17].

The paper is organized as follows. After this introduction a number of hybrid DE and evolutionary methods are referenced, leading the reader to the short recalling of EMAS basics and later presenting the DE and its hybridization with EMAS. Next the experimental results comparing EMAS-DE hybrid with the base model of EMAS are shown, and finally the paper is concluded with some remarks.

2 Differential Evolution and Its Hybrids

DE was originally developed as a real-valued parameter optimization algorithm [23], however multiple researchers successfully proven its usefulness in case of hybridization with both global optimization and local search algorithms [8]. Therefore, it is still a matter of interest for many engineers struggling with finding global optima of complex functions. This section presents firstly a high-level description of the DE algorithm, then discusses briefly hybrids of DE and other optimization methods, and finally motivates the legitimacy of this research.

2.1 Differential Evolution Algorithm

In fact, DE may be viewed as a framework defining a skeleton for computations, which must be completed with building blocks forming the specific evolutionary algorithm. A block diagram presenting all stages of any DE algorithm is presented in the Fig. 1. Every DE algorithm begins with the generation of initial population, then subsequent steps are performed. Each step consists of three succeeding actions applied to each of individuals forming the population: mutation based on difference vector(s), recombination and selection. At the end of any step, a new population is created and acts as an input for the next step. The next paragraphs introduce briefly each of aforementioned stages, however their full descriptions and mathematical formulations can be found in [8,23].

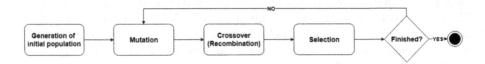

Fig. 1. Block diagram of the DE algorithm.

DE algorithm acts on a population of individuals in order to find a point in a multidimensional, generally finite, search space, which determines the global optimum of an objective function. It is naturally possible to be more than one such point, however in this paper it is assumed that we are looking for *at least* one optimal solution, not *all*. An individual is usually a multidimensional, real-valued point (also known as a genotype) belonging to the search space and acting as a solution of the problem under optimization. The population is subjected to cyclic evolution, which produces new individuals owning other (and possibly better, i.e. closer to the optimal one) solutions. In order to start the process of evolution, an initial population has to be created. It is usually achieved by uniform sampling of the search space. The size of a population (given as a control parameter NP) is constant over time.

The first stage in the process of evolution is mutation. The goal of a mutation operator is to provoke random disturbance of a parent genotype (also known as a *target*) in order to obtain a slightly different solution (also known as a *donor*). In case of DE, a donor solution is obtained by picking the base solution and perturbing it with a difference of other solutions, scaled by a *mutation parameter* (given as a control parameter F). This high-level description is the first gap which must be fulfilled in order to create a complete variant of the DE algorithm. There are multiple mutation methods, which are distinguished by strategies of choosing the base solution (e.g. random solution, current (target) solution, the best solution from current population) and the perturbing solutions (e.g. a difference of: two randomly chosen solutions, the best solution and a randomly chosen one). However it must be noted that random solutions have to be sampled from current population for each target solution independently and they must be pairwise distinct from each other and from the target solution.

Crossover (also called recombination), applied as the second stage of any evolution step, enhances diversity of the evolving population. This operator forms new, *trial* solution by exchanging parts of genotypes between the target and the donor solutions. As in the case of mutations, there are multiple recombination strategies, while two of them - binomial (also known as uniform) and exponential (also called two-point modulo) - are the most often used. Both the methods work stochastically and include an average of *crossover rate* (given as a control parameter CR) portion of the donor's genes to the trial solution (and the rest of its genes belongs to the target solution). However, the exponential crossover operator always includes a coherent fragment of donor's genotype to the trial genotype, while the binomial crossover operator decides whether to choose either target's or donor's gene (corresponding to a single real-valued coordinate of a solution point) for the trial genotype.

The last stage of the evolution step is selection. The goal of a selection operator is to preserve constant population size from one cycle to another by deciding whether the target solution should survive to the next generation, or maybe it should be replaced with a newly-created trial solution. There is rather one common strategy adopted for solution operators which promotes to the new population this individual, which genotype determines not greater (in case of minimization problems) value of the objective function.

As already presented above, a DE algorithm is built by employing specific mutation and recombination operators which highly impact on its optimization capabilities. Different variants of DE methods are classified under the *Family of Storn and Price* [8] introducing a string notation *DE/[basis]/[perturbation]/[selection]*. The *DE* prefix is common to all DE algorithms; the *[basis]* parameter represents a strategy of choosing the base solution for the mutation operator (usually it is one of: *rand* (random solution), *curr* (target solution) or *best* (best solution)); the *[perturbation]* attribute denotes the number of perturbation vectors used by the mutation operator (usually its value is one or two); finally, the *[selection]* feature indicates the selection strategy adopted by the algorithm (usually either *exp* for exponential or *bin* for binomial).

2.2 Hybrids of Differential Evolution

DE algorithm has proven efficiency in hybridization with both global optimization and local search methods [8]. Although the DE scheme is inherently designed for real-valued parameter optimization, there are multiple research works towards its adaptation to the problems of discrete and binary optimization [8]. The next paragraphs introduce some the most significant concepts for hybridization of DE with other algorithms, however a quite comprehensive overview on this topic is covered by [8].

Synergy between DE and Particle Swarm Optimization (PSO) has been examined by many researchers adopting multiple different approaches. One of the most popular methods of introducing a cooperation between these two algorithms is to integrate DE steps into the PSO evolution scheme. For example, Hendtlass [13] proposed to perform DE steps on swarm particles only at certain intervals, while Zhang and Xie [26] suggested to perform DE and PSO steps alternately. Another approach to combining the two algorithms in question is to incorporate the DE succession scheme into PSO. This idea was adopted by Das *et al.* [7], who constrained the PSO algorithm to update positions of only these particles, which improve their fitness. Finally, Kannan *et al.* [15] used DE as a meta-optimizer for adaptive improvement of PSO control parameters.

There are also numerous research works treating about hybridization of DE with other (mainly bio-inspired) global optimization algorithms, such as: Any Colony Optimization (ACO) [21], Simulated Annealing (SA) [19], Grey Wolf Optimizer (GWO) [14] and Artificial Immune Systems (AIS) [12]. In case of combining DE with local search methods (exploring small research space in the neighborhood of the solution found by DE), many strategies have also been developed, including: Hill Climbing [20], Variable Neighborhood Search [25], Random Walk and Harmony Search [18], Nelder-Mead Search and Rosenbrock Search [5]. Similarly to the hybrids of DE and global search algorithms, local search methods are usually responsible for either improving solutions found by DE or elaborating optimal values for DE control parameters [8].

Conducted review of existing solutions results in the statement that there are no hybrid solutions combining DE and agent-based evolutionary algorithms, such as EMAS. Therefore, this research tries to address the aforementioned issue by implementing three algorithms: classical EMAS, pure DE in the agent-based environment and EMAS/DE hybrid acting on a population of agents as in the case of an ordinary EMAS algorithm and performing additional DE steps in some iterations. This approach is expected to enjoy the benefits of both agent-based computing (presented in Sect. 3.1) and proven efficiency of DE and its hybrids in solving complex, high-dimensional, real-valued parameter optimization problems.

3   Hybridization of Differential Evolution and EMAS

3.1   Evolutionary Multi Agent-Systems

Evolutionary Multi Agent-System [6] can be treated as an interesting and quite efficient metaheuristic, moreover with a proper formal background proving its correctness [1]. Therefore this system has been chosen as a tool for solving the problem described in this paper.

Evolutionary processes are by nature decentralized and therefore they may be easily introduced in a multi-agent system at a population level. It means that agents are able to *reproduce* (generate new agents), which is a kind of cooperative interaction, and may *die* (be eliminated from the system), which is the result of competition (selection). A similar idea with limited autonomy of agents located in fixed positions on some lattice (like in a cellular model of parallel evolutionary algorithms) was developed by Zhong et al. [27]. The key idea of the decentralized model of evolution in EMAS [16] was to ensure full autonomy of agents.

Such a system consists of a relatively large number of rather simple (reactive), homogeneous agents, which have or work out solutions to the same problem (a common goal). Due to computational simplicity and the ability to form independent subsystems (sub-populations), these systems may be efficiently realized in distributed, large-scale environments (see, e.g. [2]).

Agents in EMAS represent solutions to a given optimization problem. They are located on islands representing distributed structure of computation. The islands constitute local environments, where direct interactions among agents may take place. In addition, agents are able to change their location, which makes it possible to exchange information and resources all over the system [16].

In EMAS, phenomena of inheritance and selection—the main components of evolutionary processes—are modeled via agent actions of *death* and *reproduction* (see Fig. 2). As in the case of classical evolutionary algorithms, inheritance is accomplished by an appropriate definition of reproduction. Core properties of the agent are encoded in its genotype and inherited from its parent(s) with the use of variation operators (mutation and recombination). Moreover, an agent may possess some knowledge acquired during its life, which is not inherited. Both inherited and acquired information (phenotype) determines the behavior of an agent. It is noteworthy that it is easy to add mechanisms of diversity enhancement, such as allopatric speciation (cf. [4]) to EMAS. It consists in introducing population decomposition and a new action of the agent based on moving from one evolutionary island to another (migration) (see Fig. 2).

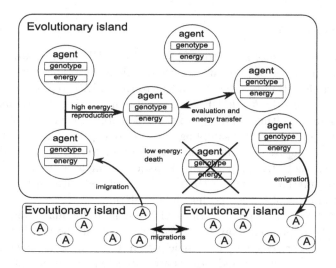

Fig. 2. Evolutionary multi-agent system (EMAS)

Assuming that no global knowledge is available, and the agents being autonomous, selection mechanism based on acquiring and exchanging non-renewable resources [6] is introduced. It means that a decisive factor of the agent's fitness is still the quality of solution it represents, but expressed by the amount of non-renewable resource it possesses. In general, the agent gains resources as a reward for "good" behavior, and looses resources as a consequence of "bad" behavior (behaviorism here may be understood as, e.g. acquiring sufficiently good solution). Selection is then realized in such a way that agents with a lot of resources are more likely to reproduce, while a low level of resources increases the possibility of death. So according to classical Franklin's and Graesser's taxonomy—agents of EMAS can be classified as Artificial Life Agents (a kind of Computational Agents) [10].

Many optimization tasks, which have already been solved with EMAS and its modifications, have yielded better results than certain classical approaches. They include, among others, optimization of neural network architecture, multi-objective optimization, multimodal optimization and financial optimization. EMAS has thus been proved to be a versatile optimization mechanism in practical situations. A summary of EMAS-related review has is given in [3].

EMAS may be held up as an example of a cultural algorithms, where evolution is performed at the level of relations among agents, and cultural knowledge is acquired from the energy-related information. This knowledge makes it possible to state which agent is better and which is worse, justifying the decision about reproduction. Therefore, the energy-related knowledge serves as situational knowledge. Memetic variants of EMAS may be easily introduced by modifying evaluation or variation operators (by adding an appropriate local-search method).

3.2 Hybrid EMAS

The adopted hybridization strategy of EMAS and DE incorporates single DE steps for each member of population with specific energy. Therefore, single step of the hybrid algorithm is either equivalent to the EMAS step or consists of the EMAS step enriched by single DE step, possibly improving the results.

The EMAS algorithm, similarly to the DE method, begins with generating initial population by uniform sampling of the search space. Then, multiple steps are performed, one by one, simulating evolution of the population. Unlike the DE scheme, single EMAS step is rather holistic, i.e. single step function manages the whole population, while the DE step is associated with an individual and replicated to all members composing the population.

The EMAS evolution process starts with mating individuals randomly. Each member of the population may be paired with at most one another individual (or stay alone). The selected pairs participate in the process of sexual reproduction, while the not selected individuals are subjected to the process of asexual reproduction.

If both of the parents satisfy given reproduction predicate (associated with acquiring sufficient amount of resources), an offspring genotype is produced by applying a recombination strategy. Then, the newly-created genotype is also mutated, and finally a proportional part of both parents' resources is transferred to the child. Conversely, if at least one of the parents does not satisfy the reproduction predicate, a fight operator is applied to them. In this case, an individual with better fitness wins the fight and a portion of loser's resources is absorbed by the winner.

In case of asexual reproduction, only a mutation operator is applied (naturally, crossover is not applicable to a single genotype). The mutated genotype forms an offspring, who acquires proportional amount of parent's energy. However, if the parent does not satisfy the reproduction predicate, the whole reproduction process is discarded.

Descendant genotypes, created during both sexual and asexual reproduction processes, are then merged and evaluated by the objective function. Finally, the whole population (parents and children) is putting together. Each of the individuals is then evaluated, whether it meets a death predicate (associated with lack of resources). If so, given member is eliminated from the population.

Once in a while, the population of living genotypes is subjected to the process of migration to other islands (workplaces). There are generally two main strategies of choosing the individuals to switch workplaces: either randomly select members of the population or the best ones are migrated. Regardless of the method chosen, chosen individuals are removed from the current population and will be added to populations living on neighboring islands before starting the next step. The remaining agents survive to the next step on the same workplace.

The hybrid algorithm, similarly to the other two methods in question, generates an initial population on start up and performs multiple evolution cycles. As briefly mentioned before, single step of the hybrid algorithm always performs an EMAS step and it sometimes enables an additional DE step for each individual satisfying certain preconditions. A block diagram presenting single iteration of the hybrid algorithm is presented in the Fig. 3.

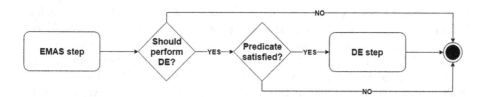

Fig. 3. Block diagram of the hybrid step.

The DE step is performed after successful completion of the EMAS step if a global predicate is satisfied. This condition is similar in spirit to the EMAS part of the hybrid algorithm, because its evaluation impacts the whole population (i.e. the DE step is totally discarded for each member of the population if this predicate evaluates to the `false` value). In case of this research, the global predicate allows to perform the DE step only for members with energy between 70 (initial energy) and 100 (threshold of reproduction), however multiple other conditions may be adopted (e.g. once in a time interval or at early/late stages of the evolution only).

When the holistic predicate is satisfied, each individual belonging to the population returned by the EMAS step (containing all non-migrated, alive agents) is tested against the DE-application predicate. In case of the implementation for this research, this predicate allows only specific genotypes to perform the DE step.

When both predicates are met, single DE step is performed for each individual satisfying given preconditions. Here, several details have to be addressed. Firstly, implemented DE and hybrid algorithms perform mutation with the population limited to current workplace, thus - in case of multiple workplaces - donor vectors are calculated on the basis of solutions belonging to individuals living on the same island (therefore the DE step is naturally not applied to agents intended for migration) as the target solution. Furthermore - because of the DE scheme, which does not consider energy parameters of agents - an offspring genotype created within the DE step replaces current genotype of the agent under mutation, while the amount of energy held by this agent remains unchanged. Finally, adopted DE selection operator promotes either trial or target (original) solution, depending on values of the objective function corresponding to them. Hence, there is no need to introduce additional selection operator choosing which genotype (i.e. the one returned by the EMAS step or the one yielded by the DE operator) should be finally included in the final population, because this issue is covered by the DE succession scheme.

4   Experimental Results

The experiments were performed taking advantage of AgE 3 platform[1], which is distributed, agent-based computational platform developed by Intelligent Information Systems Group. The platform was further developed in order to combine DE with EMAS. Finally, three evolutionary, islands-aware algorithms: classical EMAS, pure DE and EMAS hybrid with additional DE steps have been implemented and evaluated on three benchmark functions: Griewank, Rosenbrock and Sphere. This section provides a comprehensive description of adopted control parameters, algorithms' configuration, conducted tests and their results.

The experiments were conducted on a laptop with following specification: Intel Core i5-8250U CPU 1.6 GHz, 1800 MHz, 4 Cores, 8 Logical Processors; 8 GB Physical Memory (RAM); OS Microsoft Windows 10 Pro 64-bit.

4.1   Test Scenarios

Each of the three implemented algorithms was evaluated on three well-known, colorred multidimensional benchmark functions for optimization, namely Griewank, Rosenbrock and De Jong functions [9].

---

[1] http://www.age.agh.edu.pl.

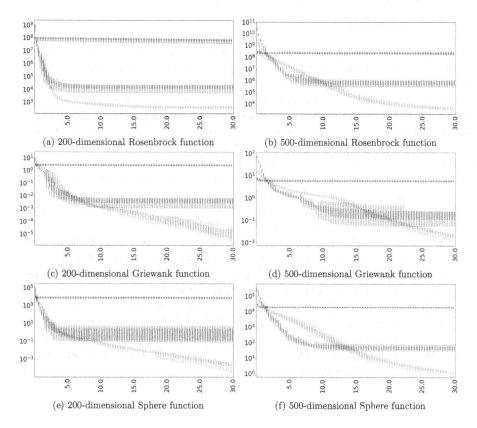

(a) 200-dimensional Rosenbrock function   (b) 500-dimensional Rosenbrock function

(c) 200-dimensional Griewank function   (d) 500-dimensional Griewank function

(e) 200-dimensional Sphere function   (f) 500-dimensional Sphere function

Fig. 4. Best fitness values in the time domain for the selected benchmark functions.

Tests were conducted as follows. Firstly, configuration of all three algorithms was elaborated separately for each test function. Values of control parameters and recombination strategies were choosing empirically, however they were based on suggestions from other researchers such as [8,11]. Afterwards, the simulations were carried out by performing ten runs of each algorithm for each test function. The optimization processes in question are stochastic, they have to be evaluated statistically by analyzing data gathered from multiple experiments. Finally, quantitative analysis of implemented algorithms was performed by plotting and analyzing charts presenting best fitness values obtained and the number of objective function evaluations depending on time.

Each of the charts presented in Fig. 4 consists a box plot describing the best values of an objective function obtained for all individuals in the population depending on the computing time. Blue color is the results of EMAS, red of DE algorithm and green of Hybrid algorithm. Mean, median, standard deviation, minimum and maximum of results can be found in Table 1.

4.2 Configuration

For each of the algorithms, an initial population consisted of 30 individuals for each workplace. Furthermore, all three algorithms were run with identical structure of workplaces and migration details. The living space of agents consisted of three fully connected (in a form of triangle) workplaces, while the migration process was conducted once in five hundred iterations. During the migration, five percent of randomly selected agents were moved to another island. Finally, both agent-based algorithms (i.e. EMAS and hybrid) were configured with the same energy parameters. Each newly-created individual received a deposit of 70 units of energy (i.e. non-renewable resources) and was intended to die when it lost all its resources. During both sexual and asexual reproduction, a parent agent transferred 30% of its energy to a child, whereas a fight loser forfeited 10 units of energy in favor of a winner.

In EMAS Gaussian mutation was used ($\mu = 0$, $\sigma = 0.1$, rate: 0.2), while in DE: *DE/rand/1/bin* with mutation factor 0.1 and crossover rate 0.1. In the case of hybrid algorithm, the DE mutation was the difference of two best individuals multiplied by mutation factor, the mutation factor $F = 1.5$, the DE crossover rate $CR = 0.1$, the DE step interval was 1. Note that the main goal of this research is to examine how DE impacts on EMAS algorithms, thus - in order to prevent the objectivity of research - the EMAS-related parameters for hybrid algorithm were configured in the same way as for the classical EMAS algorithm. Each run of the algorithm was 30 s.

4.3 Discussion of the Results

Looking at the Fig. 4, it can be seen that EMAS achieved good results, DE better, but the best results were achieved by the hybrid – they have reached the lowest values for all the tested functions. This applies to both the 200-dimensional and 500-dimensional problems. There is a chance that the hybrid would get better results if the experiment continued because it did not get stuck in the local minimum like other algorithms.

Analyzing the data on the Griewank charts, one can notice a significant impact of the DE algorithm on EMAS. The hybrid improved its score throughout the experiment, achieving a much better result than the base metaheuristics. This appears in both sizes of test function.

The next results placed on the charts also confirm the positive influence of both algorithms on themselves. For this size of problems, using a hybrid is much better than any of the component algorithms.

Table 1. Results of EMAS, DE and Hybrid algorithms for all problems.

EMAS

	Mean	Median	SD	Minimum	Maximum
Griewank 200	2.65	2.64	2.32×10^{-1}	2.31	3.09
Griewank 500	5.92	5.84	4.68×10^{-1}	5.36	7.02
Rosenbrock 200	6.72×10^7	6.72×10^7	2.05×10^7	3.51×10^7	1.12×10^8
Rosenbrock 500	2.28×10^8	2.25×10^8	3.44×10^7	1.86×10^8	2.96×10^8
Sphere 200	7.45×10^3	7.58×10^3	1.11×10^3	5.59×10^3	9.03×10^3
Sphere 500	1.90×10^4	1.89×10^4	1.28×10^3	1.77×10^4	2.23×10^4

DE

	Mean	Median	SD	Minimum	Maximum
Griewank 200	1.12×10^{-2}	4.08×10^{-3}	1.82×10^{-2}	1.09×10^{-3}	6.37×10^{-2}
Griewank 500	2.19×10^{-1}	1.47×10^{-1}	1.67×10^{-1}	5.74×10^{-2}	6.05×10^{-1}
Rosenbrock 200	2.99×10^4	1.48×10^4	4.22×10^4	5.48×10^3	1.48×10^5
Rosenbrock 500	8.68×10^5	8.30×10^5	5.60×10^5	3.10×10^5	2.32×10^6
Sphere 200	1.46	4.08×10^{-1}	1.68	8.23×10^{-2}	4.73
Sphere 500	4.98×10^1	4.80×10^1	1.62×10^1	2.91×10^1	7.70×10^1

Hybrid

	Mean	Median	SD	Minimum	Maximum
Griewank 200	1.75×10^{-3}	1.14×10^{-5}	5.21×10^{-3}	2.79×10^{-6}	1.74×10^{-2}
Griewank 500	2.18×10^{-2}	2.09×10^{-2}	6.58×10^{-3}	1.18×10^{-2}	3.65×10^{-2}
Rosenbrock 200	3.90×10^2	3.55×10^2	7.78×10^1	2.85×10^2	5.08×10^2
Rosenbrock 500	4.37×10^3	4.15×10^3	9.14×10^2	3.21×10^3	6.30×10^3
Sphere 200	3.57×10^{-4}	2.76×10^{-4}	3.29×10^{-4}	3.12×10^{-5}	1.29×10^{-3}
Sphere 500	1.24	1.17	1.73×10^{-1}	1.10	1.72

5 Conclusion

In this paper we have presented a hybrid optimization algorithm consisting in introducing differential evolution into agent-based computing system as a means for mutation. The presented experiments show that the hybrid algorithm outperforms the reference algorithms on the tested function with varying dimensionality. The presented results encourage further research, i.e. in the following steps, complex agent-based metaheuristic algorithm will be developed, putting together different nature-inspired techniques (such as complex mutations like the one presented in this paper, memetic local search and others). The general idea is to support the exploitation at the individual agent's perspective, giving it the means for improving the local search, while maintaining the exploration by controlling these individuals in the decentralized way as parts of the agent system. The presented paper may be treated as a sound step towards constructing of such metaheuristic algorithm.

Acknowledgment. The research presented in this paper was supported by the funds assigned by the Polish Minister of Science and Higher Education to AGH University of Science and Technology.

References

1. Byrski, A., Schaefer, R., Smołka, M.: Asymptotic guarantee of success for multi-agent memetic systems. Bull. Pol. Acad. Sci. Tech. Sci. **61**(1), 257–278 (2013)
2. Byrski, A., Debski, R., Kisiel-Dorohinicki, M.: Agent-based computing in an augmented cloud environment. Comput. Syst. Sci. Eng. **27**(1) (2012)
3. Byrski, A., Drezewski, R., Siwik, L., Kisiel-Dorohinicki, M.: Evolutionary multi-agent systems. Knowl. Eng. Rev. **30**(2), 171–186 (2015). https://doi.org/10.1017/S0269888914000289
4. Cantú-Paz, E.: A summary of research on parallel genetic algorithms. IlliGAL Report No. 95007. University of Illinois (1995)
5. Caponio, A., Neri, F., Tirronen, V.: Super-fit control adaptation in memetic differential evolution frameworks. Soft Comput. **13**(8), 811–831 (2009). https://doi.org/10.1007/s00500-008-0357-1
6. Cetnarowicz, K., Kisiel-Dorohinicki, M., Nawarecki, E.: The application of evolution process in multi-agent world (MAW) to the prediction system. In: Tokoro, M. (ed.) Proceedings of the 2nd International Conference on Multi-Agent Systems (ICMAS 1996). AAAI Press (1996)
7. Das, S., Konar, A., Chakraborty, U.K.: Improving particle swarm optimization with differentially perturbed velocity. In: Proceedings of the 7th Annual Conference on Genetic and Evolutionary Computation, pp. 177–184. ACM (2005)
8. Das, S., Suganthan, P.N.: Differential evolution: a survey of the state-of-the-art. IEEE Trans. Evol. Comput. **15**(1), 4–31 (2011). https://doi.org/10.1109/TEVC.2010.2059031
9. Digalakis, J., Margaritis, K.: An experimental study of benchmarking functions for evolutionary algorithms. Int. J. Comput. Math. **79**(4), 403–416 (2002). citeseer.ist.psu.edu/digalakis02experimental.html
10. Franklin, S., Graesser, A.: Is It an agent, or just a program? A taxonomy for autonomous agents. In: Müller, J.P., Wooldridge, M.J., Jennings, N.R. (eds.) ATAL 1996. LNCS, vol. 1193, pp. 21–35. Springer, Heidelberg (1997). https://doi.org/10.1007/BFb0013570
11. Gämperle, R., Müller, S.D., Koumoutsakos, P.: A parameter study for differential evolution. In: Grmela, A., Mastorakis, N. (eds.) Advances in Intelligent Systems, Fuzzy Systems, Evolutionary Computation, pp. 293–298. WSEAS Press (2002)
12. He, X., Han, L.: A novel binary differential evolution algorithm based on artificial immune system. In: Proceedings of the IEEE Congress on Evolutionary Computation, pp. 2267–2272. IEEE (2007)
13. Hendtlass, T.: A combined swarm differential evolution algorithm for optimization problems. In: Monostori, L., Váncza, J., Ali, M. (eds.) IEA/AIE 2001. LNCS (LNAI), vol. 2070, pp. 11–18. Springer, Heidelberg (2001). https://doi.org/10.1007/3-540-45517-5_2
14. Jitkongchuen, D.: A hybrid differential evolution with grey wolf optimizer for continuous global optimization. In: Proceedings of the 7th International Conference on Information Technology and Electrical Engineering, pp. 51–54. IEEE (2015)

15. Kannan, S., Slochanal, S.M.R., Subbaraj, P., Padhy, N.P.: Application of particle swarm optimization technique and its variants to generation expansion planning problem. Electr. Power Syst. Res. **70**(3), 203–210 (2004). https://doi.org/10.1016/j.epsr.2003.12.009

16. Kisiel-Dorohinicki, M.: Agent-oriented model of simulated evolution. In: Grosky, W.I., Plášil, F. (eds.) SOFSEM 2002. LNCS, vol. 2540, pp. 253–261. Springer, Heidelberg (2002). https://doi.org/10.1007/3-540-36137-5_19

17. Korczynski, W., Byrski, A., Kisiel-Dorohinicki, M.: Buffered local search for efficient memetic agent-based continuous optimization. J. Comput. Sci. **20**(Supplement C), 112–117 (2017). https://doi.org/10.1016/j.jocs.2017.02.001. http://www.sciencedirect.com/science/article/pii/S1877750317301345

18. Liao, T.W.: Two hybrid differential evolution algorithms for engineering design optimization. Appl. Soft Comput. **10**(4), 1188–1199 (2010). https://doi.org/10.1016/j.asoc.2010.05.007

19. Liu, K., Du, X., Kang, L.: Differential evolution algorithm based on simulated annealing. In: Kang, L., Liu, Y., Zeng, S. (eds.) ISICA 2007. LNCS, vol. 4683, pp. 120–126. Springer, Heidelberg (2007). https://doi.org/10.1007/978-3-540-74581-5_13

20. Noman, N., Iba, H.: Accelerating differential evolution using an adaptive local search. IEEE Trans. Evol. Comput. **12**(1), 107–125 (2008). https://doi.org/10.1109/TEVC.2007.895272

21. Rahmat, N.A., Musirin, I.: Differential Evolution Ant Colony Optimization (DEACO) technique in solving economic load dispatch problem. In: Proceedings of the IEEE International Power Engineering and Optimization Conference, pp. 263–268. IEEE (2012)

22. Sörensen, K.: Metaheuristics–the metaphor exposed. Int. Trans. Oper. Res. **22**(1), 3–18 (2015). https://doi.org/10.1111/itor.12001

23. Storn, R., Price, K.: Differential evolution: a simple and efficient adaptive scheme for global optimization over continuous spaces. Technical report TR-95-012, ICSI, USA, March 1995

24. Wolpert, D., Macready, W.: No free lunch theorems for optimization. IEEE Trans. Evol. Comput. **67**(1) (1997)

25. Yang, Z., Yao, X., He, J.: Making a difference to differential evolution. In: Siarry, P., Michalewicz, Z. (eds.) Advances in Metaheuristics for Hard Optimization, pp. 397–414. Springer, Heidelberg (2008). https://doi.org/10.1007/978-3-540-72960-0_19

26. Zhang, W.J., Xie, X.F.: DEPSO: hybrid particle swarm with differential evolution operator. In: Proceedings of the IEEE International Conference on Systems, Man and Cybernetics, pp. 3816–3821. IEEE (2003)

27. Zhong, W., Liu, J., Xue, M., Jiao, L.: A multiagent genetic algorithm for global numerical optimization. IEEE Trans. Syst. Man Cybern. Part B Cybern. **34**(2), 1128–1141 (2004)

Verifying Usefulness of Ant Colony Community for Solving Dynamic TSP

Andrzej Siemiński$^{(\boxtimes)}$ ⓘ

Department of Information Systems, Faculty of Computer Science
and Management, Wrocław University of Science and Technology,
Wybrzeże Wyspiańskiego 27, 50-370 Wrocław, Poland
Andrzej.Sieminski@pwr.edu.pl

Abstract. The paper describes Ant Colony Communities (ACC) and verifies their usefulness for Dynamic Travelling Salesman Problem (DTSP). DTSP is a version of the classical TSP in which the distance matrix change in time. The ACC consists of a set of separate ant colonies with a server that coordinates their work and sends them cargos of data for processing. The colonies could be distributed over many computers working in a LAN or even over Internet. Such a mode of operation is especially useful for dynamic tasks where solutions must catch up with the changing environment. The ACC is used for the regular ACO and its version designed for dynamic environments: PACO and Immigrant ant colonies. The experiments show that for all types Ant Colonies the introduction of the community boots the performance. The routes are far shorter than in the case of original colonies.

Keywords: Dynamic TSP · Ant Colony Community · PACO ·
Immigrant based colonies · Parallel implementation of ACO

1 Introduction

The Travelling Salesmen Problem is an ideal problem for computer science. It is extremely simple: find the shortest path that connects all cities without revisiting any one on them and yet any attempt to solve it using brute force approach is doomed to failure. Even for a set of just 50 cities the number of all possible solutions is equal to 3,04141E +64 exceeding by far the weight in kilograms of visible universe. There is no algorithmic solution to the problem and we have to resolve to heuristics solutions. An extensive survey of used here metaheuristics is in [1]. The problem is not a just mathematical puzzle, it has numerous practical applications especially in planning and logistics [2]. Adding dynamics that is allowing the distances to change over time makes this hard problem more challenging. For over 30 years this is an active area of research [3].

With the lack of algorithmic solutions to the TSP we have to resolve to heuristic approaches. Among them we have genetic algorithms, simulated annealing, tabu search, and ant colony optimization. The last approach was introduced by as early as mid-90's by Dringo [4] and still counts among the most popular of them.

The paper is organized as follows. The second section presents some basic information about the regular ACO and popular ways of adopting it to distance changes.

N. T. Nguyen et al. (Eds.): ACIIDS 2019, LNAI 11432, pp. 242–253, 2019.
https://doi.org/10.1007/978-3-030-14802-7_21

The next section is describes the architecture and operational rules of a Community of colonies. The Graph Generator which is responsible for the distance dynamics used in our study is described in the 4th Section. During the experiment the standard ACO, PACO and Immigrant based ACO are run within the ACC environment. The experiment results and their interpretation are presented in the next, 5th section. The results confirm that the introduction of the community boots the performance of all algorithms under study. The paper concludes with some ideas for future research work.

2 Related Work on Ant Based Solutions for TSP

Ant Colony Optimization (ACO) is a popular metaheuristic that is used to find approximate solutions to difficult optimization problems, especially for Travelling Salesman Problem (TSP). As in other metaheuristics the key issue is to generate a sequence of plausible solutions while mitigating the danger of premature results stagnation.

At first the basic model is described, followed by its variants useful for dynamic version of TSP.

2.1 ACO Basics

In ACO the optimization task is transformed into the problem of finding the best path on a weighted graph. The search for a solution is constructed by a number of ants that traverse the graph depositing pheromone on their way. Two values are associated with all graph edges: $\eta_{k,j}$ and $\tau_{k,j}$. The former denotes the some property of an edge connecting the nodes k and j (e.g. length of route separating the nodes or time necessary to travel from one node to another) whereas the latter ($\tau_{k,j}$) represents the pheromone level. The pheromone level matrix represents a collective knowledge of all about the usefulness of edges.

There many ACO variants. All of them share some basic functionality. The solution is the result of many iterations. Each iteration starts with placing the ants in a random manner on the nodes or vertices. Initially the memory of an ant is empty. At each step of an iteration an ant decides which node from those not yet visited to select. The usefulness of an edge leading from node r to node t is measured by a quality function $qf(r, t)$:

$$qf(r,t) = \tau(r,t)^{\alpha} * \eta(r,t)^{\beta} \qquad (1)$$

Where α and β are positive real parameters that specify the relative importance of pheromone and heuristic data.

An ant can work in deterministic or nondeterministic mode of operation. In the former one it just selects a node that maximizes $qf(r, t)$ function. In the latter one the selections is done according to a stochastic process with the following probability function:

$$pr(r,t) = \frac{qf(r,t)}{\sum_{u \in A_{ny}} qf(r,u)} \qquad (2)$$

A_{ny} is the set of all not visited nodes. Another parameter specifies what is the probability of the above modes. The ants deposit pheromone on the edges and the pheromone evaporates. The precise way in which this is done is ACO variant specific.

2.2 Ant System (AS)

Ant System is the predecessor of all ACO approaches. The node selection is done always with the probability specified by formula (2). The pheromone values are updated by all ants once they have completed the tour. To update the pheromone level we use the formula (3).

$$\tau_{r,t} = (1 - \rho)\tau_{r,t} + \sum_{k=1}^{m} \Delta\tau_{r,t}^{k} \tag{3}$$

Where:

$\rho \in (0, 1]$: is the evaporation rate
m: the number of ants
$\Delta\tau_{r,t}^{k}$ is the quantity of pheromone laid on the edge connecting vertices r and t by the k-th ant.

$$\Delta\tau_{r,t}^{k} = \begin{cases} \frac{1}{L_k} \text{ if ant k used the edge from r to t} \\ \quad 0 \text{ otherwize} \end{cases} \tag{4}$$

Where L_k denotes the length of ant's k route. Pheromone on not selected edges evaporates at the ρ rate. Additionally the pheromone levels on edges selected by an ant increase inversely proportionally to the length of its solution.

2.3 Ant Colony System ACS

The ACS node selection process is not completely probabilistic as in the case of AS. It is controlled by a parameter $q_0 \in [0, 1]$ and a randomly generated number q that also belongs to [0, 1]. If $q \leq q_0$ then the selection is non deterministic and the formula (2) is used. The former mode of operation is called exploitation whereas the latter is known as exploration. Such a selection rule could result in a rather greedy operation mode, especially for q_0 close to 1. To mitigate the danger of premature stagnation of solutions the local update rule is introduced. It is performed by all ants after each construction step on the last edge traversed.

$$\tau_{r,t} = (1 - \delta)\tau_{r,t} + \delta\tau_0 \tag{5}$$

Where τ_0 is the initial value of pheromone and $\delta \in [0, 1]$ is pheromone decay coefficient. Here the global update procedure looks much the same as for AC but only the route found the best so far ant is taken into account.

2.4   Max-Min Ant System

The Max-Min Ant System (MMAS) attempts to preserve the diversity of generated solutions by imposing limits on pheromone levels. Let us assume that at certain iteration the pheromone level is very large or small. In that case the corresponding edge has a very high or very low probability of being selected. This phenomenon could be not favorable as it may cause stagnation of paths generated in subsequent iterations. The distinguished features of MMAS are:

- The new value of pheromone update is calculated by the following formula:

$$\tau_{i,j} = (1 - \rho)\tau_{i,j} + \Delta\tau_{i,j}^{best}$$

Where $\Delta_{i,j}^{best} = 1/L_{best}$ and depending on implementation L_{best} is the length of the iteration best or best so far route.
- The resulting is further adjusted to stay within predefined range:

$$\tau_{i,j} = min\left(max\left(\tau_{min}, \tau_{i,j}\right), \tau_{max}\right)$$

- The value of τ_{min} is experimentally chosen.
- τ_{max} is set to $1/(\rho * L_B)$. L_b stands for the shortest path, in case it is not know it is the best so far solution.
- The initial value for all $\tau_{i,j}$ is set to τ_{max}.

2.5   Population Based ACO (PACO)

As its name suggest PACO uses a population of ants to update the pheromone matrix. This is a way to preserve the diversity of its search. The node selection procedure is the same as in standard ACO. For a population of K ants for the first K iterations there is no pheromone evaporation at all [5]. The iteration best ant is just added to the population. After each following iteration the best ant enters the population replacing one that is already there. Adding an ant to the population requires modifying the pheromone levels on all edges of its route using the formula (6)

$$\tau_{i,j} = \tau_0 + \Delta\{\pi \in P | (i,j) \in \pi\} \tag{6}$$

where:

$$\Delta = \frac{\tau_{max} - \tau_0}{K}$$

π is the ant added to the population P of size K
τ_{max} is maximal allowed value for pheromone level
τ_0 is the initial value for pheromone value.

Removing an ant from the population requires that the Δ amount of pheromone be removed from the edges of its route. As you can see the update procedure does not depend on the quality of the ant's route. Such an updating procedure reduces the computational complexity of PACO in comparison to ACO makes room for many optimization procedures described e.g. in [6].

There are 3 basic ways for selecting the ant to remove from the population:

(1) age-based: the oldest ant is removed, in mimics the FIFO strategy.
(2) quality-based: the iteration best ant is admitted to the population only if its route length is shorter than the length of the worst populations ant
(3) elitist-based: it is like quality-based approach but the comparison is always with the population's best ant. If the current iteration path is shorter then we replace the elitist ant with the new one. Otherwise we follow the FIFO operation mode. There is always a different way in pheromone update level. For the best ant it is $v_e *\tau_{max}$ where τ_{max} and v_e are a parameters. For the rest of ants the used formula is:

$$\Delta * \frac{(1 - \vartheta_e)}{K - 1}$$

Unlike the age-based strategy the elitist strategy quarantines that the best so far solution is always present in the population and on the other hand it eliminates older solutions quickly than the quality based rule.

2.6   Immigrant Schemes for ACO

Immigrant schemes are a modification of PACO that was designed especially for dynamic environments [7]. As in PACO we have here a population of "best" ants but it changes in a more radical way than previously. The population does not replace just one ant with another one but rather replaces a large number of worst ants with newly generated ants. The newly introduced ants could be:

- Randomly generated: the path is randomly generated.
- Based on elitism principle: the path of best ant is modified in such a way that two randomly selected nodes are switched.
- Follow hybrid operation manner in which both above approaches are combined.

3   Ant Colony Community

ACO is computing intensive. Pretty soon the researches realized, that in order to shorten processing time we have to resolve to parallel implementations. The first parallel versions of the ACO were presented just a few years after their introduction [8]. A fairly comprehensive presentation of different approaches to parallelism for AC is in [9]. This paper gives a new taxonomy for the AC parallelism.

The crucial factor that differentiates the solutions is the scope of interaction between individual ants. In non-parallel versions, the behavior of each ant impacts the operation of all other ants. For parallel implementations this calls for hardware support due to communication overhead. This, however, limits drastically the number of possible ants. Therefore other approaches reduce the level of ant interaction. They include:

- master-slave model: the multiple clients are controlled in a centralized way by a server;
- cellular model: the distance matrix is divided into small, overlapping neighborhoods and a single colony communicates directly other colonies, each one solving its part of the problem;
- parallel independent model: the ACOs run concurrently, without any communication among them;
- multi-colony model: the ACOs interact directly with each other by periodically exchanging best solutions, there is no server at all.

3.1   Ant Colony Community (ACC) Architecture

ACC is a version of the master-slave model. Its previous version was described in [10]. The measures for the approximation of communities' computational power and its scalability are given in [11]. Its two implementations are described in a follow-up paper [12]. The first one uses a number of workstations communicating via Sockets and the second is implemented using in the Hadoop environment. Both of them were applied successfully for the traditional, static TSP.

The Master of ACC sends packages of data, called in what follows cargos, to Ant Colonies, receives the results of their work and evaluates them. A cargo includes a complete specification of the optimization task: the distance and pheromone matrixes and the parameters that define ACO operation - the number of ants, iteration number e.c.t. A client processes the data and then sends to the Master the best found route and an updated pheromone matrix.

In this paper describes an extension of the ACC that aims at adopting it to DTSP. The server and clients are augmented by the environment component. The overview of the ACC architecture is presented on the Fig. 1.

Each rounded block on the Fig. 1 represents a separate process. The processes communicate via low level and efficient sockets mechanism. This make ACC highly flexible. The processes could run on a single computer, computers located on a LAN or ever spread over Internet. Once the connection is established the transfer of data via sockets is very efficient and the difference in the transfer rate between processes that run a on single computer or on a LAN is not significant. Most of the rectangles of a process run a single thread. This is not valid for: the processes Client and Server. In ACO each ant is implemented by a single thread. The Cargo Dispatcher component of the server has a dedicated thread to handle each Client.

A more detailed specification of the obligation if the components is included in Table 1.

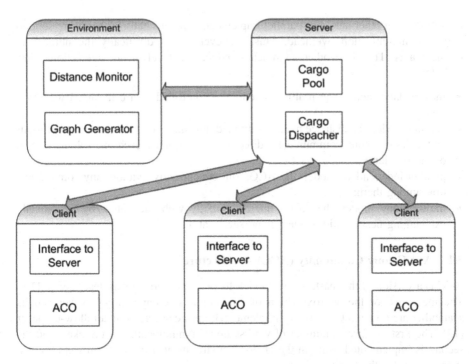

Fig. 1. An overview of ACC components

Table 1. Functionality of ACC components

Component	Tasks
Distance monitor	Every n milliseconds examines the current distance matrix
	Evaluates the route lengths for the solutions kept in the Cargo pool
	Selects the best route and calculates the Cumulative Path Length (CPL)
Graph generator	Changes the distance matrix, see Sect. 4
Cargo pool	Stores cargos with partial solutions delivered by client ACO's
	The Pool has a fixed size, new coming cargos are processed according the elitist principle that is a new cargo is inserted into the Pool if it contains a path shorter than the longest of paths included already in the Pool
Cargo dispatcher	Selects a cargo from the Cargo Pool
	Maintains a separate thread for each client ACO
	Sends cargos to the designated client and receives calculated solution
Master interface	Establishes the connection to the Master
	Receives a cargo with data to process
	Creates a new ACO to process data
	Sends results to the Master
ACO	Processes data with parameters specified by the Master Interface

4 Dynamic Environment

The Dynamic TSP was introduced for the first time by Psaraftis in [13] who up to now remains a key figure in research in the area and has recently published a review paper [3]. In most papers on DTSP, a simple model of distance changes is used. The model assumes that a certain number of nodes are deleted and new nodes are added to replace them to the distance matrix. In our opinion such a mode of operation has three shortcomings:

- The binary nature of modifications (a connection is operational or not) corresponds well to some applications like airline connections but it fails to model cases where the distance could change in a more subtle way. Consider the case when the matrix represents not distance but rather time necessary to travel from one node to another.
- The algorithm has to be omniscient in order to instantly know the edges to eliminate. This is not necessary a case of many real-life applications.
- Less drastic changes in distances make such an algorithm not useful.

For these reasons, we have adopted a more realistic approach. A distance matrix is replaced by a graph generator. In what follows we uses a slightly modified version of a graph generator described early in [14]. The number of nodes remains the same and the generator modifies an existing distance matrix. Its operation is controlled by 3 parameters:

- sleep-time (ST): the time expressed in milliseconds between distance changes.
- individual severity factor (ISF): a coefficient that specifies the scope of changes of a single distance
- total severity (TS): the sum of all distance changes.

During a modification the distances could increase or decrease but the average distance length remains the same. Not changing the average is very important as it enables us to evaluate efficiency of work by comparing the results obtained for the static version of the matrix. The code to calculate the new value of a single distance is presented on Fig. 2.

The ISF limits the scope of distance modification, the random number generator is used to mitigate the change. During the experiments, two graph generators were used. Their properties are in the Table 2. The number of nodes was equal to 50.

Table 2. Properties of graph generators used in experiment

Graph generator code	ST in seconds	ISF	TS	Description
RB (Rare Big)	10	0.7	100	Graph modification are Rare (every 10 s) but Big both in individual range and total scope
FS Frequent Small	1	0.2	10	Graph modifications Frequent (every 1 s) but rather Small

```
public synchronized double newDistance(double oldDist) {
    Random randGen= new Random();
    double res=oldDist;
    double x= randGen.nextDouble();
    if (x>totalSeverity)
        return(res);
    x= randGen.nextDouble()*individualSeverityFactor;
    if (randGen.nextDouble()>0.5) { // increasing distance
        res=oldDist+x;
        if (res>1.0) res=res-1.0; // distance normalization
    } else { // decreasing distance
        res=oldDist-x;
        if (res<0.0) res=1.0+res; // distance normalization }
    }
    return (res);
}
```

Fig. 2. Code used by the graph generator for modifying distances.

5 Analysis of Experiment Results

The aim of the experiment was to studied in what impact has the ACC when applied to the original ACO, PACO and ACO with immigrants. In all cases the client colony was the modified version of ACO as described by Chirico [15]. Their parameters were not optimized as the task was to assess the scope of impact the Community. The population size of PACO was 20 and the replacement rule was age based. The immigrants based version also used a population of 20 ants. The number of immigrants was fixed and set to 5.

The graph generators introduced in Sect. 4 were used. The number of nodes and the number of ants in each cargo was equal to 50. The change rate of the FS generator was rather fast. One second sleep-time is roughly sufficient for a regular client to process a cargo that has a graph with 50 nodes, 50 ants and 50 iterations. Standard ACO is not capable of delivering any decent results with just 50 iteration as their typical number exceeds 1000. On the other hand the distances were not much changed the ISF was equal to 0.2. The RG generator with sleep-time of 10 s gives more time for optimization but the inflected distance modification are more severe with ISF set to 0.7. The performance is measured by the Cumulative Path Length (CPL) calculated by the Distance Monitor, see Sect. 3.

During the experiments the CPL is measured over a 5 min time period. That mean that even for the slow changing RG generator had 30 times modified the distances. Therefore the data are statistically significant. The calculation of CPL took place every 10 ms.

The results are shown in the Tables 3, 4 and 5. Note that the rows with client number and pool size booth equal to 1 represent the basic version that does not use ACC.

There are several conclusion that could be drawn from the tables.

Table 3. Cumulative Path Length for standard ACS

RB (Rare Big) generator				FS Frequent Small generator			
#iter	#cls	#pool	CPL	#iter	#cls	#pool	CPL
75	15	1	1640.3	75	15	25	1485.6
75	15	25	1663.8	75	15	1	1486.7
75	1	25	2019.4	75	1	25	1573.2
200	15	25	2361.2	75	1	1	1934.5
200	15	1	2449.8	200	15	25	2100.7
75	1	1	2608.7	200	1	25	2193.1
200	1	25	3366.2	200	1	1	2540.3
200	1	1	4475.0	200	15	1	28075

Table 4. Cumulative Path Length for PACO

RB Rare and Big				FS Frequent Small			
#iter	#cls	#pool	CPL	#iter	#cls	#pool	CPL
75	15	25	1628.7	75	15	25	1328.5
75	15	1	1823.4	75	15	1	1509.2
75	1	25	2076.9	75	1	25	1861.6
200	15	25	2351.0	200	15	25	2020.8
200	15	1	2463.1	200	15	1	2035.1
75	1	1	3244.9	200	1	25	2654.5
200	1	25	3733.3	75	1	1	2711.7
200	1	1	4985.3	200	1	1	4213.2

Table 5. Cumulative Path Length for immigration based ACO

RB Rare and Big				FS Frequent Small			
#iter	#cls	#pool	CPL	#iter	#cls	#pool	CPL
75	15	25	1606.0	75	15	25	1543.4
75	15	1	1732.0	75	15	1	1633.6
75	1	25	2174.8	75	1	1	1809.3
200	15	25	2433.9	75	1	25	1891.5
200	15	1	2541.0	200	15	25	1930.4
75	1	1	3413.7	200	15	1	1939.0
200	1	25	3513.1	200	1	1	2487.3
200	1	1	4406.5	200	1	25	2704.6

- In all cases the introduction of ACC has resulted in a very significant reduction of the CPL, in most case it is at least 2 times shorter than for non-parallel version.
- The data generated by RB graph is more hard to optimize. The difference between the best and the worst result is far larger than in the case of the FS graph generator.
- It is worth noting that the most geared up version of ACC with 15 clients and a pool of 25 solutions handles RG data almost as efficiently as the data from a much more stable FS graph.
- The various tested versions of ACO achieve not much different results.
- PACO and immigrant versions of ACO are more susceptible to the community introduction than the basic version.
- Decreasing the number of iterations is necessary but not sufficient.
- Increasing the number of clients is useful. It the hardware resource do not allow for large number of clients, increasing the pool size may help.

Increasing the number of clients makes higher demands on used computer power. This is not necessary a significant disadvantage. The ACC is supposed to utilize the previously not utilized computer idle time. In the paper [11] contains a method of estimating workload of a hardware configuration and balancing it so it does not harm the responsiveness of a workstation.

6 Conclusions

The results of all reported here experiments strongly indicate that the ACC approach offers a substantially decreases the length of routs found by all tested versions of ACO. The dynamics of route lengths described in the paper is in our opinion capable of simulating many real life phenomena. These are the main contributions of the paper.

Parallelism relies on computer power but it is not a significant disadvantage. Ubiquitous of computers power and its ever decreasing cost prompt us to look at proposed methods from a slightly different angle. We have interests not only in the computational complexity of a solution but also in a degree to which its operation could be run in parallel. The paper follows that way of thinking. The distribution of tasks to computers spread over local area network or even the Internet makes it highly scalable. The communication overhead necessary for the ACC to operate is not great and the experiments show, that the lack of direct cooperation of individual ants from different colonies does not impact much the achieved results.

The optimization of Ant Colonies parameters is an hard task. It's inherent indeterminism and the complexity of interactions between ant make analytical optimization not possible. Experiments show that the higher the number of ants is the better is the quality of delivered solutions. ACC makes it possible to increase their number without increasing the processing time.

ACC was previously successfully applied to static TSP but is real power is visible when we use it for the DTPS. The paper documents an initial stage of work on the subject. The next stage will involve a more detailed analysis of the impact of the number of iterations. This should enable us to propose an adopting schema to ACO to it adjusts the iteration number to the current changeability rate of the distance matrix as well as more advanced routines for pool handling and an adaptation of these ideas to the Hadoop version.

References

1. Antosiewicz, M., Koloch, G., Kamiński, B.: Choice of best possible metaheuristic algorithm for the travelling salesman problem with limited computational time: quality, uncertainty and speed. J. Theor. Appl. Comput. Sci. **7**(1), 46–55 (2013)
2. Dorigo, M., Stuetzle, T.: Ant colony optimization: overview and recent advances, IRIDIA - Technical Report Series, Technical report No. TR/IRIDIA/2009-013 (2009)
3. Psaraftis, H.N., Wen, M., Kontovas, C.A.: Dynamic vehicle routing problems: three decades and counting. Networks **67**(1), 3–31 (2016)
4. Dorigo, M.: Optimization, Learning and Natural Algorithms, Ph.D. thesis, Politecnico di Mila-no, Italie (1992)
5. Guntsch, M., Middendorf, M.: Applying population based ACO to dynamic optimization problems. In: Dorigo, M., Di Caro, G., Sampels, M. (eds.) ANTS 2002. LNCS, vol. 2463, pp. 111–122. Springer, Heidelberg (2002). https://doi.org/10.1007/3-540-45724-0_10
6. Skinderowicz, R.: Implementing population-based ACO. In: Hwang, D., Jung, Jason J., Nguyen, N.-T. (eds.) ICCCI 2014. LNCS (LNAI), vol. 8733, pp. 603–612. Springer, Cham (2014). https://doi.org/10.1007/978-3-319-11289-3_61
7. Mavrovouniotis, M., Yang, S.: Ant colony optimization with immigrants schemes for the dynamic travelling salesman problem with traffic factors. Appl. Soft Comput. **13**(10), 4023–4037 (2013)
8. Randall, M., Lewis, A.: A parallel implementation of ant colony optimization. J. Parallel Distrib. Comput. **62**(9), 1421–1432 (2002)
9. Pedemonte, M., Nesmachnow, S., Cancela, H.: A survey on parallel ant colony optimization. Appl. Soft Comput. **11**(8), 5181–5197 (2011)
10. Siemiński, A.: Parallel implementations of the ant colony optimization metaheuristic. In: Nguyen, N.T., Trawiński, B., Fujita, H., Hong, T.-P. (eds.) ACIIDS 2016. LNCS (LNAI), vol. 9621, pp. 626–635. Springer, Heidelberg (2016). https://doi.org/10.1007/978-3-662-49381-6_60
11. Siemiński, A.: Measuring efficiency of ant colony communities. In: Zgrzywa, A., Choroś, K., Siemiński, A. (eds.) Multimedia and Network Information Systems. AISC, vol. 506, pp. 203–213. Springer, Cham (2017). https://doi.org/10.1007/978-3-319-43982-2_18
12. Andrzej, S., Marek, K.: Comparing efficiency of ACO parallel implementations. J. Intell. Fuzzy Syst. **32**(2), 1377–1388 (2017)
13. Psarafits, H.N.: Dynamic vehicle routing: Status and Prospects. National Technical Annals of Operations Research, University of Athens, Greece (1995)
14. Siemiński, A.: Using ACS for dynamic traveling salesman problem. In: Zgrzywa, A., Choroś, K., Siemiński, A. (eds.) New Research in Multimedia and Internet Systems. AISC, vol. 314, pp. 145–155. Springer, Cham (2015). https://doi.org/10.1007/978-3-319-10383-9_14
15. Chirico, U.: A Java framework for ant colony systems. In: Ants2004: Forth International Workshop on Ant Colony Optimization and Swarm Intelligence, Brussels (2004)

Physical Layer Security Cognitive Decode-and-Forward Relay Beamforming Network with Multiple Eavesdroppers

Nguyen Nhu Tuan[1(✉)] and Tran Thi Thuy[2]

[1] Department of Cryptography, Academy of Cryptography Technique, Hanoi, Vietnam
nguyennhutuan@bcy.gov.vn
[2] FPT University, Hoa Lac High Tech Park, Hanoi, Vietnam
thuytt@fpt.edu.vn

Abstract. Physical layer security (PLS) approach for wireless communication can prevent eavesdropping without upper layer data encryption. In this paper, we consider the transmission of confidential message over wireless channel with the help of multiple cooperating relays in the Decode-and-Forward cooperative scheme. Based on an information-theoretic formulation of problem we take into account a wireless network system, in which an one-way relay wireless system including a source, a legitimate destination, multiple trusted relays and multiple eavesdroppers. For decode-and-forward beamforming protocol in the presence of one eavesdropper, the closed-form optimal solution is derived for the relay weights. For another case, when the communication system has the presence of multiple eavesdroppers, it becomes more difficult to find the optimal solution of the secrecy rate maximization problem. Hence, there are several criteria were considered leading to a suboptimal but simple solution. Specifically, nulling out the message signals completely at all eavesdroppers results in a simpler form of this problem and a suboptimal solution is explicitly obtained. In this paper, we propose a new approach based on the well known method, DC programming and DCA to solve the original problem, without considering any additional constraints.

Keywords: Decode-and-Forward · Relay beamforming ·
Physical layer security · DC programming and DCA

1 Introduction

In wireless communication systems, generally security measure are implemented in the upper layers of protocol stack by using cryptographic algorithms. However, current advances in computation technology pose threats for such systems, prompting many researchers to explore alternatives a secrecy method such as PLS.

© Springer Nature Switzerland AG 2019
N. T. Nguyen et al. (Eds.): ACIIDS 2019, LNAI 11432, pp. 254–263, 2019.
https://doi.org/10.1007/978-3-030-14802-7_22

Since 1975, the PLS approach was introduced by Wyner [1] and subsequent works [2,3] have been interested in the information-theoretic aspects of PLS. There are three common cooperative models in PLS in literature, they are Decode-and-Forward (DF), Amplify-and-Forward (AF), and Cooperative Jamming (CJ). In this paper, we concentrate to deal with DF scheme.

The remaining paper is structured as follows. The state-of-the-art in PLS on DF relaying networks and short introduce DC programming and DCA are presented in the rest of Sect. 1. In Sect. 2, we showed the secrecy rate maximization problem derived from the system model. The existing work is described in Sect. 3. In Sect. 4, we showed how to apply DC programming and DCA for solving the considered problem. Finally, Sects. 5 and 6 reported the numerical results and the conclusions, respectively.

1.1   PLS with DF Beamforming Scheme

DF relaying cooperative network for improving transmission quality in the absence of any eavesdropper were presented in many papers in several years ago, but this relaying cooperative scheme for improving communication security in the present of one or more eavesdroppers were considered in recent years [10–12].

If the communication system has one eavesdropper, the achievable secrecy rate will be is

$$R_s = \max\{R_d - R_e\} \tag{1}$$

In which the maximum is taken over possible input covariance matrices, R_d and R_e are the achievable rates of the channel between the source and destination, and the channel between the source and eavesdropper, respectively.

When the total relay power constraint was considered, the maximal secrecy rate problem is completely solved by the generalized eigenvalue [10,12]. In the case of individual power constraint, the above problem becomes more difficult to solve, it is nonconvex problem and thus intractable. A general method to deal with this case is the semidefinite relaxation (SDR) technique, and in [12], we showed that it can be solved efficiently by using DC programming and DCA.

If the communication system have multiple eavesdroppers, the achievable secrecy rate will be is

$$R_s = \max \min_j \{R_d - R_e^j\}, \tag{2}$$

In which the maximum is again taken over possible input covariance matrices, R_e^j is the achievable rate of the channel between the source and the jth eavesdropper. The problem (2) always is nonconvex and difficult to solve. The existing method for solving this problem [10] is showed in the Section below.

1.2   Brief Introduction to DC Programming and DCA

DC (difference of convex functions) programming and DCA (DC algorithms) were launched by Pham Dinh in 1985 and have been well exploited by Le Thi and Pham Dinh since 1994 [5–7]. These tools are increasingly used by many researchers all over the world due to their advance and efficiency. Many works show the success of DC programming and DCA in solving many large-scale non-smooth non-convex programs in various fields such as communication system [8] and other areas [9]. Furthermore, the performance of DCA is shown to be superior to that of the other standard methods. That is reason why we choose the method based on DCA in order to find a better solution compared with the proposed one for the problem described above.

A standard DC program involves minimizing a function $f(x) = g(x) - h(x)$ on a convex set C in which $g(x)$ and $h(x)$ being lower semicontinuous proper convex functions on C. In other words, a DC program can be stated as follows.

$$m = \inf\{f(x) := g(x) - h(x) : x \in C\}$$

Where the function $f(x)$ is called a DC function, $g(x) - h(x)$ is a DC decomposition of $f(x)$, $g(x)$ and $h(x)$ are DC components of $f(x)$.

DCA is for dealing with the DC program above. The idea of DCA is simple, thus it is not difficult to implement DCA. From an initial point, the second DC components $h(x)$ is approximated by its linear minorant at that point, which results in a convex subproblem. The optimal solution of this convex subproblem is used to approximate the second DC component in the next iteration. This process is repeated until a stopping condition is satisfied. The steps of DCA can be summarized below

---

1. **Initialization.** Choose an initial point x^0.

2. **Repeat.**
   Step 1. Calculating $y^k \in \partial h(x^k)$
   Step 2. Compute x^{k+1} as the solution of the subproblem
   $\inf\{g(x) - [h(x^k) - \langle x - x^k, y^k \rangle] : x \in C\}$
   Step 3. $k \leftarrow k + 1$,
3. **Until** stopping condition.

To the best of our knowledges, DCA can efficiently solve even large-scale DC programs and not many existing algorithms can do that like DCA. The simplicity but efficiency, robustness and the scalability of DCA make it more and more popular and become one of top choices for researchers when handling optimization problems (see the list of references in [9] and those therein).

Notation: In this paper, the notations $(\Delta)^T$, $(\Delta)^\dagger$ and $(.)^*$ are transpose, conjugate transpose and conjugate, respectively; $\mathbf{I}_k$ denotes the identity matrix of size $k \times k$; $E\{.\}$ represents for expectation; $\langle .,. \rangle$ is the inner product and $\|.\|$ is the Euclidean norm. Column vectors are denoted by bold lowercase letters while matrices are denoted by bold uppercase letters. $\mathbf{D}(\mathbf{a})$ is the diagonal matrix with $\mathbf{a}$ on its main diagonal. Re(.) and Im(.) are the real and imaginary part of its argument, respectively.

2 System Model

The system consists of a single source (S), a single destination (D), M relays nodes $(i \in M = \{1, 2, ..., M\})$ and K eavesdroppers $k \in K = \{1, 2, ..., K\}$, as in Fig. 1. In this system, we assume that there is no any direct transmission from the source to the destination or to the eavesdroppers. The channel gain from a node p to a node q is denoted by a complex constant h_{pq}.

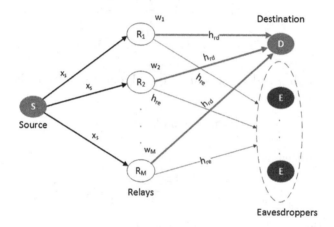

Fig. 1. Channel model

There are two phases in DF protocol. In the first phase, the source (S) broadcasts signal x_s to its trusted relays using the first transmission slot with power $E(|x_s|^2) = 1$. The received signals at the M relays, stacked in vector $\mathbf{y}(M \times 1)$, equal

$$\mathbf{y}_r = \sqrt{P_s}\mathbf{h}_{SR}x_s + \mathbf{n}_r \tag{3}$$

Where $\mathbf{n}_r$ is the background noise vector at relays that has a Gaussian distribution with zero mean and variance of σ_r^2, and P_s is the average transmit power of source. Recalling that P is the overall power for transmitting the symbol, it holds that $0 < P_s < P$.

In the second phase, we assume that all M relays successfully decode the source message, then re-encode the message and cooperatively transmit the re-encode symbols to the destination, using the second transmission slot. Specifically, each relay normalizes the decode message and then multiplies it by the weight factor, stacked in vector $\mathbf{w}(M \times 1)$, and let x_r be the re-encode symbol at relays. When transmitting the symbol x_r, the received signal at the destination and the eavesdropper j^{th} are given by

$$y_d = \Sigma_{m=1}^{M} \mathbf{h}_{rd} w_m x_r + n_d, \tag{4}$$

$$y_{e,j} = \Sigma_{m=1}^{M} \mathbf{h}_{re,j} w_m x_r + n_{e,j} \tag{5}$$

Where $\mathbf{h}_{rd} = [h_{1d}, \cdots, h_{Md}]$ and $\mathbf{h}_{re,j} = [h_{1e,j}, \cdots, h_{Me,j}]$, $j = 1, .., K$ are the channel coefficients between the relays and the destination, and the relays and the jth eavesdropper, respectively; n_d and $n_{e,j}$ are the Gaussian background noise components at the destination and the jth eavesdropper.

The transmit power budget for second phase, on relays, is $P_r = P - P_s$.

Based on the (4) and (5), the rate at the destination and the jth eavesdropper are

$$R_d = \frac{1}{2} \log(1 + \frac{\mathbf{w}^\dagger \mathbf{R}_{rd} \mathbf{w}}{\sigma^2}), \tag{6}$$

$$R_{e,j} = \frac{1}{2} \log(1 + \frac{\mathbf{w}^\dagger \mathbf{R}_{re,j} \mathbf{w}}{\sigma^2}), \tag{7}$$

where $\mathbf{R}_{rd} = \mathbf{h}_{rd} \mathbf{h}_{rd}^\dagger$ and $\mathbf{R}_{re,j} = \mathbf{h}_{re,j} \mathbf{h}_{re,j}^\dagger$. From (6) and (7), the achievable secrecy rate is given by

$$R_s = \max_{w} \min_{j} \{R_d - R_{e,j}\},$$

which can be written as

$$R_s(w) = \max_{\mathbf{w}} \min_{j=1..K} \left(\frac{1}{2} \log(1 + \frac{\mathbf{w}^\dagger \mathbf{R}_{rd} \mathbf{w}}{\sigma^2}) - \frac{1}{2} \log(1 + \frac{\mathbf{w}^\dagger \mathbf{R}_{re,j} \mathbf{w}}{\sigma^2}) \right). \tag{8}$$

The optimization problem (8) could be formulated as

$$\max_{\mathbf{w}} \min_{j=1..K} \left(\frac{1}{2} \log(1 + \frac{\mathbf{w}^\dagger \mathbf{R}_{rd} \mathbf{w}}{\sigma^2}) - \frac{1}{2} \log(1 + \frac{\mathbf{w}^\dagger \mathbf{R}_{re,j} \mathbf{w}}{\sigma^2}) \right) \tag{9}$$
$$\text{s.t. } \mathbf{w}^\dagger \mathbf{w} \le P_r.$$

The problem above is called the original secrecy rate maximization problem. Determining the weights factor ($\mathbf{w}$) for the original problem of secrecy rate maximization (9) might be difficult. In the following, we present the existing work to solve this problem and then we propose a method which uses DC programming and DCA for that problem.

3   Existing Work

In [10], the original problem (9) was transformed to an easier form and a simple closed-form solution was found.

By introducing addition constraints $\mathbf{w}^\dagger \mathbf{R}_{re,j} \mathbf{w} = 0$, $\forall j \in 1..K$, the signal is completely nulled out at all the eavesdroppers in Stage 2. Note that in this case the condition $M > K$ is needed. In the case of $M \leq K$ we cannot null out signals at all eavesdroppers.

By nulling the signals at the eavesdroppers, the capacity to the eavesdroppers in Stage 2 becomes zero, the problem (9) is reduced to

$$
\max_{\mathbf{w}} \quad \frac{1}{2}\log(1 + \frac{\mathbf{w}^\dagger \mathbf{R}_{rd}\mathbf{w}}{\sigma^2}) \tag{10}
$$
$$
\text{s.t.} \quad \mathbf{w}^\dagger\mathbf{w} \leq P_r
$$
$$
\mathbf{w}^\dagger \mathbf{R}_{re,j}\mathbf{w} = 0, \ \forall j \in 1..K.
$$

Because of the feasible set of (10) is narrower than that of (9), thus (10) only provides a suboptimal secrecy rate.

The optimization problem of maximizing the achievable secrecy rate in (10) can be equivalently formulated as

$$
\max_{\mathbf{w}} \quad \mathbf{w}^\dagger \mathbf{R}_{rd}\mathbf{w} \tag{11}
$$
$$
\text{s.t} \quad \mathbf{w}^\dagger \mathbf{R}_{re,j}\mathbf{w} = 0, \ \forall j \in 1..K
$$
$$
\mathbf{w}^\dagger\mathbf{w} \leq P_r.
$$

In [10], it is shown that the use of the equality power constraint ($\mathbf{w}^\dagger\mathbf{w} = P_r$) is equivalent to the inequality constraint ($\mathbf{w}^\dagger\mathbf{w} \leq P_r$). The optimization problem of maximizing the achievable secrecy rate in (11) is referred to as the null-steering beamforming in the array signal processing literature, and its optimal solution is given by [10]

$$
\mathbf{w} = \frac{\sqrt{P_r}}{\|(\mathbf{I}_N - \mathbf{P}_{re})\mathbf{h}_{rd}\|}(\mathbf{I}_N - \mathbf{P}_{re})\mathbf{h}_{rd}. \tag{12}
$$

Where $\mathbf{P}_{re} = \mathbf{R}_{re}(\mathbf{R}_{re}^\dagger\mathbf{R}_{re})^{-1}\mathbf{R}_{re}^\dagger$ is the orthogonal projection matrix onto the subspace spanned by the columns of $\mathbf{R}_{re}$.

4   DC Programming and DCA for Problem (9)

In this section, we offer a DC decomposition then design a DCA for solving the original secrecy rate maximization problem (9). The objective function of this problem can be rewritten as follows

$$
\begin{aligned}
R_s(\mathbf{w}) &= \max_{\mathbf{w}} \frac{1}{2}\log(1 + \frac{\mathbf{w}^\dagger \mathbf{R}_{rd}\mathbf{w}}{\sigma^2}) - \frac{1}{2}\log \max_{j=1..K}(1 + \frac{\mathbf{w}^\dagger \mathbf{R}_{re,j}\mathbf{w}}{\sigma^2}) \\
&= \frac{1}{2}\max_{\mathbf{w}}\log \frac{\sigma^2 + \mathbf{w}^\dagger \mathbf{R}_{rd}\mathbf{w}}{\max_{j=1..K}(\sigma^2 + \mathbf{w}^\dagger \mathbf{R}_{re,j}\mathbf{w})}. \tag{13}
\end{aligned}
$$

The optimization problem of maximizing the achievable secrecy rate in (13) with the total relays power constraint can be equivalently formulated as the following problem.

$$\min_{\mathbf{w},\tau} \quad -\frac{\sigma^2 + \mathbf{w}^\dagger \mathbf{R}_{rd}\mathbf{w}}{\tau} \tag{14}$$

$$\text{s.t} \quad \mathbf{w}^\dagger \mathbf{w} \le P_r$$

$$\sigma^2 + \mathbf{w}^\dagger \mathbf{R}_{re,j}\mathbf{w} \le \tau, \forall j \in 1..K$$

$$\tau > 0.$$

The problem (14) can be transformed to a real form as below.

$$\min_{\tau,\mathbf{x}} \quad -\frac{\sigma^2 + \mathbf{x}^T \mathbf{Z}\mathbf{x}}{\tau} \tag{15}$$

$$\text{s.t} \quad \mathbf{x}^T \mathbf{B}_j \mathbf{x} \le \tau - \sigma^2, \forall j \in 1..K$$

$$\mathbf{x}^T \mathbf{x} \le P_r$$

$$\tau > 0,$$

where

$$\mathbf{x} = \begin{bmatrix} \mathrm{Re}(w) \\ \mathrm{Im}(w) \end{bmatrix}, \quad \mathbf{Z} = \begin{bmatrix} \mathrm{Re}(R_{rd}) & -\mathrm{Im}(R_{rd}) \\ \mathrm{Im}(R_{rd}) & \mathrm{Re}(R_{rd}) \end{bmatrix} \quad \text{and} \quad \mathbf{B}_j = \begin{bmatrix} \mathrm{Re}(R_{re,j}) & -\mathrm{Im}(R_{re,j}) \\ \mathrm{Im}(R_{re,j}) & \mathrm{Re}(R_{re,j}) \end{bmatrix}$$

The above problem is restated as a standard DC program below:

$$\min_{\tau,\mathbf{x}} \quad \mathbb{G}(\tau,\mathbf{x}) - \mathbb{H}(\tau,\mathbf{x}) \tag{16}$$

$$\text{s.t} \quad \mathbf{x}^T \mathbf{B}_j \mathbf{x} \le \tau - \sigma^2, j = 1,..,K$$

$$\mathbf{x}^T \mathbf{x} \le P_r$$

$$\tau > 0,$$

where $\mathbb{G}(\tau,\mathbf{x}) = 0$ and $\mathbb{H}(\tau,\mathbf{x}) = \frac{\sigma^2 + \mathbf{x}^T \mathbf{Z}\mathbf{x}}{\tau}$.

The function $\mathbb{H}(\tau,\mathbf{x})$ is smooth when $\tau > 0$ and its gradient at the point $(\tau^l, \mathbf{x}^l)$ is

$$\mathbf{y}^l = \nabla \mathbb{H}(\tau^l, \mathbf{x}^l) = \begin{bmatrix} -\frac{\sigma^2}{(\tau_l)^2} - \frac{(\mathbf{x}^l)^T \mathbf{Z}\mathbf{x}^l}{(\tau^l)^2} \\ 2\frac{\mathbf{Z}\mathbf{x}^l}{\tau^l} \end{bmatrix}.$$

The DCA scheme corresponding to the DC decomposition $\mathbb{G}(\tau,\mathbf{x}) - \mathbb{H}(\tau,\mathbf{x})$ in (16) is designed as follows.

DCA Scheme:
Initialization: Choose a random initial point $\mathbf{x}^0$, $\tau^0 > 0$, and set $l = 0$, $\mathbf{v}^0 = (\tau^0, \mathbf{x}^0)$
Repeat: $l = l + 1$,   Calculating $\mathbf{v}^l = (\tau^l, \mathbf{x}^l)$ by solving this convex subproblem:

$$\min_{\mathbf{v}=(\tau,\mathbf{x})} \quad 0 - \langle \mathbf{y}^{l-1}, \mathbf{v} \rangle \tag{17}$$

$$\text{s.t} \quad \mathbf{x}^T \mathbf{B}_j \mathbf{x} \leq \tau - \sigma^2, \forall j \in K$$

$$\mathbf{x}^T \mathbf{x} \leq P_r$$

$$\tau > 0.$$

Until: ($\frac{\|\mathbf{v}^l - \mathbf{v}^{l-1}\|}{1+\|\mathbf{v}^l\|} \leq \epsilon$ or $\frac{|f(\mathbf{v}^l)-f(\mathbf{v}^{l-1})|}{1+|f(\mathbf{v}^l)|} \leq \epsilon$), where $f(\mathbf{v}^l) = \frac{\sigma^2 + (\mathbf{x}^l)^T \mathbf{Z} \mathbf{x}^l}{\tau^l}$

5 Numerical Results

In this section, the performance of DCA is compared with that of the former, Null steering algorithm. The channel coefficients are drawn from a complex Gaussian distribution with zero mean and variance $\sigma^2 = 1$. The number of relay is set to $M = 10$ while the number of eavesdroppers is set to $K = 5, 7, 9$ respectively. The reported results were taken average over 100 independent trials.

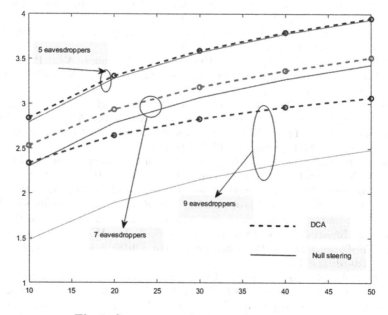

Fig. 2. Secrecy rate vs. power relay constraint

It can be observed from Fig. 2 that the secrecy rate achieved by both algorithms is increasing with the relay power while decreasing with the number of eavesdroppers. DCA always provides better secrecy rates than the Null steering does in all cases. Especially, when the number of eavesdroppers is larger, the secrecy rates produced by DCA are far superior to those of the existing one.

6 Conclusion

In this work, we have solved the secrecy rate maximization problem derived from a Decode-and-Forward relay beamforming networks having multiple eavesdroppers. We restate this problem in a form of DC program and design the DCA scheme for solving it. The numerical results show that DCA is more efficient than the existing one in terms of secrecy rate. In particular, the larger number of eavesdroppers is, the bigger gap of optimal values obtained by DCA and the existing one is. In other words, our DCA-based approach designed the beamforming coefficients that achieves better secrecy for the considered system while satisfying the relay power constraint.

Acknowledgements. We would like to thank Prof. Hoai An Le Thi, The director of LITA, University of Lorraine, France for helping us do research in this field. We also thank the referees for leading us to improve the manuscript.

References

1. Wyner, A.D.: The Wire-tap channel. Bell Sys. Tech. J. **54**, 1355–1387 (1975)
2. Csiszar, I., Korner, J.: The broadcast channels with confidential messages. IEEE Trans. Inf. Thoery **IT–24**(3), 339–348 (1978)
3. Leung-Yan-Cheong, S.K., Hellman, M.E.: The Gaussian wire-tap channel. IEEE Trans. Inf. Theory **IT–24**, 451–456 (1978)
4. Mukherjee, A., Fakoorian, S.A.A., Huang, J., Swindlehurst, A.L.: Principles of physical layer security in multiuser wireless networks: a survey. IEEE Commun. Surv. Tutorials **16**, 1550–1573 (2014)
5. Pham Dinh, T., Le Thi, H.A.: Convex analysis approach to DC programming: theory, algorithms and applications. Acta Math. Vietnamica **22**, 289–357 (1997)
6. Pham Dinh, T., Le Thi, H.A.: Optimization algorithms for solving the trust region subproblem. SIAM J. Optim. **8**, 476–505 (1998)
7. Pham Dinh, T., Le Thi, H.A.: Recent advances in DC programming and DCA. In: Nguyen, N.-T., Le-Thi, H.A. (eds.) Transactions on CCI XIII. LNCS, vol. 8342, pp. 1–37. Springer, Heidelberg (2014). https://doi.org/10.1007/978-3-642-54455-2_1
8. Tuan, N.N., Son, D.V.: DC programming and DCA for Enhancing physical layer security in amplify-and-forward relay beamforming networks based on the SNR approach. In: Le, N.-T., Van Do, T., Nguyen, N.T., Thi, H.A.L. (eds.) ICCSAMA 2017. AISC, vol. 629, pp. 23–33. Springer, Cham (2018). https://doi.org/10.1007/978-3-319-61911-8_3
9. Le Thi, H.A.: DC Programming and DCA. http://www.lita.univ-lorraine.fr/lethi/
10. Dong, L., Han, Z., Petropulu, A.P., Poor, H.V.: Improving wireless physical layer security via cooperating relays. IEEE Trans. Signal Process. **58**(3), 1875–1888 (2010)

11. Vishwakarma, S., Chockalingam, A.: Decode-and-Forward relay beamforming for security with finite-alphabet input. IEEE Commun. Lett. **17**(5), 912–915 (2013)
12. Thuy, T.T., Tuan, N.N., An, L.T.H., Gély, A.: DC programming and DCA for enhancing physical layer security via relay beamforming strategies. In: Nguyen, N.T., Trawiński, B., Fujita, H., Hong, T.-P. (eds.) ACIIDS 2016. LNCS (LNAI), vol. 9622, pp. 640–650. Springer, Heidelberg (2016). https://doi.org/10.1007/978-3-662-49390-8_62

Co-exploring a Search Space in a Group Recommender System

Dai Yodogawa[1]([⊠]) and Kazuhiro Kuwabara[2]

[1] Graduate School of Information Science and Engineering, Ritsumeikan University,
Kyoto, Japan
`is0288rr@ed.ritsumei.ac.jp`
[2] College of Information Science and Engineering, Ritsumeikan University Kusatsu,
Shiga 525-8577, Japan

Abstract. This paper presents a group recommendation system that focuses on helping users to search for an item that is agreed upon by all users in a group. We consider the case with a group of two users as a starting point and regard the search process for the recommended item as a negotiation process between the two users. More specifically, after the user's preferences are inputted in the proposed system, the system maps possible items on a two dimensional-plane according to the item's utility values for the two users. The users are expected to negotiate their preferences to reach an agreement. To examine the characteristics of the proposed system, simulation experiments are conducted with four different user models, which are created based on conflict resolution behaviors as described in human psychology literature. These four user types are represented by different parameter values to control their behaviors. The paper presents the simulation experiments and their results.

Keywords: Group recommender system · Negotiation model ·
User model · Simulation

1 Introduction

With the development of Web technology, the amount of information that users can access has increased in recent years; moreover, an increasing amount of research has focused on developing a recommendation system [12]. Many of these studies have focused on recommending things to individuals, while many recommendation scenarios involving multiple users can be considered in such domains as TV programs, music, collaborative learning support, and electronic libraries, which are promising target areas for group recommendation systems [8].

An implicit goal of a group recommendation system is to recommend an item that satisfies each group member [5]. Thus, simply aggregating results of the recommendations targeted for each group member often does not suffice. For example, for the task of finding a travel destination for a group trip, the recommendation fails if one member of the group is not satisfied with the recommended destination. It is therefore necessary to identify the item that group

© Springer Nature Switzerland AG 2019
N. T. Nguyen et al. (Eds.): ACIIDS 2019, LNAI 11432, pp. 264–274, 2019.
https://doi.org/10.1007/978-3-030-14802-7_23

members can agree on. To cope with this, we view the process of finding an item as a negotiation process [1], and we thus create a recommendation system that helps users to explore possible recommendations and adjust their preferences to find an item that each member is satisfied with.

In this paper, we use a case with two users as a starting point. After the users input their requirements, the system shows a recommended item. If both users are satisfied with the recommended item, the recommendation process ends. Otherwise, the system lets the users change their preferences so that another item can be recommended. Since a user's preferences may depend on the other user's preferences, it is difficult for a user to know his/her exact preferences from the start. Thus, it is important to allow a user to change his/her preferences to reach an agreement.

With two users, we can plot items on the two-dimensional plane where the x-axis corresponds to the item's utility value for one user and the y-axis corresponds to the item's utility value for the other user. In that way, the system can provide users with a global view of the current search space and offer a hint to update their preferences to reach a possible agreement. In a sense, the role of the proposed system is to act as a facilitator to prompt users to change their preferences.

The remainder of the paper is organized as follows: Sect. 2 describes related work, and Sect. 3 presents our proposed mechanism for a group recommendation system. Section 4 presents the simulation experiments conducted to examine the characteristics of the proposed mechanism, and Sect. 5 presents and discusses the experimental results. Section 6 concludes the paper and describes some future work.

2   Related Work

Several studies have proposed different approaches for a group recommendation system. For example, a dialog function is equipped in the system and through dialogues among multiple users, users' preferences are extracted [11]. Changing the degree of importance of the recommended items based on users' feedback or critiques and reflecting the users' preferences to the recommendation are shown to be effective approaches [9,10]. For a group recommendation method, Delic *et al.* generated a user model based on the human-specific dynamics in a group [3]. To improve the effectiveness of the item recommendation, a method to maximize the average satisfaction of group users was also proposed [2].

To solve the problem of being able to satisfy all group users, a system was proposed that urges group users to negotiate among themselves to determine their preferences [6,16]. The system assigns an intelligent agent to each group member, and the agent negotiates on behalf of the group member.

It is the user's role to judge whether the recommended item is useful or not. The recommendation system is considered to act as a neutral facilitator and support the user's decision-making process; thus, a good quality of recommendation and satisfaction of item selection can be expected [1,14]. While Rossi *et al.* considered the conflicting styles among group users in negotiation situations and used them to support users' decision-making [15], we apply conflict resolution styles among group users to construct user models for the simulation experiments in this paper.

3   Proposed Group Recommendation Mechanism

3.1   Data Model

We assume that there are m items from which a recommendation is made. Each item is characterized by n attributes. That is, item t_i ($1 \leq i \leq m$) has n attributes to describe its features. For each attribute j ($1 \leq j \leq n$), item t_i has an evaluation function $eval_j^{(t_i)}$, which accepts user's requirement value for the corresponding attribute and returns a score that represents how much the user's requirement is satisfied by the item regarding the attribute. We set the range of this evaluation function to be between 0 and 1.

The users' requirements are represented as a list of values, each of which corresponds to attribute j. The requirement of user u_k for attribute j is represented as $r_j^{(u_k)}$. Note that the range of a requirement differs from one attribute to another. For example, an attribute that corresponds to the price of an item has a different value range from an attribute that corresponds to the size of an item.

To select an item to recommend, we calculate a utility of item t_i for user u_k, $utility^{(u_k)}(t_i)$ as follows:

$$utility^{(u_k)}(t_i) = \frac{\sum_{j=1}^{n}(w_j^{(u_k)} \cdot eval_j^{(t_i)}(r_j^{(u_k)}))}{\sum_{j=1}^{n} w_j^{(u_k)}}, \tag{1}$$

where $w_j^{(u_k)}$ is a weight to represent how much the corresponding attribute j has an effect on user u_k's utility value. This weight is assumed to be in the range of $1 \ldots 5$, and its default value is set to 3. We call a combination of a user's requirement $r_j^{(u_k)}$ for an attribute and attribute weights $w_j^{(u_k)}$, *user preferences*. In the process of searching for an agreement, the user updates their *user preferences*.

3.2   Control Flow

Figure 1 shows the overall control flow of the proposed system. First, users input their initial preferences, which consist of requirement values for all attributes and weights of attributes. Then, the system calculates the utility value of each item for the two users. Here, we assume that the two users negotiate the items to reach an agreement. If the two users adopt a *monotonic concession protocol* [4] for negotiation and they follow *Zeuthen strategy* [17], it is known that they can reach an agreement that maximizes the product of both users' utility values [13]. Based on this observation, we calculate the product of both users' utility values

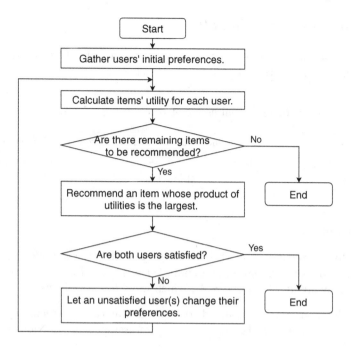

Fig. 1. Overall control flow of the proposed system

of all the possible items and select the item that has the maximum value. If there are multiple items with the highest value, one item is selected at random and presented to users as a recommended item.

If both users are satisfied with the recommended item, the recommendation process ends. If either user is unsatisfied with the recommended item, the system enters the exploring phase.

3.3 Exploring Phase

In the exploring phase, the user who is not satisfied with the recommended item is required to change their preferences. The system plots items on the two-dimensional plane, where the x-axis corresponds to the item's utility value for one user and the y-axis corresponds to that for the other user. Figure 2 in Sect. 4 shows an example of these plots.

The users are expected to change their preferences (weight or requirement) so that another item can be found that both parities agree on. Changing the weight will change the utility value of the items, leading to a change in the recommended item. We also allow the user to change their requirements as the search for an agreement continues. To avoid falling into an infinite loop, a previously recommended item is not recommended again.

4   Simulation Experiments

To examine the characteristics of our proposed recommendation system, we conducted simulation experiments as explained below.

Table 1. Initial user preferences for u_A and u_B

Attribute	price	distance	access	landscape	crowdedness	barrier free
Requirement $(r_j^{(u_k)})$	1000	10	3	4	3	4
Weight $(w_j^{(u_k)})$	3	3	3	3	3	3

4.1   Dataset

We manually created a dataset of sightseeing spots in Okinawa prefecture, Japan. The dataset contains 12 sightseeing spots $(m = 12)$, and each sightseeing spot is described by six attributes $(n = 6)$, which we define as follows:

- price: amount of money expected to be needed at the sightseeing spot.
- distance: distance from the nearby airport (assumed to be Naha airport in this dataset).
- access: how easy it is to access the sightseeing spot.
- landscape: how beautiful is its landscape.
- crowdedness: how crowded it is.
- barrier free: how easy it is for people with disabilities to visit.

Here, access, landscape, crowdedness, and barrier free are represented as the five-star scores from reviews. Thus, the value range of these attributes is between 1 and 5. Table 1 shows the initial user preferences of two users u_A and u_B. Note that a weight of an attribute is in the range 1 to 5.

4.2   User Model

To simulate the different personalities of the users, we considered four types of user models based on the interpersonal conflict-handling behavior, as described by Kilmann et al. [7], who considered personality dimension on two axes: *cooperativeness* and *assertiveness* (Table 2).

   In our simulation experiments, these user types determine the threshold value of whether the user is satisfied with the recommended item or not. Here, we assume that if the minimum value of the results of evaluation function for each attribute is equal to or higher than the *satisfaction threshold*, the user is satisfied with the recommended item. That is, $\min_j eval_j^{(t_i)}(r_j^{(u_k)}) \leq$ *satisfaction threshold*. To reflect the characteristics of each user type in the *satisfaction threshold*, it is set according to the first column of Table 3.

   Since an avoiding user type avoids any confrontation, the threshold value is set to 0, meaning that the user agrees on any item the system recommends.

Table 2. User types used in the simulation experiments (based on [7])

User type	cooperativeness	assertiveness	Description
collaborating	high	high	They retain their preferences, and they respect others' preferences
accommodating	high	low	They prioritize others' preferences and suppress their own preferences
competing	low	high	They prioritize their preferences without considering others' preferences
avoiding	low	low	They show no interest in their preferences or those of others

Table 3. Characteristics of each user type

User type	satisfaction threshold	How to change the user's preferences
collaborating	0.8	Change requirements $(r_j^{(u_k)})$
accommodating	0.7	Change requirements $(r_j^{(u_k)})$
competing	0.9	Change attribute weights $(w_j^{(u_k)})$
avoiding	0.0	—

In the exploring phase, there are two options for changing a user's preferences: requirement or attribute weights. Since collaborating and accommodating user types have higher levels of *cooperativeness*, for these user types, changing a user's preference is implemented by changing user's requirement $(r_j^{(u_k)})$. By contrast, a competing user type has lower *cooperativeness*, and their user preferences are updated by changing attribute weights $(w_j^{(u_k)})$. Note that, as the avoiding user type does not participate in the exploring phase, it is not necessary to define the method used for changing the user's preferences.

When the user's requirement needs to be changed, we select the attribute with an evaluation value that is lower than the threshold value of 0.8 and change its requirement value. The amount of change depends on the attribute; in the simulation experiments, the value of the price attribute is set to 500, meaning that the price requirement is loosened by 500; the value of distance attribute is set to 10 and the other attributes are set to −1. As for changing a weight, we select the attribute(s) with an evaluation value that is lower than the threshold of 0.9, and we increase their weight(s) by 1 and decrease the weights of the other attributes by 1.

4.3   Example Execution

Let us assume that the two users, u_A and u_B, have initial preferences as shown in Table 1. We calculate the items' utility values for u_A and u_B and plot the items on the two-dimensional plane: the x-axis represents $utility^{(u_A)}$ and the y-axis represents $utility^{(u_B)}$ as shown in Fig. 2(a). Since both users have the same preferences, all items are plotted on the line of $y = x$. The system will recommend the item that is the furthest from the origin.

Let us also assume that both users are not satisfied with the recommended item. They change their preferences, and the items' utilities are recalculated as shown in Fig. 2(b).

Figure 3 shows a snapshot of sightseeing spots during the simulation run. This chart shows a sightseeing spot's evaluation value of each attribute being plotted on the two-dimensional plane whose x-axis corresponds to u_A and whose y-axis corresponds to u_B.

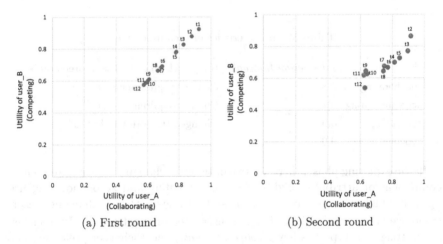

(a) First round   (b) Second round

Fig. 2. Example visualization of items based on their utility (x-axis: $utility^{(u_A)}$; y-axis: $utility^{(u_B)}$)

5   Experimental Results and Discussion

Since there are four user types, and two users are assumed to have the same user preferences, there are 10 different patterns for pairing the user types in the experiment. Table 4 presents the simulation results, showing the product of the two users' utility values when the agreement is reached, and the number of rounds needed to reach the agreement. In Table 4, *fail* means that the agreement could not be reached even if all the items were tried.

As seen in the results, out of 10 possible patterns, an agreement was reached for six patterns ({collaborating, collaborating}, {collaborating, accommodating},

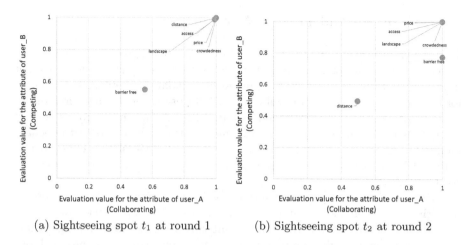

(a) Sightseeing spot t_1 at round 1   (b) Sightseeing spot t_2 at round 2

Fig. 3. Example snapshots of a sightseeing spot's evaluation value for each attribute

Table 4. Simulation results (the product of users' utility values when the agreement is reached and the number of rounds before the agreement is reached) for a pair of user types

u_A	u_B			
	collaborating	accommodating	competing	avoiding
collaborating	0.94/3	0.94/3	*fail*	0.81/3
accommodating	—	0.94/3	*fail*	0.81/3
competing	—	—	*fail*	*fail*
avoiding	—	—	—	0.85/1

{collaborating, avoiding}, {accommodating, accommodating}, {accommodating, avoiding}, and {avoiding, avoiding}).

This result indicates that it is essential to consider the preferences of the other user, because collaborating and accommodating user types could reach an agreement after just three rounds. Since there are 12 items in the dataset used in the simulation experiments, the number of possible maximum rounds is 12. In this sense, these two user types contribute well to reaching an agreement.

By contrast, the competing user type did not reach an agreement. Since the competing user type has low *collaborativeness*, we modeled it to change the attribute weights instead of changing the user's requirements when the recommended item is not satisfactory. However, this method turned out to be less effective for reaching an agreement, and the competing user type did not succeed in finding an agreement.

In addition, for the avoiding users, an agreement was reached immediately, since the avoiding user type considers any item satisfactory. Thus, the exploring phase was not required to reach an agreement.

6   Conclusion and Future Work

This paper proposed a method for exploring possible agreements in a group recommendation system, in which a user's preferences might depend on other users' preferences. We viewed a group recommendation system as a support for negotiation among group members, which let users update their preferences while exploring possible recommended items. To examine the characteristics of the proposed system, we constructed different user models based on the behavioral patterns of human personality, and we conducted simulation experiments. Although the user types greatly influenced the results, certain user models could reach an agreement.

With two users, the proposed system used a monotonic concession protocol to derive the recommended items and visualized the search space on a two-dimensional plane. Our future work will lift this limitation of being applicable for only two users. In addition to applying a monotonic concession protocol, our future work will consider various social choice functions for selecting an item. A visualization method is also needed to adapt to the method of selecting an item for recommendation.

The current recommendation mechanism visualizes items' utility values and waits for a user to change their preferences. However, the system might need to actively participate in the negotiation process. For example, when a system detects that a user is not an active participant, the system needs to give such advice as *"Please critique the recommended item,"* or *"Please change your preferences."* The timing of giving such advice also needs to be considered; one possible option would be when the system detects that a user fails to make significant changes to their user preferences.

From the viewpoint of evaluating the recommender system, another venue for future work would be to make a more elaborate user model. The current model reflects different user types in terms of the difference in the threshold values and how the user preferences are updated. Thus, we may need to consider the number of rounds in the exploring phase. For example, after a certain number of rounds have elapsed, the threshold may need to be lowered or the user preferences may need to be changed. More detailed user models would make it easier to evaluate a group recommender mechanism more realistically.

In our simulation experiments, we used virtual user models created from the theoretical user types in the negotiation process. However, we also plan to conduct evaluation experiments with real users to show the effectiveness of our proposed system and to verify how much the virtual user models we created reflect real users.

References

1. Bekkerman, P., Kraus, S., Ricci, F.: Applying cooperative negotiation methodology to group recommendation problem. In: Felfernig, A., Zanker, M. (eds.) Proceedings of the ECAI 2006 Workshop on Recommender Systems, pp. 72–75 (2006)
2. Carvalho, L.A.M.C., Macedo, H.T.: Users' satisfaction in recommendation systems for groups: an approach based on noncooperative games. In: Proceedings of the 22nd International Conference on World Wide Web, WWW 2013 Companion, pp. 951–958. ACM, New York (2013).https://doi.org/10.1145/2487788.2488090
3. Delic, A., Neidhardt, J., Nguyen, T.N., Ricci, F.: An observational user study for group recommender systems in the tourism domain. Inf. Technol. Tourism **19**(1), 87–116 (2018). https://doi.org/10.1007/s40558-018-0106-y
4. Endriss, U.: Monotonic concession protocols for multilateral negotiation. In: Proceedings of the Fifth International Joint Conference on Autonomous Agents and Multiagent Systems, AAMAS 2006, pp. 392–399. ACM, New York (2006). https://doi.org/10.1145/1160633.1160702
5. Felfernig, A., Boratto, L., Stettinger, M., Tkalčič, M.: GroupRecommender Systems - An Introduction. Springer, Cham (2018). https://doi.org/10.1007/978-3-319-75067-5
6. Garcia, I., Sebastia, L., Pajares, S., Onaindia, E.: Approaches to preference elicitation for group recommendation. In: Murgante, B., Gervasi, O., Iglesias, A., Taniar, D., Apduhan, B.O. (eds.) ICCSA 2011. LNCS, vol. 6786, pp. 547–561. Springer, Heidelberg (2011). https://doi.org/10.1007/978-3-642-21934-4_45
7. Kilmann, R.H., Thomas, K.W.: Interpersonal conflict-handling behavior asreflections of Jungian personality dimensions. Psychol. Rep. **37**(3), 971–980 (1975). https://doi.org/10.2466/pr0.1975.37.3.971
8. Kompan, M., Bielikova, M.: Group recommendations: survey and perspectives. Comput. Inf. **33**(2), 446–476 (2014). http://www.cai.sk/ojs/index.php/cai/article/viewArticle/1077
9. Álvarez Márquez, J.O., Ziegler, J.: Negotiation and reconciliation of preferences in a group recommender system. J. Inf. Process. **26**, 186–200 (2018). https://doi.org/10.2197/ipsjjip.26.186
10. McCarthy, K., McGinty, L., Smyth, B., Salamó, M.: The needs of the many: a case-based group recommender system. In: Roth-Berghofer, T.R., Göker, M.H., Güvenir, H.A. (eds.) ECCBR 2006. LNCS (LNAI), vol. 4106, pp. 196–210. Springer, Heidelberg (2006). https://doi.org/10.1007/11805816_16
11. Nguyen, T.N., Ricci, F.: Situation-dependent combination of long-term and session-based preferences in group recommendations: an experimental analysis. In: Proceedings of the 33rd Annual ACM Symposium on Applied Computing, SAC 2018, pp. 1366–1373. ACM, New York (2018).https://doi.org/10.1145/3167132.3167279
12. Ricci, F., Rokach, L., Shapira, B. (eds.): Recommender Systems Handbook, 2nd edn. Springer, New York (2015). https://doi.org/10.1007/978-1-4899-7637-6
13. Rosenschein, J.S., Zlotkin, G.: Rules of Encounter: Designing Conventions for Automated Negotiation among Computers. MIT Press, Cambridge (1994)
14. Rossi, S., Di Napoli, C., Barile, F., Liguori, L.: A multi-agent system for group decision support based on conflict resolution styles. In: Aydoğan, R., Baarslag, T., Gerding, E., Jonker, C.M., Julian, V., Sanchez-Anguix, V. (eds.) COREDEMA 2016. LNCS (LNAI), vol. 10238, pp. 134–148. Springer, Cham (2017). https://doi.org/10.1007/978-3-319-57285-7_9

15. Rossi, S., Napoli, C.D., Barile, F., Liguori, L.: Conflict resolution profiles and agent negotiation for group recommendations. In: 17th Workshop "FromObjects to Agents" (WOA 2016) (2016)
16. Villavicencio, C., Schiaffino, S., Diaz-Pace, J.A., Monteserin, A., Demazeau, Y., Adam, C.: A MAS approach for group recommendation based on negotiation techniques. In: Demazeau, Y., Ito, T., Bajo, J., Escalona, M.J. (eds.) PAAMS 2016. LNCS (LNAI), vol. 9662, pp. 219–231. Springer, Cham (2016). https://doi.org/10. 1007/978-3-319-39324-7_19
17. Zeuthen, F.: Problems of Monopoly and Economic Warfare. Routledge, Abingdon (1930)

Differential Cryptanalysis of Symmetric Block Ciphers Using Memetic Algorithms

Kamil Dworak[1,2(✉)] and Urszula Boryczka[1]

[1] University of Silesia, Sosnowiec, Poland
{kamil.dworak,urszula.boryczka}@us.edu.pl
[2] Future Processing, Gliwice, Poland
kdworak@future-processing.com

Abstract. The paper presents a new differential cryptanalysis attack based on memetic algorithms. A prepared attack is directed against the ciphertext generated by one of the most popular ciphers named Data Encryption Standard (*DES*) reduced to six rounds of an encryption algorithm. The main purpose of the proposed *MASA* attack is to indicate the last encryption subkey, which allows the cryptanalyst to find 48 from 56 bits of decrypting key. With a simple comprehensive search, it's possible to get the remaining 8 bits. The memetic attack is based on the simulated annealing algorithm, used to improve the local search process, to achieve the best possible solution. The described algorithm will be compared with a genetic algorithm attack, named *NGA*, based on an additional heuristic operator.

Keywords: Differential cryptanalysis · Memetic algorithms · DES · Cryptography · Simulated annealing

1 Introduction

The growing popularity of computers entails an increase in demand for more and more advanced security measures. This is primarily related to the improvement of security techniques currently in use. Over the last few decades, engineers in this area have designed special protocols and encryption algorithms covering such security aspects as confidentiality, integrity and undeniability. The main objective of the cryptography is to maintain the secrecy of the real image of the plaintext and of the decryption key [1]. The purpose of the encryption algorithm is not to hide the fact of the existence of information, but to transform it into a form that remains unreadable [2].

The discipline dealing with searching for weak points of cryptographic systems, consisting mainly in recreating plaintext or decryption key, is dealt with by cryptanalysis. In most cases it is assumed that the attacker knows the details of the cipher, and sometimes even its implementation. If it is not possible to break the cryptographic algorithm, knowing how it works, it is certainly not possible

© Springer Nature Switzerland AG 2019
N. T. Nguyen et al. (Eds.): ACIIDS 2019, LNAI 11432, pp. 275–286, 2019.
https://doi.org/10.1007/978-3-030-14802-7_24

to break it without knowing how the algorithm works [1]. Therefore, the security of the cipher is based only on the security of the key.

In 1990 Biham and Shamir developed a new cryptanalytic attack, based on the analysis of the difference between prepared pairs of plaintexts, namely differential cryptanalysis [3]. Until today this technique is used as one of the most popular methods of attack against symmetric block ciphers. Modern cryptographic algorithms carry out the encryption process by means of several encryption rounds, based on Feistel's network assumptions and a generalized network of substitutions and permutations S-P [2]. The high complexity of these algorithms made the verification of basic security measures, and thus the use of differential cryptanalysis too time-consuming.

In order to improve the efficiency of the attack, a combination of memetic algorithms (MA) and the differential cryptanalysis algorithm has been proposed. The developed attack enables automatic rejection of subkeys with the worst value of the fitness function. This allows for an optimal narrowing down the solutions space. Additional analytical properties of MA improve the process of local search in order to achieve the best possible solution in the shortest possible time. Good results of MA suggest their effective use when testing the immunity of other symmetric block ciphers and cryptographic systems used in industry. Breaking the cryptographic algorithm is equivalent to solving the NP-hard problem [4].

Evolutionary computation (EC) methods, such as genetic algorithms (GA) or MA, are becoming more and more popular in the field of computer security. In 2007 the first attacks using GA on a simplified version of the DES cipher were published [5,6]. In recent years, many publications have been published focusing on the optimization of cryptanalytic processes with the use of various evolutionary techniques, such as evolutionary algorithms (EA) and GA [6–8], MA [9,10], cuckoo search algorithms (CSA) [11,12] or the particle swarm optimization algorithm (PSO) [13].

The next chapter of this paper presents the DES encryption algorithm. In the third chapter the basics of differential cryptanalysis of DES cipher are discussed. The next chapter describes the proposed memetic attack. The fifth chapter contains the results of individual tests together with a comparison with the NGA algorithm mentioned in [7]. In the last section a summary and future plans are presented.

2   Symmetric Block Ciphers

The basic feature of symmetric block ciphers is the presence of one K key used simultaneously for encryption and decryption of information. These ciphers divide the message into a finite set of blocks of equal length that are processed one by one. Each block of data is transformed into a block of encryption of the same length. Most block ciphers have been designed in such a way that the avalanche effect is as large as possible and has been present from the very beginning on encryption algorithm [1].

2.1   Data Encryption Standard (*DES*)

DES is an encryption algorithm that transforms 64-bit plaintext blocks into 64-bit encryption blocks using a 64-bit encryption key [2, 14]. The encryption key is reduced to 56 bits, by removing every eighth bit, used to verify key correctness. The key is then decomposed into a set of six 48-bit subkeys dedicated to each round of the algorithm, $K_1, ..., K_n$ [1]. Figure 1 shows a 6-round *DES* encryption algorithm.

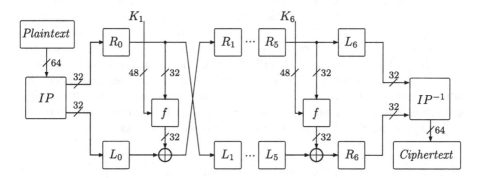

Fig. 1. *DES* encryption algorithm reduced to 6 rounds

From the point of view of differential cryptanalysis, the round function *f*, shown in Fig. 2, is quite important. At the beginning, a 32-bit part of data is passed to an extension permutation *E*. The purpose of this permutation is to equalize the block length to the subkey. The sequence generated in this way is subject to a symmetrical difference operation with bits of the K_i subkey.

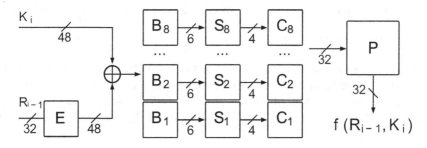

Fig. 2. Round function *f* of *DES* encryption algorithm

Generated string is divided into eight 6-bit blocks $B_1, ..., B_8$. Each of them will be passed to a special base array called S-blocks S_j. They are used to compress the input data. It is worth mentioning that these matrixes are the only non-linear element of the *DES* encryption algorithm. At the end of the *f* the generated C_j blocks are merged into one data string, which is passed to the final permutation *P* [1].

3   Differential Cryptanalysis of the *DES* Algorithm

It is assumed that a cryptanalyst has continuous access to the encryption algorithm, which allows him to select appropriate pairs of plaintext and analyze the generated cryptograms. This type of attack is known as chosen-plaintext attack [15]. The pairs indicated by the attacker may be generated in a pseudo-random manner, although it is a necessary condition they differ in a one certain way. Next, it is analyzed how the difference of the given pair changes over the course of successive rounds of the encryption algorithm. Owing to the detailed analysis of changes in the mentioned difference it is possible to determine the probabilities for each pair suggesting the correctness of some subkeys [1,3].

Each of the differences is described by a certain probability, hereinafter referred to characteristics Ω. Every characteristic determines how often the round function f will return the expected value [3]. Owing to the knowledge of the function f and assuming that $E = E(R_{i-1})$ it is possible to determine the input symmetric difference B' with the following expression:

$$B' = \prod_{j=1}^{8} B_j \oplus B_j^* = \prod_{j=1}^{8} (E_j(R_{i-1}) \oplus K_i) \oplus (E_j(R_{i-1}^*) \oplus K_i) = \prod_{j=1}^{8} E_j \oplus E_j^*. \quad (1)$$

On the basis of the above formula it can be stated that B' is completely independent from the K_i subkey. With B_j' for each j it is possible to determine the set of all ordered pairs (B_j, B_j^*) for the input symmetric difference on the basis of the expression presented in [15]:

$$\Delta(B_j') = \{(B_j, B_j \oplus B_j') : B_j \in (\mathbb{Z}_2)^6\}. \quad (2)$$

By calculating the output symmetric difference $C_j' = S_j(B_j) \oplus S_j(B_j^*)$, it is possible to determine the distribution of all possible inputs to all output differences:

$$IN_j(B_j', C_j') = \{B_j \in (\mathbb{Z}_2)^6 : S_j(B_j) \oplus S_j(B_j \oplus B_j') = C_j'\}. \quad (3)$$

In general, this distribution, described in [15], will be uniform. The main task of the attack is to find distributions with the greatest possible degree of inconsistency. Formula 3 can be used to determine the test set $test_j$ [15]. If the number of elements of this set equals the power of the IN_j set, then this set must contain bits of the K_{ij} subkey.

$$test_j(E_j, E_j^*, C_j') = \{B_j \oplus E_j : B_j \in IN_j(E_j', C_j')\}. \quad (4)$$

4   Memetic Algorithms

MA were developed by Moscato in 1989 [16]. They are a hybrid solution, combining *EA* - used to explore the current solutions space, together with local search algorithms - responsible for the process of exploitation of selected subspace solutions. *MA* have additional information acquired during the evolution process, which makes it possible to obtain much better results than these obtained e.g. with the use of classic *EA* [16,17].

MA operate on a finite set of solutions called a population - in this case a set of n-bit subkeys. In the initial phase of the algorithm, the initial population of *P(0)* is generated. Each individual is evaluated to determine its usefulness by means of a fitness function. Through successive iterations of the algorithm, under the influence of genetic operators, individuals evolve and improve the quality of their adaptation.

After prior selection, individuals can be crossed-over to form a new population of offspring. This operator is used to explore the solutions space. Moreover, each of the newly generated individuals may be subject to a mutation operator. This operator relies on the perturbation of a genotype of one offspring - small perturbations are more likely than large ones [18]. This operator is used to leave the algorithm of the local extremum in order to find an even more interesting solution.

An additional step responsible for the exploitation of a part of the solution space can be placed behind any genetic operator.

5   Proposed Memetic Attack

IP and IP^{-1} permutations can be completely omitted - they do not add any value from the attack point of view. The algorithm starts with the selection of the two most probable 3-round Ω_P^1 and Ω_P^2 characteristics mentioned in [15,19].

The probability of each of the characteristics is $P_\Omega = \frac{1}{16}$. In the fourth round of the encryption algorithm S-Blocks S_2, S_5, S_6, S_7, S_8 for Ω_P^1 and S_1, S_2, S_4, S_5, S_6 for Ω_P^2 for some input symmetric difference B_j' return the output symmetric difference C_j' equal to zero. From this observation it is possible to determine sets $I_1 = \{2, 5, 6, 7, 8\}$ for Ω_P^1 and $I_2 = \{1, 2, 4, 5, 6\}$ for Ω_P^2. The further description of the attack is identical for Ω_P^1 and Ω_P^2, therefore it was decided to generalize it by introducing a set I containing elements from set I_1 and then elements from set I_2.

The next stage is to generate a set of pairs of clear texts, whose symmetric difference corresponds to the characteristics of Ω_P^1 and Ω_P^2. On the basis of this set, a set of cryptograms is also determined. The number of pairs is determined by the signal-to-noise ratio:

$$S/N = \frac{m \cdot p}{m \cdot \alpha \cdot \beta / 2^k} = \frac{2^k \cdot p}{\alpha \cdot \beta} = \frac{2^{30} \cdot \frac{1}{16}}{4^5} = 2^{16}, \tag{5}$$

where m is the number of generated pairs not affecting S/N, p is the probability of the selected Ω characteristic, k is the number of bits of the searched subkeys, α is the average number of subkeys suggested by one pair, and β is the ratio of the pairs analyzed to all possible.

According to the suggestion described in [3] for $S/N = 2^{16}$, 7–8 correct pairs are needed for each of the characteristics. Due to the probability of P_Ω at least 120 pairs of clear text should be generated [3].

The set of generated pairs is additionally filtered. For each pair, a test set is determined from formula 4. If the power of the $test_j$ set, for at least one element from set I, is equal to 0, then the pair will be rejected:

$$\bigwedge_{j \in I} |test_j| > 0. \tag{6}$$

The purpose of the proposed attack is to guess the last encryption key K_6. When the difference C' and part of R_5 is known, it is possible to analyze different subkeys by comparing the output of S-blocks with C'. Application of brute force attack would require checking all 2^{30} possible solutions. Using MA can serve as an excellent optimization tool, finding the right solution in a shorter time and checking much smaller set of potential subkeys.

Each individual is represented by a 30-bit K_j subkey. The fitness function is defined as follows:

$$F_f = \sum_{i=0}^{n} L - H \sum_{j \in I} (S_j(B_j) \oplus S_j(B_j^*)), P^{-1}(R_6' \oplus L_3')), \tag{7}$$

where H is the Hamming distance, L is the length of the subkeys, with a probability of P_Ω. It is possible to estimate the value of L_3', and R_6' can be obtained by analyzing a pair of generated codes. F_f counts the number of overlapping bits between the difference obtained from S-blocks and the difference C'.

The algorithm uses single-point crossover operator. The intersection point is selected in pseudo-random way from 1 to 30. The newly created chromosomes can be perturbed by the mutation operator, whose purpose is to replace two randomly selected bits of the subkey. The proposed algorithm uses tournament selection to perform a correctly executed crossover operator with parents characterized by high fitness function value. Out of all subkeys, only one tournament leader is selected. The selection process is then repeated to select a second parent.

At the end of the algorithm an additional operator responsible for the local search process is activated. The classic algorithm of simulated annealing is responsible for this step. $MASA$ attack pseudocode, for Ω_P characteristics, is presented on Algorithm 1:

Algorithm 1: *MASA* attack pseudocode

---

1 $\Omega_P := find\_most\_probabilistic\_characteristic()$
2 $I := determine\_set\_of\_indexes()$
3 $set\_of\_pairs := generate\_set\_of\_plaintext\_and\_ciphertext\_pairs()$
4 **foreach** $pair \in set\_of\_pairs$ **do**
5   **foreach** $j \in I$ **do**
6          $test_j = determine\_test\_set(pair)$
7   **if** $|test_j| == 0$ **then**
8              $filter\_invalid\_pair(set\_of\_pairs, pair)$
9   **break**
10   **end**
11   **end**
12 **end**
13
14 $P(0) := create\_initial\_population()$
15 **for** $i := 0$ **to** $number\_of\_iterations$ **do**
16     $calculate\_fitness\_function\_value\_for\_each\_individual()$
17   **for** $j := 0$ **to** $population\_size$ **do**
18         $parent_A := tourney\_selection()$
19         $parent_B := tourney\_selection()$
20         $offspring := [parent_A, parent_B]$
21   **if** $random(0, 1) \geqslant crossover\_probability$ **then**
22             $child_A, child_B := crossover(parent_A, parent_B)$
23   **if** $random(0, 1) \geqslant mutation\_probability$ **then**
24                 $child_A := mutation(child_A)$
25   **end**
26
27   **if** $random(0, 1) \geqslant mutation\_probability$ **then**
28                 $child_B := mutation(child_B)$
29   **end**
30             $offspring := [child_A, child_B]$
31   **end**
32
33   **foreach** $child \in offspring$ **do**
34             $T = T_0$
35   **while** $T \geqslant T_{MIN}$ **do**
36                 $new\_child := change\_random\_bit(child)$
37                 $difference := new\_child.fitness - child.fitness$
38   **if** $difference > 0$ **or**
39                   $probability\_fun(difference, T) > random(0, 1)$ **then**
40                     $child := new\_child$
41   **end**
42                 $T = T \cdot \alpha$
43   **end**
44   **end**
45   **end**
46 **end**
47

Before running the local search algorithm, the initial T and minimal T_{MIN} temperatures are set. In subsequent iterations the temperature T is lowered by the $\alpha \in [0,1]$ constant until the T_{MIN} value is reached. In each iteration of the algorithm, a new key K_6' is generated by replacing the random bit with the opposite one. If the new individual turns out to be better than its predecessor, it is automatically replaced by it. In addition, a definition of an additional probability function shall be introduced to accept a worse solution if the algorithm cannot find a better subkeys. It is defined according to the following expression:

$$Probability = \exp\left(\frac{F_f(K_6') - F_f(K_6)}{k \cdot T}\right), \tag{8}$$

where k is the Boltzmann's constant.

Running the $MASA$ algorithm for Ω_P^1 will make it possible to guess 30 of the 48 bits of the K_6 subkey. Restarting the algorithm, this time for Ω_P^2, makes it possible to find additional 12 bits. In order to obtain the remaining 6 bits of the K_6 subkeys - coming from the S-block S_3, can be guessed using a complete overview. With the K_6 sub-key it is possible to restore 48 of the 56 bits of the decryption key by reversing the key distribution process. The remaining 8 bits can be guessed by once again using a complete overview.

6   Experiments and Results

The analysis of the proposed $MASA$ and NGA algorithm sin terms of quality and number of obtained solutions was an important element of the research. It was important to check whether the proposed algorithms make it possible to improve the time of finding solutions by checking a smaller number of subkeys. Among other things, the influence of selected parameters on the similarity of the algorithm and the quality of the obtained solutions was examined. The parameters are set as follows (Table 1):

Table 1. Parameters of the NGA and $MASA$ algorithms

Parameter	NGA	MASA
Maximum number of iterations It_{MAX}	100	100
Population size N	70	20
Number of plaintext pairs γ	200	200
Tourney size T_{SIZE}	5	10
Crossover probability P_c	0,8	0,9
Mutation probability P_m	0,01	0,2
Heuristic operator probability P_n	0,25	-
Initial temperature T_0	-	1
Minimal temperature T_{MIN}	-	0,1
Cooling rate α	-	0,9

As part of the conducted tests the values of parameters were used in various combinations, and for subsequent experiments the best values in terms of algorithm operating time were determined.

Tests were conducted on the same cryptograms and encryption keys. All experiments were carried out on the same computer equipped with a 2.1 GHz Intel i7 processor and 8 GB of RAM. Each attack was run thirty times.

Table 2. Fitness function values for *NGA* algorithm

ID	Ω_P^1						Ω_P^2					
	Min.	Med.	Avg.	Max.	Std. Dev.	It.	Min.	Med.	Avg.	Max.	Std. Dev.	It.
1	1635	1763	1773,6	1923	62,9	32	1382	1545	1561,1	1762	108,5	37
2	1391	1612	1587,2	1678	71	22	1480	1721	1701,5	1845	74,8	13
3	1703	1851	1865,8	2023	95,2	18	1298	1430	1420,9	1582	71,6	27
4	1526	1778	1764,8	1949	105,7	15	1435	1642	1647,6	1834	98,8	21
5	1590	1859	1836,3	1999	123,9	12	1343	1474	1483,3	1626	47,4	26
6	1378	1641	1583,8	1685	93,2	26	1379	1615	1591,6	1712	96,8	23
7	1457	1728	1702,1	1878	124,7	11	1419	1653	1613,8	1748	91,2	37
8	1610	1845	1815,6	1966	93	15	1456	1633	1618,7	1775	80,3	12
9	1457	1621	1623,6	1743	61,5	-	1440	1604	1606,9	1744	75,5	15
10	1446	1645	1635,6	1837	96,9	26	1330	1482	1470,7	1613	86,9	-

In case of the *DES* encryption algorithm there is no Ω characteristics fully guaranteeing the correctness of the solution. The best available characteristics are characterized by a probability of $P_\Omega = \frac{1}{16}$. As mentioned in the description of the attack, in order to correctly determine the 30-bit subkeys two independent characteristics Ω_P^1 and Ω_P^2 are required.

Table 2 presents a summary of the results for the *NGA* algorithm. It contains the first ten tests of thirty. In the last column there is a parameter *It.* which denotes the number of the iteration in which the correct decryption key was found. In most cases, the *NGA* algorithm found all 30 bits of the subkeys. In general, up to 40 iterations of the algorithm. Only in one case, successively for Ω_P^1 characteristics in test 9, and for Ω_P^2 characteristics in test 10, it was not possible to find a correct solution.

Table 3 shows the results for the *MASA* algorithm. Unlike the previous attack, the correct subkey was broken every time. In most cases the solution can be found in up to 30 iterations of algorithm work, with the help of much smaller population than in the case of the *NGA* algorithm.

Figure 3 shows the number of all verified subkeys for each experiment under consideration. On the basis of the bar graphs presented below, it can be seen that the memetic algorithm *MASA* checks a smaller number of subkeys than its genetic equivalent.

Table 3. Fitness function values for *MASA* algorithm

ID	Ω_P^1						Ω_P^2					
	Min.	Med.	Avg.	Max.	Std. Dev.	It.	Min.	Med.	Avg.	Max.	Std. Dev.	It.
1	1642	1770	1790,2	1923	78,4	9	1398	1554	1573,6	1762	120,6	11
2	1484	1619	1608,1	1678	57,7	17	1492	1716	1695,3	1845	99,8	13
3	1725	1963	1922,8	2023	85,3	20	1333	1448	1459,6	1594	90,6	11
4	1529	1725	1758,3	1949	123,9	27	1444	1680	1677	1834	111,96	30
5	1619	1907	1860,9	2012	125,5	19	1278	1485	1507,2	1626	79,4	11
6	1413	1654	1632,1	1685	73,4	30	1366	1656	1614,5	1723	105,1	22
7	1477	1773	1721,2	1878	130,3	12	1482	1577	1606,7	1773	82,1	28
8	1613	1890	1852,7	1966	116,9	8	1453	1659	1662,5	1775	89,1	4
9	1471	1597	1621,9	1755	65,5	16	1472	1607	1604,8	1744	87,7	13
10	1477	1696	1667,4	1837	83,3	15	1356	1492	1490,1	1613	85,3	16

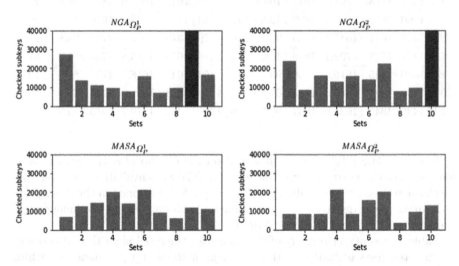

Fig. 3. Number of checked subkeys for each attack with characteristics Ω_P^1 and Ω_P^2

A more comprehensive overview reflecting the number of all proven solutions for each attack and the corresponding characteristics is presented in Table 4.

It is clearly visible that the proposed *MASA* algorithm checks on average 4000–5000 fewer subkeys than its genetic equivalent. On a scale of all running experiments, *MASA* checks 130000–140000 fewer subkeys than the previously proposed *NGA* algorithm.

Table 4. Comparison of checked subkeys between NGA and $MASA$ attacks

Attack	Total number of checked subkeys	Average number of checked subkeys
NGA algorithm		
Ω_P^1	593815	19793,8
Ω_P^2	581037	19367,9
$MASA$ algorithm		
Ω_P^1	465470	15515,6
Ω_P^2	440870	14695,6

7 Conclusions

The paper presents the results of works for the NGA genetic algorithm, based on the use of an additional heuristic negation operator and the $MASA$ memetic algorithm using a standard simulated annealing algorithm. These attacks undoubtedly improve the classic process of differential cryptanalysis. In particular, it is visible due to the acceleration of cryptanalysis process achieved by minimizing the number of reviewed subkeys. This improves the efficiency and effectiveness of the attack, which is important from the cryptanalyst's point of view.

The GA, as well as the MA, finds the correct solution already in the middle of the algorithm - up to 30–40 iterations.

On the basis of the developed research it can be seen that the $MASA$ performs slightly better, using a much smaller population, than the previously proposed NGA attack. Although it always manages to find the correct solution, the $MASA$ algorithm checks fewer possible solutions in order to guess the correct subkeys.

An adaptive version of the Memetic Algorithm is expected to be developed in the future. This would allow for automatic adjustment of parameters. It is also planned to introduce modifications for the crossover operator, which would allow for faster exploration of the space of the solutions. The algorithm should also be directed against more advanced symmetric block codes, such as AES or $Twofish$.

References

1. Schneier, B.: Applied Cryptography: Protocols, Algorithms, and Source Code in C. Wiley, New York (1996)
2. Menezes, A.J., Oorschot, P.C., Vanstone, S.A.: Handbook of Applied Cryptography. CRC Press, Boca Raton (1997)
3. Biham, E., Shamir, A.: Differential cryptanalysis of DES-like cryptosystems. J. Cryptol. **4**(1), 3–72 (1991)
4. Pieprzyk, J., Hardjono, T., Seberry, J.: Fundamentals of Computer Security. CRC Press, Boca Raton (2003)

5. Song, J., Zhang, H., Meng, Q., Zhangyi, W.: Cryptanalysis of four-round DES based on genetic algorithm. In: Wireless Communications, Networking and Mobile Computing, pp. 2326–2329. IEEE (2007)
6. Tadros, T., Hegazy, A., Badr, A.: Genetic algorithm for DES cryptanalysis. Int. J. Comput. Sci. Netw. Secur. **10**(5), 5–11 (2007)
7. Dworak, K., Boryczka, U.: Genetic algorithm as optimization tool for differential cryptanalysis of DES6. In: Nguyen, N.T., Papadopoulos, G.A., Jędrzejowicz, P., Trawiński, B., Vossen, G. (eds.) ICCCI 2017. LNCS (LNAI), vol. 10449, pp. 107–116. Springer, Cham (2017). https://doi.org/10.1007/978-3-319-67077-5_11
8. Dworak, K., Boryczka, U.: Differential cryptanalysis of FEAL4 using evolutionary algorithm. In: Nguyen, N.-T., Manolopoulos, Y., Iliadis, L., Trawiński, B. (eds.) ICCCI 2016. LNCS (LNAI), vol. 9876, pp. 102–112. Springer, Cham (2016). https://doi.org/10.1007/978-3-319-45246-3_10
9. Dworak, K., Nalepa, J., Boryczka, U., Kawulok, M.: Cryptanalysis of SDES using genetic and Memetic algorithms. In: Król, D., Madeyski, L., Nguyen, N.T. (eds.) Recent Developments in Intelligent Information and Database Systems. SCI, vol. 642, pp. 3–14. Springer, Cham (2016). https://doi.org/10.1007/978-3-319-31277-4_1
10. Garg, P.: A comparison between Memetic algorithm and genetic algorithm for the cryptanalysis of simplified data encryption standard algorithm. Int. J. Netw. Secur. Appl. (IJNSA) **1**(1), 34–42 (2009)
11. Jain, A., Chaudhari, N.S.: A new heuristic based on the cuckoo search for cryptanalysis of substitution ciphers. In: Arik, S., Huang, T., Lai, W.K., Liu, Q. (eds.) ICONIP 2015. LNCS, vol. 9490, pp. 206–215. Springer, Cham (2015). https://doi.org/10.1007/978-3-319-26535-3_24
12. Jain, A., Chaudhari, N.S.: A novel cuckoo search strategy for automated cryptanalysis: a case study on the reduced complex knapsack cryptosystem. Int. J. Syst. Assur. Eng. Manag. **9**(4), 942–961 (2017)
13. Abd-Elmonim, W.G., Ghali, N.I., Hassanien, A.E., Abraham, A.: Known-plaintext attack of des-16 using particle swarm optimization. In: Third IEEE World Congress on Nature and Biologically Inspired Computing, pp. 12–16 (2011)
14. Stallings, W.: Cryptography and Network Security: Principles and Practice. Pearson, Upper Saddle River (2011)
15. Stinson, D.R.: Cryptography: Theory and Practice. CRC Press, Boca Raton (1995)
16. Moscato, P.: On evolution, search, optimization, genetic algorithms and martial arts: towards Memetic algorithms. In: Caltech Concurrent Computation Program (1989)
17. Neri, F., Cotta, C., Moscato, P.: Handbook of Memetic Algorithms, Studies in Computational Intelligence, vol. 379. Springer, Heidelberg (2012). https://doi.org/10.1007/978-3-642-23247-3
18. Michalewicz, Z.: Genetic Algorithms + Data Structures = Evolution Programs. Springer, London (1996). https://doi.org/10.1007/978-3-662-03315-9
19. Stamp, M., Low, R.M.: Applied Cryptanalysis. Breaking Ciphers in the Real World. Wiley-Interscience, Hoboken (2007)

Intelligent Methods and Artificial Intelligence for Biomedical Decision Support Systems

Modeling of Articular Cartilage with Goal of Early Osteoarthritis Extraction Based on Local Fuzzy Thresholding Driven by Fuzzy C-Means Clustering

Jan Kubicek$^{(\boxtimes)}$, Alice Krestanova, Marek Penhaker,
Martin Augustynek, Martin Cerny, and David Oczka

FEECS, VSB-Technical University of Ostrava,
K450, 17. Listopadu 15, Ostrava-Poruba, Czech Republic
{jan.kubicek, alice.krestanova, marek.penhaker,
martin.augustynek, martin.cerny, david.oczka}@vsb.cz

Abstract. One of the routine tasks in the Orthopedics practice is the articular cartilage assessment. Proper cartilage assessment includes a precise localization, and recognition of spots indicating the cartilage loss caused by the osteoarthritis. Unfortunately, such tasks are performed manually, without the SW feedback, which leads to various clinical outputs based on the physician's experience. Based on such facts, a development of the fully automatic systems bringing automatic modeling and classification of the cartilage is clinically very important. In our paper we have proposed a local thresholding multiregional segmentation method for the cartilage segmentation from the MR (Magnetic Resonance) images. In our approach, an optimal configuration of the fuzzy triangular sets is driven by the FCM clustering to obtain an optimal segmentation model based on the thresholding. We have verified the proposed model on a sample of the 200 MR image records containing the early osteoarthritis signs.

Keywords: Articular cartilage · Image segmentation · FCM · MR ·
Local thresholding

1   Introduction

Articular cartilage assessment is one of the most challenging issues in a clinical practice of the orthopedics. This clinical procedure is standardly connected with two issues which are done by the clinical experts. Firstly, the articular cartilage should be precisely located. This crucial task ensures recognition of its morphological structure from other knee elements [1, 2]. There are plenty MR imaging modalities generating various image records with better and worse image parameters. It leads to a wide range of the image results with different recognition of the articular cartilage structure [3, 4]. Second important related to the articular cartilage is a classification of spots having clinical signs of the cartilage loss. Here, we have to mention that there are four levels of the cartilage osteoarthritis according to them the knee cartilage is more or less deteriorated. The higher state of the deterioration is found, the more severe cartilage complication it

© Springer Nature Switzerland AG 2019
N. T. Nguyen et al. (Eds.): ACIIDS 2019, LNAI 11432, pp. 289–299, 2019.
https://doi.org/10.1007/978-3-030-14802-7_25

causes. This fact points out that SW tools able to automatically recognize and classify just early state of the cartilage loss would be worth for the clinical practice [5, 6].

One of the most important aspects of the cartilage assessment is the medical imaging systems. The articular cartilage is exclusively exanimated on the MR due to a lot of its benefits, including a great spatial resolution and contrast of individual knee structures. Also, the MR offers plenty sequences allowing for different imaging. For the cartilage imaging, the Proton-dense (PD) and Fat Saturation (FS) sequences are conventionally used.

Besides the MR imaging, there are also clinical alternatives, which may be used for the cartilage assessment. Since the cartilage is located in the knee area, which complexly represent dense tissue, the X-ray imaging may be used. This method well reflects tibial and femoral bones, where the cartilage is located between them. Unfortunately, the cartilage has lower density thus, only fissure between the mentioned bones is observable. Similar problem can be found when using the ultrasound imaging. This method is also able to just imagine interface between bones. In the Fig. 1, we report the MR image of the knee area including the basic description.

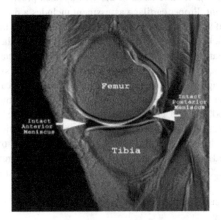

Fig. 1. MR 1.5 T image showing knee separated into tibial and femoral area.

2   Related Work

In this section, we describe important facts regarding segmentation methods for the cartilage segmentation. Generally, we require to segmentation methods would be reliable and robust even in the noisy environment. Regarding the cartilage segmentation, this task may be done, based on a level of the user's intervention, by three ways. Manual segmentation usually serves as a gold standard for evaluation and verification of the segmentation methods. Semi-automatic methods standardly require a certain level of the user's intervention by setting the initial parameters, nevertheless the entire segmentation process works automatically.

In the recent literature, the following segmentation methods have been utilized for the cartilage semi-automatic segmentation: thresholding [7, 8], watershed [9], edge detection [10], energy-minimization [2], Live Wire [11], graph-cuts [12] and active contours [2]. Contrarily, recent trends for the cartilage segmentation is focused especially to the fully autonomous methods, which should completely work automatically not regarding particular MR images, here we can mention the texture analysis [12] and supervised learning (Neural networks kNN) [13] which may be adopted for this task.

Regardless the mentioned methods, we should be aware differences, when segmenting the healthy cartilage and pathological modified cartilage structures. Mostly, the segmentation methods are tested only for the healthy cartilage, thus it is not completely clear whether such methods are same effective, when the cartilage loss is present.

3 Problem Definition

From the clinical point of the view, attention should be paid to precise identification of the cartilage boarders in order to properly specify the cartilage morphological structure for both the healthy cartilage and pathological changes. In this task we cooperate with the University hospital of Ostrava, where we were given data of 200 patients suffering from the osteoarthritis in the early stage. Particularly, we are focused on the early cartilage loss manifesting by tiny cracks impairing originally homogenous cartilage structure. Such findings are badly observable, even by the naked eyes. Therefore, a development of the sensitive segmentation methods is really challenging. In our work, we proposed a multiregional segmentation model relying on the histogram decomposition based on the triangular fuzzy functions. Based on the experimental results, this approach appears to be effective even for the MR data not having optimal parameters due to the image noise artifacts.

4 A Proposal of Local Fuzzy Thresholding for Cartilage Segmentation

In the conventional image hard thresholding, the image histogram is partitioned into a predefined number of classes on the base on the hard boarders. Such approach exclusively defines a pixel's membership in the respective class. When segmenting pixels belonging on the boarder of adjacent tissues, it is complicated to determine a unified rule defining such pixels membership. These situations represent segmentation inaccuracies, when using the hard thresholding methods [1, 2].

In our approach, we are using a histogram thresholding segmentation based on a sequence of the triangular fuzzy functions, where each of them represents one segmentation class. This classification procedure ensures that each pixel relatively belongs to respective class depending on its membership level. This soft thresholding seems to be much more robust especially when segmenting edge pixels or such pixels represented weak contrast. It is particular problem of the early cartilage loss.

4.1 Soft Thresholding for Cartilage Segmentation

In this section, we introduce a concept of the fuzzy thresholding algorithm. Supposing we have a knee MR image being represented by the histogram. We have to define a finite number of the segmentation classes. Each of the triangular fuzzy sets is defined by centroid. This centroid represents a histogram spot with a significant concentration of the pixel's intensities for respective region. When using the triangular fuzzy function, its vertex determines the class centroid. Mathematically, each tissue intensities range is approximated by the $\mu(I(r))$ fuzzy membership function, where r stands for the pixel intensity. Such configuration is perceived as a fuzzy descriptor classifying individual pixels. Here, we are getting over the main limitation of the hard thresholding because pixel can be theoretically classified into more classes with strength depending on the membership level. The final classification is done by the operator maxima. When using the triangular function, each pixel can be classified into two classes, as it is reported in the Fig. 2. Thus, this fuzzy descriptor assigns a vector of membership degrees to each pixel, as is stated:

$$\mu(I(r)) = [\mu_1(I(r))\mu_2(I(r))...\mu_L(I(r))] \tag{1}$$

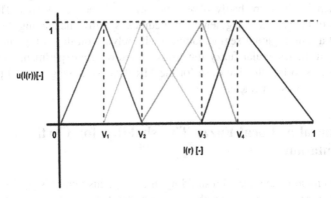

Fig. 2. Configuration of a set of the triangular membership functions approximating the image histogram.

4.2 Definition of Centroid Class

As we stated before, each of the triangular classes is determined by its centroid. The main issue is definition of a method for finding the centroid location. We also stated that such point represents a part of the knee tissue corresponding with significant concentration of the intensities having higher frequency. It means that we are seeking a cluster identifying such location. Centroid of respective class is given as center of initial cluster.

Theoretically, more clustering methods may be adopted for the centroid definition. For this task, the Fuzzy c-means (FCM) clustering appears to be sensitive.

This iterative clustering method producing an optimal c partition on the base of minimizing the weights within sum of squared error objective function J_{FCM}.

$$J_{FCM} = \sum_{k=1}^{n} \sum_{i=1}^{c} (u_{ik})^q d^2(x_k, v_i) \qquad (2)$$

where $X = \{x_1, x_2, \ldots, x_n\} \subseteq R^p$ represents the image intensities, c stands for the number of clusters, where $2 \le c < n$, u_{ik} is a membership degree of x_k belonging to the ith cluster, q is the weighted exponent of membership, $d^2(x_k, v_i)$ is a distance between object x_k and cluster center v_i. An objective function J_{FCM} is figured out based on iterative process, which is described by the following way:

- Setting values for c, q, and ε.
- Initialize the fuzzy partition matrix.
- Calculate the cluster centers $c \left\{ v_i^{(b)} \right\}$ with $U^{(b)}$

$$v_i^{(b)} = \frac{\sum_{k=1}^{n} \left(u_{ik}^{(b)} \right)^q x_k}{\sum_{k=1}^{n} \left(u_{ik}^{(b)} \right)^q} \qquad (3)$$

- Calculation of membership $U^{(b+1)}$. For $k=1$ to n calculate the following procedure: $I_k = \{i | 1 < i \le c|, d_{ik} = \|x_k - v_i\| = 0\}$, $\tilde{I}_k = \{1, 2, \ldots, c\} - I_k$ for the kth column of the matrix. The new membership values are computed:

$$u_{ik}^{(b+1)} = \frac{1}{\sum_{j=1}^{c} \left(\frac{d_{ik}}{d_{jk}} \right)^{2/(q-1)}} \qquad (4)$$

5 Segmentation of Real MR Image Records

In this section, we report results of the articular cartilage modeling. In this task we cooperate with the University hospital in Ostrava, Czech Republic where we were given patient's MR results. The major issue is recognition of the early cartilage loss to quantify a level of the osteoarthritis severity. Since we suppose that cartilage takes relatively small part of the knee area we are using the Region of Interest (RoI) extraction (Fig. 3). Since we extract relatively small part of the image area containing fewer pixels, we have to count with worse image parameters. This unfavorable fact is partially compensated by the cubic image interpolation. For our analysis, we have used the cubic interpolation of the fourth degree.

The Soft thresholding requires definition of number of the segmentation classes, as an input. There is not exact method would doing this task. We have empirically set six

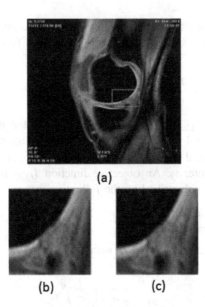

(a)

(b)   (c)

Fig. 3. Native MR 1.5 T image taken from fat suppression sequence (a), native RoI (b) and RoI completed by cubic interpolation.

segmentation classes, which appearing as a relevant compromise for the cartilage segmentation. As we reported on the Fig. 3, the image interpolation has favorable effect on the image quality. Besides that, it influences the segmentation quality as well. When segmenting low-contrast images, we come across lower segmentation effectivity. This unfavorable fact is also compensated by the image interpolation. When processing the non-interpolated RoI we obtain less accurate segmentation results, as we report on the Fig. 4.

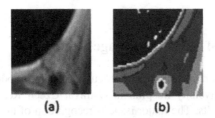

(a)   (b)

Fig. 4. Native RoI (a), Result of the cartilage Soft segmentation (six classes) applied on native RoI (b).

Contrarily, when using the interpolation procedure, individual image structures appear smoothly in the segmentation model (Fig. 5). The segmentation model performs decomposition of the knee structure into segmentation classes via the fuzzy functions. Mathematically, we obtain segmentation results in a form of the index matrix, where each of the classes is represented by a unique number denoting respective class. Such matrix may be shown via sequence of the artificial colors, as we report on the Fig. 5.

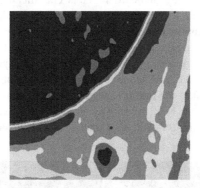

Fig. 5. Soft segmentation with six classes applied on the interpolated MR image.

In the segmentation model, the articular cartilage is identified as red contour. When comparing with the native MR image record (Fig. 3), the early cartilage loss may be reliably identified. It is apparent that the soft thresholding is able to identify even tiny intensity variations in the cartilage structures. The Fig. 6 shows complex situation of the model building. The red class represents the cartilage structure. This structure is apparently interrupted, which is caused by the early cartilage loss well manifesting in the cartilage model. Since we are using the index matrix for the segmentation, classes not representing the cartilage structure may be suppressed, when extracting the cartilage.

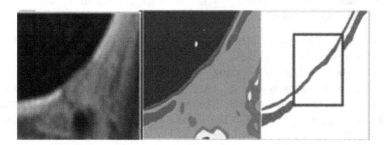

Fig. 6. Example of native MR 1.5 T data, where RoI indicates the early cartilage loss (left), Soft segmentation model (middle), and extraction of the cartilage morphological structure, where the early cartilage loss indicating osteoarthritis is marked blue (right). (Color figure online)

6 Quantitative Testing and Comparison

In the last section, the quantitative comparison and testing is carried out. Previously, we have mentioned an inaccuracy of boarder pixels classification when using the hard intensity thresholding. Therefore, we are focused on the articular cartilage edge pixels, regarding their proper classification.

We have carried out quantitative assessment in a contrast of gold standard images. These reference images have been generated by the clinical experts tracing the cartilage edges manually (Fig. 7). We have done a comparative evaluation of the Soft thresholding against three state of the art segmentation methods: Otsu thresholding (thresholding segmentation), K-means clustering and Region growing method. For the quantitative comparison, the following parameters are considered:

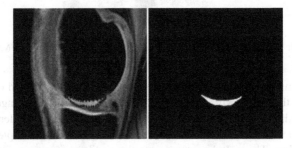

Fig. 7. Gold standard image generated by the clinical expert by the manual tracing of the cartilage boarder (left) and binary mask of its contour.

- **Correlation (*Corr*)** representing a strength of the linear dependence between the gold standard and tested model.
- **Mean Square Error (*MSE*)** represents an average value of squared differences between the gold standard and tested image.

In order to evaluate the segmentation method's robustness, we have incorporated the artificial Gaussian image noise generators deteriorating quality of the image records. We should be aware working with various MR images taken in different quality. Thus, reproducibility of such results is wide. Noise generators can simulate variable quality of the image records. The Fig. 8 shows four different segmentations, where the Gaussian noise is applied. The Table 1 reports of the averaged segmentation results for 200 images.

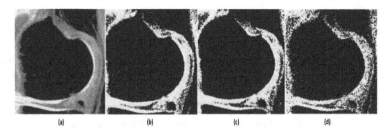

Fig. 8. Articular cartilage segmentation with the Gaussian noise: (a) $\mu = 0$, $\sigma = 0.01$, (b) $\mu = 0$, $\sigma = 0.08$, (c) $\mu = 0$, $\sigma = 0.1$ a (d) $\mu = 0$, $\sigma = 0.8$.

Table 1. Quantitative comparison of correlation coefficient (*Corr*) for the segmentation model against the gold standard.

	Native data	Gaussian noise (0, 0.01)	Gaussian noise (0, 0.08)	Gaussian noise (0, 0.1)	Gaussian noise (0, 0.8)	Gaussian noise (0, 0.2)
Soft thresholding	0.96	0.91	0.91	0.89	0.87	0.81
K – means	0.78	0.76	0.71	0.65	0.65	0.62
Region growing	0.79	0.77	0.74	0.71	0.69	0.66

Table 2. Quantitative comparison of *MSE* for the segmentation model against the gold standard.

	Native data	Gaussian noise (0, 0.01)	Gaussian noise (0, 0.08)	Gaussian noise (0, 0.1)	Gaussian noise (0, 0.8)	Gaussian noise (0, 0.2)
Soft thresholding	32.25	33.16	35.63	36.87	39.87	42.15
K – means	44.12	44.89	45.78	46.81	46.99	53.55
Region growing	41.87	42.69	43.98	44.54	45.78	47.96

7 Conclusion

In this paper, we present the multiregional soft thresholding for the articular cartilage segmentation, and objectification of the early cartilage loss. Due to many advantages of the MR imaging, we use the MR knee images.

Soft segmentation, contrarily from the hard thresholding methods, classifies pixels to respective classes based on membership degree. Thus, each pixel may be classified into more regions. Consequently, higher membership denotes higher probability of assignment. This soft segmentation model is optimized based on the fuzzy c-means clustering performing initial clustering. Consequently, we take centers of individual clusters as centroids of the individual segmentation classes of the segmentation model.

We have experimentally set six classes in the soft segmentation model. This configuration appears as optimal when segmenting the cartilage from the MR. The proposed model is able to identify the articular cartilage morphological structure, and pathological spots representing the early cartilage loss, where the cartilage is missing. Extraction of the cartilage is done based on the index matrix, where other classes, not corresponding with the cartilage, are suppressed.

Lastly, we have compared the segmentation results against the gold standard represented by manually segmented images by the clinical experts. Based on the objective evaluation reported in the Tables 1 and 2, the Soft thresholding appears as the most robust segmentation, when comparing with alternative segmentations utilizing the hard thresholding for pixels classification even in the noisy environment.

Acknowledgment. The work and the contributions were supported by the project SV4508811/ 2101 Biomedical Engineering Systems XIV'. This study was also supported by the research project The Czech Science Foundation (GACR) 2017 No. 17-03037S Investment evaluation of medical device development run at the Faculty of Informatics and Management, University of Hradec Kralove, Czech Republic. This study was supported by the research project The Czech Science Foundation (TACR) ETA No. TL01000302 Medical Devices development as an effective investment for public and private entities.

References

1. Kubicek, J., Penhaker, M., Augustynek, M., Cerny, M., Oczka, D.: Multiregional soft segmentation driven by modified ABC algorithm and completed by spatial aggregation: volumetric, spatial modelling and features extraction of articular cartilage early loss. In: Nguyen, N.T., Hoang, D.H., Hong, T.-P., Pham, H., Trawiński, B. (eds.) ACIIDS 2018. LNCS (LNAI), vol. 10752, pp. 385–394. Springer, Cham (2018). https://doi.org/10.1007/ 978-3-319-75420-8_37
2. Kubicek, J., Vicianova, V., Penhaker, M., Augustynek, M.: Time deformable segmentation model based on the active contour driven by Gaussian energy distribution: extraction and modeling of early articular cartilage pathological interuptions. Front. Artif. Intell. Appl. **297**, 242–255 (2017)
3. Kubicek, J., Valosek, J., Penhaker, M., Bryjova, I.: Extraction of chondromalacia knee cartilage using multi slice thresholding method. In: Vinh, P.C., Alagar, V. (eds.) ICCASA 2015. LNICST, vol. 165, pp. 395–403. Springer, Cham (2016). https://doi.org/10.1007/978-3-319-29236-6_37
4. Kubicek, J., Penhaker, M., Bryjova, I., Kodaj, M.: Articular cartilage defect detection based on image segmentation with colour mapping. In: Hwang, D., Jung, Jason J., Nguyen, N.-T. (eds.) ICCCI 2014. LNCS (LNAI), vol. 8733, pp. 214–222. Springer, Cham (2014). https:// doi.org/10.1007/978-3-319-11289-3_22
5. Kim, J.J., Nam, J., Jang, I.G.: Fully automated segmentation of a hip joint using the patient-specific optimal thresholding and watershed algorithm. Comput. Methods Programs Biomed. **154**, 161–171 (2018)
6. Pitikakis, M., et al.: Automatic measurement and visualization of focal femoral cartilage thickness in stress-based regions of interest using three-dimensional knee models. Int. J. Comput. Assist. Radiol. Surg. **11**(5), 721–732 (2016)

7. Kumarv, A., Jayanthy, A.K.: Classification of MRI images in 2D coronal view and measurement of articular cartilage thickness for early detection of knee osteoarthritis. In: 2016 IEEE International Conference on Recent Trends in Electronics, Information and Communication Technology, RTEICT 2016 - Proceedings, art. no. 7808167, pp. 1907–1911 (2017)
8. Mallikarjuna Swamy, M.S., Holi, M.S.: Knee joint cartilage visualization and quantification in normal and osteoarthritis. In: International Conference on Systems in Medicine and Biology, ICSMB 2010 - Proceedings, art. no. 5735360, pp. 138–142 (2010)
9. Fripp, J., Crozier, S., Warfield, S.K., Ourselin, S.: Automatic segmentation and quantitative analysis of the articular cartilages from magnetic resonance images of the knee. IEEE Trans. Med. Imaging 29(1), 55–64 (2010). art. no. 5071225
10. Wang, P., He, X., Lyu, Y., Li, Y.-M., Qiu, M.-G., Liu, S.-J.: Automatic segmentation of articular cartilages using multi-feature SVM and elastic region growing. Jilin Daxue Xuebao (Gongxueban)/J. Jilin Univ. (Eng. Technol. Ed.) 46(5), 1688–1696 (2016)
11. Gougoutas, A.J., et al.: Cartilage volume quantification via Live Wire segmentation. Acad. Radiol. 11(12), 1389–1395 (2004)
12. Dodin, P., Pelletier, J.P., Martel-Pelletier, J., Abram, F.: Automatic human knee cartilage segmentation from 3D magnetic resonance images. IEEE Trans. Bio-Med. Eng. 57(11), 2699–2711 (2010)
13. Xia, Y., Manjon, J.V., Engstrom, C., Crozier, S., Salvado, O., Fripp, J.: Automated cartilage segmentation from 3D MR images of hip joint using an ensemble of neural networks. In: Proceedings - International Symposium on Biomedical Imaging, art. no. 7950701, pp. 1070–1073 (2017)

Modeling and Features Extraction of Heel Bone Fracture Reparation Dynamical Process from X-Ray Images Based on Time Iteration Segmentation Model Driven by Gaussian Energy

Jan Kubicek[✉], Alice Krestanova, Iveta Bryjova, Marek Penhaker, Martin Cerny, Martin Augustynek, David Oczka, and Jan Vanus

FEECS, VSB-Technical University of Ostrava,
K450, 17. Listopadu 15, Ostrava-Poruba, Czech Republic
{jan.kubicek,alice.krestanova,iveta.bryjova,
marek.penhaker,martin.cerny,martin.augustynek,
david.oczka,jan.vanus}@vsb.cz

Abstract. Tracking of the bone reparation is one of the crucial task in the clinical traumatology. Such reparation period is conventionally subjectively observed by the clinicians. Such procedure leads to subjective errors. Therefore, mathematical model would autonomously classify respective stage of the bone healing would have significant impact to clinical practice of the traumatology. We have proposed a time deformation segmentation model based on the fitting Gaussian energy for detection and modeling of the periosteal callus which is clinically perceived as one of the dominant features determining stage of the heel bone fracture, as well as speed of the heeling. In our analysis we have compared two groups of the patients: controlled and granted group where each of them was differently loaded after placing heel bone fixator. This analysis leads to objective classification of such therapeutic procedure corresponding with the most optimal healing process.

Keywords: Heel bone · Fracture · Active contour · Periosteal callus · Fixator

1 Introduction

The bone fracture period, including the heel bone is composed from sequence of the events. Such events include initial hemorrhage, thrombotic factors, tissue's breakdown release mediators causing the migration of the blood cells. In the secondary stage we start observing an indirect bone healing, where the external periosteal callus is being formed. This element stabilizes the fracture fragments over the healing period [1–4].

For proper fracture healing, adequate blood supply and mechanical stability represent necessary elements. Next important aspect is the oxygen. This element is important for many aspects of cell metabolism. Most of these biological aspects being responsible for the heel bone fracture healing are badly observable during the healing period. On the

© Springer Nature Switzerland AG 2019
N. T. Nguyen et al. (Eds.): ACIIDS 2019, LNAI 11432, pp. 300–310, 2019.
https://doi.org/10.1007/978-3-030-14802-7_26

other hand, periosteal callus is clinically perceived as an object observable from the clinical image records well reflecting a process of the bone healing [5–7].

Secondary reparation of the bone healing is accompanied with forming the periosteal callus. In the clinical practice, they recognize several stages over which the periosteal callus is being formed:

- Stage of proliferation (0–7 days).
- Stage of differentiation (8–21 days).
- Stage of the ossification (since 4th week).
- Modeling and remodeling phase (8–12 week).

The Fig. 1 shows situation of the healing period. As it is apparent, the periosteal callus is badly observable from the native X-ray records due to insufficient contrast. Periosteal callus recognition is subjective with inter-physician variability of 20–25%. In this regards it is important to automatically track the periosteal callus evolution over the time. Such procedure would be able to confirm hypothesis whether the periosteal callus growth is dependent on the physical load after heel bone surgery. Based on these facts we would obtain predictive mathematical model would autonomously estimate a progress of stage of the healing based on the automatic modeling of the periosteal callus. It would represent a significant improvement for the clinical practice [8–10].

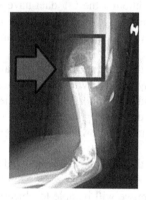

Fig. 1. Manifestation of the periosteal callus (in blue RoI) during bone healing. (Color figure online)

2 Clinical Notes of Periosteal Callus

Periosteal callus may be tracked by several ways. One of the alternatives is the X-ray imaging. This clinical method well shows hard tissues having a great density. On the other hand, soft tissues are partially suppressed depending on their density. When speaking about the periosteal callus it represents a tissue with a middle density. This fact is bit complication because it is imagined under weaker contrast. It leads to worse observation. Therefore, a robust and sensitive segmentation method should be considered [11, 12].

Regarding to tracking the periosteal callus, we are going to extract geometrical features well reflecting its dynamical development. Based on the clinical assumptions, the size and diameter of the periosteal callus should be considered for development tracking.

By the clinical assumptions we can theoretically predict a direction of the development. Immediately after the surgery we can observe tiny structure of the periosteal callus nearly merging with bone structure. Over the time, its structure is getting expanded. This development is supposed to be tracked over the time to obtain predictive model reflecting the healing period [13–15].

Another challenging area deals with adjacent factors being more or less responsible for the periosteal callus growth. In this regard a process of the heel bone loading is very important. Besides the automatic modeling of the periosteal callus, we have done a comparative analysis between two significantly different physical loading in order to evaluate prediction of the most suitable heel bone loading based on the dynamical callus features [16–18].

3   Materials and Methods

In the task of the dynamical modelling of the periosteal callus we cooperate with the University hospital of Ostrava, department of traumatology. We were given both X-ray and CT clinical image data. Although the CT data have significantly higher contrast, when comparing with the X-ray images, CT images are deteriorated by the artefacts, especially hard beaming artefact. Based on this fact we are using the X-ray images for the periosteal callus modeling.

According to the needs of the clinical practice, we have developed fully automatic segmentation model completely working without user intervention. User only specifies an initial curve being consequently modified during the segmentation process. We assume to obtain a binary model classify the periosteal callus area from other X-ray structures. Such model consequently allows for features extraction which are related to healing procedure.

We analyze two groups of the patients: controlled and granted group. Each of them was differently physically loaded, and we expect it would appear on the callus model. By this experimental procedure we will be able to objectively select the most suitable loading of the heel bone fracture based on the callus evolution.

4   Design of Segmentation Algorithm

For the periosteal callus segmentation, we are using the Active contour model driven by local Gaussian distribution fitting energy by using the local means and variables. In this model, we describe local image intensities by the Gaussian distributions having different parameters (means and variances). Firstly, the energy functional is defined:

$$E^{LGDF} = \int_{\Omega} \left(\sum_{i=1}^{N} \int_{\Omega_i} -\omega(x-y) \log p_{i,x}(I(y)) dy \right) dx \qquad (1)$$

In this term, each region Ω_i is represented by the probability density $(p_{i,x}(I(y)))$. Conversion from the maximization to the minimization is ensured by the term $-log$. The method supposes classification the whole image area into two binary classes characterizing the periosteal callus.

We need to achieve a precise and smooth evolution of the active contour. This assumption is done by incorporating a regularization term of the level set function Φ. This procedure is done by the penalization procedure. Penalization is based on the Φ deviation from the sign distance function (Eq. 2).

$$P(\phi) = \int \frac{1}{2}(|\nabla\phi(x)| - 1)^2 dx \qquad (2)$$

The penalization procedure (Eq. 3) utilizes the Heaviside function H.

$$L(\phi) = \int |\nabla H(\phi(x))| dx \qquad (3)$$

In the next step, the entire energy functional is defined:

$$\left(\phi, u_1, u_2, \sigma_1^2, \sigma_2^2\right) = E^{LGDF}\left(\phi, u_1, u_2, \sigma_1^2, \sigma_2^2\right) + vL(\phi) + \mu P(\phi) \qquad (4)$$

Weight constants are represented by: $v, \mu > 0$. The time evolution of the level-set function $\frac{\partial\phi}{\partial t}$ is defined by the Eq. 5.

$$\frac{\partial\phi}{\partial t} = -\delta_\varepsilon(\phi)(e_1 - e_2) + v\delta_\varepsilon(\phi)div\left(\frac{\nabla\phi}{|\nabla\phi|}\right) + \mu\left(\nabla^2(\phi) - div\left(\frac{\nabla\phi}{|\nabla\phi|}\right)\right) \qquad (5)$$

where terms $e_1(x)$ and $e_2(x)$ are defined by the Eq. 6, respectively 7.

$$e_1(x) = \int_\Omega \omega(y - x)\left[\log(\sigma_1(y)) + \frac{(u_1(y) - I(x))^2}{2\sigma_1(y)^2}\right] dy \qquad (6)$$

$$e_2(x) = \int_\Omega \omega(y - x)\left[\log(\sigma_2(y)) + \frac{(u_2(y) - I(x))^2}{2\sigma_2(y)^2}\right] dy \qquad (7)$$

Smoothness of the segmentation process is driven by the parameters reported in the Table 1.

Table 1. Parameters setting for active contour model

Parameter	Description
Δt	Time step
ε	Width of Dirac impulse
σ	Kernel size
ν	Weighted constant
μ	Weighted constant
n	Number of iterations
λ_1	Outer weight
λ_2	Inner weight

5   Results of Periosteal Callus Model

In this analysis, we do modeling of the periosteal callus time-dynamical progress indicating a process of the heel bone healing. During this analysis, two patients groups are being compared. In the granted patient's group, the physical heel bone loading starts since from the third week. It leads to assumption of faster periosteal callus growth. In the controlled group, the heel bone is loaded from the sixth week. It leads to assumption to a slower callus growth.

Based on the segmentation model we take advantage the fact that the segmentation curve is well propagated in such environment not containing rapid intensity changes, it means that the intensity spectrum is nearly focused. Steep intensity changes, as if an interface bone-periosteal callus reliably terminates the segmentation curve thereby the segmentation model is determined. In the initial phase of the segmentation, we are using the initial circle with a radius lower than the periosteal callus.

During the segmentation process this curve is being gradually formed, and such way adopts the periosteal callus shape within a predefined number of the iterations. We have experimentally set 500 iterations appearing as optimal settings. The Fig. 2 shows an application of the active contour on three X-ray image records of same patient taken in the third, sixth and ninth week after surgery.

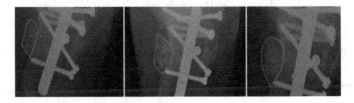

Fig. 2. Segmentation curve, indicated blue, for periosteal callus: 3 weeks (left), 6 weeks (middle) and 9 weeks after surgery. (Color figure online)

In the consequent step, we have built a segmentation model classifying the periosteal callus area (inside the segmentation curve), and adjacent heel bone structures (area outside from the curve). For this task we are using the energy map of the active contour. When the active contour is spreading, its energy is being formed. Inside the active contour curve there is a negative energy, contrarily outside the curve is indicated by the positive energy. Based simple energy thresholding the periosteal callus can be formed (Fig. 3).

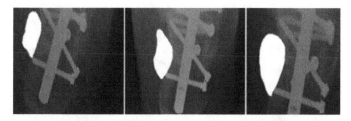

Fig. 3. Binary segmentation model overlapped with the native X-ray images of periosteal callus for period: 3 weeks (left), 6 weeks (middle) and 9 weeks after surgery.

The proposed segmentation model, besides the visualization, allows for the features extraction reliably evaluate dynamical progress of the periosteal callus. We have calculated area, horizontal and vertical diameter of the respective model. The Table 2 represents periosteal callus area computed for seven anonymized patients from the granted group (Fig. 4).

Table 2. Results of the periosteal callus (PC) size calculated in pixels for individual tracked weeks in the granted group

	PC size – 3 weeks	PC size – 6 weeks	PC size – 9 weeks	PC size – 12 weeks
Patient 1	13057	17611	21566	29909
Patient 2	11439	15688	19998	23981
Patient 3	11987	14998	20145	23991
Patient 4	13884	15889	19882	22115
Patient 5	13962	15992	19445	23186
Patient 6	10897	14854	16992	19981
Patient 7	11864	15667	18754	24551

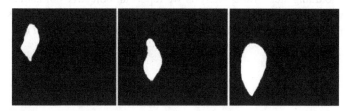

Fig. 4. Binary segmentation model of periosteal callus for period: 3 weeks (left), 6 weeks (middle) and 9 weeks after surgery.

In the Fig. 5, we report the regression model making a prediction trend for the granted group.

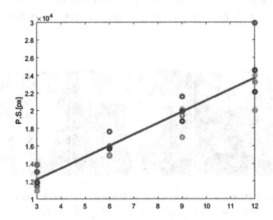

Fig. 5. Prediction model for the granted group based on the linear regression.

As a contrast to the granted group, we have also analyzed the controlled group of the patients. In this group of the patients, we have expected a slower periosteal callus development due to a lower loading. From the clinical point of view they suppose that bone tissue has delayed reactions, in the result this phenomenon should be observable on the periosteal callus geometrical features. The Table 3 represents extract of three patients having heel bone fracture which are assigned to the controlled group.

Table 3. Results of the periosteal callus (PC) size calculated in pixels for individual tracked weeks in the controlled group

	PC size – 3 weeks	PC size – 6 weeks	PC size – 9 weeks	PC size – 12 weeks
Patient 1	13113	14556	17991	18099
Patient 2	11441	13441	15442	16991
Patient 3	10911	12422	13991	15845

As same as in the previous case we report the linear regression model reflecting trend of the periosteal callus based on the proposed model (Fig. 6).

In the concluding step of this analysis we report an objective comparison between both groups. We have measured a slope of the linear regression models. As it is expected, this parameter is significantly different for both cases (Fig. 7).

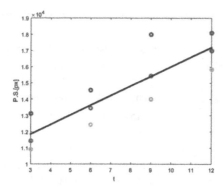

Fig. 6. Prediction model for the controlled group based on the linear regression.

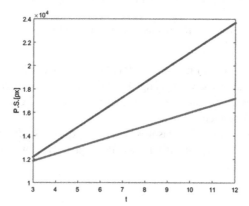

Fig. 7. Comparative analysis for linear regression model for controlled group (red) and granted group (blue). (Color figure online)

6 Quantitative Comparison and Evaluation

In the last part of our analysis we have carried out the objective evaluation of the proposed model against some state of the art segmentation techniques including the Otsu multiregional segmentation, K-means and FCM clustering and Region growing segmentation method. In the all cases we have compared individual method against the ground truth images representing a gold standard for the periosteal callus given manually by the clinical ophthalmologists. Table 4 brings average results for 50 X-ray images where the best result for each test is highlighted. For this objective comparison the following parameters are considered:

- **Rand Index (*RI*)** measures a similarity level between two regions. This parameter compares pair of the elements based on the correctly classified against all the elements. *RI* is a normalized parameter where 0 indicates no similarity, and 1 stands for totally same areas.

Table 4. Quantitative comparison for the proposed model and alternative segmentations against the gold standard

	Proposed model	Otsu	K-means	FCM	Region growing
RI	**0.96**	0.88	0.74	0.91	0.86
VI	3.12	3.92	4.54	**3.11**	4.21
Corr	**0.94**	0.91	0.65	0.88	0.45
MSE	**33.12**	35.15	36.85	33.95	44.15

- **Variation of Information (*VI*)** measures a distance between two regions regarding their conditional entropy. Lower values of *VI* stand for higher similarity. This parameter is given by the Eq. 8.

$$VI(C_1, C_2) = \mathcal{H}(C_1) + \mathcal{H}(C_2) - 2I(C_1, C_2) \tag{8}$$

$\mathcal{H}(*)$ represents entropy of the respective region, and $I(*, *)$ mutual information of two regions.

- **Linear correlation (*Corr*)** measures a level of the linear dependency between two vectors. In our analysis we are using the Pearson correlation coefficient in a normalized range where 0 stands for no linear dependence, contrarily 1 stands for complete linear dependence.
- **Mean Square Error (*MSE*)** measures an average quadratic difference between two regions. More are two regions dissimilar, the higher *MSE* we obtain.

7   Conclusion

In this paper we have proposed a segmentation model for periosteal callus modeling and tracking. Such procedure allows for prediction of the heel bone fracture development which is crucially important for the clinical practice. The proposed segmentation model iteratively approximates area of the periosteal callus based on the Gaussian energy fitting. The model has been utilized for a classification of the callus development regarding different physical load represented by controlled and granted group which have been clinically observed. We have objectively evaluated different steepness of the time predictive model on different heel bone fracture load. Other word speaking such analysis also serves as a prove that the healing period is dependent on a physical load. As an important part of the analysis, we have compared efficiency of the model against some other state of the art segmentation methods with satisfactory results, reported in the Table 4.

We are aware certain limitations. We must mention the X-ray image parameters may influence the image quality. In this regard it would be worth focusing on the periosteal callus segmentation, when the image noise and artefact are present. Such analysis would give robustness of the method even in the noisy environment. Also, we are going to focus on parameters may influence the dynamical model, as if age or alcohol drinking – generally parameters are responsible for bone density.

Acknowledgment. The work and the contributions were supported by the project SV4508811/ 2101Biomedical Engineering Systems XIV'. This study was also supported by the research project The Czech Science Foundation (GACR) 2017 No. 17-03037S Investment evaluation of medical device development run at the Faculty of Informatics and Management, University of Hradec Kralove, Czech Republic. This study was supported by the research project The Czech Science Foundation (TACR) ETA No. TL01000302 Medical Devices development as an effective investment for public and private entities. This article was supported by the Ministry of Education of the Czech Republic (Project No. SP2018/170). This work was supported by the European Regional Development Fund in the Research Centre of Advanced Mechatronic Systems project, project number CZ.02.1.01/0.0/0.0/16_019/0000867 within the Operational Programme Research, Development and Education.

References

1. Chikazu, D., et al.: Cyclooxygenase-2 activity is important in craniofacial fracture repair. Int. J. Oral Maxillof. Surg. **40**(3), 322–326 (2011)
2. Lujan, T.J., Madey, S.M., Fitzpatrick, D.C., Byrd, G.D., Sanderson, J.M., Bottlang, M.: A computational technique to measure fracture callus in radiographs. J. Biomech. **43**(4), 792–795 (2010)
3. Kubicek, J., Penhaker, M., Oczka, D., Augustynek, M., Cerny, M., Maresova, P.: Analysis and modelling of heel bone fracture with using of active contour driven by Gaussian energy and features extraction. In: Nguyen, N.T., Hoang, D.H., Hong, T.-P., Pham, H., Trawiński, B. (eds.) ACIIDS 2018. LNCS (LNAI), vol. 10752, pp. 405–414. Springer, Cham (2018). https://doi.org/10.1007/978-3-319-75420-8_39
4. Kubicek, J., Penhaker, M., Bryjova, I., Augustynek, M., Zapletal, T., Kasik, V.: Tracking of bone reparation process with using of periosteal callus extraction based on fuzzy C-means algorithm. In: Król, D., Nguyen, N.T., Shirai, K. (eds.) ACIIDS 2017. SCI, vol. 710, pp. 271–280. Springer, Cham (2017). https://doi.org/10.1007/978-3-319-56660-3_24
5. Bottlang, M., et al.: Dynamic stabilization of simple fractures with active plates delivers stronger healing than conventional compression plating. J. Orthop. Trauma **31**(2), 71–77 (2017)
6. Hagiwara, Y., et al.: Fixation stability dictates the differentiation pathway of periosteal progenitor cells in fracture repair. J. Orthop. Res. **33**(7), 948–956 (2015)
7. Salonen, A., Lahdes-Vasama, T., Mattila, V.M., Välipakka, J., Pajulo, O.: Pitfalls of femoral titanium elastic nailing. Scand. J Surg. **104**(2), 121–126 (2015)
8. Gangwar, T., Calder, J., Takahashi, T., Bechtold, J.E., Schillinger, D.: Robust variational segmentation of 3D bone CT data with thin cartilage interfaces. Med. Image Anal. **47**, 95–110 (2018)
9. Pandey, P., Guy, P., Hodgson, A.J., Abugharbieh, R.: Fast and automatic bone segmentation and registration of 3D ultrasound to CT for the full pelvic anatomy: a comparative study. Int. J. Comput. Assist. Radiol. Surg., 1–10 (2018)
10. Pérez-Carrasco, J.-A., Acha, B., Suárez-Mejías, C., López-Guerra, J.-L., Serrano, C.: Joint segmentation of bones and muscles using an intensity and histogram-based energy minimization approach. Comput. Methods Programs Biomed. **156**, 85–95 (2018)

11. Aouache, M., Hussain, A., Zulkifley, M.A., Wan Zaki, D.W.M., Husain, H., Abdul Hamid, H.B.: Anterior osteoporosis classification in cervical vertebrae using fuzzy decision tree. Multimed. Tools Appl. **77**(3), 4011–4045 (2018)
12. Shah, R., Sharma, P.: Bone segmentation from X-Ray images: challenges and techniques. In: Bhateja, V., Nguyen, B.L., Nguyen, N.G., Satapathy, S.C., Le, D.-N. (eds.) Information Systems Design and Intelligent Applications. AISC, vol. 672, pp. 853–862. Springer, Singapore (2018). https://doi.org/10.1007/978-981-10-7512-4_84
13. Liu, S., et al.: Fully automated bone mineral density assessment from low-dose chest CT. In: Progress in Biomedical Optics and Imaging - Proceedings of SPIE, vol. 10575, art. no. 105750 M (2018)
14. Engelke, K.: Quantitative computed tomography—current status and new developments. J. Clin. Densitometry **20**(3), 309–321 (2017)
15. Luan, K., Liang, C., Liu, X., Li, J.: Point extraction from cross sections of fractured long bones for registration. In: 2016 IEEE International Conference on Information and Automation, IEEE ICIA 2016, art. no. 7831955, pp. 947–951 (2017)
16. Liu, S., Xie, Y., Reeves, A.P.: Individual bone structure segmentation and labelling from low-dose chest CT. In: Progress in Biomedical Optics and Imaging - Proceedings of SPIE, vol. 10134, art. no. 1013444 (2017)
17. Klintström, B., Klintström, E., Smedby, Ö., Moreno, R.: Feature space clustering for trabecular bone segmentation. In: Sharma, P., Bianchi, F.M. (eds.) SCIA 2017. LNCS, vol. 10270, pp. 65–75. Springer, Cham (2017). https://doi.org/10.1007/978-3-319-59129-2_6
18. Grepl, J., Penhaker, M., Kubicek, J., Liberda, A., Mashinchi, R.: Real time signal detection and computer visualization of the patient respiration. In: Sulaiman, H.A., Othman, M.A., Othman, M.F.I., Rahim, Y.A., Pee, N.C. (eds.) Advanced Computer and Communication Engineering Technology. LNEE, vol. 362, pp. 783–793. Springer, Cham (2016). https://doi.org/10.1007/978-3-319-24584-3_66

A Semi-Supervised Learning Approach for Automatic Segmentation of Retinal Lesions Using SURF Blob Detector and Locally Adaptive Binarization

Tathagata Bandyopadhyay[1], Jan Kubicek[1(✉)], Marek Penhaker[1],
Juraj Timkovic[2], David Oczka[1], and Ondrej Krejcar[3]

[1] FEECS, VSB-Technical University of Ostrava,
K450, 17. Listopadu 15, Ostrava-Poruba, Czech Republic
gata.tatha14@gmail.com, {jan.kubicek,marek.penhaker,
david.oczka}@vsb.cz
[2] Clinic of Ophthalmology,
University Hospital Ostrava, Ostrava, Czech Republic
timkovic.j@bluepoint.sk
[3] Faculty of Informatics and Management, Center for Basic
and Applied Research, University of Hradec Kralove, Rokitanskeho 62,
50003 Hradec Kralove, Czech Republic
ondrej.krejcar@uhk.cz

Abstract. In the clinical ophthalmology, the retinal area is routinely investigated from the retinal images, by the naked eyes. Such subjective assessment may be apparently influenced by ineligible inaccuracies. Therefore, objective assessment of the retinal image records plays an important role for the clinical evaluation and treatment planning. Retinal lesions in premature born children represent one of the most frequent retinal findings which may endanger their vison. These findings are mostly connected with the Retinopathy of Prematurity (RoP). In this paper, we have proposed a novel segmentation model utilizing the SURF blob detector and locally adaptive binarization. The proposed model is able to autonomously detect, and consequently classify retinal lesions. In the result, we obtain a segmentation model of the retinal lesions, where the retinal posterior is effectively separated. As a part of the proposed analysis, we have done objectification and quantitative comparison of the proposed method against some of the state of the art segmentation models by selected evaluating parameters. The proposed method has a potential to be used in the clinical practice as a feedback for the automatic evaluation of the retinal lesions, and also for dynamic retinal lesion's features extraction.

Keywords: Semi-Supervised learning · Image segmentation · Retinal lesions ·
RetCam 3 · Optical disc

© Springer Nature Switzerland AG 2019
N. T. Nguyen et al. (Eds.): ACIIDS 2019, LNAI 11432, pp. 311–323, 2019.
https://doi.org/10.1007/978-3-030-14802-7_27

1   Introduction

The retinal area represents a thin multilayer structure which is composed mostly from the neuronal cells. The main aim of the retina is receiving, modulating and transmitting external visual stimuli. These are led to the optical disc (nerve) and to the visual cortex of the brain. The retinal area is composed from two essential parts, comprising the optical disc (OD) and blood vessels. Our research is exclusively focused on the Retinopathy of Prematurity (RoP). This disease is connected with two significant clinical, pathological signs. Firstly, the retinal blood vessels may be affected. No blood vessel is narrow, but each of them is curved along its length. Within the RoP, some blood vessels exhibit steep oscillations. Such phenomenon is clinically called as the pathological tortuosity. Second sign of the RoP deals with the retinal lesions. Retinal lesions are usually represented by the circular shape, and they are filled by blood [12–14].

From the clinical point of the view, the RoP is perceived as a vasoproliferative disease, especially affecting the prematurely-born infants having a low birth weight. The main aim of the RoP screening is the early detection of the first signs of this disease. This procedure allows for ophthalmologists an optimal scheduling the examination plan. Thus, ophthalmologist experts can prevent a development of the severe vision deterioration. As we have stated before, the most significant clinical signs of the RoP are the oscillating retinal blood vessels and retinal lesion [15, 16]. These objects are imagined in the retinal images from the RetCam 3 (Fig. 1).

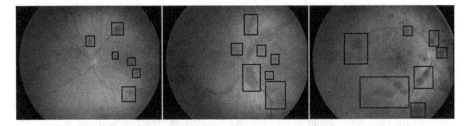

Fig. 1. Example of the retinal records containing the bloody retinal lesions. The most significant lesions are marked, as black.

2   RetCam 3

The premature born infants suffering from the RoP are standardly exanimated by the system RetCam 3. This medical imaging device represents a powerful diagnostic system equipped by the retinal fixation. RetCam 3 is a retinal probe with the image resolution 480×640 px. Besides the single retinal images, system also offers acquiring the retinal video sequences, lasting up to the two minutes. The RetCam 3 also enables changing the lens in order to achieve the retinal images with an optimal wide-angle. In the case of the RoP screening, lens with $130°$ are conventionally used. $120°$ lens are also sometimes used for achieving a higher contrast, but less accuracy. There is also the high contrast lens with the scale $80°$, suitable for the children and adult

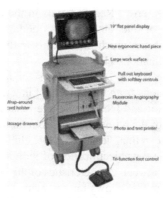

Fig. 2. Complex system RetCam 3 for retinal examination.

patients, and 30° with a high focusing and resolution for the yellow spot examination or the optical nerve (Fig. 2).

The imaging system RetCam 3 is mainly focused to be used for the retinopathy prematurity diagnostic due to the early screening possibility from two weeks after born. Based on the international standards, each newborn having a born-weight lower than 1500 g should be exanimated on the RoP presence. On the other hand, the RetCam 3 can be also used for other retinal diseases, like is the retinal cancer (Retinoblastoma) or further retinal findings (Fig. 3).

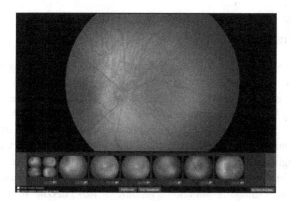

Fig. 3. SW environment of RetCam 3.

3   Problem Definition

Retinal lesions represent a significant issue in the clinical ophthalmologic. Unfortunately, conventional imaging systems, including the RetCam 3 contain only basic visualizing functions for the retinal image processing. Therefore, all the clinical evaluations are done manually, by the naked eyes. Such procedure is of course accompanied by the subjective error. Recent trends in the clinical ophthalmology are focused on a development of the autonomous models enabling retinal features extraction. These procedures bring much more accurate diagnosis results, when comparing with the manual examinations [17].

A proposed model of the retinal lesions allows for automatic segmentation and classification of the retinal lesions from the RetCam 3 image data. The proposed model utilizes the SURF blob detector and locally adaptive binarization. The retinal lesion's model brings a classification of the retinal lesions from the retinal image posterior. The most important aspect of the retinal lesions in the clinical practice is the features extraction. Such features serve for objectification of individual retinal features. One of the clinical features is a ratio between characteristics of the optical disc to the respective retinal lesions. The proposed model allows for automatic calculation of such geometric features, which can be compared with the optical disc [17].

4 Proposed Method for Retinal Lesions Segmentation

The RGB retinal images used in the current study, are observed to contain more visual contrast in the green channel as compared to the other two. Therefore, proposed algorithm first extracts the green channel image (I_{green}) from the original RGB input (I_{RGB}) to ease the subsequent processing [1, 2]. Following this step, the rest of the algorithm can be partitioned into three logical groups as: (i) Adaptive Binarization and Morphological Operations, (ii) Blob detection using SURF and (iii) Semi-Supervised learning to identify the ambiguous lesion regions. Each of these three stages are discussed as follows.

4.1 Adaptive Binarization and Morphological Operations

As the retinal lesions regions are supposed to be darker in intensity than the background, in the first stage our algorithm tries to simply identify the dark patches in the I_{green} image. The easiest way to do so is to convert the grayscale green channel image (I_{green}) to a binary image by making pixel values as 0 or 1 based on whether their green channel intensity is below or above a predetermined threshold respectively. However, choosing a particular global threshold for an image is difficult due to wide variation in intensity and brightness in the image background [3, 4]. In such case, local threshold based adaptive binarization [3, 5] comes into rescue and so here we used that. I_{green} is passed through 'adaptthresh()' function available in MATLAB® R2018a with 'ForegroundPolarity' set to 'dark' and 'Statistic' set to 'median' to compute a local threshold for each pixel as median of its neighborhood. The neighborhood size is calculated using the Eq. 1.

$$Neighborhood\ Size = 2 * floor(size(I)/16) + 1 \tag{1}$$

Where I is the input image i.e. I_{green} in this case. With that local threshold map, the green channel image is converted to a binary image (I_{Bin}).

I_{Bin} captures all the dark regions in I_{green}. However, in that process it also includes the blood vessel areas which are not regions of interest for the current study. Therefore, to exclude those regions some morphological operations are performed on I_{Bin}. To be specific, I_{Bin} is passed through median filtering, hole filling, erosion with square structuring element and finally image complementation to obtain the initial mask ($I_{InitMask}$) in which the white patches are likely to correspond to lesion regions.

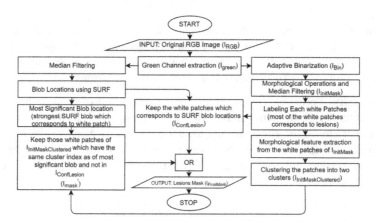

Fig. 4. The flowchart of the Proposed Segmentation Algorithm. (Values inside parenthesis denotes the output of the corresponding operation).

4.2 Blob Detection Using SURF

SURF or Speed Up Robust Feature serving as a local feature detector and descriptor [6] is widely used in Computer Vision to primarily detect blobs in scale and transformation invariant manner. It uses box-filters and Difference of Gaussian (DoG) to approximate Laplacian of Gaussian (LoG) to identify the blob regions in an image [6, 7]. In this section of the proposed algorithm, I_{green}, followed by a median filtering based smoothing, is passed through SURF algorithm ('detectSURFFeatures()') with 'NumOctaves', 'NumScaleLevels' and 'MetricThreshold' parameters empirically set to 4, 4 and 500 respectively to detect the blobs.

As this blob detector detects both bright and dark blobs, a further filtering is done to keep only those blobs which correspond to some white patches in $I_{InitMask}$. Thus, conceptually we are taking an intersection between the regions in $I_{InitMask}$ and the blobs detected by SURF to generate a new binary image ($I_{ConfLesion}$) which contains only the regions detected by both SURF and Adaptive Binarization methods. White patches in $I_{ConfLesion}$ are assumed to be confirmed Lesion regions, as those are detected by both the methods.

4.3   Semi-Supervised Learning to Identify the Ambiguous Lesion Regions

A set difference of $I_{InitMask}$ and $I_{ConfLesion}$ yields $I_{Ambiguous}$. Regions in $I_{Ambiguous}$ are ambiguous i.e. not all are lesions and not all are non-lesions either. So, to do further filtering among the ambiguous regions, a Semi-Supervised learning approach is used. A Semi-Supervised learning is a mix of Supervised and Unsupervised Machine Learning approach, where majority of the data is unlabeled and a small subset is labeled [8, 9].

To adapt the Semi-Supervised strategy in the present work, first, the regions in $I_{InitMask}$ are clustered into two groups based on their morphological properties. For each region in $I_{InitMask}$, 'Eccentricity', 'EquivDiameter', 'Solidity', 'PerimeterPerArea', 'Circularity' and 'MinorToMajorAxisRatio' are taken as features:

- **Eccentricity:** Eccentricity of an ellipse is the proportion of distance between the foci and the length of its major axis. In our current context, 'Eccentricity' refers to the eccentricity of the ellipse having the same second moments as the region.
- **Equivalent Diameter:** 'EquivDiameter' refers to the diameter of a circle having same area as that of the region patch. The Eq. 2 is used to calculate this feature for each region.

$$EquivDiameter = \sqrt{4 * Area / \pi} \tag{2}$$

- **Solidity:** 'Solidity' is the fraction of area of the region to the area of the convex hull of the region. It is calculated as:

$$Solidity = \frac{RegionArea}{ConvexHullArea} \tag{3}$$

- **Perimeter per Area:** 'PerimeterPerArea' refers to the ratio of perimeter of the region to its area. It is calculated in the same way as of [10].
- **Circularity:** It is one of the shape factors and calculated as:

$$Circularity = \frac{Perimeter^2}{4 * \pi * Area} \tag{4}$$

- **Minor to Major Axis Ratio:** This is the ratio of Minor Axis Length to Major Axis Length of the ellipse having the same normalized second moments as the region.

The features, as discussed above, are used to congregate the regions into two groups using well known k-means clustering algorithm [11]. Thus, we obtain

$I_{InitMaskClustered}$, in which each region corresponds to either one of the two cluster indices.

Even though $I_{InitMaskClustered}$ contains two clusters of regions, the cluster indices do not have any information on which index corresponds to Lesions and which index corresponds to non-Lesions and hence comes the concept of Semi-Supervised Learning to solve the issue further. Here, let us first define a term called 'MostSignificantBlob'. It refers to the strongest SURF blob which corresponds to a white patch in $I_{InitMask}$. Once the 'MostSignificantBlob' is defined, corresponding location and cluster index can be obtained from $I_{InitMaskClustered}$. Quite logically, the cluster index of the 'MostSignificantBlob' will correspond to the Lesion regions and another cluster index will correspond to non-Lesion regions. Now, to produce I'_{mask}, only those regions are kept from $I_{InitMaskClustered}$, which have same cluster indices as that of 'MostSignificantBlob' and not already taken in $I_{ConfLesion}$. Finally, an 'OR' operation is performed between $I_{ConfLesion}$ and I'_{mask}, or in other words 'Set Union' is done on those two images to create the final mask ($I_{FinalMask}$) which can be used to segment out the Lesions from the retinal images. The Fig. 4 summarizes the whole algorithm in a flowchart form.

5   Testing and Quantitative Comparison

In this section, we introduce the testing results of the retinal lesions modeling. We generally suppose that the resulting model is represented by the binary classification of the retinal area, where the surrounding retinal structures are suppressed. Such model reflects the retinal lesions features, which can be used for clinical evaluation of a state of the respective retinal lesions in the particular time. Clinically, the optical disc geometrical features are utilized as a reference for the retinal lesions objectification and quantification.

In this analysis, we cooperate with the University hospital in Ostrava, Czech Republic, particularly with the Ophthalmology ward. We are currently processing retinal image data from the RetCam 3, containing 2800 images, from 80 patients having clinical signs of the RoP. All the images have been acquired in the standardized resolution 480×640 px. Each patient record is also accompanied with the clinical diagnosis, containing particular signs of the RoP (a level of pathological tortuosity and presence of the retinal lesions).

One of the limitations is a lower retinal image resolution, which is connected with a worse recognition of the retinal structures, including the retinal lesions. Therefore, we firstly report the image preprocessing results and the SURF features (Fig. 5). Note that the native retinal data are provided in the RGB format. By the decomposition of the RGB model, we have found out that the G channel the best reflects the retinal lesion's features.

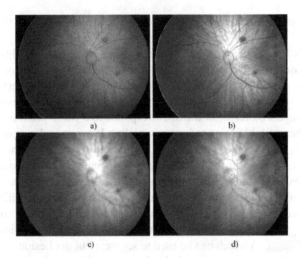

Fig. 5. Processing of the retinal images: (a) native retinal image from the RetCam 3, (b) green channel extraction and contrast enhancement, (c) image smoothing by using the median filtration and (d) SURF features location. (Color figure online)

Finally, we report the final detection and classification of the retinal lesions. We report two stage of the modeling. Firstly, the binary model is done based on the binary classification, which suppresses other retinal structures. Consequently, the binary mask is used for localization of the retinal lesions in the native retinal images. The experimental results of the retinal lesions modeling is reported in the Fig. 6.

Lastly, we have done a quantitative comparison of the proposed method against selected state of the art segmentation methods. We have done a quantitative comparison on a sample of twenty patients, which have been diagnosed the retinal lesions. For these patients, the ground truth images have been generated by the ophthalmologist experts. These images serve as a gold standard for the objective evaluation. In order to test the method robustness, we have applied the artificial Gaussian noise ($G(\mu, \sigma^2)$), which gradually increasingly deteriorate the retinal image quality (Fig. 7).

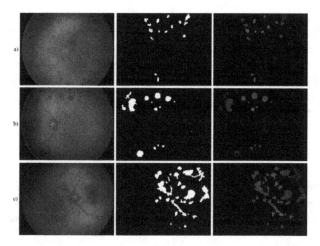

Fig. 6. Process of the retinal lesions modeling for three cases (a), (b), (c): first column: native retinal images, second column: binary model of the retinal lesions and third column: retinal lesions modeling in the native image area.

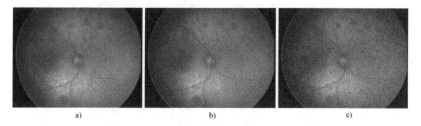

Fig. 7. Noise analysis for the quantitative comparison: (a) native retinal image, (b) G(0,0.01), (c) G(0,0.05)

The objective comparison is carried out based on a set of the evaluating parameters enabling measurement of a difference level between the gold standard images and the proposed model. The following parameters are considered for the comparison:

- **Rand Index (*RI*)** measures a level of the similarity between two segmentation regions. *RI* evaluates a compatibility of the assignment between pairs of the elements in two segmentation regions. *RI* is given by the following expression:

$$\mathrm{RI}(C_1, C_2) = \frac{2(n_{11} + n_{00})}{N(N-1)} \tag{5}$$

where N stands for the total number of the pixels, n_{11} stands for a number of pairs belonging to the C_1 and C_2 and n_{00} stands for a number of the pairs belonging to different segmentation regions.

- **Correlation coefficient (*Corr*)** measures the linear dependence between two regions. This parameter gives numbers in the range: $[0; 1]$, where 0 indicates no linear similarity, contrarily 1 stands for the full linear similarity.
- **Mean Squared Error (*MSE*)** measures differences between two segmentation regions. *MSE* computes a squared average difference of the pixels in the same positions. *MSE* is given by the formulation:

$$\text{MSE}(C_1, C_2) = \frac{1}{N} \sum_{k=1}^{N} (C_1(k) - C_2(k))^2 \qquad (6)$$

The following Tables 1, 2 and 3 bring the averaged results of the quantitative comparison for twenty patients. The quantitative comparison is done against the regional segmentation methods, including K-means, FCM, Active contours and Region growing.

Table 1. Quantitative comparison of the retinal modeling for *RI*

	Native retinal data	Gaussian noise (0, 0.01)	Gaussian noise (0, 0.03)	Gaussian noise (0, 0.05)	Gaussian noise (0, 0.07)	Gaussian noise (0, 0.2)
Proposed method	0.94	0.88	0.84	0.82	0.69	0.58
K-means	0.69	0.64	0.68	0.67	0.44	0.32
FCM	0.83	0.81	0.78	0.74	0.63	0.41
Active contours	0.91	0.88	0.81	0.81	0.65	0.61
Region growing	0.66	0.63	0.53	0.48	0.41	0.33

Table 2. Quantitative comparison of the retinal modeling for *Corr*

	Native retinal data	Gaussian noise (0, 0.01)	Gaussian noise (0, 0.03)	Gaussian noise (0, 0.05)	Gaussian noise (0, 0.07)	Gaussian noise (0, 0.2)
Proposed method	0.95	0.92	0.87	0.82	0.74	0.64
K-means	0.88	0.64	0.61	0.61	0.58	0.32
FCM	0.91	0.86	0.81	0.76	0.71	0.53
Active contours	0.89	0.91	0.85	0.82	0.78	0.68
Region growing	0.64	0.55	0.51	0.43	0.43	0.34

Table 3. Quantitative comparison of the retinal modeling for *MSE*

	Native retinal data	Gaussian noise (0, 0.01)	Gaussian noise (0, 0.03)	Gaussian noise (0, 0.05)	Gaussian noise (0, 0.07)	Gaussian noise (0, 0.2)
Proposed method	29.54	31.74	33.95	34.54	35.18	38.78
K-means	34.15	34.98	36.74	36.89	37.55	40.82
FCM	29.89	31.99	33.99	36.54	37.14	40.11
Active contours	30.55	32.12	34.81	35.12	35.59	39.78
Region growing	33.97	34.72	35.12	37.84	41.54	44.74

6 Discussion

In this paper, we present a novel method for the retinal lesions modeling. Retinal lesions represent significant issue for the clinical practice. The retinal lesions are one of the most significant signs of the RoP. From the clinical point of the view, in a comparison with the optical disc, the retinal lesions exhibit dynamical features over the time, mostly they are expanded. Therefore, clinicians need to objectively track their development, regarding the optical disc features, which are supposed to be stable over the time.

The proposed method is focused on the autonomous extraction and classification of the retinal lesions from the retinal records. Binary model reflect the geometrical features and manifestation of the retinal lesions. Such features will be clinically applicable for tracking their development. As a part of the proposed analysis, we report the quantitative comparison against some well-known state of the art methods, where the proposed method seems to be very sensitive, even in the noisy environment.

In this analysis, we primarily devote to a proposal of the segmentation method. Clinically, when detecting the multiple lesions, these lesions should be recognized from one another due to its individual tracking. Another way of the future research will be focused on the classification procedure for individual detected lesions. Second part of the future research will be a clinical evaluation of the detected lesions based on the optical disc features.

Acknowledgment. The work and the contributions were supported by the project SV45088 11/2101Biomedical Engineering Systems XIV'. This study was also supported by the research project The Czech Science Foundation (GACR) 2017 No. 17-03037S Investment evaluation of medical device development run at the Faculty of Informatics and Management, University of Hradec Kralove, Czech Republic. This study was supported by the research project The Czech

Science Foundation (TACR) ETA No. TL01000302 Medical Devices development as an effective investment for public and private entities.

References

1. Adal, K.M., et al.: An automated system for the detection and classification of retinal changes due to red lesions in longitudinal fundus images. IEEE Trans. Biomed. Eng. **65**(6), 1382–1390 (2018)
2. Kubicek, J., et al.: Optical nerve segmentation using the active shape method. Lékař a technika-Clinician Technol. **46**(1), 13–20 (2016)
3. Sezgin, M., Sankur, B.: Survey over image thresholding techniques and quantitative performance evaluation. J. Electron. Imaging **13**(1), 146–166 (2004)
4. Zhang, J., Hu, J.: Image segmentation based on 2D Otsu method with histogram analysis. In: 2008 International Conference on Computer Science and Software Engineering. IEEE (2008)
5. Sauvola, J., Pietikäinen, M.: Adaptive document image binarization. Pattern Recogn. **33**(2), 225–236 (2000)
6. Bay, H., Tuytelaars, T., Van Gool, L.: SURF: speeded up robust features. In: Leonardis, A., Bischof, H., Pinz, A. (eds.) ECCV 2006. LNCS, vol. 3951, pp. 404–417. Springer, Heidelberg (2006). https://doi.org/10.1007/11744023_32
7. Tuytelaars, T., Mikolajczyk, K.: Local invariant feature detectors: a survey. Found. Trends Comput. Graph. Vis. **3**(3), 177–280 (2008)
8. Chapelle, O., Scholkopf, B., Zien, A.: Semi-supervised learning. IEEE Trans. Neural Networks **20**(3), 542 (2009). (Chapelle, O., et al. (eds.) 2006) [book reviews]
9. Zhu, X.: Semi-supervised learning literature survey. Comput. Sci. Univ. Wis. Madison **2**(3), 4 (2006)
10. Bandyopadhyay, T., Mitra, S., Mitra, S., Rato, L.M., Das, N.: Analysis of pancreas histological images for glucose intolerance identification using wavelet decomposition. In: Satapathy, S.C., Bhateja, V., Udgata, Siba K., Pattnaik, P.K. (eds.) Proceedings of the 5th International Conference on Frontiers in Intelligent Computing: Theory and Applications. AISC, vol. 515, pp. 653–661. Springer, Singapore (2017). https://doi.org/10.1007/978-981-10-3153-3_65
11. Hartigan, J.A., Wong, M.A.: Algorithm AS 136: a k-means clustering algorithm. J. Roy. Stat. Soc. Ser. C (Appl. Stat.) **28**(1), 100–108 (1979)
12. Colomer, A., Naranjo, V., Janvier, T., Mossi, J.M.: Evaluation of fractal dimension effectiveness for damage detection in retinal background. J. Comput. Appl. Math. **337**, 341–353 (2018)
13. Sreng, S., Maneerat, N., Hamamoto, K., Panjaphongse, R.: Automated diabetic retinopathy screening system using hybrid simulated annealing and ensemble bagging classifier. Appl. Sci. **8**(7), 1198 (2018)
14. Baltatescu, A., et al.: Detection of perimacular red dots and blots when screening for diabetic retinopathy: refer or not refer? Diab. Vasc. Dis. Res. **15**(4), 356–359 (2018)

15. Kubicek, J., Timkovic, J., Krestanova, A., Augustynek, M., Penhaker, M., Bryjova, I.: Morphological segmentation of retinal blood vessels and consequent tortuosity extraction. J. Telecommun. Electron. Comput. Eng. **10**(1–4), 73–77 (2018)
16. Kubicek, J., Kosturikova, J., Penhaker, M., Augustynek, M., Kuca, K.: Segmentation based on gabor transformation with machine learning: Modeling of retinal blood vessels system from retcam images and tortuosity extraction. Front. Artif. Intell. Appl. **297**, 270–283 (2017)
17. Hu, J., Chen, Y., Zhong, J., Ju, R., Yi, Z.: Automated analysis for retinopathy of prematurity by deep neural networks. IEEE Trans. Med. Imaging **38**, 269–279 (2018)

Autonomous Segmentation and Modeling of Brain Pathological Findings Based on Iterative Segmentation from MR Images

Jan Kubicek[1(✉)], Alice Krestanova[1], Tereza Muchova[1],
David Oczka[1], Marek Penhaker[1], Martin Cerny[1],
Martin Augustynek[1], and Ondrej Krejcar[2]

[1] FEECS, VSB-Technical University of Ostrava,
K450, 17. Listopadu 15, Ostrava-Poruba, Czech Republic
{jan.kubicek,alice.krestanova,tereza.muchova.st,
david.oczka,marek.penhaker,martin.cerny,
martin.augustynek}@vsb.cz
[2] Faculty of Informatics and Management, Center for Basic
and Applied Research, University of Hradec Kralove,
Rokitanskeho 62, 50003 Hradec Kralove, Czech Republic
ondrej.krejcar@uhk.cz

Abstract. This paper deals with the design of an automated algorithm for segmentation and modeling pathological areas of MR brain imaging data. For segmentation purposes was used namely active contouring method in MATLAB. The proposed algorithm was tested on a dataset of 21 MR frames.

This work also deals with the comparison efficiency of preprocessing image to improve segmentation results and subsequently testing and verifying the proposed algorithm for real image data.

Keywords: Brain · Pathological area of the brain · Magnetic resonance · Detection · Image processing · Active contours · MATLAB

1 Introduction

Pathological areas of the brain are manifested in a different bright spectrum from physiological structures. In the brain, these pathological areas are malignant or benign tumors. Malignant tumors grow very fast, they are unlimited and metastasized. Benign tumors grow very slowly, they are bordered and non-metastasis. Yet benign brain tumors can kill the affected person as they grow in the closed space of the skull [1].

One type of brain tumor is glioblastoma multiforme, which affects predominantly elderly people. The average survival time is about one year. This type grows unbounded, it is soft consistency, bloody and necrotic. In most cases is benign but threatens man with his growth. Further benign meningiomas is well defined with spherical or ovoid shape. Glioblastoma is malignant tumor; its limitlessness and irregularity cannot be exploited, and it reacts badly to other healing methods. In addition to the tumor, at the CT and MR images are observable also brain abscesses [2].

N. T. Nguyen et al. (Eds.): ACIIDS 2019, LNAI 11432, pp. 324–335, 2019.
https://doi.org/10.1007/978-3-030-14802-7_28

From this reason, it is aim analysis and genesis of algorithms for segmentation pathological areas from MR images.

Brain magnetic resonance, CT, PET and more are used to visualize brain tissue and brain pathologies. Magnetic resonance uses a very strong magnetic field and physical properties of hydrogen's nucleus. Hydrogen nucleus are exposed to strong magnetic field and they are source of radio frequency wave. This signal is captured by the receiving coils. Images give to doctors very important information about all organs in the body. The advantage is that the patient is not exposed to X-rays, unlike computed tomography. The disadvantage is the relatively long examination, usually takes 20–50 min. Magnetic resonance is the most important imaging method for imaging brain tissues and spinal cord. T2-weighted sequences with suppressing signal of cerebrospinal fluid (FLAIR) are used to assess the structure tissue and fat suppression (STIR) that are used to detect spinal cord injury [3].

Another imaging method for visualize the brain is CT. The examination is performed without the use of contrast medium and assessing the state of the brain tissue, the width of the ventricular system or presence of bleeding. Principle of computed tomography is the mathematical reconstruction of the transverse cut of the body patient. Using functional magnetic resonance (fMRI) are displayed areas of brain, which actively working, they consume more oxygen. Therefore, they are more blooded than areas at rest. Examination of fMRI is time-consuming, usually takes 30–60 min [4, 5].

Sonography is used for assessment brain tissue of fetuses, newborns and infants until a large fontanel has closed them [4].

2   Related Work

This paper deals with image processing of brain from MR. In related works are used many segmentation methods for detection pathological areas of brain. Exists algorithms can classificate to three groups: manual, semi-automatic and fully automatic.

One of methods is watershed transformation, which it is based on similar principle as the method dividing and merging areas. The method is used to extract area of tumor from the brain's MRI image. The image is understood as a relief that is gradually filled with water. Water floods the relief from the local image minima [6–8].

Another method is region growing. Firstly, starting points of areas are manually selected. In the next steps pixels in the area are examined seed and are then incorporated into particular region, if they satisfy a certain homogeneity criterion [9].

Algorithm K-means is used to find tumors by detecting edges and ranks between nonparametric cluster analysis methods. The principle of the algorithm is the classification of objects into the final count clusters. The centers of the clusters are the center of mass. Initial values of the center of mass must be set manually [10–12].

Algorithm BIANCA serves for detection WHM (white hyperintensity) with assumed vascular origin. This is, for example, multiple sclerosis, acute infarctions or brain tumors. The BIANCA algorithm uses k-NN algorithm [13].

Algorithm Fuzzy C-Means allows clustering one object into multiple clusters. The Fuzzy algorithm works with a degree of truth, which determines membership to a particular cluster. The algorithm divides the final number of points into clusters according to the criterion, such as color, distance and more [14, 15].

3 Description of Dataset and Image Preprocessing

Images of patients with cancers were chosen from free available database http://www. cancerimagingarchive.net/. These images represent a large archives of medical images Data are broken down by disease area (Fig. 1). In this work were selected groups of pathological findings in brain (low grade malignancy gliomas, glioblastoma, glioblastoma multiforme, gliomas of low and high degree of malignancy). Image resolution is 256×256 pixels, except group Glioblastoma with resolution of 512×512. Several test images were selected from each group. The DICOM format from database is not suitable for segmentation purposes, so it needs to be converted to PNG.

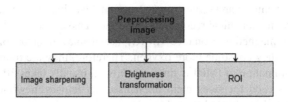

Fig. 1. Block diagram of preprocessing image.

Appropriate image preprocessing will increase the success of further lesion detection steps. Algorithm of image preprocessing, were composed from image sharpening, bright transformation and specification of ROI (Region Of Interest). Image sharpening is based on subtracting the blurred image of original, so it is created new image with high frequency components (edges) (see Fig. 2). Second step is bright transformation, specifically, point transformation that means new pixel value is calculated only from the value of the same element (see Fig. 3). Creating ROI can reduce the count of pixels in the image. Object of interest are lesion area, they are selected manually (see Fig. 4).

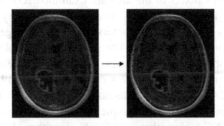

Fig. 2. Native image (left), image after application sharpening (right).

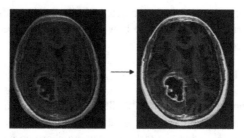

Fig. 3. Image after brightness transformation (left) and after brightness transformation (right).

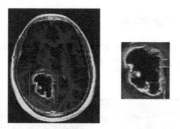

Fig. 4. Native image with selected lesion (left), extracted ROI (right).

4   Segmentation of MR Images with Active Contour

Active contours are called "snakes", is one of the more advanced image segmentation methods. It is a curve inside the image, deformed by energy. Internal energy influences the smoothness of the waveform, the image energy deals with the deformation of the contours to the edges, and the external energy draws the curve to the local minima. Total energy can be expressed as the sum of the internal and external energy and energy of image. Deformation of its shape takes place until it reaches the boundary of segmented interest object [16].

Active contour is defined as parametrical equation:

$$v(s) = [x(s), y(s)], s \in [0, 1] \tag{1}$$

where, $s = 0$ means start of curve and, $s = 1$ means end of curve. $[x(s), y(s)]$ are coordinates of point, which is at curve. Total energy of active contour is:

$$\int_0^1 E_{snake}(v(s))ds = \int_0^1 \left[E_{internal}(v(s)) + E_{image}(v(s)) + E_{con}(v(s)) \right]ds \tag{2}$$

where $E_{internal}$ is internal energy, E_{con} is energy of contour's starting points, E_{image} is energy of image. Internal energy is defined by equation:

$$E_{int} = \alpha(s)\left|\frac{\partial s}{\partial s}\right|^2 + \beta(s)\left|\frac{\partial s}{\partial s}\right|^2 \tag{3}$$

where α represents curve strength and parameter β indicates the stiffness of the contour. Using these parameters, it can determine how the contour will be shaped. Energy of image is defined by equation:

$$E_{image} = w_{tine}E_{tine} + w_{edge}E_{edge} + w_{term}E_{term} \qquad (4)$$

E_{tine} determines how the contour will be tilted to light or dark areas of the image of choice w_{tine}. Function E_{edge} the contour is attracted to locations with high gradient value (local minimum). E_{term} has the task of detecting sharp corners and end of edges by examining curvature. Blurred image is used here due to possible noise [16].

4.1   Implementation of Algorithm

In the first step it is necessary to set the following optimal parameters for detection the object of interest (Fig. 5):

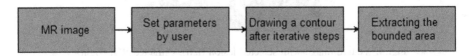

Fig. 5. Block diagram of algorithm.

- NumIter – the number of iteration steps.
- Timestep – time step of the curve shift. The larger value of the parameter is, the less the exact curve, but the faster it moves.
- mu – the constant providing the curve so that it does not deviate far from its position during the development.
- sigma – the Kernel's function parameter, examining the luminous intensity inhomogenity. The higher is this parameter, the curve moves away from the actual position.
- epsilon – determines the Dirac pulse width, which accelerates the initial movement of the initialized contour. The higher value is, the lower accurate is the contour at the end.
- c_0 – the constant multiplies the pixel values inside the contour (negative value) and outside (positive value). The higher constant is, the faster is the initial contour motion.
- lambda1 – the constant, the weight of the area of the inner contour when moving the curve.
- lambda2 – the constant, the weight of the area of the outer contour in the curve movement. It also depends on the ratio between lambda1 and lambda2. If lambda1 > lambda2, the curve shrinks. If lambda1 < lambda2, the curve expands.
- nu – the constant determining the length of the contour. The higher value is, the shorter is the contour and the attraction of any undesirable artifacts in the image.
- alf – a parameter that can set an emphasis on image energy, drawing the curve to the edges [16].

The next step is to place the initialization contour. For each image, it needs to set the size of the initialization circle individually. At Fig. 6, you can see the individual steps of the algorithm. Once all the iterative steps have been taken, a contour is created. It borders the object, which we choose. Evolution of the contour is accompanied by the energy function E (ϕ) differentiating the image area to:

- outside: area outside the segmented object (positive energy)
- inner: area outside the segmented object (negative energy)
- border: zero energy representing the edges of the object

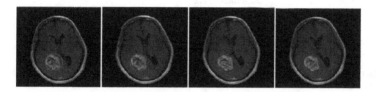

Fig. 6. Evolve contour from initialization to final step, number of iteration steps = 300.

At Fig. 7 shows an energy map, which shows the energy distribution during contour development. The pathological tissue is depicted in blue and has got negative value, the surrounding yellow area has got positive value. In the right window, the interest object is extracted by thresholding, where the negative area is white and the positive area is black.

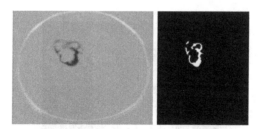

Fig. 7. Energetic map (left), binary model of object (right). (Color figure online)

5   Quantitative Testing and Comparison

Algorithm was tested on native images, preprocessed image without specifying the ROI and the ROI from the images. For testing was chosen 21 images. In the figures below, the upper image always features images with unclear boundary of object, the image in the middle is well-contrasting and the lower image represents images, where it is not entirely clear.

Only parameters of algorithm NumIter, Timestep and sigma were changed. The other parameters were set same at all testing. Values of these parameters show Table 1 below.

Table 1. Set parameters for testing images

Parameter	Values
epsilon	1
c_0	2
lambda1	1,02
lambda2	1,2
Nu	0,0005*255*255
Alf	20

5.1   Testing on Native Images

For testing on native images were used parameters with values according to the Table 2.

Table 2. Set parameters for testing image below

Parameter	Top image	Middle image	Bottom image
NumIter	300	400	200
Timestep	0.01	0.1	0.01
sigma	5	10	12

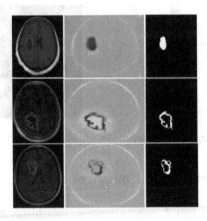

Fig. 8. Testing on native images: native MR images (left), energetic map of active contour (middle) and binary model of object of interest (right).

Figure 8 shows in the left window final contours on the native images, map of energy is displayed in the middle window and the object of interest is extracted in the right window. Segmentation for these images was successful.

5.2   Testing on Preprocessed Images Without ROI

Segmentation of preprocessed images without ROI was complicated, due by using image sharpening and bright transformation. For efficient segmentation, it was necessary to increase the sigma value. Average sigma value for native images was 9.63, while for images from this section average value was 23.05. Increasing sigma value had a negative impact on computational demands. For testing on native images were used parameters with values according to the Table 3.

Table 3. Set parameters for testing preprocessed images

Parameter	Top image	Middle image	Bottom image
NumIter	1100	100	400
Timestep	0.001	0.01	0.01
sigma	15	25	20

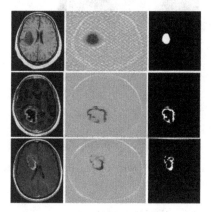

Fig. 9. Testing on preprocessed images: native MR images (left), energetic map of active contour (middle) and binary model of object of interest (right).

Top image of Fig. 9 shows inaccurate extraction. This is due to the unclear boundary between the object and the surrounding tissues. The image in the middle shows degraded results compared to the native image. The contour does not break open to the folds and the result is inaccurate. Much worse results have the bottom of the image, where the resulting extraction is incomplete.

5.3   Testing on Image with Defined ROI

ROI from images were testing and it shows, that often it occurs to distort the results by adding pixels to unwanted objects. If unwanted objects with a similar shade of gray appear in the ROI, unwanted pixels are added to the undesirable parts. Parameters of setting are showing below in the Table 4.

Table 4. Set parameters for testing ROI image

Parameter	Top image	Middle image	Bottom image
NumIter	300	400	800
Timestep	0.01	0.1	0.01
sigma	10	10	8

In the image below (see Fig. 10), the contour does not approximate the entire object. This is again caused by an inaccurate boundary between the subject and the background. The middle image is comparable to preprocessed images without ROI. In these images, glioblastoma was successfully detected. The image below does not restrict the entire object again and the resulting extraction is incomplete.

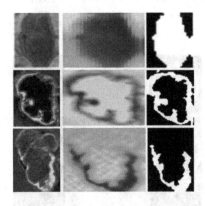

Fig. 10. Testing on ROI images: ROI MR images (left), energetic map of active contour (middle), binary model of object of interest (right).

5.4   Definition of Objectification Parameters

Evaluation of segmentation's effectivity is based on correlation coefficient. Correlation coefficient expresses the degree of "linear bond tightness," since the correlation analysis describes the linear relationships between the variables. The correlation coefficient R is from −1 to +1. The second power of the correlation R_2, called the coefficient of determination, takes values from 0 to +1, where 0 means no dependency and 1 means full dependence [17].

5.5   Results of Quantitative Testing

Testing of the segmentation model contrasted with the gold standard, which is given by manual segmentation. Gold standard is given for each image, which is compared with the binary model (Fig. 11). This is determined by the method of active contours. The correlation coefficient is calculated, and then this coefficient is converted to percentages (see Table 5). Testing was done only on native preprocessed images, as ROI images did not achieve satisfactory results.

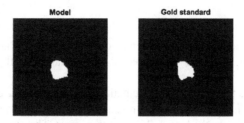

Fig. 11. Binary model of image (left), gold standard (right).

Table 5. Objective evaluation of active contour

Image	Image resolution [px]	Native images [%]		Preprocessed images [%]	
1.	512 × 512	1.	95.63	1.	92.24
		2.	94.11	2.	92.76
		3.	95.71	3.	92.64
		Average	95.15	Average	92.54
2.	512 × 512	1.	89.05	1.	93.07
		2.	88.77	2.	92.74
		3.	87.19	3.	92.92
		Average	88.33	Average	92.91
3.	512 × 512	1.	92.88	1.	90.80
		2.	94.39	2.	91.72
		3.	94.96	3.	92.24
		Average	94.07	Average	91.58
4.	256 × 256	1.	92.87	1.	87.97
		2.	95.96	2.	88.87
		3.	95.85	3	86.54
		Average	94.89	Average	87.79
5.	256 × 256	1.	87.98	1.	93.99
		2.	89.07	2.	94.98
		3.	90.90	3.	94.41
		Average	89.31	Average	94.46
6.	256 × 256	1.	93.36	1.	89.34
		2.	92.49	2.	87.57
		3.	92.22	3.	87.08
		Average	92.69	Average	87.99

The resulting average value of native images is 90.92% and the average value of preprocessed images is 90.52%. The highest percentage difference is recorded for the tenth image, where it is 8.01%. This image has a dark background and the lesions are lighter, so the contrast is enhanced by preprocessing and lesion is better bordered and visible.

6   Conclusion

The Active Contour method was implemented on test images. The optimal combination of parameters for each image was selected individually for the detection of the respective object.

A great effect on the resulting segmentation has the location of the initialization contour. If the object has unclear border, the contour has nothing to attach. Two images failed to detect at all due to the low visibility of the object. Therefore, the clearly visible boundaries of the tumors and surrounding tissues are very important. The next point is the size of the initialization contour. If a small contour is used, the curve decreases until it disappears. In the case of large contours, it can attach itself to other structures.

The best results are native data, on the other hand the worst results have been achieved with ROI images. This is evidenced by the fact that 19 images from 21 native data were detected, and only 15 images from 21 images with ROIs were detected. If the subject is brighter than the background of the native image, it will help to preprocessing of image. There is highlighting light colors and better contrast between the subject and the background.

However, if the tumor is darker than background, the background will be highlighted by preprocessing. The subject is not highlighting. The result is inaccurate segmentation.

The Active Contour method was evaluated to verify the success of this method for segmentation and extraction of the pathological regions of the brain and also to determine whether better native or preprocessed images are best for this method. The resulting values are very similar, it cannot determine exactly whether better native or preprocessed images are for the active contour method.

Acknowledgment. The work and the contributions were supported by the project SV4508811/2101Biomedical Engineering Systems XIV'. This study was also supported by the research project The Czech Science Foundation (GACR) 2017 No. 17-03037S Investment evaluation of medical device development run at the Faculty of Informatics and Management, University of Hradec Kralove, Czech Republic. This study was supported by the research project The Czech Science Foundation (TACR) ETA No. TL01000302 Medical Devices development as an effective investment for public and private entities.

References

1. Stříteský, J.: Patologie: [učebnice pro zdravotnické školy a bakalářské studium]. Epava, Olomouc (2001)
2. Ferda, J., Mírka, H., Baxa, J., Malán, A.: Základy zobrazovacích metod. Galén, Praha (2015)
3. Seidl, Z., Vaněčková, M.: Magnetická rezonance hlavy, mozku a páteře. Grada, Praha (2007)
4. Meningeom. Neurologienet. http://www.neurologienetz.de/fachliches/neuroradiologie/meningeom/. Accessed 20 Nov 2017
5. Heřman, M.: Radiologické zobrazování mozku: Magnetická rezonance (MR). E-learningová podpora mezioborové integrace výuky tématu vědomí na UP Olomouc, Olomouc (2012)

6. Subudhi, A., Rajendra, A., Manasa, D., Subhransu, J., Sukanta, S.: Automated approach for detection of ischemic stroke Delaunay Triangulation in brain MRI images. Comput. Biol. Med. **103**, 116–129 (2018)
7. Cerrolaza, J.J., Villanueva, A., Cabeza, R.: Hierarchical statistical shape models of multiobject anatomical structures: application to brain MRI. IEEE Trans. Med. Imaging **31**(3), 713–724 (2012)
8. Benson, C.C., Lajish, V.L., Rajamani, K.: Brain tumor extraction from MRI brain images using marker based watershed algorithm. In: 2015 International Conference on Advances in Computing, Communications and Informatics (ICACCI), pp. 318–323. IEEE, Kochi (2015)
9. Despotovic, I., Goossens, B., Philips, W.: Segmentation of the human brain: challenges, methods, and applications. Comput. Math. Methods Med. **2015**, 1–23 (2015)
10. Bernal, J., et al.: Deep convolutional neural networks for brain image analysis on magnetic resonance imaging: a review. Artif. Intell. Med. pii, S0933–3657(16)30520–6 (2018)
11. Naveen, A., Velmurugan, T.: Identification of calcification in MRI brain images by k-Means algorithm. Indian J. Sci. Technol. **8**(29) (2015). https://doi.org/10.17485/ijst/2015/v8i29/83379
12. Bal, A.., Banerjee, M., Chakrabarti, A., Sharma, P.: MRI brain tumor segmentation and analysis using rough-fuzzy C-Means and shape based properties. J. King Saud Univ. Comput. Inf. **30**(4) (2018)
13. Griffanti, L., Zamboni, G., Khan, A., et al.: BIANCA (Brain Intensity Abnormality Classification Algorithm): a new tool for automated segmentation of white matter hyperintensities. NeuroImage **141**, 191–205 (2016)
14. Jipkate, B.R., Gohokar, V.V.: A comparative analysis of fuzzy C-Means clustering and K means clustering algorithms. Int. J. Comput. Eng. Res. **2**, 737–739 (2012)
15. Kotte, S., Pullakura, R.K., Injeti, S.K.: Optimal multilevel thresholding selection for brain MRI image segmentation based on adaptive wind driven optimization. Measurement **130**, 340–361 (2018)
16. Vicianová, V.: Extrakce objektů z medicínských obrazů na základě metody aktivních kontur. Bakalářská práce. VŠB – Technická univerzita Ostrava. Ostrava (2017)
17. Milde, D.: Korelace. Univerzita Palackého v Olomouci, Olomouc (2011)

Design and Analysis of LMMSE
Filter for MR Image Data

Jan Kubicek[1]([✉]), Alice Krestanova[1], Martina Polachova[1],
David Oczka[1], Marek Penhaker[1], Martin Cerny[1],
Martin Augustynek[1], and Ondrej Krejcar[2]

[1] FEECS, VSB-Technical University of Ostrava,
K450, 17. Listopadu 15, Ostrava-Poruba, Czech Republic
{jan.kubicek,alice.krestanova,martina.polachova,
david.oczka,marek.penhaker,martin.cerny,
martin.augustynek}@vsb.cz
[2] Faculty of Informatics and Management,
Center for Basic and Applied Research, University of Hradec Kralove,
Rokitanskeho 62, 50003 Hradec Kralove, Czech Republic
ondrej.krejcar@uhk.cz

Abstract. This paper deals with the method of removing the noise in MRI images- During data capture and transmission, the data is disturbed by a noise component that cannot be completely reproduced exclude. Noise is defined in signal theory as additive information that was added to the original purchasing equipment or during the transport. Study of noise models is a very important part of image processing. On the images are applied noise generators, and design LMMSE filter, which is used for shaded images. They were tested salt and pepper noise, Gaussian noise and Rican noise. For each noise, more than one level of this noise. Another task was objective and subjective evaluation of the success of the filtration.

Keywords: Noise · Gauss distribution · Rican distribution · LMMSE filter

1 Introduction

Many imaging applications for MRI (magnetic resonance) images are based on stochastic methods, which are based on knowledge of noise statistics. This work deals models of noise and filters that play an important role in the noise reduction process. To eliminate noise can also lead the signal estimate by removing the noise, which must be preceded by a well-defined noise statistical data model (usually Gaussian distribution). Magnetic resonance contains noise from various sources (including noise from stochastic variation and noise from eddy currents and many physiological processes) and artefacts caused by magnetic sensitivity between adjacent ones tissues, rigid and unstable movement of the body and other sources.

Electrically conductive tissue in the body the patient creates thermal noise, which is the main source of MRI noise. Noise is also the result errors that occur during the

© Springer Nature Switzerland AG 2019
N. T. Nguyen et al. (Eds.): ACIIDS 2019, LNAI 11432, pp. 336–348, 2019.
https://doi.org/10.1007/978-3-030-14802-7_29

acquisition. Random noise that enters the display system from external sources, typically has a Gaussian or normal distribution.

If we want to compare more models of noise and discuss their use, we must accept a reasonable one a compromise between the accuracy of the model and its generalization capability. The data are obtained in k-space 1 using regular Cartesian sampling. Noise contributions are independent, total noise is the sum of the noise from each single source. For noise modeling, we also take into assumption the power of noise, whose value is constant [1].

Noise can be removed by various techniques such as spatial and time filters, anisotropic diffusion filtering, filter of nonlinear dimeter (NLM; non local means), bilateral an trilateral filter, wavelet transform, linear minimal estimates quadratic errors, approaches with maximum probability, or statistical estimates nonparametric analysis of neighboring and singular functions [1].

2 Related Work

Noise filtering methods in MRI can be divided into three groups: methods defined in spatial domain, methods working in the transformed domain and methods using statistical properties signals. Heat noise occurs during signal acquisition and is one of the most common causes of image degradation.

There are many ways to reconstruct an image or set of data. An important property of good filter is that it should completely remove the noise and keep the edges. Traditionally, there are two types of models – linear and nonlinear. They are more used linear filters. Their advantage is the speed, and the disadvantage insufficient ability effectively to protect the borders of images that are blurred. We can divide filters for high frequency and low frequency [2, 3].

After applying these types of filters, the edges in the image are deleted, and the image is smoothed. The smoothing effect depends on the size of the mask used. These filters are used to remove noise and blur image. This group includes, for example, average, Gaussian and median filter.

The median filter is used to remove the noise known as salt&pepper. Which representing the impulsive noise model [4].

Wiener's filter based on statistical approach. The filter principle is based on minimizing the mean quadratic error between the original matrix and the reconstructed matrix.

Gaussian filter represents an extension of the filter by averaging, namely Gaussian layout [5].

High-frequency filters include Laplacian filter, Sobel's filter. This filter is used for edge detection in the image. However, the method is susceptible to noise, so it is good use to any filter to remove the noise from image [14].

The Sobel's filter highlights the edges either in the horizontal and vertical plane. Alternatively, it is possible combine both possibilities and obtain Sobel's overall gradient. However, this is a filter becomes nonlinear.

Prewitt's, Robertson's or Kirch's filter work on the same principle as the Sobel's filter [6, 7, 12, 13].

3 Methods

3.1 Analysis of Test MR Data

Describing MR image structures according to signal strength of a given sequence is used to indicate hypersensitive for a tissue that is represented by a bright part of the brightness spectrum, isosignal or also hyposignal for the tissue, which is represented by the dark part of the brightness spectrum, and for tissues the black color is the term assigned. It is also important to indicate the sequence in which the image was generated because the signal variability analyzed in the different sequences occurs tissues.

MR examination provides an advantage in showing the structures of both soft tissues (as for example ligaments, muscles, tendons, etc.), as well as bones and meniscus, and is therefore widely used in the diagnosis of pathologies musculoskeletal system.

The knee is one of the most investigated joints by just this one MR method. MR images of knee and CT images of knee areas were tested in this work. Figure 1 shows two knee images that have been tested [1].

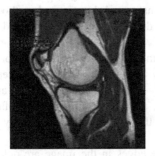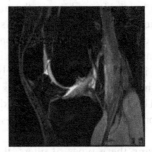

Fig. 1. Input dataset (images of knee) – CT image (left), MRI image (right).

The MRI image shows knee, where the cartilage is represented as the lightest area of the image. Can be recognized two cartilages around the knuckles of the femur and the tibia and the third is in the area of the patella. Visible gray areas belong to the muscle tissues, among which in the knee area belong m. gastrocnemius and m. vastus medialis. Bone tissue is presented with the large black areas, which are femur and tibia bones.

Due to the absence of ionizing radiation, MRI is the ideal imaging method for repeated examination in the context of long-term monitoring, which can be used in angiography. This method is suitable, for the examination of calcifications blood vessels, in which a gadolinium contrast agent is applicated. During projection the area under investigation belongs to the significant light regions, while another vessel appears only in gray. Selected image MR image is used in tested dataset (see Fig. 2).

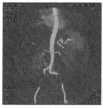 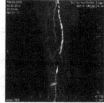 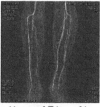 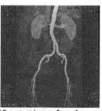

(a) contrast MR image of blood vessels system of descending part of the aorta   (b) calcification of the femoral blood vessels   (c) contrast MR image of the vascular system of the calf   (d) contrast image of vascular system of division of a descending aorta

Fig. 2. Input dataset (vascular systems).

The tested dataset includes next MR images as pathological image of human brain and cavity image abdominal (see Fig. 3). The resolution of all images is shown in Table 1.

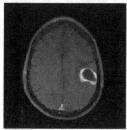 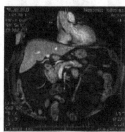

(a) pathological image of brain   (b) MR image of belly walls

Fig. 3. Input dataset (brain and belly walls).

Table 1. Resolution of input data

Input image	Resolution
Knee (CT)	600×600
Knee (MR)	600×600
Contrast MR image of blood vessels of descending aorta	640×640
Calcification of the femoral vascular system	960×960
Contrast MR image of the vascular system of the calf	640×640
Contrast MR image of the vascular system of division of the descending aorta	640×640
Brain (MR)	256×256
Belly (MR)	480×480

3.2   Used Synthetic Generators of Image Noise

In this work are used generators for three types of image noise salt&pepper, Gauss noise and Rician noise, which are applied to MR images.

Salt&pepper noise is often named as impulsive noise. Impulsive noise is very common in digital images. It is always independent of the pixels of the image and is distributed over the image. This type of noise generally damages the digital image by disturbing the pixel elements in camera sensors, insufficient memory space, errors in the digitalization process, and many others. Pixels randomly acquire three values. The pixels will gain values for the color white, black or does not change [5, 7] (Fig. 4).

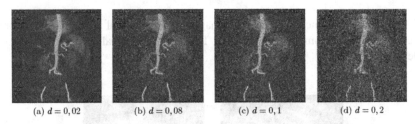

(a) $d = 0,02$ (b) $d = 0,08$ (c) $d = 0,1$ (d) $d = 0,2$

Fig. 4. Analysis of vascular system: contrast MRI image with superposed salt&pepper noise of different levels.

Gaussian noise is also called electronic noise, because it arises in amplifiers or detectors. This noise is caused by natural sources such as thermal vibrations of atoms and discrete disposition of radiation warm objects. Gaussian white noise affects the image by adding a value from normal distribution to each pixel, the normal distribution has a certain scattering, zero mean value, and intensity typically less than the maximum intensity in the image. The noise spectrum is uniform, which means that all image frequencies are equally affected. This is broadband noise [4, 7] (Fig. 5).

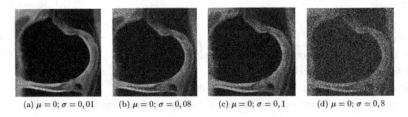

(a) $\mu = 0; \sigma = 0,01$ (b) $\mu = 0; \sigma = 0,08$ (c) $\mu = 0; \sigma = 0,1$ (d) $\mu = 0; \sigma = 0,8$

Fig. 5. Cartilage analysis: image with superposed Gaussian noise with different values μ (mean value) and σ (standard deviation).

Probability distribution of the Rician noise is given by the equation:

$$p_M(M) = \frac{M}{\sigma^2} e^{-\frac{M^2 + A^2}{2\sigma^2}} I_0 \left(\frac{A \cdot M}{\sigma^2} \right) \tag{1}$$

where M is measured intensity of pixels, σ is standard deviation of Gaussian noise in real and imaginary image, A is intensity of pixels in the absence of noise, I_0 is modified zero order Bessel's function of the first type [7, 9, 10].

3.3 Objectivization Parameters of Filtration Efficiency

To evaluate the effectiveness of the filters were used the parameter SNR, which indicate the signal-to-noise ratio, and MSE expressing a mean quadratic error. The SSIM (structural similarity index) is also used to compare output quality, which is based on local variance QILV (quality index) and covariance.

SSIM expresses the structural similarity of two images. The index takes values from 0 to 1, when 1 expresses the same images. For color images, it is usually only calculated in the brightness folder.

QILV is based on comparing the local distribution of scattering between the image and the "gold standard". The index is suitable for better image instability evaluation, so it explicitly focuses on image structure.

The simplest relationship of two metric variables is a linear relationship whose rate can be ascertained correlation coefficient. Filtration was considered based on 2D correlation. Pearson's correlation coefficient (r) takes values from -1 to $+1$, which indicates the perfect linear relationship (negative or positive) [11].

3.4 Design of LMMSE Filter

In statistics and signal processing is estimation of minimal mean quadratic error (MMSE) which minimizes the central quadratic error (MSE), which is a common measure of quality estimation of the measured values of the dependent variable.

LMMSE filter is based on minimizing MSE, works on noise estimation (by the method of statistical moments). Perhaps the best filtration results is achieved when it is assumed, that the signal and the noise are statistically dependent on Rican distribution. Estimation noise of single MRI image is usually done from background pixels (binarization), where the signal is assumed to be zero. The bright pixel component un that background is zero. LMMSE estimate for 2D signal with Rican distribution can be written as:

$$\hat{A}_{ij}^2 = \mathrm{E}\left\{ A_{ij}^2 \right\} + C_{A_{ij}^2 M_{ij}^2} C_{M_{ij}^2 M_{ij}^2}^{-1} \left(M_{ij}^2 - \mathrm{E}\left\{ M_{ij}^2 \right\} \right) \tag{2}$$

where: A_{ij} is an unknown pixel intensity value (i, j), M_{ij} is brightness magnitude of signal, $C_{A_{ij}^2 M_{ij}^2}$ is vector of cross covariance, $C_{M_{ij}^2 M_{ij}^2}$ is covariant matrix.

After simplifying estimation to point with vectors and matrices become scalar values. Modification we can get the shape:

$$\hat{A}_{ij}^2 = E\left\{A_{ij}^2\right\} + \frac{E\left\{A_{ij}^4\right\} + 2E\left\{A_{ij}^2\right\}\sigma_n^2 - E\left\{A_{ij}^2\right\}E\left\{M_{ij}^2\right\}}{E\left\{M_{ij}^4\right\} - E\left\{M_{ij}^2\right\}^2} \left(M_{ij}^2 - E\left\{M_{ij}^2\right\}\right) \quad (3)$$

Assuming local ergodicity, the final listing for LMMSE can be defined as:

$$\hat{A}_{ij}^2 = \left\langle M_{ij}^2 \right\rangle - 2\sigma_n^2 + K_{ij}\left(M_{ij}^2 - \left\langle M_{ij}^2 \right\rangle\right) \quad (4)$$

where:

$$K_{ij} = 1 - \frac{4\sigma_n^2\left(\left\langle M_{ij}^2 \right\rangle - \sigma_n^2\right)}{\left\langle M_{ij}^4 \right\rangle - \left\langle M_{ij}^2 \right\rangle^2} \quad (5)$$

It follows from this equation, that the value σ_n^2 must be properly estimated. This is mostly done from the selected area from the pixels in the background. The performance of the overall is therefore highly dependent on quality estimation of noise variance. However, it does not just work with such a selection, it can also be used as distribution a local second order moment [1, 9].

3.5   Testing on Real Data

Data testing phase can be divided into several steps in MatLab. First, the above types of noise were applied to the loaded data. For each type of noise, it was selected more that one parameter of level noise. Work with cell fields has been selected to simplified the code. Into cell fields were saved individual images. The next phase after was application of LMMSE filter. After filtering, the new images were also stored in a new cellular field, totally two fields were created.

The first field contain images with noise. The field has a size of 8×11 – it contains eight rows and eleven columns. The first column consists of input images, the other columns are images with added noise. Algorithm continues with the cell field always working so that the entire first column is input variable for noise-adding command.

The main part is the analysis and testing of the efficiency of the LMMSE filter for the selected MR image data. For all images is applicated filtration with several values of the parameter W_s, which determines size of mask filter. Selected values W_s were 5, 7, 9, 13, 15 and 17. For all of these filter levels were calculated objectivization parameters SNR, MSE, SSIM, QILV, correlation and their values have been entered into tables for easier evaluation of the level of filtration.

3.6   Filtration for Noise Salt&pepper

Filtration of impulsive noise of type salt&pepper based on the LMMSE filter appears to be ineffective in terms of test results. This filter is ineffective for all mask size values W_s. From this testing, it can be assumed, that the LMMSE filter can not be applied to impulsive noise, as shown in Fig. 6.

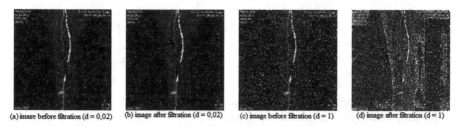

(a) image before filtration (d = 0,02)   (b) image after filtration (d = 0,02)   (c) image before filtration (d = 1)   (d) image after filtration (d = 1)

Fig. 6. Images of the vascular system with the application salt&pepper noise at different levels after filtering with mask W_s

The result corresponds to expectation, because many sources state that for this type of noise is available to use median filter. This is not necessary in more detail to deal with parameter values whose averages are shown in Table 2.

Table 2. Average parameter values for the filter salt&pepper

Parameter	Value
SNR	0,9270 dB
MSE	0,0275
SSIM	0,0710
QILV	0,0175
r	0,1590

3.7   Filtration for Gaussian Noise

The shape of the Rician and Gaussian noise distribution curve is very similar (see Fig. 7) and manifestation of noise LMMSE filter in MR images assumes Rician distribution.

Gaussian noise can be expected to be more successful than salt&pepper. Four levels of Gaussian noise were tested, namely:

1. Mean value 0,1; variance 0,1
2. Mean value 0,1; variance 0,01
3. Mean value 0,5; variance 0,01
4. Mean value 0,05; variance 0,01

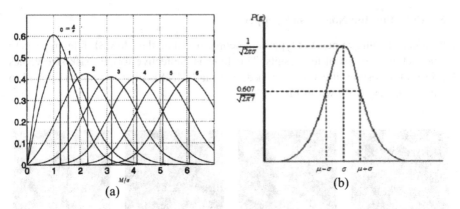

Fig. 7. (a) Rician distribution for different values SNR and $\frac{A}{\sigma}$, (b) PDF (function of the distribution probability) of gaussian noise [4, 10].

Various mask size values have also been tested. Table 4 contains average values of monitored parameters for different noise levels. These specific values were calculated as the averages of the eight filtered images on which they were applied filter with value $W_s = 5$. Parameter SNR is significantly higher than in the salt&pepper. MSE is low, which is as the error is desirable. QILV, SSIM and correlation as parameters with maximal value 1 can also be rate as higher (Table 3).

Table 3. Average parameters for filtration of Gaussian noise

Parameter	Value			
	$\mu = 0,1$ $\sigma^2 = 0,1$	$\mu = 0,1$ $\sigma^2 = 0,01$	$\mu = 0,5$ $\sigma^2 = 0,01$	$\mu = 0,05$ $\sigma^2 = 0,01$
SNR	3,625 dB	8,507 dB	2,036 dB	9,295 dB
MSE	0,039	0,019	0,173	0,019
SSIM	0,135	0,411	0,232	0,396
QILV	0,182	0,661	0,250	0,666
R	0,543	0,822	0,519	0,823

Effective can be defined as filtering with parameters $\mu = 0,1$ and $\sigma^2 = 0,01$ (third column of the table) and $\mu = 0,05$ and $\sigma^2 = 0,01$ (fifth column of the table). The image output of one of the efficient filters is shown in Fig. 8. On the other hand, inefficient filtration is shown in Fig. 9.

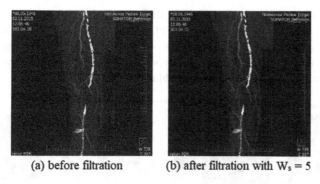

| (a) before filtration | (b) after filtration with $W_s = 5$ |

Fig. 8. Images with noise (Gauss; $\mu = 0,1$ and $\sigma^2 = 0,01$).

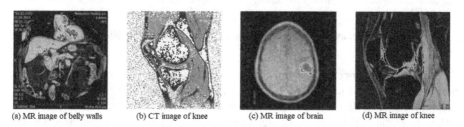

| (a) MR image of belly walls | (b) CT image of knee | (c) MR image of brain | (d) MR image of knee |

Fig. 9. Ineffective filtration of Gaussian noise (average 0,5; variance 0,01).

3.8 Filtration for Rician Noise

MRI images with Rician noise were filtered with values of variance 0,05; 0,08; 0,1 and 0,5. For testing were used different mask sizes of filter (W_s 5, 7, 9, 13, 15 and 17). The resulting parameters are in Table 4. Size of the filter mask were changed and did not make any major changes in average values, therefore only the lowest and highest values are listed [8].

Table 4. Average values of parameters for filtration Rician noise

Parameter		Value			
		$\mu = 0,1$ $\sigma^2 = 0,1$	$\mu = 0,1$ $\sigma^2 = 0,01$	$\mu = 0,5$ $\sigma^2 = 0,01$	$\mu = 0,05$ $\sigma^2 = 0,01$
$W_s = 5$	SNR	11,739 dB	9,982 dB	9,373 dB	1,037 dB
	MSE	0,019	0,019	0,019	0,078
	SSIM	0,538	0,461	0,424	0,116
	QILV	0,647	0,604	0,571	0,172
	r	0,842	0,853	0,857	0,646
$W_s = 17$	SNR	11,299 dB	9,509 dB	8,791 dB	1,750 dB
	MSE	0,019	0,019	0,020	0,067
	SSIM	0,518	0,467	0,452	0,247
	QILV	0,543	0,467	0,415	0,103
	r	0,869	0,857	0,849	0,699

The most efficient filtration was achieved for variance 0,05 and $W_s = 5$, as insufficient on the other hand is filtration with set variance 0.5 and $W_s = 17$, which can be objectively to judge based on parameters above and subjectively according to Fig. 10. Still it is obvious the grain areas, which should be homogeneous, and transitions between different levels of brightness are blurred.

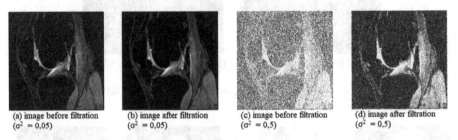

(a) image before filtration ($\sigma^2 = 0.05$)   (b) image after filtration ($\sigma^2 = 0.05$)   (c) image before filtration ($\sigma^2 = 0.5$)   (d) image after filtration ($\sigma^2 = 0.5$)

Fig. 10. Comparison of effective and ineffective filtration of Rician noise.

Increasing variance of noise changes the calculated parameter values. At Rician noise and variance from 0,5 to 1, there are dependencies, which are demonstrated graphs in Fig. 11. With increasingly variance of noise have parameters SNR, SSIM and QILV decreasing character. Parameter values are plotted on the vertical axis. On the other hand MSE value rises, which corresponds to expectation, because MSE is parameter specifying average quadratic error.

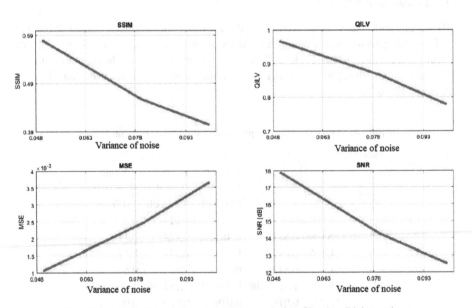

Fig. 11. Independence of each parameters during filtration Rician noise.

Parameters SSIM, SNR, QILV have the most efficient filtration, because their values are the highest and parameter MSE has the lowest value. The charts were crated from the output for the mask size 5. Size of parameters are changed slightly with change of mask size (see Table 4), but the behavior is always the same, so graphs for the other mask size looked very similar.

4 Conclusion

Eight MRI images were tested, each image was infiltrated for ten different levels of noise. Size of mask filter were tested for six different values. The total set obtained after filtration was 480 images of MRI. These images have been evaluated for filtration efficiency.

Since alone the mathematical basis of the filter is based on Rician distribution, it can be assumed that for Rician and Gaussian noise is the best application for the filter. In many sources is also presented the fact that for impulsive noise type of salt&pepper is more convenient to select low frequency filter such as median filter or diameter filter. For Rician noise, also applies, that the quality of filtration decreases with increasing variance.

If variance above 0,1, filtering is significant and the images are already too degraded. For Gaussian noise was filtration is achievable at variance of 0,01 and mean [0,05; 0,1].

Acknowledgment. The work and the contributions were supported by the project SV4508811/2101Biomedical Engineering Systems XIV'. This study was also supported by the research project The Czech Science Foundation (GACR) 2017 No. 17-03037S Investment evaluation of medical device development run at the Faculty of Informatics and Management, University of Hradec Kralove, Czech Republic. This study was supported by the research project The Czech Science Foundation (TACR) ETA No. TL01000302 Medical Devices development as an effective investment for public and private entities.

References

1. Aja-Fernandez, S.: Selected papers on statistical noise analysis in MRI: LPI. Universidad de Valladolid, Springer International Publishing, Valladolid (2015)
2. Novozámský, A.: NPGR032 – CVIČENÍ III: Šum a jeho odstranění – teorie & praxe. Department of Image Processing. Institute of Information Theory and Automation of the ASCR, Praha (2015)
3. Patidar, P., Gupta, M., Sprivastava, S., Nagawat, A.K.: Image de-noising by various filters for different noise. Int. J. Comput. Appl. **9**(4), 45–50 (2010)
4. Pikora, J.: Implementace grafických filtrů pro zpracování rastrového obrazu. Bakalářská práce. Masarykova univerzita, Brno (2008)
5. Roy, V.: Spatial and transform domain filtering method for image de-noising: a review. Int. J. Mod. Educ. Comput. Sci. **5**(7), 41–49 (2013)

6. Varghese, J., Khan, M.S., Siddappa, M., Subash, S., Ghouse, M., Hussain, O.B.: Efficient adaptive fuzzy-based switching weighted average filter for the restoration of impulse corrupted digital images. IET Image Process. **8**(4), 199–206 (2014)
7. Kaur, G., Kumar, R., Kainth, K.: A review paper on different noise types and digital image processing. Int. J. Adv. Res. Comput. Sci. Softw. Eng. **6**(6), 562–565 (2016)
8. Roy, V.: Spatial and transform domain filtering method for image de-noising: a review. Int. J. Mod. Educ. Comput. Sci. **5**(7), 41–49 (2013)
9. Aja-Fernandez, S., Alberola-Lopez, C., Westin, F.: Noise and signal estimation in magnitude MRI and Rician distributed images: a LMMSE approach. IEEE Trans. Image Process. **17**(8), 1383–1398 (2008)
10. Gudbjartsson, H., Patz, S.: The Rician distribution of noisy MRI data. Magn. Reson. Med. **36**(2), 332–333 (1996)
11. Walek, P., Lamoš, M., Jan, J.: Analýza biomedicínských obrazů. VUT, Brno (2013)
12. Kubicek, J., Augustynek, M., Vodakova, A., Penhaker, M., Cerny, M., Oczka, D.: Segmentation and modeling of scattered RTG irradiation on quality of skiagraphy images in clinical conditions. Paper presented at the 2017 IEEE conference on big data and analytics, ICBDA 2017, 2018-January, pp. 105–110. The Rician distribution of noisy MRI data (2018)
13. Bryjova, I., Kubicek, J., Molnarova, K., Peter, L., Penhaker, M., Kuca, K.: Multiregional segmentation modeling in medical ultrasonography: extraction, modeling and quantification of skin layers and hypertrophic scars. In: Nguyen, N.T., Papadopoulos, G.A., Jędrzejowicz, P., Trawiński, B., Vossen, G. (eds.) ICCCI 2017. LNCS (LNAI), vol. 10449, pp. 182–192. Springer, Cham (2017). https://doi.org/10.1007/978-3-319-67077-5_18
14. Kubicek, J., et al.: Automated extraction and modeling of calcifications and blood stream volumetric parameters. Front. Artif. Intell. Appl. **297**, 256–269 (2017)

A Nearest Neighbour-Based Analysis to Identify Patients from Continuous Glucose Monitor Data

Michael Mayo$^{(\boxtimes)}$ and Vithya Yogarajan

Department of Computer Science, University of Waikato, Hamilton, New Zealand
{michael.mayo,vyogaraj}@waikato.ac.nz

Abstract. Continuous glucose monitors (CGMs) are minimally invasive sensors that detect blood glucose levels (usually in patients with diabetes) at high frequency. The devices produce considerable volumes of sensor data when used for weeks and months. We consider the following research question: is it possible to uniquely identify a patient from a fragment of their CGM data? That is, supposing a patient's medical records are stored in a database along with a large sample of their CGM data, could an attacker with a much smaller sample of data from a different time period match the two time series and positively identify the patient? If the answer is yes, then significant patient privacy concerns are raised since many health records are now stored online. Our investigations using existing public CGM datasets reveal that many subjects can be uniquely identified using a simple nearest neighbour-based analysis approach.

Keywords: Continuous glucose monitors · Diabetes ·
Nearest neighbour analysis · Time series data · Privacy ·
Medical Internet of Things · Data security

1 Introduction

In this paper we use techniques from machine learning, information retrieval and statistics to investigate the question of whether is it possible to uniquely identify a patient with diabetes from a sample of his or her blood glucose data recorded using a continuous glucose monitor (CGM) device.

A sample of data from such a device is given in Fig. 1. The figure outlines in red the desired target zone for a healthy individual's blood glucose, the point of least risk (given by the central red line), and boundaries indicating moderate hypo- or hyperglycemia [6]. This figure shows blood glucose fluctuations characteristic of diabetes sampled by a CGM every five minutes.

We consider the scenario of an attacker gaining access (authorized or unauthorized) to a fragment of CGM data and using it to retrieve other medical records in a database by matching the attacker's CGM data fragment to larger

© Springer Nature Switzerland AG 2019
N. T. Nguyen et al. (Eds.): ACIIDS 2019, LNAI 11432, pp. 349–360, 2019.
https://doi.org/10.1007/978-3-030-14802-7_30

samples of the same type of data stored in the database. The database may be public and de-identified or it may be private and have been compromised by the attacker.

If the attacker knows the identity of the victim, then patient privacy could potentially be compromised since CGM data is likely linked to other electronic health record (EHR) data in a comprehensive medical database.

The investigation presented in this paper is novel and as far as we are aware this issue has not been considered to date. CGM technology is a relatively recent development, and explorations of the potential misuse of data generated by these devices has not been explored in the literature.

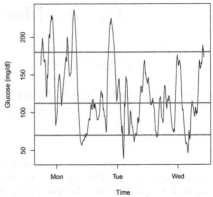

Fig. 1. Fragment of the blood glucose data from a subject consisting of 700 continuous samples over 2–3 days. (Color figure online)

2 Background

2.1 Wearable CGM

Wearable CGM devices provide non-invasive or minimally invasive blood glucose readings every 1–5 min [4,5]. This is in distinct contrast to the self-monitoring of blood glucose (SMBG) approach (involving a lancet device for drawing a blood sample) where samples are typically taken only a few times a day. Most current commercial CGMs use a needle sensor inserted into the subcutaneous tissue that measures the electric signal generated by the glucose-oxidase reaction. This signal is converted into an estimate of the patient's interstitial glucose concentration [4], which in turn is a reasonable estimate of the blood glucose concentration (subject to a varying time lag).

First generation commercial CGMs have been available since 2005 but suffered from lower accuracy than SMBG [4]. Recent developments in the technology include greater accuracy, greater user comfort, longer wear time, and the reduced need for regular calibration [1,4,10]. Such advances have led to CGM devices being more widely accepted by patients and doctors as a means of making clinical decisions. The most accurate devices may now potentially be used safely as the sole means of glucose monitoring [10]. Other recent developments include efforts to develop non-invasive wearable devices that can continuously monitor glucose concentrations via the skin, sweat, breath, saliva and ocular fluid [2].

Medically, use of CGM devices has been shown to improve glycemic (i.e. blood glucose concentration) control in both patients utilising continuous insulin pumps as well as those following a multiple daily insulin injection regime. These devices can be used by patients with both type 1 and type 2 diabetes, as well as pregnant and hospitalized patients [10].

CGM has been shown to reduce the number of hypoglycemic episodes in patients at high risk of hypoglycemia, but not reduce the risk of severe hypo-glycemic episodes that may lead to seizure, coma or hospital admissions. The reasons for this are currently not clear, but this may in part be because the frequency of severe hypoglycemic episodes are linked to factors other than blood glucose levels alone [10]. Therefore linking CGM data to a larger dataset of more general EHR data may be invaluable for research, despite the security risks.

CGM has other potential applications in diabetes treatment and research. Masked (or professional mode) CGM provides data to doctors so they can mon-itor compliance or response to therapy, but not to patients (who might alter their normal behavior based on real-time feedback) [10]. A minimum of 14 days of monitoring appears needed to establish a good profile of the patients' activities and blood glucose profiles [10].

CGM is also used synergistically with other technologies to improve glycemic control. For example, CGM can be one of the components of an "artificial pan-creas" system that actively regulates blood glucose through the controlled deliv-ery of insulin and/or glucagon [10]. CGM data combined with an intelligent driven decision support system (and a telemedicine system that allows medical supervision) can allow patients to modify their insulin doses based on glucose predictions [9].

In contrast to HbA1c (which shows long term glycemic control) and regular SMBG (which shows the level at the time of monitoring), CGM can also pro-vide data on nocturnal patterns, and the duration, severity and frequency of previously undetected hypo- and hyper-glycemic episodes [10]. This makes these devices extremely useful in evaluating the effects of new drugs in clinical trials.

2.2 Interpreting CGM Measurements of Blood Glucose Control

The standard blood glucose range (approximately 20 to 600 mg/dl) is not symmetrical. Moderate hypoglycemia is defined as $b_t < 70$ mg/dl where b_t is the blood glucose level at time t, while moderate hyperglycemia is defined as $b_t > 180$ mg/dl. The range considered clinically normal (around 112.5 mg/dl) is therefore not the same as the middle of the interval of possible blood glucose readings, which is approximately 310 mg/dl. This fact potentially makes para-metric statistics inappropriate, which may have clinical implications if statistical methods are used that make normality assumptions about the data [6,10].

In order to account for this, therefore, [6] proposed an appropriate logarithmic transform mapping blood glucose sensor readings to the so-called *risk domain*. The risk domain transform [7] is defined as follows, where r_t denotes the risk domain value and b_t is a raw blood glucose sensor reading in mg/dl:

$$r_t = 1.509(\ln(b_t)^{1.084} - 5.381) \tag{1}$$

Following the risk domain transform, the point of lowest risk is defined as $r_t = 0$, and the target range for healthy blood glucose levels is transformed to approximately $[-0.9, 0.9]$.

Research shows that low risk domain values predict the incidence of severe hypoglycemia better than low blood glucose readings and HbA1c. Similarly, high risk values correlate to HbA1c, which can be used to assess hyperglycemic control [6].

2.3   Datasets

This study uses datasets from the following three sources: Mauras *et al.* (2011) [8], Buckingham *et al.* (2013) [3] and The Nightscout Foundation (2014) [11]. The Mauras *et al.* [8] dataset consists of CGM data from young children aged 4 to 9 years with Type 1 diabetes who were part of a randomised clinical trial. The Buckingham *et al.* [3] dataset consists of CGM data from a group of carefully chosen individuals who went through a clinical trial. Inclusion criteria for Buckingham *et al.* [3] included age 6 to <46 years, clinical diagnosis of type 1 diabetes, and initiation of insulin therapy within seven days prior to the trial. The Nightscout [11] dataset is a set of CGM data from type 1 diabetic subjects who have shared their data for the purposes of research. We use the version of the Nightscout dataset dated 14th November 2017.

Table 1. Datasets used in the evaluation study, following removal of subjects with less than 5,000 samples.

Dataset	#Subjects	Samples/subject (min; median; max)
Mauras *et al.* [8]	127	(5,478; 21,827; 71,359)
Buckingham *et al.* [3]	63	(5,246; 66,062; 237,114)
Nightscout [11]	18	(13,693; 54,644; 208,410)

Table 1 gives a summary of the datasets. The CGM monitor readings for all three datasets have a frequency of five minutes. However, the length of the each time series varies between datasets and individual subjects within datasets due to differences in device usage. For this study, each sensor reading from a CGM is referred to as one "sample". The length of a CGM trace will be reported as a number of samples, since CGM device usage is generally intermittent and therefore two different time series with the same number of samples will likely cover different amounts of absolute time. We have excluded data from the patients with less than 5,000 samples. Table 1 provides the minimum, maximum and median of the samples per subject of the datasets after removal of the excluded subjects.

3   Methodology

3.1   Pre-processing of Dataset

As explained in Sect. 2.2, the standard glucose range is not symmetric. Figure 2 depicts a frequency histogram of raw blood glucose samples for the same subject whose data is depicted in Fig. 1. It is evident that the data is left-skewed with unevenly spread hypo- and hyperglycemia readings.

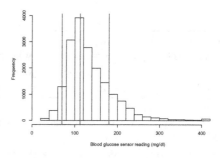

Fig. 2. Blood glucose readings from a CGM for one subject from Nightscout [11] dataset, $n = 17{,}892$.

Figure 3 presents the same data, but after the risk domain transform defined by Eq. 1. Clearly the transformed data more closely approximates a normal distribution. We therefore applied this risk domain transform to all of the data before continuing.

3.2   Experimental Methodology

We set out to show that if an attacker holds a fragment of a victim's CGM data disjoint from the main portion of the victim's CGM data (which might be stored in an online EHR database), then it is possible to unmask the specific victim's identity or at very least retrieve the victim's other medical records associated with their CGM data.

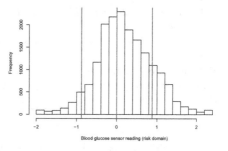

Fig. 3. Same data as shown in Fig. 2, but following the risk domain transformation given in Eq. 1.

To achieve our goal, we use a nearest neighbor analysis approach inspired by the K-Nearest Neighbour algorithm from machine learning. In our approach, a privacy attack is "simulated" by iteratively picking one subject in the current database to be the victim. On each iteration, the attacker attempts to correctly retrieve the victim's full CGM time series by matching the query fragment with every other CGM trace in the database. The other CGM traces are then ranked by similarity to the attacker's fragment. Clearly, if the attack is successful, then the victim's CGM data will be the top ranked match; on the other hand, if the attack is only partially successful or unsuccessful, then the victim's CGM data will be poorly ranked.

Statistically, the results of such an approach (i.e. the averaging matching rank across all subjects in a database) should give an expected estimate of how difficult it is to match disjoint CGM data from the same individual using a nearest neighbour technique. More formally, our experimental methodology is outlined in Algorithm 1.

Algorithm 1. Experimental Methodology. Experimental parameters are (i) a database $DB = \{S^{(1)}, S^{(2)}, \ldots S^{(n)}\}$ where $S^{(i)}$ is the CGM time series in the risk domain belonging to the ith subject; (ii) a distance function $Dist$ designed for fixed-length vectors; and (iii) a time delay gap.

---

for all $S^{(i)} \in DB$ **do**

　Let $s_1^{(i)}, s_2^{(i)}, \ldots s_t^{(i)}$ be individual CGM samples belonging to the current subject $S^{(i)}$.

　$T^{(i)} = s_1^{(i)}, s_2^{(i)}, \ldots s_{t-1000-gap}^{(i)}$ is the database version of the current subject's CGM data.

　$Q = s_{t-999}^{(i)}, s_{t-998}^{(i)}, \ldots s_t^{(i)}$ is the attacker's fragment of the current subject's CGM data

　Temporarily replace $S^{(i)}$ in DB with $T^{(i)}$.

　for all $S^{(j)} \in DB, j \neq i$ **do**

　　Set $d^{(j)} = Dist(f(S^{(j)}), f(Q))$, where f is a function to extract a fixed number of features from a time series of sensor readings.

　end for

　Rank elements in DB by distance to Q (where rank 1 is best, and rank $|DB|$ is worst).

　Record the ranking $rank^{(i)}$ of the target time series $T^{(i)}$ according to the given distance function.

　Delete $T^{(i)}$ from DB and put back into DB the original $S^{(i)}$ for the next round.

end for

Return the list of rankings $rank^{(1)}, rank^{(2)}, \ldots rank^{(|DB|)}$, where each individual's rank will be a reflection of how easy/difficult is was to match that individual's CGM data.

The basic idea is to iteratively perform the following for each subject in each database: firstly, extract their CGM data. Next, split the time series into a database portion and a portion held by the attacker, after accounting for a time delay. We assume that the attacker holds only the last 1,000 samples from the time series, so this data and the samples occurring during the time delay are temporarily deleted from the database, but they will be put back for the next iteration. We then match the attacker's portion of the time series with each subject's time series in the database. When this is done, the subjects in the database can then be ranked by similarity of their data to the attacker's fragment. The rank of the victim (i.e. the true owner of the attacker's fragment) can then be determined. This rank will either be 1, indicating a perfect match, or a number between 2 and the number of subjects in the database, indicating an increasingly imperfect match.

To illustrate the methodology, consider a simple database with CGM data from three hypothetical subjects. The database is $DB = \{S^{(1)}, S^{(2)}, S^{(3)}\}$, with lengths in samples of 5,000, 10,000 and 15,000 respectively. Suppose we wish to determine if an attacker can unmask the second subject. Assuming a time delay of 3,000 samples, we take the second subject's original data and split it such that the first 6,000 samples are assigned to $T^{(2)}$, the next 3,000 samples are discarded, and the last 1,000 samples are assigned to Q, the attacker's fragment. The database DB is temporarily changed to $DB = \{S^{(1)}, T^{(2)}, S^{(3)}\}$ so that Q does not exist in DB. Upon matching Q to each element in DB, a set of distances will be calculated. The distances, for example, may be 0.15, 0.1 and 0.09. Thus, the attacker's query Q is a best match for subject 3's CGM data, which is incorrect. The actual owner of Q is subject 2, whose full CGM time series data has rank 2. Further analysis by the attacker (e.g. filtering the rankings by demographic information such as gender) may improve the ranking of the victim, perhaps even enabling the attacker to go from nearly a perfect match to a perfect match if such information is available.

Table 2. Features computed from risk domain time series. For features x_5–x_{14}, the percentiles used are 0.9, 0.75, 0.5, 0.25 and 0.1.

Feature	Description
x_1	Median risk domain value
x_2	Risk domain interquartile range
x_3	Fraction of samples with $r_t \leq -0.9$
x_4	Fraction of samples with $r_t \geq 0.9$
$x_{5,6,7,8,9}$	Percentiles from sample distribution where $r_t < 0$
$x_{10,11,12,13,14}$	Percentiles from sample distribution where $r_t > 0$
x_{15}	Fraction of samples s.t. $r_t \geq 0.9$ but $r_{t-1} < 0.9$
x_{16}	Fraction of samples s.t. $r_t \leq -0.9$ but $r_{t-1} > -0.9$

In order to implement our methodology, we require a function f for computing a fixed number of features from a varying-length time series of sensor readings, and a distance function $Dist$ for matching features. Table 2 gives details of the features that we compute. Briefly, feature x_1 gives the median risk domain value over the entire time series and can be considered a rough proxy of HbA1c after the risk transform. Feature x_2 is a measure of the general volatility of the blood glucose levels. Features x_3 and x_4 capture the general amount of time spent in the hypo/hyperglycemic ranges while features x_5 to x_{14} are an attempt to be more specific about which parts of the ranges are visited. For example, if a patient has relatively frequent and intense excursions into the hyperglycemic range, then the value for x_{14} will be high compared to other patients. Finally, features x_{15} and x_{16} are an attempt to measure the number of excursions into the hyper- or hypoglycemic ranges, and exploit the notion that -0.9 and 0.9 are boundaries in the risk domain and therefore crossing them is significant.

These features can therefore be thought of as another means of measuring the volatility of the blood glucose levels in addition to x_2.

After computing the raw features, we standardized them across all subjects in the dataset (excluding the attacker's fragment Q) such that all features have zero mean and unit standard deviation. Two distance metrics were then employed to measure the similarity between attacker and database fragments: Euclidean $(Dist(\mathbf{x}, \mathbf{y}) = \sqrt{\sum_i (x_i - y_i)^2})$ and Manhattan $(Dist(\mathbf{x}, \mathbf{y}) = \sum_i |x_i - y_i|)$, both of which are widely known distance metrics.

In our initial set of experiments, we carried out the methodology described in Algorithm 1 with (i) the three different CGM datasets described in the previous section; (ii) the two distance metrics defined above; and (iii) two different time delays: 0 samples and 3,000 samples. Varying the time delay was of interest to use so that we could assess the impact of the most recent samples on the matching precision. These three experimental parameters with their different settings led to $3 \times 2 \times 2 = 12$ different configurations. We did not mix patients from different datasets as we did not want to inadvertently end up exploiting general differences between the datasets.

After these initial experiments, we further performed experiments to explore the impact of the attacker having demographic information about the victim as well, and we also considered a longer time delay scenario.

In all experiments, we recorded the median rank of the victim's CGM data after the attack. Since in many cases the victim's CGM data is not the top ranked time series, we also considered the precision@K metric for three values of K, namely 1, 5 and 10. Differences between the distance metrics, and the effects of the time delay, were assessed using a Wilcoxon signed rank test for statistical significance.

4   Evaluation

4.1   Initial Experiments

Following execution of our first round of experiments, Table 3 provides the median rank over all subjects in a given dataset. For each dataset, results are presented for both no time delay and with a time delay. It is evident that, even with only a small fragment of the CGM data held by an attacker, the median rank across all datasets is reasonably high. This indicates that patient glycemic variability is reasonably unique, a fact that a privacy attacker could take advantage of.

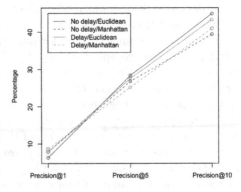

Fig. 4. Precision results for the Maurus et al. [8] dataset.

A statistical test was performed comparing the rankings across all subjects against those that would be expected if the rankings were random. The p values as shown in the table are all extremely small indicating a strongly significant effect.

Table 3. Median rank over all subjects for a target subject's CGM data following a query by an attacker who holds a disjoint query fragment from the same subject. Results are given by time delay (i.e. number of samples between the database copy and the query) and the distance metric used for matching. The p value is computed using a one sample Wilcoxon signed rank test comparing the ranking against the expected rank obtained if the query results were random.

Dataset	Delay (samples)	Distance metric	Median rank	p
Mauras *et al.* [8]	0	Euclidean	12/127	9.65E−20
	0	Manhattan	13/127	2.38E−20
	3,000	Euclidean	12/127	1.49E−17
	3,000	Manhattan	13/127	1.44E−18
Buckingham *et al.* [3]	0	Euclidean	17/63	1.91E−06
	0	Manhattan	13/63	4.37E−07
	3,000	Euclidean	17/63	4.09E−05
	3,000	Manhattan	15/63	7.73E−06
Nightscout [11]	0	Euclidean	1.5/18	3.21E−04
	0	Manhattan	3/18	2.93E−04
	3,000	Euclidean	1.5/18	3.60E−04
	3,000	Manhattan	3/18	3.18E−04

Figures 4, 5 and 6 present the precision results for the three datasets. It is evident from these more detailed plots that for a small number of subjects, their CGM data is so distinctive that they can be easily uniquely identified. For example, for the smaller Nightscout foundation [11] datasets (see Fig. 6), 33% or 39% of subjects were perfectly matched depending on the matching method. For the larger datasets, the perfect match rate drops to between 6% to 16%. However, if we consider whether or not the correct match occurs in the top 10 ranked places then even for the large datasets the precision improves considerably to 35–45%.

Table 4. Result of six Wilcoxon signed rank tests comparing distance metrics (Euclidean vs. Manhattan) with and without the 3,000 sample time delay.

Dataset	p (no time delay)	p (time delay)
Maurus et al.	0.86	0.57
Buckingham et al.	0.48	0.33
Nighscout	0.48	0.48

Table 5. Result of three Wilcoxon signed rank tests comparing time delay (0 vs. 3,000 samples) using the Euclidean distance metric.

Dataset	p
Maurus et al.	0.10
Buckingham et al.	8.69E−4
Nightscout	0.16

We next compared the differences in rankings due to the choice of distance metric. Table 4 presents the p values of the Wilcoxon signed rank tests comparing the Euclidean distance metric-produced rankings to that of the Manhattan metric-produced rankings, both with and without a time delay. The table shows no significant difference between distance metrics when the time delay is constant, since all p values are >0.05.

Table 5 presents the p values when comparing the time delay with no time delay for Euclidean distance metric. It is evident that there is an effect on the rankings due to time delay for the Buckingham *et al.* [3] dataset, but no effect (if $p \leq 0.05$ is considered the criteria for significance) for the other datasets.

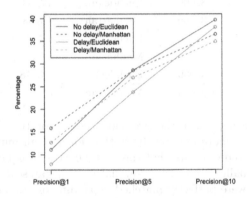

Fig. 5. Precision results for the Buckingham [3] dataset.

4.2   Further Experiments

In our next round of experiments, we explored the effects of longer time delays and filtering using demographic information, in the event that the attacker has access to it. In these experiments, we focus on the single largest dataset with 127 subjects.

A longer time delay of 21,000 samples was considered for subjects with more than 21,000 samples. Table 6 presents median rank over all subjects in the Mauras *et al.* [8] dataset, following a query by an attacker with a fragment of 1,000 samples but a time delay from the database copy of 21,000 samples instead of 3,000 samples, which we considered previously. A statistical test comparing the rankings produced after using the two distance metrics did yield a significant difference between the metrics this time, with a p value of 4.15E−3. Therefore we can conclude that when the time delay is long, using Manhattan distance rather then Euclidean distance allows a more effective attack.

We next explored the effects on the ranking precision by filtering the attacker's query results by gender. Table 6 also presents the median rank over all subjects in the Mauras et al. [8] dataset, following an attack where the results are filtered by gender, for both male and female subjects separately. Compared to the results presented in Table 3 for the same dataset, it is evident the median rank improves dramatically. As previously, the p values comparing the mean rank against the expected rank if the query results were random are very small.

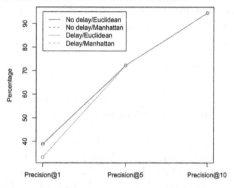

Fig. 6. Precision results for the Nightscout [11] dataset.

Table 6. Further results additional to those in Table 3 but for three subsets of the Mauras et al. [8] dataset instead.

Dataset	Delay (samples)	Distance metric	Median rank	p
Subjects with >21,000 samples	21,000	Euclidean	12/64	1.46E−04
	21,000	Manhattan	9.5/64	1.01E−05
Males subjects	0	Euclidean	6/69	4.93E−12
	0	Manhattan	7/69	5.36E−12
	3,000	Euclidean	8/69	1.90E−11
	3,000	Manhattan	6/69	1.49E−11
Female subjects	0	Euclidean	5/58	4.35E−09
	0	Manhattan	6/58	1.30E−09
	3,000	Euclidean	5/58	1.24E−07
	3,000	Manhattan	5.5/58	1.45E−08

Finally, we also examined the precision@K values and found similar trends of precision increase as K is increased. The increases under gender filtering were quite striking especially for the precision@10 metric which reached values in the 50–60% range. The plots for these experiments, however, are omitted due to space restrictions.

5 Conclusion

We have shown that an attacker possessing even a relatively small fragment of CGM data can unmask some patients' identities in a larger set of EHRs with relatively high precision. The accidental or intentional release of even a minor amount of CGM data (half a week's worth of recording could yield the 1,000 CGM sensor used in this paper) may pose privacy and identity theft risks, given a sufficiently motivated attacker.

References

1. Bailey, T.S., Chang, A., Christiansen, M.: Clinical accuracy of a continuous glucose monitoring system with an advanced algorithm. J. Diabetes Sci. Technol. **9**(2), 209–214 (2014)
2. Bruen, D., Delaney, C., Florea, L., Diamond, D.: Glucose sensing for diabetes monitoring: recent developments. Sensors **17**(8), 1866 (2017)
3. Buckingham, B., et al.: Effectiveness of early intensive therapy on β-cell preservation in type 1 diabetes. Diabetes Care **36**, 4030 (2013)
4. Cappon, G., Acciaroli, G., Vettoretti, M., Facchinetti, A., Sparacino, G.: Wearable continuous glucose monitoring sensors: a revolution in diabetes treatment. Electronics **6**(3), 65 (2017)
5. Hassanalieragh, M., et al.: Health monitoring and management using Internet-of-Things (IoT) sensing with cloud-based processing: opportunities and challenges. In: 2015 IEEE International Conference on Services Computing (SCC), pp. 285–292. IEEE (2015)
6. Kovatchev, B.P., Cox, D.J., Gonder-Frederick, L.A., Clarke, W.: Symmetrization of the blood glucose measurement scale and its applications. Diabetes Care **20**(11), 1655–1658 (1997)
7. Kovatchev, B.P., Otto, E., Cox, D., Gonder-Frederick, L., Clarke, W.: Evaluation of a new measure of blood glucose variability in diabetes. Diabetes Care **29**(11), 2433–2438 (2006)
8. Mauras, N., et al.: A randomized clinical trial to assess the efficacy and safety of real-time continuous glucose monitoring in the management of type 1 diabetes in young children aged 4 to <10 years. Diabetes Care **35**, 204 (2011)
9. Pérez-Gandía, C., et al.: Decision support in diabetes care: the challenge of supporting patients in their daily living using a mobile glucose predictor. J. Diabetes Sci. Technol. **12**(2), 243–250 (2018)
10. Rodbard, D.: Continuous glucose monitoring: a review of recent studies demonstrating improved glycemic outcomes. Diabetes Technol. Ther. **19**(S3), S25 (2017)
11. The Nightscout Foundation: About the Nightscout Data Commons on Open Humans (2014). http://www.nightscoutfoundation.org/data-commons/

PCA Kernel Based Extreme Learning Machine Model for Detection of NS1 from Salivary SERS Spectra

Nur Hainani Othman[1], Khuan Y. Lee[1,2(✉)],
Afaf Rozan Mohd Radzol[1,2], Wahidah Mansor[1,2],
and N. A. Z. M. Zulkimi[1]

[1] Faculty of Electrical Engineering,
Universiti Teknologi MARA, Shah Alam, Selangor, Malaysia
leeyootkhuan@salam.uitm.edu.my
[2] Computational Intelligence Detection RIG,
Pharmaceutical and Lifesciences Communities of Research,
Universiti Teknologi MARA, Shah Alam, Selangor, Malaysia

Abstract. Use of NS1 as a biomarker in saliva has led to the non-invasive and early detection of Flaviviridae related diseases. Saliva is preferred as medium of detection because of its advantages such as non-invasive, painless and easy to collect. Work here intends to compare the performance of KELM classifier with linear and RBF kernels for classification of NS1 from salivary SERS spectra. Prior to KELM, PCA with different termination criteria (Cattle Scree test, CPV and EOC) are used to extract important features and reduce the dimension of SERS spectra dataset. Regularization coefficient (C-value) for linear kernel and Regularization coefficient (C-value) and γ-value for RBF kernel are varied to find the optimum KELM classifier model. For linear kernel, 100% accuracy, precision, sensitivity, specificity is achieved for Linear model with EOC criterion and C-value set to 0.1, 0.2, 0.5, 1 and 2. For RBF kernel, 100% performance of accuracy, precision, sensitivity and specificity is achieved with RBF model with EOC criterion and values of 0.04, 0.06, 0.08, 0.1, 0.2, 0.4, 0.6, 0.8 and 1. The C-value is fixed to 1. The best Kappa value of 1 is obtained when all performance indicators scored 100%. For both Linear-KELM and RBF-KELM, EOC termination criterion gives the highest performance. It also observed that KELM classifier is data dependent.

Keywords: NS1 · Saliva · SERS · PCA · KELM ·
Linear kernel and RBF kernel

1 Introduction

Dengue is known as arboviruses (arthropod-borne viruses) that mostly found in countries that have tropical and sub-tropical climate. It is spread by two types of Aedes mosquitoes; Aedes aegypti and Aedes albopictus (Asian tiger mosquito) [1]. Dengue virus (DENV) is classified as genus of Flavivirus in the family of Flaviviridae, which also includes the West Nile, Japanese encephalitis, Yellow fever and Tick-borne

© Springer Nature Switzerland AG 2019
N. T. Nguyen et al. (Eds.): ACIIDS 2019, LNAI 11432, pp. 361–372, 2019.
https://doi.org/10.1007/978-3-030-14802-7_31

encephalitis virus [2]. DENV consists of four serotypes namely DENV1, DENV2, DENV3 and DENV4. Malaysia is placed as ninth most dengue-infected country as reported by WHO. Under Ministry of Science, Technology and Innovation (MOSTI), Malaysian Remote Sensing Agency (MRSA) [3] Malaysia has recorded 22, 564 cases from the January until May 2018 with 49 fatal cases.

There are several methods used to detect the dengue virus, yet the most common is using enzyme-linked immunosorbent assay (ELISA) based on detection of dengue-specific antibodies; Immunoglobulin M (IgM) or Immunoglobulin G (IgG). Yet, this method has several limitations such as prolonged time to seroconversion (3 to 7 days) thus it decreases the feasibility to detect dengue in acute phase.

In recent years, many studies have been conducted to acknowledge the detection of non-structural protein 1 (NS1) antigen as a potential alternative biomarker using various methods such as ELISA [4], RT-PCR [5], lab-on-a-disk [6], circular dichroism [7] and immunochromatographic rapid test [8]. NS1 is encoded in ribonucleic acid (RNA) of Flavivirus genome and made up of structural and nonstructural proteins. During acute phase (day 1 up to day 9), NS1 is found in the blood serum of dengue patients and therefore, NS1 has been categorized as an early biomarker for Flavivirus infection diseases. By looking on the advantages of saliva which have been described in [9], saliva is seems suitable to be used as a medium of detection that circumvent the disadvantages of blood.

Raman spectroscopy is a specific spectroscopic technique which is capable to produce unique molecular fingerprint of a molecule based on the vibration of molecular structure. The Raman signal is noticed to be very weak since Raman scattering phenomenon occurrence is very rare comparative to Rayleigh scattering [10]. Surface-Enhanced Raman Spectroscopy (SERS) is an enhanced technique of Raman spectroscopy with ability to produce stronger Raman scattering. By using noble metals e.g. gold (Au), silver (Ag) and copper (Cu) as substrate, normally, SERS can amplify the Raman signal with amplification from 10^4 to 10^6. With special substrate, the amplification can be high as 10^8 to 10^{14}. The use of SERS has been proved in other diseases such as urinary tract infections [11], acquired immune deficiency syndrome (AIDS) [12], lung cancer [13], bacterial meningitis [14] and hepatocellular carcinoma (HCC) [15].

Feature extraction method known as Principal Component Analysis (PCA) is a technique introduced by Karl Pearson in 1901 that used to decompose data with large dimension into smaller dimension while extracting important features from the data [16]. The new uncorrelated with lower dimension variables produced by PCA is known as principal components (PCs). Extreme Learning Machine (ELM) classifier has been successfully applied in detection of diseases such as breast cancer [17], heart disease [18], Parkinson's disease [19], pathological brain disease [20], diabetic molecular edema disease [21] and, Amyotrophic Lateral Sclerosis (ALS) and myopathy disease [22]. However, no attempt has been reported on classification of dengue fever based on the presence of NS1 feature from salivary Raman spectra.

In this study, PCA and ELM are used as feature extraction and classification method for classification of dengue fever from salivary Raman spectra. Simulated saliva samples of dengue infected patients are produced by mixing NS1 protein with saliva of healthy volunteer. The mixtures are named as NS1 adulterated saliva samples and are used to evaluate the performance of PCA-ELM algorithm developed for this study. To the best of our knowledge, the methodology proposed in this study is novel. Section 2 elaborates in detail on theoretical background of PCA and ELM. Section 3 presents the methodology which includes dataset, termination criteria of PCA and selection of the kernel parameters. Section 4 presents the performance of KELM classifiers for each termination criteria in term of accuracy, sensitivity, specificity, precision and Kappa value.

2 Theoretical Background

2.1 Principal Component Analysis

It is very important to estimate the number of PCs to be retained because it will summarize important information resides within the dataset. Underestimation will cause the loss of information while overestimation will include the unnecessary information with bigger data dimension. Thus, termination criteria can be used as guidelines to select the number of PCs suitable for the analysis. There are many termination criteria available in literature, amongst the popular are Kaiser, Cattle scree test and Cumulative Percent of Variance (CPV) [23, 24]. Kaiser's stopping rule or known as Eigenvalue-One-Criterion (EOC) is a procedure that includes all PCs with eigenvalues larger than 1 and discards PCs with smaller eigenvalues [23]. Cattell's scree test uses Scree plot to estimate the non-trivial PCs to be retained. Scree plot is a plot PCs versus its' eigenvalues. The number of non-trivial PCs to be retained is corresponding to PCs where a sharp changes in slope (known as elbow) is observed on the Scree plot [24]. CPV is a criterion that preserves all PCs within the appointed threshold or higher. Typically, the threshold value for CPV is in the range of 70% to 99% of cumulative variance [16].

2.2 Kernel Extreme Learning Machine

ELM has been developed by Huang et al. in 2004 [25]. It uses single hidden layer feed-forward network (SLFN) and with advantages of better generalization, faster learning speed and less human intervention. Unlike the conventional algorithm such as Back Propagation (BP) which requires initial setting of learning parameters, the only parameter that need to be specified in ELM is the number of hidden nodes. The other learning parameters of the hidden nodes such as input weight and biases are automatically specified by the ELM algorithm [26]. Kernel based ELM (KELM) are introduced with several variants to overcome the shortcomings of standard ELM. ELM algorithm can be summarized as follows:

ELM Inputs: Dataset for training, N, activation function f and the number of hidden nodes L.

ELM Output: Weight of output layer, β.

Step 1: Assign randomly a_i and $b_i, i = 1, \ldots, L$ for parameters of hidden nodes (a_i, b_i) define by Eq. (1),

$$o_j = \sum_{i=1}^{L} G(a_i, b_i, x)\beta_i \tag{1}$$

Where $G(a_i, b_i, x)$ is the activation function and L hidden nodes for standard SLFN is defined by Eq. (1), whereas a_i is the input weight vector connecting the ith hidden node and the input nodes; β_i is the weight vector connecting the ith hidden node and the output node; b_i is the threshold or bias of the ith hidden node.

Step 2: Calculate H using Eq. (2).

$$H = \begin{bmatrix} h(x_1) \\ \ldots \\ h(x_N) \end{bmatrix} = \begin{bmatrix} G(a_1, b_1, x_1) & \ldots & G(a_L, b_L, x_1) \\ \ldots & \ldots & \ldots \\ G(a_1, b_1, x_N) & \ldots & G(a_L, b_L, x_N) \end{bmatrix} \tag{2}$$

Step 3: Calculate β using Eq. (3), where the Moore-Penrose generalized inverse of H is represented as $H^\dagger$.

$$\beta = H^\dagger T \tag{3}$$

Where,

$$T = \begin{bmatrix} t_1^T \\ \ldots \\ t_N^T \end{bmatrix} = \begin{bmatrix} t_{11} & \ldots & t_{1m} \\ \ldots & \ldots & \ldots \\ t_{N1} & \ldots & t_{Nm} \end{bmatrix} \tag{4}$$

3   Methodology

3.1   Salivary Raman Spectra Dataset

The datasets used in this study are obtained from UiTM-NMRR-12-1278-12868-NS1-DENV database. It consists of Raman spectra of saliva acquired from the saliva of healthy volunteers and saliva adulterated with NS1protein. The datasets are named as control group saliva and NS1 adulterated saliva, respectively. NS1 adulterated saliva is a mixture of saliva and NS1 protein (ab64456) purchased from Abcam. NS1 protein was diluted to concentrations less than 1 ppm and mixed with the saliva of the healthy volunteers to simulate the saliva of dengue infected patients. The control group saliva were collected from healthy volunteers with range of 22 to 34 years old. Details on the Raman spectra acquisition procedure was described in our previous study [27]. The dimension of each dataset is [64 × 1801].

3.2 Classification of NS1 Based on Salivary Raman Spectra

Figure 1 shows the overall algorithms developed for classification NS1 based on Raman spectra of saliva. First, the datasets were pre-processed to remove the unwanted features and improve it SNR. The pre-processing algorithm has four stages which are background subtraction, baseline removal, smoothing and normalization [28]. Then, the pre-processed spectra were analyzed using PCA for feature extraction and dimension reduction. According to the termination criteria discussed in Sect. 2.2, the number of estimated PCs based on Cattle Scree test, CPV and EOC are 7, 70 and 115, respectively.

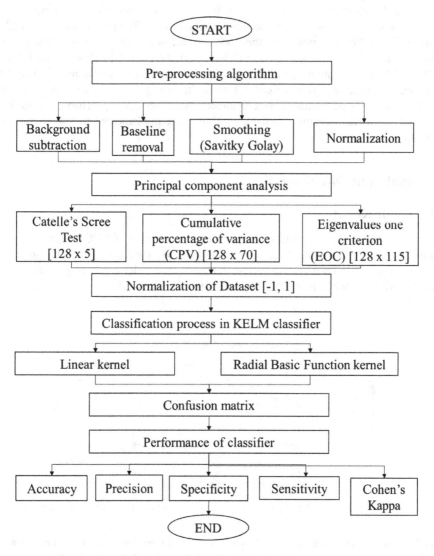

Fig. 1. Overall flowchart of classification of NS1 adulterated saliva.

Next the estimated PCs are normalized into $[-1, 1]$ using Eq. (5) as recommended in [29].

$$Y = \frac{(y_{max} - y_{min}) \cdot (x - x_{min})}{(x - x_{min}) + y_{min}} \quad (5)$$

where, Y is the normalized value of every row; y_{max} is equals to $+1$; y_{min} is equals to -1; x is real finite value to normalize; x_{min} is minimum value of every row x and x_{max} is maximum value of each row.

Finally, the normalized PCs are fed into KELM algorithm as input. For every set of PCs proposed by the termination criteria, the datasets were divided for the training and testing of the algorithm with ratio of 70% to 30%. The control group outputs are labelled as '0' and the NS1 adulterated are labelled as '1'. The performance of KELM classifier is highly dependent on the variable parameters of the used kernels. For Linear-KELM classifier, the values of C were varied from 0.1 to 10000. And for the RBF-KELM, the variation of the C value does not show any significant effect to the classifier performance thus was fixed at 1 while the value of γ were varied from 0.01 to 10. The classifiers performance were evaluated based accuracy, precision, sensitivity, specificity, and Kappa value. All the algorithms were implemented in Matlab version R2014a environment.

4   Result and Discussion

4.1   Linear-KELM

Figure 2 shows the performance of Linear KELM with input of 5 PCs as proposed by Cattle Scree test. It is observed that all the performance parameters do not change as the C-value varies. The accuracy, precision, sensitivity, specificity and Kappa are maintained at 97.37%, 95%, 100%, 94.74% and 0.9474 respectively.

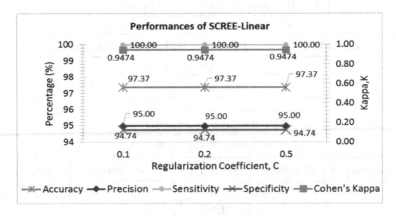

Fig. 2. Accuracy, precision, sensitivity, specificity, Kappa-value for Linear-KELM model with Cattle Scree test criterion.

With reference to Fig. 3, increasing the number of PCs to 70 as proposed by CPV criterion, it is observed that for all values of C; sensitivity is maintained at 100%. A decreasing trend is observed on specificity performance thus effecting the accuracy, precision and Kappa value with the similar trend. However for C-values of 0.1, 0.2 and 0.5, all the performance parameters scored 100% which gives better performance than the Cattle Scree test criterion.

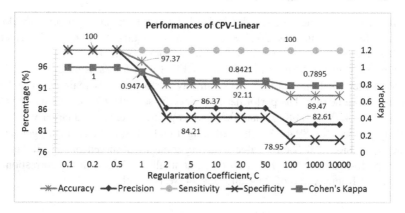

Fig. 3. Accuracy, precision, sensitivity, specificity, Kappa-value for Linear-KELM model with CPV criterion as C-value varies.

The performance of Linear KELM with 115 PCs as proposed by EOC criterion is shown in Fig. 4. At C value of 0.1, 0.2, 0.5, 1 and 2, all the performance parameters scored 100%, then as C-value increases to 5 the sensitivity performance drops to 94.74% thus reducing the accuracy performance to 97.37%. Increasing the C-value further to 50, causing the specificity and precision performance to drop from 100% to 94.74%. This reduces the accuracy performance to 94.74%.

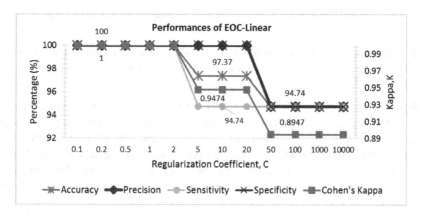

Fig. 4. Accuracy, precision, sensitivity, specificity, Kappa-value for Linear-KELM model with EOC criterion as C-value varies.

In summary, the performance of Linear-KELM classifiers increases as the number of PCs increase. However, as more PCs are included, the performance of the classifier model decrease as C-value increase. From findings, EOC termination criterion is selected as the optimal classifier model for Linear-KELM because it gives better performance for wider range of C-value.

4.2   Radial Basis Function (RBF)-KELM

For RBF-KELM classifier, the C-value is fixed at 1 as the classifier performance do not shows significant variance as C-value varies. Thus, for this study the performance of RBF-KELM classifier is observed as γ is varied from 0.01 to 10.

Figure 5 shows the performance of RBF-KELM with 5 PCs as proposed by Cattle Scree test criterion. It shows that the sensitivity performance is maintained at 100% regardless the value of γ. The performance of specific scored 94.74% for small γ-values (0.01, 0.02, 0.04, 0.06, 0.08, 0.1 and 0.2) and as γ-value increases to larger values, the specificity reduces to 89.47%. The reducing trend of specificity is triggering the accuracy, precision and Kappa value reducing trend as observed in Fig. 5. The best accuracy performance at 97.37% is observed at small γ-values with the corresponding precision and Kappa-value of 95% and 0.9474, respectively.

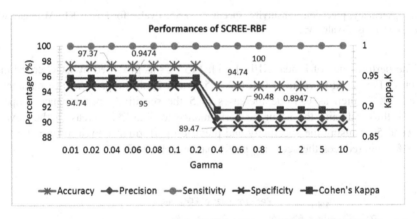

Fig. 5. Accuracy, precision, sensitivity, specificity, Kappa-value for RBF-KELM model with Cattle Scree test criterion as γ-value varies and C-value is fixed at 1.

Figure 6 shows the performance of RBF-KELM with 70 PCs proposed by CPV criterion. Initially, at γ-value of 0.01, the sensitivity, specificity, accuracy, precision and Kappa value start at 89.47%, 100%, 94.74%, 100% and 0.8947, respectively. As γ-value increases to 0.02, the only performance parameter showing an increasing trend is sensitivity while the others parameter are decreases in performance. However as γ-value increases to 0.04, the other parameters performance starts to increase and scored 100% as γ-value set to 2.

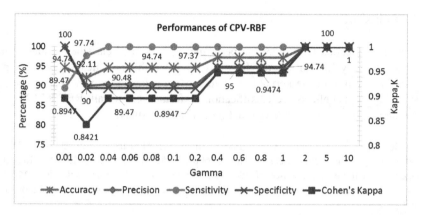

Fig. 6. Accuracy, precision, sensitivity, specificity, Kappa-value for RBF-KELM model with CPV criterion as γ-value varies and C-value is fixed at 1.

Figure 7 shows the classification performance of RBF-KELM classifier with 115 PCs proposed by EOC criterion. It is observed that the specificity performance scored 100% for all γ-values. The sensitivity performance shows some variation as γ-values varies where it starts at 68.42% and increase to 100% when γ-values is 0.04. It remained at 100% until γ-values increases to 1. For larger γ-values at 2, 5 and 10, the sensitivity performance reduces slightly to 94.74%. The variation of sensitivity causing similar variation trend in accuracy and precision.

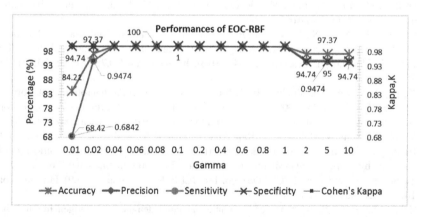

Fig. 7. Accuracy, precision, sensitivity, specificity, Kappa-value for RBF-KELM model with EOC criterion as γ-value varies and C-value is fixed at 1.

Overall, it is observed that the performance of the RBF-KELM is increase as the number of retained PCs increase. RBF-KELM with EOC criterion is considered as the optimal model since 100% of accuracy can be achieved with wider range of γ-values.

5   Conclusion

This paper is intended to explore the performance of KELM classifier in classifying NS1 from salivary Raman spectra. It is observed that the classification performance varies with variation of the kernel parameters. Moreover, the number of PCs used as classifier inputs also effects the classification performance. Overall, RBF-KELM with EOC criterion is selected as the optimal model for this study.

Acknowledgment.   The author would like to thank the Ministry of Education (MOE) of Malaysia, for providing the research funding 600-RMI/ERGS 5/3 (20/2013); the Research Management Institute and Faculty of Electrical Engineering, Universiti Teknologi MARA, Selangor, Malaysia, for the support and assistance given to the authors in carrying out this research.

References

1. Gubler, D.J.: Dengue and dengue hemorrhagic fever. Clin. Microbiol. Rev. **11**(3), 480–496 (1998). https://doi.org/10.1201/9780203752463
2. Darwish, N.T., Alias, Y.B., Khor, S.M.: An introduction to dengue-disease diagnostics. TrAC Trends Anal. Chem. **67**, 45–55 (2015). https://doi.org/10.1016/j.trac.2015.01.005
3. Laman Utama iDengue. http://idengue.remotesensing.gov.my/idengue. Accessed 21 May 2018
4. Radzol, A.R.M., Lee, K.Y., Mansor, W., Ariffin, N.: Biostatistical analysis of principle component of salivary Raman spectra for NS1 infection. In: 2016 IEEE EMBS Conference on Biomedical Engineering and Sciences, pp. 13–18. IEEE, Kuala Lumpur (2016). https://doi.org/10.1109/iecbes.2016.7843406
5. Anand, A.M., Sistla, S., Dhodapkar, R., Hamide, A., Biswal, N., Srinivasan, B.: Evaluation of NS1 antigen detection for early diagnosis of dengue in a tertiary hospital in Southern India. J. Clin. Diagn. Res. **10**(4), 1–4 (2016). https://doi.org/10.7860/JCDR/2016/15758.7562
6. Yusoff, N.A., Soin, N., Ibrahim, F.: Lab-on-a-disk as a potential microfluidic platform for dengue NS1-ELISA. In: 2009 IEEE Symposium on Industrial Electronics & Applications, pp. 946–950. IEEE, Kuala Lumpur (2009). https://doi.org/10.1109/isiea.2009.5356330
7. Yusoff, N., Soin, N., Rahman, N., Raja Abd Rahman, R.N., Abd Kahar, M.A., Ibrahim, F.: Binding characteristics study for dengue virus non-structural protein 1 of Antigen and its antibody by using circular dichroism technique. In: 2009 International Conference for Technical Postgraduates (TECHPOS), pp. 1–6. IEEE, Kuala Lumpur (2009). https://doi.org/10.1109/techpos.2009.5412060
8. Zainah, S., et al.: Performance of a commercial rapid dengue NS1 antigen immunochromatography test with reference to dengue NS1 antigen-capture ELISA. J. Virol. Methods **155**(2), 157–160 (2009). https://doi.org/10.1016/j.jviromet.2008.10.016
9. Anders, K.L., et al.: An evaluation of dried blood spots and oral swabs as alternative specimens for the diagnosis of dengue and screening for past dengue virus exposure. Am. J. Trop. Med. Hyg. **87**(1), 165–170 (2012). https://doi.org/10.4269/ajtmh.2012.11-0713
10. Fleischmann, M., Hendra, P.J., McQuillan, A.J.: Raman spectra of pyridine adsorbed at a silver electrode. Chem. Phys. Lett. **26**, 163–166 (1974). https://doi.org/10.1016/0009-2614(74)85388-1

11. Hadjigeorgiou, K., Kastanos, E., Pitris, C.: Surface enhanced Raman spectroscopy as a tool for rapid and inexpensive diagnosis and antibiotic susceptibility testing for urinary tract infections. In: 2016 IEEE Healthcare Innovation Point-Of-Care Technologies Conference (HI-POCT), pp. 158–161. IEEE, Cancun (2016). https://doi.org/10.1109/hic.2016.7797721

12. Yan, W., et al.: Preliminary study on the quick detection of acquired immure deficiency syndrome by saliva analysis using surface enhanced Raman spectroscopic technique. In: 2009 Annual International Conference of the IEEE Engineering in Medicine and Biology Society, pp. 885–887. IEEE, Minneapolis (2009). https://doi.org/10.1109/iembs.2009.5333131

13. Wang, Y., et al.: A feasibility study of early detection of lung cancer by saliva test using surface enhanced Raman scattering. In: 5th International Conference on BioMedical Engineering and Informatics, pp. 135–139. IEEE, Chongqing (2012). https://doi.org/10.1109/bmei.2012.6513160

14. Huang, C.Y., et al.: Hybrid SVM/CART classification of pathogenic species of bacterial meningitis with surface-enhanced Raman scattering. In: 2010 IEEE International Conference on Bioinformatics and Biomedicine (BIBM), pp. 406–409. IEEE, Hong Kong (2010). https://doi.org/10.1109/bibm.2010.5706600

15. Santiago-Cordoba, M.A., Romano, P.R., Mackay, A., Demirel, M.C.: Raman based hepatocellular carcinoma biomarker detection. In: 2011 Annual International Conference of the IEEE Engineering in Medicine and Biology Society, pp. 3672–3675. IEEE, Boston (2011). https://doi.org/10.1109/iembs.2011.6090620

16. Joliffer, I.T.: Principal Component Analysis. Springer Series of Statistic, 2nd edn. Springer, Heidelberg (2002). https://doi.org/10.2307/1270093

17. Salma, M.U.: BAT-ELM: a bio inspired model for prediction of breast cancer data. In: 2015 International Conference on Applied and Theoretical Computing and Communication Technology (iCATccT), pp. 501–506. IEEE, Davangere (2015). https://doi.org/10.1109/icatcct.2015.7456936

18. Ismaeel, S., Miri, A., Chourishi, D.: Using the extreme learning machine (ELM) technique for heart disease diagnosis. In: 2015 IEEE Canada International Humanitarian Technology Conference (IHTC2015), pp. 1–3. IEEE, Ottawa (2015). https://doi.org/10.1109/ihtc.2015.7238043

19. Shahsavari, M.K., Rashidi, H., Bakhsh, H.R.: Efficient classification of Parkinsons disease using extreme learning machine and hybrid particle swarm optimization. In: 2016 4th International Conference on Control, Instrumentation, and Automation (ICCIA), pp. 148–154. IEEE, Qazvin (2016). https://doi.org/10.1109/icciautom.2016.7483152

20. Lu, S., Wang, H., Wu, X., Wang, S.: Pathological brain detection based on online sequential extreme learning machine. In: 2016 International Conference on Progress in Informatics and Computing (PIC), pp. 219–223. IEEE, Shanghai (2016). https://doi.org/10.1109/pic.2016.7949498

21. Kumar, S.J.J., Ravichandran, C.G.: Macular Edema severity detection in colour fundus images based on ELM classifier. In: 2017 International Conference on I-SMAC (IoT in Social, Mobile, Analytics and Cloud) (I-SMAC), pp. 926–933. IEEE, Palladam (2017). https://doi.org/10.1109/i-smac.2017.8058316

22. Mishra, V.K., Bajaj, V., Kumar, A.: Classification of normal, ALS, and myopathy EMG signals using ELM classifier. In: 2016 2nd International Conference on Advances in Electrical, Electronics, Information, Communication and Bio-Informatics (AEEICB), pp. 455–459. IEEE, Chennai (2016). https://doi.org/10.1109/aeeicb.2016.7538330

23. Kaiser, H.F.: The application of electronic computers to factor analysis. Educ. Psychol. Meas. **20**(1), 141–151 (1960). https://doi.org/10.1177/001316446002000116

24. Cattell, R.B.: The scree test for the number of factors. Multivariate Behav. Res. 1(2), 245–276 (1966). https://doi.org/10.1207/s15327906mbr0102
25. Huang, G.B., Zhu, Q.Y., Siew, C.K.: Extreme learning machine: a new learning scheme of feed forward neural networks. In: Proceedings of International Joint Conference on Neural Networks (IJCNN2004), pp. 985–990. IEEE, Budapest (2004). https://doi.org/10.1109/ijcnn.2004.1380068
26. Ding, S., Xu, X., Nie, R.: Extreme learning machine and its applications. Neural Comput. Appl. 25(3–4), 549–556 (2014). https://doi.org/10.1007/s00521-013-1522-8
27. Radzol, A.R.M., Lee, K.Y., Mansor, W.: Nonstructural protein 1 characteristic peak from NS1-saliva mixture with surface-enhanced Raman spectroscopy. In: 2013 35th Annual International Conference of the IEEE Engineering in Medicine and Biology Society (EMBC), pp. 2396–2399. IEEE, Osaka (2013). https://doi.org/10.1109/embc.2013.6610021
28. Radzol, A.R.M., Lee, K.Y., Mansor, W., Othman, N.H.: Principal component analysis for detection of NS1 molecules from Raman spectra of saliva. In: 11th International Colloquium on Signal Processing & Its Applications (CSPA), pp. 168–173. IEEE, Kuala Lumpur (2015). https://doi.org/10.1109/cspa.2015.7225640
29. Extreme Learning Machines. http://www.ntu.edu.sg/home/egbhuang/elm_kernel.html

Improving the Robustness of the Glycemic Variability Percentage Metric to Sensor Dropouts in Continuous Glucose Monitor Data

Michael Mayo[(⊠)]

Department of Computer Science, University of Waikato, Hamilton, New Zealand
michael.mayo@waikato.ac.nz

Abstract. Continuous glucose monitors generate significant volumes of high frequency blood glucose data. Analysis of this data by a physician may entail the calculation of various glycemic variability metrics. In this paper, we consider the problem of metric robustness to sensor dropouts. We show that the standard metrics for glycemic variability are unreliable with missing data. A more recent metric, glycemic variability percentage, is shown to consistently underestimate glycemic variability as the amount of missing data increases. We therefore propose a new algorithm based on random sampling combined with linear regression to correct this underestimation, and show that the metric's accuracy is significantly increased with our correction.

1 Introduction

Glycemic variability (GV) is an important issue in the management of patients with diabetes. Normal (healthy) patients usually exhibit low GV compared to patients with type 1 and type 2 diabetes. High GV implies that fluctuations in blood glucose concentration are significant. Patients with high GV are at risk of complications arising from hypoglycemia if blood glucose levels fall too low or hyperglycemia if glucose levels rise too far. Both conditions, if left untreated, are generally dangerous.

To assess a patient's GV, a number of approaches are used by doctors. Self monitoring of blood glucose (SMBG) [1] is one such approach, and currently the predominant one. Under an SMBG regime, a patient receives lancet device for obtaining blood samples (typically from the finger) along with a blood glucose meter that measure glucose levels from blood droplets that are applied to a reagent strip and inserted into the meter. The patient will typically measure his or her own glucose levels 3–4 times per day (hence the "self" in SMBG), recording the values for subsequent analysis.

More recently, continuous glucose monitoring (CGM) systems [1] have become popular. A CGM consists of two parts: a sensor and a receiver. The sensor consists of a small needle inserted into the subcutaneous region around

ⓒ Springer Nature Switzerland AG 2019
N. T. Nguyen et al. (Eds.): ACIIDS 2019, LNAI 11432, pp. 373–384, 2019.
https://doi.org/10.1007/978-3-030-14802-7_32

the abdomen or the upper arm. When the needle is inserted, the sensor measures blood glucose levels at a specified temporal frequency (e.g. every five minutes) and sends the data wirelessly to the receiver, either in real time (these are the so-called "real time" CGMs, or rtCGMs [4]) or whenever a patient "scans" or "swipes" the receiver past the sensor (these type of CGMs are known as "flash" or "intermittent" CGMs, or iCGMs [4]). A clear difference between CGMs and SMBG is the amount of data acquired: with a sampling frequency of five minutes, a CGM can be used to obtain up to 288 blood glucose readings in 24 h compared to the small handful of readings that is typical with SMBG.

An example of four days of CGM data for one patient is shown in Fig. 1. If the point of least risk in the chart is 112.5 mg/dl, then that chart shows significant and frequent excursions above and below this safe value. This quality of data is clearly not obtainable using SMBG. Since CGMs record date- and time-stamps with each blood glucose sensor reading, the data forms a natural time series.

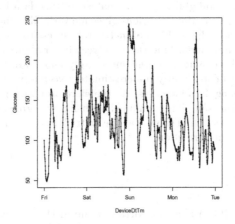

Fig. 1. Four days of CGM data for one real subject sampled at a frequency of five minutes. Points denote actual samples. Glucose is measured in mg/dl.

A flip side to the much greater volume of data obtainable using CGMs is that the data needs to be handled and analysed correctly. Several approaches are available to a doctor in this respect. The obvious one is that the data can be visualised, as is done in the figure above.

GV metrics may also be computed from the data. The purpose of a metric is to assess in a single number the glycemic state of a patient. This metric can then be compared to a reference population in order to estimate the severity of the patient's condition. Similarly, metrics taken from the same patient at different points in time can be used to assess how well the patient is self-managing the condition.

An example of a GV metric is the mean glucose level over the past month, which can give a doctor an idea of whether a patient's average glucose levels are safe or not. However, the mean gives no indication of the size of fluctuations in glucose levels nor their frequency, and therefore other metrics may also be used.

To date, many GV metrics have been proposed in the literature. A recent paper discussing the international consensus among physicians treating patients with diabetes using CGMs recommends a handful of GV metrics [4] and provides guidelines for the minimum amount of data needed before metrics should be calculated. This minimum level of data is two weeks, of which data must exist for at least 10 days.

In this paper, we consider the effect of missing data on these metrics. With CGM devices, missing data is prevalent because patients do not wear the devices for every hour of the day. There are many reasons why a CGM cannot sometimes be worn, e.g. swimming, skin irritation caused by the sensor, running out of calibration strips etc. Moreover, iCGMs have the added limitation that sensors can only store a limited number of hours worth of data: therefore if the subject forgets to scan the sensor in time, valuable data may be lost.

We show that most of the standard GV metrics are not robust to missing data. We also examine a very recently proposed metric (glycemic variability percentage, GVP) and show that this metric is also not robust to missing data. The main novel contribution of this paper is the observation that GVP responds in a unique way to missing data compared to the other metrics: in the case of other metrics, missing data causes an increase in error magnitude but the direction of the error is more or less random; in the case of GVP, the metric is consistently underestimated when data is missing. Therefore we propose a new method for correcting the GVP metric according to the amount of missingness in the data. We show using data from ten patients that our new approach better approximates the "true" gold standard GVP compared to simply ignoring the missing data and calculating GVP according to the published definition.

2   Background

A single sensor reading from a CGM can be denoted by the tuple (x_i, y_i) where x_i is the time of the sensor reading, y_i is the blood glucose reading, and i is an index. An analysis frame $F = \{(x_1, y_1), (x_2, y_2), \ldots (x_n, y_n)\}$ consists of n sensor readings occurring during a determined period of time.

Two common metrics for assessing GV are the mean and the coefficient of variation (CV), both of which are straightforward. The mean is defined as:

$$\text{mean}(F) = \frac{1}{n} \sum_{i=1}^{n} y_i \tag{1}$$

and the CV metric can be written as:

$$\text{cv}(F) = \frac{\sqrt{\frac{1}{n} \sum_{i=1}^{n} (y_i - \text{mean}(F))^2}}{\text{mean}(F)} \tag{2}$$

The mean blood glucose value clearly assesses the average glucose level over the entire analysis frame (which clinically speaking may be high, low, or in the target range) while the CV metric measures the size of the fluctuations relative to the mean. A high CV even if the mean glucose is reasonable may indicate extreme excursions into the hypo- or hyperglycemic ranges.

One issue with simple metrics calculated directly from raw glucose sensor readings is that of balance between the hypoglycemic and hyperglycemic ranges. For example, the level 2 hypoglycemic range is defined as a blood glucose concentration of <54 mg/dL (3.0 mmol/L) while the level 2 hyperglycemia range is defined as >250 mg/dL (13.9 mmol/L) [4]. Since the maximum possible blood glucose concentration is 600 mg/dl (33.33 mmol/L), and most CGMs have a maximum sensor reading of 400 or 500 mg/dl, then the ranges are skewed and distributions of blood glucose values therefore tend not to be normally distributed [7].

Recognising that this could lead to problems with statistical analyses that make the assumption of normality, [7] proposed the risk domain as an alternative space in which analysis of blood glucose concentrations could be carried out. The risk domain transform is defined as:

$$r_i = 10 \times (1.509(\ln(y_i)^{1.084} - 5.381))^2 \tag{3}$$

where y_i are blood glucose concentrations in mg/dl units, and r_i is a non-linear transformation.

An advantage of the risk domain transform is that it equalises the hypo- and hyperglycemic ranges. The point of least risk on the raw scale (defined by [7] as 112.5 mg/dl [6.25 mmol/L]) now has a risk value of zero, while the extremes of the ranges (approx. 20 mg/dl at the low end and 600 mg/dl at the high end) have risk values close to 100.

From the values produced by applying this risk transform, two further metrics can be defined: the low blood glucose index (LBGI) and the high blood glucose index (HBGI), both of which were introduced in [8]. The LBGI has been demonstrated to be a good predictor for the occurrence of severe hypoglycemia [3] while the HBGI characterises time spent in the hyperglycemic range, which in turn can be a risk predictor for diabetic complications such as diabetic nephropathy.

Both of these metrics are defined simply once the risk transform has been applied to the y values in the current analysis frame:

$$\text{LBGI}(F) = \frac{1}{n} \sum_{i=1}^{n} \begin{cases} r_i & \text{if } y_i < 112.5 \text{ mg/dl} \\ 0 & \text{otherwise} \end{cases} \tag{4}$$

$$\text{HBGI}(F) = \frac{1}{n} \sum_{i=1}^{n} \begin{cases} r_i & \text{if } y_i > 112.5 \text{ mg/dl} \\ 0 & \text{otherwise} \end{cases} \tag{5}$$

According to [5], key low/medium/high risk zones for the LBGI and HBGI metrics are LBGI < 2.5, LBGI between 2.5 and 5, and LBGI > 5; and HBGI < 4.5, HBGI between 4.5 and 9, and HBGI > 9. Since both metrics were originally defined with SMBG in mind, [5] conducted a study to determine if the same metrics could be used for CGM data as well. It was concluded that LGBI computed from CGM measurements tends to underestimate risk, and therefore the following linear correction was suggested for computing LBGI from CGM data:

$$\text{LBGI}'(F) = 1.0199(\text{LBGI}(F)) + 0.6521 \tag{6}$$

HBGI, however, was found to require no changes for CGM data.

The final metric considered in this paper is the glycemic variability percentage (GVP). The metric was very recently proposed in 2018 by [10]. Unlike other metrics computed solely from the blood glucose levels, a key feature of this metric is that it incorporates time into the metric calculations.

The definition of GVP is as follows:

$$\text{GVP}(F) = 100.0(\frac{\text{L}}{\text{L0}} - 1) \tag{7}$$

where

$$\text{L}(F) = \sum_{i=2}^{n} \sqrt{(\triangle x_i^2 + \triangle y_i^2)} \tag{8}$$

and

$$\text{L0}(F) = \sum_{i=2}^{n} \triangle x_i \tag{9}$$

The $\triangle x_i$ and $\triangle y_i$ terms are defined as the consecutive changes in value of x and y: that is, $\triangle x_i = x_i - x_{i-1}$ in minutes and $\triangle y_i = y_i - y_{i-1}$ in mg/dl. Thus, the metric sums over the Euclidean distances between consecutive points on a 2D plot of the blood glucose concentrations, and divides by the time taken. Therefore a patient with many high amplitude/rapid fluctuations in blood glucose will have a high GVP because of the higher degree of "distance travelled" on the 2D plot, while a patient with small or slower moving fluctuations will have lower GVP. Analysis in [10] showed that healthy patients have a median GVP of 18.4 while type 1 diabetic adults have a median GVP of 42.3, which is a significant difference.

The mean, CV, LBGI and HBGI are all recommended risk metrics to be used when assessing a patient from CGM data [4]. The GVP metric, being more recent, is not currently "recommended" but it none-the-less has been shown to be more effective at capturing frequency variations in blood glucose concentration than mean, CV and other metrics not described here [10]. GVP is also one component of a larger system aiming to better characterise a patient's risk [6].

In terms of the amount of data that is required before computing these metrics, a recent international consensus [4] was that two weeks worth of data with a minimum coverage of 70–80% or 10 days is the minimum size for the analysis frame.

3 Sensitivity Analysis of Existing Metrics to Sensor Dropouts

In this section, we introduce mathematically the notion of sensor reading coverage and use it to define an objective metric for measuring the missingness of sensor readings in an analysis frame. In turn, this will enable us to evaluate the effect of missing data on the risk metrics described in the previous section.

Let the total amount of time (in minutes) between the first and last sensor readings for an analysis frame F be denoted by mins (F). If the sampling frequency f (also in minutes) of the CGM device used to produce F is known then we can easily calculate the maximum possible number of sensor readings that could be made in the given time period by dividing mins (F) by f.

The coverage is therefore defined as the fraction of actual sensor readings taken. That is, if n is the number of actual sensor readings taken, then the coverage is defined as:

$$\text{cov}(F) = \frac{n}{\left(\frac{\text{mins}(F)}{f}\right)} = \frac{nf}{\text{mins}(F)} \tag{10}$$

This quantity ranges between 0 and 1 with a coverage of 1 indicating full coverage and 0 indicating no coverage at all.

Next, we define our notion of missingness in the context of CGM data. Let a "sensor dropout" be a block of time during which the CGM sensor is not recording data. This could be for a variety of reasons as previously mentioned.

Due to the way in which CGMs are used, it makes sense to model missing data as missing contiguous blocks of sensor readings, as opposed to randomly missing sensor readings. The randomly missing approach works best when the samples are statistically independent, but in our case they are not, because the sensor readings are ordered in time and in practice would be "missing in blocks".

We define a simulated sensor dropout, then, as the random removal of a contiguous sequence of sensor readings spanning a fixed amount of time, e.g. four hours. The coverage prior to and after the simulated dropout can be recorded, and GV metrics prior to and after the removal can likewise be calculated. Moreover, we can perform multiple random sensor dropouts to simulate data being lost on multiple occasions over a longer period such as two weeks.

An example of this is given by Fig. 2. This figure depicts the same data as shown in Fig. 1, but random four hour blocks have been removed to reduce the coverage from 0.99 in Fig. 1 to 0.60 in Fig. 2.

To conduct our experiments, we used publicly available CGM data sources. The first source is from a clinical trial (CT) involving 68 patients with type 1 diabetes who used a CGM for varying amounts of time [2]. In some cases, over one year's worth of data is available for one patient. The second source of data was from the Night Scout (NS) Data Commons project [11], a repository where individuals experimenting with closed loop artificial pancreas systems (of which CGMs are integral parts) can deposit and share their data.

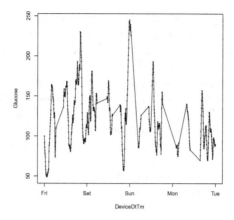

Fig. 2. The same data as shown in Fig. 1 but with several four-hour blocks of sensor readings randomly removed to reduce the coverage.

From the two different sources of CGM data, we selected ten random subjects, six from the CT dataset and four from the NS dataset. Inclusion criteria were that the frequency of the sensor readings was five minutes, and there was sufficient data to obtain at least one analysis frame of length two weeks with coverage greater than 0.9. One such frame was then chosen at random for each subject. Statistical details of each analysis frame and the corresponding gold standard metrics are shown in Table 1. Characteristics of each patient in the analysis frame's time period can be deduced from the table. For example, patient CT6 has a very high value for HBGI and GVP, indicating significant fluctuations into the hyperglycemic range. Conversely, subject NS3 scores highly with the LBGI metric implying high risk of hypoglycemia-related complications.

Table 1. Gold standard analysis frames chosen randomly from ten random subjects. Each analysis frame spans two weeks.

Subject	n	cov	mean	cv	LBGI	HBGI	GVP
CT1	3,661	0.91	128.84	0.30	1.22	2.35	33.75
CT2	3,679	0.91	125.01	0.25	0.86	1.63	22.84
CT3	3,848	0.95	109.14	0.31	2.50	1.13	28.74
CT4	3,654	0.91	140.17	0.25	0.38	3.07	28.48
CT5	3,768	0.94	101.57	0.27	3.05	0.51	17.41
CT6	3,678	0.91	204.28	0.48	1.08	16.50	57.45
NS1	3,647	0.92	195.81	0.33	0.32	12.78	35.63
NS2	3,887	0.96	105.01	0.24	2.06	0.54	25.03
NS3	3,613	0.93	145.72	0.39	6.55	4.80	31.54
NS4	3,874	0.96	128.21	0.33	1.22	2.67	41.04

Our first analysis considered the robustness of GV metrics to missing data. To achieve this, we took each of the analysis frames F in turn and copied the frame 80 times obtain copies $F_1, F_2, \ldots, F_{80}$. Each copy F_i was then "corrupted" by randomly removing i four hour blocks of data. Blocks were allowed to overlap, but each block was required to be completely enclosed inside the bounds of the analysis frame and to consist of at least one sensor reading that would be removed. This gave us, for each gold standard analysis frame F, a set of realistic subsets of F with coverage values ranging from very low all the up to $\text{cov}(F)$.

We then computed our five metrics of interest for each corrupted analysis frame and plotted coverage vs. metric value. For gold standard analysis frame CT1, the results are depicted in Fig. 3. All of the remaining gold standard analysis frames shows similar trends, so we omit those figures for reasons of space.

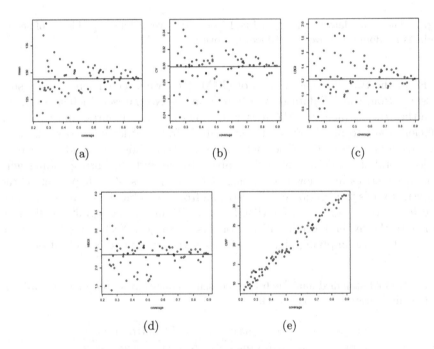

Fig. 3. Plots showing various risk metrics vs. coverage for subject CT1's analysis frame. The horizontal lines indicate the gold standard for the same analysis frame without any missing data, values of which are given in Table 1.

Figure 3 shows that for the "standard" metrics (mean, CV, LBGI and HBGI) listed in [4] the error magnitude clearly has a linearly increasing trend as coverage decreases. For coverage values of 0.7, the error on the metrics is approximately 10–20% in most cases. At lower coverage, the LBGI metric appears to overestimate the risk slightly (see Fig. 3c) while HBGI appears to underestimated slightly (Fig. 3d). Beyond that, there is no clear trend in error direction with decreasing coverage.

GVP, on the other hand, shows different behaviour: as coverage decreases, GVP consistently underestimates the gold standard. This relationship appears to be strongly linear. Furthermore, unlike the other metrics, the variance of the error does not increase with decreased coverage. Instead, it appears to increase very slowly initially and then remain constant as coverage decreases below 0.8.

This property of the GVP follows intuitively from its definition in Eqs. 7, 8 and 9. Since the L quantity in the equations measures distanced travelled in the time vs. mg/dl space 2D plot, then removing points from this space must naturally decrease L since "zig/zag" patterns in the data are being replaced by straight lines that have a shorter distance. Since for most individuals, the patterns of GV are usually constant over a two week time frame, the decrease in L is linear with the decrease in coverage.

4   GVSD: A New Metric Robust to Sensor Dropouts

This curious property of the GVP metric suggests the following possibility: to better account for sensor dropouts, we could (for each patient) fit a linear model that can be used to estimate a patient's true GVP from his or her estimated GVP given the available data. For example, suppose a patient presents with two weeks of CGM data but that data has a coverage of only 0.6. According to our previous analysis and the guidelines [4], this is an insufficient amount of data for computing reliable metrics. However, if we can generate a set of points similar to those shown in Fig. 3e from the patient's data, then we could in turn fit a trend line to the data and therefore estimate the "true" GVP that would have been calculated if the coverage was 0.9 or 1.0.

The main question is how to achieve this. The answer is to use a similar process to that described in the previous section: start with the initial low-coverage data, iteratively create corrupted copies of the data by simulating sensor dropouts of varying degrees, then compute the coverage and GVP metric for each corrupted copy of the original dataset. If the original dataset has low coverage, then the corrupted versions of the dataset will have even lower coverage. However, if the trend is still linear at these lower levels, then it should be possible to estimate what the GVP would have been with coverage at higher levels using simple univariate linear regression.

An example of this algorithm in action will now be described for the dataset shown in Fig. 2. Recall that this CGM dataset has coverage 0.6, and it is a corrupted version of the gold standard analysis frame CT1 shown in Fig. 1 which has coverage 0.9. The GVP metric for the data shown in Fig. 2 is 23.90 which is clearly a significant underestimate compared to the "true" GVP at coverage 0.9 which is 33.75.

Let us create twenty corrupted versions of the 0.6 coverage dataset. Each corrupted version of the data has a different set of random sensor dropouts, with sensor readings being randomly removed in blocks of size four hours. As a consequence, each corrupted version of the data will have a different coverage and a different calculated GVP score.

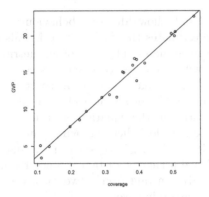

Fig. 4. Example of the algorithm applied to a dataset with coverage 0.6. Each point is a corrupted version of the original dataset with coverage less than 0.6. The figure shows the GVP computed from each dataset along with the line of best fit.

Figure 4 gives an example of this approach. The figure shows twenty points, one for each corrupted version of the original dataset. It can be observed clearly that there is a wide range of coverages. Linear regression from the machine learning toolkit `scikit-learn` [9] with data normalisation turned on was then applied to this data to produce a linear model relating coverage and GVP. The slope of the linear model is 41.33 and the intercept is −0.45. The coefficient of determination (R^2) value is 0.97, indicating a strong fit. Using this model to estimate the patient's GVP with coverage of 0.9 or 1.0 yields GVP values of 36.75 and 40.88 respectively – both estimates are considerably higher than the estimate derived from the incomplete version of the data presented by the patient.

To test this algorithm, we repeated this experiment for all nine remaining analysis frames. Random four hour blocks of data were removed from each frame until coverage was below 0.6. It was assumed that this version of the data with missing blocks was the only version of the data available, although for experimental purposes we also retained the complete (at least, with coverage of 0.9) versions of the data for use as gold standards. We then computed the GVP metric from the sample with missing data.

Next, we applied the algorithm outlined above to compute new estimates of the GVP. Even though the coverage of each original sample is approximately 0.6, once the linear model was fitted to the available low-coverage data, we were able to estimate the GVP for coverages of 0.9 (close to coverage of our gold standard) and 1.0 (probably the more realistic estimate).

Table 2 summarises our results. The columns labelled GVSD ("**G**ylcemic **V**ariability Percentage **R**obust to **S**ensor **D**ropouts") give our algorithm's estimates of the GVP. The gold standard GVP values are the same as Table 1. Comparing the sample and gold GVPs to the GVSD values, it is clear that the algorithm effectively estimates the gold standard GVP values better than a direct calculation of GVP from the available data. In all cases, the GVP(gold)/GVSD error is 10% or less, in stark contrast to the GVP(gold)/GVP(sample) error which is much greater.

Table 2. Results showing the estimation of GVP using our new algorithm compared to the standard method.

Subject	n	cov	GVP (sample)	GVP (gold)	GVSD (0.9)	GVSD (1.0)
CT1	2411	0.60	23.90	33.75	36.75	40.88
CT2	2392	0.59	14.23	22.84	22.07	24.50
CT3	2400	0.60	17.57	28.74	26.52	29.44
CT4	2401	0.60	19.34	28.48	29.20	32.41
CT5	2385	0.60	11.25	17.41	16.92	18.74
CT6	2410	0.60	41.49	57.44	60.29	66.47
NS1	2366	0.60	24.38	35.63	35.10	38.60
NS2	2378	0.60	15.29	25.03	23.77	26.44
NS3	2287	0.59	20.55	31.54	30.07	33.18
NS4	2375	0.59	25.53	41.04	38.69	42.76

5 Conclusion

To summarise, this paper describes a new metric for measuring glycemic variability. This new metric is a variant of GVP, but one that takes into account the effect of sensor dropouts on the metric and corrects for such dropouts. An open question for future research is whether or not this approach could also be applied to other metrics. Clearly, as Fig. 3 shows, the method would not be suitable for metrics such as HBGI where the sign of the error is nearly random and the variance of the errors increase as coverage decreases. There are several other "distance travelled"-type metrics in the literature however which may benefit from our approach. Another question is under what conditions the correction proposed in this paper is accurate. We have considered only two week analysis frames so far. However, we could also consider larger frames, e.g. three months. Finally, we have modelled missing data as the random removal of "blocks" of sensor readings of size four hours. Is this a realistic assumption? A future investigation could look at patterns of real missing data in iCGM and rtCGM devices and develop a more realistic approach for modelling sensor dropouts.

References

1. Bruen, D., Delaney, C., Florea, L., Diamond, D.: Glucose sensing for diabetes monitoring: recent developments. Sensors **17**(8), 1866 (2017)
2. Buckingham, B., et al.: Effectiveness of early intensive therapy on β-cell preservation in type 1 diabetes. Diabetes Care **36**, 4030 (2013)
3. Cox, D.J., Gonder-Frederick, L., Ritterband, L., Clarke, W., Kovatchev, B.P.: Prediction of severe hypoglycemia. Diabetes Care **30**(6), 1370–1373 (2007). https://doi.org/10.2337/dc06-1386. http://care.diabetesjournals.org/content/30/6/1370

4. Danne, T., et al.: International consensus on use of continuous glucose monitoring. Diabetes Care **40**(12), 1631–1640 (2017). https://doi.org/10.2337/dc17-1600. http://care.diabetesjournals.org/content/40/12/1631
5. Fabris, C., Patek, S.D., Breton, M.D.: Are risk indices derived from CGM interchangeable with SMBG-based indices? J. Diabetes Sci. Technol. **10**(1), 50–59 (2016). https://doi.org/10.1177/1932296815599177
6. Hirsch, I.B., Balo, A.K., Sayer, K., Garcia, A., Buckingham, B.A., Peyser, T.A.: A simple composite metric for the assessment of glycemic status from continuous glucose monitoring data: implications for clinical practice and the artificial pancreas. Diabetes Technol. Ther. **19**(S3), S-38–S-48 (2017). https://doi.org/10.1089/dia.2017.0080
7. Kovatchev, B.P., Cox, D.J., Gonder-Frederick, L.A., Clarke, W.: Symmetrization of the blood glucose measurement scale and its applications. Diabetes Care **20**(11), 1655–1658 (1997)
8. Kovatchev, B.P., Straume, M., Cox, D.J., Farhy, L.S.: Risk analysis of blood glucose data: a quantitative approach to optimizing the control of insulin dependent diabetes. J. Theor. Med. **3**(1), 1–10 (2000). https://doi.org/10.1080/10273660008833060
9. Pedregosa, F., et al.: Scikit-learn: machine learning in Python. J. Mach. Learn. Res. **12**, 2825–2830 (2011)
10. Peyser, T.A., Balo, A.K., Buckingham, B.A., Hirsch, I.B., Garcia, A.: Glycemic variability percentage: a novel method for assessing glycemic variability from continuous glucose monitor data. Diabetes Technol. Ther. **20**(1), 6–16 (2018). https://doi.org/10.1089/dia.2017.0187
11. The Nightscout Foundation: About the Nightscout Data Commons on Open Humans (2014). http://www.nightscoutfoundation.org/data-commons/

GenPress: A Novel Dictionary Based Method to Compress DNA Data of Various Species

Péter Lehotay-Kéry[(✉)] and Attila Kiss

Faculty of Informatics, Department of Information Systems,
ELTE Eötvös Loránd University, Budapest, Hungary
lkp@caesar.elte.hu, kissae@ujs.sk

Abstract. There can be a data boom in the near future, due to cheaper methods make possible for everyone to keep their own DNA on their own device or on a central medical cloud. With the development of sequencing methods, we are able to get the sequences of more and more species. However the size of the human genome is about 3 GB for each person. And for other species it can be more.

The need is growing for the efficient compression of these data and general compressors can not reach a satisfying result. These are not aware of the special structure of these data. There are already some algorithms tried to reach smaller and smaller rates. In this paper, we would like to present our new method to accomplish this task.

Keywords: Bioinformatics · Biology · Compression · Genetics

1 Storing of the DNA

Deoxyribonucleic acid (DNA) is a complex molecule which contains the genetic information. These are built up by nucleotides. Each nucleotide is built up by 3 components: nucleobases (adenine - A, guanine - G, cytosine - C, thymine - T), a sugar called deoxyribose and a phosphate group.

In bioinformatics, we store DNA sequences as strings, which are composed by the 4 characters for the 4 nucleobases: 'C', 'G', 'A' and 'T'. Similarly, we can store RNA and protein sequence data too.

Because we have only 4 types of possible characters, we can store every base on 2 bits. However it is not enough, so far DNABit had its worst compression rate at 1.58 bits/base and COMRAD even reached 0.25 bits/base with small datasets. These are dictionary based methods, so we aimed to start our research with methods like these.

[1] presents 2 scenarios about how one should choose the compression algorithm. The first case is when the data are transferred over a network and then

---

Dr. Kiss was also with J. Selye University, Komárno, Slovakia.

© Springer Nature Switzerland AG 2019
N. T. Nguyen et al. (Eds.): ACIIDS 2019, LNAI 11432, pp. 385–394, 2019.
https://doi.org/10.1007/978-3-030-14802-7_33

decompressed. The second case is when the are accessed in real-time. Our algorithm gives an alternative for the first case.

We know that similar species have much in common in their genomes, like the different kind of viruses have more in common than a virus and a fungus for example. Our method uses this, builds different dictionaries based on different classifications of species, using the same dictionaries for the same classes. Different classes have different frequencies of subsequences in their genomes, but in a given class, species have a lot in common.

Our method greatly depends on how we choose our parameters, so in our worst case we can get great compression rate, but our best case can produce really small compression rate.

2   Related Works

As we mentioned, there are some algorithms that have already reached below 2 bits/base rate.

Biocompress-2 1994 searches for exact repeats and complement palindromes and encodes them with the length of the repeat and the position of the previous repeat. If there's no repeat, then it uses order-2 arithmetic encoding [2].

Cfact 1996 is searching for the longest exact matching repeat. It uses suffix tree to find it [3].

Gencompress 1999 has better compression rate than Biocompress. It searches for approximate matches, that are satisfy the C condition. The algorithm finds the optimal prefix and uses order-2 arithmetic encoding. It also finds the approximate complement palindromes. Their compression rate is 1.7428 average [4].

CTW+LZ 2000 is the combination of GenCompress and CTW. Uses local heuristics for problem of greedy choice. It has slow running time [5].

DNA compress 2002 reaches a little better compression rate than Gencompress. Finds all approximate matches, including the complement palindromes and encodes the approximate repeating regions and the non-repeating regions. It has 1.7254 average compression rate. It uses PatternHunter to preprocess. It has good running time [6].

DNASequitor 2004 is a language based compressor. The modified version of Sequitor (1997) to DNA [7].

DNAPack 2005 searches for the repeats with dynamic programming and uses context tree and 2 bit encoding for the non-repeating parts [8].

Exploring Three-Base Periodicity for DNA Compression 2006 uses the 3 base periodicity in protein coding zones with the help of a DNS model that has 3 deterministic states [9].

Relative Lempel-Ziv Compression 2010 stores the index of base sequences and compresses every further sequences relative to the bases. Uses suffix arrays, self-indexes, relative Lempel-Ziv factorization and compressed integer sets [10].

GenBit 2010 segments the input to parts of 4 characters, so 256 combinations exists. Every 4 segments of 4 bases are replaced by a binary number of 8 bits. If parts following each other are equals, then 1 bit comes as the 9th, else 0 [11].

DNABit 2011 assigns binary bits to smaller DNA base segments to compress repetitive and non-repetitive DNA sequences. It has 1.58 bits/base compression rate at worst case [12].

COMRAD 2012 reaches a really impressive rate: 0.25 bits/base as an example for smaller datasets, also it gives random accesses to subsequences without decompressing. Finds exact repeats and uses expensive process of multiple passes [13].

In 2016 researchers implemented and compared 4 different algorithms to compress DNA: LZW algorithm, run length encoding algorithm, Arithmetic coding and Substitution method with various species. They found that different organisms can be encoded effectively with different algorithms [14].

In 2017 an algorithm presented using multiple dictionaries with LZW, but it have not been implemented and tested yet [15].

3   The Algorithm

The main idea is that we have prebuilt, efficient dictionaries for the different biological classes that have similarities. Based on these similarities we can have the dictionary of the most frequent sequences in these species. In these dictionaries we assign much shorter bit words, than the others.

In this way, when encoding, first we must select our dictionary (e.g. virus, fungus, ...), then break the sequence into equivalent sized parts. If we found a part that appears in the dictionary of the frequent parts, then we start with a 0 bit, then the bits from the dictionary, else we start with a 1 bit, then the bits from the non-frequent dictionary.

If there is a k-mer which can not be found in neither dictionaries, then the non-frequent dictionary can be extended dynamically.

Sample code snippet to compress genome, where genome is the genome to compress, k is the size of subsequences in the dictionary, d1 is the dictionary for frequent subsequences and d2 is the dictionary for non-frequent subsequences:

```
FUNCTION genomeToBits(genome,n,d1,d2):
    bits=''
    i=0
    while i<len(genome):
        IF genome[i:i+n] in d1:
            bits+='0'
            bits+=d1[genome[i:i+n]]
        ELSEIF genome[i:i+n] in d2:
            bits+='1'
            bits+=d2[genome[i:i+n]]
        ENDIF
        i+=n
    ENDWHILE
    RETURN bits
ENDFUNCTION
```

Decoding is very simple. If we see 0, we get our subsequence back from the dictionary of frequent subsequences, otherwise if it is 1, we get our subsequence from the non-frequent dictionary.

Sample code snippet to decode the compressed genome, where bits is the encoded genome, n is the length of the bitwords for the frequent subsequences, k is the length of the bitwords for the non-frequent subsequences, r1 is the reversed dictionary for the dictionary of the frequent subsequences and r2 is the reversed dictionary for the non-frequent subsequences:

```
FUNCTION bitsToGenome(bits,r1,r2,n,k):
    genome=''
    i=0
    while i<len(bits):
        IF bits[i]=='0':
            i+=1
            genome+=r1[bits[i:i+k]]
            i+=k
        ELSE:
            i+=1
            genome+=r2[bits[i:i+n]]
            i+=n
        ENDIF
    ENDWHILE
    RETURN genome
ENDFUNCTION
```

When we first built up the dictionaries, for a dictionary with sequences of length k, we ran through species of that specific family, took out every k-mer of their genome, ordered them by frequencies, and put some of the beginning of this list into the more frequent dictionary with smaller bit words and the others into the less frequent dictionary.

So the first step is building substring indexes for the given family, like in [16], except we do not store the positions of the occurrences, but the frequencies instead.

If we would simply store all possible k-mers, it would be 4^k entry and we would have to generate bit word for every one of these. But if we chose k big enough, not all possibilities will appear as k-mers in the genomes and we can spare 1 or 2 bits for each k on this too.

Sample code snippet to build dictionaries, where ind is an index is a dictionary of subsequences with their frequencies, in descending order by frequencies:

```
FUNCTION generateDicts(ind):
    d1,d2={},{}
    n=math.sqrt(len(ind))
    j=0
    for i in ind:
        IF j<n:
            bits=bin(j)[2:]
            while len(bits)<(math.sqrt(n)/2)+1:
                bits='0'+bits
            ENDWHILE
            d1[i[0]]=bits
        ELSE:
            bits=bin(j)[2:]
            while len(bits)<math.sqrt(math.sqrt(len(ind)))+1:
                bits='0'+bits
            ENDWHILE
            d2[i[0]]=bits
        ENDIF
        j+=1
    ENDFOR
    RETURN d1,d2
ENDFUNCTION
```

4   Compression Rates

For the first measurements, we chose to classify species based on the biological kingdoms defined by Cavalier-Smith in 1998: bacteria, protozoa, chromista, plantae, fungi, animalia [17].

It is not yet decided if viruses can be included in the tree of life [18,19], now we have put them into a class too.

Name	Bases
Herpes	211518
Lambda	48502
E coli	5277676
Abiotrophia defectiva	2041839
Zymomonas mobilis	2061413
Candida tenuis	10747050
Absidia glauca excerpt	2999927
Mortierella verticillata excerpt	2999885

For the measurements, we got the viruses from the NCBI database [20], bacteria and fungus from the Ensembl Genome databases [21]. For some species, we only used excerpts. On this table you can see each genome with its length:

As our first measurement (Fig. 1), we checked the compress rates of bacteria with different ks. We can see that with the increasing of k, we can reach lower rates. But inside this biological kingdom, with same k, there is not much difference between the compression rates of the species.

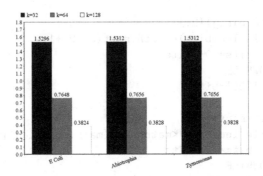

Fig. 1. Compression rates of bacteria with different ks. y = bits/base

As our second measurement (Fig. 2), we checked the compress rates of fungus with different ks. These are longer sequences, so with the same ks we get greater compression rates, but the result is similar in that perspective, that bigger k gives lower compression rates and the there is little difference between the species of the biological kingdom.

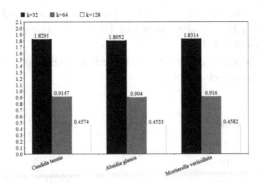

Fig. 2. Compression rates of fungus with different ks. y = bits/base

So we can conclude that we can reach lower rates with greater k inside a biological class, but with longer sequences with the same k we get bigger rates. Let us see how much we can compress really small sequences: viruses (Fig. 3).

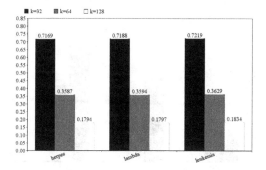

Fig. 3. Compression rates of virus with different ks. y = bits/base

We can reach really small rates with greater ks and smaller sequences, as we suspected. Now let us check the connection between the choice of k and the running times.

5   Running Times

We can see that, there's not much connection between the decompression time and k. It seems decompression time is only affected by the length of the genomes (Fig. 4).

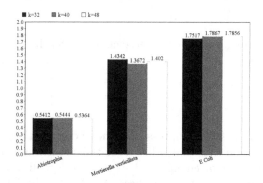

Fig. 4. Average decompression time of genomes with different lengths and ks. y = sec

However as we can see that, with greater k, the compression time decreases (Fig. 5). But then what is the cost of increasing k?

Generating the indexes and the dictionaries takes much longer time. Good thing is the user will not have to do this, the dictionaries will be ready for the user. But with increasing k, the generation time increases too (Fig. 6).

We made our measurements with Intel i5-8350U CPU @ 1.70 GHz 1.90 GHz, 16.0 GB RAM. We used python as the implementation language.

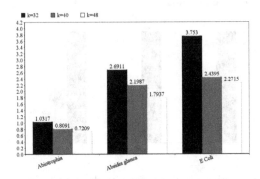

Fig. 5. Average compression time of genomes with different lengths and ks. y = sec

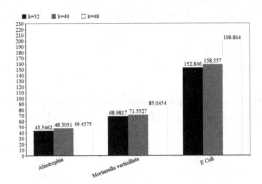

Fig. 6. Average time to generate index and dictionaries for different species with different lengths and ks. y = sec

6   Conclusion

Our algorithm works well to compress the genetic data, with good choice of k it reaches the compression rate that other algorithms reach, or even better rates.

General compression mechanisms does not work well with DNA strings, because of the special structures of these. We have only 4 types of possible characters, so we can store every base on 2 bits. We build up the indexes of different biological classes, then make two dictionaries for each race based on the occurrence frequencies of the subsequences. We assign shorter bit words to the more frequent subsequences and longer to the less frequent ones.

We must consider that not the user will have to set k, but the dictionaries will be prepared with an optimized k. Most ideally these dictionaries would be available online as a web service.

7   Future Work

We are planning to extend the functionalities of our secured bioinformatic database library with our compression and making efficient processing functions on these compressed formats.

We are also planning to increase the efficiency in memory usage and running time. For example it would be easy to make the compression parallel.

Moreover we will develop our compressing dictionaries as a web service.

Acknowledgment. The project was supported by the European Union, co-financed by the European Social Fund (EFOP-3.6.3-VEKOP-16-2017-00002).

References

1. Deorowicz, S., Grabowski, S.: Data compression for sequencing data. Algorithms Mol. Biol. **8**(1), 25 (2013)
2. Grumbach, S., Tahi, F.: A new challenge for compression algorithms: genetic sequences. Inf. Process. Manag. **30**(6), 875–886 (1994)
3. Rivals, E., Delahaye, J.-P., Dauchet, M., Delgrange, O.: A guaranteed compression scheme for repetitive DNA sequences. In: Proceedings of Data Compression Conference, DCC 1996, p. 453. IEEE (1996)
4. Chen, X., Kwong, S., Li, M.: A compression algorithm for DNA sequences and its applications in genome comparison. Genome Inform. **10**, 51–61 (1999)
5. Matsumoto, T., Sadakane, K., Imai, H.: Biological sequence compression algorithms. Genome Inform. **11**, 43–52 (2000)
6. Chen, X., Li, M., Ma, B., Tromp, J.: DNACompress: fast and effective DNA sequence compression. Bioinformatics **18**(12), 1696–1698 (2002)
7. Cherniavsky, N., Ladner, R.: Grammar-based compression of DNA sequences. DIMACS Working Group on The Burrows-Wheeler Transform, 21 (2004)
8. Behzadi, B., Le Fessant, F.: DNA compression challenge revisited: a dynamic programming approach. In: Apostolico, A., Crochemore, M., Park, K. (eds.) CPM 2005. LNCS, vol. 3537, pp. 190–200. Springer, Heidelberg (2005). https://doi.org/10.1007/11496656_17
9. Ferreira, P.J.S.G., Neves, A.J.R., Afreixo, V., Pinho, A.J.: Exploring three-base periodicity for DNA compression and modeling. In: Proceedings of the 2006 IEEE International Conference on Acoustics, Speech and Signal Processing, ICASSP 2006, vol. 5, p. V. IEEE (2006)
10. Kuruppu, S., Puglisi, S.J., Zobel, J.: Relative Lempel-Ziv compression of genomes for large-scale storage and retrieval. In: Chavez, E., Lonardi, S. (eds.) SPIRE 2010. LNCS, vol. 6393, pp. 201–206. Springer, Heidelberg (2010). https://doi.org/10.1007/978-3-642-16321-0_20
11. Rajeswari, P.R., Apparo, A., Kumar, V.K.: Genbit Compress Tool (GBC): a Java-based tool to compress DNA sequences and compute compression ratio (bits/base) of genomes. arXiv preprint arXiv:1006.1193 (2010)
12. Rajarajeswari, P., Apparao, A.: DNABit compress-genome compression algorithm. Bioinformation **5**(8), 350 (2011)
13. Kuruppu, S., Beresford-Smith, B., Conway, T., Zobel, J.: Iterative dictionary construction for compression of large DNA data sets. IEEE/ACM Trans. Comput. Biol. Bioinform. (TCBB) **9**(1), 137–149 (2012)

14. Machhi, V., Patel, M.S.: Compression techniques applied to DNA data of various species. DNA Seq. **8**(3) (2016)
15. Keerthy, A.S., Priya, S.M.: Lempel-Ziv-Welch compression of DNA sequence data with indexed multiple dictionaries. Int. J. Appl. Eng. Res. **12**(16), 5610–5615 (2017)
16. Bockenhauer, H.-J., Bongartz, D.: Algorithmic Aspects of Bioinformatics. Springer, Heidelberg (2007). https://doi.org/10.1007/978-3-540-71913-7
17. Cavalier-Smith, T.: A revised six-kingdom system of life. Biol. Rev. **73**(3), 203–266 (1998)
18. Moreira, D., López-García, P.: Ten reasons to exclude viruses from the tree of life. Nat. Rev. Microbiol. **7**(4), 306 (2009)
19. Hegde, N.R., Maddur, M.S., Kaveri, S.V., Bayry, J.: Reasons to include viruses in the tree of life. Nat. Rev. Microbiol. **7**(8), 615 (2009)
20. NCBI National Center for Biotechnology Information. https://www.ncbi.nlm.nih.gov/
21. Ensembl genomes. http://ensemblgenomes.org/

Utilizing Pretrained Deep Learning Models for Automated Pulmonary Tuberculosis Detection Using Chest Radiography

Thi Kieu Khanh Ho[1], Jeonghwan Gwak[2,3,4(✉)], Om Prakash[5],
Jong-In Song[1], and Chang Min Park[2,4,6,7]

[1] School of Electrical Engineering and Computer Science, Gwangju Institute
of Science and Technology, Gwangju 61005, Korea
[2] Department of Radiology, Seoul National University Hospital,
Seoul 03080, Korea
james.han.gwak@gmail.com
[3] Biomedical Research Institute, Seoul National University Hospital,
Seoul 03080, Korea
[4] Department of Radiology, Seoul National University College of Medicine,
101, Daehak-ro, Jongno-gu, Seoul 03080, Korea
[5] Computer Engineering Department, Institute of Technology, Nirma University,
Ahmedabad 382481, India
[6] Institute of Radiation Medicine, Seoul National University Medical Research
Center, Seoul 03080, Korea
[7] Seoul National University Cancer Research Institute, Seoul 03080, Korea

Abstract. Tuberculosis (TB) is determined as a major health threat resulting in approximately 1.8 million people died in 2015 in most of the low and middle income countries. Many of those deaths could have been prevented if TB had been diagnosed and treated at an earlier stage. Nevertheless, recent advanced diagnosis techniques such as the methods of frontal thoracic radiographs have been still cost prohibitive for mass adoption due to the need of individual analysis of each radiograph by properly experienced radiologists. In addition, current outperformances of deep learning accomplish significant results for classification tasks on diverse domains, but its capability remains limited for tuberculosis detection. Therefore, in this study, we examine the efficiency of deep convolutional neural networks (DCNNs) for detecting TB on chest radiographs using public ChestXray14 as training dataset and Montgomery and Shenzhen as two external testing datasets. Multiple preprocessing techniques, tSNE visualization and data augmentation are first performed. Three different pre-trained DCNNs, namely ResNet152, Inception-ResNet and DenseNet121 models are then used to classify X-ray images as having manifestations of pulmonary TB or as healthy. We observe that appropriate data augmentation techniques are able to further increase accuracies of DCNNs. We achieve the best classifier having an average AUC of 0.95 with DenseNet121 while 0.91 and 0.77 with Inception-ResNet and ResNet121, respectively.

Keywords: Chest X-ray · TB detection · ChestX-ray14 ·
Deep learning · Pre-trained CNNs

© Springer Nature Switzerland AG 2019
N. T. Nguyen et al. (Eds.): ACIIDS 2019, LNAI 11432, pp. 395–403, 2019.
https://doi.org/10.1007/978-3-030-14802-7_34

1 Introduction

As an infectious disease caused by the bacillus Mycobacterium tuberculosis, Tuberculosis (TB) has been leading cause of death worldwide, alongside human immunodeficiency virus–acquired immune deficiency syndrome (known as HIV-AIDS). TB mainly affects the lung region and typically manifests near the clavicles showing different pathological patterns or manifestations in the lungs corresponding to various factors and no single or specific sign can confirm its presence. However, TB, as a curable disease, can be effectively detected and treated under powerful techniques.

It was reported that there is an insufficient resource of radiology interpretation expertise with diagnosis agreements among skilled radiologists in detection of TB-prevalent locations. Thus, developments of computer-aided diagnosis (CAD) systems for automatic detection of TB from chest radiographs (CXRs) has been a great functional tool for under-developed and developing countries where high burden of patients and poor medical services are currently challenging in TB diagnosis. Traditionally, CAD systems consist of four phases, including pre-processing, segmentation, feature extraction and classification. In these systems, the region-of-interest of TB is segmented and thereafter important texture and features are manually extracted. However, there is a need to discover more strategies in TB diagnosis and treatments that promising deep learning techniques in radiology has been outperformed conventional CADs. By building the end-to-end architecture, best features are automatically extracted and can be usable for further purposes.

Deep learning techniques have recently earned outstanding results in a broad range of machine learning tasks. Convolutional Neural Networks (CNNs), known as one of the powerful architectures, handle four different manners: training the weights from scratch done on very large available datasets, fine-tuning the weights of an existing pretrained CNN with much smaller datasets, using the unsupervised pre-training to initialize the weights before putting inputs into CNN models and using pre-trained CNN called an off-the-shelf or out-of-box CNN as a feature extractor. CNNs have especially proved it powers in image classifications and being successfully applied in disease diagnosis either on detection of pleural effusion and cardiomegaly at chest radiography, mediastinal lymph nodes at computed tomography (CT), lung nodules at CT, pancreatic and brain segmentation or others [1–5], but its application on TB detection still remains limited. Therefore, in this study, we evaluate the efficacy of different DCNN models for TB detection on Chest X-ray14 and two external Montgomery and Shenzhen datasets.

The rest of the paper is organized as follows. Section 2 briefly presents the related works done till date. Visualization of the complexity of our chosen dataset, data augmentation techniques and proposed DCNNs models are described in Sect. 3. Section 4 summarizes experimental results using proposed models with three public datasets. The paper concludes in Sect. 5 with a short future works.

2 Related Works

Chest radiographs typically have the presence of different types of manifestations, including cavities, opacities, consolidation, focal lesions and nodules that lead to the difficulty to detect solely TB. Among the first proper methods proposed for automatic TB detection, [6] extracted and analyzed the local texture of CXRs to detect different abnormalities. Lung regions were first divided into several overlapping areas. K-nearest neighbor was thereafter used at region level to facilitate classification tasks based on the combination of each region with the weighted multiplier. [7] proposed a hybrid method that suspected cavities to be detected using adaptive thresholding and active contour model. Bayesian classifier was thereafter used to confirm or discard the suspected cavities. A dual scale method to detected TB was proposed by [8] with three major steps. Cavity candidates at the coarse level was first identified using Gaussian template matching, local binary pattern and histogram of oriented gradient features. Then, these cavity candidates were segmented using active contour-based method. Lastly, at the finer level, Support Vector Machine (SVM) was performed to remove false positives based on different features. This works gave 82.6% of accuracy on testing 35 CXRs. [9] composed a proposal to detect TB using a MIL technique. Each CXR was divided into unlabeled sub-regions called instances and features based on pixel intensity distributions were extracted. In this approach, the label of each sub-region was unknown while label of whole images was known that leading to not capable to classify abnormalities on CXR at instance levels using standard SVM. Authors, hence, proposed the usage of MiSVM which enabled the classification of groups of samples. By utilizing three private proprietary datasets (Gambia, Tanzania and Zambia), they gave a competitive result with current literation with AUC of 0.86 to 0.91.

Regarding deep learning-based approaches for CXR abnormality detections, a survey of automated disease prediction literature revealed the usage of pre-trained CNNs, known as feature extractors. [10] reported the acceptable pre-trained models in identifying pleural effusion and cardiomegaly on frontal CXRs. By combining features extracted from pre-trained CNN and Pico-Descriptors, their performance achieved promising results with AUC of 0.93 and 0.89 for the detection of the right pleural effusion and cardiomegaly, respectively. Toward TB detection, in particular, the capability of CNN was demonstrated by [11] using a customized CNN model with AlexNet framework and transfer learning trained on private 10K CXR images. Although their customized model gave fair results when trained with initializations of random weights, they obtained competitive results on the publicly available Montgomery and Shenzhen datasets with AUC of 0.884 and 0.926, respectively. [13] also used the same fine-tuned deep learning models with the shuffle sampling methods and cross validation on the unbalanced datasets. They achieved 85.56% of classification accuracy. [12] aimed to evaluate five pre-trained CNNs such as AlexNet, VGG-16, VGG-19, Xception and ResNet towards improving the accuracy of TB screening from CXRs. Based on the learning from task-specific features from the posterior-anterior (PA) CXRs, authors tried to proposed models that could be optimized for hyperparameters in minimizing the classification error during training process. As a result, VGG19 gave the best AUC of 0.956.

In addition, another proposed DCNNs-based model to classify CXR images into different categories of TB manifestations was presented by [14]. By modifying the GoogleNet model from Caffe as the pre-training network and developing a new machine learning and mobile health screening system to minimize the wait-time for diagnosis, they achieved 89.6% of binary classification after 100,000 iterations, whereas their accuracy for multi-class classification (in five classes, including normal, cavity, lymphadenopathy, infiltration and pleural effusion) was 67.07% after 10,000 iterations. [15] increased the number of original TB images using data augmentation such as 90, 188 and 270° rotations, mirror images and histogram equalization. They then put the augmented images into AlexNet and VGGNet for desire classification purposes. [16] proposed a potential CNN architecture comprising of seven convolutional layers and three fully connected layers. With three different optimizers, including Adam, momentum, and stochastic gradient descent (SGD), they evaluated that the Adam optimizer gave the best accuracy. Their further works have been expanded by applying a ResNet model for the same classification objective [17]. Here, they increased the datasets with more data augmentation techniques resulting in the best features which could be automatically extracted. [18] aimed to present three different approaches to the usages of CNN such as simple and direct feature extraction, multiple instance learning and ensembles of classifiers to create s complementary screener of TB diagnosis CXR. Each of those proposed approaches used three pre-trained CNN architectures (GoogleNet, ResNet and VGGNet) as feature extractors, Support Vector Machine was later applied to identify whether the images contain tuberculosis. Their obtained results were fairly competitive with other published works demonstrating the potential of pre-trained CNN in medical image extractors.

3 Materials and Methodology

3.1 Datasets

Our experiments use three publicly available datasets from ChestX-ray14, Montgomery, Maryland and Shenzhen, China, maintained by the National Library of Medicine, National Institute of Health. Details pertaining to the origin and characteristics of each dataset are illustrated by Table 1.

3.2 Preprocessing

All CXR images are initially resized to specific range pixels based on deep models being applied due to the different sizes of three datasets. As mentioned previously, ChestX-ray14 is used for our training dataset, but two classes were unequally represented, called imbalanced datasets. Often real-world datasets are predominately consisted of normal samples with a small percentage of abnormal samples. Hence, it is the case that the cost of abnormal sample misclassification from normal ones is much higher than the cost of the reverse error. We therefore propose the data augmentation techniques to balance the number images of two classes. While under-sampling of the majority (CXRs from healthy control patients) is applied as a good classifier of

Table 1. CXR datasets and characteristics

Origin	No. of positive cases for TB	No. of healthy control patients	File type	Bit depth	Resolution
Chest-Xray14 [19]	5689	48922	PNG	8-bit	1024 × 1024
Montgomery [20]	58	80	PNG	8-bit	4020 × 4892
Shenzhen [20]	336	326	PNG	8-bit	948-3001 × 1130-3001

increasing sensitivity to the minority class (CXRs from TB patients), we also apply over-sampling approach in which the minority class is over-sampled by data augmentation techniques, including rotation 25°, flipping left-to-right, random cropping with 0.8 and 0.9% areas, histogram equalization and 25° shearing.

As a technique to visualize the variant distribution and its complexity, we perform the combination of Principle Component Analysis (PCA) and T-Distributed Stochastic Neighboring Entities (t-SNE) to first reduce dimensions in the training dataset whilst retaining most information, then to visually explore the two-class ChestX-ray14 data distributions before and after augmenting techniques. Particularly, PCA uses the correlation between some dimensions and tries to offer a minimum amount of variables keeping the maximum amount of variation or information about the original dataset is distributed. Thus, we first create new dataset containing 50 dimensions generated by PCA algorithms and it roughly hold around 87.71% and 88.16% of total variation in the data before and after augmenting, respectively. t-SNE thereby minimizes the divergence between two distributions: one measures pairwise similarities of inputs and one measures pairwise similarities of the corresponding low dimension and tries to present our data using less dimensions by matching both distributions. Here, PCA inputs can be fed into t-SNE algorithm. We use 10,000 samples out of 54611 samples (before augmenting) and 10,000 samples out of 97844 samples (after augmenting) (see Fig. 1).

3.3 Proposed Deep Convolutional Neural Networks (DCNNs)

Since CNN models become increasingly deep, a new research problem emerges. For example, the input's information or gradients pass through many layers, it can vanish by the time when it reaches the end or beginning of the network. Various recent publications have addressed this problem to ensure maximum information flow between layers in the network. One of an orthogonal approach for making networks deeper is to increase the network's width. The Inception module, known as GoogleNet concatenates feature maps produced by filters of different sizes. A variant of ResNet, on the other hand, is proposed with the wide generalized residual blocks. In fact, ResNets can improve its performance provided the sufficient depth by simply increasing the number of filters in each layers. In comparison, instead of exploring the representation of extremely deep and wide architectures like Inceptions and ResNets. DenseNets exploit the network's potential through feature reuse, yielding condensed models to be easier in training with highly parameter-efficient. By concatenating feature maps

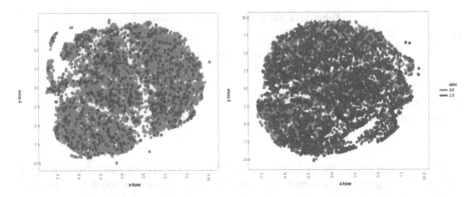

Fig. 1. PCA – t-SNE visualization of two categories (0: Healthy Control Patients and 1: TB Patients) from ChestX-ray14 datasets before (left) and after (right) augmenting data.

learned by different layers, DensNets also can increase variation of inputs of subsequent layers and training efficiency compared to ResNets. Compared to Inception networks which also concatenate features from different layers, DenseNets are much simpler and effective. Based on their outperformances compared to existing models proposed previously such as AlexNet, VGG, MobileNet, etc., we adopt three different DCNN architectures to our TB study, ResNet152 [21], Inception-ResNet [22] and DenseNet [23] including pre-trained models that were already trained on 1.2 million color images for ImageNet consisting of 1000 categories.

Images are down-sampled to 299 × 299 and 224 × 224 pixel resolutions to match the input sizes for pre-trained ResNets, Inception-ResNet and DenseNet. They are then loaded onto a Linux operating System (Ubuntu 16.04) with Tensorflow deep learning framework with CUDA 9.0/cuDNN 7.5 dependencies for graphics processing unit acceleration. Each layer of pre-trained CNNs produces an activation map that match primitive features such as blobs, edges, colors from earlier layers. The deeper layers then capture the previous information and formulate higher level features to present more affluent representation. The architecture and weights are downloaded from GitHub repository. We also experimentally determine the optimal layers from three pre-trained DCNNs for improving usable feature extractions and TB detection accuracy right after.

4   Experimental Results

Of over 10,000 patients from ChesX-ray14 datasets, 1266 (12.7%) random patients are selected for internal testing. Of these 1266 patients, 633 are positive and 633 are healthy. Among the remaining 8734 patients, it is randomly split 80:20 ratios into 6987 patients used for training whilst 1747 patient for validation. Data augmentation techniques are then applied for training and validation datasets, as mentioned in Subsect. 3.2. Moreover, two external testing sets, Montgomery and Shenzhen, are also used for evaluating our model efficacies.

A summary of our results are reported by AUC results and plots (see Table 2 and Fig. 2). The best performing model with AUC of 0.9872 on internal ChestX-ray14, 0.9139 on Montgomery testing set and 0.9384 on Shenzhen testing set is DenseNet121. It is significantly greater than the performances of two remaining pre-trained models, ResNet152 and Inception-ResNet. Table 3 also compares our results obtained with other available studies in the literature on TB detection using two public datasets. It shows that our proposed pre-trained DenseNet121 achieved a competitive result in comparison with a customized DCNN from Hwang et al. [11], pre-trained AlexNet and VGG19 from Sivaramakrishnan et al. [12].

Table 2. AUC results of three different testing sets

Pre-trained model	Internal ChestX-ray14	Montgomery	Shenzhen
ResNet152	0.8675	0.7002	0.7496
Inception-ResNet	0.9606	0.8552	0.9179
DenseNet121	0.9872	0.9139	0.9384

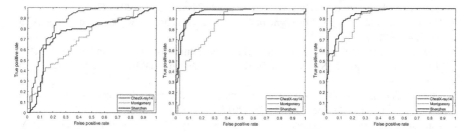

Fig. 2. Comparison of ROC curves for the pre-trained ResNet152, Inception-ResNet and DenseNet121 models (from left to right).

Table 3. Comparison AUC results with literature on Montgomery and Shenzhen datasets

Datasets	Proposed	Hwang et al. [11]	Sivaramakrishnan et al. [12]
Montgomery	**0.939**	0.926	0.926
Shenzhen	**0.914**	0.884	0.833

5 Conclusions

In this study, we compare the performances of three different pre-trained DCNNs model toward improving the accuracy of TB detection from frontal CXRs. Due to the scarcity of data and highly imbalanced distribution across two classes form ChestX-ray14 dataset which directly impact to the performance of each model, we first demonstrate the appropriate data augmentation techniques. We then observe that pre-trained DenseNet121 model significantly yields better results compared to pre-trained ResNet and Inception-ResNet models on three categories of testing datasets, including internal ChestX-ray14, Montgomery and Shenzhen sets. As our future works, further improvements can be evaluated in terms of using different settings of DNNs' configurations and different existing standard pre-trained CNN models to determine capabilities of transfer learning and fine-tuning schemes. Moreover, classifying CXRs images into distinguished TB manifestations, for which a larger dataset is needed, is indispensable topic to address.

Acknowledgements. This work was supported by the Basic Science Research Program through the NRF funded by the Ministry of Education (NRF-2017R1D1A1B03036423, NRF-2017R1A2B4008517) and the Brain Research Program through the National Research Foundation of Korea (NRF) funded by the Ministry of Science, ICT & Future Planning (NRF-2016M3C7A1905477, NRF-2019M3C7A1020406).

References

1. Apou, G., et al.: Detection of lobular structures in normal breast tissue. Comput. Biol. Med. **74**, 91–102 (2016)
2. Hua, K.-L., et al.: Computer-aided classification of lung nodules on computed tomography images via deep learning technique. Onco Targets Ther. **8**, 2015–2022 (2015)
3. Havaei, M., et al.: Brain tumor segmentation with deep neural networks. Med. Image Anal. **35**, 18–31 (2017)
4. Kawahara, J., BenTaieb, A., Hamarneh, G.: Deep features to classify skin lesions. In: 2016 IEEE 13th International Symposium on Biomedical Imaging (ISBI), pp. 1397–1400. IEEE (2016)
5. Miki, Y., et al.: Classification of teeth in cone-beam CT using deep convolutional neural network. Comput. Biol. Med. **80**, 24–29 (2017)
6. van Ginneken, B., Katsuragawa, S., ter Haar Romeny, B.M., Doi, K., Viergever, M.A.: Automatic detection of abnormalities in chest radiographs using local texture analysis. IEEE Trans. Med. Imaging **21**(2), 139–149 (2002)
7. Shen, R., Cheng, I., Basu, A.: A hybrid knowledge-guided detection technique for screening of infectious pulmonary tuberculosis from chest radiographs. IEEE Trans. Biomed. Eng. **57** (11), 2646–2656 (2010)
8. Xu, T., Cheng, I., Long, R., Mandal, M.: Novel coarse-to-fine dual scale technique for tuberculosis cavity detection in chest radiographs. EURASIP J. Image Video Process. **2013** (1), 1 (2013)
9. Melendez, J., et al.: A novel multiple-instance learning-based approach to computer-aided detection of tuberculosis on chest X-rays. IEEE Trans. Med. Imaging **34**(1), 179–192 (2015)
10. Bar, Y., Diamant, I., Wolf, L., Lieberman, S., Konen, E., Greenspan, H.: Chest pathology detection using deep learning with non-medical training. In: Proceedings of the ISBI, pp. 294–297 (2015)
11. Hwang, S., Kim, H.-E., Jeong, J., Kim, H.-J.: A novel approach for tuberculosis screening based on deep convolutional neural networks. In: Proceedings of the SPIE, vol. 9785, p. 97852W (2016)
12. Sivaramakrishnan, R., et al.: Comparing deep learning models for population screening using chest radiography. In: Medical Imaging 2018: Computer-Aided Diagnosis, vol. 10575. International Society for Optics and Photonics (2018)
13. Liu, C., et al.: TX-CNN: detecting tuberculosis in chest X-Ray images using convolutional neural network. In: IEEE Proceeding of the ICIP, pp. 1–5 (2017)

14. Cao, Y., et al.: Improving tuberculosis diagnostics using deep learning and mobile health technologies among resource-poor and marginalized communities. In: IEEE First International Conference on Connected Health: Applications, Systems and Engineering Technologies (CHASE), pp. 274–281. IEEE (2016)
15. Rohilla, A., et al.: TB detection in chest radiograph using deep learning architecture. In: ICETETSM-17, pp. 136–147 (2017)
16. Hooda, R., et al.: Deep-learning: a potential method for tuberculosis detection using chest radiography. In: IEEE ICSIPA, pp. 497–502 (2017)
17. Hooda, R., et al.: Automated tuberculosis classification of chest radiographs by using convolutional neural networks. In: IJETSR, pp. 310–317 (2018)
18. Lopes, U.K., Valiati, J.F.: Pre-trained convolutional neural networks as feature extractors for tuberculosis detection. Comput. Biol. Med. **89**, 135–143 (2017)
19. Wang, X., et al.: Chestx-ray8: hospital-scale chest x-ray database and benchmarks on weakly-supervised classification and localization of common thorax diseases. arXiv preprint arXiv:1705.02315 (2017)
20. Jaeger, S., Candemir, S., Antani, S., Wáng, Y.X.J., Lu, P.X., Thoma, G.: Two public chest X-ray datasets for computer-aided screening of pulmonary diseases. Quant. Imaging Med. Surg. **4**(6), 475–477 (2014)
21. He, K., Zhang, X., Ren, S., Sun, J.: Identity mappings in deep residual networks. In: Leibe, B., Matas, J., Sebe, N., Welling, M. (eds.) ECCV 2016. LNCS, vol. 9908, pp. 630–645. Springer, Cham (2016). https://doi.org/10.1007/978-3-319-46493-0_38
22. Szegedy, C., Ioffe, S., Vanhoucke, V., Alemi, A.A.: Inception-v4, inception-resnet and the impact of residual connections on learning. In: AAAI, vol. 4, p. 12 (2017)
23. Huang, G., Liu, Z., Van Der Maaten, L., Weinberger, K.Q.: Densely connected convolutional networks. In: CVPR, vol. 1, no. 2, p. 3 (2017)

Intelligent and Contextual Systems

Advanced Neural Network Approach, Its Explanation with LIME for Credit Scoring Application

Lkhagvadorj Munkhdalai[1] , Ling Wang[2] , Hyun Woo Park[1] ,
and Keun Ho Ryu[3,4(✉)]

[1] Database/Bioinformatics Laboratory,
School of Electrical and Computer Engineering,
Chungbuk National University, Cheongju 28644, Republic of Korea
{lhagii,hwpark}@dblab.chungbuk.ac.kr
[2] Department of Computer Technology, School of Information Engineering,
Northeast Electric Power University, Jilin City 132012, China
smile2867ling@neepu.edu.cn
[3] Faculty of Information Technology, Ton Duc Thang University,
Ho Chi Minh City 700000, Vietnam
khryu@tdtu.edu.vn
[4] Department of Computer Science, Chungbuk National University,
Cheongju 28644, Republic of Korea
khryu@chungbuk.ac.kr

Abstract. Neural network models have achieved a human-level performance in many application domains, including image classification, speech recognition and machine translation. However, in credit scoring application, neural network approach has been useless because of its black box nature that the relationship between contextual input and output cannot be completely understood. In this study, we investigate the advanced neural network approach and its' explanation for credit scoring. We use the LIME technique to interpret the black box of such neural network and verify its' trustworthiness by comparing a high interpretable logistic model. The results show that neural network models give higher accuracy and equivalent explanation with the logistic model.

Keywords: Neural network · LIME · Credit scoring

1 Introduction

Credit scoring model accurately estimates the borrowers' credit risk as well as explains the association between borrowers' characteristics and their credit scores. Although neural network models achieve a higher predictive accuracy of the borrowers' credit-worthiness, their decision-making process is rarely understood because of the models' black box nature that they are currently neither interpretable nor explainable. Without explanations, neural network approach cannot meet regulatory requirements, and thus cannot be adopted by financial institutions and would likely not be accepted by consumers [1].

© Springer Nature Switzerland AG 2019
N. T. Nguyen et al. (Eds.): ACIIDS 2019, LNAI 11432, pp. 407–419, 2019.
https://doi.org/10.1007/978-3-030-14802-7_35

Therefore, this study investigates the advanced neural network models and their explanations for credit scoring in order to fill the gap between experimental studies from the literature and the demanding needs of the lending institutions [2].

More recently, Ribeiro et al. [3] proposed the LIME technique, short for Local Interpretable Model-agnostic Explanations, in an attempt to explain any decision process performed by a black box model. We adopt this technique to interpret our models as well as explain the black box of neural networks. However, LIME could explain the black box of neural network models, but it is essential to verify its trustworthiness. In order to evaluate the trustworthiness of LIME, we compare our result to the logistic model constructed by Logistic regression [4]. This is the most popular and powerful white-boxing method that commonly used on credit scoring application. Here are some properties of logistic regression that make it a benchmark - good predictive accuracy, high-level of interpretability and the modeling process is faster, easier and makes more sense [5]. Therefore, we can utilize it to verify the trustworthiness of LIME by comparing the unbiased logistic regression coefficients with contextual input data and its significance.

In the experimental part, we apply the advanced neural network approach with LIME explanation to over three real-world credit scoring datasets. Then we perform an extensive comparison between the neural network and logistic models for both simple and complex neural network architectures. The performance of model interpretability is measured by R-squared and the Wald chi-squared test is used to determine significance of variables. In addition, the model's predictive performance of test set is evaluated against four theoretical measures, an area under the curve (AUC), AUC-H, H-measure, and accuracy [6].

This paper is organized as follows. Section 2 briefly presents advanced neural network approach and LIME. Section 3 indicates the performance of neural network and logistic models, explanation of black box model and its trustworthiness. Finally, Sect. 4 concludes and discusses the general findings from this study.

2 Methods

2.1 Advanced Neural Networks

Neural networks are a set of algorithms that can learn to approximate an unknown function between any input and output. Nowadays, neural networks are being popular because these methods have dramatically improved the state-of-the-art in visual object recognition, speech recognition, object detection, genomics, energy consumption as well as financial domains [7]. In addition, Multi-Layer Perceptron (MLP) is widely used approach in neural network applications [8–10]. This study utilizes MLP for credit scoring application.

MLP is a general architecture in an artificial neural network that has been developed similar with human brain function, the basic concept of a single perceptron was introduced by Rosenblatt, (1958) [11]. MLP consists of three layers with completely different roles called input, hidden and output layers. Each layer contains a given number of nodes with the activation function and nodes in neighbor layers are linked by weights. MLP achieves the optimal weights by optimizing objective function using backpropagation algorithm to construct a model as:

$$\arg\min_{\omega} \frac{1}{T} \sum_t l(f(\omega x + b); y) + \lambda \Omega(\omega) \tag{1}$$

where ω denotes the vector of weights, b is the bias and $f(*)$ is the activation function, x is the dependent variables, y is the independent variable and $\Omega(*)$ is a regularizer. There are several parameters that need to be determined in advance for the training model such as number of hidden layers, number of their nodes, learning rate, batch size and epoch number. In addition, there are several methods to avoid overfitting problem.

Early Stopping. An Early Stopping algorithm for finding the optimal epoch number based on given other hyper-parameters. This algorithm is to prematurely stop the training at the optimal epoch number when the validation error starts to increase. It also helps to avoid overfitting [12]. However, overfitting is still a challenging issue when the training neural networks are extremely large or working in domains which offer very small amounts of data. If the training neural networks are extremely large, the model will be too complex and it would be transformed into an untrustworthy model.

Dropout. The dropout technique was proposed for addressing this problem [13]. This method is able to efficiently prevent overfitting by randomly dropping out nodes in the network. Dropping nodes creates thinned networks during training. At test time the results of different thinned networks are combined using an approximate model averaging procedure. Dropout has significantly reduced overfitting and gave such improvements in many applications including image classification and automatic speech recognition [14].

Optimizer. The choice of optimization algorithm in neural network has a significant impact on the training dynamics and task performance. There are many techniques to improve the gradient descent optimization and one of the best optimizer is Adam [15]. Adam computes adaptive learning rates for different parameters from estimates of first and second moments of the gradients and realizes the benefits of both Adaptive Gradient Algorithm (AdaGrad) and Root Mean Square Propagation (RMSProp). Therefore, Adam is considered one of the best gradient descent optimization algorithms in the field of deep learning because it achieves good results and faster [16].

2.2   LIME

Mostly machine learning algorithms are notoriously difficult to interpret because of the model's black box nature. Recently, in order to overcome this issue, Ribeiro et al., proposed the LIME technique that explains the predictions of any classifier in an interpretable and faithful manner, by learning an interpretable model locally around the prediction [3]. The key idea is that the simple model can be used to explain the predictions of the more complex model locally. Accordingly, LIME attempts to approximate an interpretable model based on a specific combination of the original representation of an instance being explained, x, made by the uninterpretable model (neural network) $f(x)$. This explanation model is defined as a $g \in G$, where G is a class of potentially interpretable simple models such as simple regressions, decision trees, etc. But not every g is simple enough to be interpretable, thus the complexity of g is measured with $\Omega(g)$. The locality around x is defined as the distance measure π_x and the local unfaithfulness between uninterpretable model $f(x)$ and $g(x)$ is expressed with the function $\mathcal{L}(f, g, \pi_x)$. The explanation obtained from LIME can be expressed as:

$$\xi(x) = \arg\min_{g \in G} \mathcal{L}(f, g, \pi_x) + \Omega(g) \tag{2}$$

$\xi(x)$ is the explanation model g that minimizes the locality-aware loss $\mathcal{L}(f, g, \pi_x)$, without making any assumptions about $f(x)$ since the explainer is model-agnostic. Then it is suggested to let G be a class of linear models in which case the minimization search can be implemented using perturbed samples of x. LIME uses the local behavior of $f(x)$ by sampling from the instances around x to obtain the interpretable representations based on slightly different inputs, x'. Thus x is treated as context used to express x'. Distance-based context creation allows for the creation of an accurate representation of neural networks expressed with a simple linear model.

Ribeiro et al. [3] used the locally weighted square loss function as $\mathcal{L}$:

$$\mathcal{L}(f, g, \pi_x) = \sum_{x, x'} \pi_x(x')(f(x) - g(x'))^2 \tag{3}$$

where $\pi_x(x') = \exp\left(-D(x, x')^2/\sigma^2\right)$ is an exponential kernel defined on some distance function D, with width σ, which is picked depending on the classification problem. Moreover, LIME also provides the explanation fit that allows us to understand how well that model explains the local region and the weighted importance for each variable that best describes the local relationship.

3 Results

3.1 Experimental Set-Up

In this section, MLP neural network method and its explanation with LIME is compared with the white-boxing logistic model in terms of three real-world credit datasets. Two datasets from UCI repository [17], namely German and Taiwan, and other one dataset from FICO's explanation machine learning challenge [18], namely FICO. A summary of all the datasets is presented in Table 1.

Table 1. Summary of the three datasets.

Dataset	Instances	Variables	Training set	Validation set	Test set	Good/bad
German	1000	21	640	160	200	700/300
Taiwan	30000	23	19200	4800	6000	23364/6636
FICO	9871	24	6318	1579	1974	5136/4735

Regarding logistic regression, in order to create unbiased estimators, we have to detect severe multicollinearity because it provides insignificant estimates or estimates with wrong signs for important contextual variables [19]. A popular technique to correct multicollinearity is to drop one or more of the highly correlated explanatory variables. Therefore, in data preprocessing, correlation and random forest variable importance were used to identify the most important ones from highly correlated variables – for example: in German dataset, credit amount and duration are contextually correlated (0.64), and credit duration variable is more important than credit amount, thus we dropped credit amount variable according to random forest feature importance. In a similar way as with the mentioned previously, 1, 9 and 6 variables were dropped from German, Taiwan and FICO datasets, respectively.

In the advanced neural network, we compared 12 neural networks that consist of different number of hidden layers, nodes and with or without dropout. We also set the learning rate to 0.001, maximum epoch number for training to 1000 and use a mini-batch with 32, 128 and 64 instances at each iteration for the German, Taiwan and FICO datasets, respectively.

All experiments are performed using R programming language, 3.4.0 version, on a PC with 3.4 GHz, Intel Core i7, and 32 GB RAM, using Microsoft Windows 10 operating system. Particularly, this study used several libraries such as 'Fselector', 'Keras' and 'lime' in R [3, 20, 21].

3.2 Predictive Performances

Our aim in this empirical evaluation is to display that the advanced neural network approach can lead to better performance than the industry-benchmark logistic regression in terms of different evaluation metrics. To validate neural network approach and to make a reliable conclusion, Tables 2, 3 and 4 compared the performance metrics of the 12 neural network classifiers that consist of a different number of hidden layers, nodes and with or without dropout, and logistic regression on the three datasets. The experiments are looped 5 times to enhance their robustness and to avoid sample dependency, and the evaluation measures are averaged in the comparison of results.

For the German dataset (see Table 2), logistic model indicates the best performance in terms of the all evaluation metrics. Logistic model achieves 76.9% AUC, 79.3% AUC-H, 0.304 H-measure, and 78.5% accuracy, which are 0.00%, 0.003%, 0.08 and 0.01% better than neural network model that consists of 3 hidden layers, 64 nodes in each hidden layer and without dropout, respectively. The AUC and AUC-H indicate classifying ability between borrowers as good and bad. Whereas, H-measure is better at dealing with cost assumptions among credit classes. In general, it is found that with the German dataset, logistic model shows promising predictive performances over the most evaluation metrics, indicating that logistic regression is an appropriate approach with the small dataset in credit scoring and results for logistic model consistent with conclusion by Lessman et al. [5].

In the Taiwan dataset (see Table 3), the neural network that consists of 5 hidden layers, 64 nodes in each layer, and without dropout model provides improvement over logistic model by 2.8% AUC. The AUC of network model achieves 75.4%, which indicates its classification ability between bad and good borrowers. In addition, the H-measure proves that the network model is better at dealing with cost assumptions between classes, as it scores 0.257, better than logistic model by 0.36. Finally, for the accuracy, the neural network that has 3 hidden layers, 64 nodes at each hidden layer and without dropout model achieves the best accuracy of the probabilities, at 82%. The reason why the AUC, H-measure and accuracy give better results than logistic model in the German dataset might be that the Taiwan dataset contains more instances.

Regarding the FICO dataset (see Table 4), the neural network that has 3 hidden layers, 16 nodes at each layer and without dropout model improves the predictive performance over the logistic model by 0.01% AUC, 0.01% AUC-H and 0.04% accuracy respectively. In terms of the AUC, the neural network model achieves 80.9%. The H-measure of the advanced neural network approach achieves 0.332, which is very nearly to the performance of logistic regression. Lastly, the accuracy of neural network model succeeds the best accuracy of the probabilities with 74.3%. This again provides evidence that the neural network models constructed on the datasets, which contain more instances, are better than the logistic model. Overall, regarding the FICO dataset, the logistic regression's results indicate that it is a close rival to the neural network models for all evaluation metrics. Since it has proven that the advanced neural network approach is an efficient and promising classifier for developing credit scoring models when dataset contains a large number of instances, we will consider its explanation ability with LIME by comparing it to the white-boxing logistic model in the next section.

Table 2. Predictive performances for the German dataset over the different evaluation metrics.

Models	AUC	AUC-H	H measure	Accuracy
Logistic	**0.769 ± 0.021**	**0.793 ± 0.022**	**0.304 ± 0.053**	**0.785 ± 0.022**
MLP (1-16)	0.759 ± 0.015	0.781 ± 0.017	0.275 ± 0.038	0.773 ± 0.027
MLP (1-64)	0.767 ± 0.020	0.789 ± 0.022	0.294 ± 0.050	0.768 ± 0.027
MLP with dropout (1-16)	0.752 ± 0.012	0.780 ± 0.012	0.278 ± 0.033	0.769 ± 0.017
MLP with dropout (1-64)	0.763 ± 0.016	0.786 ± 0.016	0.289 ± 0.042	0.772 ± 0.019
MLP (3-16)	0.759 ± 0.017	0.784 ± 0.019	0.280 ± 0.039	0.773 ± 0.028
MLP (3-64)	0.769 ± 0.022	0.790 ± 0.020	0.296 ± 0.044	0.775 ± 0.020
MLP with dropout (3-16)	0.752 ± 0.006	0.776 ± 0.004	0.264 ± 0.021	0.731 ± 0.007
MLP with dropout (3-64)	0.762 ± 0.012	0.786 ± 0.011	0.284 ± 0.031	0.750 ± 0.016
MLP (5-16)	0.759 ± 0.012	0.785 ± 0.010	0.288 ± 0.033	0.742 ± 0.014
MLP (5-64)	0.767 ± 0.018	0.789 ± 0.015	0.293 ± 0.034	0.761 ± 0.024
MLP with dropout (5-16)	0.752 ± 0.004	0.775 ± 0.004	0.266 ± 0.020	0.724 ± 0.007
MLP with dropout (5-64)	0.762 ± 0.015	0.785 ± 0.014	0.285 ± 0.035	0.757 ± 0.010

Table 3. Predictive performances for the Taiwan dataset over the different evaluation metrics

Models	AUC	AUC-H	H measure	Accuracy
Logistic	0.726 ± 0.001	0.731 ± 0.000	0.221 ± 0.000	0.805 ± 0.002
MLP (1-16)	0.747 ± 0.002	0.751 ± 0.002	0.251 ± 0.003	0.819 ± 0.000
MLP (1-64)	0.75 ± 0.002	0.754 ± 0.002	0.255 ± 0.002	0.818 ± 0.001
MLP with dropout (1-16)	0.741 ± 0.002	0.746 ± 0.002	0.245 ± 0.002	0.817 ± 0.001
MLP with dropout (1-64)	0.741 ± 0.002	0.745 ± 0.002	0.245 ± 0.001	0.818 ± 0.001
MLP (3-16)	0.753 ± 0.001	0.757 ± 0.001	0.256 ± 0.002	0.819 ± 0.001
MLP (3-64)	0.752 ± 0.001	0.756 ± 0.001	0.255 ± 0.002	$\mathbf{0.820 \pm 0.001}$
MLP with dropout (3-16)	0.744 ± 0.003	0.748 ± 0.004	0.248 ± 0.002	0.818 ± 0.001
MLP with dropout (3-64)	0.751 ± 0.001	0.755 ± 0.002	0.254 ± 0.002	0.819 ± 0.001
MLP (5-16)	0.746 ± 0.016	0.751 ± 0.013	0.252 ± 0.008	0.818 ± 0.002
MLP (5-64)	$\mathbf{0.754 \pm 0.002}$	$\mathbf{0.758 \pm 0.002}$	$\mathbf{0.257 \pm 0.002}$	0.819 ± 0.001
MLP with dropout (5-16)	0.739 ± 0.011	0.745 ± 0.008	0.247 ± 0.003	0.818 ± 0.001
MLP with dropout (5-64)	0.751 ± 0.001	0.754 ± 0.001	0.253 ± 0.002	0.819 ± 0.001

Table 4. Predictive performances for the FICO dataset over the different evaluation metrics

Models	AUC	AUC-H	H measure	Accuracy
Logistic	0.808 ± 0.007	0.813 ± 0.007	0.332 ± 0.013	0.739 ± 0.007
MLP (1-16)	0.809 ± 0.006	0.814 ± 0.006	0.332 ± 0.012	0.738 ± 0.005
MLP (1-64)	0.81 ± 0.005	0.815 ± 0.006	0.332 ± 0.01	0.74 ± 0.005
MLP with dropout (1-16)	0.804 ± 0.008	0.808 ± 0.008	0.322 ± 0.016	0.737 ± 0.008
MLP with dropout (1-64)	0.807 ± 0.008	0.812 ± 0.008	0.329 ± 0.016	0.74 ± 0.009
MLP (3-16)	$\mathbf{0.809 \pm 0.006}$	$\mathbf{0.814 \pm 0.005}$	$\mathbf{0.332 \pm 0.01}$	$\mathbf{0.743 \pm 0.006}$
MLP (3-64)	0.808 ± 0.006	0.813 ± 0.006	0.33 ± 0.011	0.737 ± 0.007
MLP with dropout (3-16)	0.803 ± 0.01	0.808 ± 0.01	0.321 ± 0.019	0.737 ± 0.007
MLP with dropout (3-64)	0.808 ± 0.007	0.813 ± 0.008	0.332 ± 0.015	0.738 ± 0.01
MLP (5-16)	0.809 ± 0.005	0.814 ± 0.005	0.332 ± 0.01	0.74 ± 0.006
MLP (5-64)	0.808 ± 0.007	0.813 ± 0.006	0.332 ± 0.013	0.739 ± 0.006
MLP with dropout (5-16)	0.805 ± 0.008	0.81 ± 0.008	0.324 ± 0.015	0.737 ± 0.009
MLP with dropout (5-64)	0.808 ± 0.007	0.814 ± 0.007	0.333 ± 0.014	0.737 ± 0.009

3.3 Model Explanation with LIME

In this section, firstly, we choose the best model which provides the trade-off between prediction and explanation. As mentioned above, LIME provides the model explanation fit and the interpretation for why a borrower has been assigned a specific model score by giving each input variable an importance weight.

Figures 1, 2 and 3 illustrate the trade-off between prediction and explanation of the neural network models for the German, Taiwan and FICO datasets, respectively. Although the neural network models cannot show the best predictive performance on the German dataset, we compared their explanation using LIME with the logistic model. From Fig. 1, there are two models that we can choose, the first neural network model is the one that has 3 hidden layers, 64 nodes at each layer and without dropout model achieves the highest predictive performance, but its explanation fit is lower than the neural network that has a hidden layer with 64 nodes and without dropout model. In addition, we compared the interpretable models' variable coefficients produced by LIME to the regression coefficients of the logistic model as shown in Tables 5, 6 and 7. The results show that the interpretable models' all variable coefficients are consistent with the white-boxing logistic model in terms of signs and significance of the regression coefficient, and there is no critical difference between two interpretable models' variable coefficients, thus we can choose the model with higher predictive performance. Moreover, it can be observed that if the significance level of the regression coefficients in the logistic model is lower, the absolute values of variable coefficients of the interpretable model are close to zero. This indicates that those variables are not associated with the borrower's creditworthiness.

Since the variable coefficients of the interpretable model produced by LIME are consistent with regression coefficients in the logistic model, we can make some explanations – for instance, if an account balance (Acc.Bal) held by a borrower increase, the borrower's creditworthiness would be higher and this can be the most important variable to explain the borrower's credit score. For the other two datasets, we can designate the conclusions same as the German dataset.

To summarize, the following conclusions can be drawn according to the interpretable model results produced by LIME:

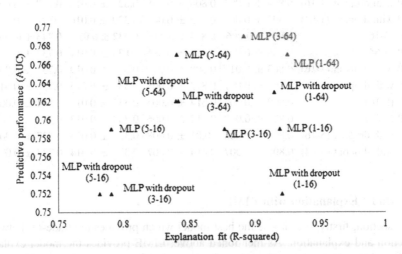

Fig. 1. The trade-off between prediction and explanation for the neural network models on the German dataset. There are two models which can be used to predict and to explain borrower's creditworthiness as expressed by red color. (Color figure online)

1. LIME produces a simple and easily understandable explanation for the neural network as a black box.
2. Those local interpretable models produced by LIME are consistent with white-boxing logistic regression and logically related to the real world.
3. The explanation fit of the interpretable model does not depend on the number of nodes in the neural network. However if the number of the hidden layers increases, the explanation fit of the interpretable model deteriorates.
4. As mentioned above, if our dataset can be predicted well by a simpler neural network architecture, we can obtain good performances for both prediction and explanation as the results of the German and FICO datasets.

Table 5. The comparison between the simple model of the neural network produced by LIME and logistic model for the German dataset. The ***, **, * and $'$ denote significance of p-value at 0.001, 0.01, 0.05, and 0.1 respectively.

Variables	Logistic	MLP (3-64)	MLP (1-64)
	Regression coefficients	Coefficients of the simple models produced by LIME	
Acc.Bal	0.579***	0.321 ± 0.041	0.327 ± 0.024
Savings.Stocks	0.232***	0.193 ± 0.052	0.19 ± 0.046
Pay.Status.Prev	0.386***	0.156 ± 0.026	0.147 ± 0.04
Marriage	0.250*	0.124 ± 0.036	0.115 ± 0.049
Guarantors	0.342	0.118 ± 0.018	0.106 ± 0.013
Type.apartment	0.284$'$	0.111 ± 0.01	0.095 ± 0.019
Concur.Credits	0.244*	0.091 ± 0.021	0.083 ± 0.025
Foreign	1.052$'$	0.081 ± 0.028	0.082 ± 0.025
Length.employ	0.160*	0.069 ± 0.022	0.066 ± 0.027
Age.years	0.008	0.066 ± 0.019	0.076 ± 0.016
Telephone	0.214	0.044 ± 0.003	0.047 ± 0.007
Purpose	0.033	0.002 ± 0.022	−0.003 ± 0.027
Dur.address	−0.012	0.005 ± 0.003	−0.006 ± 0.004
Dependents	−0.165	−0.022 ± 0.004	−0.017 ± 0.004
Occupation	−0.039	−0.048 ± 0.051	−0.04 ± 0.049
Credits.Bank	−0.261	−0.062 ± 0.006	−0.054 ± 0.009
Inst.percent	−0.218**	−0.089 ± 0.024	−0.098 ± 0.032
Valuable.asset	−0.203*	−0.106 ± 0.01	−0.117 ± 0.012
Dur.Credit.M	−0.037***	−0.328 ± 0.038	−0.326 ± 0.065

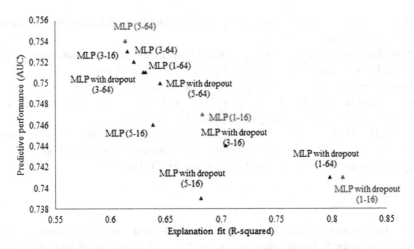

Fig. 2. The trade-off between prediction and explanation for the neural network models on the Taiwan dataset. There are three models which can be used to predict and to explain borrower's creditworthiness as expressed by red color. (Color figure online)

Table 6. The comparison between the simple model of the neural network model produced by LIME and logistic model for the Taiwan dataset. ***, **, * and ′ denote significance of p-value at 0.001, 0.01, 0.05, and 0.1 respectively.

Variables	Logistic	MLP (5-64)	MLP (1-64)	MLP with dropout (1-16)
	Regression coefficients	Coefficients of the simple models produced by LIME		
Limit_Bal	0.18***	0.162 ± 0.011	0.156 ± 0.012	0.113 ± 0.002
Pay_Amt2	0.048***	0.104 ± 0.006	0.109 ± 0.009	0.12 ± 0.008
Pay_Amt4	0.048***	0.086 ± 0.005	0.088 ± 0.005	0.092 ± 0.006
Pay_Amt1	0.055***	0.077 ± 0.02	0.095 ± 0.021	0.129 ± 0.01
Marriage	0.158***	0.05 ± 0.008	0.066 ± 0.011	0.056 ± 0.007
Education	0.082***	0.048 ± 0.014	0.073 ± 0.005	0.057 ± 0.002
Pay_Amt6	0.014**	0.028 ± 0.007	0.02 ± 0.009	0.025 ± 0.004
Pay_Amt5	0.012*	0.029 ± 0.003	0.013 ± 0.01	0.027 ± 0.002
Sex	0.093**	0.019 ± 0.002	0.018 ± 0.002	0.019 ± 0.002
Bill_Amt4	−0.000	−0.03 ± 0.014	−0.034 ± 0.009	−0.021 ± 0.005
Bill_Amt1	−0.006	−0.043 ± 0.046	−0.074 ± 0.012	−0.033 ± 0.01
Age	−0.007***	−0.056 ± 0.011	−0.027 ± 0.011	−0.053 ± 0.009
Pay_6	−0.224***	−0.334 ± 0.02	−0.337 ± 0.015	−0.293 ± 0.008
Pay_0	−0.554***	−1.078 ± 0.058	−1.042 ± 0.018	−1.019 ± 0.019

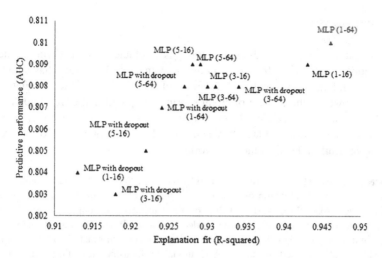

Fig. 3. The trade-off between prediction and explanation for the neural network models on the FICO dataset. There is only one model which can be used to predict and to explain borrower's creditworthiness as expressed by red color. (Color figure online)

Table 7. The comparison between the simple model of the neural network model produced by LIME and logistic model for the FICO dataset. ***, **, * and ' denote significance of p-value at 0.001, 0.01, 0.05, and 0.1 respectively.

Variables	Logistic	MLP (1-64)
	Regression coefficients	Coefficients of the simple model produced by LIME
Num.Inq.Last6 M	0.092***	1.208 ± 0.058
Net.Frac.RevBurd	0.01***	0.636 ± 0.009
Num.RevTr.wBal	0.055***	0.542 ± 0.012
Percent.Install.Tr	0.011***	0.212 ± 0.029
Net.Frac.InstBurd	0.000	0.058 ± 0.004
Num.InstTr.wBal	−0.001	−0.024 ± 0.011
Most.Recent.Delq	−0.027**	−0.032 ± 0.034
Percent.Tr.wBal	−0.001	−0.08 ± 0.021
Max.Delq.Last12M	0.007	−0.123 ± 0.073
Most.Recent.Inql7D	−0.038	−0.158 ± 0.117
Num.Tr.Open.Last12M	−0.065**	−0.157 ± 0.01
Num.Trades.60Ever	−0.03***	−0.217 ± 0.002
Num.Satisf.Trades	0.005*	−0.436 ± 0.289
Percent.Never.Delq	−0.028***	−0.455 ± 0.013
Avg.Months	−0.018***	−0.462 ± 0.009
MSince.Recent	−0.01***	−0.772 ± 0.033
External.Risk	−0.064***	−1.133 ± 0.03

4   Conclusions

One of the main focuses of lending institutions is an efficient credit scoring model that provides good ability of both prediction and explanation. This study has shown that advanced neural networks with LIME approach can provide better predictive and explanatory power in the context of credit scoring compared to the well-known white-boxing logistic model. LIME produces simple and easily understandable explanations for the neural network as a black box and those local interpretable models are consistent and logically related to the real world.

Acknowledgements. This work was supported by the Basic Science Research Program through the National Research Foundation of Korea (NRF) funded by the Ministry of Science, ICT & Future Planning (No. 2017R1A2B4010826), by the Business for Cooperative R&D between Industry, Academy, and Research Institute funded Korea Small and Medium Business Administration (Grants No. C0541451), by the Private Intelligence Information Service Expansion (No. C0511-18-1001) funded by the NIPA (National IT Industry Promotion Agency) and by National Natural Science Foundation of China (Grant No. 61701104).

References

1. Vellido, A., Martín-Guerrero, J.D., Lisboa, P.J.G.: Making machine learning models interpretable. In: ESANN, vol. 12, pp. 163–172 (2012)
2. Louzada, F., Ara, A., Fernandes, G.B.: Classification methods applied to credit scoring: systematic review and overall comparison. Surv. Oper. Res. Manage. Sci. **21**(2), 117–134 (2016)
3. Ribeiro, M.T., Singh, S., Guestrin, C.: Why should i trust you? Explaining the predictions of any classifier. In: Proceedings of the 22nd ACM SIGKDD International Conference on Knowledge Discovery and Data Mining, pp. 1135–1144. ACM (2016)
4. Cox, D.R.: The regression analysis of binary sequences. J. R. Stat. Soc. Series B (Methodological) **20**, 215–242 (1958)
5. Lessmann, S., Baesens, B., Seow, H.-V., Thomas, L.C.: Benchmarking state-of-the-art classification algorithms for credit scoring: an update of research. Eur. J. Oper. Res. **247**(1), 124–136 (2015)
6. Hand, D.J., Anagnostopoulos, C.: A better Beta for the H measure of classification performance. Pattern Recogn. Lett. **40**, 41–46 (2014)
7. LeCun, Y., Bengio, Y., Hinton, G.: Deep learning. Nature **521**(7553), 436 (2015)
8. West, D.: Neural network credit scoring models. Comput. Oper. Res. **27**(11–12), 1131–1152 (2000)
9. Lee, T.-S., Chen, I.-F.: A two-stage hybrid credit scoring model using artificial neural networks and multivariate adaptive regression splines. Expert Syst. Appl. **28**(4), 743–752 (2005)
10. Wong, B.K., Selvi, Y.: Neural network applications in finance: a review and analysis of literature (1990–1996). Inf. Manage. **34**(3), 129–139 (1998)
11. Rosenblatt, F.: The perceptron: a probabilistic model for information storage and organization in the brain. Psychol. Rev. **65**(6), 386 (1958)
12. Girosi, F., Jones, M., Poggio, T.: Regularization theory and neural networks architectures. Neural Comput. **7**(2), 219–269 (1995)

13. Srivastava, N., Hinton, G., Krizhevsky, A., Sutskever, I., Salakhutdinov, R.: Dropout: a simple way to prevent neural networks from overfitting. J. Mach. Learn. Res. **15**(1), 1929–1958 (2014)
14. Krizhevsky, A., Sutskever, I., Hinton, G.E.: ImageNet classification with deep convolutional neural networks. In: Advances in Neural Information Processing Systems, pp. 1097–1105 (2012)
15. Kingma, D.P., Ba, J.: Adam: a method for stochastic optimization. arXiv preprint arXiv: 1412.6980 (2014)
16. Ruder, S.: An overview of gradient descent optimization algorithms. arXiv preprint arXiv: 1609.04747 (2016)
17. Asuncion, A., Newman, D.: UCI machine learning repository (2007)
18. FICO, Xml challenge. https://community.fico.com/s/explainable-machine-learning-challenge. Accessed 01 Oct 2018
19. Farrar, D.E., Glauber, R.R.: Multicollinearity in regression analysis: the problem revisited. Rev. Econ. Stat. **49**, 92–107 (1967)
20. Amadoz, A., Sebastian-Leon, P., Vidal, E., Salavert, F., Dopazo, J.: Using activation status of signaling pathways as mechanism-based biomarkers to predict drug sensitivity. Sci. Rep. **5**, 18494 (2015)
21. Arnold, T.B.: KerasR: R interface to the Keras deep learning library. J. Open Source Softw. **2**, 296 (2017)

Non-uniform Initialization of Inputs Groupings in Contextual Neural Networks

Maciej Huk[(⊠)] [iD]

Faculty of Computer Science and Management,
Wroclaw University of Science and Technology, Wroclaw, Poland
`maciej.huk@pwr.edu.pl`

Abstract. Contextual neural networks which are using neurons with conditional aggregation functions were found to be efficient and useful generalizations of classical multilayer perceptrons. They allow to generate neural classification models with good generalization and low activity of connections between neurons in hidden layers. The key factor to build such solutions is achieving self-consistency between continuous values of weights of neurons' connections and their mutually related non-continuous aggregation priorities. This allows to optimize neuron inputs aggregation priorities by simultaneous gradient-based optimization of connections' weights with generalized BP algorithm. But such method additionally needs initial setting of connections groupings (scan-paths) to define priorities of signals during first ω epochs of training. In earlier studies all connections were initially assigned to a single group to give neurons access to all input signals at the beginning of training. We found out that such uniform solution not always is the best one. Thus within this text we compare efficiency of training of contextual neural networks with uniform and non-uniform, random initialization of connections groupings. On this basis we also discuss the properties of analyzed training algorithm which are related to characteristics of used scan-paths initialization methods.

Keywords: Classification · Self-consistency · Scan-paths initialization ·
Aggregation functions

1 Introduction

Contextual neural networks (CxNNs) were proposed as generalization of feedforward neural models. They are using conditional, multi-step aggregation functions to express nonlinear relations between inputs of neurons [1, 2]. Both for benchmark and real-life problems such solution allows to limit internal activity of connections between neurons without decreasing accuracy of classification [2, 3]. It was shown that CxNNs can be effective for detection of fingerprints for crime-related analyses [4]. They were also used with success for spectrum prediction in cognitive radio as well as for research related to measuring awareness of computational systems [5].

The structures of contextual neural networks can be used to build classifiers with properties better than their non-contextual versions including MLP [2]. But what is more important they possess also ability to strongly limit activity of connections

© Springer Nature Switzerland AG 2019
N. T. Nguyen et al. (Eds.): ACIIDS 2019, LNAI 11432, pp. 420–428, 2019.
https://doi.org/10.1007/978-3-030-14802-7_36

between neurons without lowering accuracy of network outputs – during as well as after training process [3]. In highly constrained applications it can help to reduce time and energy costs of usage of trained neural networks.

Moreover, limiting of activity of neurons' inputs in CxNN is done adaptively to processed data. Neurons that build contextual neural network are aggregating input data in multiple steps instead of one. During each step of aggregation given subset of input signals is read-in to check if already analyzed information is enough to calculate the output value of the neuron with acceptable precision. The composition and order of groups of inputs are built for each neuron during the training process. The ordered list of groups of input connections is realization of Starks' scan-path theory due to multi–step character of conditional aggregation functions [2, 6]. Such functions aggregate signals from groups of inputs in following steps until cumulated activation of the neuron is lower than given constant threshold. RFA, CFA, OCFA and Sigma-if are examples of such functions used to build contextual neural networks [3].

Finally, usage of contextual neurons modifies the character of the neural network from black-box to grey-box model. This is because in CxNN it can be checked which data attributes are needed to calculate output values of the network for each input vector. It can be done by analyzing activity of inputs of neurons in the first hidden layer of the neural network. Finally, this allows to order data attributes by their importance found by the CxNN during processing of vectors that define given problem [1].

The construction of CxNNs can be done by using generalized error backpropagation algorithm (GBP) [2]. What is interesting, during training of contextual neural network the GBP algorithm initially assumes that for every neuron with conditional aggregation function all connections belong to the same single group. Thus for first ω epochs of training the network is treated as classical MLP neural network trained with error backpropagation algorithm. When number of epochs becomes equal to the interval of inputs grouping ω connections groupings are updated according to weights of related connections. This forces conditional aggregation functions to prioritize inputs of neurons and this process is repeated after every ω epochs.

In this paper we explore properties of different version of generalized error back-propagation algorithm. The proposed modification makes it to assume that for first ω epochs of training the connections of neurons do not belong to the same group. Instead, for every neuron with conditional aggregation function connections belong to random groups. This assignment to groups is made before the first epoch and is not related with weights of connections.

We have observed that for many analyzed benchmark problems proposed modi-fication of GBP algorithm has two beneficial influences on the process of training and properties of best constructed neural networks. First, it allows to achieve neural models of lower average internal activity of connections. And second - it helps to finish the training much faster, both in terms of number of training epochs as well as of time. While the latter benefit can be viewed as a direct result of the decrease of internal connections activity, the former is more interesting. As it was shown earlier [1, 2], the decrease o connections activity of CxNNs can be also related with additional increase of their classification accuracy.

The rest of the paper is constructed as follows. After the introduction, in the second section the GBP algorithm and contextual neural networks are presented with their most important properties. Next, in the section three, we define the proposed solution of randomized initialization of connections grouping for GBP algorithm. Fourth section includes results of experiments conducted to verify expected properties of modified GBP algorithm. Finally in section five we give conclusions with description of further research plans.

2   Generalized Backpropagation Algorithm

The GBP algorithm is a well known error backpropagation modified to be able to train contextual neural networks [2, 3]. The key modification is based on using self-consistency paradigm frequently used in physics [7]. It allows to use gradient based method to optimize both continuous (connection weights) and non-continuous, non-differentiable parameters of neuron's aggregation functions. The source of such properties of the GBP algorithm is keeping mutual relation of those both groups of parameters. It is done by defining non continuous behavior of multi-step aggregation as function Ω of continuous weights. The Ω relation can be formulated as a join of two operations: sorting N dendrites of neuron according to their weights and then dividing those ordered inputs into list of K equally-sized groups. N/K connections with highest weights are assigned to the first, most important group, next N/K inputs with highest weights are included into the second group, etc. In the effect a scan-path can be defined as the list of groups from first to last. It can then be used within aggregation function to read-in groups of inputs one after another until given condition is met. It was formally proven that GBP can train neural networks using various types of multi-step conditional aggregation functions [3]. The block diagram of the modified form of GBP algorithm without details of aggregation function can be found on the Fig. 1.

It is interesting to notice how the value of interval ω sets the strength of self consistency between weights of neuron inputs and parameters of aggregation functions. During the first ω training epochs all neurons have all input connections assigned to the first group of their scan-paths. The idea behind such solution is to not make any assumptions about the importance of inputs of neurons and to not limit their access to possibly crucial information. Then, after each ω epochs scan-paths are updated with use of Ω function according to the values of weights of connections. In the effect, the value of interval ω can significantly influence the training process. For $\omega = \infty$ (or number of groups K = 1) the GBP algorithm behaves like the classical error backpropagation, and CxNN works like MLP. In the case when K > 1 and the value of ω is close to one the parameters space of the CxNN is reconfigured frequently due to the changes of scan-paths of neurons after each ω epochs. Such way of operation can decrease the effectiveness of the training. But it was shown for many problems that for K > 1 and $5 < \omega \ll \infty$, the output error of the CxNN can increase after the update of the scan-path, but during following epochs it typically decreases again, together with the activity of neural network connections.

In the effect of described construction of the GBP algorithm, CxNNs built with it can show better classification properties as well as lower average activity of neurons' inputs than their non-contextual versions trained with error backpropagation method. This can be caused by the fact, that cyclic changes of scan-paths of the neurons during the GBP together with evolution of decision space of each neuron during conditional signals aggregation, create mechanism similar to the dropout technique. But one can easily find also important differences. While dropout does not depend on the data processed by the neural network, the decision spaces of neurons of contextual neural network change accordingly to processed data vector and during GBP take into account what the whole model has learned before given epoch. Dropout decreases the internal activity of the network connections only during the training, and CxNNs limit it both during and after the training. This makes GBP algorithm and contextual neural models valuable solutions to be used in place of feedforward neural networks such as MLP as well as within convolutional neural networks.

3 Randomized Initialization of GBP Connections Grouping

From the previous section one can find out that one of the most important phases of the GBP method is the modification of scan–paths of neurons done by inputs sorting. The reason for that is that scan–paths update merges gradient-based search of the error backpropagation algorithm with self-consistency paradigm. But one can also notice that while results of BP strongly depend on initial connection weights selection, the behavior of GBP algorithm should depend not only on initial values of connection weights, but also on initial selection of scan-paths (inputs groupings) of neurons.

All previously described versions of GBP were setting initial scan-paths to single step aggregation of the whole input space (all connections in single group). This was making CxNNs to behave as MLP during first epochs of training. The rationale behind such solution was to not block GBP from finding data important for solving given problem by improper selection of scan-paths. We assumed, that it will be better to give CxNN access to all signals at the beginning of training, and then let every neuron to filter out less important inputs at later stages of GBP.

But during our analyses of context distribution within various data sets [5] it turned out that for many real life data sets the information needed to solve the problem is distributed between many data attributes. At the same time CxNNs can achieve good accuracy of classification with final connections activity much lower than 100%. Thus we formulated hypothesis that initial setting of scan-paths of hidden neurons to single step grouping (all inputs belong to one group) is not needed to train contextual neural network. Moreover, in such case it can be expected, that when all inputs of neurons will not belong to single group prior to first scan-paths update, the average time of training should be lower than when initially CxNN neurons behave like in MLP. This would be the result of decreased activity of connections between neurons which should emerge from the very beginning of CxNN training with GBP. Such effect could be very valuable especially for problems described by large data sets with big number of input attributes.

Using the above as a guideline, we have modified the standard version of Generalized error Backpropagation method. This formed a solution in which the initialization sequence of GBP uses random assignments of neurons' inputs to groups with non uniform numbers of inputs within each group. This means that some groups initially can be larger than others and some of them can include only one connection or be empty. But even empty group is not a problem for the GBP algorithm – in such case it just skips the group during signals aggregation because such group does not contribute to the neuron activity. For clarity we present the modified GBP algorithm on the Fig. 1.

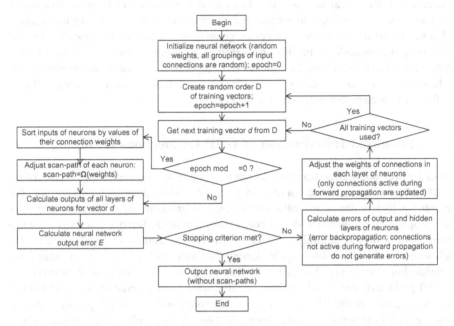

Fig. 1. General flowchart of the modified generalized error backpropagation algorithm (GBP) for given scan–path creation function Ω. Initial grouping of connections is set by function R as random with non-uniform numbers of connections within groups. Scan-paths update interval ω is controlling the strength of self–consistency between weights of connections and parameters of aggregation functions of neurons. Temporary neuron scan-paths are not needed after the end of the training because can be re-created from values of weights with use of function Ω.

In the effect of proposed modification even before the first scan-paths update, i.e. during training epochs from 1 to ω, the average activity of hidden connections (*hca*) can be lower than 100%. It is good to note that in this period of epochs connections grouping can be highly not related with connections weights – such relation is established during first scan-paths update after ω epoch. But this also should not be a problem for GBP algorithm, because in its' previously analyzed form with initial inclusion of all input connections to a single group such assignment was also not related with connections weights.

4   Results of Experiments

Proposed modification of the GBP algorithm changes values of initial parameters of the training process. As it was suggested earlier, this can have considerable influence on the results of this gradient-based method. Thus it was needed to check experimentally if such effect exists and if obtained results of training justify usage of analyzed change of GBP initialization. To do that we have performed a set of experiments of training contextual neural networks with GBP algorithm with uniform and non–uniform, random initialization of scan-paths. As a point of reference we have used selected UCI ML repository benchmarks [8] as well as three real-life data sets with cancer gene expression microarray data, namely ALL-AML Leukemia (Golub), ALL-MLL Leukemia (Armstrong) and Small Round Blue Cell Tumors (SRBCT) [9–11]. Basic properties of considered classification problems and architectures of trained artificial neural networks are given in Table 1.

Table 1. Basic properties of data sets and neural networks used within experiments

Data set	Number of attributes	Number of classes	Number of data vectors	Number of hidden neurons	Number of hidden connections	Number of groups
Iris	4	3	150	3	21	4
Wine	13	3	178	10	160	5
Sonar	60	2	208	10	620	7
Golub	7129	2	72	10	71310	10
Armstrong	12582	2	72	10	125840	10
SRBCT	2308	4	83	10	23120	10

Attributes were represented with one-hot encoding. Values of parameters of training method were: SGD training step $\alpha = 0.1$ (0.01 for Armstrong data set) with batch size = 1, connection weights are from range $(-0.2, 0.2)$, interval of groups actualization $\omega = 25$, aggregation function threshold $\varphi^* = 0.6$. Activation functions of neurons was bipolar sigmoid. Aggregation function of neurons in CxNNs was Sigma-if. Momentum was not used. The stopping criterion was: perfect classification of training data or maximal number of 300 training epochs without decrease of training error.

All results were obtained as averages of measurements from 10-fold cross validation. For each type of analyzed initialization of connections grouping and MLP, the same architectures and random weights were used within neural networks. Experiments were done with use of Pentium 4 CPU working at 2597.2 ± 0.2 MHz and pseudo–random generator built in the Delphi 7 environment. The statistical differences of groups of measurements were checked with two-sample T-Test (confidence above 90%). The normality of series was analyzed with Shapiro-Wilk normality test. Results for MLP are presented in Table 2. Results for contextual neural networks with uniform and random initialization of connections groupings are given in Tables 3 and 4, respectively.

The results show what was observed in previous experiments: CxNNs can decrease average activity of hidden connections without decrease of classification accuracy for test data. At the same time CxNNs can use the contextual processing of signals to solve selected problems with higher accuracy than their non-contextual counterparts (Table 3, Wine and Armstrong). Both effects are statistically significant. But it can be

Table 2. Average results with standard deviations for MLP neural networks trained with BP algorithm. Highlighted values are statistically better than related values measured for CxNNs.

Data set	Training epochs [1]	Training time [s]	Test error [%]	Hidden conn. activity [%]
Iris	219.6 ± 93.8	**2.06 ± 0.82**	5.3 ± 6.9	100 ± 0
Wine	67.3 ± 103.7	0.25 ± 0.38	3.9 ± 4.6	100 ± 0
Sonar	97.7 ± 56.6	**6.4 ± 3.7**	16.0 ± 11.7	100 ± 0
Golub	12.7 ± 5.4	52.6 ± 20.9	8.4 ± 9.9	100 ± 0
Armstrong	35.7 ± 43.8	301.3 ± 351.0	5.7 ± 7.3	100 ± 0
SRBCT	6.5 ± 1.4	1.9 ± 0.4	5.0 ± 8.7	100 ± 0

Table 3. Average results with standard deviations for CxNN with GBP algorithm and uniform initialization of connections groups. Highlighted values are statistically better than related values measured for MLP.

Data set	Training epochs [1]	Training time [s]	Test error [%]	Hidden conn. activity [%]
Iris	292.0 ± 104.1	5.3 ± 1.9	6.6 ± 8.3	**82.3 ± 3.9**
Wine	**8.3 ± 2.2**	**0.04 ± 0.01**	**1.1 ± 2.3**	100 ± 0
Sonar	282.9 ± 172.6	15.2 ± 8.5	17.3 ± 8.8	**50.2 ± 11.4**
Golub	13.2 ± 3.9	51.0 ± 12.9	4.1 ± 6.6	100 ± 0
Armstrong	18.1 ± 9.3	**128.9 ± 51.8**	**2.9 ± 6.0**	96.1 ± 12.3
SRBCT	7.2 ± 2.4	2.2 ± 0.7	7.5 ± 12.1	100 ± 0

Table 4. Average results with standard deviations for CxNN with GBP algorithm and random initialization of connections groups. Highlighted values are statistically better than related values measured for MLP.

Data set	Training epochs [1]	Training time [s]	Test error [%]	Hidden conn. activity [%]
Iris	296.8 ± 114.3	5.4 ± 2.1	4.7 ± 5.5	**82.3 ± 3.9**
Wine	142.1 ± 87.1	0.6 ± 1.1	5.0 ± 6.7	**68.3 ± 8.0**
Sonar	340.3 ± 163.0	18.5 ± 8.4	21.2 ± 8.6	**51.2 ± 5.3**
Golub	32.4 ± 2.0	**41.2 ± 3.1**	8.2 ± 9.8	**35.4 ± 6.3**
Armstrong	35.1 ± 5.1	**94.6 ± 15.1**	8.2 ± 11.4	**34.6 ± 4.5**
SRBCT	33.0 ± 4.2	4.2 ± 0.3	13.2 ± 14.8	**39.6 ± 6.1**

also noticed that with uniform initialization of connections groupings the hidden connections activity of CxNN models for most of considered data sets is close to 100%. In such case *hca* dropped significantly in relation to MLP only for Sonar and Iris problems (Table 3). Moreover, this doesn't guarantee decrease of the training time: for Sonar and Iris the average training times increased without change of classification accuracy for test data.

On the other hand, when random connections initialization was used average *hca* of best CxNN models measured for test data dropped for all considered problems (Table 4). And what is more important this modification of properties happened without statistically significant change of classification accuracy for test data in relation to MLP. Also the training time has not changed in most cases. And when training time changed the result was positive - it significantly decreased for two data sets: Golub and Armstrong (by 22% and 69%, respectively). This is the type of effect which was expected from the random initialization of scan-paths during proposed, modified GBP algorithm.

The source of presented behavior is as follows. In the case of uniform initialization of scan-paths all inputs of each neuron belong to single group and at least for first ω epochs the CxNN is working as MLP. The value of ω controls the strength of the bond of gradient descent and self–consistency methods used by GBP training algorithm. Thus its' low value can cause too frequent changes of evolving error space of CxNN making training too difficult. And it was found in earlier experiments that safe values of ω are in the range from 5 to 40. In the effect, for big contextual neural networks before e.g. ω = 25 training epoch one can expect enormous amount of computation identical as in the case of MLP. On the other hand, with proposed non-uniform, random initialization of connections groupings in hidden layers makes CxNN to process information with number of active connections reduced from the very first epoch of training. The level of initial activity reduction is proportional to assumed number of inputs groups K, but can change during training.

Even if the general mechanism causing the above effect is clear, at the current stage of experiments it is hard to explain the detailed reasons why the proposed modification of GBP algorithm decreased its' time of training only for Golub and Armstrong problems. But one can suppose that this is related to the number of hidden connections within neural networks. The number of hidden connections in CxNNs for Golub and Armstrong problems is at least 3 and 5 times greater than in the next biggest considered neural network for SRBCT data, and over 100 and 200 times greater than in the case of Sonar benchmark (see Table 1). And earlier experiments suggest that decrease of *hca* is most effective in neural networks with higher number of hidden connections. Still, this should be the subject of further analyses.

5   Conclusions

In this paper we have presented new, random method of connections groupings initialization for use at the beginning of the GBP algorithm. As expected, the method allowed for statistically significant reduction of hidden connections activity of CxNNs without decrease of classification accuracy on test data. At the same time for contextual networks with big number of hidden connections such initialization can considerably reduce the time of training and amount of related computations. This in turn can be used in energy limited applications of CxNNs to save energy and extend potential battery life of related devices.

To evaluate proposed method we have used a set of benchmark problems from UCI ML repository as well as three real-life data sets with cancer gene expression microarray data. To extend presented analysis it would be needed to perform analogous experiments for wider set of benchmarks and for various values of ω. Next interesting question to ask would be if CxNNs created with use of the GBP algorithm with random scan-paths initialization also express Miller' effect (decrease of hidden connections activity and increase accuracy of classification of training data for number of groups K between 5 and 9) [2, 12]. It would be also valuable to find out if proposed solution can be successfully applied also for problems other than classification (e.g. regression).

The continuation of above work can include analysis of changes of activity not only of hidden connections but also activity of neural network inputs and of data attributes. In relation to the research on context distribution within training data, this can shed light on the sources of the observed phenomenon that limiting *hca* with random initial scan-paths does not decrease classification accuracy of contextual neural networks for test data. Finally, in further studies we plan additionally to explore properties of other possible scan-paths initialization methods that can be valuable modifications of GBP method.

References

1. Huk, M.: Learning distributed selective attention strategies with the Sigma-if neural network. In: Akbar, M., Hussain, D. (eds.) Advances in Computer Science and IT, pp. 209–232. InTech, Vukovar (2009)
2. Huk, M.: Backpropagation generalized delta rule for the selective attention Sigma-if artificial neural network. Int. J. Appl. Math. Comput. Sci. **22**, 449–459 (2012)
3. Huk, M.: Notes on the generalized backpropagation algorithm for contextual neural networks with conditional aggregation functions. J. Intell. Fuzzy Syst. **32**, 1365–1376 (2017)
4. Szczepanik, M., Jóźwiak, I.: Data management for fingerprint recognition algorithm based on characteristic points' groups. In: Pechenizkiy, M., Wojciechowski, M. (eds.) New Trends in Databases and Information Systems. AISC, vol. 185, pp. 425–432. Springer, Heidelberg (2013). https://doi.org/10.1007/978-3-642-32518-2_40
5. Huk, M.: Measuring the effectiveness of hidden context usage by machine learning methods under conditions of increased entropy of noise. In: 2017 3rd IEEE International Conference on Cybernetics (CYBCONF), pp. 1–6. IEEE (2017)
6. Privitera, C.M., Azzariti, M., Stark, L.W.: Locating regions-of-interest for the Mars Rover expedition. Int. J. Remote Sens. **21**, 3327–3347 (2000)
7. Raczkowski, D., Canning, A.: Thomas-Fermi charge mixing for obtaining self-consistency in density functional calculations. Phys. Rev. B **64**, 121101–121105 (2001)
8. UCI Machine Learning Repository. http://archive.ics.uci.edu/ml
9. Golub, T.R., et al.: Molecular classification of cancer: class discovery and class prediction by gene expression monitoring. Science **286**, 531–537 (1999)
10. Armstrong, S.A.: MLL translocations specify a distinct gene expression profile that distinguishes a unique leukemia. Nat. Genet. **30**, 41–47 (2002)
11. Khan, J., et al.: Classification and diagnostic prediction of cancers using gene expression profiling and artificial neural networks. Nat. Med. **7**(6), 673–679 (2001)
12. Miller, G.A.: The magical number seven, plus or minus two: some limits on our capacity for processing information. Psychol. Rev. **63**(2), 81–97 (1956)

Implementation and Analysis of Contextual Neural Networks in H2O Framework

Krzysztof Wołk[(⊠)] and Erik Burnell

Wroclaw University of Science and Technology, Wroclaw, Poland
krzysztof.mateusz.wolk@gmail.com,
erik.d.burnell@gmail.com

Abstract. Contextual neural networks utilizing conditional multi-step aggregation functions have many useful properties. For example, their ability to decrease the activity between internal neuron connections may decrease computational costs, whereas their built-in automatic selection of attributes required for proper classification can simplify problem setup. The research of contextual neural networks was motivated by a limited number of satisfactory machine learning solutions providing these features. An implementation of the CxNN model in the H2O.ai machine learning framework was also developed to validate the method. In this article we explain relevant terms and the implementation of contextual neural networks as well as conditional multi-step aggregation functions. To validate the solution, experiments and their results are presented for selected UCI benchmarks and Cancer Gene Expression Microarray data.

Keywords: Aggregation functions · H2O.ai · GBP · Scan-paths · Sigma-if

1 Introduction

The implementation of contextual neural networks and conditional multi-step aggregation functions could have been done from scratch. However, many parts of machine learning software are reusable and have been collected into frameworks, allowing developers and scientists to focus on the new solutions, which increases effectiveness of ML and related research [1, 2]. Thus, for this research project, the H2O.ai framework was chosen. It contains many classifier models, selected by the H2O.ai developers, and supports training them and analyzing their properties. This selection includes well known machine learning algorithms like deep learning networks (DNN), distributed random decision forests (DRF), generalized linear model (GLM) and Naive Bayes [3, 4]. The H2O.ai environment is used in over 9,000 reputable organizations such as PayPal Holdings Inc. and Cisco Systems indicating its reputability and reliability [5, 6]. Thanks to its' open-source license and very efficient data processing system based on distributed processing and other technologies, H2O.ai is a freely accessible, powerful platform for data processing and analysis [4, 7, 8]; this is also why it was chosen as the framework for the implementation of contextual neural networks and conditional multi-step aggregation functions. Promoting automation, existing software integration and research, the H2O.ai developers included an extensive server-side API as well as interfaces to Python, Scala or Java languages [9]. Measurements can be collected and

© Springer Nature Switzerland AG 2019
N. T. Nguyen et al. (Eds.): ACIIDS 2019, LNAI 11432, pp. 429–440, 2019.
https://doi.org/10.1007/978-3-030-14802-7_37

exported into single or separate files to speed up solution analysis. Finally, H2O.ai Flow is the user-friendly graphical web interface to the H2O.ai framework [5, 9].

The H2O.ai framework was built to be simple to use, where more complex tasks are partitioned into individual activities. Starting the H2O.ai Flow application server is done by launching the executable Java jar file, which turns on a web server accessible at the address http://localhost:54321 unless otherwise configured. Even though one of its strengths lies in distributed processing, H2O.ai can also be run on a single node, allowing individual research or undertaking smaller problems. The developers' focus on the excellent optimization of the H2O.ai engine took time away from implementing actual machine learning algorithms, thus only some of the most common ones have been included so far. This means that some problems won't be covered by the existing software, requiring additional development prior to solving them. Nonetheless, they will still benefit from the efficient H2O.ai engine once this initial work is completed. Contextual neural networks are one such algorithm and their implementation is described in this article.

This article has been organized as follows. Section 2 contains a simple description of contextual neural networks and the Generalized Error Backpropagation algorithm. Section 3 describes the required changes and additions made in the H2O.ai core architecture, the H2O.ai API as well as the H2O.ai Flow web application. Utilizing the newly added functionality, Sect. 4 presents results of experiments performed on data from the UCI ML Repository as well as on Cancer Gene Expression Microarray data. Finally, Sect. 5 wraps up with conclusions about the results as well as suggested directions for further research on contextual neural networks and connected topics.

2 Contextual Neural Networks and Generalized Error Backpropagation

Contextual neural networks (CxNNs) can solve many well-known machine learning benchmarks [10, 11, 13, 14]. They can also be used to solve more complex real-life problems, such as fingerprint detection [12] and problems related to rehabilitation [13]. This proves that contextual neural networks are a viable machine learning algorithm. What makes them stand out are some of their documented properties [11–13]. The use of conditional multi-step aggregation functions in CxNNs automates the selection of attributes preferred during classification; the selections are easily accessible and thus can be analyzed. This attribute selection allows CxNNs to solve problems using low number of internal neuron connections while still providing satisfactory results. The benefits can be seen during the training process and classification process, the most important of which is the decrease in computational needs and in effect power costs of computer utilization during machine learning.

For many problems, the activity of internal neuron connections can be effectively decreased during training by up to ten times compared to standard MLP networks [11]. This effect can be achieved by either CxNNs or by using dropout with standard MLP networks [14–16]. The difference between these two approaches is that only CxNNs do achieve this effect by using knowledge already gathered by the neural network and are associated with the data processing step. Furthermore, both approaches can decrease internal neuron connections during the training process, but only CxNNs can decrease internal neuron connections activity also during the classification process after the training [14].

The standard choice of aggregation function in neural networks, the sum of products, can be replaced with conditional multi-step aggregation functions. In these functions, the aggregation of input signals takes place in multiple steps rather than one [14]. At each step, a subset of the neuron's input connections is read, and a decision is made whether the aggregated information is sufficient to calculate the neuron's output. The level of signals that is sufficient to stop aggregation is specified by the aggregation threshold parameter φ^*. Thus, the aggregation ends when neuron activation reaches φ^* or when all the neuron's inputs are aggregated. Conditional aggregation functions include e.g. CFA and Sigma-if [14, 17].

Contextual neural networks do not require many more parameters than standard MLP networks. All that needs to be added is information about the composition and propensity of input groups for each neuron [13]. The parameters are as follows: the already mentioned aggregation threshold φ^*, the K parameter, which specifies the number of groups into which neurons' inputs should be divided, and the "scan-paths" [12, 14], which are simply the lists of the order of groups and are stored inside the connections themselves [12]. These scan-paths can be trained together with the neural network weights by using Generalized Error Backpropagation (GBP) [10, 18]. GBP is a generalized version of the standard neural network backwards error propagation algorithm, further extended for use with contextual neural networks with conditional multi-step aggregation functions. To optimize the output error of the network and the scan-paths defined by the non-differentiable, non-continuous conditional multi-step aggregation functions, the GPB algorithm applies the self-consistency paradigm known from physics [19]. Thanks to GBP, the dependencies between weights and discontinuous virtual parameters, which define groups of neuron inputs and their order, are preserved by the algorithm. These virtual parameters are calculated by the Ω function using connection weights. When training the network with the GBP algorithm, the scan-paths are updated only once per ω epochs, to control these dependencies [10]. Therefore, during training the scan-paths must be kept in separate structures from the weights of the connections, but after training they can be recreated from values of connections' weights and number of groups K.

Neurons utilizing conditional multi-step aggregation function use the Ω function to define scan-paths. This function sorts all inputs of a given neuron according to their weights and then divides them into K groups. In each group there are N/K neuron inputs selected according to weight values. Collected into the first group are N/K inputs with the largest weights, into the second group N/K inputs with the largest weights out of the inputs not assigned to any group yet and so forth. This division of inputs into groups defines the scan-paths. The conditional multi-step aggregation function uses these paths by reading subsequent groups until the specified condition is met [10]. The following Fig. 1 shows a simplified GPB algorithm.

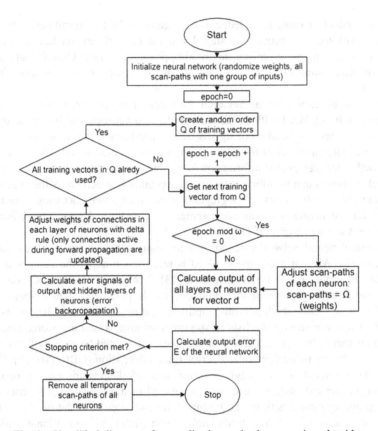

Fig. 1. Simplified diagram of generalized error backpropagation algorithm

Due to the randomness of connection weights between neurons, for the first ω learning epochs, scan-paths have only one group. After the first and each subsequent ω learning epochs, the scan-paths is updated using the Ω function.

3 Implementation

The GBP and BP algorithm work identically when the group update interval is infinite and contextual neural networks work the same as standard MLP networks when the number of groups equals 1 [10]. This suggests that the extension of the base implementation of neural networks and the back-propagation algorithm in the H2O.ai framework should be possible. A deeper analysis of the code revealed the complexity of modifying parts of H2O.ai required for the implementation of contextual neural networks due to the high optimization in the H2O.ai framework.

3.1 Implementation of Aggregation Functions

The third version of the H2O.ai framework implements activation functions as classes that inherit from the basic class "Neurons" which allows overloading the "fprob" and "bprob" methods. This approach complicates the development of additional aggregation functions. Maintaining this structure in the implementation of conditional multi-step aggregation functions would result in the necessity of creating many combinations of sub-classes connecting all activation and aggregation functions such as "SigmoidBipolar_SigmaIf", "Rectifier_SigmaIf" etc. Developing software in such a way would lead to unnecessary expansion of the source code, hindering code maintenance and slowing code integration with newer H2O.ai versions. Thus, the focus during the implementation was to keep modifications to the base H2O.ai source code as small and concise as possible.

With this focus in mind, a satisfactory solution would be to extend the transfer function class, rather than to replace it. When the new class, called "NeuronsExt", is used, activation and aggregation functions can be selected separately without developing a new class for each combination. The original H2O.ai functionality is available just as it was before, but now there is an additional, optional operation method. A welcome side effect of leaving the H2O.ai base functionality intact is the simplicity of comparing it with the new functionality for analysis. Conveniently, support for the generalized error backpropagation (GBP) algorithm could also be included in this extended class. GBP was implemented in the "conditionalBprop" method, which overrides the previous "bprop" method. The "NeuronConnectionGroups" and "Layer ConnectionGroups" classes are also new additions and are related to the GBP algorithm and new transfer functions. Using these new classes, these aggregation functions were implemented: CFA, Sigma-if and OCFA. Two new activation functions were also added for comparison purposes: Leaky Rectifier and Bipolar Sigmoid.

3.2 Implementation of Analysis Tools

More effort had to be devoted to the implementation of methods for analyzing the new functionality, especially for measuring the activity of connections between neurons. Since the base version of H2O.ai did not offer a satisfactory extension point for such measurements a new solution was required. The method in which distributed processing was developed in H2O.ai complicated adding an analysis tool for measuring neuron connection activity. Regardless of the number of processors available for calculations, the H2O.ai environment always utilizes more than one thread, adapted to the computing power of the computer it is run on. To perform measurements in this architecture, the tool for measuring the activity of neuron connections has been implemented in a Map/Reduce manner. For simplicity it cannot be used together with cross-validation in H2O, as the H2O.ai implementation of cross-validation trains different neural networks in each computing node. Therefore, at the time of this writing, the tool can be used during the training only when cross-validation is disabled. Still, the activity of all models built during the cross-validation can be measured after the training.

The distributed processing implementation in H2O.ai maintains several independent local copies of the model. Once the distributed processing is complete, these models are combined to create a single shared model and returned as the result of the training process. For this reason, activity of neuron connections is measured only on local models after each epoch. Disjoint data vectors are processed by computing nodes to which the shared model has been delivered. Active neuron inputs are counted when calculating results for data vectors in each of the mentioned models. The number of active connections for a given local model is obtained by adding the active inputs of each neuron. To obtain the average activity of internal connections, the number of active connections of a given model is divided by the number of data vectors. A similar process is performed after completing training for the shared model.

All activity measurements have been implemented as part of the "LayerNeuronsGroup" class. They were placed in the method responsible for forward propagation, "fprop". This function was modified to return a value of type long rather than void so that it could return the number of active connections for the processed data vector. The results of the "fprop" method are aggregated and once training ends, averaged. To facilitate testing, the results of activity measurements are saved in a text file. To perform the described measurements, the "train_samples_per_iteration" parameter must be set to zero. This parameter value guarantees that regardless of the number of nodes, the data vector will be processed exactly once per epoch.

3.3   H2O.ai Flow and API Modifications

H2O.ai Flow and the H2O.ai API also had to be modified to enable the new functionality. Changes were done in the "DeepLearningModel" and "DeepLearningModelv3" classes to allow specifying new parameters in the H2O.ai API and interactively selecting the parameters in the H2O.ai Flow web interface. Introduction of the new "Ext" transfer function to the GUI is associated with the creation of a new "DeepLearningParameter.ExtActivation" parameter. The modified GUI is shown in Fig. 2.

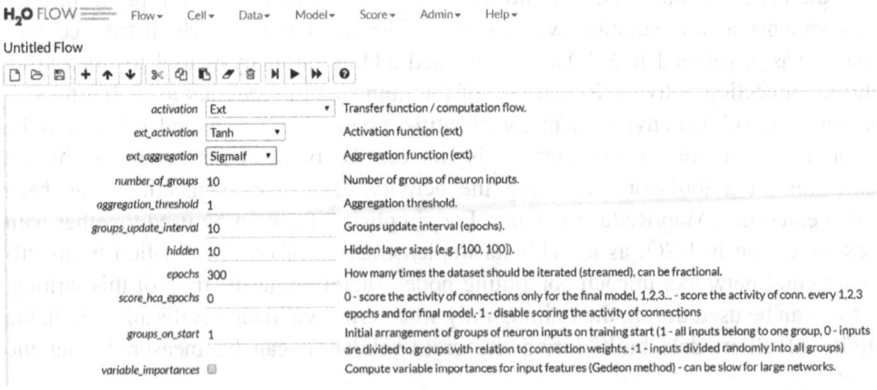

Fig. 2. Fragment of modified H2O Flow application.

One can notice that the name of the original "activation" parameter was not changed. The reason for leaving the name of this parameter was to keep the solution in sync with newer H2O.ai versions. Still, to indicate additional use-case of the "activation" parameter, its' description was renamed from "Activation function" to "Transfer function/computation flow". New parameter controls such as "ext_aggregation" appear only after selecting the new "Ext" transfer function value of the "activation" parameter. Other added parameters include "groups_update_interval", "number_of_groups" and "aggregation_treshold", corresponding respectively to Ω, K, φ^*.

4 Results of Experiments

To ensure that the CxNN implementation is correct, we have performed experiments on data sets from the UCI Machine Learning Repository Internet Advertisements, Qualitative_Bankruptcy and Census Income [20] as well as Armstrong' Cancer Gene Expression Microarray data [21]. For good measure, we deliberately chose two small collections, namely Armstrong and Qualitative_Bankruptcy, as well as two larger collections, Internet Advertisements and Census Income. The parameters that were analyzed were the average validation classification error and the average activity of connections between neurons "avg_hca".

For the experiments, we left most parameters at their default values and changed only a few of them. Cross-validation was not used, to analyze changes of connections activity during training. The "score_training_samples" and "train_samples_per_iteration" parameters have been set to zero, which means that all vectors in the training data set are used. When it comes to other parameters, "score_each_iteration" and "reproducible" have been set to true. For sets Qualitative_Bankruptcy and Armstrong values of parameters: "ext_activation", "ext_aggregation, "groups_update_interval", "aggregation_treshold" and "score_hca_epochs" were set successively to: Tanh, Sigma-if, 20, 0.1, 1. For bigger collections, Census Income and Internet Advertisements, only "aggregation_treshold" was changed to 0.7, while the rest of CxNN parameters were set to the same value as in small collections. The values of other parameters are shown in Table 1.

Table 1. Structure of CxNN used in experiments for chosen problems

Training data set	Number of inputs	Number of hidden neurons	Number of connection between neurons	Number of groups of neuron inputs (K)	Number of classes	Number of data vectors
Qualitative_Bankruptcy	7 (17)	10	260	10	2	250
Armstrong	12582 (0)	10	125840	10	3	72
Internet Advertisements	1558 (0)	20	31200	10	2	3279
Census Income	14 (423)	20	8780	10	2	48842

A single binary input for each value is a representation of nominal attributes, continuous attributes represent single inputs. In addition, H2O.ai introduces additional attributes to the list of network inputs whose value is determined by analyzing the interaction between the attributes of the data set. Table 1 shows the number of additional attributes in round brackets. Figures 3 and 4 show the results of experiments for the Qualitative_Bankruptcy and Armstrong problems. There is a visible decrease in the average classification error in successive epochs using the GPB algorithm. Those figures also show that in both cases "avg_hca" drops semi-logarithmically as the average classification error decreases. In the case of Qualitative_Bankruptcy, "avg-hca" decreased from 100% to 29% and in the case of Armstrong from 100% to 10%.

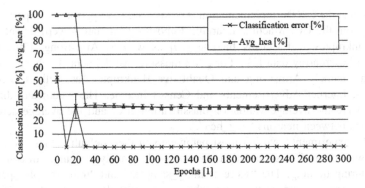

Fig. 3. Average activity of hidden connections and average classification error of CxNN when solving Qualitative_Bankruptcy benchmark problem.

The decrease in the activity of connections between neurons was expected because it is characteristic of neural networks using conditional multi-step aggregation functions. The decrease of "avg_hca" is dependent on both the values of aggregation threshold and number of groups K. We selected K = 10 due to suggestions in [10, 17] that for many problems it is useful to set K from 3 to 11.

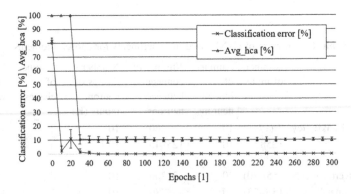

Fig. 4. Average activity of hidden connections and average classification error of CxNN when solving Armstrong benchmark problem.

When the number of neuron input groups K is equal to 1, a CxNN acts as a MLP network and in both cases "avg_hca" is 100% [10, 18]. Due to the implementation of the GBP algorithm, in the first ω epochs, the number of groups K is set to a value of 1, which is visible in Figs. 3 and 4, where the activity is equal to 100% at the beginning of training. After the experiments were conducted, it turns out that for the Qualitative_Bankruptcy and Armstrong problems, the accuracy of CxNN classification and MLP networks are comparable. This may be due to dynamic changes in the CxNN architecture observable when training the datasets with the GBP algorithm, which prevents overfitting. To more accurately evaluate the new functionality, additional experiments were performed on the larger data sets available in the UCI ML repository, Internet Advertisements and Census Income, for which results are shown in Figs. 5 and 6.

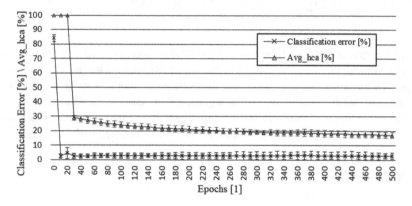

Fig. 5. Average activity of hidden connections and average classification error of CxNN when solving Internet Advertisements benchmark problem.

In these larger datasets, the classification error stays low while "avg_hca" continues to decrease down to around 20% for the Internet Advertisements problem and 10% for the Census Income problem. It is especially in larger problems that such a decrease in network connection activity could be highly beneficial.

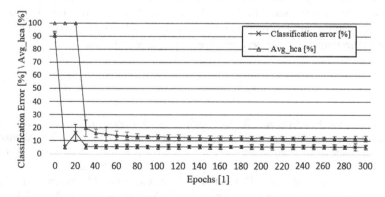

Fig. 6. Average activity of hidden connections and average classification error of CxNN when solving Census Income benchmark problem

CxNN measurements were also collected with the number of groups K set to 1 so that it would perform as a standard MLP network (MLP2). The results of its' training were compared with MLP (MLP1) network implemented in the H2O.ai version 3.10.0.3 (Table 2).

Table 2. Comparison of training results between MLP and CxNN neural networks

Training data set	MLP2 avg_hca [%]	CxNN avg_hca [%]	MLP1 classification error [%]	MLP2 classification error [%]	CxNN classification error [%]
Internet Advertisements	100 ± 0	17,1 ± 1,98	3,02 ± 1,25	3,22 ± 2,1	2,75 ± 2
Census Income	100 ± 0	12,1 ± 1,15	5,71 ± 1,61	5,59 ± 2,03	5,55 ± 1,75

Experiments indicate that the implemented GBP algorithm and conditional multi-step aggregation functions work correctly. When the parameter K, the number of groups, is set to a value of 10, "avg_hca" drops by more than four times in comparison to a standard MLP network without a significant increase in classification error. In addition, when the number of groups is set to 1, activity of hidden connections is equal to 100% by the learning process. These properties are characteristic of CxNN and GBP.

5 Conclusions

The results of experiments, performed on considered data sets, validate the methodologies presented. The decrease of hidden connections activity was observed also for benchmark data sets such as Qualitative_Bankruptcy, Internet Advertisements and Census Income. This set of new results is extension of earlier experiments [22] done with CxNNs on other data sets from UCI Machine Learning repository [20]. Additional experiments for these data sets have confirmed that in special cases, when the number of groups K is set to 1 CxNN works in the same way as a standard MLP network. Thus presented results suggest, that prepared software can be used for further research on CxNNs and their properties. The topics for next experiments include analyzing the properties of various multi-step aggregation functions and using CxNNs with large data sets to validate computational performance of prepared software.

References

1. Richter, A.N., Khoshgoftaar, T.M., Landset, S., Hasanin, T.: A multi-dimensional comparison of toolkits for machine learning with Big Data. In: IEEE International Conference on Information Reuse and Integration, pp. 1–8, IEEE, San Francisco (2015)
2. Ng, S.S.Y., Zhu, W., Tang, W.W.S., Wan, L.C.H., Wat, A.Y.W.: An independent study of two deep learning platforms - H2O and SINGA. In: International Conference on Industrial Engineering and Engineering Management, IEEM 2016, pp 1–5. IEEE Press, Bali (2016)

3. Cook, D.: Practical Machine Learning with H2O. Powerful, Scalable Techniques for Deep Learning and AI. O'Reilly Media, Beijing (2016)
4. Liang, M., Trejo, C., Muthu, L., Ngo, L.B., Luckow A., Apon, A.W.: Evaluating R-based Big Data analytic frameworks. In: IEEE International Conference on Cluster Computing (CLUSTER), pp. 1–2, IEEE, Chicago (2015)
5. H2O.ai homepage. https://www.h2o.ai/company/h2o-ai-partners-with-ibm-to-bring-enterprise-ai-to-ibm-power-systems/. Accessed 31 Oct 2018
6. Domingos, S.L., Carvalho, R.N., Carvalho, R.S., Ramos, G.N.: Identifying IT purchases anomalies in the Brazilian government procurement system using deep learning. In: 15th IEEE International Conference on Machine Learning and Applications (ICMLA) (2016)
7. Grolinger, K., Capretz, M.A.M., Seewald, L.: Energy consumption prediction with Big Data: balancing prediction accuracy and computational resources. In: IEEE International Congress on Big Data (BigData Congress), pp. 1–8 (2016)
8. Suleiman, D., Al-Naymat, G.: SMS spam detection using H2O framework. In: Procedia Computer Science, vol. 113, pp. 154–161 (2017)
9. H2O.ai 3.10.0.3 documentation. https://h2o-release.s3.amazonaws.com/h2o/rel-turing/3/docs-website/h2o-docs/index.html
10. Huk, M.: Backpropagation generalized delta rule for the selective attention Sigma-if artificial neural network. Int. J. Appl. Math. Comput. Sci. **22**, 449–459 (2012)
11. Huk, M.: Learning distributed selective attention strategies with the Sigma-if neural network. In: Akbar, M., Hussain, D. (eds.) Advances in Computer Science and IT, pp. 209–232. InTech, Vukovar (2009)
12. Szczepanik, M., Jóźwiak, I.: Data management for fingerprint recognition algorithm based on characteristic points' groups. In: Pechenizkiy, M., Wojciechowski, M. (eds.) New Trends in Databases and Information Systems. Advances in Intelligent Systems and Computing, vol. 185, pp. 425–432. Springer, Heidelberg (2013). https://doi.org/10.1007/978-3-642-32518-2_40
13. Huk, M.: Context-related data processing with artificial neural networks for higher reliability of telerehabilitation systems. In: 17th International Conference on E-health Networking, Application & Services (HealthCom), pp. 217–221. IEEE Computer Society, Boston (2015)
14. Huk, M., Kwasnicka, H.: The concept and properties of Sigma-if neural network. In: Ribeiro, B., Albrecht, R.F., Dobnikar, A., Pearson, D.W., Steele, N.C. (eds.) Adaptive and Natural Computing Algorithms, pp. 13–17. Springer, Vienna (2005). https://doi.org/10.1007/3-211-27389-1_4
15. Srivastava, N., Hinton, G., Krizhevsky, A., Sutskever, I., Salakhutdinov, R.: Dropout: a simple way to prevent neural networks from overfitting. J. Mach. Learn. Res. **15**, 1929–1958 (2014)
16. Huk, M.: Sigma-if neural network as the use of selective attention technique in classification and knowledge discovery problems solving. Annales UMCS Sectio AI - Informatica **4**(2), 121–131 (2006)
17. Huk, M.: Notes on the generalized backpropagation algorithm for contextual neural networks with conditional aggregation functions. J. Intell. Fuzzy Syst. **32**, 1365–1376 (2017)
18. Huk, M.: Manifestation of selective attention in Sigma-if neural network. In: 2nd International Symposium Advances in Artificial Intelligence and Applications, International Multiconference on Computer Science and Information Technology, IMCSIT/AAIA 2007, vol. 2, pp. 225–236 (2007)
19. Raczkowski, D., Canning, A.: Thomas-Fermi charge mixing for obtaining self-consistency in density functional calculations. Phys. Rev. B **64**, 121101–121105 (2001)
20. UCI Machine Learning Repository. http://archive.ics.uci.edu/ml

21. Armstrong, S.A.: MLL translocations specify a distinct gene expression profile that distinguishes a unique leukemia. Nat. Genet. **30**, 41–47 (2002)
22. Janusz, B.J., Wołk, K.: Implementing contextual neural networks in distributed machine learning framework. In: Nguyen, N.T., Hoang, D.H., Hong, T.-P., Pham, H., Trawiński, B. (eds.) ACIIDS 2018. LNCS (LNAI), vol. 10752, pp. 212–223. Springer, Cham (2018). https://doi.org/10.1007/978-3-319-75420-8_20

Low-Level Greyscale Image Descriptors Applied for Intelligent and Contextual Approaches

Dariusz Frejlichowski$^{(\boxtimes)}$ (iD)

Faculty of Computer Science and Information Technology, West Pomeranian University of Technology, Szczecin, Żołnierska 52, 71-210 Szczecin, Poland
dfrejlichowski@wi.zut.edu.pl

Abstract. The process of image recognition and understanding is not always a trivial task. The automatic analysis of the image content can be difficult and not obvious. Usually, it requires the identification of particular objects visible in a scene, however, this assumption not always provides the expected results. In many cases, the whole context of an image or relations between objects provide important information about an image and can lead to other conclusions than in case of the analysis of single objects separately. Hence, the obtained result can be considered more 'intelligent'. The contextual analysis of images can be based on various features. Amongst them the low-level descriptors are successfully applied in the problem of image analysis and recognition. Using the obtained representations of objects one can conclude the context of an image as a whole. In the paper the possibility of applying selected greyscale descriptors in the intelligent systems is analytically and experimentally analyzed. The works have been performed by means of algorithms employing the transformation of pixels from Cartesian into polar co-ordinates.

Keywords: Image recognition · Object identification · Greyscale descriptors · Polar co-ordinates

1 Introduction and Motivation

The automatic digital image recognition systems became something natural in our everyday life for good, becoming its component improving particular activities, giving handy tools, sometimes only entertaining, but usually significantly increasing our safety or comfort. Usually, we do not wonder, how an algorithm works, which localizes a face when we are taking a picture using digital camera or modifies our look in a funny way, when we already have taken this picture. It surprises us, how quick is the automatic search in Internet for many images similar to the one indicated by us as an example. Comfort and safety when driving a modern car is improved by automatic systems for traffic signs recognition, analysis of the driver's fatigue or detecting change of the trajectory.

Above-mentioned systems are based on the recognition of images as a whole or objects placed on them. In the second case one can apply two different approaches. The most popular and less complicated method is to recognize each object separately.

© Springer Nature Switzerland AG 2019
N. T. Nguyen et al. (Eds.): ACIIDS 2019, LNAI 11432, pp. 441–451, 2019.
https://doi.org/10.1007/978-3-030-14802-7_38

However, when considering contextual or intelligent algorithms it is sometimes better to take into account not only an object but also some more sophisticated data, e.g. the relation between objects or context of the information. The last problem is especially tempting in case of intelligent CBIR (Content Based Image Retrieval) systems, in which one tries to automatically identify the content of images based on pre-assumed conditions and query details. It is a popular example that an orange ball visible on an image can be identified as an orange fruit, basketball or the Sun, depending on the context of the rest of the image under analysis.

Regardless of the applied general approach, the selection of appropriate and effective features representing objects is crucial. For this purpose, one can apply shape, color, texture or greyscale. The descriptors of low-level features became very popular mainly because of their fast and easy derivation [1]. These features have different properties, usually related to particular applications. The greyscale descriptors can be very effective, since in real world, greyscale can bring important and useful information, sometimes better than in case of other features. However, the greyscale descriptors are less popular than for example shape or color ones. That is why in this paper the application of greyscale descriptors for the intelligent approaches designed for image recognition is considered and analyzed. It is assumed that development of the algorithms for greyscale feature will lead to the achievement of more efficient methods for intelligent recognition of objects placed on digital images. The effective method for object description is an important and useful task. However, this efficiency can become higher when using additionally some particular classification algorithms, as one of the stages of the final developed approach applied after the object description. Hence, this issue is also addressed in the paper.

In order to obtain the assumed goals, the paper was divided into the following sections. The second section describes the state of the art in the area of greyscale description from various points of view. The third section provides the description of the selected and applied greyscale descriptors for an exemplary real problem. The next section is devoted to the discussion of the possibility of improving the results by means of some modern classification methods. Finally, the last section concludes the paper.

2   Related Works

Greyscale descriptors are less popular than the other ones, as it was already mentioned. However, there are some applications and works based on them made so far, e.g. the descriptor based on the moment theory [2]. However, in many cases, the used approaches are utilized for the image as a whole, above all for the object localization. The SIFT (Scale-Invariant Feature Transform) algorithm is particularly popular in this application [3]. It is based on searching for local characteristic points and is especially useful in the process of searching for objects of interest (OOI) in an image. The extended SIFT approach was also proposed [4]. The SURF interesting point descriptor works efficiently for the same problem [5]. Similarly, corner detection is performed in order to localize features of interest on an image [6]. The detection is a goal for the authors of the Scale-Invariant Shape Features [7]. Greyscale images were applied for human detection on images using Histograms of Oriented Gradients [8]. Also, the histograms derived for the greyscale can be used as the object descriptor [9], however this representation is strongly limited. Slightly different approach was described in [10], where the analysis of scene in greyscale was based on histogram of distances between characteristic points. The completely another approach is the transformation to the gradient image and consecutive work on the level of edges in binary image. Such solution was for example applied in [11]. For the analysis of greyscale feature some local descriptors were also proposed [12].

The greyscale analysis is in many approaches limited to the textures and not applied for the description of objects extracted from digital images. Nevertheless, in some cases the same algorithms could be applied both for textures and segmented objects of interest, hence, texture descriptors should be mentioned here. Gabor features [13] are very popular in texture representation, however some other methods were also proposed for this task, e.g. Gaussian Markov Random Fields [14], non-uniform patterns [15], Local Binary Pattern with Local Phase Quantization [16], connectivity indexes in local neighborhoods [17], Fisher tensors with 'bag-of-words' [18], genetic algorithms [19], morphological approach [20], and local fractal dimensions [21].

3 Description of the Selected Algorithms

The algorithms selected for the experimental analysis have the same property of applying the transformation from Cartesian into polar co-ordinates for pixels represented in greyscale. They were designed for object representation regardless the planned application. Hence, they can be used for example in the recognition, identification or retrieval of extracted objects, but their usage is not limited only to the enumerated and that is why they can be applied in other intelligent and contextual systems.

The selection of the transformation to polar co-ordinates as the basis of the analyzed algorithms is not accidental. Above all, the obtained representation is invariant to translation and scaling. If combined with histogram or Fourier transform, it is also invariant to rotation. In case of the second mentioned transform, the descriptor can be additionally robust to noise.

The Polar-Fourier Greyscale Descriptor is the first algorithm selected for the experiments. It is based on the usage of simple algorithmic approaches – transformation of the pixels into polar co-ordinates and Fourier transform during the object description, and Euclidean distance at the stage of template matching. The whole algorithm includes several auxiliary stages, e.g. median filtering, low-pass filtering, the construction of the constant rectangle containing the object, filling the gaps with the background color, and resizing the polar image to the constant size. So far, this approach was applied to several problems: recognition of erythrocytes for automatic or semi-automatic diagnosis of some diseases [22], biometric identification of persons based on ear images [23], recognition of butterfly species [24] and traffic signs [25].

The second approach is based on polar transform and vertical and horizontal projections. The first part of the algorithm is similar to the above-mentioned approach, i.e. the pre-processing and transformation of co-ordinates system is performed. However, later instead of the Fourier transform applied at the end, the horizontal and vertical projections are obtained. In result, the representation is invariant to scaling and translation, but not to the rotation of the object in the image plane. The described approach was so far applied only to one real problem – identification of persons based on ear images in greyscale [26].

As it was already mentioned, the most important element of the selected algorithms is the transformation of the pixels from Cartesian into polar coordinates. It results in a significant change in the object's appearance. As an example, in Fig. 1 some objects and their polar-transformed representations are given [26]. For this example, two different applications were selected. On the left, the erythrocytes are shown. The recognition of red blood cells can be applied for example in automatic or semi-automatic diagnosis of some diseases (e.g. anemia or malaria), because they significantly change the cell's appearance. On the right, the butterflies are presented. In this case the automatic species recognition would be performed.

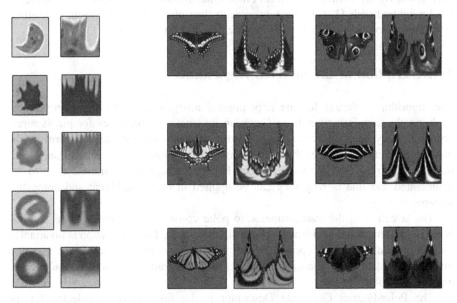

Fig. 1. Examples of polar transformed greyscale objects, belonging to two different classes and applications – erythrocytes in the process of automatic diagnosis and butterflies in the species recognition.

The details of both selected for the experiments and analysis algorithms are provided below.

3.1   The Polar-Fourier Greyscale Descriptor

Step 1. Median filtering of the input sub-image I, with the kernel size 3.
Step 2. Low-pass filtering, realized through the convolution with mask 3×3 pixels, and normalization parameter equal to 9.

Step 3. Calculation of the centroid by means of the moments [27]. Firstly, m_{00}, m_{10}, m_{01} are derived, and then the centroid $O = (x_c, y_c)$:

$$m_{pq} = \sum_x \sum_y x^p y^q I(x, y), \tag{1}$$

$$x_c = \frac{m_{10}}{m_{00}}, y_c = \frac{m_{01}}{m_{00}}. \tag{2}$$

Step 4. Finding the maximal distances d_{maxX}, d_{maxY} for X- and Y-axis respectively from the boundaries of I to O.

Step 5. Expanding the image into both directions by $d_{maxX} - x_c$ and $d_{maxY} - y_c$ and filling in the new parts using constant greyscale level, e.g. 127.

Step 6. Derivation of the polar coordinates and insertion in the image P:

$$\rho_i = \sqrt{(x_i - x_c)^2 + (y_i - y_c)^2}, \quad \theta_i = \operatorname{atan}\left(\frac{y_i - y_c}{x_i - x_c}\right). \tag{3}$$

Step 7. Resizing P to the constant rectangular size, $n \times n$, e.g. $n = 128$.

Step 8. Derivation of the two-dimensional Fourier transform:

$$C(k, l) = \frac{1}{HW} \left| \sum_{h=1}^{H} \sum_{w=1}^{W} P(h, w) \cdot e^{\left(-i\frac{2\pi}{H}(k-1)(h-1)\right)} \cdot e^{\left(-i\frac{2\pi}{W}(l-1)(w-1)\right)} \right|, \tag{4}$$

where:

H, W — height and width of P,

k — sampling rate in vertical direction ($k \geq 1$ and $k \leq H$),

l — sampling rate in horizontal direction ($l \geq 1$ and $l \leq W$),

$C(k, l)$ — the coefficient of discrete Fourier transform in k-th row and l-th column,

$P(h, w)$ — value in the resultant image plane with coordinates h, w.

Step 9. Selection of the spectrum sub-part, e.g. $10 \ldots 10$ size and concatenation into vector V.

3.2 The Approach Based on Polar Transform and Vertical and Horizontal Projections

Step 1. Median filtering of the input image I, with the kernel size 3.

Step 2. Low-pass filtering, realized through the convolution with mask 3×3 pixels, and normalization parameter equal to 9.

Step 3. Calculation of the centroid by means of the moments [27]. Firstly m_{00}, m_{10}, m_{01} are derived, and later the centroid $O = (x_c, y_c)$:

$$m_{pq} = \sum_x \sum_y x^p y^q I(x, y), \tag{5}$$

$$x_c = \frac{m_{10}}{m_{00}}, y_c = \frac{m_{01}}{m_{00}}. \tag{6}$$

Step 4. Transforming I into polar coordinates (resultant image is denotes as P), by means of the formulas:

$$\rho_i = \sqrt{(x_i - x_c)^2 + (y_i - y_c)^2}, \quad \theta_i = \operatorname{atan}\left(\frac{y_i - y_c}{x_i - x_c}\right). \tag{7}$$

Step 5. Resizing P to the constant rectangular size, $n \times n$, e.g. $n = 128$.
Step 6. Deriving the horizontal and vertical projections of P:

$$H_i = \sum_{j=1}^{n} P_{i,j}, \quad V_j = \sum_{i=1}^{n} P_{i,j}. \tag{8}$$

Step 7. Concatenating the obtained vectors H and V into one, $C = HV$, representing an object.

4   Conditions and Results of the Experiment

The goal of the performed experiment was the comparison of the two selected algorithms for greyscale object representation in an exemplary real application. For this purpose, the problem of traffic signs recognition was selected by means of The German Traffic Sign Recognition Benchmark [28], one of the most popular publicly available benchmark databases. Some examples of applied images are provided in Fig. 2.

Fig. 2. Examples of traffic signs images applied for performed experiment.

The extracted road signs were used. In total, 10000 images for 20 different classes were taken from the database in order to constitute the test data. For each class 50 instances were randomly selected from 500 images and used as the learning examples (i.e. they were the templates) and 200 random images were employed as the test data. This procedure was repeated ten times and the average recognition rate was obtained. The same test procedure was applied for both analyzed descriptors. The results for the Polar-Fourier Greyscale Descriptor were taken from [25], but the second algorithm was applied for the analyzed problem for the first time. By means of the obtained representation and the Euclidean distance the template closest to a test object was selected resulting in the identification of the recognized class. The average recognition rate for particular classes in case of the first greyscale descriptor is provided in Table 1, and for the second one – in Table 2.

Table 1. The experimental results obtained for the Polar-Fourier Greyscale Descriptor

Class	Correct results	Wrong results	Recognition rate	Class	Correct results	Wrong results	Recognition rate
1	1630	370	81.50%	11	1821	179	91.05%
2	1832	168	91.60%	12	1875	125	93.75%
3	1728	272	86.40%	13	1750	250	87.50%
4	1743	257	87.15%	14	1876	124	93.80%
5	1715	285	85.75%	15	1412	588	70.60%
6	1910	90	95.50%	16	1694	306	84.70%
7	1863	137	93.15%	17	1945	55	97.25%
8	1535	465	76.75%	18	1837	163	91.85%
9	1901	99	95.05%	19	1712	288	85.60%
10	1959	41	97.95%	20	1954	46	97.70%
				TOTAL	**35692**	**4308**	**89.23%**

The obtained results are different for investigated greyscale representation methods. The Polar-Fourier Greyscale Descriptor worked significantly better in the analyzed application. The worse results of the second approach are connected with less sophisticated steps of the algorithm. It is noticeably faster, but less effective. That brings the conclusion that for the full approach, taking the classification stage into account, the first descriptor works better.

5 Discussion on the Possibility of Improving the Results by Means of Properly Selected Classification Methods

The results provided in the previous section are not ideal, especially in case of the second descriptor. However, one has to take into account that the efficiency of a feature representation algorithm can become considerably higher when applying after the object description particular, efficient classification methods. It has to be emphasized that every intelligent approach or system designed for image recognition depends not only on the quality of information coded in feature space, but also on properly selected learning algorithm. The above elements have to work effectively together in order to give good results.

The proper selection of classification algorithms is crucial, since they would work well only if the data constituting the input are appropriately prepared. On the other hand, for particular data a specific classifier would work better than the other ones.

Table 2. The experimental results obtained for the greyscale descriptor based on polar transform and vertical and horizontal projections

Class	Correct results	Wrong results	Recognition rate	Class	Correct results	Wrong results	Recognition rate
1	1432	568	71.60%	11	1645	355	82.25%
2	1635	365	81.75%	12	1690	310	84.50%
3	1518	482	75.90%	13	1538	462	76.90%
4	1559	441	77.95%	14	1582	418	79.10%
5	1518	482	75.90%	15	1138	862	56.90%
6	1835	165	91.75%	16	1308	692	65.40%
7	1724	276	86.20%	17	1691	309	84.55%
8	1317	683	65.85%	18	1615	385	80.75%
9	1809	191	90.45%	19	1527	473	76.35%
10	1718	282	85.90%	20	1709	291	85.45%
				TOTAL	**31508**	**8492**	**78.77%**

Therefore, the research on the selection of classifiers appropriate for the analyzed greyscale descriptors should be conducted broadly. According to this assumption the future search for efficient machine learning approach should consider at least testing accuracy of various simple classifiers (the experiment described in this paper was limited only to the Euclidean distance) as well as advanced classifiers belonging to various groups, e.g. Support Vector Machines, perceptron networks, Radial Basis Function based networks, Hidden Markov Models, decision trees. The application of ensemble classifiers (such as AdaBoost, RealBoost, random forests) is also taken into consideration. Some additional approaches, complementary to the above, would also be taken into account, for example by means of model complexity selection or regularization techniques.

The usefulness of ensemble classifiers would be particularly utilized in order to compose an effective approach. Above all, the property of high resistance to overfitting [29] shall be experimentally investigated according to the features given by the grayscale descriptors. Similarly, the regularization methods as well as the model complexity selection should significantly improve the future results. In the second case, the possibility of testing successive supersets of classifiers would result in selection of an appropriate complexity of a model, with high probabilistic confidence.

Finally, the deep learning algorithms, which won affection and popularity lately, stimulate hopes for particularly effective approaches in intelligent and contextual systems designed for image representation and recognition in various applications. Many results obtained for the Convolutional Neural Networks have proven high efficiency of this approach, when considering the classification results. Thanks to this, they were successfully applied in various problems of computer vision – in video analysis [30], traffic sign recognition [31], food recognition [32] and many others. Considering the way the CNNs are applied, additionally it would be interesting to compare the usage of CNNs with and without description algorithms.

6 Conclusions

In the paper two greyscale descriptors were experimentally compared in an exemplary practical application from the domain of computer vision. The Polar-Fourier Greyscale Descriptor obtained better results in the traffic signs recognition problem. The algorithms are planned to be applied in intelligent image recognition systems. However, in order to do so, better approaches for classification are needed. In the experiment, simple Euclidean distance was applied. It is obvious that usage of more sophisticated algorithms for classification would result in more effective approaches and consequently better results. In this paper the problem of object representation was only analyzed, however the stress on the second part of the approach should result in higher efficiency. That is why, initially, the possibility of applying particular algorithms for this purpose was discussed in the paper. It is planned to verify the efficiency of other simple classifiers as well as advanced and ensemble ones. Moreover, model complexity selection and regularization techniques should give additional benefit to intelligent approach under development. Also, the deep learning algorithms, very popular lately, should be applied and analyzed during future works. Finally, the main goal of the described work was the comparison of approaches based on polar transform. However, in order to compare the efficiency of the algorithms (by means of the recognition rate obtained for particular tests) with other related works in this specific problem, more experimental results will be performed.

References

1. Verma, M., Raman, B.: Center symmetric local binary co-occurrence pattern for texture, face and bio-medical image retrieval. J. Vis. Commun. Image Represent. **32**, 224–236 (2015)
2. Shu, H., Zhang, H., Chen, B., Haigron, P., Luo, L.: Fast computation of Tchebichef moments for binary and grayscale images. IEEE Trans. Image Process **19**(12), 3171–3180 (2010)
3. Lowe, D.G.: Distinctive image features from scale-invariant keypoints. Int. J. Comput. Vis. **60**(2), 91–110 (2004)
4. Ke, Y., Sukthankar, R.: PCA-SIFT: a more distinctive representation for local image descriptors. In: Proceedings of the 2004 IEEE Computer Society Conference on Computer Vision and Pattern Recognition (CVPR 2004), vol. 2, pp. II-506–II-513 (2004)
5. Bay, H., Ess, A., Tuytelaars, T., Van Gool, L.: SURF: speeded up robust features. Comput. Vis. Image Underst. **110**(3), 346–359 (2008)
6. Shui, P., Zhang, W.: Corner detection and classification using anisotropic directional derivative representations. IEEE Trans. Image Process. **22**(8), 3204–3218 (2013)
7. Jurie, F., Schmid, C.: Scale-invariant shape features for recognition of object categories. In: Proceedings of the 2004 IEEE Computer Society Conference on Computer Vision and Pattern Recognition, (CVPR 2004), vol. 2, pp. II-90–II-96 (2004)
8. Dalal, N., Triggs, B.: Histograms of oriented gradients for human detection. In: Proceedings of the IEEE Computer Society Conference on Computer Vision and Pattern Recognition, CVPR 2005, San Diego, vol. 1, pp. 886–893 (2005)

9. Gbèhounou, S., Lecellier, F., Fernandez-Maloigne, C.: Evaluation of local and global descriptors for emotional impact recognition. J. Vis. Commun. Image Represent. **38**, 276–283 (2016)

10. Chin, T.J., Suter, D., Wang, H.: Boosting histograms of descriptor distances for scalable multiclass specific scene recognition. Image Vis. Comput. **29**(4), 241–250 (2011)

11. Vu, N.S., Dee, H.M., Caplier, A.: Face recognition using the POEM descriptor. Pattern Recogn. **45**(7), 2478–2488 (2012)

12. Castrillón-Santana, M., de Marsico, M., Nappi, M., Riccio, D.: MEG: texture operators for multi-expert gender classification. Comput. Vis. Image Underst. **156**, 4–18 (2017)

13. Kumar, M., Singh, Kh.M: Retrieval of head–neck medical images using Gabor filter based on power-law transformation method and rank BHMT. Signal Image Video Process. **12**(5), 827–833 (2018)

14. Dharmagunawardhana, C., Mahmoodi, S., Bennett, M., Niranjan, M.: Gaussian Markov random field based improved texture descriptor for image segmentation. Image Vis. Comput. **32**(11), 884–895 (2014)

15. Nanni, L., Brahnam, S., Lumini, A.: A simple method for improving local binary patterns by considering non-uniform patterns. Pattern Recogn. **45**(10), 3844–3852 (2012)

16. Nanni, L., Melucci, M.: Combination of projectors, standard texture descriptors and bag of features for classifying images. Neurocomputing **173**, 1602–1614 (2016)

17. Florindo, J.B., Landini, G., Bruno, O.M.: Three-dimensional connectivity index for texture recognition. Pattern Recogn. Lett. **84**, 239–244 (2016)

18. Faraki, M., Harandi, M.T., Wiliem, A., Lovell, B.C.: Fisher tensors for classifying human epithelial cells. Pattern Recogn. **47**(7), 2348–2359 (2014)

19. Wang, S., et al.: Texture analysis method based on fractional Fourier entropy and fitness-scaling adaptive genetic algorithm for detecting left-sided and right-sided sensorineural hearing loss. Fundamenta Informaticae **151**(1–4), 505–521 (2017)

20. Aptoula, E., Lefèvre, S.: Morphological texture description of grey-scale and color images. Adv. Imaging Electron Phys. **169**, 1–74 (2011)

21. Florindo, J.B., Bruno, O.M.: Local fractal dimension and binary patterns in texture recognition. Pattern Recogn. Lett. **78**, 22–27 (2016)

22. Frejlichowski, D.: Identification of erythrocyte types in greyscale MGG images for computer-assisted diagnosis. In: Vitrià, J., Sanches, J.M., Hernández, M. (eds.) IbPRIA 2011. LNCS, vol. 6669, pp. 636–643. Springer, Heidelberg (2011). https://doi.org/10.1007/978-3-642-21257-4_79

23. Frejlichowski, D.: Application of the Polar-Fourier Greyscale Descriptor to the problem of identification of persons based on ear images. In: Choraś, R.S. (ed.) Image Processing and Communications Challenges 3. AISC, vol. 102, pp. 5–12. Springer, Heidelberg (2011). https://doi.org/10.1007/978-3-642-23154-4_1

24. Frejlichowski, D.: An experimental evaluation of the Polar-Fourier Greyscale Descriptor in the recognition of objects with similar silhouettes. In: Bolc, L., Tadeusiewicz, R., Chmielewski, L.J., Wojciechowski, K. (eds.) ICCVG 2012. LNCS, vol. 7594, pp. 363–370. Springer, Heidelberg (2012). https://doi.org/10.1007/978-3-642-33564-8_44

25. Frejlichowski, D.: Application of the Polar-Fourier Greyscale Descriptor to the automatic traffic sign recognition. In: Kamel, M., Campilho, A. (eds.) ICIAR 2015. LNCS, vol. 9164, pp. 506–513. Springer, Cham (2015). https://doi.org/10.1007/978-3-319-20801-5_56

26. Frejlichowski, D.: A new algorithm for greyscale objects representation by means of the polar transform and vertical and horizontal projections. In: Nguyen, N.T., Hoang, D.H., Hong, T.-P., Pham, H., Trawiński, B. (eds.) ACIIDS 2018. LNCS (LNAI), vol. 10752, pp. 617–625. Springer, Cham (2018). https://doi.org/10.1007/978-3-319-75420-8_58

27. Hupkens, T.M., de Clippeleir, J.: Noise and intensity invariant moments. Pattern Recogn. Lett. **16**(4), 371–376 (1995)
28. Stallkamp, J., Schlipsing, M., Salmen, J., Igel, C.: The German traffic sign recognition benchmark: a multi-class classification competition. In: Proceedings of the IEEE International Joint Conference on Neural Networks, pp. 1453–1460 (2011)
29. Friedman, J., Hastie, T., Tibshirani, R.: Additive logistic regression: a statistical view of boosting. Ann. Stat. **28**(2), 337–407 (2000)
30. Burney, A., Syed, T.Q.: Crowd video classification using convolutional neural networks. In: International Conference on Frontiers of Information Technology (FIT), Islamabad, pp. 247–251 (2016)
31. Sermanet, P., LeCun, Y.: Traffic sign recognition with multi-scale convolutional networks. In: Proceedings of the International Joint Conference on Neural Networks, San Jose, pp. 2809–2813 (2011)
32. Kagaya, H., Aizawa, K., Ogawa, M.: Food detection and recognition using convolutional neural network. In: Proceedings of the 22nd ACM International Conference on Multimedia, pp. 1085–1088 (2014)

Intelligent Systems and Algorithms in Information Sciences

Motion Controlling Using Finite-State Automata

Michal Jaluvka, Eva Volna$^{(\boxtimes)}$, and Martin Kotyrba

Department of Informatics and Computers, University of Ostrava,
30. dubna 22, 70103 Ostrava, Czech Republic
{michal.jaluvka,eva.volna,martin.kotyrba}@osu.cz

Abstract. This paper, a discussion of fundamental finite-state algorithms, constitutes an approach from the perspective of dynamic system control. First, we describe fundamental properties of deterministic finite-state automata. We propose an algorithm focused on correct finite-state automaton state switching. This approach will then be used to propose a finite-state automaton to solve real-time movement in a maze. We also illustrate the use of such a proposed approach to control movement in a robotic system which requires correct states to achieve correct movement. Evaluation of achieved outputs is described in the conclusion, which also includes the future focus of the proposed way of finite-state automaton state switching.

Keywords: Deterministic finite-state automaton ·
Finite-state automaton state switching · Movement in a maze ·
Robotics motion controlling

1 Introduction

Finite-state automata were first studied in the 1950's by Stephen Kleene [6] and found a number of important applications in computer science: for example, in the design of computer circuits, and in the lexical analysers of compilers. In the 1960's and 1970's, mathematicians such as Samuel Eilenberg, Marcel-Paul Schützenberger, and John Rhodes [10] pioneered the mathematics of finite-state automata. More recently, other mathematicians have come to appreciate the usefulness of automata in such areas as combinatorial group theory and symbolic dynamics [2].

The theory of finite-state machine is rich. Finite-state devices, such as finite-state automata have been present since the emergence of computer science and are extensively used in areas as various as program compilation, hardware modelling, database management, cellular automata, beverage automata, traffic systems, robotic systems, etc. Finite-state automata techniques are used in a wide range of domains, including switching theory, pattern matching, pattern recognition, speech processing, handwriting recognition, optical character recognition, encryption algorithm, data compression, operating system analysis (e.g. Petri-net), electronic dictionaries, natural language processing [9], etc. In a sum: systems where states changing in time occur can include a finite-state automaton.

© Springer Nature Switzerland AG 2019
N. T. Nguyen et al. (Eds.): ACIIDS 2019, LNAI 11432, pp. 455–464, 2019.
https://doi.org/10.1007/978-3-030-14802-7_39

The paper is organised as follows. Section 1 is an introduction. Section 2 introduces terminology concerning finite-state automata, especially deterministic finite-state automaton. Section 3 describes the proposed way of state switching for a finite-state automaton. Sections 4 and 5 present experimental verification of the proposed algorithm. Section 4 shows a proposal of a finite-state automaton designed for real-time movement in a maze. Section 5 shows a proposal of a finite-state automaton designed for a robotic system. Finally, Sect. 6 presents our conclusions.

2 Finite-State Automata

A *finite-state machine* or *finite-state automaton* (plural: *automata*), *finite automaton*, or simply a *state machine*, is a mathematical model of computation [11]. It is an abstract machine that can be in exactly one of a finite number of states at any given time. There are two types of finite-state automata (FA): deterministic finite-state automaton (DFA), and nondeterministic finite-state automaton (NFA). There are slight variations in ways that state machines are represented visually, but the ideas behind them stem from the same computational ideas. In DFA, for each input symbol, one can determine the state to which the machine will move. Hence, it is called Deterministic Automaton. As it has a finite number of states. DFA can recognize or accept regular languages, and a language is regular if a deterministic finite-state automaton accepts it. Unlike a DFA, for some state and input symbol, the next state may be nothing or one or two or more possible states, i.e. the next state is an element of the power set of the states, which is a set of states to be considered at once. Both types of finite-state machines are usually taught using languages made up of binary strings that follow a particular pattern.

As our proposed system is based on deterministic finite-state automata, so only it is introduced in the following text. A deterministic finite-state automata M is a 5-tuple, $(Q, \sum, \delta, q_0, F)$, consisting of [9]

- a finite set of states Q
- a finite set of input symbols called the alphabet $\sum$
- a transition function $\delta \colon Q \times \sum \to Q$
- an initial or start state $q_0 \in Q$
- a set of accept states $F \subseteq Q$

Let $w = a_1, a_2, \ldots, a_n$ be a string over the alphabet Σ. The automaton M accepts the string w if a sequence of states, $r_0, r_1, \ldots, r_n$, exists in Q with the following conditions:

- $r_0 = q_0$
- $r_{i+1} = \delta \ (r_i, a_{i+1})$, for $i = 0, \ldots, n - 1$
- $r_n \in F$.

In words, the first condition says that the machine starts in the start state q_0. The second condition says that given each character of string w, the machine will transition from state to state according to the transition function δ. The last condition says that the machine accepts w if the last input of w causes the machine to halt in one of the accepting states. The set of strings that M accepts is the language recognised by M and this language is denoted by $L(M)$.

There are four possibilities how to represent DFA:

1. To list all elements from the pentad of a finite-state automaton.
2. A table
3. A state diagram (a graph).
4. A state tree.

In the following text, a deterministic finite-state automaton will be used.

3 Switching the States for a Finite-State Automaton

In [7], there are algorithms for searching through a graph, if we stem from a state diagram. However, the trouble is that most algorithms (search into depth and width) generate a sequence of states having presented the input and output state. Other algorithms are mostly for oriented graphs which have valued edges. A solution how to achieve a sequence – *queue*, so that a dynamic system switched to a new state, is by using the following algorithm:

1. Initialisation of the initial state $q_0 = S_p$
2. Initialisation of the final state $\sum = S_K$
3. Initialisation of a modified matrix M_{MS}
4. Initialisation of a queue (empty set)
5. Initialisation of the current state $S = S_p$
6. While($S! = S_K$) do
   (a) Read edge from $M_{MS}[index_S][index_{S_K}]$
   (b) Store edge (no sign) into queue
   (c) Switch the state according to edge $S \xrightarrow{[edge]} S$
7. End while

A modified matrix M_{MS} differs from an adjacency matrix MS as follows:

- instead of ones, cells contain a type of edges leading directly to a certain state
- instead of zeros, cells contain a type of edges beginning with a minus
- instead of zeros, cells on the main diagonal have empty spaces

A modified matrix contains information on the type of edges which are to be followed from the initial state $q_0 = S_p$ into the final state $\sum = S_K$. If an edge has a minus sign, it means that the final sate is achieved through one or more states. If we take into consideration the configuration of the finite-state automaton, a modified matrix is created using an algorithm of findings ways, Dijkstra's algorithm. According to a definition in [7] and one of its implementation in [3], the shortest path from the initial to the final state can be found as well as a list of individual states it passes through. The list then can serve to identify the way (a set of edges) along which one can move from the initial to the final state (although the final state can be switched through other states).

4 Proposal of a Finite-State Automaton Designed for a Real-Time Movement in a Maze

This chapter describes an approach of a finite-state automaton for moving in a maze in real time. In the maze, the starting position (location of a penguin figure) and the final destination (in a form of *chequered flag*) are defined at the beginning. The main task of the moving figure is to get to the final destination, if possible, by the shortest path. However, in the running experiment, the location of the final destination dynamically changes, it means that the path of the moving figure is optimised according to the actual location of the final destination (see Fig. 1).

In this experiment, a finite-state automaton (see Fig. 2) is used, in which individual states represent individual squares in the maze (Fig. 2b) defined in Fig. 2a. State switching occurs according to the transfer function, where its inputs are the actual state (marked as upper-case letter, this is the position of penguin) and an edge (marked as lower-case letter) and its output is a new position (a new location of the penguin). The choice of the relevant edge is solved by the proposed algorithm allowing the way how to switch states for the finite-state automaton (see Sect. 3).

For this experiment, the parameters of the finite-state automaton according to definition [9] are set:

- $Q = \{AA, AB, AC, AD, \ldots, US, UT, UU\}$
- $\sum = \{a, b, cd\}$
- $q_0 = AA$
- $F = \{AA\}$
- $\delta = \{AA \xrightarrow{c} BA, AB \xrightarrow{b} AC, AB \xrightarrow{c} BB, \ldots, UT \xrightarrow{b} UU, UU \xrightarrow{d} UT\}$

The total number of elements of set Q is 441 and the number of transfer functions is 885. Modified matrix M_{MS} is generated from finite-state automaton settings (its state diagram is shown in Fig. 2b). A part of matrix M_{MS} is shown in Table 1. In fact, this matrix has its size of 441×441 according to the total number of elements of set Q.

Table 1. Modified matrix M_{MS} for the maze.

	AA	AB	AC	...	UR	US	UT	UU
AA		c	c		c	c	c	c
AB	c		b		c	c	c	c
AC	d	d			d	d	d	d
⋮				⋱				
UR	a	a	a			b	b	b
US	d	d	d		d		a	a
UT	a	a	a		a	a		b
UU	d	d	d		d	d	d	

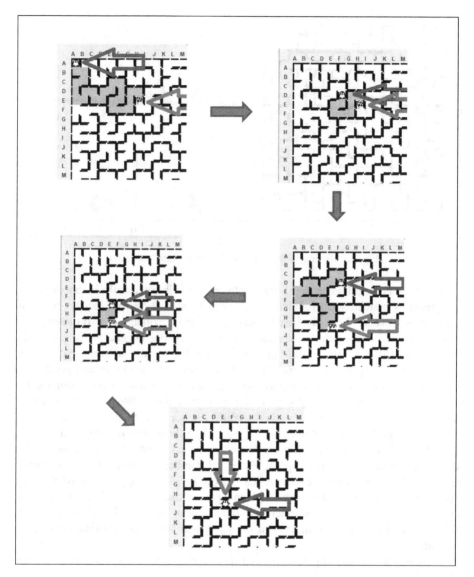

Fig. 1. Experimental environment for the maze (cut-out). Start is the position of *the Penguin* 🐧. Destination is the position of *chequered flag* 🏁.

The experiment progress can be found on YouTube channel https://www.youtube.com/watch?v=gWK3Kl2rNsQ. During running the experiment, it is not necessary to perform calculations for generating the shortest path according to the change its final destination, because the propose approach of state switching is resistant to changes of the final destination.

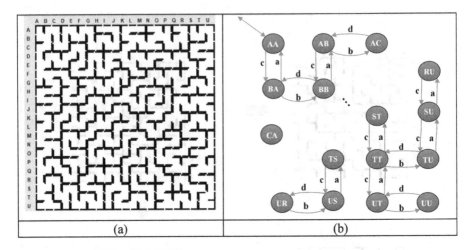

Fig. 2. Moving in the maze with a finite-state automaton.

Maze solving and finding the shortest path or all possible exit paths in mazes can be interpreted as mathematical problems which can be solved algorithmically. In [4] is summarized several chemistry-based concepts for maze solving in two-dimensional standard mazes which rely on surface tension driven phenomena at the air-liquid interface. In [5] a genetic algorithm is used to solve a class of maze pathfinding problems, where agents find a complete set of paths directing them from any position in the maze towards a single goal. Behaviour of these experimental agents is suitable for games, because they do not always find the shortest paths. Work [1] proposes an intelligent maze solving robot that can determine its shortest path on a line maze based on image processing and artificial intelligence algorithms. The developed algorithm solves the maze by examining all possible paths exist in the maze that could convey the robot to the required destination point. After that, the best shortest path is determined and then the instructions are sent to the robot. In [8] the application of Cellular Automata to the problem of robot path planning in a maze is presented. It is shown that a Cellular Automata allows the efficient computation of an optimal collision-free path from an initial to a goal configuration on a physical space for real-time robot path planning.

The benefits of the proposed solution based on the finite-state automata are the following:

- Modeling of agent behavior using finite-state automata.
- Determining how to switch states without programming.
- The ability to switch from one state to another one without the use of any planning algorithms.
- Only one simple operation is performed in each state of the finite-state automata.
- Changes of the state of the finite-state automata do not recalculate the path.

5   Finite-State Automaton Proposal for a Robotic System

A finite-state automaton proposal for a robotic system is another possibility where the proposed algorithm for state switching can be used. A finite-state automaton becomes a powerful tool for modelling of robotic system behaviour. It makes us understand what actual state the system is in and how it is possible to change the state. State changing is performed by another system or human interaction.

Let's have a representative of a robotic system, a humanoid robot, and we want to teach it several activities. Such several activities include basic movements like sitting, lying, walking, and running. An example of modelled behaviour can be a finite-state automaton shown in Fig. 3. This finite-state automaton consists of states (position or robot movement) and transfers (edges representing commands called by the user or system itself). In every state is a set of individual movements for a given activity. Every state has a unique edge to preserve the property of determinism.

Initial state q_0 is defined as state "Staying", in which the robotic system waits for command or command sequence. After commands are inserted, the finite-state automaton is switched to the state and then the automaton waits for next commands. In case the robot is in the state "Lying" and we want to get it to state "Sitting", it is simple to generate command "lie". Of course, there is a variant, when we want the robot to run although the robot is sitting. It means that the robot must pass 5 states to get from state "Lying" to new state "Running".

The proposed model of robotic system behaviour is shown in a state diagram in Fig. 3. The meanings of individual used states and edges are shown in Table 2.

Table 2. Legend for finite-state automaton from Fig. 3

State legend		Transfer legend	
State	Meaning	Transfer (edge)	Meaning
A	Staying	a	Go
B	Squatting	b	Stop
C	Sitting	c	Go back
D	Lying	d	Squat
E	Walking	e	Sit down
F	Running	f	Lie down
G	Stepping back	g	Stand up
H	Turning right	h	Jump
I	Turning left	i	Bend your knees
J	Knees bending	j	Run
K	Jumping	k	Turn left
		l	Turn right

Modified matrix M_{MS}, which is generated from finite-state automaton settings in Fig. 3, is shown in Table 3.

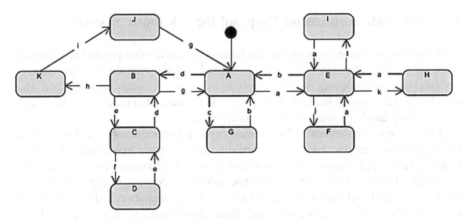

Fig. 3. Robotic system state diagram

Table 3. Modified matrix M_{MS} for robot system

	A	B	C	D	E	F	G	H	I	J	K
A		d	-d	-d	a	-a	c	-a	-a	-d	-d
B	g		e	-e	-g	-g	-g	-g	-g	-h	h
C	-d	d		f	-d	-d	-d	-d	-d	-d	-d
D	-e	-e	e		-e	-e	-e	-e	-e	-e	-e
E	b	-b	-b	-b		j	-b	k	l	-b	-b
F	-a	-a	-a	-a	a		-a	-a	-a	-a	-a
G	b	-b	-b	-b	-b	-b		-b	-b	-b	-b
H	-a	-a	-a	-a	a	-a	-a		-a	-a	-a
I	-a	-a	-a	-a	a	-a	-a	-a		-a	-a
J	g	-g	-g	-g	-g	-g	-g	-g	-g		-g
K	-i	-i	-i	-i	-i	-i	-i	-i	-i	i	

The proposed algorithm for state switching in a finite-state automaton is presented as a scenario form. A scenario creation (command generating) can be applied in case the robotic system is influenced by its surrounding. An example may be obstacle detection before the robotic system. There are two solutions how the robot could proceed. The robot either stops, waits for user's instruction (see the 1st scenario in Fig. 4), or tries to overcome the obstacle (see the 2nd scenario in Fig. 5). Next use depends on who generated the command sequence. The proposed algorithm is integrated either into the robotic system or another system which controls the robotic system.

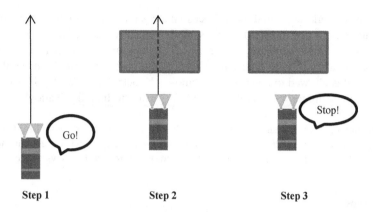

Fig. 4. Obstacle detected – 1st scenario

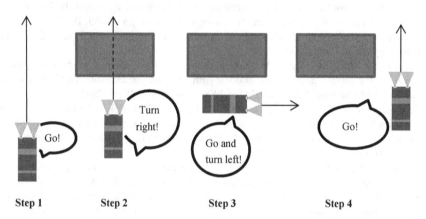

Fig. 5. Obstacle detected – 2nd scenario

6   Conclusions

This paper, a discussion of fundamental finite-state algorithms, constitutes an approach from the perspective of dynamic system control. We propose an algorithm focused on states switching for a finite-state automaton. This approach was experimentally verified in a proposal of a finite-state automaton designed for real-time movement solution in a maze, see https://www.youtube.com/watch?v=gWK3Kl2rNsQ. We also illustrate the use of this proposed approach for movement control in a robot.

A great benefit of the purposed algorithm is simple and secure state switching of the finite-state automaton which controls a dynamic system. It eliminated a system collapse. With a continuous state switching with commands, which are provided by the algorithm, we achieve the right control system function. Possible shortcomings in system activities are fixed only at the state level (penguin moving in the maze, robot joints). This state switching was demonstrated in the performed experiments.

In our next work, we would like to focus on a design of finite finite-state automaton whose state settings and transfer functions are dynamically changed in real time. A part of this focus is implementation of the existing proposal how to switch states in a dynamically changing finite-state automaton. In the maze experiment, it would be a maze where it is allowed to generate and remove obstacles (maze walls). However, the task of moving a figure remains the same: "To get to the final destination".

Acknowledgments. The research described here has been financially supported by University of Ostrava grant SGS04/PřF/2018. Any opinions, findings and conclusions or recommendations expressed in this material are those of the authors and do not reflect the views of the sponsors.

References

1. Aqel, M.O., Issa, A., Khdair, M., ElHabbash, M., AbuBaker, M., Massoud, M.: Intelligent maze solving robot based on image processing and graph theory algorithms. In: International Conference on Promising Electronic Technologies (ICPET), pp. 48–53. IEEE (2017)
2. Berthe, V., Rigo, M.: Combinatorics, Words and Symbolic Dynamics. Cambridge University Press, Cambridge (2016)
3. Broumi, S., Bakal, A., Talea, M., Smarandache, F., Vladareanu, L.: Applying Dijkstra algorithm for solving neutrosophic shortest path problem. In: International Conference on Advanced Mechatronic Systems (ICAMechS), pp. 412–416. IEEE (2016)
4. Čejková, J., Tóth, R., Braun, A., Branicki, M., Ueyama, D., Lagzi, I.: Shortest path finding in mazes by active and passive particles. In: Adamatzky, A. (ed.) Shortest Path Solvers. From Software to Wetware. ECC, vol. 32, pp. 401–408. Springer, Cham (2018). https://doi.org/10.1007/978-3-319-77510-4_15
5. Gordon, V.S., Matley, Z.: Evolving sparse direction maps for maze pathfinding. In: Congress on Evolutionary Computation, CEC 2004, vol. 1, pp. 835–838. IEEE (2004)
6. Kleene, S.C.: Representation of events in nerve nets and finite automate. In: Automata Studies, Annals of Mathematics Studies, pp. 2–42. Princeton University Press (1956)
7. Lafore, R.: Data Structures and Algorithms in Java. Sams Publishing (2017)
8. Martins, L.G.A., Cândido, R.P., Escarpinati, M.C., Vargas, P.A., de Oliveira, G.M.B.: An improved robot path planning model using cellular automata. In: Giuliani, M., Assaf, T., Giannaccini, M.E. (eds.) TAROS 2018. LNCS (LNAI), vol. 10965, pp. 183–194. Springer, Cham (2018). https://doi.org/10.1007/978-3-319-96728-8_16
9. Roche, E., Schabes, Y.: Finite-State Language Processing. MIT Press, Cambridge (1997)
10. Salomaa, A.: Theory of Automata. Elsevier (2014)
11. Straubing, H.: Finite Automata, Formal Logic, and Circuit Complexity. Springer, New York (2012). https://doi.org/10.1007/978-1-4612-0289-9

Automatic Recognition of Kazakh Speech Using Deep Neural Networks

Orken Mamyrbayev[1]([✉]) [iD], Mussa Turdalyuly[1]([✉]) [iD],
Nurbapa Mekebayev[2] [iD], Keylan Alimhan[1] [iD],
Aizat Kydyrbekova[2] [iD], and Tolganay Turdalykyzy[1] [iD]

[1] Institute of Information and Computational Technology,
050010 Almaty, Kazakhstan
morkenj@mail.ru, mkt_001@mail.ru
[2] al-Farabi Kazakh National University, 050040 Almaty, Kazakhstan

Abstract. This article presents a deep neural network (DNN) system based on automatic speech recognition for Kazakh language, developed using the Kaldi speech recognition tool. DNNs are initialized using the restricted Boltzmann machines (RBM) and are trained using cross-entropy as the objective function and the standard back propagation of error. In order to achieve optimal results, the training has been modified based on peculiarities of Kazakh language. A 76 hours-corpus has been used in training. Results are compared for two different sets of values between classical models and various DNN settings.

Keywords: DNN · ASR · Kazakh speech recognition · LM

1 Introduction

The creation of natural-language man-machine interfaces and, in particular, automatic speech recognition systems has recently become one of the main areas and tasks in the field of artificial intelligence. Speech technologies provide a more natural user interaction with computing and telecommunications complexes compared to the standard graphical interface.

With the development of personal computers and a wide range of public information and entertainment services, speech and then multimodal interfaces are now more focused on application in social intelligent services, which imposes its conditions on speech processing systems. In particular, the vocabulary of lexical units increases, the variability of speech increases, and processing should be carried out in real time to maintain a natural dialogue with the user. The development of a compact way of presenting the dictionary is especially relevant for agglutinative languages with a relatively rich morphology. To take into account the variability and learning models of phonemes and words require huge text and speech materials, the preparation of which requires meticulous expert work.

In [1], three types of speech disruptions that are most characteristic of spontaneous speech were analyzed: (1) voiced pause, (2) repetition of words, (3) modification of the sentence from the very beginning. As a material, speech corps Spoken Dutch Corpus (CGN) and Switchboard-1 were used. The number of voiced pauses made up 3% of all

© Springer Nature Switzerland AG 2019
N. T. Nguyen et al. (Eds.): ACIIDS 2019, LNAI 11432, pp. 465–474, 2019.
https://doi.org/10.1007/978-3-030-14802-7_40

lexical units in these corpora. Most often these were interjections, and they were located in all parts of sentences. The relative number of repetitions was approximately 1%. And the twenty most frequent repetitions are short words consisting of one syllable.

In [2], an audiovisual detector of voiced pauses was used to filter unwanted speech failures in multimedia recordings of lectures. The recorded multimedia corpus of lectures lasting about 7 h contained an image of the tablet computer screen, on which the lecturer made handwritten notes displayed to the audience on a multimedia projector, as well as a sound stream with the lecturer's speech and background noise. The analysis of the body showed that the vast majority of hesitations occurs when the lecturer does not use a tablet (or pad, data tablet), so a two-stage algorithm was used to filter the pauses. First of all, the moments of time were determined when the image on the monitor screen did not change, and then only during these periods of time the search for filled pauses in the audio stream was carried out. In the analysis, voiced pauses with a duration of more than 120 ms, pronounced in isolation (i.e., those containing segments with silence before and after hesitation), as well as within a word, were considered. The use of preliminary segmentation of the audio segments and the video analysis from a tablet helped to increase the recognition accuracy of the hesitations up to 85%.

In the rapid development of speech technology is associated with the development of artificial neural networks and is now becoming increasingly popular research on the use of DNN for recognition of Kazakh speech. At the same time, there are no effective systems of automatic recognition of Kazakh speech at the moment and the development of ASR is relevant.

In this article, we consider the method of creating an automatic speech recognition system using DNN using Kaldi tools. In this study, the existing speech corpus was expanded, the speech and text corpus for the Kazakh language was assembled, and acoustic and language models were created on the basis of neural network (NN), which allowed to increase the accuracy of recognition of Kazakh speech.

For speech preprocessing, we used the following algorithms: Mel-cepstral coefficients (MFCC) and Perceptual Linear Prediction Coefficients (PLP). For acoustic modeling, it uses the hidden Markov model (HMM), the Gaussian distribution mixture model (GMM), Subspace Gaussian mixture model (SGMM) and deep neural networks (DNN). Language modeling is performed using finite transducers (FSTs) with linear algebra support — the BLAS and LAPACK libraries.

The paper is organized as follows. Section 2 describes the work on the relevant scientific research area. Section 3 discusses data preprocessing methods Sect. 4 describes the methodology for automatic speech recognition. Section 5 describes the DNN architecture and Sects. 6, 7 discuss the results of the experiment and the conclusion.

2   Related Works

Currently DNN is often used in speech research for speech recognition and the results of the research show good results. For example, studies [3] present a system of recognition of spontaneous Czech, Slovak and Russian speech for processing interviews of Holocaust witnesses. In this paper, basic transcriptions were created automatically using a specific set of rules, and for many words several transcription variants were generated to take into account the phonetic phenomena of continuous speech (for example, assimilation of consonants at the word boundary). Then transcriptions describing spoken pronunciation variants were created, as well as for the Russian language and accent, as interviews were taken not only from the residents of Russia, but also from Russians living in Ukraine, Israel, and the USA. In addition, non-speech phenomena were modeled. The size of the corpus used to create acoustic models for the Russian language was 100 h and DNN was used. The language model was a bigram model using the return method (Katz's backing-off scheme). With a dictionary size of 79 thousand transcriptions, the percentage of incorrectly recognized words was 38.57%.

Another class of speech recognition applications is shorthand.

Most often, with such a task, some monologue is carried out, recorded in fairly good acoustic conditions using a headset microphone. Therefore, in contrast to systems of mass service, where speech comes through telephone channels and/or is recorded on the street, automatic transcribing systems receive a speech signal with a much better recording quality. Since there are softer requirements for recognition speed, the system can process the speech signal in several passes, using the methods of adaptation to the voice of the speaker and the applied problem [4].

Scientists from Russia conducted a study on the recognition of continuous Russian speech using DNN confidence), described in [5]. A method using finite state machine-based converters was used for speech recognition. It was shown that the proposed method allows to increase the accuracy of speech recognition in comparison with hidden Markov models.

The study [6] compares the language models constructed using a feedforward neural network and a recurrent neural network. Three different implementations of the language model on neural networks were used: (1) LIMSI software tools for creating a feedforward neural network in which the output layer is limited to the most frequent words; (2) feedforward neural network with clustering (the entire dictionary is used); (3) recurrent neural network with clustering. Experimental results showed that language models built using a feedforward neural network, work worse than recurrent neural networks. On the test data, the recurrent network showed an improvement of 0.4% compared to the use of a feedforward neural network.

3 Speech Preprocessing

The conversion of input data into a set of features is called feature extraction. Speech recognition efficiency deteriorates dramatically in the presence of noise due to spectral mismatch between training and testing data. With conventional MFCC feature extraction, the logarithm function is used for the energy of the Mel filter Bank to reduce their dynamic range. Root cepstral analysis replaces the logarithmic function with a constant root function and gives the RCC coefficients. The coefficients of the RCC showed the best resistance to noise. In the RCC method, the compressed speech spectrum is calculated as shown in (1):

$$L_d(n) = L(n)^\tau, 0 \leq \tau \leq 1 \tag{1}$$

where $L_d(n)$ - compressed spectrum, $L(n)$ is the original spectrum, τ is the compression ratio, and m - filter bank index. Feature extraction involves simplifying the amount of resources required to accurately describe a large dataset. The feature extraction was performed using 13 MFCC coefficients [7].

Therefore, relation (1) expands, as shown in (2):

$$L_d(n) = L(n)^{\tau(m)}, 0 \leq \tau(m) \leq 1 \qquad (2)$$

where the compression ratio depends on the frequency band and is called non-uniform spectral compression. We show that by incorporating a speech recognition system into the process of adjusting the compression ratio, the recognition rate is further improved.

4   The Proposed Automatic Speech Recognition System

The methodology of our work is as follows:

4.1   Construction of Experimental Speech Corpus

Over the past ten years, a number of speech corpuses have been created in the world, containing up to a thousand speakers recorded in various environmental conditions. The recording of acoustic data for the creation of an acoustic corpus of the language was carried out at the Institute of Information and Computational Technologies of the Scientific Committee of the Ministry of Education and Science of the Republic of Kazakhstan in Almaty. For this, a sound-proofing, professional recording studio from Vocalbooth.com was used (Fig. 1). The cabin for recording audio data consists of two noise insulation layers, with the same hermetic door. The interior design consists of a pyramid-shaped sound-absorbing acoustic material of red color and the cabin is equipped with a silent air exchange system. The studio is designed to record high quality audio data.

Fig. 1. Soundproof professional recording studio of the company Vocalbooth.com (Color figure online)

The recorded audio materials have been preserved with expansion .wav. Each sentence was saved as a separate file, and the name consisted of the following identifiers:

<Region_code> + <gender> + <birth year> + <initials_name> + <education_code> + <text_number> + <sentence_number_in_text>

For example, speaker from the Almaty region, named Mamyrbayev Orken, male, born in 1979, with a higher education, voiced the text number 5 and sentence 82, will be identified as 05M79OM3_T005_S082.

All audio materials have the same characteristics:

- file extension: .wav;
- method of converting to digital form: PCM;
- discrete frequency: 8 kHz;
- digit capacity: 16 bits;
- number of audio channels: one (mono).

As speakers, people were selected without any problems with the pronunciation of speech. For research purposes and further use of the data, the speakers were surveyed according to a previously created template (Fig. 1). 200 speakers of different ages (age from 18 to 50 years) and genders were used for recording. It took an average of 40–50 min to sound and record one speaker. For each speaker, a text consisting of 100 sentences was prepared. The sentences were recorded in separate files. Each sentence consists of an average of 6–8 words. Sentences are selected with the most rich phoneme of words. Text data were collected from news sites in the Kazakh language, and other materials were used in electronic form. In total 76 h of audio data were recorded. During recording, transcriptions were created – a description of each audio file in a text file. The corpus created gives us, firstly, work with large volumes of databases, checking the proposed system characteristics and, secondly, studying the effect of database expansion on the recognition rate.

4.2 Acoustic Model

The acoustic model p(x|w) provides conditional probability of a sequence of feature vectors x, given a sequence of words w has occurred. This can be thought of as a measure of acoustic similarity of the input features to a sequence of words, regardless of the grammatical correctness of that word sequence. For an ASR system each word may be represented by a sequence of sub-word units called acoustic states. During acoustic model training, the statistics of each state are calculated from the occurrences of feature vectors corresponding to that state. For ASR with very large vocabulary sizes of thousands of words, due to data sparsity it is not feasible to accumulate sufficient statistics for each word separately. We would like to recognize even those words which may have few or no occurrences in the training data. To alleviate this problem, the words are defined as sequences of phonetic units called phonemes, just like the word pronunciations are represented in language dictionaries. Such a representation based on sub-word units is called a pronunciation lexicon. Each word in the lexicon may have one or more pronunciations.

4.3   Kazakh Language Model

Language model - allows determining the most likely word sequences. The complexity of constructing a language model depends largely on the specific language. So, for the English language, it is enough to use statistical models (the so-called N-grams). For agglutinative languages with a relatively rich morphology, statistical models are not suitable and hybrid models are used.

The language model p(w) gives the prior probability of the word sequence w. Basically it shows how likely a word sequence is to be uttered, based on grammatical rules of a language. Since this model only depends on the text and is independent of acoustic data, therefore large amounts of text available on the books, journal, articles etc., can be used as input source. Additionally, we want the language model to capture the topic-specific information for special ASR systems. To capture certain character-istics associated with human speech e.g. certain grammatical errors common in speaking, repetitions, hesitations etc., transcriptions of spoken text are also a useful input data source. Since the total number of possible word sequences is unlimited, simplifying assumptions have to be made to have reliable non-sparse estimates. The standard way to calculate the language model probabilities is through accumulation of counts of neighbouring words. It assumes that the probability of current word w_n depends only on the previous m−1 words w_n−1 … n−m+1.

A speech recognition involves a number of different components such as feature extraction, acoustic modeling, language modeling and DNN, as shown in the Fig. 2.

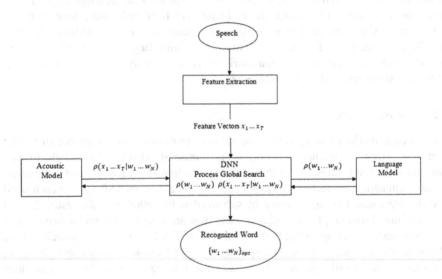

Fig. 2. Overview of an ASR system

5 The DNN Architecture and Training

For the development of ASR, we used the Kaldi tool and the DNN library in it, the modified Karel Vesely setting on the CUDA graphics processor was used for training.

We consider the DNN model:

First layers L

$$v^l = f(z^l) = f(W^l v^{l-1} + b^l), \text{ for } 0 < 1 < L,$$

where $z^l = W^l v^{l-1} + b^l \in R^{N_l \times 1}, v^l \in R^{N_l \times 1}, W^l \in R^{N_l \times N_{l-1}}, b^l \in R^{N_l \times 1}$, and $N_l \in R$ respectively, the excitation vector, the activation vector, weight matrix, displacement vectors and the number of neurons in layer 1. $v^0 = 0 \in R^{N_0 \times 1}$ - observation vector, $N_0 = D$ that is the element size and $f(\cdot) : R^{N_l \times 1} \rightarrow R^{N_l \times 1}$ activation function in relation to the excitation vector elementwise. In most applications, the sigmoid function

$$\sigma(z) = \frac{1}{1 + e^{-z}}$$

or hyperbolic tangent function

$$\tanh(z) = \frac{e^z - e^{-z}}{e^z + e^{-z}}$$

is used as an activation function. Next, we consider the algorithm for this model [8].

Algorithm Direct DNN calculation.

```
1: procedure ForwardComputation(O)
           > Each column O is an observation vector.
2:    v⁰ ← O
3:    for l ←1; l<L; l←l+1 do      > L total number of
layers
4:       Zˡ ← WˡVˡ⁻¹ + Bˡ    > Each column  Bˡ is  bˡ
5:       Vˡ ← f(Zˡ)   > f (.)may be sigmoidal  tanh, ReLU,
other functions
6:    end for
7:    Zᴸ ← WᴸVᴸ⁻¹ + Bᴸ
8:    if regression then      > regression task
9:       Vᴸ ← Zᴸ
10: else
11:    Vᴸ ← softmax(Zᴸ)
12: enf if
13: Return Vᴸ
14: end procedure
```

During training, we use the algorithm of single-stage selection by the Monte-Carlo method in the Markov chain. RBM have Gauss-Bernoulli units and is trained at an initial learning rate of 0.01 and other RBM have Bernoulli-Bernoulli units. Training was not controlled, the number of iterations was 4, the number of hidden layers was up to 6 and the number of units per layer was up to 2048.

6 Experimental Results

In the course of this work, feature extraction methods such as MFCC and acoustic, language model, DNN were investigated. The results were evaluated by the word error rate (WER) for classical models. The results indicating the vertical axis are percentages, and the horizontal axis: training monophonic models (Mono), the passage of the first (Tri1) and second (Tri2) and third (Tri3) thyryphon (Fig. 3). The best result is 36.76% WER for SAT Training.

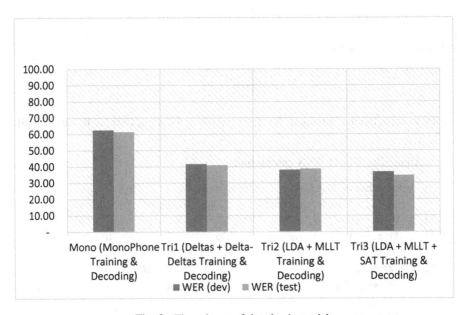

Fig. 3. The rule set of the classic model.

The results obtained with the application of DNN using from 0 to 6 hidden layers. The optimal result of 32.72% WER was obtained for 6 hidden layers, and this was an improvement over the classical models (Fig. 4).

It is important to note that performance is improved when the volume of the corpus for training is large. The best results have been obtained using DNN and SMBR algorithm.

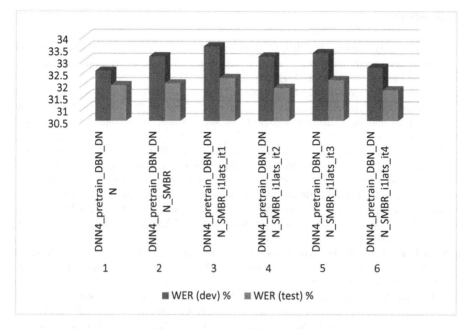

Fig. 4. Results of different DNNs.

7 Conclusion and Future Work

In this article we have developed and implemented a system of automatic speech recognition of Kazakh speech, which works on the basis of DNN. According to the results of the study we can see that it is better to use DNN for automatic speech recognition than classical algorithms. The paper analyzed the existing models and methods and considered a speech compression algorithm using the MFCC algorithm and gave an example of the ASR architecture. In this regard, it was stated that the MFCC and DNN methods provided the best results. The test error rate reached 0.56% for the corpus with 76 h of speech.

Future work will focus on improving the training corpus and exploring different optimization approaches for designing and implementing ASR for real-time applications such as voice-controlled robots.

Acknowledgements. This work was supported by the Ministry of Education and Science of the Republic of Kazakhstan. IRN AP05131207 Development of technologies for multilingual automatic speech recognition using deep neural networks.

References

1. Stouten, F., Duchateau, J., Martens, J.-P., Wambacq, P.: Coping with disfluencies in spontaneous speech recognition: acoustic detection and linguistic context manipulation. Speech Commun. **48**, 1590–1606 (2006)
2. Tsiaras, V., Panagiotakis, C., Stylianou, Y.: Video and audio based detection of filled hesitation pauses in classroom lectures. In: Proceedings of the 17th European Signal Processing Conference (EUSIPCO 2009), Glasgow, Scotland, 24–28 August 2009, pp. 834–838 (2009)
3. Psutka, J., Ircing, P., Psutka, J.V., Hajič, J., Byrne, W.J., Mirovsky, J.: Automatic transcription of Czech, Russian, and Slovak spontaneous speech in the MALACH project. In: Proceedings of Eurospeech, Portugal, Lisboa, 4–8 September 2005, pp. 1349–1352 (2005)
4. Young, S., et al.: The HTK Book (for HTK Version 3.4), Cambridge, UK, 375 p. (2009)
5. Karpov, A., Kipyatkova, I., Ronzhin, A.: Very large vocabulary ASR for spoken Russian with syntactic and morphemic analysis. In: Proceedings INTERSPEECH-2011, Florence, Italy, pp. 3161–3164 (2011)
6. Serizel, R., Giuliani, D.: Vocal tract length normalization approaches to DNN-Based children's and adults' speech recognition. In: IEEE Workshop on Spoken Language Technology, pp. 135–140 (2014)
7. Behbahani, Y.M., Babaali, B., Turdalyuly, M.: Persian sentences to phoneme sequences conversion based on recurrent neural networks. Open Comput. Sci. **6**, 219–225 (2016)
8. Yu, D., Deng, L.: Automatic Speech Recognition, p. 315. Springer, London (2014). https://doi.org/10.1007/978-1-4471-5779-3

Feedback Shift Registers Evolutionary Design Using Reverse Polish Notation

Miłosław Chodacki$^{(\boxtimes)}$

Institute of Computer Science, Silesian University in Katowice, Katowice, Poland
miloslaw.chodacki@us.edu.pl

Abstract. The paper presents the use of Genetic Algorithm in the design of registers with linear and non-linear feedback. Such registers are used, among others, in the diagnostics of digital circuits as Test Pattern Generators, and Test Response Compactors. Of particular importance are the registers that generate the maximum cycle, and in practice, respectively long. The length of the test generator cycle is important to Fault Coverage. The selection of the register feedback structure to achieve the maximum cycle is a difficult task, especially for the register with a non-linear feedback function. It is a novelty to propose coding solutions by means of Reverse Polish Notation, thanks to which the simple mechanism of a stack with automation, realizing a context-free grammar of logical expressions can be used to evaluate these solutions. This form of representation of the genotype of solutions is a certain generalization and gives greater possibilities to search the space of acceptable solutions. Such solutions must be minimized due to the limitation of area overhead on the silicon implementation of the tester. The obtained results indicate that the proposed approach gives positive solutions.

1 Introduction

In the Built-In Self-Test (BIST) technique, it dominates the approach based on the use of Linear Feedback Shift Registers (LFSR) as Test Pattern Generator (TPG) and Test Response Compactor (TRC). TRC are realised by use of Multi-Input Signature Register (MISR) or Single-Input Signature Register (SISR). This approach is classical and industrially used [1]. Nonlinear constructions are also known in the form of a Self-Test Path (STP) and Circular Self-Test Path (CSTP), using the Circuit Under Test (CUT) as a feedback. STP and CSTP structures can be modeled using the Non-Linear Feedback Shift Register (NLFSR) [2]. NLFSR registers are more difficult to analyze due to the non-linear nature of feedback, however they can potentially generate longer pseudo-random sequences of binary test vectors in compared to LFSR registers and Cellular Automata (CA). An important however, the size of the system implementing the specified feedback logic of the register is a limitation NLFSR. In self-testing BIST, the Area Overhead (AO) on the tester's implementation is limited due to the available silicon surface of the digital system.

© Springer Nature Switzerland AG 2019
N. T. Nguyen et al. (Eds.): ACIIDS 2019, LNAI 11432, pp. 475–485, 2019.
https://doi.org/10.1007/978-3-030-14802-7_41

Registers, especially LFSR, have many applications outside the domain of digital circuits testing, for example they are used in encrypting information on SIM cards, in the coding of car pilot signals and in the hardware implementation of selected cryptological algorithms [3]. Some CA, especially those generating the maximum cycle, e.g. with the rule 90/150, can be also used in similar areas, e.g. in wireless communication [4]. There are no simple mechanisms to select such feedbacks to get maximum cycle, especially for NLFSR registers [5]. For LFSR registers, one should indicate the primitive polynomial for obtaining a cycle of maximum length $2^p - 1$, where p is a length of register. This difficult task of designing the feedback was entrusted to the Genetic Algorithm (GA).

GA has some useful features, such as the ability to deliver multiple point solutions, and so the lack of concentration of solutions around a certain subclass of LFSR and NLFSR configuration. The algorithm mimics natural evolutionary processes, and therefore there exists the possibility of self-control calculations in such a way that a solution better adapted to a greater extent affects the entire population of solutions. The GA directs the search in the space of feasible solutions by environmental evaluation of the fitness function of each solution (Individual) [6]. The new approach, described in this paper, is based on coding solutions (representation of feedback function) using text expressions.

Typically, matrix notation, also a characteristic polynomial for LFSR, is used to describe the LFSR or NLFSR register function. The essence of the proposed approach is the use of Reverse Polish Notation (RPN) expressions to generate strings, symbolizing the feedback structure of registers, and which are the genotype of individuals for the GA. With the help of RPN, it can effectively process strings of characters, operands and operators without the use of parentheses. This solution is very important for simplifying the grammar of the language of such expressions, as well as for the automaton with the stack, which is the implementation of this grammar. Of course, there is no doubt about the priorities of the logic activities carried out despite the lack of parentheses. The RPN allows writing logical and arithmetic expressions without the use of parentheses, but itself the order of performed operations is strictly defined and unambiguous. Following the authors of RPN are considered to be Burks, Warren and Wright (1954), as well as independently F. L. Bauer and E. W. Dijkstra (1960s), who used the concept stack PDA (Pushdown Automata) for evaluation of arithmetic expressions and minimization necessary computer resources to accomplish this task [7]. Improved procedure RPN calculations were presented by the Australian, Charles Hamblin [8]. One can also use RPN to encrypt information [9]. In RPN, the record of arithmetic expressions, operands (arguments) and operators, and thus arithmetic functions operating on operands, it has a post-fixed character, i.e. that in the instruction after operands are placed *AAOp* operators (Argument, Argument, Operator) for binary operators. No need to use parentheses to modification of priorities of arithmetic operations and algorithm, based on implementation stack LIFO (called Last In, First Out) allows one to perform faster calculations in a digital machine. Some additional information on variations of PDAs can be found in [10]. Therefore, the RPN method was used in calculators Hewlett-Packard, National

Semiconductor, due to minimizing the length of arithmetic expressions (no parentheses) and use stack. Up to now RPN is used in the evaluation of arithmetic expressions in programming languages as well as in PostScript, BibTEX, as well as in some software calculators, expressions parsing and translating [11,12]. Thing obvious, RPN results directly from the Polish Notation, which was developed by the outstanding Polish mathematician of the Lwow-Warsaw school of mathematics in the 1920s, Jan Łukasiewicz; the Lwow school gave many to the world eminent scholars, including Stefan Banach, Stanisław Ulam, and Hugo Steinhaus, Stanisław Mazur (Scottish Book [13]).

The paper is organized as follows. In Sect. 2 basic information on LFSR and NLFSR is presented. Section 3 includes some description of RPN notation and in Sect. 4. the Genetic Algorithm and its using to create feedback structures is shown. Next, in Sect. 5. some results of evolutionary searching for register feedback functions are presented, and finally Sect. 6 concludes the paper.

2   Feedback Shift Register

The feedback register is created by means of linear or nonlinear feedback. In the digital technology, linear feedback is carried out only with ExOR logical functors, and then such a register can be described by means of a characteristic polynomial that uniquely identifies the feedback structure. In turn, non-linear feedback occurs when at least one logical functor other than ExOR participates in its implementation.

2.1   Linear Feedback Shift Register

The classic approach to designing test generators is based on the use of LFSR registers, usually those that generate a maximum cycle of 2^{p-1}, where p is the length of the register (the number of its cells). The register that generates the maximal cycle is described by the primitive polynomial $g(x)$.

A feedback of register can be realized by set of ExOR logic gates. The equation of the next state of the register is expressed by the statement (1) in general and for a register with a length p by (2):

$$Q(t+1) = T * Q(t), \tag{1}$$

where t is discrete clock-time, $Q(t)$ - state of register in an actual t clock-time, $Q(t+1)$ - state of register in next $t+1$ clock-time and $Q = \{q_0, q1, q_2, ..., q_{p-1}\}$ is a set of register p-number flip-flops (register's cells), and T is a connection square matrix.

$$
\begin{vmatrix}
q_0(t+1) \\
q_1(t+1) \\
\vdots \\
q_m(t+1) \\
\vdots \\
q_{p-1}(t+1)
\end{vmatrix}
=
\begin{vmatrix}
g_0 & g_1 & \cdots & g_{m-1} & \cdots & g_{p-2} & g_{p-1} = 1 \\
1 & 0 & \cdots & 0 & \cdots & 0 & 0 \\
\vdots & \vdots & \ddots & \vdots & \cdots & \vdots & \vdots \\
0 & 0 & \cdots & 1 & \cdots & 0 & 0 \\
\vdots & \vdots & \cdots & \vdots & \ddots & \vdots & \vdots \\
0 & 0 & \cdots & 0 & \cdots & 1 & 0
\end{vmatrix}
*
\begin{vmatrix}
q_0(t) \\
q_1(t) \\
\vdots \\
q_m(t) \\
\vdots \\
q_{p-1}(t)
\end{vmatrix}
. \tag{2}
$$

LFSR registers can be classified due to the type of feedback connection schema. And so one can distinguish standard LFSR registers (Fibonacci type) with external (Fig. 1), modular LFSR (Galois type) with internal (Fig. 2) and external/internal (Fig. 3) feedback, and the corresponding matrices of connections are presented in (3), (4) and (5) respectively.

$$
T_g =
\begin{vmatrix}
g_0 & g_1 & \cdots & g_{m-1} & \cdots & g_{p-2} & g_{p-1} = 1 \\
1 & 0 & \cdots & 0 & \cdots & 0 & 0 \\
\vdots & \vdots & \ddots & \vdots & \cdots & \vdots & \vdots \\
0 & 0 & \cdots & 1 & \cdots & 0 & 0 \\
\vdots & \vdots & \cdots & \vdots & \ddots & \vdots & \vdots \\
0 & 0 & \cdots & 0 & \cdots & 1 & 0
\end{vmatrix}, \tag{3}
$$

$$
T_h =
\begin{vmatrix}
0 & 0 & \cdots & 0 & \cdots & 0 & h_0 = 1 \\
1 & 0 & \cdots & 0 & \cdots & 0 & h_1 \\
\vdots & \vdots & \ddots & \vdots & \cdots & \vdots & \vdots \\
0 & 0 & \cdots & 1 & \cdots & 0 & h_{m-1} \\
\vdots & \vdots & \cdots & \vdots & \ddots & \vdots & \vdots \\
0 & 0 & \cdots & 0 & \cdots & 1 & h_{p-1}
\end{vmatrix}, \tag{4}
$$

$$
T_{gh} =
\begin{vmatrix}
g_0 & g_1 & \cdots & g_{m-1} & \cdots & g_{p-2} & g_{p-1} = h_0 \\
1 & a_{1,1} & \cdots & a_{1,m-1} & \cdots & a_{1,p-2} & h_1 \\
\vdots & \vdots & \ddots & \vdots & \cdots & \vdots & \vdots \\
0 & 0 & \cdots & 1 & \cdots & a_{m-1,p-2} & h_{m-1} \\
\vdots & \vdots & \cdots & \vdots & \ddots & \vdots & \vdots \\
0 & 0 & \cdots & 0 & \cdots & 1 & h_{p-1}
\end{vmatrix}, \tag{5}
$$

where
$$
a_{i,j} = \begin{cases} 1 \text{ if } h_i = g_j = 1 , \\ 0 \text{ if } h_i \neq g_j \text{ or } h_i = g_j = 0 , \end{cases} \tag{6}
$$

and $g_{p-1} = h_0 = 1$, $g_i, h_j, a_{i,j} \in GF(2)$.

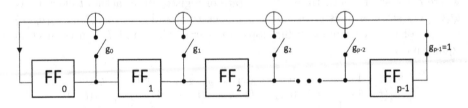

Fig. 1. Standard LFSR with external ExOR feedback.

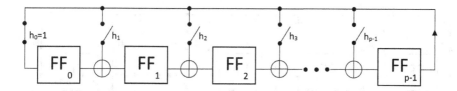

Fig. 2. Modular LFSR with internal ExOR feedback.

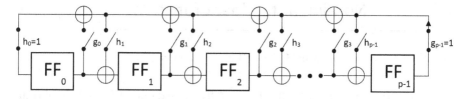

Fig. 3. LFSR with external/internal ExOR feedback.

2.2 Non-linear Feedback Shift Register

By choosing the right feedback, the NLFSR register (Fig. 4) can generate a cycle longer than LFSR, i.e. 2^p. The feedback of such a register is carried out by logical gates like OR, AND etc. There is no universal method for designing a nonlinear feedback to obtain the maximum cycle of such a register [5]. The NLFSR register can simplify modeling the behavior of the STP and CSTP, and the feedback is realized by the Circuit Under Test (CUT) [2]. A very important issue of the evaluation of the feedback structure design is an area overhead for its silicon implementation in BIST, in simplified terms it can be expressed by the number of elements realizing the designed logic function (logic gates numbers). The description of the NLFSR register can be described as follows (7), and for a register with a length p by (8):

$$Q(t+1) = T * Q(t) \oplus F(Q(t)), \tag{7}$$

where $F(Q(t)) = (f_0(Q(t)), f_1(Q(t)), ..., f_{p-1}(Q(t)))$ is non-linear feedback function and the remaining symbols are as previously explained in (1) and the operator $\oplus$ is an addition in GF(2). For a register with the length p, the expression (7) takes the detailed form presented in (8).

$$
\begin{vmatrix}
q_0(t+1) \\
q_1(t+1) \\
\vdots \\
q_m(t+1) \\
\vdots \\
q_{p-1}(t+1)
\end{vmatrix}
= T *
\begin{vmatrix}
q_0(t) \\
q_1(t) \\
\vdots \\
q_m(t) \\
\vdots \\
q_{p-1}(t)
\end{vmatrix}
\oplus
\begin{vmatrix}
f_0(q_0(t), q_1(t), ..., q_{p-1}(t)) \\
f_1(q_0(t), q_1(t), ..., q_{p-1}(t)) \\
\vdots \\
f_m(q_0(t), q_1(t), ..., q_{p-1}(t)) \\
\vdots \\
f_{p-1}(q_0(t), q_1(t), ..., q_{p-1}(t))
\end{vmatrix}
. \tag{8}
$$

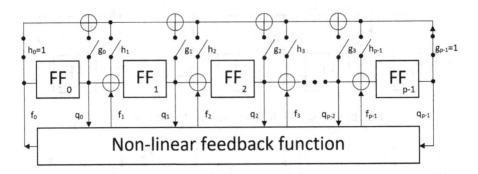

Fig. 4. NLFSR general model.

The selection of NLFSR feedback is usually carried out with the help of computer simulation and heuristic methods [15]. There is a lack of a well-developed theory of such registers to this day, in contrast to the LFSR registers [16].

3   Reverse Polish Notation

In Fig. 5 an activity diagram is shown illustrating the course of the RPN algorithm. The construction of this algorithm requires using the stack PDA, and the infix expressions are converted to postfix representation including the value of operators (Table 1) without using brackets (the symbol '!' will represent a unary negation operator). In this notation, the interpretation of performing arithmetic-logic operations is unambiguous.

Table 1. RPN priorities of operators.

Operator	Priority
$+$	2
$*$	1
$!$	0

Figure 6 shows a combinational circuit that performs a sample feedback function. The description of this circuit function in the traditional infix notation is presented in (9), and its equivalent in RPN (10); the same RPN algorithm for this translation is presented in Table 2.

$$y = !\,(!(a*b)*!(c*d) + !(!(c*d) + !(e*f)))\,. \tag{9}$$

$$y = ab*!cd*!*cd*!ef*!+!+!\,. \tag{10}$$

Encoding solutions using RPN can reduce the size of expressions (compare expressions (9) and (10)) and gives the possibility of unambiguous interpretation of the text entry by a simple PDA machine.

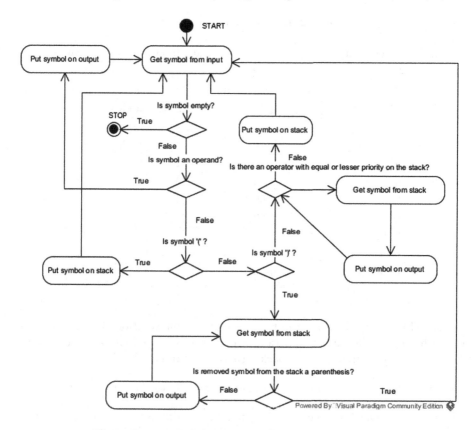

Fig. 5. Reverse Polish Notation UML activity diagram.

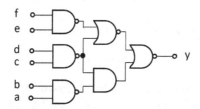

Fig. 6. Example of more complicated non-linear feedback.

4   Genetic Algorithm as Method of FSR Design

Genetic Algorithm has some useful features, such as the ability to deliver multiple point solutions, and so the lack of concentration of solutions around a certain subclass of LFSR/NLFSR feedback structure and configuration. The algorithm mimics natural evolutionary processes, and therefore there exists the possibility of self-control calculations in such a way that a solution better adapted to a greater extent affects the entire population of solutions (Selective Pressure).

Table 2. RPN stack automata (PDA)

Input	Stack	Output	Input	Stack	Output	Input	Stack	Output
!	!		*	!(*!(*		+	!(+!(+	!
(	!(		d		d	!	!(+!(+!	
!	!(!		)	!(*!	*	(	!(+!(+!(	
(	!(!(		+	!(+	!	e		e
a	!(!(	a	!	!(+!	*	*	!(+!(+!(*	
*	!(!(*		(	!(+!(		f	!(+!(+!(*	f
b	!(!(*	b	!	!(+!(!		)	!(+!(+!	*
)	!(!	*	(	!(+!(!(		)	!(+!	!
*	!(*	!	c	!(+!(!(	c	)	!(	+
!	!(*!		*	!(+!(!(*				!
(	!(*!(		d	!(+!(!(*	d			+
c	!(*!(	c	)	!(+!(!	*			!

The GA directs the search in the space of feasible solutions by environmental evaluation of the fitness function of each solution (Individual). The fitness function for the search of a feedback structure is presented in (11).

$$Fit(x \in V(p)) = w_0 x_0 + w_1(x_{1max} - x_1) + w_2(x_{2max} - x_2) + w_3(x_{3max} - x_3) \ , \ (11)$$

where x_0 is a length of cycle, x_1 is a number of ExORs used to create linear feedback, x_2 is a number of T-type flip-flops used to design LFSR/NLFSR, x_3 is a number of logic gates used to create non-linear feedback function, and $\sum_{i=0}^{n=3} w_i = 1$.

GA is searching for such $x \in V(p)$ that maximizes the following formula:

$$f(x^*) = max_{x \in V(p)} fit(x) \qquad (12)$$

The genotype of the solution (individual, feedback function structure) consists of chromosomes formed in the RPN notation. For a Fibonacci type register, genotype contains a single chromosome encoding the function of element q_0, and for the Galois type register, chromosomes describing all q_i for $i = 0, \ldots, p-1$ cells functions. The rest of the genotype describes initial state of the register and schema of used D-type or T-type flip-flops as cell of register in ordered sequence like DTDTD (q_0 is D-type, q_1 is T-type flip-flops and so on).

5   Results

Figure 7 shows an example generated schema solution of the NLFSR register. The set of non-linear functions $F(Q(t))$ is presented in (13), then the equation describing the operation of the register is showed in (14). If one assumes that in this way $L(Q) = \sum_0^2 q_i * 2^i$ a decimal number is created, then transitions graph of this register is presented in Fig. 8.

$$\begin{aligned}
f_0(t) &= !q_0(t)!q_2(t) + q_0(t)q_2(t) \\
f_1(t) &= q_1(t) \\
f_2(t) &= !q_1(t)q_2(t) + q_0(t)q_2(t) + !q_0(t)q_1(t)!q_2(t)
\end{aligned} \tag{13}$$

$$\begin{aligned}
q_0(t) &= q_2(t) \oplus f_0(t) \\
q_1(t) &= q_0(t) \oplus f_1(t) \\
q_2(t) &= q_1(t) \oplus f_2(t)
\end{aligned} \tag{14}$$

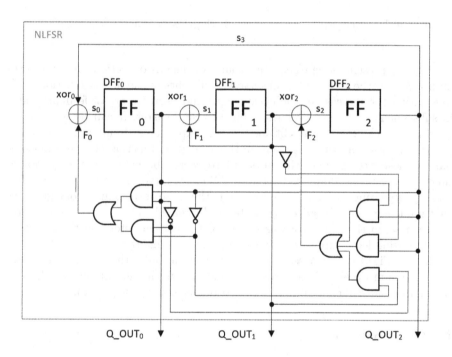

Fig. 7. NLFSR example solution.

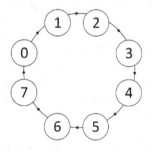

Fig. 8. NLFSR transition graph.

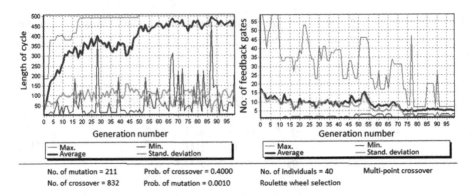

Fig. 9. Length of cycle and no. of feedback gates vs. number of generations.

Figure 9 presents a representative course of the GA operation (in the sense of an increase in the value of the observed coefficients of the fitness function (11)) illustrating the increase of the cycle length and decrease no. of feedback gates in subsequent generations for the 9-bit NLFSR register. It can be seen that a maximum cycle $2^9 - 1$ was obtained in this experiment. Similar charts can be presented for other criteria present in the evaluation function (11). Results similar to those presented in the work [14,16] were obtained, but the generated functions are more complex. Therefore GA must, however, be pre-adapted, frequently appearing solutions are excessive in the sense of the number of logic gates that implement non-linear feedback. It can be remedied by applying the appropriate value of the weight of this factor, or the chosen deterministic algorithm of minimizing logic expressions (local optimization algorithm).

In the Fig. 9 one can see a curiosity that although the register generates the maximum cycle, it is an ordered sequence, which can not be considered a pseudorandom sequence (it is an ordered binary counter sequence).

6   Conclusions

The paper presents the use of RPN for genetic coding solutions, i.e. the structure of register feedback function. Instead of operating on matrices, sequences of text expressions describing the feedback logic function are processed. It gives more freedom of expression - the statements can be as long as possible - which contributes to better coverage of the search space. In turn, too long phrases can introduce an unacceptable excess to the implementation of the feedback function. RPN allows the use of a simple mechanism of a PDA machine (LIFO automata). This simplicity of evaluation of expressions is crucial for machine processing (PDA). GA requires proper tuning in the selection of the parameters controlling the evolution of the population of solutions. It also seems necessary to have the function of penalty for too large trees of cutting the grammatical sentences describing the feedback. Some solutions are too large trees, hence the need to apply the penalty function or the classic algorithm of minimizing logical expressions, i.e. local optimization. The carried out research indicates the usefulness of processing the text as representation of the feedback structure of registers and evolutionary design.

References

1. Wunderlich, H.J.: BIST for systems-on-a-chip. Integration VLSI J. **26**, 55–78 (1998)
2. Chodacki, M.: Genetic algorithm as self-test path and circular self-test path design method. Vietnam J. Comput. Sci. **5**(3–4), 263–278 (2018)
3. Bhal, A.S., Dhillon, Z.: LFSR based stream cipher (enhanced A5/1). Int. J. Adv. Comput. Eng. Network. **2**(12) (2014)
4. Roy, S., Karjee, J., Rawat, U.S., Dayama Pratik, N., Dey, N.: Symmetric key encryption technique: a cellular automata based approach in wireless sensor networks. Procedia Comput. Sci. **78**, 408–414 (2016)
5. Dubrova, E.: Generation of full cycles by a composition of NLFSRs. Des. Codes Crypt. **73** (2014). https://doi.org/10.1007/s10623-014-9947-3
6. Michalewicz, Z., Fogel, D.B.: How to Solve It: Modern Heuristics. Springer, Heidelberg (2004). https://doi.org/10.1007/978-3-662-07807-5
7. Burks, A.W., Warren, D.W., Wright, J.B.: An analysis of a logical machine using parenthesis-free notation. Math. Tables Other Aids Comput. **8**(46), 53–57 (1954)
8. Hamblin, C.L.: Translation to and from Polish Notation. Comput. J. **5**(3), 210–213 (1962)
9. Dhenakaran, S.S.: Employing reverse Polish Notation in encryption. Adv. Comput. Int. J. (ACIJ) **2**(2) (2011)
10. Chaudhary, N.K., Karmacharya, R., Ghimire, B., Srinivasu, N.: Stack variation in push down automata (PDA). Int. J. Eng. Trends Technol. (IJETT) **4**(5) (2013)
11. Jackson, D.: Parsing and translation of expressions by genetic programming. In: Genetic and Evolutionary Computation Conference, GECCO 2005, Proceedings, Washington DC, USA, 25–29 June 2005
12. Rastogi, R., Mondal, P., Agarwal, K.: An exhaustive review for infix to postfix conversion with applications and benefits. In: 2nd International Conference on Computing for Sustainable Global Development (INDIACom), New Delhi, pp. 95–100 (2015)
13. Mauldin, R.D.: The Scottish Book: Mathematics from the Scottish Cafe, 1st edn. Birkhauser, 1 April 1982
14. Dubrova, E.: A list of maximum period NLFSRs (2018)
15. Janicka-Lipska, I., Stokłosa, J.: Boolean feedback functions for full-length nonlinear shift registers. J. Telecommun. Inf. Technol. (JTIT) **4** (2004)
16. Dubrova, E., Teslenko, M., Tenhunen, H.: On analysis and synthesis of (n, k)-nonlinear feedback shift registers. In: Design, Automation & Test in Europe (DATE08) (2008)

Intelligent Supply Chains
and e-Commerce

IT Value for Customer: Its Influence on Satisfaction and Loyalty in E-commerce

Arkadiusz Kawa[✉] and Justyna Światowiec-Szczepańska

Poznan University of Economics and Business,
al. Niepodległości 10, 61-875 Poznań, Poland
{arkadiusz.kawa,
justyna.swiatowiec-szczepanska}@ue.poznan.pl

Abstract. Today, in the time of a technological revolution, which has a great influence on the economy, we can say that the world is smaller than ever. IT solutions have significantly changed business activities, which can be developed in almost any place. A specific beneficiary of this amendment is e-commerce. The customer does shopping online because of lower prices, convenience, a wider range of products, etc. Beside logistics and marketing reasons, they pay more and more attention to other aspects, too. One of them is IT solutions which support the majority of processes in online retailers, especially marketing communication and website usability. They can have influence on their satisfaction, and, consequently, on their loyalty. These aspects are relatively rarely studied. The goals of this paper are to identify the components of value for customers related to IT, and to present their influence on satisfaction and loyalty in e-commerce. We administered our survey to managers representing e-tailers, suppliers, complementors, and to customers.

Keywords: E-commerce · Value for customer · Satisfaction · Loyalty

1 Introduction

E-commerce has quite a short history and initially its meaning was marginal. Nowadays, however, e-commerce is recognized as one of the most important sectors of the economies of many countries, including Poland. The dynamic development of e-commerce is driven by rapidly expanding Internet access, but also by growing mobility and popularity of portable devices, via which customers order goods and services at a convenient time and place more and more frequently. They do not only order things of greater value, but, more and more often, everyday products to which they want to have very fast access.

According to the research carried out by Gemius [7], the main reason behind the fact that we buy online is the convenience of twenty-four-hour access to stores, lower prices, greater assortment selection than in brick-and-mortar stores and the ease of comparing different offers available on the market. Sales aspects seem to be very significant. However, it must stressed that logistics issues, such as a convenient form of delivery and return of the goods purchased are also taken into account by customers. That is why logistics plays an increasingly important role in e-commerce. It not only

© Springer Nature Switzerland AG 2019
N. T. Nguyen et al. (Eds.): ACIIDS 2019, LNAI 11432, pp. 489–498, 2019.
https://doi.org/10.1007/978-3-030-14802-7_42

attracts new customers, but also helps to retain the existing ones by punctuality, conformity of the delivered products with the order, no damage. These two aspects are extensively described in the literature, in particular the one related to sales. However, there is a deficit of empirical studies on the IT solutions which support the majority of the processes in e-tailers, especially website usability and marketing communication. Thanks to them, customers can have access to information about companies and their products at any time on the e-tailer website, in social media or blogs provided by experts. They can easily find offers, compare them and read other users' opinions. They may also purchase new products in a simple way by registering at online store or even without registration. All the positive experience in all the processes of online shopping is positively related to customer satisfaction.

Many different entities are involved in e-commerce. For this reason, e-commerce should be treated as a system which consists of e-tailers, their customers, suppliers, and complementors. E-tailers are mostly companies that have their own online shops or cooperate with other intermediaries such as marketplaces, auction platforms. The customers can be almost any individual or institutional persons who have access to the Internet and financial resources. Suppliers include suppliers of products sold through electronic channels. However, complementors are understood as subcontractors supporting e-commerce, e.g. providers of logistics services, financial services, IT solutions, etc.

The subject of value in e-commerce, which is analyzed from different perspectives by different entities, is still little recognized, even though the issue of value has been discussed in the literature for many years. In particular, there is a lack of empirical research that would address the issue of value creation for customers from different perspectives. Therefore, the aim of the article is to identify customer value factors related to IT created and delivered by e-tailers, suppliers, complementors, and customers. The second goal is to present the influence of this value on satisfaction, and, subsequently, on loyalty of customers in e-commerce.

2   Value for Customer

Originally introduced in 1954 by P. Drucker, the notion of value is defined in the literature in various ways. Broadly speaking, value can be defined as an evaluation of the utility of a product understood as a relationship between what has been received and what has been given – value represents a compromise between what can be obtained and what should be given. The notion of value in management sciences often refers to the customer. In that case it is described as a value for customer [12].

The existing body of literature on e-commerce is primarily devoted to explaining the essence of e-commerce, e-customer characteristics and e-marketing concepts. To a large extent, it is also focused on the B2B market. Moreover, the literature provides a little insight into the area of value creation despite the fact that the issue of value and value creation has been investigated for many years. The research work of Zott et al. [24], in which the authors focus on relationships between firms operating in the Internet environment to create a value, can serve as an example. On the other hand Piy-athasanan et al. [17] in their article analyze how customers perceive value during contacts with e-tailers.

The theoretical considerations as well as the research carried out are very fragmented, which is manifested in analyzing value creation from an individual market participant's point of view. The customers and companies in most cases serve as focal points. On the other hand, the attempt to simultaneously include the point of view of both customers and companies is reduced to analyzing value creation in dyadic relationships. A simplified perspective is adopted in which a single company creates value for a single customer without taking into account other entities directly or indirectly involved in the provision of the value.

With the abovementioned characteristics of the research into value creation in mind, namely fragmentation of the research and the dyadic approach to the problem, it was postulated to investigate this issue (value creation) in a broad and holistic way. This requires to include an analysis of all (if possible) market entities involved in value creation. A reaction to this need is a concept of value networks that can be defined as spontaneous combinations of actors interacting to co-produce service offerings, exchange them, and co-create value [14]. The value network includes not only companies but also customers.

3 Value for Customer Related to IT in E-commerce

3.1 Satisfaction and Loyalty

Olivier [16] claims that satisfaction is the customer's fulfillment response and notices that satisfaction is not merely about the extent of being pleased, but can be described as a process as well [15]. According to Kotler [12], satisfaction is understood as the degree to which the perceived features of the product meet the expectations of the buyer. Satisfaction is, then, a gradable feeling. In the case of e-commerce, satisfaction means contentment with the transaction and a sense of understanding the customer's needs by the entities of network value.

In turn, loyalty means the customer's willingness to buy the product of a given brand or to use a service again. It leads to repeatability of purchases, despite the situation impacts and marketing efforts to promote competing brands [16]. This means that the customer will continue to buy even if the products and services offered by other sellers are more competitive. Loyalty is a determinant of the customer's trust in the company and is evident primarily in the customer's emotional attachment to a given entity and maintenance of a special type of ties. In the case of e-commerce, loyalty does not only refer to the product itself, but also to the place where one re-purchases, such as marketplaces.

Satisfaction and loyalty are well-known constructs widely discussed in the literature in various industries. Along with the dynamic development of e-commerce, researchers have started to study e-satisfaction and e-loyalty. This applies in particular to the impact of satisfaction on loyalty, where most studies show a positive effect of satisfaction on loyalty [4] or repurchase intention [2, 13]. Moreover, researchers have found out that the link between satisfaction and customer spending is positive when higher satisfaction results in more spending in e-commerce [15].

The above observations are the basis for the research hypothesis, which is as follows: **H1.** *Satisfaction has positive impact on loyalty in e-commerce.*

3.2 Moderation Effect Between Satisfaction and Loyalty

In most studies, many factors, such as trust and commitment, are positively linked with satisfaction or loyalty [20]. They are increasingly associated with customer value. For example, Chiou and Pana studied the indirect impact of service quality on customer loyalty among online bookshop customers. The authors confirmed a positive correlation between perceived quality and customer satisfaction and customer trust, which in turn affect customer loyalty [2].

Nisar and Prabhakar [15] supported the hypothesis that customer satisfaction has a positive impact on consumer spending in American-based e-commerce retailers. It is very important to note that the received value should be conceptualized and measured as a cognitive construct, but both satisfaction and loyalty should be conceptualized as affective constructs [16]. However, satisfaction can be a moderator between value and loyalty in e-commerce. That is why in this paper, apart from exploring the direct relationship between value for customers and loyalty, the moderation effect between these two constructs is studied.

3.3 IT-Assisted Marketing Communication

Information technologies are widely used in marketing activities, especially those related to the Internet, which is known as e-marketing. The researchers describe it as company activities undertaken to inform customers, communicate, promote and sell by means of the Internet [19]. Here, the focus is on marketing communication which has been used for many years with the aim of developing relationships with organizational stakeholders. Thanks to the Internet this communication has become more interactive. One of excellent examples of new marketing communication tools is social media which have provided yet another platform for a multi-step theory of communication [15]. Social media are based on word of mouth, which has been treated as a major influence on what people know, feel and do. The rapid development of social media has led to a transformation of e-commerce from a product-oriented environment into a social and customer-centered one.

Thanks to media tools such as forums and blogs collective intelligence is provided, which allows to deepen knowledge about products and services or to solve specific problems [10]. Blogs are a specific marketing tool which is based on leaders' opinion. Consumers often consider opinions of leaders in their purchase decision-making processes, not only after buying a product or in order to check how to use it, but also before the purchase to know more about it [22]. Research shows that many customers depend on ratings and reviews given by others who have already purchased the products and are using them [18].

IT-assisted marketing communication brings customers a lot of benefits. One of them is that more customized information can be communicated in the same unit of time, which may lead to a reduction of the transaction and coordination costs. Moreover, customers are able to easily access firms and get to know their brands at any time

[15]. The beneficiaries are also customers. If they have more appropriate information about e-tailers, their products and services, they can be more satisfied with their shopping. Electronic word-of-mouth recommendations have significant importance in shaping customer behavior. According to Hsu and Lin [8], social media based on social norms and identification are included as additional beliefs and have influence on customer intentions to use apps and make in-app purchases.

On the basis of the above considerations, we formulated the following hypotheses:

H2. *IT-assisted marketing communication has positive impact on loyalty in e-commerce.*

H2a. *IT-assisted marketing communication positively moderate the link between satisfaction and loyalty in e-commerce.*

3.4 Website Customization

Around 25 years ago, the first online shops started to be established. Amazon was founded in 1994 and eBay was opened in 1995. Initially, very simple websites were used, where orders were placed through a special form. Over time, virtual baskets and other amenities imitating shopping at a traditional shop appeared. Currently, an e-tailer's website is equipped with many tools facilitating the whole shopping process.

The website of a shop, apart from the previously mentioned adverts, social media, blogs, is an essential form of communication with the customer in e-commerce. Due to its complexity and lack of occurrence in traditional marketing communications, it has been treated in this article as a separate construct.

The homepage of the online store is very important, because it is visited most frequently by customers in the first place. It plays the role of a business imagine of the shop, represents its strategy and distinguishes it from the competition, so it must contain information tailored to the profile of the company and its products. In addition to the homepage, the online shop should have other website categories with various specific objectives: product category pages that allow customers to browse and compare products freely; a product card that contains detailed information about the product and a landing page that is a dedicated website, e.g. with a promotional campaign.

The website of a shop should also include a tool that enables to compare products. It makes the user spend more time on the website, analyze information about several products and search for a product most suitable for their needs. Companies also run cross-selling and up-selling activities, which consist in presenting complementary or similar products.

Buyers should be offered different options of searching for a product, e.g. by product category, using a search engine (without a need to leave the store and search, for example, via Google). The facilitating search for product information [23] is very important because customers can easily change e-retailers – it is enough to find another website with the given product. Well-customized websites attract customers and make them attached to e-tailer brands or Internet platforms. An example is seeking products by customers who use a search engine of a marketplace or an online retail service (e.g. Amazon, AliExpress) as their first choice [11].

Another very important aspect of websites is personalization, which allows to keep your customers by matching products or services to their individuality or preferences.

Companies have an opportunity to build closer relationships with customers by using the information that the customer gives when registering or by agreeing to the cookies policy. According to Fernandez-Lanvin [6], personalization enhances customer retention and increases sales by improving customer perception of the site quality.

The access to customer information, storage, processing and sharing thereof must be transparent. E-tailers cannot fail to obey the privacy policy [23]. Research shows that insufficient, incomplete and wrong information in purchase process decision makes customers less satisfied, and, subsequently, results in refusal to do shopping. Zhou et al. [23] suggest that improving the transparency attracts consumers and increases the purchase rate.

Website customization is related to ease of use as well as shopping directions, navigation, interaction, personality and review swapping. According to researchers, website design and the detail extent with respect to product information provided are some of the main elements influencing customer satisfaction in e-commerce. Moreover, they proved that website design positively affects the purchase intention [15].

The presented observations lead to another research hypotheses:

H3. *Website customization has positive impact on loyalty in e-commerce.*

H3a. *Website customization positively moderate the link between satisfaction and loyalty in e-commerce.*

4 Methodology

The research presented in this paper was carried out for the needs of a project whose aim is to develop a model of value network creation in e-commerce. Achievement of this goal will be possible thanks to conducting three-stage research, which consists of: an in-depth literature review, empirical research and multi-level modeling. The empirical research was divided into two parts: research using the qualitative method and research using the quantitative method. This article presents the results of the second type of research.

The research assumed that the respondent (representing e-tailers, suppliers, complementors) was to look at the value through the final customer's "eyes", regardless of their role in e-commerce. The main reason of this was that the value network is created around its customers. The central point of the e-commerce system is the customer who ultimately evaluates the value and converts it into a monetary equivalent for the other network members [11].

CATI (computer-assisted telephone interview) was selected as a technique of information collection, which had been preceded by FGIs (focus group interviews). The research with the use of qualitative methods was aimed at a preliminary analysis of the problems of value creation and providing information necessary for the proper organization of the research by the quantitative method, including, most importantly, the design of a measuring instrument.

The database of companies operating in e-commerce, suppliers and customers in Poland was used as the sample. It included, inter alia, data from the Regon database kept by the Central Statistical Office in Poland and Polish commercial databases, such as DBMS, Bisnode. Approximately 10 thousand respondents took part. In all four general populations, non-random purposeful sampling was applied. The sample was selected from those entities that had relevant experience in selling (e-tailers) or purchasing (customers) products via the Internet, cooperating with e-tailers (suppliers and complementors).

The research was conducted between November 2017 and May 2018 by an external entity – a research and expert agency with extensive experience in empirical research. In total, 800 correctly completed questionnaires were received (200 records in each group – customers, e-tailers, suppliers, and complementors). There were not any errors and incomplete information; that is why no surveys were rejected. All questionnaires were qualified for further analysis, which, assuming the size of the fraction of 50% and the confidence level of 95%, gives an acceptable measurement error of 7% in each group [9].

In the study 5 measures, which correspond to the previously presented theoretical considerations, were distinguished. These are respectively: IT-assisted marketing communication, website customization, satisfaction, and loyalty. Since relationship related variables were latent, we adopted multi-item scale approach in this research to increase item reliability. We used all items for measuring variables of interest from existing literature. The respondents were asked to indicate the extent to which they agreed with a given description using a five-point Likert-type scale. In all constructs the scale was as follows: 1 = strongly disagree to 5 = strongly agree. The quality of the results were verified using validity and reliability measures (all Cronbach's alpha coefficients of constructs were higher than 0.75).

4.1 Analysis and Results

We analysed the data using hierarchical moderated multiple regression, which is particularly appropriate for testing of moderation effects [1] proposed in hypotheses H2a and H3a. The empirical models include satisfaction, IT-assisted marketing communication and website customization variables and the interaction between satisfaction and IT-assisted marketing communication and satisfaction and website customization as explanatory variables. To avoid the problem of collinearity of variables in the regression model with interaction terms all continuous predictors (including both independent variables and moderators) were mean-centered (i.e., the mean of the variable was subtracted from it, so the new version has a mean of zero) [5]. We also include control variables of regions and three dichotomous variables indicating a role in value network, i.e. e-tailer, supplier or complementor. Table 1 reports means, standard deviation, and Pearson's correlation coefficients for all variables used in this study.

Table 2 presents the hypothesised results of main interaction effects. For all the regressions, we report robust standard error to address the heteroscedasticity.

Model 1 is the baseline model with all control variables included. Model 2 and 3 includes main effects put forward in hypotheses 1, 2 and 3. Model 4 includes the interaction terms needed to test hypotheses H2a and H3a. Corroborating H1, the effect of satisfaction on loyalty is found to be significant and positive ($\beta = 0.527$, $p < 0.001$) as shown in Model 2. The results of Model 3 strongly support H2, which proposes a positive impact of IT-assisted marketing communication on loyalty ($\beta = 0.215$, $p < .001$). However, the results of Model 3 do not support H3, which proposes a

Table 1. Descriptive statistics

	Variable	Mean	S.D.	1	2	3	4	5	6	7
1.	Loyalty	3.472	0.960							
2.	Satisfaction	3.901	0.678	0.555**						
3.	Web customization	3.655	0.705	0.240**	0.231**					
4.	Communication	3.370	0.818	0.366**	0.167**	0.564**				
5.	Regions	8.100	4.467	0.004	−0.101**	−0.084*	−0.080*			
6.	Role: e-tailer	0.250	0.433	−0.306**	−0.410**	−0.026	−0.025	0.057		
7.	Role: supplier	0.250	0.433	0.249**	0.087*	0.108**	0.234**	−0.062	−0.333**	
8.	Role: complementor	0.250	0.433	0.235**	0.186**	−0.043	0.080*	−0.035	−0.333**	−0.333**

*$p < .05$, **$p < .01$; N = 800

Table 2. Hierarchical regression analysis results

	Y = Loyality			
	Model 1	Model 2	Model 3	Model 4
Control variables				
Regions	0.040	0.082**	0.091**	0.088**
	(0.007)	(0.006)	(0.006)	(0.006)
Role: e-tailer	−0.097**	0.119**	0.059*	0.072*
	(0.087)	(0.079)	(0.078)	(0.078)
Role: supplier	0.323***	0.346***	0.261***	0.266***
	(0.087)	(0.074)	(0.076)	(0.076)
Role: complementor	0.311***	0.294***	0.237***	0.248***
	(0.087)	(0.074)	(0.075)	(0.075)
Main effects				
Satisfaction		0.527***	0.486***	0.490***
		(0.043)	(0.042)	(0.043)
Communication			0.215***	0.218***
			(0.040)	(0.040)
Web customization			−0.002 (0.045)	0.004
				(0.045)
Interaction effects				
Communication × satisfaction				0.037
				(0.054)
Web customization × satisfaction				0.065**
				(0.062)
R^2	0.183	0.412	0.451	0.459
Adjusted R^2	0.179	0.408	0.446	0.453
Δ Adjusted R^2	-	0.228	0,039	0,008
ΔF-statistics	44.556***	308.292***	28.199***	5.865***

Note: Standardised beta coefficients reported. Standard errors reported in parentheses. Significance levels: *$p < 0.05$, **$p < 0.01$, ***$p < 0.001$.

positive impact of website customization on loyalty. The coefficient for website customization is non-significant ($\beta = -0.002$, $p > 0.05$). Compared with the baseline model, Model 2 and 3 explain a significantly greater variance of the firm loyalty (respectively $\Delta R^2 = 0.228$, $p < 0.001$ and $\Delta R^2 = 0.267$, $p < 0.001$). H2a proposes that IT-assisted marketing communication positively moderates the link between satisfaction and loyalty. In Model 4 this interaction term ($\beta = 0.037$, $p > .05$) is not statistically significant. However, the interaction term between website customization and loyalty is positive and statistically significant ($\beta = 0.065$, $p < 0.05$). These findings support H3a. Compared with Model 3, Model 4 does not explain a significantly greater variance of firm loyalty ($\Delta R^2 = 0.008$). Hence, the interaction effects were supported only partly.

5 Conclusion

Trade with the use of Internet technology enables expansion on a larger scale for existing businesses and offers prospects for rapid development of new companies. This is possible thanks to low barriers to market entry which encourage the sales of products. Customers who have access at all times to information about sellers, and thus their products, also benefit. They can easily find offers, compare them, read other users' opinions and, as a result, buy cheaper.

Value in e-commerce is created and delivered not only by e-tailers, but also by many other entities, such as: product manufacturers, sales platforms, financial institutions, logistics companies, warehouse service providers and customers themselves. The research carried out for the purpose of this study confirmed a strong relationship between customer satisfaction and loyalty in the value network in e-commerce. Moreover, our analyses lead to the conclusion that communication directly influences loyalty, but is not a moderator of the satisfaction-loyalty relationship. On the other hand, website customization does not directly affect loyalty, but acts as a moderator in the satisfaction-loyalty relationship.

The research results have managerial implications. Entities of the value network in e-commerce, i.e. e-tailers, suppliers and complementors, should not only take into account marketing and logistics issues, but also those related to information technologies.

The limitation of this research is the methodological nature of this study. It is based on an empirical model which simplifies the economic reality and thus reduces the complexity of the actual state. Therefore, further research work on the development of the variable measurement reflecting the value network should be carried out.

Acknowledgements. This paper has been written with financial support of the National Center of Science [Narodowe Centrum Nauki] – grant number DEC-2015/19/B/HS4/02287.

References

1. Carte, T.A., Russell, C.J.: In pursuit of moderation: nine common errors and their solutions. MIS Q. **27**, 479–501 (2003)
2. Chiou, J.-S., Pan, L.-Y.: Antecedents of internet retailing loyalty: differences between heavy versus light shoppers. J. Bus. Psychol. **24**, 327–339 (2009)
3. Chiu, C.-M., Lin, H.-Y., Sun, S.-Y., Hsu, M.-H.: Understanding customers' loyalty intentions towards online shopping: an integration of technology acceptance model and fairness theory. Behav. Inf. Technol. **28**(4), 347–360 (2009)

4. Cyr, D.: Modeling Web site design across cultures: relationships to trust, satisfaction, and E-loyalty. J. Manag. Inf. Syst. **24**(4), 47–72 (2008)
5. Dawson, J.F.: Moderation in management research: what, why, when and how. J. Bus. Psychol. **29**(1), 1–19 (2014)
6. Fernandez-Lanvin, D., de Andres-Suarez, J., Gonzalez-Rodriguez, M., Pariente-Martinez, B.: The dimension of age and gender as user model demographic factors for automatic personalization in e-commerce sites. Comput. Stan. Interfaces **59**, 1–9 (2018)
7. Gemius: E-commerce w Polsce 2018 (E-commerce in Poland 2018), Warszawa (2018)
8. Hsu, C.L., Lin, J.C.C.: Effect of perceived value and social influences on mobile app stickiness and in-app purchase intention. Technol. Forecast. Soc. Chang. **108**, 42–53 (2016)
9. http://www.raosoft.com/samplesize.html
10. Huang, Z., Benyoucef, M.: From e-commerce to social commerce: a close look at design features. Electron. Commer. Res. Appl. **12**(4), 246–259 (2013)
11. Kawa, A., Światowiec-Szczepańska, J.: Value network creation and value appropriation in e-commerce. Przedsiębiorczość i Zarządzanie **19**(6), 9–21 (2018)
12. Kotler, P.: Marketing Management: Analysis Planning Implementation and Control. Prentice Hall, Upper Saddle River (1994)
13. Kwon, W.-S., Lennon, S.J.: What induces online loyalty? Online versus offline brand images. J. Bus. Res. **62**, 557–564 (2008)
14. Lusch, R., Vargo, S., Tanniru, M.: Service, value networks and learning. J. Acad. Mark. Sci. **38**(1), 19–31 (2010)
15. Nisar, T.M., Prabhakar, G.: What factors determine e-satisfaction and consumer spending in e-commerce retailing? J. Retail. Consum. Serv. **39**, 135–144 (2017)
16. Olivier, R.I.: Whence consumer loyalty. J. Mark. **63**, 33–44 (1999)
17. Piyathasanan, B., Mathies, Ch., Wetzels, M., Patterson, P.G., Ruyter, K.: A hierarchical model of virtual experience and its influences on the perceived value and loyalty of customers. Int. J. Electron. Commer. **19**(2), 126–158 (2015)
18. Raja, K., Pushpa, S.: Feature level review table generation for e-commerce websites to produce qualitative rating of the products. Future Comput. Inf. J. **2**(2), 118–124 (2017)
19. Sharma, A., Rishi, O.P.: A study on e-marketing and e-commerce for tourism development in Hadoti Region of Rajasthan. In: Satapathy, S.C., Joshi, A. (eds.) ICTIS 2017. SIST, vol. 83, pp. 128–136. Springer, Cham (2018). https://doi.org/10.1007/978-3-319-63673-3_16
20. Sullivan, Y.W., Kim, D.J.: Assessing the effects of consumers' product evaluations and trust on repurchase intention in e-commerce environments. Int. J. Inf. Manage. **39**, 199–219 (2018)
21. Yan, Q., Wu, S., Wang, L., Wu, P., Chen, H., Wei, G.: E-WOM from e-commerce websites and social media: which will consumers adopt? Electron. Commer. Res. Appl. **17**, 62–73 (2016)
22. Zhao, Y., Kou, G., Peng, Y., Chen, Y.: Understanding influence power of opinion leaders in e-commerce networks: an opinion dynamics theory perspective. Inf. Sci. **426**, 131–147 (2018)
23. Zhou, L., Wang, W., Xu, J.D., Liu, T., Gu, J.: Perceived information transparency in B2C e-commerce: an empirical investigation. Inf. Manag. **55**(7), 912–927 (2018)
24. Zott, Ch., Amit, R., Donlevy, J.: Strategies for value creation in e-commerce: best practice in Europe. Eur. Manag. J. **18**(5), 463–475 (2000)

Multicriteria Selection of Online Advertising Content for the Habituation Effect Reduction

Anna Lewandowska[1], Jarosław Jankowski[1]([⊠]), Wojciech Sałabun[1],
and Jarosław Wątróbski[2]

[1] Faculty of Computer Science and Information Technology,
West Pomeranian University of Technology in Szczecin,
ul. Żołnierska 49, 71-210 Szczecin, Poland
{atomaszewska,jjankowski,wsalabun}@wi.zut.edu.pl
[2] Faculty of Economics and Management, University of Szczecin,
Mickiewicza 64, 71-101 Szczecin, Poland
jaroslaw.watrobski@wneiz.pl

Abstract. While extant research has examined the various areas related to interactive marketing, techniques to overcome habituation effects without negative impact on users are still less explored. Usually effects such as vividness are used and they have an influence on consumer skepticism toward websites and brands, which negatively influences their attitudes. This research contributes to the field through exploration and identification of consumer responses to multimedia content focused on reduction of habituation effects through visual elements with high intensity used. The results were obtained from the natural responses of web users and subjective experiments in which a group of observers rated differently arranged banners. Proposed decision support model based on the COMET method helps to validate proposed scenarios towards compromise solutions.

1 Introduction

Due to the high intensity of marketing activities, they are performed as a continuous balance between consumers avoiding advertisements and companies trying different methods to get advertisements delivered and noticed. Increasing effectiveness is often connected with the growth of intrusiveness. This leads to a negative impact on the customer and consequent attempts to disable such multimedia content [8]. Research related to intrusiveness is based mainly on controlled experiments and deals with selected factors like frequency and size, or the ability to remove the content [14].

Most research is conducted in artificial and simulated environments and the results are not based on responses in real web environments, however field experiments have been used as well [9]. Our approach, where we understand an advertisement as multimedia component, combined two sources of knowledge based on

© Springer Nature Switzerland AG 2019
N. T. Nguyen et al. (Eds.): ACIIDS 2019, LNAI 11432, pp. 499–509, 2019.
https://doi.org/10.1007/978-3-030-14802-7_43

perceptual and online experiments in a real environment. Experimental advertising tests were created by using several components that affected the evaluation of intrusiveness along with measuring the effects online. This approach makes it possible to identify the level to which it is worth increasing the level of intrusiveness and how it is related to results from a real environment. Subjective experiments are based on the forced-choice pairwise comparison method, which results in the smallest measurement variance and thus produces the most accurate results [10]. This approach may be useful for website services owners to prevent losing customers through the use of excessively intrusive banners, and for marketers in searching for trade-off solutions.

Paper is organized as follows: literature review is presented in Sect. 2. The conceptual framework is illustrated in Sect. 3. Section 4 discusses results from subjective study and online experiment. Procedure of searching for compromise solutions is presented in Sect. 5 and concluded in Sect. 6.

2   Related Work

The increasing importance of interactive technologies and especially the Internet in marketing has been brought about by the development of new media and is affected by information technology. One of the most important features is the possibility to interact with the receiver, while the traditional mass media use a one to many communication model. Such an approach makes it impossible to analyze the direct effects and influence on consumer behavior [12]. The main distinguishing feature of electronic media (and the Internet) is two-way communication, which allows information to be sent and feedback to be received [1]. New stages of interactive communication create areas related to the measurement of interactions and effectiveness, design of multimedia content [5] and call to action messages, or changing the structure of advertisements using data about consumer behavior [15].

Because of the extensive use of similar techniques and the high number of advertisements visible online, a very low fraction of advertisements is clicked on. Due to habituation effects and limited ability to process information online users miss online marketing content [2,3]. To increase performance and visibility of marketing content invasive techniques are often used. They lead to higher click rates but they can affect brand perception negatively and lead to advertisement avoidance [11]. Reduction of the overall effectiveness because of limited cognitive capacity and negative affective response, such as irritation and annoyance, was also reported [13].

Therefore, the subjective preferences should be analyzed. The goal of these experimental procedures is to find a scalar valued impression correlate that would express the level of impairment (in the case of video compression) or overall impression [4]. In this work we discuss how to interpret such impression correlates in the context of rating the negative impact of an advertisement. To design the experiment efficiently we utilized procedures described in [10].

Most of the presented in literature research is based on controlled experiments within artificial environments. The approach based on real advertisements were proposed in [9], however the data were gathered only based on perceptual experiments. Our intention in the paper is to analyze advertisement' intrusive elements not only in artificial environment but integrate them with data from an online campaign and a perceptual experiment.

3   Conceptual Framework

The research presented in this paper uses different approaches to analyze intrusiveness of multimedia content and extends available solutions to subjective metrics with the possibility of evaluating the intrusiveness of advertising content. The analysis combines knowledge obtained from online data and the results of a subjective experiment. It shows how increasing intrusiveness affects online results and enables the selection of a compromise design. The basis for this research is interactive marketing and the decomposition of the multimedia content into several sections with a scalable influence on the user. The presented approach extends our earlier work related to online subjective experiments [10] and adjustable influence levels [6,9].

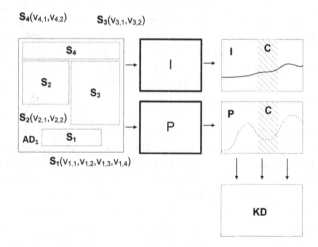

Fig. 1. The process of gathering knowledge (KD) from subjective (P) and online (I) experiments for ADi test advertisement.

Following [4], we called the parameter modified in the advertisement a condition. For every tested advertisement there is a set of n sections $S = \{S_1, S_2, \ldots, S_n\}$ and a number of m possible conditions for each of them defined as follows: $V_i = \{v_{i,1}, v_{i,2}, \ldots, v_{i,m}\}$. The research assumes that for each condition $v_{i,j}$, the level of persuasion $p_{i,j}$, can be defined, depending on the number and size of flushing elements, their frequency, type of animation, and vividness. Figure 1 depicts an example test advertisement used in the research process.

Four main sections, S_1, S_2, S_3, S_4, can be distinguished within a test advertisement AD_1, and for each section there are assigned available conditions and intensity levels $S_1(v_{1,1}, v_{1,2}, v_{1,3}, v_{1,4})$, $S_2(v_{2,1}, v_{2,2})$, $S_3(v_{3,1}, v_{3,2})$, $S_4(v_{4,1}, v_{4,2})$. Constructed in this way, the test advertisements can be verified as part of real campaigns conducted over the Internet (denoted as I), and with the perceptual experiment (denoted as P). The experiments were conducted for every condition of the test advertisement. During the tests over the Internet, design variants with the different conditions were generated and presented to web users. The response expected by the advertiser was gathered in the form of the number of clicks. The perceptual experiment, with the use of the forced-choice metrics, was based on the advertisement pairs-comparison toward intrusiveness. The results from both parts were gathered and analyzed to select the compromise level C, at which intrusiveness is on a moderate level and response is at an acceptable level. The data from a series of experiments were gathered in the knowledge database KD.

4   Experimental Study

Our goal was to compare the level of negative impact of selected parts of an advertisement. Negative impact was assessed for ten advertisements, each depicting a different content and prepared at several conditions. The experiment was run for many observers through the Internet. We collected data through the pairwise comparison experiment. Forced-choice pairwise comparison is an ordering method in which observers decide which of the two displayed images has a higher negative impact. The method is popular, but is very tedious if a large number of conditions need to be compared. However, as reported in [10], it results in the smallest measurement variance and thus produces the most accurate results. The advertisements were assessed through the Internet by naïve observers who were confirmed to have normal or corrected to normal vision. The age of the observers ranged between 20 and 68. Fourteen observers completed the pairwise comparison experiment. For additional reliability, all observers repeated each experiment three times, but no two repetitions took place on the same day in order to reduce the learning effect. According to [10], collecting 30–60 repetitions per condition is sufficient. The experiments were run through the Internet; therefore, the conditions were not stabilized. We established that a properly designed advertisement should have a similar impact on observers despite the display conditions. The observers were free to adjust the viewing distance to their preference. In real-world applications images are seen from varying distances on screens of different resolutions. Therefore, the data is more representative of real-world conditions if the variability due to uncontrolled viewing conditions is included in the measurements. Five campaigns were designed for the experiments. Each contained a different layout and content type, including fitness, games, travel agency, virtual reality and social site. We selected four main elements that may negatively impact on perceptions of an advertisement: vividness, animation (vertical and horizontal), different frequency (3 level: static, medium and high frequency), the size of the animated area (10, 20 and 40% of the whole advertisement area) and the number of flashing elements (1, 2 and 3). Including all modifications, we had 10 conditions per single advertisement.

Once we collected the experimental data, our goal was to find a scalar measure for each test advertisement that would rate its intrusiveness on a continuous interval scale. The following sections describe how this can be done. An online experiment was conducted in a real environment and natural responses from web users were gathered based on their clicking on advertisements. The advertisements were the same as in the subjective experiment and were shown with the same frequency, which resulted in a similar number of impressions; however, the number of impressions was different for each category. Advertisements related to games were shown 79998 times and received 105 clicks; advertisements related to travel were shown 44235 times with 113 clicks; advertisements related to the virtual world had 64097 impressions and 11 clicks. Advertisements related to fitness were shown 130751 times and resulted in 107 clicks, while advertisements related to social platforms were shown 60671 times and got 106 clicks. For each advertising banner a click-through ratio was computed, representing the number of clicks in relation to the number of impressions; this is shown in Table 1.

Table 1. Click-through ratios from the online experiment.

Type	Games	Travel	Virtual	Fitness	Social	Avg
Static	0.10%	0.15%	0.16%	0.09%	0.17%	0.13%
Flashing 25 ms	0.12%	0.17%	0.20%	0.05%	0.23%	0.16%
Flashing 50 ms	0.25%	0.36%	0.21%	0.07%	0.19%	0.22%
Vividness	0.12%	0.18%	0.22%	0.07%	0.14%	0.15%
20% area flashing	0.12%	0.27%	0.11%	0.12%	0.22%	0.17%
40% area flashing	0.08%	0.28%	0.18%	0.06%	0.11%	0.14%
2 elements flashing	0.15%	0.26%	0.13%	0.09%	0.21%	0.17%
3 elements flashing	0.16%	0.23%	0.11%	0.08%	0.13%	0.14%
Horizontal animation	0.13%	0.35%	0.14%	0.12%	0.20%	0.19%
Vertical animation	0.14%	0.36%	0.27%	0.08%	0.16%	0.20%
Avg	0.14%	0.26%	0.17%	0.08%	0.18%	0.17%
SD	0.05%	0.08%	0.05%	0.02%	0.04%	0.03%

The observers may have reported implausible impression scores because they misunderstood the experiment instruction or did not engage in the task and gave random answers. If the number of participants is low, it is easy to spot unreliable observers by inspecting the plots. However, when the number of observers is very high or it is difficult to scrutinize the plots, the [4] standard, Annex 2.3.1, provides a numerical screening procedure. We performed this procedure on our data and found no participants whose data needed to be removed.

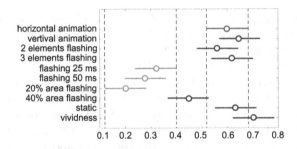

Fig. 2. Comparison of invasiveness from fixed mean scores computed for all test advertisements. Fixed scores are scores shifted about the absolute score minimum. The thin black error bars visualize forced-choice statistical testing. If the two thin bars from two different conditions overlap at any point, the difference between them is too small to be statistically significant.

Horizontal bars, such as those shown in Fig. 2, are a common way to visualize rating experiment results. In addition to the mean scores, most studies are expected to also report the 95% confidence interval for the mean, i.e. the range of values in which the true mean score resides with 95% probability [4]. There is no significant difference between the vividness and the 40% flashing area. This means that flashing elements for a small area (about 10% of the whole advertisement area - that was established in our tests) are acceptable to the subjects. With the growing size of the area and an increasing number of elements, the invasiveness is reported as being more irritating. Animated elements also decrease the positive impression of an advertisement.

The results show that the highest average click-through ratio was obtained for flashing with 50 ms while the lowest was reported for a static condition. Surprisingly, the elements with the highest potential intrusiveness, like vividness, 40% space flashing and three flashing elements, did not increase the response rate. These elements are usually used in online advertising with the intention of increasing the click rates, but this experiment showed the opposite results. The response rates varied between categories. The lowest average response at the level of 0.08% was obtained for advertisements related to fitness, while the highest response at the level of 0.26% was for the travel category.

5   Searching for Compromise Solutions

Presented results showed different performance for objects used. Then the question is what strategy should be selected to deliver results at acceptable level in terms of clicks and overcome habituation effects without negative influence on web users. The problem of assessing web content based on its intrusiveness and conversion is a classic MCDA (Multi-Criteria Decision-Analysis) problem, where two opposing criteria are involved (intrusiveness is cost criterion, and conversion is profit criterion). Therefore, the appropriate MCDA method should be selected carefully [7], what can be effectively done using the generalized framework for multi-criteria method selection proposed by Wątróbski et al. [16]. In that way, the COMET (Characteristic Objects METhod) method has been chosen, which was first presented by Sałabun [17]. This method is characterized by unique properties that are rare in the MCDA field. First of all, the COMET is free of the rank reversal paradox, which is one of the most significant shortcomings of MCDA methods [18]. This property is the result of fact that the COMET method provides an alternatives assessment using the model identified on the basis of characteristic objects. These objects depend on the set of assessed decision variants. It means that in contrast to many other MCDA methods, there is no comparison here of the decision-making variants of each other. The assessment of the decision variants is obtained solely from the obtained model. Therefore, if we use the same model decision-making, the values of assessments for alternatives will not change regardless of the number, so the mentioned paradox will never occur [19]. The COMET method additionally allows relatively easy identification of linear and non-linear expert decision functions. We use the COMET algorithm, which was extensively presented in [17]. The three decision models have been identified by using the result of presented experiments and expert knowledge, where in each case the domain of the problem has been presented by two normalized criteria presented with five triangular fuzzy numbers (presented in Figs. 3 and 4).

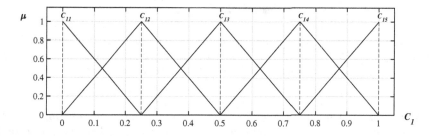

Fig. 3. The set of five triangular numbers for the normalized conversion criterion (C_1).

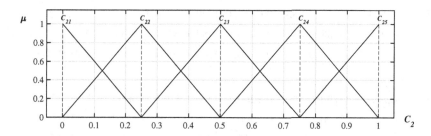

Fig. 4. The set of five triangular numbers for the normalized intrusiveness criterion (C_2).

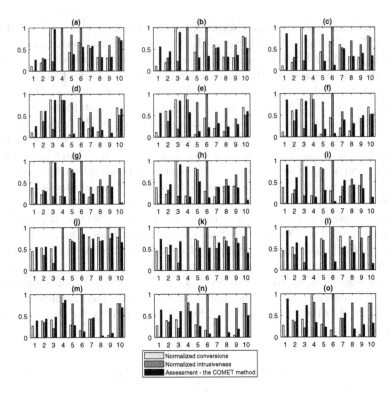

Fig. 5. The set of considered alternatives and their assessment, where (a–c) represents travel; (d–f) social platform; (g–i) games; (j–l) fittness; (m–o) virtual world.; (a, d, g, j, m) model P_1; (b, e, h, k, n) model P_2; (c, f, i, l, o) model P_3.

We assume that in the first model the most important is maximizing conversion and intensive reduction of habituation due to high visibility (1), the second model is sustainable model with limited negative impact of user and limited ability to decrease habituation effect (2), and the third one is minimizing intrusiveness and visual intensity as well (3). The presented preference vectors allow to recreate rules base (the lower index means the number of the model) (Fig. 5).

$$P_1 = [0.1667, 0.1250, 0.0833, 0.0417, 0.0000, 0.3750, 0.3333, 0.2917, 0.2500,$$
$$0.2083, 0.5833, 0.5417, 0.5000, 0.4583, 0.4167, 0.7917, 0.7500, 0.7083, \quad (1)$$
$$0.6667, 0.6250, 1.0000, 0.9583, 0.9167, 0.8750, 0.8333]$$

$$P_2 = [0.5000, 0.3750, 0.2500, 0.1250, 0.0000, 0.6250, 0.5000, 0.3750, 0.2500,$$
$$0.1250, 0.7500, 0.6250, 0.5000, 0.3750, 0.2500, 0.8750, 0.7500, 0.6250, \quad (2)$$
$$0.5000, 0.3750, 1.0000, 0.8750, 0.7500, 0.6250, 0.5000]$$

$$P_3 = [0.8333, 0.6250, 0.4167, 0.2083, 0.0000, 0.8750, 0.6667, 0.4583, 0.2500,$$
$$0.0417, 0.9167, 0.7083, 0.5000, 0.2917, 0.0833, 0.9583, 0.7500, 0.5417, \quad (3)$$
$$0.3333, 0.1250, 1.0000, 0.7917, 0.5833, 0.3750, 0.1667]$$

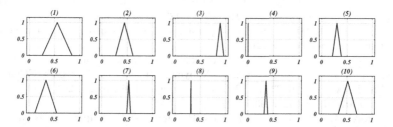

Fig. 6. The assessment for ten travel experiments presented as triangular fuzzy numbers

Based on the obtained results, we can identify triangular fuzzy numbers that evaluate the generalized assessment of advertising campaigns. From the set of 3 obtained assessment for each campaign, the two extreme ones are the support of the fuzzy set and the middle value is its core. For travel, this is shown in Fig. 6.

6   Conclusions

The comparison of the results with the online experiment showed a new approach within interactive marketing to evaluating online advertisements regarding their effectiveness in terms of habituation and limiting their negative impact on users. The results, based on integrated knowledge, showed that the elements used to attract visual attention such as vividness effects, do not necessarily bring better results and can negatively affect users; this was confirmed by the perceptual experiment. The main contribution of this paper is the hybrid approach to visual elements selection towards reduced habituation and negative effect on target users. Integrated knowledge from perceptual and online experiment was treated as input to COMET method, which was used for searching for compromise solutions. It also provides another step toward making the online environment friendlier. Future work should include making more extended experiments with various forms of advertising and building metrics on higher sets of advertisements. Another direction could be to implement the perceptual experiment within the online environment and conduct the analysis among target users.

Acknowledgments. The work was supported by the National Science Centre of Poland, the decisions no. 2017/27/B/HS4/01216 (JJ), 2016/23/N/HS4/01931 (WS) and by the Faculty of Computer Science and Information Technology, West Pomeranian University of Technology, Szczecin statutory funds.

References

1. Hoffman, D.L., Novak, T.P.: Marketing in hipermedia computer-mediated enviroments: conceptual foundations. J. Mark. **60**(3), 50–68 (1996)
2. Benway, J.P.: Banner blindness: the irony of attention grabbing on the World Wide Web. Proc. Hum. Factors Ergon. Soc. Annu. Meet. **42**(5), 463–467 (1998)
3. Burke, M., Hornof, A., Nilsen, E., Gorman, N.: High-cost banner blindness: ads increase perceived workload, hinder visual search, and are forgotten. ACM Trans. Comput. Hum. Interact. (TOCHI) **12**(4), 423–445 (2005)
4. ITU-R.REC.BT.500-11. Methodology for the subjective assessment of the quality for television pictures (2002)
5. Jankowski, J.: Integration of collective knowledge in fuzzy models supporting web design process. In: Jędrzejowicz, P., Nguyen, N.T., Hoang, K. (eds.) ICCCI 2011. LNCS (LNAI), vol. 6923, pp. 395–404. Springer, Heidelberg (2011). https://doi.org/10.1007/978-3-642-23938-0_40
6. Jankowski, J., Wątróbski, J., Ziemba, P.: Modeling the impact of visual components on verbal communication in online advertising. In: Núñez, M., Nguyen, N.T., Camacho, D., Trawiński, B. (eds.) ICCCI 2015. LNCS (LNAI), vol. 9330, pp. 44–53. Springer, Cham (2015). https://doi.org/10.1007/978-3-319-24306-1_5
7. Wątróbski, J., Jankowski, J., Piotrowski, Z.: The selection of multicriteria method based on unstructured decision problem description. In: Hwang, D., Jung, J.J., Nguyen, N.-T. (eds.) ICCCI 2014. LNCS (LNAI), vol. 8733, pp. 454–465. Springer, Cham (2014). https://doi.org/10.1007/978-3-319-11289-3_46
8. Krammer, V.: An effective defense against intrusive web advertising. In: Proceedings of the 2008 Sixth Annual Conference on Privacy, Security and Trust (PST 2008), pp. 3–14. IEEE Computer Society, Washington, DC (2008)
9. Lewandowska (Tomaszewska), A., Jankowski, J.: The negative impact of visual web advertising content on cognitive process: towards quantitative evaluation. Int. J. Hum. Comput. Stud. **108**, 41–49 (2017)
10. Mantiuk, R.K., Tomaszewska, A., Mantiuk, R.: Comparison of four subjective methods for image quality assessment. Comput. Graph. Forum **31**(8), 2478–2491 (2012)
11. McCoy, S., Everard, A., Polak, P., Galletta, D.F.: The effects of online advertising. Commun. ACM **50**(3), 84–88 (2007)
12. Rust, T.: Advertising Media Models. Lexington Books, Lexington (1989)
13. Yoo, C.Y., Kim, K., Stout, P.A.: Assessing the effects of animation in online banner advertising: hierarchy of effects model. J. Interact. Advertising **4**(2), 49–60 (2004)
14. Zha, W., Wu, H.D.: The impact of online disruptive ads on users? Comprehension, evaluation of site credibility, and sentiment of intrusiveness. Am. Commun. J. **16**(2), 15–28 (2014)
15. Zorn, S., Olaru, D., Veheim, T., Zhao, S., Murphy, J.: Impact of animation and language on banner click-through rates. J. Electron. Commer. Res. **13**(2), 173–183 (2012)
16. Wątróbski, J., Jankowski, J., Ziemba, P., Karczmarczyk, A., Zioło, M.: Generalised framework for multi-criteria method selection. Omega (2018)
17. Sałabun, W.: The characteristic objects method: a new distance-based approach to multicriteria decision-making problems. J. Multi Criteria Decis. Anal. **22**(1–2), 37–50 (2015)

18. Wang, X., Triantaphyllou, E.: Ranking irregularities when evaluating alternatives by using some ELECTRE methods. Omega **36**(1), 45–63 (2008)
19. Piegat, A., Sałabun, W.: Identification of a multicriteria decision-making model using the characteristic objects method. Appl. Comput. Intell. Soft Comput. **2014**, 14 (2014)

Dark Side of Digital Transformation in Tourism

Meghdad Abbasian Fereidouni[1] and Arkadiusz Kawa[2(✉)]

[1] Center of Post Graduate Studies,
Limkokwing University of Creative Technology,
Inovasi 1-1, Jalan Teknokrat 1/1, 63000 Cyberjaya, Malaysia
mgabbasian@gmail.com
[2] Poznan University of Economics and Business,
al. Niepodległości 10, 61-875 Poznań, Poland
arkadiusz.kawa@ue.poznan.pl

Abstract. Research on technology in tourism has mostly investigated the benefits and the applications of digitalization, while the risk of structural dependency and data control has mostly neglected in smart tourism research. This study aims to investigate the digital transformation changes in tourism. This research underscores the potential dark side of five digital transformation drivers in tourism through in depth analysis of four activity system elements. The findings highlight the long-term economic, political, and social consequences of digital transformation that may lead to digital colonialism in tourist destination. The research underlines three main digital transformation gaps (productivity, technology, regulation) that may lock the tourist destinations into digital colonialism and dependency.

Keywords: E-commerce · Colonialism · Digitalization · Tourism ·
Platform economy

1 Introduction

From digitalization to age of acceleration, research on information technology in tourism has been almost exclusively emphasizing on the benefits and the applications of technology [3, 5, 9, 18] and rarely on the drawbacks [6, 12]. Today tourism witnesses' new entrants such as online travel agencies (OTAs), meta-search engines, and travel service platforms have been reshaping whole tourism value chain [15, 20]. In most developed and many developing countries, travel and tourism have been heavily dependent on dominant platforms that providing travel information search, reservation and booking, accommodation, transportation, and financial services. Consequently, a powerful elite of private software, Internet, hardware, financial, logistics, and infrastructure enterprises have been able to establish control over destination marketing, operation, and management through dominant travel service platforms such as Google in travel information search, Booking and Expedia as online travel agencies (OTAs), TripAdvisor in travel reviews, Uber and Grab in Urban transportation, and AirBnB in lodging and accommodation sector. Some of these companies go further and start to

© Springer Nature Switzerland AG 2019
N. T. Nguyen et al. (Eds.): ACIIDS 2019, LNAI 11432, pp. 510–518, 2019.
https://doi.org/10.1007/978-3-030-14802-7_44

cooperate. For example, Booking has recently invested $200 million in Grab. Thanks to that those two companies will team up to offer reciprocal services. On the one hand, transportation services will be integrated into Booking's system, and on the other hand, travel accommodation services will be the part of Grab's application [16]. It is worth to emphasize that besides Booking.com and Agoda, Booking Holding also is owner of Kayak, Priceline.com, Rentacars.com and OpenTable.

Meantime, the new regime around global digital economy reveals that the powerful high tech corporations are aiming to secure the dominancy of the early mover in digital domain by systematic norm creation and restricting government's ability to regulate their technologies and services [13, 14]. Hence, the dominant travel service providers by having the intellectual property right and dominating both digital ecosystem and major non-state multi-stakeholder Internet governance forums aim to entrench the power of incumbent and shape a new era of digital colonialism in tourism destinations [14]. The term "digital colonialism" is referred to the structural dependency of tourism destination to technology and knowledge of digital service providers.

There are different studies that have highlighted the potential threat of the digitalization and their dominant platforms in various terms such as on user privacy and ethical concern [7], digital discrimination [4], lodging service [1, 2, 8], housing market [10], and regulatory challenges [19, 21]. Nevertheless, tourism research has widely neglected to address the risk of structural dependency and data control, per se, digital colonialism in destination infrastructure and travel services. The emergence of filling this gap in tourism research is indispensible and timely. Therefore, this study aims to address the potential risk of digital colonialism in tourism by elaborating on main digital transformation drivers.

The paper employs a narrative interpretation to the extant theoretical and empirical research in order to provide a steal overview of dark side of digital transformation in tourism. This approach was taken to the quality, scarcity and diversity of the literature retrieved with less emphasis on evaluation criteria and methodological matters than other forms of review. Therefore, framing on activity theory, the study proposes a research model to track the digital transformation impact on tourism activity system. Figure 1 illustrates how five digital transformation drivers (ecommerce and platform technology, big data and intelligent system, artificial intelligence and Internet of Things, policy and governance system, and fiscal and technical architecture) may unfavorably influence four main elements of tourism activity systems (subject, rule, community, division of labor) and provoke digital colonialism in extant of digital transformation gaps (technology, productivity, and policy).

It has to be justified that this study refers to the methodological contribution of Jørgensen [11] and underlines a subject (tourist, supplier, intermediary) can be affected by the nature of the subject, subject characteristics, and destination or market characteristics. Similarly, the rules are influenced by explicit rules such norms, legislations, and cultural values. Likewise, the community can be impacted by individuals, institutions, companies, or non-human actors. The division of labor is exposed to change by relational characteristics, decision makers, and influencers.

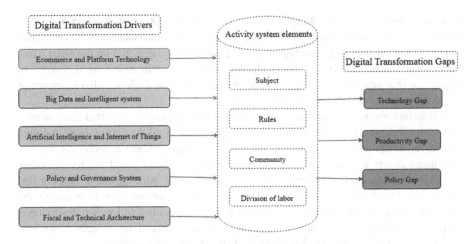

Fig. 1. Proposed research model of digital transformation in tourism

The structure of the paper is as follows. Section 2 describes five digital transformation drivers: ecommerce and platform technology, big data and intelligent system, artificial intelligence and Internet of Things, policy and governance system, and fiscal and technical architecture. The discussions elaborate how digital transformation drivers can exert negative forces on a destination by influencing on subject (tourist, supplier, and intermediary), rules (norms, legislations, and cultural values), community (individuals, institutions, companies, and non-human actors), and division of labor (relational characteristics, decision makers, and influencers). Section 3 provides the reasons behind current threats and highlights the extend digitalization gaps in tourism. Section 4 summarizes the paper, identifies the research limitations, and points to future directions of the research.

2   Digital Transformation Drivers

2.1   E-commerce and Digital Platform

E-commerce as centerpiece of global digital economy has primarily induced businesses, households, communities, and governments to harness centralized networking platforms and interact with billions of online users across the world. Digital technology and the Internet soon transformed both service and manufacturing industries and reformed supply chains that allow giant digital enterprises to run their platform and algorithm anywhere in the world. Consequently, few mostly US-lead giant tech companies as early movers have been able to dominate their platform in each sector. In travel and tourism content, market capitalization comparison of OTAs, Airlines, and hotels demonstrates the new errant players run some aggregative digital platforms like Booking Holding (owner of Booking.com, Priceline.com, and Agoda.com) with $70 billion and AirBnB with $30 billion that outpace the largest hotel chains and airline companies [20]. The report shows that this money migrates from traditional travel booking intermediaries and undifferentiated hotels to dominant OTAs and short-term rentals service providers respectively.

In an increasingly personalization demand for service and product, each company has been entailed to sense, analyze, and integrate location and spatial information and knowledge in order to effective and efficient marketing strategy and increase competitive advantage, for instance, Google in travel search, TripAdvisor in online travel review, Booking Holding group in reservation service mostly aggregating the information available to both travelers and tourism firms. Moreover, those seeking to access international platforms or e-marketplaces to sell products must negotiate with those aggregators that have already pursing their place in travel and tourism value chain. Small, medium, and sometimes even large tour and travel companies, hotels, and airlines lack negotiating power with the aggregators and have to comply with mega-digital platforms due to their popularity, bigness, and user-friendly features, or lose the market.

2.2 Big Data and Intelligent System

Digital platforms serve consumers and business to communicate, buy and sell online. They provide free service in exchange for data. The machine learning and analytic systems are fed by data using algorithm and artificial intelligence to create required intelligent controlling the representative sector. In an exponentially cumulative phenomenon, intelligent platforms aggregate information of personalized travel experiences, destination infrastructure and environment, and soon plays as brain of travel and tourism sector. On the one hand, big data centralized in dominant platforms improves travel experience by personalization and benefit businesses by growth in online sale, service revenue and operational profit, but, on the other hand, commodification of big data enables a business model that surpasses any other in profitability and durability that cementing the platform dominance. The real costs involved are twofold. First, on an unbeatable network, dominant digital enterprises begin to leverage their position in service providers, curbing competition, and in long-term monopolizing tourism infrastructure. The big data enable the platform owners to block access of other competitors, prioritize content, lower the speed, and extract rent from competitors [13]. Secondly, key national resources of personal and social data, as well as information of buildings, vehicles, natural resources are siphoned off and exploited by foreign firms and governments that can lead to geo-economic and geo-political insecurity and instability [6, 13]. For instance, big data aggregated from short-term accommodation and share homes can be fed to intelligent systems to influencing the housing market [10].

2.3 Artificial Intelligent and Internet of Things

Artificial intelligence (AI) and Internet of Things (IoTs) facilitate businesses to provide automated location-based personalized services to travelers in a fully integrated ecosystem, and monitor, analysis, and control the destination [6, 18]. On the other hand, there are long-term economic and social consequences. At initial cost, the owners of customer network platforms are enabled to access local computers, devices, and IT facility to enrich their database. Secondly, the network created from IoTs acts as nerves centers for brain of global network, the platforms. The value data captured by these nerve centers is being stored at the foreign centers for creating the digital intelligence required to exercise control over the country's various sectors and social systems. Thirdly, most of software, hardware, sensor, and automated device have been patenting in few developed countries that monopolizing and controlling the technology market. A multiplicity of rules in the mega-agreement, from source code secrecy and bans on technology transfer to unrestricted cross-border supply of services and financial flows, would constrain the development of local AI and its regulation. Fourthly, automation eliminates many low wages jobs that are critical to developing countries with growing population, while substitutes new jobs that demands offshoring skilled workers and training services. Studies show that up to 7 800 000 traditional jobs in hospitality industry are eliminated, while potential jobs are created outside the industry's value chain in the digital ecosystem of IoT, robotics, connectivity, and data analytics [20].

2.4   Digital Industrialization Policy and Governance System

There are currently very few restrictions that govern the Internet and digital ecosystem [17]. However, the powerful elite of private dominant digital enterprises try to hege-monize the digital domain and every service sector by submitting mega-trade and investment agreements and forming strategic norms in respective industry [14]. With scrutinizing mega-agreements on e-commerce chapter, Kelsey reveals how the group of giant digital enterprise looking for restricting governments' ability to regulate the technologies and services. Under provision of agreements like TiSA (Trade in Services Agreement), a small group of private mega-corporations are able to hegemony most valuable global economy's infrastructure of 4th industrial revolution; technology, digital platforms, search engine, and big data, and restrict the power of governments [14].

The vast economic and social implications of the mega-trade agreement and the strategic norm creation on future of tourism industry are yet blurred. Under provision of ecommerce and economic development, some regulatory factors of digital colonialism forces can be deciphered: prohibitions on national requirements to store or process data locally, weak provisions on the protection of consumers and personal information, promoting actual cross-border commerce by eliminating customs duties on electronic transmissions [14].

2.5   Fiscal and Technical Architecture

The technical architecture of cross-border e-commerce requires countries to comply with free flow of data and rules such data source secrecy, non-mandatory of local presence, and data storage. It means that dominant platforms as major travel and tourism service providers with blurred and complex corporate structure are able to circumvent from paying taxes to local governments. Framing from Kelsey [13] study, with new rules of tax on cross-border transaction, profit earned in foreign country, custom duty, and tariffs, it will be even more problematic for tax authorities. With more challenging mega-agreement tax exemption, prohibition requirements of local presence and data storage let more tax revenue goes out of country for cross-border transactions and profits earned in a country when local presence is absent in that country [13].

Moreover, the new global regime around digital would permanently ban the customs duties on electronic transmissions that WTO members agreed to provisional moratorium in 1998. The loss of these revenues or damage to domestic competitors would be huge. Falling revenue would deprive developing countries of the investment required for effective digital industrialization strategies and to address the impacts on local businesses, employment, and communities.

Meanwhile, having the intellectual property right and dominating both digital ecosystem and major non-state multi-stakeholder Internet governance forums have brought the technological advantage to dominant digital service providers [14]. Therefore, the giant tech enterprises have required competition rules at the global level with developing countries that unfairly lead to losing competition for small and undifferentiated companies in travel and tourism industry. Moreover, dominant OTAs and travel service provider platforms that are mainly located in the US and few developed harvest from local tourism revenues while paying taxes to their parent countries. The cross-border services might provide better quality to local developing tourism, but lack of commitment to maintain supply will intensify the destination long-term dependency.

3   Digital Transformation Gaps

3.1   Technology Gap

The technological advancement has surged destination revenue, reduced operational costs, efficient marketing, and enhance management. However, the dominancy of digital service providers has been bigger and stronger every day deepening the destination dependency on technology and knowledge of their providers. Subsequently, this dependency has exerted a technology gap in digital infrastructure and provision of technology in tourism destination. The ubiquitous presence of IoTs around destination infrastructure and environment together with structural dependency of tourists, residence, businesses, communities, and government on dominant and resourceful digital platforms expand the technological gap in the destination. Simultaneously, unrestricted supply of services and uncompetitive activity of dominant players erect extremely difficult barriers to entry of major tourism stakeholders in digital service ecosystem. This gap can justify the severe impact of digital colonialism in the destination. For instance, the dominant digital enterprises that having technology advantage are able to halt a payment system, cease rival airline operation, limit travel agencies access to real-time data, disrupt recreational sites, and international hotel chains activities, all by a faulty software, ISP outage or even a single malware.

3.2   Productivity Gap

The dominant platforms that aggregated destination data for years now act as brain for travelers and businesses in which no one can technically compete with the big data that they have. These platform cumulatively optimizes information for both travelers and businesses using massive content that created by tourist experience and business data such as information of accommodation, transportation, and excursions. Therefore, tourism destinations that have dug more on using these dominant platforms are technically reliant more on provision of service by them. The dependency on knowledge and soft skills such as advanced digital content, language, and data analytics can be referred to productivity gap in the destination. According to World Economic Forum [20] report on impact of digitalization on aviation, travel and tourism, within next decade, 940 000 jobs will be eliminated due to demand shift from hotels to short-term rentals that migrate $55 billion out of hotel industry. However, digitalization has created new jobs in travel and tourism industry that demand new skills and knowledge.

3.3 Regulatory Gap

E-commerce rules such as prohibition of requirements for local presence, using local facility that the country invested in establishing, employing local people and train them, transfer knowledge and technology, and intellectual property right exacerbates this dependency [14]. This can be referred to extant of regulatory gaps in digitalization of tourism. As tradeoff between market access and losing the profit, businesses have to sale their products through dominant channels and implement digitalization rules and norms. Likewise, e-commerce rules on free data flow and priority of financial service in fast and secure transaction have underpinned the proliferation of using electronic payment methods such as major international electronic payment systems (e.g. Visa, MasterCard), specialist online exchange (e.g. PayPal, Alipay), and integrated platforms (e.g. Google Wallet, ApplePay) [13]. These payment methods not only shift direct destination income to platform owners' parent countries but also they are highly profitable for dominant digital service providers and challenging for tax authority due to ever more creative and non-transparent features to circumvent national regulation [14].

4 Conclusions

Digital transformation has been enabling tourism destinations to employ IoTs and AI in order to monitor, analysis, and optimize the resource utilization, that, in turn, benefits tourism ecosystem in different ways. Nevertheless the digitalization has engendered some digital enterprises as early mover have been able to dominant in the travel services and control over the destination infrastructure. The main contribution of this study is the conceptualization of digital colonialism through elaborating on digital transformation changes in four main tourism activity systems (subject, rules, community, and division of labor). Based on the narrative interpretation, the study found that three gaps in technology, productivity, and regulation are erecting the risk of digital colonialism in tourism. It is very important for further studies on technology in tourism because the risk of structural dependency and data control has so far been neglected in tourism research. This approach should change the perception and comprehension of tourism value chain which is reshaping by OTAs, meta-search engines, and travel service platforms.

   In our paper, we have discussed that the socio-economic consequences of digital colonialism are more dramatic as current regime around the Internet let few monopolistic digital enterprises to run their platforms everywhere and soak up information of buildings, vehicles, natural resources and exploit the key national resources of personal and social data. Technological elements such as big data and intelligent systems can be employed to analyze and influence social trends, shape public opinion, and mislead to harm a destination image by propagating negative comments and manipulating the information different individuals see. These sustainability issues will be structurally deepened if the digital transformation gaps in productivity, technology, and regulation do not be puzzled out and settled by tourism researcher, practitioners, and policy makers. However, the limitation of this research should not be dismissed.

This study has not concentrated on generalizing the model; it has rather limited to draw the framework to interpret the dark side of digital transformation in tourism. Therefore, the further studies are indispensible. The next step will be to conduct empirical research in which interviews with experts (from digital platforms, universities, e-commerce companies, e-commerce service providers, etc.) will be applied. Thank to that, we will deepen the phenomena of digital colonialism. In order to check the theoretical assumptions, especially to expand the awareness of digital colonialism threat in tourist destinations, we will implement a quantitative method based on questionnaire which would allow to test the hypotheses on a larger sample, and thus to identify and explain the poorly recognized phenomena and relationships between them. They will allow to carry out a statistical description and explanation of the needs emerging in larger populations on the basis of the representative samples.

Acknowledgements. This paper has been written with financial support of the National Center of Science [Narodowe Centrum Nauki] – grant number DEC-2015/19/B/HS4/02287.

References

1. Aznar, J.P., Sayeras, J.M., Rocafort, A., Galiana, J.: The irruption of Airbnb and its effects on hotel profitability: an analysis of Barcelona's hotel sector. Intangible Capital **13**(1), 147–159 (2017)
2. Blal, I., Singal, M., Templin, J.: Airbnb's effect on hotel sales growth. Int. J. Hospitality Manage. **73**, 85–92 (2018)
3. Buhalis, D., Law, R.: Progress in information technology and tourism management: 20 years on and 10 years after the Internet. State eTourism Res. **29**, 609–623 (2008)
4. Cheng, M., Foley, C.: The sharing economy and digital discrimination: the case of Airbnb. Int. J. Hospitality Manage. **70**, 95–98 (2018)
5. Del, G., Baggio, R.: Knowledge transfer in smart tourism destinations: analyzing the effects of a network structure. J. Destination Mark. Manage. **4**(3), 145–150 (2015)
6. Gretzel, U.: Intelligent systems in tourism: a social science perspective. Ann. Tourism Res. **38**(3), 757–779 (2011)
7. Gretzel, U., Werthner, H., Koo, C., Lamsfus, C.: Computers in human behavior conceptual foundations for understanding smart tourism ecosystems. Comput. Hum. Behav. **50**, 558–563 (2015)
8. Hajibaba, H., Dolnicar, S.: Substitutable by peer-to-peer accommodation networks? Ann. Tourism Res. **66**, 185–188 (2017)

9. Huang, C.D., Goo, J., Nam, K., Woo, C.: Smart tourism technologies in travel planning: the role of exploration and exploitation. Inf. Manage. **54**, 757–770 (2017)

10. Horn, K., Merante, M.: Is home sharing driving up rents? Evidence from Airbnb in Boston. J. Hous. Econ. **38**, 14–24 (2017)

11. Jørgensen, M.T.: Reframing tourism distribution - activity theory and actor-network theory. Tour. Manage. **62**, 312–321 (2017)

12. Kang, B., Brewer, K.P., Bai, B.: Biometrics for hospitality and tourism: a new wave of information technology. FIU Hospitality Tourism Rev. **25**(1), 1–9 (2007)

13. Kelsey, J.: The Risks for ASEAN of New Mega-Agreements that Promote the Wrong Model of e-Commerce (2017)

14. Kelsey, J.: TiSA Foul play. UNI Global Union (2017)

15. Möller, K., Halinen, A.: Managing business and innovation networks from strategic nets to business fields and ecosystems. Ind. Mark. Manage. **67**, 5–22 (2017)

16. Russel, J.: Southeast Asia's Grab pulls in $200M from travel giant Booking, TechCrunch (2018). https://techcrunch.com/2018/10/29/grab-raises-200m-booking/

17. Singh, P.J.: Report on developing countries in the emerging global digital order table of contents digital re-ordering of society geopolitics of data-based intelligence, pp. 1–8 (2017)

18. Wang, D., Li (Robert), X., Li, Y.: China "smart tourism destination" initiative: a taste of the service-dominant logic. J. Destination Mark. Manage. **2**(2), 59–61 (2013)

19. Williams, C.C., Horodnic, I.A.: Regulating the sharing economy to prevent the growth of the informal sector in the hospitality industry. Int. J. Contemp. Hospitality Manage. **29**(9), 2261–2278 (2017)

20. World Economic Forum. Digital Transformation Initiative Aviation, Travel and Tourism Industry, Accenture (2017)

21. Zale, K.: When everything is small: the regulatory challenge of scale in the sharing economy. San Diego Law Rev. **53**(4), 949–1016 (2016)

Increasing User Engagement and Virtual Goods Life Span Through Products Diversity and Intensity of Content Updates

Kamil Bortko, Piotr Bartków, Patryk Pazura, and Jarosław Jankowski[✉]

Faculty of Computer Science and Information Technology,
West Pomeranian University of Technology in Szczecin,
ul. Żołnierska 49, 71-210 Szczecin, Poland
{kbortko,pbartkow,ppazura,jjankowski}@wi.zut.edu.pl

Abstract. The virtual goods sector has become one of the main business models for social platforms, games and virtual worlds. While online systems are under continuous development, their users require frequent updates of virtual goods, new digital content and functionality. System developers face dilemmas concerning the frequency of updates or content drops to decrease the habituation effect and increase the life span of digital products. The presented research shows how the diversity and number of virtual products can increase user engagement and interest in new products. Apart from the empirical study, a multi-criteria decision support model is proposed for the evaluation of strategies used in system updates and virtual goods distribution.

1 Introduction

Nowadays, virtual worlds live in symbiosis with the real world. They can be implemented within games, social platforms or dedicated systems with complex social and economic phenomena observed [3]. They enable the possibility of correlating real-time communication with respect to economic activity between users. Together with their evolution, new business models were developed based on the distribution of virtual goods [4] such as avatars [7]. The sale of virtual goods has become an important business model for online games and virtual worlds [1,2]. Virtual goods usually refer to virtual items such as avatar clothing, weapons, pets, coins, characters and tokens [10,12]. While there is substantial research on the motivation to purchase virtual goods [11], the research related to identifying the role of content characteristics for product life span and factors affecting user engagement is limited. Online systems are continuously being updated with new game content [20] and then content spreads within the system [16]. Adjusting the frequency, volume and quality of this new content requires analysis and planning. The presented research shows how strategies of content updates affect user engagement and product life span. The empirical study is

© Springer Nature Switzerland AG 2019
N. T. Nguyen et al. (Eds.): ACIIDS 2019, LNAI 11432, pp. 519–530, 2019.
https://doi.org/10.1007/978-3-030-14802-7_45

followed by a multi-criteria evaluation of these strategies. The rest of the paper is organized as follows: Sect. 2 includes the literature review and Sect. 3 presents the conceptual framework and assumptions for the proposed approach. In the next section, the empirical results are presented, followed by a multi-criteria strategy evaluation presented in Sect. 5 and summary in Sect. 6.

2 Related Work

Virtual goods are the basic source of income for many internet ventures and platforms [10]. There has been a substantial increase in research on the purchases of virtual goods over the last decade [11]. The literature focuses primarily on the relationship between game design and the business model of selling content within games and virtual worlds [11] and their role in creating a positive user experience [2]. Players motivations to acquire various virtual goods are analyzed, as well as developing advantages over the competition. An important aspect of this is related to expressing yourself through, for example, a special outfit, theme or avatar. The motivation shown by users in participating in activities and using the services they engage in is related to their attitudes towards virtual goods [12].

Virtual goods are usually closely related to the specific game in which customers buy them. Therefore, game developer should foresee the continuity and benefits of using and purchasing virtual goods. However, before taking advantage of this, the intention of further use must be assessed [9]. An important aspect of this is the subjective assessment of the player based on the psychological assessment of motivation and exploring the mechanisms of the decision-making process. This is measured by the level of user involvement when buying virtual goods.

From the perspective of MMO (massively multiplayer online) games, we can see a relationship between revenue and the motivation of players [24]. Motivation itself can result from various aspects such as maintaining distance from other players, acquiring new things or even identifying with a virtual character. It was discovered that revenue is strongly correlated with the user's motivations. Therefore, we must analyze how these aspects, such as usefulness and attitude, prompt the user to purchase virtual goods.

Often, because it costs nothing to get the game itself, free online games are strongly focused on the sale of virtual items. The publishers, however, encourage the purchase of virtual goods to improve the character's abilities and collect unique items such as armor or clothing. They allow the user to emphasize their individuality. However, users can still use the website's services free of charge. This enables the publishers to build a growing base of new customers [13].

There is currently a strong emphasis on customer engagement in the usage and distribution of digital content. This is achieved with the help of various mechanisms of content propagation. People are more willing to promote and distribute different types of information [8]. People who trust the content themselves are more likely to recommend it to others whilst also being more susceptible to receiving this type of information [22].

Developers face dilemmas related to optimal frequency of virtual goods updates, their volumes and intensity with focus on continuous development [20]. A low frequency of updates can result in user churn, while frequent development of new content increases operational costs. From another point of view, users may have a limited ability for digital content consumption when content is updated frequently. This may be considered as an unwise budget allocation when content production is significantly higher than demand. Another problem is taking into account limited availability of users within the system [17].

The life span of online gaming products is usually shorter than that of traditional products, and users are constantly expecting new content and system updates [18,19]. Another problem is the habituation effect resulting from the short life span of virtual goods and limited time in which the product can attract web users. This opens up new directions for research since so far it has mainly only been studied for traditional markets [23].

3   Conceptual Framework

To gather knowledge about virtual products usage and behavior the system is monitored towards engagement, content life span and gathered knowledge is used during platform development. Generalization of presented problem assumes n content updates $U_1, U_2, ..., U_n$ with the use of m content categories $C_1, C_2, ..., C_m$ (Fig. 1). Within each update U_i and category C_j new elements $E_{i,j,k}$ are introduced, where k represents the number of element from category j assigned to content update i, where $i = 1, 2, ..., n$, $j = 1, 2, ..., m$ and $k = 1, 2, ..., N(U_i, C_j)$.

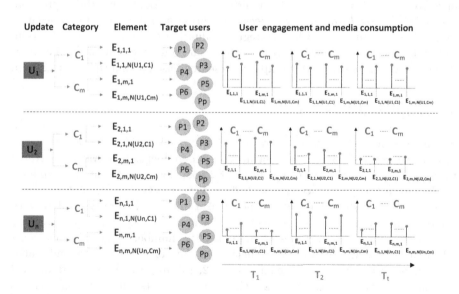

Fig. 1. Evaluation of digital content distribution strategies based on product diversification and scale of updates

Value $N(U_i, C_j)$ represents the number of elements within category C_j used during update U_i. Content is delivered to target platform's users $P = P_1, P_2, ..., P_p$. In the following time periods $T_1, T_2,..., T_t$ user activity and consumption of each content unit is monitored. Regarding used strategies and content elements assigned to each update U_i different effect can be observed. For example update U_1 resulted continuous usage of delivered virtual products with long life span. For U_2 high intensity of usage is observed only at the beginning period and then users loose interest in new product. Update U_n resulted growing usage till saturation point and then dropping user interest observed. Different strategies for new content introduction can be considered. The complexity of the problem leads to the need for decision support tools and analytical methods for better planning and the optimization of system development strategies. It is assumed that several factors are monitored, such as total product usage or consumption dynamics, over monitored time periods. If updates are based on new elements, content production costs should be considered alongside the number of elements in each update and their diversity. A multi-criteria evaluation of results can deliver guidelines for future development planning. This leads to multi-criteria problems with preferences assigned to evaluation measures. Various methods can be used for strategies evaluation and results ranking. The PROMETHEE method was selected to create the presented research [5,6,21]. It delivers the ability of building of an outranking between different strategies. It uses be a set of solutions, in our case possible strategies A, each $a \in A$, $f_j(a)$ represents the evaluation of a solution a, to a given criterion f_j. The preference function $P_j(a, b)$ represents the degree of preference of solution a over solution b for a given evaluation criterion f_j. A multi-criteria preference index $\pi(a, b)$ of a over b is used. It takes all the criteria into account with the expression: $\pi(a, b) = \sum_{j=1}^{k} w_j P_j(a, b)$. In the decision process positive $\phi + (a)$ and negative outranking $\phi - (a)$ is used for strategies evaluation and final ranking creation. In the next sections PROMETHEE method is used for the evaluation of content development strategies used in empirical research based on multiplayer platform.

4   Empirical Results

Empirical study is based on behavior within virtual world and usage of newly introduced avatars [14,15]. Content updates where focused on avatars delivered to users with four types of elements E1, E2, E3, E4. Total twenty one updates U_1–U_{21} were taken into account with data of user activity gathered after new content was introduced. Result is shown in Table 1 each content update with showed number of times each element of avatar was used. The table shows the division into various types of data, i.e. the number of types of elements (their diversity), the number of elements, the sum of total changes and statistics for individual types of elements: E1, E2, E3 and E4. Also results divided into four weeks are presented as well as in aggregated form for all periods (Figs. 2 and 3). Comprehensive analysis of the dynamics of the division for specific weeks is presented as a reversed geometric function graph. In first week 59%

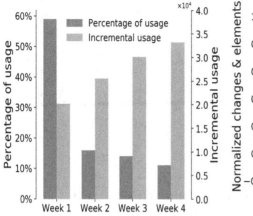

Fig. 2. Percentage of usage and incremental usage

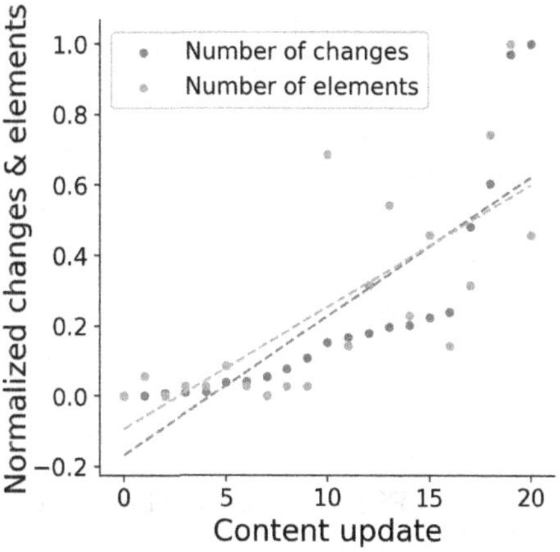

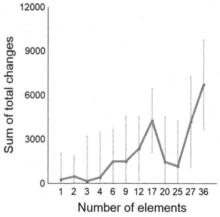

Fig. 3. Number of changes and number of elements

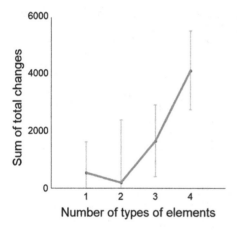

Fig. 4. Relation between total usage and product diversity represented by the number of introduced types of elements

Fig. 5. Relation between total usage and the number of elements used for content update

of all usages take place. This value in the following weeks decreases accordingly: second week to 16%, third week to 14% and last week to 11%. The values of changes in individual weeks have changed as follows: Week I equal to 20060, Week II equal to 25484, Week III equal to 30083 and 33797 in Week IV. The first week had the largest share in the number of changes within analyzed four weeks. In the next step ANOVA analysis was performed. We analysed number of types of elements in relation to the sum of changes within 28 days (Fig. 4). Each content updates could have a maximum of 4 types of elements. The factor F is greater than one (6.9586), it is close to the dependent variable, i.e. the number

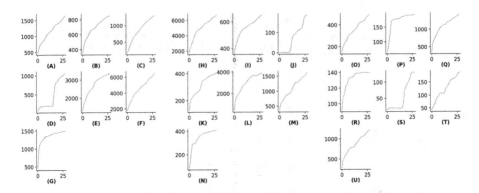

Fig. 6. Incremental usage after each content update U1–U21

Table 1. Content updates with the division into various types of data, i.e. the number of types of elements, the number of elements, the sum of total changes and statistics for individual types of elements: E1, E2, E3 and E4

Content update	Types of elements	E1	E2	E3	E4	Number of elements	Sum of changes	Week I	Week II	Week III	Week IV
U1	3	0	2	2	2	6	1750	861	300	238	246
U2	1	0	0	0	2	2	871	592	100	100	60
U3	1	0	12	0	0	12	1340	639	266	235	163
U4	4	5	6	6	8	25	1176	190	11	675	180
U5	3	7	3	0	2	12	3394	2312	611	257	184
U6	4	15	8	8	5	36	6721	3620	1026	847	998
U7	4	2	5	1	1	9	1499	1270	115	64	43
U8	4	7	3	3	4	17	6914	4142	997	751	759
U9	1	0	0	0	2	2	665	517	49	36	47
U10	2	1	0	1	0	2	198	2	79	34	72
U11	3	1	2	0	1	4	407	249	94	37	24
U12	4	8	6	9	4	27	4224	2215	953	483	264
U13	3	10	2	0	5	17	1649	844	280	246	245
U14	1	0	0	0	2	2	411	295	80	19	11
U15	1	0	1	0	0	1	507	285	65	77	64
U16	1	1	0	0	0	1	192	175	7	8	2
U17	3	10	6	4	0	20	1463	923	214	138	138
U18	1	0	0	0	3	3	140	128	10	2	0
U19	1	0	0	1	0	1	133	19	0	79	33
U20	2	1	1	0	0	2	201	94	20	48	26
U21	3	2	3	1	0	6	1259	688	147	225	155
Sum:		70	60	36	41	207	35114	20060	5424	4599	3714

of types of elements (6.95). A significance level $p <= 0.0029$ confirms the dependence of number of types of elements on the sum of changes. Another goal was to examine whether a number of changes depends on the number of elements used during update. Our dependent variable was the sum of changes within 28 days (Fig. 5). The sum of the items ranged from 1 to 36. The factor F is greater than 1 (3.16890), it is close to the dependent variable, i.e. the number of elements.

Thanks to this we can determine if the test is statistically significant. A significance level p $<=0.047058$ obtained confirms the dependence of sum of changes within 28 days on the number of used elements. The graph shows the spread of normalized data taking into account the number of elements and the sum of changes for four weeks. We can see a clear coverage of the number of items versus the sum of changes within 28 days. Both trend lines running through the graph almost overlap. Incremental usage after each content update is presented in Fig. 6.

5   Multicriteria Strategy Evaluation

Empirical study delivered data from content updates within the real system. Various approaches were used with different number of elements, their diversity and different number of elements in each category. Analysis showed that results represented by user engagement where dependent on a number of used elements and their diversity. Used strategies can be evaluated from the perspective of costs, number of elements and results represented by user engagement depending on preferences of decision maker.

In first scenario analysis was performed from the perspective of four criteria: content production cost, number of types of elements used, number of elements and the total usage in analyzed period. Three variants were analyzed with different weights assigned to criteria. For the first variant (Cost variant I) the same weights to all the criteria were assigned. Results for this variant shows the ranking of the strategies with best results achieved by eight content update (U8).

Analyzing the result from PROMETHEE we can conclude that U8 is preferred to all the other actions in the PROMETHEE ranking. They have cost 47, number of types of elements 4, number of elements 17 and total of changes 6914. U12 is on top of U6 but they are very close to each other. U6 and U12 is incomparable with the U8, because they have a worse score on $\Phi+$, but still above 0,4. U16, U19 and U15 have the worst position in the ranking with $\Phi+$ below 0.3 (Table 2).

In diagram A in Fig. 7, we see that U2, U9, U14 and U19 are clearly the cheapest options as it project completely to the left side. U15 and U18 the second best choices with respect to cost. They are very close to each other. U6 and U12 are very close to each other and are the most expensive options. This information is of course highly dependent on the localization on the GAIA plane. For lower level one can expect more distortions with respect to actual evaluations. U4 is the best one in number of types of elements. U8 is the best option on sum of changes and number of elements and not so bad on number of type of elements but it is weak on cost. If our determining criterion will be sum of changes, with weight 70% (Cost variant III), we should chose U8 with cost equal to 47, the number of types of elements equal to 4, number of elements 17, total number of changes equal to 6914. If our determining criterion will be cost with weight 70% (Cost variant II), we should chose U2 with low cost equal to 2, one type of elements, two elements used and total 871 changes. In second variant

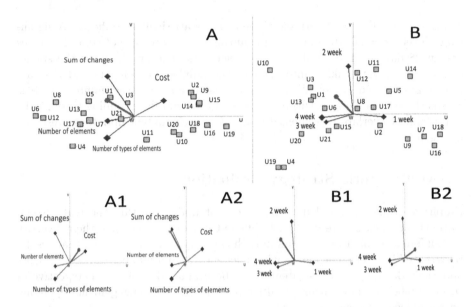

Fig. 7. GAIA analysis diagrams from scenario I cost - A, A1, A2 and scenario II dynamics B, B1, B2.

criteria have different weight: cost = 70%, number of types of elements = 10%, number of elements = 10% and the total of changes = 10%. Results for Cost variant II show the ranking of the actions according to $\Phi+$ with content update number 2 (U2) with the best result, followed by others. Analyzing data from a different side the ranking according to $\Phi-$: U2 is still on top, but it is followed by U9 and U14. We can conclude that U2 is preferred to all the other actions in the PROMETHEE ranking. They have cost equal to 2, number of types of elements equal to 1, number of elements equal to 2 and total of changes at the level of 871. U9 is on top of U14 but they are very close to each other. U9 and U14 is incomparable with the U2, because they has a worse score on $\Phi+$, but above 0.64 updates U6 and U17 have the worst position in the ranking with $\Phi+$ below 0.3.

In third variant criteria have different weights: cost = 10%, number of types of elements = 10%, number of elements = 10%, total number of changes = 70%. Results tab for cost variant III shows the ranking of the actions according to $\Phi+$: content update number 8 (U8) with best result, followed by other. Results shows the ranking according to $\Phi-$: U8 is still on top, but it is followed by U6 and U12. We can conclude that U8 is preferred to all the other actions in the PROMETHEE ranking.

U8 is on top of U6 but they are very close to each other. U6 and U12 is incomparable with the U8, because they has a worse score on $\Phi+$, but above 0.4. U18, U16 and U19 have the worst position in the ranking with $\Phi+$ below 0.3.

Table 2. Scenario I: cost variants for three variants: cost variant I, cost variant II, cost variant III with MCDA analysis

Scenario I	Cost variant I				Cost variant II				Cost variant III			
Rank	Update	Φ	$\Phi+$	$\Phi-$	Update	Φ	$\Phi+$	$\Phi-$	Update	Φ	$\Phi+$	$\Phi-$
1	U8	0,46	0,70	0,24	U2	0,47	0,66	0,19	U8	0,79	0,88	0,09
2	U6	0,43	0,69	0,26	U9	0,46	0,65	0,19	U6	0,71	0,85	0,14
3	U12	0,40	0,68	0,28	U14	0,44	0,64	0,20	U12	0,64	0,81	0,17
4	U7	0,30	0,63	0,33	U19	0,34	0,60	0,26	U5	0,51	0,74	0,23
5	U4	0,23	0,59	0,36	U18	0,21	0,57	0,36	U1	0,45	0,71	0,26
6	U1	0,23	0,58	0,35	U15	0,20	0,56	0,36	U13	0,37	0,67	0,30
7	U5	0,23	0,58	0,35	U1	0,09	0,53	0,44	U7	0,36	0,67	0,31
8	U13	0,18	0,55	0,38	U10	0,08	0,53	0,45	U17	0,23	0,60	0,38
9	U17	0,11	0,53	0,41	U16	0,05	0,50	0,46	U4	0,09	0,54	0,45
10	U21	0,08	0,50	0,43	U11	0,04	0,51	0,47	U21	0,09	0,53	0,44
11	U11	−0,06	0,44	0,50	U20	0,02	0,50	0,48	U3	0,08	0,52	0,44
12	U3	−0,10	0,40	0,50	U7	0,00	0,49	0,49	U2	−0,10	0,42	0,52
13	U2	−0,10	0,36	0,46	U21	−0,03	0,47	0,50	U9	−0,17	0,38	0,55
14	U9	−0,13	0,35	0,48	U8	−0,12	0,43	0,55	U14	−0,31	0,31	0,62
15	U14	−0,18	0,33	0,50	U5	−0,15	0,41	0,56	U15	−0,31	0,32	0,63
16	U10	−0,26	0,34	0,60	U3	−0,22	0,37	0,59	U11	−0,33	0,33	0,65
17	U20	−0,26	0,34	0,60	U13	−0,29	0,34	0,63	U20	−0,47	0,26	0,72
18	U18	−0,30	0,30	0,60	U4	−0,33	0,33	0,66	U10	−0,53	0,23	0,75
19	U15	−0,33	0,28	0,60	U12	−0,38	0,30	0,68	U18	−0,66	0,15	0,81
20	U19	−0,43	0,21	0,64	U6	−0,43	0,28	0,71	U16	−0,68	0,14	0,82
21	U16	−0,49	0,20	0,69	U17	−0,44	0,27	0,71	U19	−0,77	0,09	0,86

In second scenario user engagement and dynamics in analyzed time periods was taken into account. It represents situations when decision maker is interested in specific results, for example high usage in short time after content update, or longer product life span. Four criterias were used at this stage of analysis: dynamics of the first week as percentage of total usage in all periods, dynamics in the second week, dynamics in the third week and dynamics of the fourth last analysed week.

Results for used scenarios in the linear variant shows the ranking of the actions according to $\Phi+$: content update number ten (U10) is on top, followed by other. Analyzing data from a different side the ranking according to $\Phi-$: U10 is still on top, but it is followed by U13 and exequo U1 and U3.

We can conclude that U10 is preferred to all other update in PROMETHEE ranking with $\Phi+$ equal to 0.6875, and $\Phi-$ equal to 0.2875. Overall score Φ is 0.4. It is almost a double advantage over the previous one (U20). They have dynamics of the first week in percent 1%, dynamics of the second week as a percentage 42%, dynamics of the third week as a percentage 18%, dynamics of the fourth last week as a percentage 39%. U1 and U3 they are very close to each other. They have the same scores in Φ (0.1875), $\Phi+$ (0.55) and $\Phi-$ (0.3625). U7, U9, U16 and U18 have the worst position in the ranking with $\Phi+$ values within limits −0.2375 to −0.4 (Table 3).

Table 3. Scenario II: dynamics for three variants: linear, reversed geometric and Gaussian with MCDA analysis

Scenario II	Linear				Reversed geometric				Gaussian			
Rank	Update	Φ	$\Phi+$	$\Phi-$	Update	Φ	$\Phi+$	$\Phi-$	Update	Φ	$\Phi+$	$\Phi-$
1	U10	0,40	0,69	0,29	U14	0,41	0,69	0,28	U10	0,64	0,80	0,16
2	U13	0,20	0,55	0,35	U18	0,35	0,65	0,30	U3	0,45	0,69	0,24
3	U1	0,19	0,55	0,36	U16	0,33	0,64	0,31	U13	0,28	0,58	0,31
4	U3	0,19	0,55	0,36	U5	0,31	0,62	0,31	U12	0,27	0,58	0,31
5	U6	0,16	0,51	0,35	U7	0,28	0,63	0,34	U1	0,26	0,58	0,33
6	U21	0,09	0,48	0,39	U9	0,25	0,61	0,36	U20	0,18	0,57	0,39
7	U20	0,06	0,51	0,45	U2	0,22	0,57	0,35	U11	0,16	0,54	0,38
8	U8	0,06	0,48	0,41	U11	0,21	0,57	0,37	U21	0,14	0,51	0,37
9	U12	0,06	0,46	0,40	U17	0,18	0,58	0,40	U6	0,13	0,49	0,37
10	U11	0,05	0,48	0,43	U8	0,11	0,52	0,41	U15	0,07	0,49	0,43
11	U15	0,05	0,48	0,43	U12	0,05	0,46	0,41	U4	0,02	0,50	0,48
12	U17	0,00	0,48	0,48	U6	−0,06	0,41	0,47	U8	0,01	0,45	0,44
13	U4	0,00	0,49	0,49	U15	−0,06	0,42	0,48	U5	−0,02	0,45	0,47
14	U5	0,00	0,45	0,45	U21	−0,08	0,39	0,47	U19	−0,02	0,48	0,50
15	U19	−0,01	0,49	0,50	U1	−0,16	0,39	0,55	U14	−0,06	0,43	0,49
16	U14	−0,06	0,44	0,50	U13	−0,16	0,38	0,54	U17	−0,06	0,43	0,49
17	U2	−0,06	0,40	0,46	U3	−0,23	0,36	0,58	U2	−0,12	0,37	0,49
18	U9	−0,24	0,34	0,58	U10	−0,34	0,33	0,67	U9	−0,47	0,23	0,70
19	U7	−0,34	0,30	0,64	U20	−0,38	0,28	0,67	U7	−0,56	0,18	0,74
20	U16	−0,40	0,28	0,68	U4	−0,57	0,21	0,78	U16	−0,64	0,16	0,80
21	U18	−0,40	0,28	0,68	U19	−0,64	0,18	0,82	U18	−0,64	0,16	0,80

In diagram B in Fig. 7 in two out of three variants (Linear and Gaussian) except reversed geometric variant we could see that U10 was the best options. In Week I we could see the highest growth rate for U16 and U18 (91%). The lowest growth rate have, our best option U10 (1%). Next options with lowest rate is U19 and U4 with 15% and 18%. Then next have dynamics over 45% and even higher. Average of dynamics in Week I is 57.71%. The case is different in the case of the dynamics of the Week II U10 has rate of dynamics 42%. It was the best rate of all options. In Week II U10 has rate 18% with (average for all is equal to 16.43%). Then the last period had dynamics of 40%, which was again the highest value. In linear and gaussian options, U10 is the most desirable option.

In reversed geometric variant the best is U14. In Week I value is equal to 73%, in Week II is equal to 20% in Week II 5% and Week IV 3%. Φ is equal to 0.4075, $\Phi+$ equal to 0.69 and $\Phi-$ equal to 0.2825. These are the best values, but slightly relative to the next three options: U18 (Week I 91%, Week II 7%, Week III 1%, Week IV 0%) and U16 (Week I 91%, Week II 4%, Week III 4%, Week IV 1%). Both options showed a similar relationship each week. In revered geometric variant U14 is the best option.

Performed analysis shows how evaluation of results from content updates is dependent on preferences of decision maker and strategical goals. Different strategies can be considered as successfully when main target is high coverage with.

6 Conclusions

The increased importance of digital environments and the role of virtual goods in online business models is creating the need for new analytical tools and methods. Phenomena typical to offline markets are also often observed within electronic systems and are related to product life cycles, consumer habituation and strategies of new products development. The presented research shows how content update strategies can affect user engagement and the life span of virtual products. High diversification of products within a single content drop influences user interest and the products' propagation within the system. Other factors analyzed include the number of elements within a single content drop and the dynamics of product usage after introduction. The proposed conceptual framework based on multi-criterial model makes an evaluation of the used strategies possible. Two main approaches were discussed based on implementation costs and usage dynamics. Future work will focus on a more detailed analysis of propagation within social networks and use behavior prediction based on earlier behaviors.

Acknowledgments. The work was supported by the National Science Centre of Poland, the decision no. 2017/27/B/HS4/01216.

References

1. Alha, K., Koskinen, E., Paavilainen, J., Hamari, J.: Critical acclaim and commercial success in mobile free-to-play games. In: DiGRA/FDG (2016)
2. Alha, K., Koskinen, E., Paavilainen, J., Hamari, J., Kinnunen, J.: Free-to-play games: professionals' perspectives. In: Proceedings of Nordic DiGRA 2014 (2014)
3. Bainbridge, W.S.: The scientific research potential of virtual worlds. Science **317**(5837), 472–476 (2007)
4. Bakshy, E., Karrer, B., Adamic, L.A.: Social influence and the diffusion of user-created content. In: Proceedings of the 10th ACM Conference on Electronic Commerce, pp. 325–334. ACM (2009)
5. Brans, J.P., Mareschal, B.: The PROMCALC & GAIA decision support system for multicriteria decision aid. Decis. Support Syst. **12**(4–5), 297–310 (1994)
6. Brans, J.P., Vincke, P., Mareschal, B.: How to select and how to rank projects: the Promethee method. Eur. J. Oper. Res. **24**(2), 228–238 (1986)
7. Castronova, E.: Virtual worlds: a first-hand account of market and society on the Cyberian frontier (2001)
8. Chiu, H.C., Hsieh, Y.C., Kao, Y.H., Lee, M.: The determinants of email receivers' disseminating behaviors on the internet. J. Advertising Res. **47**(4), 524–534 (2007)
9. Hamari, J.: Why do people buy virtual goods? Attitude toward virtual good purchases versus game enjoyment. Int. J. Inf. Manage. **35**(3), 299–308 (2015)

10. Hamari, J., Järvinen, A.: Building customer relationship through game mechanics in social games. In: Business, Technological, and Social Dimensions of Computer Games: Multidisciplinary Developments, pp. 348–365. IGI Global (2011)

11. Hamari, J., Keronen, L.: Why do people buy virtual goods? A literature review. In: 2016 49th Hawaii International Conference on System Sciences (HICSS), pp. 1358–1367. IEEE (2016)

12. Hamari, J., Lehdonvirta, V.: Game design as marketing: how game mechanics create demand for virtual goods (2010)

13. Hanner, N., Zarnekow, R.: Purchasing behavior in free to play games: concepts and empirical validation. In: 2015 48th Hawaii International Conference on System Sciences (HICSS), pp. 3326–3335. IEEE (2015)

14. Jankowski, J., Bródka, P., Hamari, J.: A picture is worth a thousand words: an empirical study on the influence of content visibility on diffusion processes within a virtual world. Behav. Inf. Technol. **35**(11), 926–945 (2016)

15. Jankowski, J., Michalski, R., Bródka, P.: A multilayer network dataset of interaction and influence spreading in a virtual world. Sci. Data **4**, 170144 (2017)

16. Jankowski, J., Michalski, R., Kazienko, P.: The multidimensional study of viral campaigns as branching processes. In: Aberer, K., Flache, A., Jager, W., Liu, L., Tang, J., Guéret, C. (eds.) SocInfo 2012. LNCS, vol. 7710, pp. 462–474. Springer, Heidelberg (2012). https://doi.org/10.1007/978-3-642-35386-4_34

17. Jankowski, J., Michalski, R., Kazienko, P.: Compensatory seeding in networks with varying avaliability of nodes. In: 2013 IEEE/ACM International Conference on Advances in Social Networks Analysis and Mining (ASONAM), pp. 1242–1249. IEEE (2013)

18. Kaplan, A.M., Haenlein, M.: Consumer use and business potential of virtual worlds: the case of second life. Int. J. Media Manage. **11**(3–4), 93–101 (2009)

19. Kwong, J.A.: Getting the goods on virtual items: a fresh look at transactions in multi-user online environments. Wm. Mitchell L. Rev. **37**, 1805 (2010)

20. Lu, H.-P., Wang, S.-m.: The role of internet addiction in online game loyalty: an exploratory study. Internet Res **18**(5), 499–519 (2008)

21. Mareschal, B., Brans, J.P., Vincke, P., et al.: Promethee: a new family of outranking methods in multicriteria analysis. ULB-Universite Libre de Bruxelles, Technical report (1984)

22. McPherson, M., Smith-Lovin, L., Cook, J.M.: Birds of a feather: homophily in social networks. Annu. Rev. Sociol. **27**(1), 415–444 (2001)

23. Wathieu, L.: Consumer habituation. Manage. Sci. **50**(5), 587–596 (2004)

24. Wohn, D.Y.: Spending real money: purchasing patterns of virtual goods in an online social game. In: Proceedings of the SIGCHI Conference on Human Factors in Computing Systems, pp. 3359–3368. ACM (2014)

Sensor Networks and Internet of Things

Analysis of the Error Rate in Electrometers for Smart Grid Metering

Josef Horalek and Vladimir Sobeslav[(✉)]

Faculty of Informatics and Management, University of Hradec Kralove,
Hradec Kralove, Czech Republic
{josef.horalek,vladimir.sobeslav}@uhk.cz

Abstract. The following article deals with the analysis of electrometers used for remote data acquisition in dependence on the protocols used and their error rate. The analysis is based on the real data from the specific sample of smart meters, which proportionally represent actual population on the device side. During the research, real communication of the smart grid control center with chosen meter types using different protocols for the given reading was surveyed. Types of the stored data were described, as well as the system of their collection and the analysis results from several viewpoints. The analysis focused on error rates of meters according to used transfer protocols VDEW, DLMS, and SCTM, and the measuring was underway continuously for a 4-month period. The results of the analysis, including detailed measuring process throughout the whole period, are important leads in AMR system development and its gradual integration into Smart Grid networks.

Keywords: Smart metering · Smart grid · Power distribution ·
Control systems · GPRS · SCTM · VDEW · DLMS

1 Introduction

Remote measuring in the distribution network is a highly topical issue, which is, among others, accentuated by current employment of remote metering in tens of thousands receiving points all over the Czech Republic. Despite significant technological advance and smart electrometer employment in Smart Grid networks pilot projects, data collection using automated electrometer measuring technology remains the standard, as can be seen in [1, 2] (AMR), which belongs to the group of Smart metering [3–5]. Not only because of the significant economic burden on the distributors, which a replacement of the current technologies with newer ones would entail, it is necessary to extensively analyze behavior, reliability, and security of the current solutions, and their maximum possible utilization in the Smart Grids as the authors in [6–9] have pointed out. AMR system itself only allows for sending measured data from the electrometers to the smart grid control center [10] and it is therefore only one-way communication, which, despite allowing for any reading and lowering the expenses on manual data reading, does not allow for remote settings of individual meters. For the data transfer between the meters and the data center, GPRS technology and GSM backup connection are used. An integral part of the data collection are the communication protocols

© Springer Nature Switzerland AG 2019
N. T. Nguyen et al. (Eds.): ACIIDS 2019, LNAI 11432, pp. 533–542, 2019.
https://doi.org/10.1007/978-3-030-14802-7_46

defining the transfer rules between the electrometers and the reading central. Among the longest used protocols belongs Serial Code Tele Metering (SCTM) protocol, which is used exclusively in meters with older manufacturing date. Details about this protocol can be found e.g. in [11]. The most frequently used are two protocols that are completely different. One is VDEW protocol, which is based on three-layer architecture EPA, which is described in [12], and the other is DLMS protocol, which has become the international standard working on the principle of server – client, with is specifics being described in [13]. Some electrometers can operate with both of the aforementioned protocols, and some with only one. This, therefore, points at the importance of the current solution analysis from the viewpoint of reliability and efficiency of reading during real operation, which can help to determine potential flaws of the technical equipment and the individual protocols in use. Potential high unreliability and error rate cannot only negatively affect the technical solution itself, but can also have impact on the economic side with excessive expenses spent on human resources and inefficient resource expenditure on the communications and technical equipment.

2   Chosen Meter Testing Process

The testing contained several meters including the types most frequently used for HV and EHV voltage in distribution for the Czech Republic. 60 locations with sufficient signal provided by the mobile network operator or failure-free data link were chosen. Data readings through the mobile network operator were realized using GSM or GSM/GPRS technology. Data collection was realized using Converge data center manufactured by Landis&Gyr. The data for the subsequent analysis were obtained from the data/reading center Converge version 3.9.7 by Landis&Gyr and ran by CEZ Distribuce – the national power distribution company. Individual meter series chosen for the testing deviated only in insignificant details (accuracy class differences, tariff number, use of the meter in primary/secondary measuring). ZxD (ZxD3xx/4xx), SL7000, FAG and FBC, LZQ, EKM647, and DC3 meters were used during the testing.

2.1   Data Reading Process

Data reading from the meters was performed using GSM/GPRS and PSTN technologies. In Converge data center, virtual electrometer was set up. This electrometer had pre-defined protocol type and number of data reading profiles. The data reading was carried out once a day from 1 September 2016 to 31 December 2016. During the data collection, individual telegram values (profile and register values) were downloaded, and at the same time, reading center time was compared with the meter time. If the times difference is under 2 s, automatic synchronization with the reading center time is performed. If the difference is above 2 s, synchronization is not performed and the times must be adjusted manually. Profile values had 15-min periods, and in order to prevent their overlapping and overwriting of saved profiles, value reading started at the latest data stamp. However, during register reading, a part of the data is overwritten (immediate values in particular), phase voltage can be changed, or maximum values from the previous periods can be overwritten. Along with reading the data, reading center and meter times were checked. In GSM mode, communication with the electrometer is similar to a regular phone call. In the reading center, so-called virtual electrometer with data number is set up, using its data number to call the corresponding electrometer. The communication via GPRS works similarly to network connection between two devices. Every location with remote reading via GPRS has allocated one SIM with an IP address, using which two-sided identification with the IP address registered in the reading central system Converge, which uses only static IP addresses during remote reading. Besides the IP addresses, data phone number is assigned to the communication as well in order to allow for choosing between GSM and GPRS modes. Multi-master communication is used during cascade connection, when all the electrometers connect using single data number. Every electrometer in the cascade has allocated one unique identification number (Password, Physical address), which usually is given electrometer's serial number, and therefore, the meters in the cascade cannot be interchanged. PSTN communication is a parallel to GSM. All tested devices also used SCTM protocol.

3 Meter Error Rate According to Protocols

In order to compare response times via the data center, testing focused on the rate of exceeding the maximum waiting time for a telegram. In case of failed attempt to collect data, time limit for this process runs out and the reading ends up in *timeout* state. One of the factors related to the failed readings was the inability of all of the used communication protocols to continue after an interrupted telegram. If an interruption during the reading occurred, new calling was initialized up to five attempts. During the testing period, this limit was exceeded only at rare occasions. Telegram readings usually succeeded at the first attempt. In Figs. 1 and 2, numbers of attempts of individual meters are depicted with reattempted readings standing out. Such readings usually followed an identical scenario. One or two telegrams were transferred during the first attempt and the rest during further attempts. If the data were read only partially or not at all, the readings ended up in error states (*Erroneous*, *Unreachable*, or *Not processed*). Reading of such data was usually finished the following day, without finding the cause of the reading failure. Nevertheless, there was no necessity for an intervention of a technician, which would be performed should the collection point remain unavailable for three consecutive days.

LZQJ, SL7000, and ZxD electrometers in average read the data on the second or third attempt maximum. Such devices use VDEW (LZQJ and ZxD) and DLMS (SL7000 and ZxD) protocols. There was a minimum number of connection interruptions and they occurred randomly. Despite having high time requirements for register reading, DLMS communication protocol usually keeps the established connection during communication with the center. On the contrary, VDEW protocol has on average a higher rate of interrupted telegrams during data reading. The highest number of instances exceeding the maximum waiting time for the telegram was registered by the DC3 meter. It reached the maximum attempt number five times, and in one occasion this limit was exceeded.

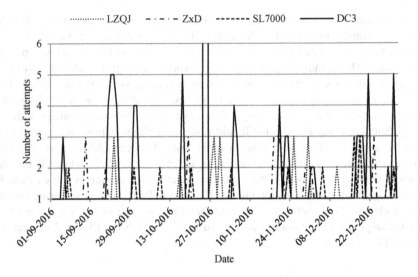

Fig. 1. Timeout of VDEW and DLMS gauges

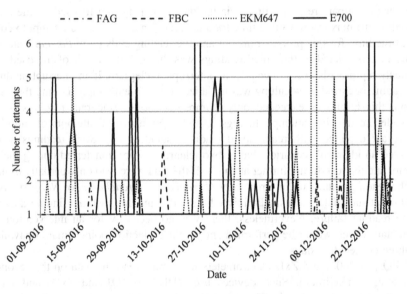

Fig. 2. Timeout of SCTM meters

Devices communicating using SCTM protocol had significantly higher rate of exceeding telegram time. That was not true for FAG and FBC encoders, which in turn have completed most readings successfully at first attempt. Both of the aforementioned meters are read via public telephone network (PSTN). Directly at the receiving point, both encoders are connected to *Stelco* device. A regular phone apparatus is connected to this device as well. According to the type of call, it distinguishes whether the incoming data is a regular phone call, or a data call initiated by the center. A disadvantage of *Stelco* device is irregular and unprompted blocking. Such issues, however, did not occur during the testing. E700 electrometer was among the devices with the highest number of timeouts. In two cases the limit of five attempts to connect was exceeded and reading ended up in complete *timeout*.

4 Specific Meter Type Error Rate

The analysis of the gathered data was evaluation of the meters, including their communication units, with analysis focusing on their overall error rate in connection to the protocols and communication technologies used. The analysis was performed using the following parameters: **without response** (as a connection error between the data center and the electrometer), **timeout** (error that occurs when maximum waiting time for the telegram is exceeded), and **check code** (defined as an error in checksum in the telegram).

LZQJ Error Rate
There were not any significant deviations in monitored values of this electrometer model. E.g. timeout had values between 10% and 11%, which manifested mainly in reading success, which was in average over 85% successfully read data in case of this electrometer. Telegram checksum (check code) values were negligible – mostly below 0.20%.

ZxD Error Rate
Compared to LZQJ, this electrometer model had a higher rate of exceeding the maximum telegram waiting time, as well as somewhat higher values of telegram checksum errors. The exception was telegram check code error values. During the everyday data collection in one electrometer, two quarter-hours were not read. This manifested mainly during the final monthly audit, when a two-sided check of the register and profile data is performed. A significant result confirmed in practice is that this was not the only instance of this problem and that it appears repeatedly in this device. The cause behind the data not being recorded at the given moment has not yet been unequivocally explained and further analysis is in progress. Error rate record in time is depicted in Fig. 3.

SL7000 Error Rate
This model has error rates similar to the previously mentioned electrometers. *Timeout* error rate oscillated around 12% and response around 3%. Telegram check code was under 0.15%.

FAG SL7000 Error Rate
Throughout the whole testing, FAG encoders had the lowest error rate during communication with the center and exceeding of the maximum telegram waiting time. *Timeout* did not exceed the threshold of 5% once and response usually oscillated around 2.5%. *Check code* entry always remained near zero and was very balanced.

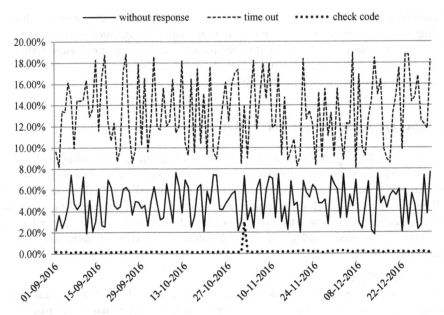

Fig. 3. ErrorZxD

FBC Error Rate

In certain aspects, FBC encoder is similar to FAG model. Depicted results are, however, different. Compared to the previous model, the values are somewhat higher, especially the entry of exceeding the maximum telegram waiting time from 27 September to 19 November 2016 reached values several times higher. During this period, one of the FBC encoders had an issue with time unit battery, which is why time synchronization occurred more often. The unfinished call error rate is also slightly increased. After exchange of the battery, everything returned back to normal state. Therefore, the increase in error rate can be attributed to the time unit's faulty battery. Error rate record in time is depicted in Fig. 4.

DC3 Error Rate

These electrometers had higher number of instances of having exceeded maximum telegram waiting time (*timeout*). Error rate oscillated between 10% and 30%. Connection between the data center and the electrometer was not established at all in one instance. Despite having far higher error rate, during data readings, DC3 electrometers did not belong among the meters with the highest time delay. *Check code* telegram percentage was also very low. This, however, has no practical effect on the overall error rate.

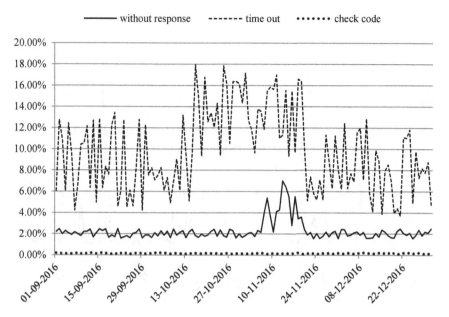

Fig. 4. Error FBC

EKM647 Error Rate

During the testing period, these meters have also had a higher error rate. Analogously to DC3, encoders had more problems with establishing connection. There was one unsuccessful reading here as well, when the limit of five connection attempts was exceeded and remote data collection ended up in complete *timeout*. *Timeout* error rate often exceeded the threshold of 20%. Another problem with EKM is regular monthly data accumulation, which, however, does not occur with all the meters of this model, which results in data inhomogeneity and the subsequent error in telegram checksum. As the calculations are on a monthly basis, this entry's graph has high error rate only in several sections. Error rate record in time is depicted in Fig. 4.

E700 Error Rate

E700 model had significant problems with the number of time limit expirations, as well as with higher number of interrupted connections that all the aforementioned models. In two instances the limit of five connection attempts was exceeded. Error rate when exceeding the telegram waiting time often exceeded the threshold of 30% and error during unsuccessful calling had rate over 20% in average. As for *check code* telegrams, values were negligible (Fig. 5).

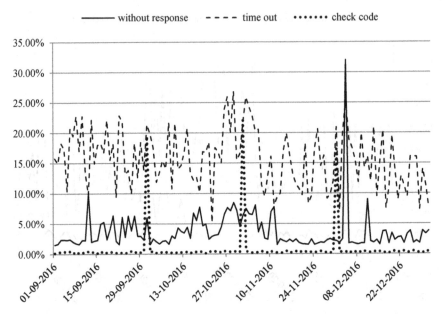

Fig. 5. Error EKM647

5 Conclusion

The aim of the research was to analyze error rate in technology used in remote electrometer reading using AMR system. During the analysis of real operation from 1 September to 31 December 2016, performance of selected AMR electrometers was tested and evaluated with the emphasis on reliability, reading efficiency, and communication in relation to used protocol and transferred data capacity. At the beginning of the research, a selection of theoretical criteria has been made based upon internal data from CEZ Distribuce, a.s. Afterwards, the practical testing between the meters selected in the theoretical part and the data (reading) center Converge manufactured by Landis&Gyr.

As for the encoder models FAG, FBC, and EKM, it can be noted that they no longer completely meet the selected criteria and that they should not be used to perform billing measuring. Despite being able to meet certain criteria for the time being, the fact that these electrometers are still employed at some collection or transfer points is caused only by the reluctance of the dealers to change the contract between the customers and the electricity distributors. Current technical solution is that these encoders gather data from the electrometers that meet all the requirements for the current billing meters. DC3 and E700 electrometers, as opposed to the aforementioned encoders, meet the current technical requirements, but they are technically obsolete and their further employment cannot be recommended, mainly because of E700's unbalanced performance and relatively high error rate. The aforementioned devices also do not fulfill Smart Metering requirements, which will most likely replace current remote data collection in the future.

The only devices that are compatible with Smart Metering are LZQJ, SL7000, and ZxD electrometers, which meet the basic standards. In general evaluation, it can be stated that all the electrometers have shown similar results during the practical testing. Individual differences were induced by the communication protocols used. During the testing, DLMS communication protocol was affected by the missing drivers in Converge center, which was made by the same manufacturer as ZxD electrometers, despite the fact that DLMS is considered to be the standard in the world, while its older counterpart, VDEW, is generally ignored and can be found only in older devices or electrometers manufactured and still supported by Landis&Gyr. During the testing, older meters with the exception in FAG had a higher error rate in response time and waiting time exceeding error which resulted in longer connection establishment and termination. According to the analysis, E700 electrometer had unequivocally the worst results in all of the performed tests. On the contrary, FAG encoder, which is considered to be rather older meter, reached the best results. Along with FBC model, FAG also has Load profile without registers at disposal, which is, nevertheless, fully in accord with the current law, according to which billing values are determined by the Load profile. It is paradoxical that only the register values go through any extensive testing in a metrological laboratory. Compared to a 15-min time interval profile, measuring progress cannot be monitored so well from the register values. However, it is principal that the testing results have not been influenced in any way.

The use of GPRS technology, which lacks certain qualities and advantages of remote data collection, remains an open question. During the analysis, problems with the availability of this service were encountered and reading had to be completed using a costly GSM technology. Therefore, it seems logical for the electrometer manufacturers to request for LTE support. Another variant being tested and considered by the distributors is the use of PLC technology (or BPL as well), which is taken into consideration in the area of Smart Grid, and respectively the whole Smart Metering. The chief reason for its use is the full control that energy distributors (such as CEZ power distributing company in Czech Republic) can assume over the most of the energy network as well as minimal expenses, as there will be no need to build a new network. One of the possibilities of replacing the whole current AMR technology is to merge it or integrate it into AMM (Advanced Metering Management), or AMI (Advanced Metering Infrastructure). Yet, in order to realize this, it is necessary to have at disposal extensive and long-term analyses of the usability of current widely diversified technical instruments for remote electrometer reading, as their mass replacement in short time is technically and financially impossible. For this reason, this real performance analysis was aimed at currently used technical means for remote electrometer reading focusing on the possibility of integrating them into AMM and AMI systems.

This work and the contribution were supported by a Specific Research Project, Faculty of Informatics and Management, University of Hradec Kralove, Czech Republic. We would like to thank to Lubos Mercl – Ph.D. candidate at Faculty of Informatics and Management, University of Hradec Kralove.

References

1. Tan, H.G.R., Lee, C.H., Mok, V.H.: Automatic power meter reading system using GSM network. In: 2007 International Power Engineering Conference (IPEC 2007), Singapore, 2007, pp. 465–469 (2007). ISSN: 1947-1262
2. Khalifa, T., Naik, K., Nayak, A.: A survey of communication protocols for automatic meter reading applications. IEEE Commun. Surv. Tutorials 13(2), 168–182 (2011). https://doi.org/10.1109/surv.2011.041110.00058. Accessed 16 Mar 2018, ISSN 1553-877X
3. Geetha, A., Jamuna, K.: Smart metering system. In: 2013 International Conference on Information Communication and Embedded Systems (ICICES), pp. 1047–1051. IEEE (2013). https://doi.org/10.1109/icices.2013.6508368. Accessed 16 Mar 2018. ISBN 978-1-4673-5788-3
4. Lin, H.-Y., Tzeng, W.-G., Shen, S.-T., Lin, B.-S.P.: A practical smart metering system supporting privacy preserving billing and load monitoring. In: Bao, F., Samarati, P., Zhou, J. (eds.) ACNS 2012. LNCS, vol. 7341, pp. 544–560. Springer, Heidelberg (2012). https://doi.org/10.1007/978-3-642-31284-7_32
5. Horalek, J., Sobeslav, V., Krejcar, O., Balik, L.: Communications and security aspects of smart grid networks design. In: Dregvaite, G., Damasevicius, R. (eds.) ICIST 2014. CCIS, vol. 465, pp. 35–46. Springer, Cham (2014). https://doi.org/10.1007/978-3-319-11958-8_4
6. Rial, A., Danezis, G.: Privacy-preserving smart metering. In: Proceedings of the 10th Annual ACM Workshop on Privacy in the Electronic Society - WPES 2011, p. 49. ACM Press, New York (2011). https://doi.org/10.1145/2046556.2046564. Accessed 16 Mar 2018. ISBN 9781450310024
7. Verbong, G.P.J., Beemsterboer, S., Sengers, F.: Smart grids or smart users? involving users in developing a low carbon electricity economy. Energy Pol. 52, 117–125 (2013). https://doi.org/10.1016/j.enpol.2012.05.003. Accessed 16 Mar 2018. ISSN 03014215
8. Roberts, B.P., Sandberg, C.: The role of energy storage in development of smart grids. Proc. IEEE 99(6), 1139–1144 (2011). https://doi.org/10.1109/jproc.2011.2116752. Accessed 16 Mar 2018, ISSN 0018-9219
9. Wolsink, M.: The research agenda on social acceptance of distributed generation in smart grids: renewable as common pool resources. Renew. Sustain. Energy Rev. 16(1), 822–835 (2012). https://doi.org/10.1016/j.rser.2011.09.006. Accessed 16 Mar 2018, ISSN 13640321
10. Horalek, J., Matyska, J., Sobeslav, V.: Communication protocols in substation automation and IEC 61850 based proposal. In: 2013 IEEE 14th International Symposium on Computational Intelligence and Informatics (CINTI), pp. 321–326. IEEE (2013). https://doi.org/10.1109/cinti.2013.6705214. Accessed 16 Mar 2018, ISBN 978-1-4799-0197-5
11. Manual LIAN 98(en). LIAN98. Erlangen: Mayr (2011). http://manuals.lian98.biz/doc.en/manual98en.html. Accessed 01 Jan 2017
12. Protocols. IPCOMM. Nuremberg: IPCOMM, c2004-2017. http://www.ipcomm.de/protocols_en.html. Accessed 02 Feb 2017
13. Microchip Makes Global Smart Meter Interoperability Easy with DLMS User Association Certified Stack for PIC® Microcontrollers. Business Wire. Lawson (2011). http://www.businesswire.com/news/home/20110719005651/en/Microchip-Global-Smart-Meter-Interoperability-Easy-DLMS. Accessed 08 Feb 2017

Multi-cell Based UWB Indoor Positioning System

JaeMin Hong[1], ShinHeon Kim[1], KyuJin Kim[2],
and ChongGun Kim[1(✉)]

[1] Department of Computer Engineering, Yeungnam University,
Gyeongsan 38541, Korea
hjm4606@naver.com, rubymix80@ynu.ac.kr, cgkim@yu.ac.kr
[2] College of Nursing, Kyungpook National University, Daegu 41944, Korea
kayjay6t@naver.com

Abstract. Indoor positioning is a big issue to trace or navigate for a moving object. Methods using Wi-Fi have been widely studied even though Wi-Fi has a lot of problems. UWB is an alternative indoor positioning method. But indoor positioning using UWB has lots of problems to solve. At an indoor environment like hospitals, many pathways like a maze is an important target for positioning. For monitoring an active moving object in some period, continuous tracking is needed. A cell system is proposed to overcome limitation of UWB modules on continuous indoor position tracking during some period. The proposed system can be used for analyzing staff workload at hospitals and for real time monitoring. Measuring nursing activities is critical steps to understand relationship between nursing practices, workload and patient safety.

Keywords: Cells · Wi-Fi · UWB · BLE · Indoor positioning ·
Workload monitoring

1 Introduction

Recently, as smart mobile devices are widely activated, location-based services (LBS) have attracted attention. With the success of LBS using Global-positioning system (GPS), service providers begin to pay attention to LBS for indoor environment. Various studies of indoor positioning is conducted, and indoor LBS are also increasing. Most widely known indoor positioning method is a method using Wi-Fi. However, indoor positioning using Wi-Fi has a lot of problems. [1] Therefore, UWB is getting attention. UWB is suitable for indoor positioning because of its high positioning of resolution and good performance on obstacle transmittance.

As a practical indoor positioning application, this system can be used for workload analysis.

Hospital staffs can be analyzed as one practical target to see their workload. Factors that influenced to nurses' workload are nurse-patient ratio, patients' average length of stay, patient turnover, job satisfaction and etc. [9]. With these criteria, nurses' workload is measured. However, these beliefs are largely unsupported by data. Most workload measurement is taken based on subjective criteria by nurses. With subjective tools, it is

© Springer Nature Switzerland AG 2019
N. T. Nguyen et al. (Eds.): ACIIDS 2019, LNAI 11432, pp. 543–554, 2019.
https://doi.org/10.1007/978-3-030-14802-7_47

limited to measure accurate nursing activities. Moreover, The Korean nurses' turnover rate was 32% in 2008 (Korea Health Industry Development Institute 2008). Many nurses are presumed to have a better condition job. Also, 58% of nurses remain as idle manpower (Ministry of Health & Welfare 2009). For reducing excessive workload and work more efficiently, alternative measurements should be needed.

In this paper, we propose an indoor tracing based on multi-cells positioning system, which can measure nurses' work activities using UWB module and monitor on the web client.

2   Related Researches

2.1   Indoor Positioning Methods

GPS (Global Positioning System) is the most widely used outdoor positioning system. But in the indoor environment, GPS cannot work properly. Therefore, many research of indoor positioning are in progress. For example, visible light communications [13], BLE signal strength [14], Wi-Fi signal strength and so on can be used.

Typical positioning methods include fingerprint and triangulation method. In this study, triangulation is used. Triangulation is a method of measuring the position from three nodes, which know the distance. The more accurate the distance value, the more accurate point can be calculated. [4] Methods such as AOA, TOA, and TDOA also use triangulation methods [5].

2.2   Indoor Positioning Based on Wi-Fi

Wi-Fi signal strength can be used for calculating indoor position using the fingerprint method and triangulation [11, 12].

The triangulation method is a geometric method of computing a coordinate by calculating each distance from three reference points. It is necessary to be able to obtain the correct distance from the wireless AP as reference point so that the correct position can be traced.

Fingerprint method is RSSI (Received Signal Strength Indication)-based, but it simply relies on the previously recorded data of the signal strength from several reference access points in the proper range. Storing this information in a database along with the known coordinates of the tracking device is clone in an offline phase. During the online position decision phase, the current RSSI vector at an unknown location is compared to those stored in the fingerprint DB and the closest matching position is returned as the estimated user location.

2.3   Ultra Wide Band (UWB)

It is usually called UWB, and it is one of the recently attracted communication methods. UWB is less impacted by obstacles and radio interference, and is more energy efficient. [3] Because of these advantages, it is suitable for indoor positioning and error is less than 1 m, which is better than other wireless communication methods. The distance is calculated using the Round Trip Time of the signal, not the RSSI value, and the position is measured by triangulation method with the distance.

Figure 1 shows the method for getting a distance between two UWB nodes.

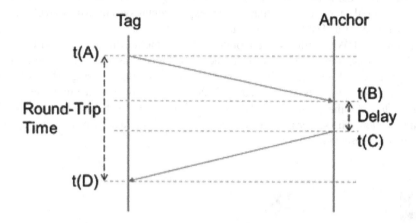

Distance = Signal Speed * (Round-Trip Time – Delay)/2

Fig. 1. Distance calculation using Round Trip Time

2.4   Moving Object Monitoring in Indoor Environment

Nursing activities and patterns of movement across a shift, department, each individual and the types of tasks that they engage which is poorly understood in hospital. With these data, it helps to estimate direct nursing activities (vital signs measurement, circulating system nursing, nervous system nursing, respiratory system, digestive system nursing, medications and etc.) and indirect activities (clinical environment management, clinical asset management, communication, information management and etc.) accurately to increase nursing practice efficiently [8].

Missed nursing care, nursing care left undone, unmet patient needs refers to necessary nursing care that is delayed partially completed, or missing in a clinical for any number of possible reasons is studied. Missed nursing care reduces the quality of nursing, causes accidents and inpatient complications, and ultimately results in negative patient outcomes such as dissatisfaction and readmission [9, 10].

It will give date on nurses' occupational physical activity which is associated with nurses' health. Sleep, diet and physical activity are the three pillars essential for good health that may combat the adverse effects of shift work. Researched with these factors can help prolonged healthy work environment for nurses [6, 7].

3 Design of Position Tracking System

3.1 Indoor Positioning on a UWB Cell

Figure 2 shows Communication concept of the indoor monitoring system on a cell. A UWB cell is composed four anchors. Moving tags can receive the distance information form only most near anchors. As an experimental system, MDEK-1001 kit of Decawave is used as UWB module is used.

Among the UWB modules, the mobile node that the user has is tag, and the fixed node in the indoor environment is called anchor. tag measures the distances to anchors using UWB signal and transmits distance data to Smartphone using BLE one-way communication. Smartphone application calculates coordinates using triangulation method, displays the position on the map, and transmits to the server using Wi-Fi communication. The server reserve coordinate values in the databases and displays the position via web client.

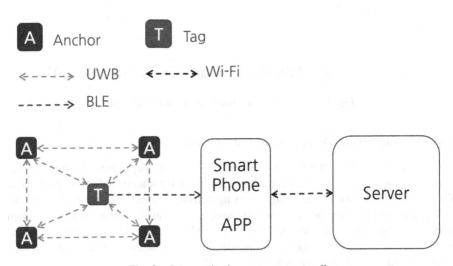

Fig. 2. Communication concept on a cell

The protocol of distance data that tag sends to the smartphone is shown in Fig. 3.

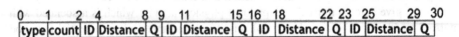

Fig. 3. The protocol of distance data

'Type' is protocol type number, and it has '1' in this protocol. 'Count' is a number of distance values, it has 1 ~ 4 value, and total protocol length can be varied from 9 to

30 bytes according to this values. 'ID' is Anchor's ID that tag measured distance. 'Distance' is the distance between the tag and the Anchor in millimeters. 'Q' is quality factor, and it typically has a value of 100.

Communication between tag and smartphone is one-way communication using BLE. Smartphone can receive distance data from multiple tags without pairing [2].

3.2 Interference Effects on UWB

When there are obstacles or people between tag and anchor, there is a little interference. Depending on the location of the person between the tag and the anchor, the error occurs differently. Figure 4 shows the experimental environment for interference by person.

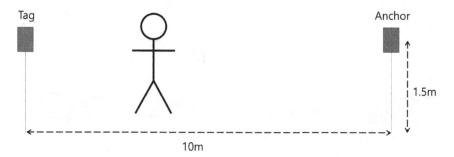

Fig. 4. Experimental environment for interference by person

A person moves from tag to anchor at 1 m intervals to check distance measurement error. Figure 5 shows the distance errors for each person's position.

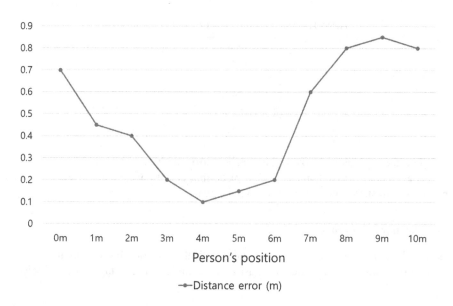

Fig. 5. Distance error for each person's position

The error is largest when a person stand in front of the anchor. And when person stand in the center, the error is the smallest.

In addition, a distance error also occurs along the antenna direction of UWB nodes. Figure 6 shows the experimental environment for measuring the distance error along the antenna direction. Experiments are conducted in 4 cases for each antenna direction.

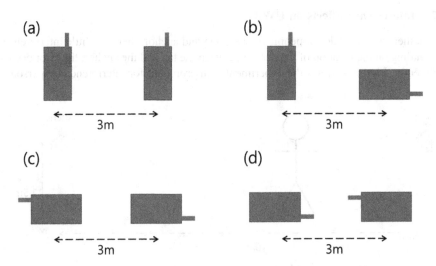

Fig. 6. 4 cases of experimental environment for each antenna direction

Table 1 shows distance error in 4 cases. (a) is best, and (c) is worst.

Table 1. Distance error in 4 cases of experiments

Case	Distance error
(a)	0.12 m
(b)	0.43 m
(c)	0.76 m
(d)	0.74 m

When the node is laid down, the distance error increases.

So, we installed an anchor at a high place that is not covered by person. And the anchors was installed upright without lying down.

3.3   Positioning Errors

The tag does not communicate sequentially with the anchors, but communicate randomly. For triangulation, 4 anchors must be communicated. The tag randomly selects

4 anchors among many anchors. The selected anchor has no specificity, and the anchor selection is changed at any time. Because of this, the position coordinate calculation also change, and a position error is occurred. Especially there is larger error in sensitive areas. The sensitive area is shown in Fig. 7.

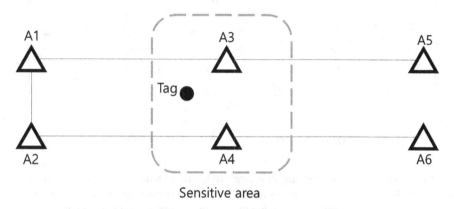

Fig. 7. Sensitive area in experimental environment

3.4   Introduction of Multi-cell

Figure 8 shows overview of multi-cell and Fig. 9 shows continuous cells in a building. A moving object must be traced by using continuous positioning. Without cell concept, sometimes selecting most nearest three anchors has problem. By introducing cells at the corridor in the building, correct nearest anchors can be selected.

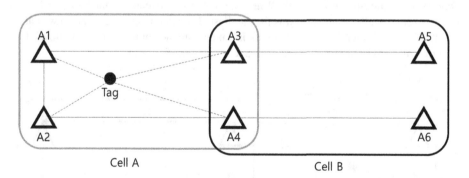

Fig. 8. Overview of multi-cell

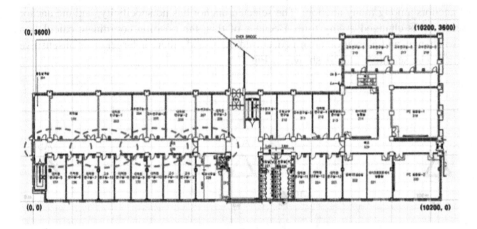

Fig. 9. Position metric of a floor of whole building

In the single cell, positioning in a large space is difficult. Since the effective measurement distance of this module is about 20 m, the positioning range is 20 m or less. However, in the multi-cell, it is possible to expand the positioning range by adding cells.

Even in the case of positioning in a large space, the multi-cell method is more advantageous. In the no cell method, the tag receives the anchor in the range at random. And it is difficult to calculate the position using the triangulation and the position error is large. However, in the multi-cell method, the tag receives anchor of a specific cell fixedly. Therefore, it is easy to calculate the position and the position error is small.

3.5　Position Error Reduction

Determining the Size of the Cell. When the distance between the anchors was 18 m, the number of anchor confusion was the smallest. The anchor spacing of 20 m was also valid, but the shorter interval of 18 m is taken into account. Figure 10 shows the experimental environment in which the anchor spacing is 18 m.

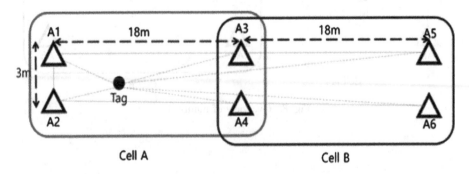

Fig. 10. Environment in which the anchor spacing is 18 m

Confirmation Effect of Cell. To reduce cell confusion, cell filtering is also used. Since the cell which the tag is located is recognized as a majority and the confused cell is recognized as a minority, the filter ignore minority cell. The filter stores consecutive frequent incoming cells in the buffer, ignoring occasional confused cells. Figure 11 shows concept of cell filtering.

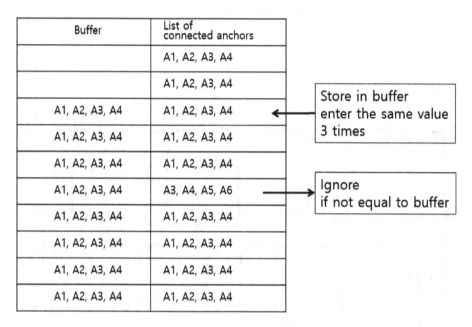

Fig. 11. Concept of cell filtering

Table 2 shows the error ranges depending on the presence or absence of filtering.

Table 2. Error range depending on the presence or absence of filtering

	Location of tag	Min error	Max error
No filter	Center of cell	0.1 m	0.4 m
	Edge of cell	0.5 m	0.7 m
Use filter	Center of cell	0.1 m	0.3 m
	Edge of cell	0.2 m	0.4 m

The effect of the filter is valid.

4 Operating Environment and Data Utilization

Figure 12 shows the overall flow of the system. Tag measures distances from each anchors using UWB signal, and smart device connected with tag calculates coordinates by triangulation method. Coordinates values are transmitted from the smart device to the server via Wi-Fi, and the server stores the position data in DB. Position data stored in the DB can be output on the map by monitoring device.

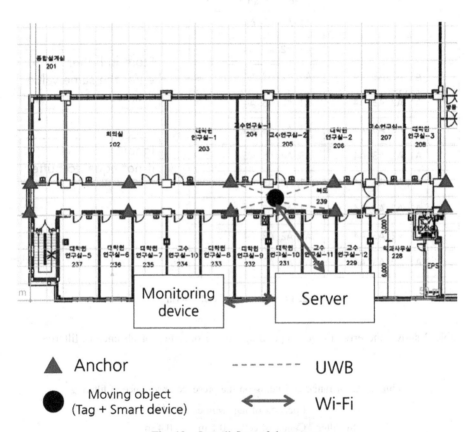

Fig. 12. Overall flow of the system

These position data can be used in various ways. It is used for real time tracing of multiple moving objects. It is also possible to monitoring of important objects. It is also used for analyzing history of movement. And it can ring an alarm when an user's abnormal activity is detected. And then these position data can be base of indoor self-driving.

5 Conclusions

In this paper, we designed an indoor position monitoring system using UWB module based on multi-cells in a building. Multi UWB cells are very suitable for position trace on several moving objects. A database is designed for deciding positions and tracing moving activities depend on time.

For the future research, more precise and applicable system has to be designed for special purpose building such as hospital and sanatorium. In addition, qualitative and quantitative improvement is needed.

Acknowledgement. This research was supported by The Leading Human Resource Training Program of Regional Neo industry through the National Research Foundation of Korea (NRF) funded by the Ministry of Science, ICT and future Planning, and the BK21 Plus Program funded by the Ministry of Education (MOE, Korea) and National Research Foundation of Korea (NRF).

References

1. Hong, J., Kim, K., Kim, C.: Comparison of indoor positioning system using Wi-Fi and UWB. In: Nguyen, N.T., Hoang, D.H., Hong, T.-P., Pham, H., Trawiński, B. (eds.) ACIIDS 2018. LNCS (LNAI), vol. 10751, pp. 623–632. Springer, Cham (2018). https://doi.org/10.1007/978-3-319-75417-8_58

2. Hong, J., Kim, S., Lee, J., Kim, K., Kim, C.: Indoor position monitoring system using UWB module. In: 2018 International Conference on Applied Mechanics, Mechatronics and Materials (2018)

3. Di Benedetto, M.-G.: UWB communication systems: a comprehensive over-view, vol. 5. Hindawi Publishing Corporation, New York (2006)

4. Dong, Q., Dargie, W.: Evaluation of the reliability of RSSI for indoor localization. In: 2012 International Conference on Wireless Communications in Unusual and Confined Areas (ICWCUCA), pp. 1–6. IEEE (2012)

5. Zhou, Y., Law, C.L., Chin, F.: Construction of local anchor map for indoor position measurement system. IEEE Trans. Instrum. Measur. **59**(7), 1986–1988 (2010)

6. Kim, K., Hong, J., Kim, C.: An alarm and response tracking system for the patient-centric perspective. J. Med. Imaging Health Inform. **8**, 190–195 (2018)

7. Ingram, S.J., Harmer, D., Quinlan, M.: UltraWideBand indoor positioning systems and their use in emergencies. In: Position Location and Navigation Symposium, PLANS 2004, pp. 706–715. IEEE (2004)

8. Saksvik-Lehouillier, I., et al.: Individual, situational and lifestyle factors related to shift work tolerance among nurses who are new to and experienced in night work. J. Adv. Nurs. **69**(5), 1136–1146 (2013)

9. Duffield, C., et al.: Nursing staffing, nursing workload, the work environment and patient outcomes. Appl. Nurs. Res. **24**, 244–255 (2011)

10. Myny, D., et al.: Determining a set of measurable and relevant factors affecting nursing workload in the acute care hospital setting: a cross-sectional study. Int. J. Nurs. Stud. **49**, 427–436 (2012)

11. OnkarPathak, P.P., Palkar, R., Tawari, M.: Wi-Fi indoor positioning system based on RSSI measurements from Wi-Fi access points –a trilateration approach. Int. J. Sci. Eng. Res. **5**(4), 1234–1238 (2014)
12. Vasisht, D., Kumar, S., Katabi, D.: Decimeter-level localization with a single WiFi access point. In: 13th USENIX Symposium on Networked Systems Design and Implementation (NSDI 2016) (2016)
13. Kong, I.-Y., Kim, H.-J.: Experiments and its analysis on the Identification of indoor location by visible light communication using LED lights. J. Korea Inst. Inf. Commun. Eng. **15**(5), 1045–1052 (2011)
14. Bluetooth Indoor Localization System: Gunter FISCHER, Burkhart DIETRICH, Frank WINKLER

An Improved Resampling Scheme for Particle Filtering in Inertial Navigation System

Wan Mohd Yaakob Wan Bejuri[1,3]([⊠]), Mohd Murtadha Mohamad[1],
Raja Zahilah Raja Mohd Radzi[1], and Sheikh Hussain Shaikh Salleh[2]

[1] School of Computing, Universiti Teknologi Malaysia, Johor Bahru, Malaysia
wanmohdyaakob@gmail.com
[2] Next Tech PLT., Skudai, Malaysia
[3] Faculty of Information and Communication Technology,
Universiti Teknikal Malaysia Melaka, Durian Tunggal, Malaysia

Abstract. The particle filter provides numerical approximation to the nonlinear filtering problem in inertial navigation system. In the heterogeneous environment, reliable state estimation is the critical issue. The state estimation will increase the positioning error in the overall system. To address such problem, the sequential implementation resampling (SIR) considers cause and environment for every specific resampling task decision in particle filtering. However, by only considering the cause and environment in a specific situation, SIR cannot generate reliable state estimation during their process. This paper proposes an improved resampling scheme to particle filtering for different sample impoverishment environment. Adaptations relating to noise measurement and number of particles need to be made to the resampling scheme to make the resampling more intelligent, reliable and robust. Simulation results show that proposed resampling scheme achieved improved performance in term of positioning error in inertial navigation system In conclusion, the proposed scheme of sequential implementation resampling proves to be valuable solution for different sample impoverishment environment.

Keywords: Resampling · Particle filter ·
Inertial navigation system and indoor positioning

1 Introduction

Nowadays, current technologies gain ground (such as: [1]). This includes also location determination technologies [2]. The knowledge of location position in a shopping mall is a common requirement for many people during shopping activities [3–5]. Considerable research and development has taken place over the recent years with regards to Location-Based Services (LBS), which can now be supplemented and expanded with the help of ubiquitous methods, and perhaps even replaced in the future.

The retrieved positioning data however must be analysed, manipulated, and interpreted before it can be used for computer processing purposes. There are many examples of area that requires this task including; target tracking [6–10], pollution monitoring [11, 12], communications [13, 14], audio engineering [15, 16], finance [17, 18] and

© Springer Nature Switzerland AG 2019
N. T. Nguyen et al. (Eds.): ACIIDS 2019, LNAI 11432, pp. 555–563, 2019.
https://doi.org/10.1007/978-3-030-14802-7_48

econometrics [19, 20] to list but a few (more examples can be found in [21–23]). It is, however, often the case that by the time these data are observed or obtained, they are 'contaminated' by the presence of 'noise', which makes it difficult to analyse the true data and retrieve relevant information. This raises important questions about the inferences and conclusions that can be drawn from the data. The practice of particle filtering attempts to understand and answer these questions. The main aim of this method is to smooth or approximate data or particles to allow the data to be smoothly read by the end user. Without smoothing method, the filtering output might be inconsistent and difficult to recognized. A typical problem arising in particle filtering is the different kinds of measurement noise and interference that inherent in the environment. Thus, the particle filter often uses a resampling algorithm to suppress noise interference. In the ubiquitous computing environment, the particle filter is required to process data or particles from different kinds (in term of specification) of sensor or real time platform. This can lead to the corrupted data or particles being retrieved, and this interruption effects the particles' value (for example; particle weight or particle state) or numbers (particle sample size). However, using resampling algorithm solution in particle filter just considering the environment in a specific case, which fails to generate reliable positioning accuracy. The work in this paper is aimed at reducing the positioning error during particle filtering process as it may interfere navigation process in mobile devices. The structure of the paper is as follows. Section 2 presents the basic concepts related to location-aware shopping assistance and puts forward the proposed method. Section 3 presents the experiment setup and the details of the preliminary results. In conclusion, a discussion and the future direction of the project are provided in Sect. 5.

2 Particle Filter in Inertial Navigation System

This section discusses the fundamental of particle filtering in inertial navigation system. The concept of inertial navigation system (refer Fig. 1 for fundamental inertial navigation system architecture) is regards to positioning determination across all environments [24–30]. Usually, it requires multi-sensor approach, augmenting standalone positioning with other signals, motion sensors, and environmental features [31–33]. This may be enhanced by using three-dimensional (3D) mapping, context awareness and cooperation between users. The navigation process of the inertial navigation system will be determined and processed by CPU by lookup the recorded positioning data which stored in database. In the system, the particle filter is used to optimize the inertial data sensed by the sensor, to help the system to determine positioning and navigation information correctly. Initially, the data (or particle) that obtained by sensor will be processed by particle filter (in CPU). The process of particle filtering will be executed by two (2) main components; which are sequential import sampling (SIS) and resampling. First, the obtained particles will be propagated and computed (weight) by SIS. In this part, the new particle set $\tilde{x}_{0:t}^{(i)}$ will be assigned according to time; the importance weight will be evaluated; and the importance weight will be normalised. Then, the particles will be processed by resampling component. Here, the particles will be replaced with new particles according to the importance weight which is in detail, the low weighted particles will be discarded and replaced with new particles. Finally, the process of resampling will move on to the SIS, and this process continues repeatedly until the particle process reaches the determined time. In the following section, we will discuss about improved resampling scheme proposed in this paper.

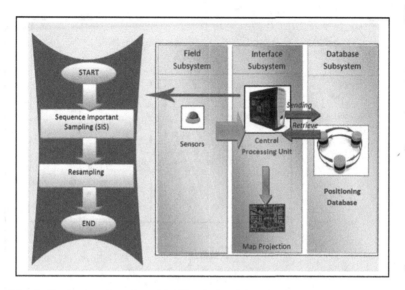

Fig. 1. Implementation of particle filter in inertial navigation system architecture

3   Adaptive Sequential Implementation Resampling

Previous section has discussed the basic concept of resampling scheme in inertial positioning architecture system. In this section, we will discuss the proposed resampling scheme. To restate, our aim is to propose a new sequential implementation resampling, based on the adaptation of noise measurement and particle number. This proposed scheme handles various sample impoverishment phenomenon and unbalance resampling's memory. The proposed resampling scheme can be seen in Fig. 2. In our proposed method, there are three (3) types of situation that will be focused, which are; the situation during inertial-based mobile IPS system using internal inertial sensor, using external inertial sensor and real time application.

The justification is, these situations are causing sample impoverishment for positioning determination. Figure 2 shows the block diagram of adaptive noise and particle based special strategies resampling. Initially, the system will receive the inertial data signal from sensor. At first, the noise will be measured. If the noise measurement is high, the algorithm will perform roughening resampling. If the noise measurement is low, the particle number will be measured. Finally, the algorithm will perform modified resampling (if the particle is high) or variable resampling (if the particle is low). Next section will discuss the special strategies resampling algorithm that will be employed and manipulated in our proposed method.

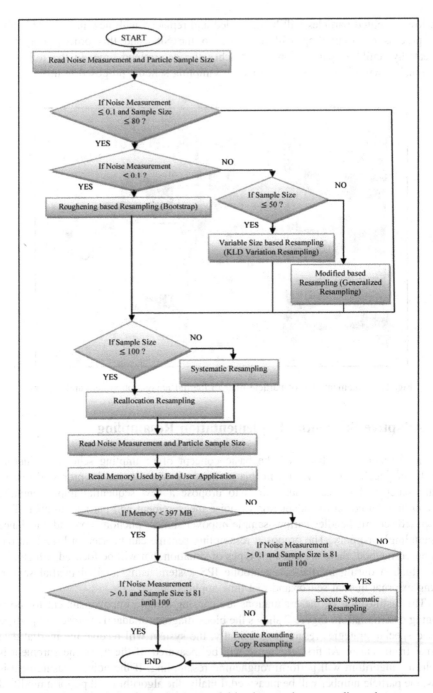

Fig. 2. Flowchart of adaptive sequential implementation resampling scheme

4 Performance Evaluation

Previous section has discussed about the adaptive sequential implementation resampling. In this section, the performance result of proposed method implementation in case study will be discussed. The case study is referring to car positioning application [34]. Basically, the digital road map is used to constrain the possible positions, where a dead reckoning of wheel speeds is the main external input to the algorithm. By matching the driven path to a road map, a vague initial position (order of kilometers) can be improved to a meter unit accuracy. To perform this, the experiment has been conducted by using MATLAB simulation. The data initially obtained in the field before simulated in the MATLAB environment (see Fig. 3 for view of during data collection process). The data of car speed has been obtained by using wheel speed sensors in ABS which is available as standard components in the test car (Volvo V40). From this, yaw rate and speed information are computed, as described in [22].

Fig. 3. View of IMU data collection task

Therefore, the velocity vector is considered available as an input signal, and the motion model is thus appropriate. The initial position is either marked by the driver or given from a different system, where crude position information is available today [16] or GPS. The initial area should cover an area not extending more than a couple of kilometers to limit the number of particles to a realizable number. With infinite memory and computation time, no initialization would be necessary. The car positioning with map matching has been implemented in a car, and the particle filter runs in real time with sampling frequency 2 Hz on a modern laptop with a commercial digital road map.

The driven path consists of a number of 90 turns. Initially, the particles are spread uniformly over all admissible positions, that is, on the roads, covering an area of about 1 km. This principle can be used as a supplement to, or even replacement for, a global positioning system (GPS). In this case study the proposed method is used to operate as resampling in the particle filter. The result of predicted value can be seen in Figs. 4 and 5. Generally, it shown that the value prediction is a near to true value, but after 50 time

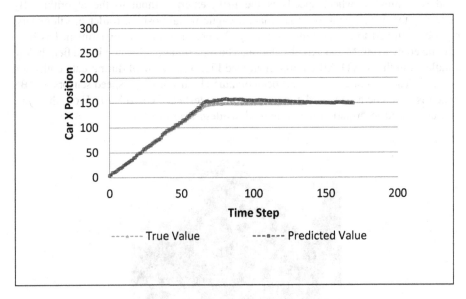

Fig. 4. Predicted value of X position

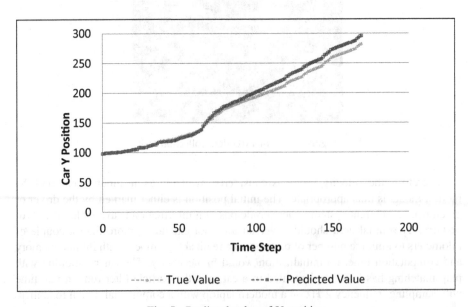

Fig. 5. Predicted value of Y position

step, the predicted value seems far more than true value. This is because the nature of dead reckoning based positioning, that generating accumulated error. As a final result, the proposed method can achieve RMSE of 8.8857 m.

5 Conclusion and Future Works

This paper discussed the problem and solutions regarding the development of a resampling scheme for particle filter, specifically in terms of its unreliable state estimation, since this may lead to large positioning error in inertial navigation system. To perform resampling scheme for particle filter, the improvement of sequential implementation resampling scheme in particle filter during inertial-based location determination is carried out. Towards this end, an adaptation of sequential implementation resampling scheme for particle filtering has been proposed. The final result shows that the large positioning error by achieving RMSE of 8.8857 m. For future works, a continuation of experiments to ascertain the extent to which the approach can be more ubiquitous in other environments will be carried out, using these results combined with other mobile sensors, e.g. cameras. This part will undoubtedly be undertaken by the next researcher.

References

1. Othmane, L.B., Weffers, H., Ranchal, R., Angin, P., Bhargava, B., Mohamad, M.M.: A case for societal digital security culture. In: Janczewski, L.J., Wolfe, H.B., Shenoi, S. (eds.) SEC 2013. IAICT, vol. 405, pp. 391–404. Springer, Heidelberg (2013). https://doi.org/10.1007/978-3-642-39218-4_29
2. Bejuri, W.M.Y.W., Mohamad, M.M., Sapri, M.: Ubiquitous positioning: a taxonomy for location determination on mobile navigation system. Signal Image Process. Int. J. SIPIJ 2(1), 24–34 (2011)
3. Schougaard, K.R., Grønbæk, K., Scharling, T.: Indoor pedestrian navigation based on hybrid route planning and location modeling. In: Kay, J., Lukowicz, P., Tokuda, H., Olivier, P., Krüger, A. (eds.) Pervasive 2012. LNCS, vol. 7319, pp. 289–306. Springer, Heidelberg (2012). https://doi.org/10.1007/978-3-642-31205-2_18
4. Bejuri, W.M.Y.W., Mohamad, M.M., Radzi, R.Z.R.M., Salleh, M., Yusof, A.F.: A proposal of location aware shopping assistance using memory-based resampling. In: Kim, K.J., Joukov, N. (eds.) ICMWT 2017. LNEE, vol. 425, pp. 482–486. Springer, Singapore (2018). https://doi.org/10.1007/978-981-10-5281-1_54
5. Bejuri, W.M.Y.W., Mohamad, M.M., Sapri, M., Rosly, M.A.: Performance evaluation of mobile U-navigation based on GPS/WLAN hybridization. J. Converg. Inf. Technol. JCIT 7(12), 235–246 (2012)
6. Ahmadi, K., Salari, E.: Social-spider optimised particle filtering for tracking of targets with discontinuous measurement data. IET Comput. Vis. 11(3), 246–254 (2017)
7. Wang, H., Nguang, S.K.: Multi-target video tracking based on improved data association and mixed Kalman/H∞ filtering. IEEE Sens. J. 16(21), 7693–7704 (2016)

562   W. M. Y. Wan Bejuri et al.

8. Bejuri, W.M.Y.W., Saidin, W.M.N.W.M., Bin Mohamad, M.M., Sapri, M., Lim, K.S.: Ubiquitous positioning: integrated GPS/Wireless LAN positioning for wheelchair navigation system. In: Selamat, A., Nguyen, N.T., Haron, H. (eds.) ACIIDS 2013. LNCS (LNAI), vol. 7802, pp. 394–403. Springer, Heidelberg (2013). https://doi.org/10.1007/978-3-642-36546-1_41

9. Bejuri, W.M.Y.W., Mohamad, M.M., Zahilah, R.: Optimization of Rao-Blackwellized particle filter in activity pedestrian simultaneously localization and mapping (SLAM): an initial proposal. Int. J. Secur. Appl. 9(11), 377–390 (2015)

10. Bejuri, W.M.Y.W., Mohamad, M.M., Raja Mohd Radzi, R.Z.: Optimisation of emergency rescue location (ERL) using KLD resampling: an initial proposal. Int. J. U- E- Serv. Sci. Technol. 9(2), 249–262 (2016)

11. Metia, S., Ha, Q.P., Duc, H.N., Azzi, M.: Estimation of power plant emissions with unscented Kalman filter. IEEE J. Sel. Top. Appl. Earth Obs. Remote Sens., 1–10 (2018)

12. Metia, S., Oduro, S.D., Duc, H.N., Ha, Q.: Inverse air-pollutant emission and prediction using extended fractional Kalman filtering. IEEE J. Sel. Top. Appl. Earth Obs. Remote Sens. 9(5), 2051–2063 (2016)

13. Yao, S., Wang, Y., Niu, B.: An efficient cascaded filtering retrieval method for big audio data. IEEE Trans. Multimed. 17(9), 1450–1459 (2015)

14. Zhou, J., Gu, G., Chen, X.: Distributed Kalman filtering over wireless sensor networks in the presence of data packet drops. IEEE Trans. Autom. Control, 1 (2018)

15. Ding, Y., Huang, J., Pelachaud, C.: Audio-driven laughter behavior controller. IEEE Trans. Affect. Comput. 8(4), 546–558 (2017)

16. Muñoz-Romero, S., Arenas-García, J., Gómez-Verdejo, V.: Nonnegative OPLS for supervised design of filter banks: application to image and audio feature extraction. IEEE Trans. Multimed. 20(7), 1751–1766 (2018)

17. Chauhan, G.S., Huseynov, F.: Corporate financing and target behavior: new tests and evidence. J. Corp. Finance 48, 840–856 (2018)

18. Finlay, W., Marshall, A., McColgan, P.: Financing, fire sales, and the stockholder wealth effects of asset divestiture announcements. J. Corp. Finance 50, 323–348 (2018)

19. González-Fernández, M., González-Velasco, C.: Can Google econometrics predict unemployment? Evidence from Spain. Econ. Lett. 170, 42–45 (2018)

20. Wilcox, B.A., Hamano, F.: Kalman's expanding influence in the econometrics discipline. IFAC-Pap. 50(1), 637–644 (2017)

21. Cappe, O., Godsill, S.J., Moulines, E.: An overview of existing methods and recent advances in sequential Monte Carlo. Proc. IEEE 95(5), 899–924 (2007)

22. Doucet, A., de Freitas, N., Gordon, N.: An introduction to sequential Monte Carlo methods. In: Doucet, A., de Freitas, N., Gordon, N. (eds.) Sequential Monte Carlo Methods in Practice. Springer, New York (2001). https://doi.org/10.1007/978-1-4757-3437-9_1

23. Doucet, A., Godsill, S., Andrieu, C.: On sequential Monte Carlo sampling methods for Bayesian filtering. Stat. Comput. 10(3), 197–208 (2000)

24. Jung, S.-H., Lee, G., Han, D.: Methods and tools to construct a global indoor positioning system. IEEE Trans. Syst. Man Cybern. Syst. (2017)

25. Li, X., Wei, D., Lai, Q., Xu, Y., Yuan, H.: Smartphone-based integrated PDR/GPS/Bluetooth pedestrian location. Adv. Space Res. 59(3), 877–887 (2017)

26. Li, Y., Zhuang, Y., Zhang, P., Lan, H., Niu, X., El-Sheimy, N.: An improved inertial/wifi/magnetic fusion structure for indoor navigation. Inf. Fusion 34, 101–119 (2017)

27. Liu, Q., Qiu, J., Chen, Y.: Research and development of indoor positioning. China Commun. 13(Supplement 2), 67–79 (2016)

28. Bejuri, W.M.Y.W., Mohamad, M.M.: Wireless LAN/FM radio-based robust mobile indoor positioning: an initial outcome. Int. J. Softw. Eng. Its Appl. 8(2), 313–324 (2014)

29. Bejuri, W.M.Y.W., Mohamad, M.M., Sapri, M., Rahim, M.S.M., Chaudry, J.A.: Performance evaluation of spatial correlation-based feature detection and matching for automated wheelchair navigation system. Int. J. Intell. Transp. Syst. Res. **12**(1), 9–19 (2014)
30. Bejuri, W.M.Y.W., Mohamad, M.M.: Performance analysis of grey-world-based feature detection and matching for mobile positioning systems. Sens. Imaging **15**(1), 1–24 (2014)
31. Ren, H., Chai, P., Zhang, Y., Xu, D., Xu, T., Li, X.: Semiautomatic indoor positioning and navigation with mobile devices. Ann. GIS, 1–12 (2017)
32. Albert, M.V., Shparii, I., Zhao, X.: The applicability of inertial motion sensors for locomotion and posture. In: Barbieri, F.A., Vitório, R. (eds.) Locomotion and Posture in Older Adults, pp. 417–426. Springer, Cham (2017). https://doi.org/10.1007/978-3-319-48980-3_26
33. Retscher, G., Roth, F.: Wi-Fi fingerprinting with reduced signal strength observations from long-time measurements. In: Gartner, G., Huang, H. (eds.) Progress in Location-Based Services 2016. LNGC, pp. 3–25. Springer, Cham (2017). https://doi.org/10.1007/978-3-319-47289-8_1
34. Gustafsson, F., et al.: Particle filters for positioning, navigation, and tracking. IEEE Trans. Signal Process. **50**(2), 425–437 (2002)

Analysis of Image, Video, Movements and Brain Intelligence in Life Sciences

Quaternion Watershed Transform in Segmentation of Motion Capture Data

Adam Świtoński[1]([✉]), Agnieszka Michalczuk[1], Henryk Josiński[1],
and Konrad Wojciechowski[2]

[1] Institute of Informatics, Silesian University of Technology,
Akademicka 16, 44-100 Gliwice, Poland
{adam.switonski,agnieszka.michalczuk,henryk.josinski}@polsl.pl
[2] Research and Development Center, Polish-Japanese Academy of Information
Technology, Aleja Legionow 2, 41-902 Bytom, Poland
kwojciechowski@pja.edu.pl
http://www.polsl.pl, http://bytom.pja.edu.pl

Abstract. The novel approach for segmentation of motion capture data is proposed. It utilizes hierarchical watershed transform for time series containing unit quaternions which describe rotations of human skeleton joints as well as their angular velocities. To approximate gradient magnitudes of subsequent time instants, aggregated geodesic distance on hypersphere S^3 for preceding and following quaternions is computed. The introduced segmentation method was applied to gait analysis. The highly precise motion capture data, registered in a human motion lab (HML), were used. A fully automatic watershed transform with detection of catchment basins as well as a marker controlled one were investigated. The obtained results are promising. By selecting proper hierarchy of a segmentation or by specifying adequate markers, it is even feasible to divide gait cycle into consecutive steps. The segmentation can be improved in respect to a considered problem, if only selected joints are taken into account by watershed transform.

Keywords: Motion capture · Watershed transform ·
Motion segmentation · Gait analysis · Unit quaternions

1 Introduction

The most precise measurements of human motion are obtained by motion capture systems. During an acquisition markers attached on human body are tracked by set of calibrated cameras. Their 3D coordinates are reconstructed, which allows to determine orientations of bone segments. Thus, skeleton based motion capture data are represented by time series of poses described by joints rotations as well as global location and orientation of a skeleton.

In baseline approach of motion analysis Euler angles are utilized for coding 3D rotations of human joints. They contain data of three successive basic

© Springer Nature Switzerland AG 2019
N. T. Nguyen et al. (Eds.): ACIIDS 2019, LNAI 11432, pp. 567–578, 2019.
https://doi.org/10.1007/978-3-030-14802-7_49

rotations performed around axes of local coordinate systems. Euler angles are intuitive, especially in case of medical diagnosis. They can correspond to flexion/extention, abduction/adduction and rotation of bone segments which are commonly analyzed in medical assessment of motion data. However, processing of rotations described by three clearly uncorrelated values is also troublesome. It can lead to improper results, because there is much more difficult to take into account mutual relationships between basic rotations in such the case. Furthermore, Euler angles suffer gimbal lock problem.

It justifies application of unit quaternions which are broadly used in computer graphics, also in motion processing and analysis. Quaternions provide compact angle-axis representation of 3D rotations performed by α angle about fixed axis described by vector $\boldsymbol{u} = (u_1, u_2, u_3)$.

$$q = cos\left(\frac{\alpha}{2}\right) + (u_1 \cdot i + u_2 \cdot j + u_3 \cdot k) \cdot sin\left(\frac{\alpha}{2}\right) \tag{1}$$

There is an algebra defined for quaternions with selected operators corresponding to transformations of 3D rotations. Moreover, they are efficient in integration over time. However, the most important advantage of unit quaternions over Euler angles is related to the simultaneous and coherent processing of complete rotational data. It can have an impact on obtained results.

Segmentation is process of data division into disjoint multiply partitions which collectively cover them. It simplifies the representation – data became easier for interpretation, further processing and recognition. In case of motion data segmentation can be useful in detection of short lasting specific movements or phases as well as in generic feature extraction for the sake of classification problems.

The current paper concerns issues of segmentation of motion data. On the basis of our previous experiences on gait identification [13,14] a new method which utilizes unit quaternions and watershed transform is introduced.

The following sections are devoted to: a short review of state of the arts methods focused on segmentation of motion data, explanation of different variants of watershed transform used in computer vision as well as descriptions of the proposed methods and obtained preliminary results.

2   Related Work

There are numerous studies on segmentation of motion capture data. Barbič et al. [4] introduced three highly cited approaches. In the first one, on the basis of selected number of principal components reconstruction of consecutive motion frames is carried on. Boundaries between detected segments are denoted by time instants of sudden increase of reconstruction error. In another variant Probabilistic Principal Components Analysis is used to model the first K frames of time series. It allows to assess fitness of frames K+1 through K+T to estimated Gaussian distribution. Mahalanobis distance is employed to distinguish new partitions. In the third approach Gaussian Mixture Models are utilized to cluster entire sequence.

Time series are divided into time instants of consecutive frames belonging to different Gaussian distributions. In all methods unit quaternion are employed to represent rotational data. However in fact they only form four dimensional vectors. Thus, it is difficult to point any advantages over the case of Euler angle usage.

Technique proposed by Bouchard and Badler [8] employs neural networks trained on the basis manual segmentation and input attributes corresponding to specified kinematic features of selected markers. To reduce oversegmentation, temporary aggregation of detected boundaries in local windows of 25 s width is carried on. The peaks are determined and they point proper segments. Supervised machine learning is also applied by Arikan et al. [2] – poses represented by coordinate vector are classified as belonging to selected simple motions e.g. running, walking. Support Vector Machines are used. The coordinate vector contains joints positions in one second local windows centered in reference to the classified pose.

The method which utilizes unit quaternions and unsupervised learning is proposed by Wu et al. [16]. Joint rotations are transformed into tangent spaces on the basis of logarithm operator. It allows to calculate similarity matrix by using baseline Euclidean metric. Further, poses are clustered by the normalized cut model and the weighted kernel k-means. In the postprocessing filtering which removes some isolated short segments is carried on. Finally, so called category strings containing sequences of subsequent segments are formed and analyzed to reduce number of obtained partitions.

Xie et al. [17] transforms motion capture data into low dimensional embedding on the basis of Laplacian Eigenmap algorithm. The proper segmentation is realized by extreme tracking of amplitudes of characteristic curve. Quite similar method is presented by Yang et al. [18]. In successive steps dimensionality of pose space is reduced by Principal Component Analysis, velocity vectors are calculated and clustered according to cosine distance. In recently proposed approach Arn et al. [3] employ generalized curvature analysis for curves residing in n-dimensional pose space.

Lin and Kulic [12] for initial partitioning analyze velocity features such as velocity peaks and zero velocity crossings. In the second stage Hidden Markov Models are employed to precisely modify segments from the initially identified candidates. In another segmentation approach Yang et al. [19] utilize Fisher's optimal partitioning algorithm for multidimensional time series transformed into matrix form.

There are also examples of proposed segmentation methods targeted to motion data registered by inertial measurement units [1,11].

Summarizing, the problem of motion data segmentation is broadly studied in the literature, there are plenty of techniques introduced. The most of the applications are related to detection of selected activities in long lasting recordings. However, there is no one general method applicable to all problems. Moreover, some of them require training data and other are only semiautomatic. Furthermore, there is no study strictly focused on gait analysis. Therefore, conducting research on gait segmentation is justified. Still new techniques and applications can be proposed.

3   Watershed Transform

The watershed transform originally introduced by Bouchard and Badler [5] has still numerous applications in 2D and 3D image segmentation [9,10,15]. The transform determines boundaries between segments in locations with the highest gradient magnitude. The markers of successive segments refer to homogeneous regions which are pointed by local minima of the gradient magnitude. It is analogous to flooding of 3D surface – barriers are built in places where water of different catchment basins is met.

In the implementation of watershed transform priority queues are utilized [7]. The priorities relate to gradient magnitudes of image pixels. In the initial step markers are extracted, labeled and inserted to the queue. In a baseline approach markers correspond to regional minima of the gradient. Further, the process is iterated until the queue is not empty. In the single iteration, the first element x from the queue is extracted and every, so far unlabelled neighbor of x is assigned with the same label as x and inserted to the queue.

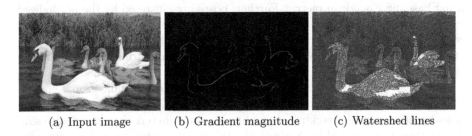

(a) Input image   (b) Gradient magnitude   (c) Watershed lines

Fig. 1. Watershed segmentation (Color figure online)

The example segmentation is visualized in Fig. 1. For the input image from Fig. 1(a) gradient magnitude on the basis on Sobel masks is calculated (Fig. 1(b)). The boundaries between obtained regions are marked with blue lines in Fig. 1(c). The watershed transformation suffers oversegmentation which means numerous, insignificant and small obtained regions. They do not correspond to real objects of the image. Oversegmentation is caused by slight, invisible changes of pixel values. It results in frequent, insignificant local minima of the gradient image and numerous regions extracted.

To reduce oversegmentation effect, some kind of region merging can be carried on. It can be realized by hierarchical watershed approaches. The most well known is the waterfall algorithm [6]. Its key idea is to apply geodesic reconstruction of gradient magnitude image from watershed lines of the previous segmentation. The process is iterated through successive hierarchies till regions are large enough. The working of hierarchical watershed segmentation is shown in Fig. 2 – for hierarchy 4 (Fig. 2(b)) still oversegmentation occurs, it is better for hierarchy 6 (Fig. 2(c)), but fitting regions into real image objects is not very precise.

| (a) Hierarchy 2 | (b) Hierarchy 4 | (c) Hierarchy 6 |

Fig. 2. Hierarchical watershed segmentation (Color figure online)

In another variant to reduce oversegmentation, marker controlled watershed transformation can be applied [7]. It means, instead of local minima, manually specified markers are utilized to initially label image points. In this case segmentation is only semiautomatic. Example results are presented in Fig. 3. Markers are labeled with red and green colors. Similarly to previous visualizations from Figs. 1 and 2 contours of extracted segments are drawn with blue color. If markers are sufficient as in Fig. 3(a) and in Fig. 3(b), obtained segments accurately correspond to real image objects – the selected swan and its surroundings. However, if they do not indicate areas with different color tones of the detected objects as in Fig. 3(c), the achieved results are unsatisfactory.

| (a) | (b) | (c) |

Fig. 3. Marker controlled watershed segmentation (Color figure online)

4   The Proposed Segmentation Method

The watershed transformation is directly applicable for any n-dimensional data. The dimensionality has an impact on the way how neighborhood is determined for analyzed points. In case of motion capture data, segmentation is carried on along time domain. Thus, instead of two dimensional images, time series are analyzed. To establish neighborhood for successive poses, preceding and following time instants are taken.

However, there is one more challenge which has to be faced. It is the way how gradient magnitudes $GMag$ in time instant t is approximated. In similar approach to commonly used in image processing Sobel or Prewitt operators, absolute difference d_P between preceding and following poses can be calculated.

$$GMag(t) = d_P(P_{t+1}, P_{t-1}) \qquad (2)$$

As mentioned before, time instants of motion capture data are described by rotations of subsequent joints as well as global orientation and translation of a skeleton. The translation is strictly related to the place of acquisition, it does not give valuable information about human movements and it is not usually taken into account during analysis. Thus, dissimilarity d_P, stated to be absolute difference, of poses P_1 and P_2 can be calculated as total distance between their corresponding rotations:

$$d(P_1, P_2) = \sum_{rotation} d_R(P_1(rotation), P_2(rotation)) \qquad (3)$$

The method d_R which compares two rotations has to be established. If Euler angles (α, β, γ) are applied to represent rotations baseline Euclidean (d_{Euc}) or Manhattan metrics which take into account periodicity of angle range can be used.

$$d_{Euc}((\alpha_1, \beta_1, \gamma_1), (\alpha_2, \beta_2, \gamma_2)) = \sqrt{\Delta(\alpha_1, \alpha_2)^2 + \Delta(\beta_1, \beta_2)^2 + \Delta(\gamma_1, \gamma_2)^2} \qquad (4)$$

where $\Delta(angle_1, angle_2) = \pi - ||angle_1 - angle_2| - \pi|$

It is not so straightforward in case of unit quaternions. The most often used dissimilarity function calculates geodesic distance $d_{geodesic}$ on hypersphere S^3, which covers all possible rotations. It utilizes dot product $\langle q_1, q_2 \rangle$ of quaternion vectors:

$$d_{geodesic}(q_1, q_2) = \frac{1}{\pi} \cdot arccos(\langle q_1, q_2 \rangle) \qquad (5)$$

In another variant quaternions transformed into tangent space at the identity on the basis of logarithm operator can be compared by Euclidean or Manhattan metrics [13]. The distance can also be approximated by the scalar part of product of the first quaternion and conjugate second quaternion as presented in [13].

Moreover, it is worthwhile to consider to analyze only the most significant joints for segmented activities. In such a case pose description is reduced to selected rotations. It not only decreases computational expensiveness, but primarily it focuses on joints performing movements in analyzed activities and reduces influence of the remaining joints. Finally, it may improve the performance of segmentation.

In motion analysis dynamics of movements expressed by angular velocities is frequently assessed. Thus, it seems that also segmentation which utilizes dynamics can lead to satisfactory results. It would extract partitions obtained on the basis of different rules. In case of watershed transform barriers would be built in sudden changes of angular velocities corresponding to local maxima of angular accelerations.

To segment motion data on the basis of dynamics of joint and body movements a method to approximate angular velocities has to be applied. Similarly to [13] differential filtering of time domain is chosen. In case of Euler angles $\Delta(angles_t, angle_{t-1})$ defined in Eq. (4) for rotations from consecutive time instants t and $t-1$ is calculated:

$$v_t = \Delta(angle_t, angle_{t-1}) \qquad (6)$$

For unit quaternions multiplication with the preceding conjugate quaternion is computed:

$$v_t^{quat} = q_t \cdot \overline{q_{t-1}} \qquad (7)$$

The further segmentation is analogous as in case of raw rotational data. Time series described by unit quaternions or n-dimensional vectors with scalar attributes are processed.

The introduced segmentation approaches utilize both baseline watershed transformation which detects catchment basins as well as hierarchical and marker controlled ones.

5 Results and Conclusions

The proposed segmentation methods were applied to human gait analysis. Motion capture data were used. The acquisition took place in Human Motion Laboratory of Polish-Japanese Academy of Information Technology (HML of PJAIT http://bytom.pja.edu.pl) equipped with Vicon hardware and software. The default skeleton model visualized in Fig. 4 of Vicon Blade was chosen. It means poses are described by 23 3D rotations of successive joints and skeleton orientation as well as global translation.

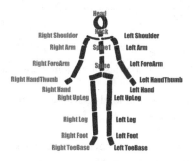

Fig. 4. Skeleton model

Segmentation concerns main cycles containing two middle adjacent steps performed by left and right lower limbs and analyzes only rotational data. To extract main cycles from gait sequences tracking of extremes of interaankle distances is carried on, as described in [13,14]. The quaternion geodesic distance function from Eq. (5) is taken to compute gradient magnitudes.

In the implementation watershed lines utilized in hierarchical segmentation are marked only at the border of segments with higher value of gradient magnitude than value at the border of neighboring segment. Moreover, it is assumed that if there is no local minima of gradient magnitude, single segment covering complete time series is extracted.

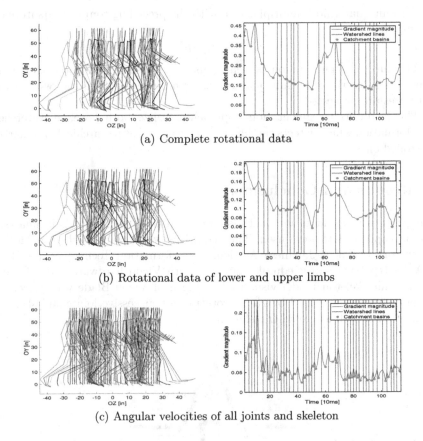

(a) Complete rotational data

(b) Rotational data of lower and upper limbs

(c) Angular velocities of all joints and skeleton

Fig. 5. Watershed segmentation of randomly selected gait instance (Color figure online)

The working of watershed segmentation is depicted in Figs. 5, 6 and 7. On the left sides gait main cycles are visualized by 20 poses arranged uniformly in time domain. YZ view is chosen, where OZ axis corresponds to the gait direction and OY to up and down orientation. The poses of successive partitions are labeled with different colors. The diagrams on the right side represent gradient magnitudes in consecutive time instants with determined catchment basins or manually pointed markers. In both visualizations watershed lines separating extracted segments are drawn with a blue color.

Experiments were preformed in respect to data of all joints and orientations of a skeleton as well as for selected parameters of lower (LeftUpLeg, RightUpLeg, LeftLeg and RightLeg) and upper (LeftShoulder, RightShoulder, LeftArm and RightArm) limbs, that are mainly responsible for gait locomotion and stability. Both raw rotational data and angular velocities computed on the basis of formula (7) are utilized in the segmentation process. In the validation stage set of one hundred randomly selected gait instances is used.

Similarly to image data (Fig. 1), in case of baseline transformation which separates every extracted catchment basin (Fig. 5), oversegmentation occurs. The resultant partitions are numerous, irrelevant and they do not correspond to interpretable gait phases. The oversegmentation is reduced for successive hierarchies of watershed segmentation as shown in Fig. 6. This is summarized in Table 1, which presents average values and standard deviations of number of extracted partitions. Hierarchical segmentation not only decreases number of extracted partitions, but primarily it makes them more meaningful.

The differential filtering which is used to approximate angular velocities, enhances acquisition noise. It increases number of insignificant minima of gradient magnitude and boosts oversegmentation effect. Therefore, greater number of hierarchies is required to obtain relevant partitions.

Table 1. Statistics of number of extracted segments

Hierarchy	Rotations		Angular velocities	
	All rotations	Limbs	All rotations	Limbs
0	23.61 ± 6.44	18.91 ± 6.35	35.85 ± 7.93	34.28 ± 8.38
1	5.67 ± 1.87	4.87 ± 1.75	9.55 ± 3.04	9.53 ± 2.93
2	1.87 ± 0.59	1.51 ± 0.71	2.64 ± 1.05	2.84 ± 1.19
3	1.01 ± 0.10	1.01 ± 0.10	1.08 ± 0.27	1.10 ± 0.33

For marker controlled watershed transformation which is visualized in Fig. 7, region boundaries are located in places of maximum values of gradient magnitude between successive markers. In case of gait data gradient is usually strongest in time instants which separate consecutive steps performed by left and right lower limbs.

Therefore, in final evaluation watershed transform was applied to separate following steps of gait main cycle. To validate performance of steps detection, manually prepared segments of aforementioned gait set were matched to extracted watershed partitions. It allows to calculate true positives (TP), true negatives (TN), false positives (FP), false negatives (FN) and accuracy (ACC) of detection for best matching segments representing consecutive steps:

$$ACC = \frac{TP + TN}{TP + TN + FP + FN} \tag{8}$$

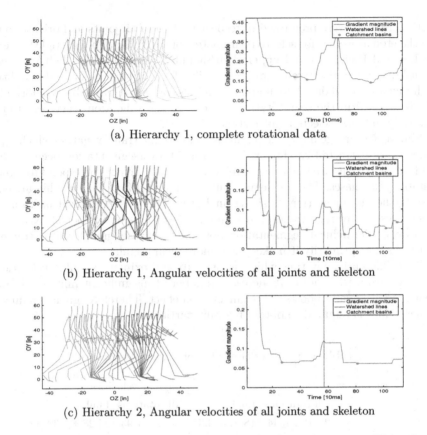

(a) Hierarchy 1, complete rotational data

(b) Hierarchy 1, Angular velocities of all joints and skeleton

(c) Hierarchy 2, Angular velocities of all joints and skeleton

Fig. 6. Hierarchical watershed segmentation of gait instance from Fig. 5 (Color figure online)

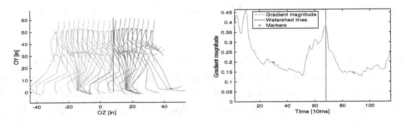

Fig. 7. Marker controlled watershed gait segmentation of gait instance from Fig. 5 represented by complete rotational data (Color figure online)

In the segmentation hierarchical as well as marker controlled (MC) watershed transforms were utilized. The markers are extracted automatically – they correspond to poses 25% and 75% of gait cycle as shown in Fig. 7. The assessment concerns both raw rotational data and angular velocities, describing movements of all joints and orientation of the skeleton as well as lower and upper

Table 2. Average accuracies of steps detection

	Rotations				Angular velocities			
	H1	H2	H3	MC	H1	H2	H3	MC
All segments	77.03%	83.91%	49.97%	95.35%	69.52%	89.34%	53.25%	94.51%
Limbs	81.47%	68.16%	49.97%	97.87%	69.73%	89.41%	53.77%	97.12%
Lower limbs	85.67%	57.65%	49.53%	98.16%	71.62%	91.30%	54.12%	97.51%

limbs. Furthermore, different hierarchies H are examined. The obtained results are depicted in Table 2. It contains average accuracies calculated according to formula (8). Because of manual coarse reference segmentation 30 ms margin surrounding partition boundaries is not taken into account in the assessment.

The best performance is achieved by marker controlled segmentation. It is consistent with our expectations – separation of successive steps corresponds to maxima of gradient magnitude and simple proposed method of marker extraction is sufficient to initially label partitions. In case of hierarchical watershed, only neighboring segments are merged, even if they are strongly separated. Thus, if oversegmentation differs for detecting steps, their segments are not precisely determined. Moreover, the best hierarchy level is not the same for every gait instance. However, hierarchical watershed has still acceptable accuracy, which exceeds 91%.

In general, selection of the most significant joints of gait, improves the segmentation. It reduces a noise and removes an influence of casual free movements of body segments not directly involved in gait locomotion and stability. Angular velocities give bit worse results if marker controlled segmentation is carried out, but substantially better in other cases.

Acknowledgments. The work was supported by Silesian University of Technology, Institute of Informatics under statute project BK/RAU2/2018.

References

1. Aoki, T., Venture, G., Kulic, D.: Segmentation of human body movement using inertial measurement unit. In: IEEE International Conference on Systems, Man, and Cybernetics (SMC), pp. 1181–1186. IEEE (2013)
2. Arikan, O., Forsyth, D.A., O'Brien, J.F.: Motion synthesis from annotations. In: ACM Transactions on Graphics (TOG), vol. 22, pp. 402–408. ACM (2003)
3. Arn, R., Narayana, P., Draper, B., Emerson, T., Kirby, M., Peterson, C.: Motion segmentation via generalized curvatures. IEEE Trans. Pattern Anal. Mach. Intell. (2018)
4. Barbič, J., Safonova, A., Pan, J.Y., Faloutsos, C., Hodgins, J.K., Pollard, N.S.: Segmenting motion capture data into distinct behaviors. In: Proceedings of Graphics Interface 2004, pp. 185–194. Canadian Human-Computer Communications Society (2004)

5. Beucher, S.: Use of watersheds in contour detection. In: Proceedings of the International Workshop on Image Processing. CCETT (1979)
6. Beucher, S.: Watershed, hierarchical segmentation and waterfall algorithm. In: Serra, J., Soille, P. (eds.) Mathematical Morphology and Its Applications to Image Processing, pp. 69–76. Springer, Dordrecht (1994). https://doi.org/10.1007/978-94-011-1040-2_10
7. Beucher, S., Meyer, F.: The Morphological Approach to Segmentation: The Watershed Transformation, Mathematical Morphology in Image Processing, pp. 433–481. Marcel Dekker Inc., New York (1992)
8. Bouchard, D., Badler, N.: Semantic segmentation of motion capture using Laban Movement Analysis. In: Pelachaud, C., Martin, J.-C., André, E., Chollet, G., Karpouzis, K., Pelé, D. (eds.) IVA 2007. LNCS (LNAI), vol. 4722, pp. 37–44. Springer, Heidelberg (2007). https://doi.org/10.1007/978-3-540-74997-4_4
9. Ciecholewski, M.: River channel segmentation in polarimetric SAR images: watershed transform combined with average contrast maximisation. Expert Syst. Appl. **82**, 196–215 (2017)
10. Huynh, H.T., Le-Trong, N., Oto, A., Suzuki, K., et al.: Fully automated MR liver volumetry using watershed segmentation coupled with active contouring. Int. J. Comput. Assist. Radiol. Surg. **12**(2), 235–243 (2017)
11. Lin, J.F.S., Joukov, V., Kulic, D.: Human motion segmentation by data point classification. In: 36th Annual International Conference of the IEEE Engineering in Medicine and Biology Society (EMBC), pp. 9–13. IEEE (2014)
12. Lin, J.F.S., Kulic, D.: Online segmentation of human motion for automated rehabilitation exercise analysis. IEEE Trans. Neural Syst. Rehabil. Eng. **22**(1), 168–180 (2014)
13. Switonski, A., Josinski, H., Wojciechowski, K.: Dynamic time warping in classification and selection of motion capture data. In: Multidimensional Systems and Signal Processing, pp. 1–32 (2018)
14. Szczesna, A., Switonski, A., Slupik, J., Zghidi, H., Josinski, H., Wojciechowski, K.: Quaternion lifting scheme applied to the classification of motion data. Inf. Sci. (2018)
15. Wisaeng, K., Sa-ngiamvibool, W.: Automatic detection and recognition of optic disk with maker-controlled watershed segmentation and mathematical morphology in color retinal images. Soft Comput. **22**, 6329 (2018). https://doi.org/10.1007/s00500-017-2681-9
16. Wu, Z., Liu, W., Xing, W.: A novel method for human motion capture data segmentation. In: IEEE 15th International Conference on Dependable, Autonomic and Secure Computing, 15th International Conference on Pervasive Intelligence & Computing, 3rd International Conference on Big Data Intelligence and Computing and Cyber Science and Technology Congress (DASC/PiCom/DataCom/CyberSciTech), pp. 780–787. IEEE (2017)
17. Xie, X., Liu, R., Zhou, D., Wei, X., Zhang, Q.: Segmentation of human motion capture data based on Laplasse eigenmaps. In: Chen, H., Zeng, D.D., Karahanna, E., Bardhan, I. (eds.) ICSH 2017. LNCS, vol. 10347, pp. 134–145. Springer, Cham (2017). https://doi.org/10.1007/978-3-319-67964-8_13
18. Yang, Y., Chen, J., Liu, Z., Zhan, Y., Wang, X.: Low level segmentation of motion capture data based on hierarchical clustering with cosine distance. Int. J. Database Theor. Appl. **8**(4), 231–240 (2015)
19. Yang, Y., Shum, H.P., Aslam, N., Zeng, L.: Temporal clustering of motion capture data with optimal partitioning. In: Proceedings of the 15th ACM SIGGRAPH Conference on Virtual-Reality Continuum and Its Applications in Industry-Volume 1, pp. 479–482. ACM (2016)

How Does State Space Definition Influence the Measure of Chaotic Behavior?

Henryk Josiński[1]([✉]), Adam Świtoński[1], Agnieszka Michalczuk[1],
Marzena Wojciechowska[2], and Konrad Wojciechowski[2]

[1] Silesian University of Technology, Akademicka 16, 44-100 Gliwice, Poland
henryk.josinski@polsl.pl
[2] Polish-Japanese Academy of Information Technology,
Koszykowa 86, 02-008 Warsaw, Poland

Abstract. In the case of experimental data the largest Lyapunov exponent is a measure which is used to quantify the amount of chaos in a time series on the basis of a trajectory reconstructed in a phase (state) space. The authors' goal was to analyze the influence of a state space definition on the measure of chaos. The time series which represent the joint angles of hip, knee and ankle joints were recorded using the motion capture technique in the CAREN Extended environment. Fourteen elderly subjects ('65+') participated in the experiments. Six state spaces based on univariate or multivariate time series describing a movement at individual joints were taken into consideration. The authors proposed a modified version of the False Nearest Neighbors algorithm adjusted for determining the embedding dimension in the case of a multivariate time series representing gait data (*MultiFNN*). The largest short-term Lyapunov exponent was computed in two variants for six scenarios of trials based on different assumptions regarding walking speed, platform inclination, and optional external perturbation. A statistical analysis confirmed a significant difference between values of the Lyapunov exponent for different state spaces. In addition, computation time was measured and averaged across the spaces.

Keywords: Nonlinear time series analysis · State space ·
Largest Lyapunov exponent · Human motion analysis ·
CAREN Extended system

1 Introduction

A basis for research was the concept of healthy flexibility and adaptability [11], according to which a human being is capable to adapt flexibly to changing and unpredictable environmental conditions, which allows the locomotor system to maintain a stable walking pattern despite very small disturbances – little variations in the walking surface and/or natural noise in the neuromuscular system. Such tiny perturbations are linked to slightly different initial conditions of successive gait cycles (strides). Theory of dynamical systems delivers a measure of the system's sensitivity to initial conditions, which in honour of Russian mathematician Aleksandr Mikhailovich Lyapunov (1857–1918) was named *the largest Lyapunov exponent* (λ_1). As concerns

© Springer Nature Switzerland AG 2019
N. T. Nguyen et al. (Eds.): ACIIDS 2019, LNAI 11432, pp. 579–590, 2019.
https://doi.org/10.1007/978-3-030-14802-7_50

human gait, the property of an extreme sensitivity to initial conditions is associated with the heel-strike events and it means the presence of deterministic chaos phenomenon in motion of some important components of human locomotor system. The capability of the locomotor system to maintain continuous walking by weakening effects of the abovementioned disturbances is called *Local Dynamic Stability* (LDS) [4].

The presence of chaotic behavior in biomedical signals (e.g. in ECG, EEG and gait kinematic data) is not in doubt and should be treated as a positive symptom – not without reason one of the subchapters of the famous book "Chaos. Making a New Science" by James Gleick is entitled "Chaos as Health".

Central to the computation of λ_1 as the measure of chaotic behavior is the notion of phase space (state space). According to the definition given by Baker and Gollub [1], a phase space is "a mathematical space with orthogonal coordinate directions representing each of the variables needed to specify the instantaneous state of the system". Such a state corresponds to a point in the phase space. Consecutive points form a phase trajectory which describes the dynamics of the system. If a system is chaotic, initially adjacent points evolve rapidly into more and more distant states, as in Fig. 1.

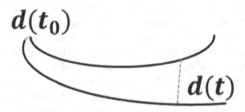

Fig. 1. A divergence of initially (i.e. at time instant t_0) nearby trajectories in the phase space.

However this behavior is only locally unstable – the trajectories diverge over short times but the distance between them cannot grow unlimitedly. The largest Lyapunov exponent defined by the following formula $\lambda_1 = \lim_{t \to \infty} \frac{1}{t} \cdot \ln \frac{d(t)}{d(t_0)}$, where $d(t)$ is a distance between points on adjacent trajectories at any time instant t, represents the average exponential rate of divergence of initially nearby phase trajectories. Its positive value means chaotic behavior. As it was mentioned before, as concerns human gait, this short-term instability is weakened shortly by the locomotor system.

The authors' goal was to analyze the influence of a state space definition on a value of λ_1. Variety of ways of constructing a state space offers a large space for interpretation (pun intended). Therefore, six variants of a state space, which are presented in next section, were tested by the authors. Experimental data were collected in cooperation with the University of the Third Age in Bytom, Poland, which brings together the elderly to help them develop their interests and remain active in their senior years.

Showing interest in elderly people is absolutely justified – we are facing a global phenomenon of ageing population, also in Poland. In 2016 over 16% of Polish people were aged 65 and older ('65+'). Many health issues affect the elderly, e.g. the threat of neurodegenerative diseases and the risk of falling. In order to avoid falling, elderly walkers decrease walking speed and step length, at the same extending the double support phase of a gait cycle [3].

Consequently, the authors decided to invite the 65+ students to participate in the experiments carried out using a powerful research environment – the multisensory CAREN Extended system (www.motekforcelink.com/product/caren/).

A basic component of the CAREN (the acronym stands for Computer Aided Rehabilitation ENvironment) Extended system is a motion platform with 6 DOF including an embedded instrumented dual belt treadmill (Fig. 2). A virtual reality environment produces visual and audio stimuli which create a fully immersive scenery of an experiment for a practicing subject (e.g. performing a rehabilitation exercise). As far as gait analysis is concerned, he/she performs a self-paced walk or adapts itself to a forced constant treadmill speed. A subject in a safety harness can be also exposed to previously defined external perturbations (e.g. platform distortions). A very important CAREN's component is an integrated motion capture system. Experimental data can be recorded using GRF and EMG measurement devices as well, giving, *inter alia*, the opportunity to assess a progress of rehabilitation. It is worthwhile to mention that the CAREN environment provides also a tool for creating own scenarios of experiments.

Fig. 2. Recordings in the centre for research and development of the Polish-Japanese Academy of Information Technology, Bytom, Poland, using the CAREN Extended system.

The current research was focused on the movement at hip, knee and ankle joints which constitute the kinetic chain for lower limb. Six scenarios of trials based on different assumptions regarding walking speed, platform inclination and optional vertical platform distortion were carried out using the CAREN Extended system. The authors state that the study has been approved by Ethical Committee and all subjects gave informed written consent to participate in the research.

2 State Spaces Structures

In the case of experimental data a phase trajectory is constructed based on (one or more) time series $x_i = x(i \cdot \tau_s)$ of a length equal to N ($i = 1, 2, \ldots, N$) where τ_s is the sampling time. The time series used in computations represent by means of angles (*joint angles*) a specific type of movement at the considered joints in sagittal (S), frontal (F) and transverse (T) planes, which divide a human body into left/right, front/back and upper/lower parts, respectively.

For experimental data a mathematical model of a system in the form of state equations as well as a phase space are often unknown. However, according to the Takens' Embedding Theorem [12] the latter can be reconstructed using time-delayed measurements of a single observed signal (a univariate time series).

Approximately 85% of the work during gait occurs in the sagittal plane. On the one hand, researchers prefer using a single time series with values of a joint angle in the sagittal plane which is mostly the only subject of analysis of movement at individual joints (e.g. [4, 9]), on the other, for instance, during a normal gait cycle the movement at a hip joint is not limited solely to this specific plane. This is the reason why benefiting from the diversity of available experimental data the authors analysed movement at a given joint based on a multivariate time series composed of three-dimensional data collected for the respective joint.

Each point on the trajectory is described by its coordinates in the reconstructed phase space. If the reconstruction process is based only on one univariate time series, two parameters: time delay τ and embedding dimension m must be determined first – m is the phase space dimensionality and τ indirectly defines an m-element vector of coordinates of a point x_i on the reconstructed phase trajectory as $[x_i, x_{i+\tau}, x_{i+2\cdot\tau}, \ldots, x_{i+(m-1)\cdot\tau}]$, where $i = 1, 2, \ldots, M = N - (m - 1) \cdot \tau$ (the reconstructed trajectory is made up of M points).

However, as regards a multivariate time series which consists of K univariate times series of equal length N, the pair (m_k, τ_k) of the abovementioned reconstruction parameters must be determined independently for each of the component time series. Consequently, dimensionality m of the reconstructed phase space is a sum of individual m_k ($k = 1, 2, \ldots, K$) parameters. The vector of m coordinates of a point x_i ($i = 1, 2, \ldots$, $M = N - \max_k[(m_k - 1) \cdot \tau_k])$ can be subdivided into K subvectors $[x_{k,i}, x_{k,i+\tau k}, x_{k,i+2\cdot\tau k}, \ldots, x_{k,i+(mk-1)\cdot\tau k}]$ of a length equal to m_k.

Discussion of state space structures in the context of motion data was initiated in [5]. The authors – Gates and Dingwell – started with the famous model of the chaotic Lorenz system and next they focused on univariate time series based on Euler angles describing rotational motion of a shoulder, using optionally the PCA method with the purpose of dimensionality reduction of state spaces. In conclusion the aforementioned authors suggest that numerical comparison between results for different state spaces should be made with caution. Nevertheless, identified trends seem to remain relevant. In addition, as far as a state space construction is concerned, the authors do not recommend the PCA reduction. In [8] a multivariate time series which comprised movement at 3 joints (hip, knee, ankle) in sagittal plane was applied in analysis of quiet standing balance. When it comes to determination of embedding dimension of a multivariate time series, the authors of [14] proposed a criterion of maximal joint entropy H, whereas Vlachos and Kugiumtzis [13] focused on adjusting the False Nearest Neighbors method [7] to a multivariate time series.

Inspired by reports in the literature the authors decided to incorporate six different state space structures into research. A "canonical" phase space for a particle moving in 3-D is a 6-dimensional structure based on three position and three velocity (or momentum) directions. Consequently, the first space considered in the research (*Canonical*) comprises three joint angles and their derivatives which are three angular velocities. This space does not require reconstruction.

Next three spaces (*UniS, UniF, UniT*) are reconstructed on the basis of one of the three univariate time series describing a movement at a joint in one of the considered planes (*S, F, T*).

Last two spaces – *MultiFull* and *MultiFNN* – are reconstructed on the basis of a multivariate time series which is composed of three abovementioned univariate time series related to the given joint. The former is a space which can be interpreted as a sum of *UniS, UniF* and *UniT* spaces, while dimensionality of the latter is determined as follows: starting from 1 the vector of coordinates is gradually expanded (and hence the embedding dimension increased) by including successive coordinates taken with appropriate time delay $\tau_i, i = 1, 2, 3$ from cyclically changed component time series x_i. This algorithm is a version of the False Nearest Neighbors method adjusted to multivariate time series, having (to some extent) regard to the domination of the sagittal plane in gait (the number of coordinates taken from the time series which describes an anterior-posterior movement cannot be smaller than contribution of other series). The stop criterion is met when the percentage of the false nearest neighbors falls below a given threshold (e.g. 1%).

Table 1 includes vectors of coordinates for each considered state space.

Regardless of the state space structure the largest Lyapunov exponent is calculated on the basis of the reconstructed phase trajectory using the Rosenstein algorithm [10]. The idea behind this method is that pairs of segments of the trajectory pretend repeatedly two initially adjacent trajectories, which makes it possible to compute the divergence of them.

The parameters of a phase trajectory reconstruction – time delay and embedding dimension – were determined by the Average Mutual Information [6] and the False Nearest Neighbors methods, respectively. However, as mentioned before, a modified version of the latter algorithm was used in the case of *MultiFNN* space. Computations were performed in MATLAB.

Table 1. Coordinates of a point on a phase space trajectory.

Method	Parameters	Vector of coordinates
Canonical	$m = 6$ (constant)	$[x_i^S, x_i^F, x_i^T, \dot{x}_i^S, \dot{x}_i^F, \dot{x}_i^T]$
UniS	τ, m	$[x_i^S, x_{i+\tau}^S, x_{i+2\cdot\tau}^S, \ldots, x_{i+(m-1)\cdot\tau}^S]$
UniF	τ, m	$[x_i^F, x_{i+\tau}^F, x_{i+2\cdot\tau}^F, \ldots, x_{i+(m-1)\cdot\tau}^F]$
UniT	τ, m	$[x_i^T, x_{i+\tau}^T, x_{i+2\cdot\tau}^T, \ldots, x_{i+(m-1)\cdot\tau}^T]$
MultiFull	$\tau_S, \tau_F, \tau_T, m_S, m_F, m_T$ $m = m_S + m_F + m_T$	$[x_i^S, x_{i+\tau S}^S, x_{i+2\cdot\tau S}^S, \ldots, x_{i+(mS-1)\cdot\tau S}^S,$ $x_i^F, x_{i+\tau F}^F, x_{i+2\cdot\tau F}^F, \ldots, x_{i+(mF-1)\cdot\tau F}^F,$ $x_i^T, x_{i+\tau T}^T, x_{i+2\cdot\tau T}^T, \ldots, x_{i+(mT-1)\cdot\tau T}^T]$
MultiFNN	$\tau_S, \tau_F, \tau_T, m_S, m_F, m_T$ $m = m_S + m_F + m_T$ $m_S \geq m_F \geq m_T$ they differ by 1 at most	$[x_i^S, x_{i+\tau S}^S, x_{i+2\cdot\tau S}^S, \ldots, x_{i+(mS-1)\cdot\tau S}^S,$ $x_i^F, x_{i+\tau F}^F, x_{i+2\cdot\tau F}^F, \ldots, x_{i+(mF-1)\cdot\tau F}^F,$ $x_i^T, x_{i+\tau T}^T, x_{i+2\cdot\tau T}^T, \ldots, x_{i+(mT-1)\cdot\tau T}^T]$

3   Experimental Research

The motion capture recordings were carried out in the Human Dynamics and Multi-modal Interaction Lab (HDMIL) of the Centre for Research and Development of the Polish-Japanese Academy of Information Technology (http://bytom.pja.edu.pl/) in Bytom, Poland. Table 2 includes mean values of age, height, weight and the body mass index (BMI) for 14 elderly participants (12 women and 2 men).

Table 2. Mean values of subjects' age, height, weight and the body mass index.

	Age [years]	Height [cm]	Weight [kg]	BMI [kg/m^2]
Mean	70.64	166	76.12	27.66
SD	3.52	7	15.57	5.18

Taking the advantage of the CAREN Extended capabilities the subjects performed first thrice a self-paced walk through a virtual forest on level ground (*Normal* scenario). Subsequently, the walk was enriched with an unexpected vertical distortion of the platform (*Perturbation* scenario). A mean value of the preferred walking speed (PWS), which was determined for each subject separately using the values measured during the *Normal* trials, after increasing or decreasing by 20% was forced as constant treadmill speed in two next scenarios (*Faster* and *Slower*). Two last scenarios (*Up* and *Down*) were based again on a self-paced walk, however they differed in platform inclination. The summary of all six scenarios is included in Table 3.

A proper analysis of a chaotic behavior by means of λ_1 requires that every time series should contain the equal number of strides and the equal number of data points [2]. Values of 50 strides and 100 points/stride were assumed, imposing in this way requirements regarding the length of recorded gait sequences.

The recordings have taken place under constant medical supervision. The participants could take a rest at any time. Starting with a new scenario every subject practiced

Table 3. Scenarios of experiments.

Scenario	Normal	Perturbation	Faster	Slower	Up	Down
Walking speed mode	Self-paced	Self-paced	120% of the subject's mean PWS	80% of the subject's mean PWS	Self-paced	Self-paced
Platform inclination [deg]	0	0	0	0	+5	−5

until he/she was ready to walk comfortably. Next, three gait sequences long enough to include the assumed 50 strides were recorded with a frequency of 100 Hz. Given the six scenarios, the total number of sequences for every subject was equal to 18 (with a few exceptions caused by fatigue).

The recorded motion data were filtered and repaired (e.g. due to occluded markers). Next, three time series related to each one of the considered joints were extracted from every gait sequence. Each of the series includes values of joint angles which represent a specific type of movement at the joint in one of the considered planes (S, F, T). The heel-strike events were identified in order to determine the limits of each stride. In the light of the aforementioned assumption the time series were cropped to 50 strides, where necessary, and every stride was separately normalized using linear interpolation to contain 100 points. It is worth noting that outcomes of additional tests based on cubic spline interpolation indicate that applied interpolation method is not relevant with regard to computed λ_1 value.

It is an established practice to compute the largest Lyapunov exponent in several variants – in authors' opinion the most important are those computed over the first step (i.e. the half of a stride, thus denoted by $\lambda_{S0.5}$) or over the first stride (λ_{S1}) immediately after a potential perturbation, which justifies calling them both "*short-term* Lyapunov exponents".

The pre-processed time series were subject to estimation of the phase trajectory reconstruction parameters (with the exception of the *Canonical* space) leading to the trajectory reconstruction. Exemplary 3-D projections of a phase trajectory for a left hip joint in *Canonical*, *UniS*, *UniF*, and *UniT* spaces are presented in Fig. 3a–d, respectively (the time series were recorded while performing the *Normal* scenario). A hip joint is a triaxial joint, so it allows for the following types of movement in three aforementioned planes: flexion/extension (S), ad-/abduction (F), and internal/external rotation (T). All these behaviors seem to be reflected in the trajectory shapes. It is necessary to mention that appropriate 3-D projections of *MultiFull* space (in the presented case it is 12-dimensional state space: $m_S = 3, m_F = 4, m_T = 5$) lead to the same figures. To some extent this remark also applies to *MultiFNN* space (6-dimensional state space: $m_S = m_F = m_T = 2$).

Both short-term Lyapunov exponents were computed using each of considered state spaces and their values were aggregated across all six scenarios and individual joints. The average values of $\lambda_{S0.5}$ and λ_{S1} for a left hip joint are presented in Fig. 4a and b, respectively.

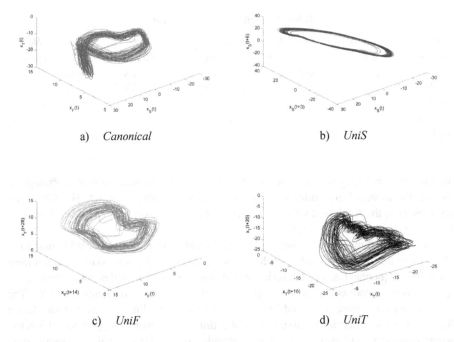

a)   *Canonical*   b)   *UniS*

c)   *UniF*   d)   *UniT*

Fig. 3. Exemplary 3-D projections of a phase trajectory for a left hip joint in different state spaces.

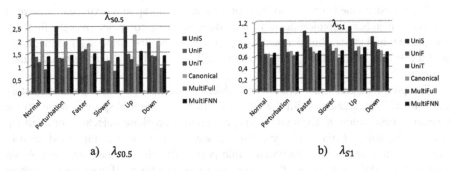

a) $\lambda_{S0.5}$ b) λ_{S1}

Fig. 4. Values of two variants of the short-term Lyapunov exponent for a left hip joint in different state spaces averaged across scenarios.

Every state space retains the effect of gradual attenuation of implications resulting from slightly different initial conditions of successive strides (generally, $\lambda_{S0.5} > \lambda_{S1} > 0$). Reconstructions based on different univariate time series lead to (significantly) different values of the short-term Lyapunov exponents (however they are always positive), which is a consequence of focusing on a single movement plane. As concerns a multivariate time series, incorporation of all available dimensions (*MultiFull*) results in values lower in comparison to other variants. Conclusion about whether there are statistically significant differences was made based on statistical tests performed by means of the Real Statistics Resource Pack (www.real-statistics.com/) for each chaos measure separately. All tests were conducted at significance level of 5%.

Firstly, the Shapiro-Wilk test was performed for each of six datasets including values of the short-term Lyapunov exponent computed in the appropriate state space. The test revealed that each dataset deviated from normal distribution. In addition, the Levene test uncovered the lack of homogeneity of datasets variances. Consequently, the non-parametric Kruskal-Wallis test was used to assess a significant difference between the six datasets.

Since the result of the Kruskal-Wallis test turned out to be positive, the Nemenyi test and the pairwise Mann-Whitney test were used to indicate which pairs of datasets are significantly different. The vast majority of them was found to be dissimilar. Table 4 includes a few exceptions.

Table 4. Pairs of datasets which are not significantly different.

Measure	Dataset 1	Dataset 2	p-value of the Nemenyi test	p-value of the pairwise Mann-Whitney test
$\lambda_{S0.5}$	UniS	MultiFNN	0.44	0.50
$\lambda_{S0.5}$	UniF	UniT	0.99	0.86
λ_{S1}	Canonical	MultiFull	0.66	0.69

A very small number of datasets pairs, which are not significantly different, indicates that each state space has its own characteristics. However, it should be emphasized that:

- each state space reveals the presence of chaos,
- each state space reveals that the effect of very little changes in initial conditions for the current stride is weakened within it.

Nevertheless, a direct comparison of Lyapunov exponents computed in different state spaces bears the risk of a wrong conclusion.

It appears that the results of the statistical analysis are a consequence of the domination of movement in the sagittal plane during gait. In particular, owing to the fact that MultiFNN – the space which is based on a multivariate time series – is not substantially different from UniS. And the reduced dimensionality which is a hallmark of MultiFNN avoids a redundant information. MultiFull space probably includes a redundant information which can have a negative impact on the computation results and according to Table 4 it would entail a critical judgment against Canonical space. It should be emphasized that Table 4 includes different pairs of datasets for each chaos measure – it can be an advantage of λ_1 that this measure differentiates UniF from UniT.

However, it is important to remember that the statistical analysis was conducted for all joints and scenarios together.

Attention should also be paid to a computation time for individual state spaces. Values of the computation time which comprises the following operations on a univariate or multivariate time series: interpolation, estimation of reconstruction parameters (if needed), trajectory reconstruction, and computation of the largest Lyapunov exponent, averaged across individual state spaces are presented in Fig. 5.

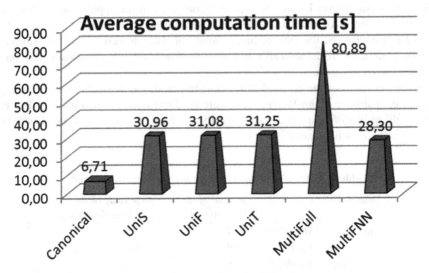

Fig. 5. Average computation time for different state spaces.

According to the authors' expectations the average computation time is:

- minimal for the *Canonical* space which does not require reconstruction,
- maximal for the *MultiFull* space (because of the 3-fold computation of embedding dimension for component time series),
- significantly smaller for the *MultiFNN* space (due to only a single computation of embedding dimension),
- roughly equal for all spaces based on a univariate time series (owing to the same set of operations).

4 Conclusion

The research presented in the paper is a part of a project aimed at analysing gait and posture of the elderly. The authors focused here on an important aspect of the computational method – an influence of a state space on the measure of chaos. The relation between both factors is indisputable, but nevertheless it is desirable to continue exploring this direction of the research. It is important to investigate if a given state space is capable of representing certain trends that will be examined (e.g. a potential difference between a patient's state before and after a rehabilitation process). The analysis on the level of individual joints is also advisable.

It is necessary to mention that a state space is not a sole factor which influences the largest Lyapunov exponent. Firstly, it was found that interpolation method is not relevant in this context. But the parameters of a state space reconstruction can be determined using various methods. Hence, focusing on a state space effect the authors confined themselves only to the Average Mutual Information (in respect of m) and the False Nearest Neighbors methods (in relation to τ). The only exception was the method proposed by the authors for determining the dimensionality of the *MultiFNN* space. The method is adjusted to a multivariate time series, eliminates a redundant information, and requires a smaller computation time compared to operations needed for other considered spaces.

Notwithstanding the aspect of a state space construction, it is necessary to mention that the positive values of the largest Lyapunov exponents imply the presence of chaotic behavior in gait of elderly subjects, however without losing sight of the fact that the participants were the students of the University of the Third Age that take care of their health and try to keep in shape.

Acknowledgments. Data used in this project were obtained from the Centre for Research and Development of the Polish-Japanese Academy of Information Technology (PJAIT) (http://bytom.pja.edu.pl/). This work was supported by Statutory Research funds of Institute of Informatics, Silesian University of Technology, Gliwice, Poland (grant No BK-213/RAU2/2018).

References

1. Baker, G.B., Gollub, J.P.: Chaotic Dynamics: An Introduction. Cambridge University Press, Cambridge (1990)
2. Bruijn, S.M., Meijer, O.G., Beek, P.J., van Dieën, J.H.: Assessing the stability of human locomotion: a review of current measures. J. Royal Soc. Interface **10**, 20120999 (2013)
3. Buzzi, U.H., Stergiou, N., Kurz, M.J., Hageman, P.A., Heidel, J.: Nonlinear dynamics indicates aging affects variability during gait. Clin. Biomech. (Bristol, Avon) **18**(5), 435–443 (2003)
4. Dingwell, J.B., Cusumano, J.P.: Nonlinear time series analysis of normal and pathological human walking. Chaos **10**(4), 848–863 (2000)
5. Gates, D.H., Dingwell, J.B.: Comparison of different state space definitions for local dynamic stability analyses. J. Biomech. **42**(9), 1345–1349 (2009)
6. Henry, B., Lovell, N., Camacho, F.: Nonlinear dynamics time series analysis. In: Akay, M. (ed.) Nonlinear Biomedical Signal Processing, Volume 2: Dynamic Analysis and Modeling, pp. 1–39. Wiley/IEEE Press (2000)
7. Kennel, M.B., Brown, R., Abarbanel, H.D.I.: Determining embedding dimension for phase-space reconstruction using a geometrical construction. Phys. Rev. A **45**(6), 3403–3411 (1992)
8. Liu, K., Wang, H., Xiao, J.: The multivariate largest lyapunov exponent as an age-related metric of quiet standing balance. Comput. Math. Methods Med. **2**, 1–11 (2015)
9. Piórek, M., Josiński, H., Michalczuk, A., Świtoński, A., Szczęsna, A.: Quaternions and joint angles in an analysis of local stability of gait for different variants of walking speed and treadmill slope. Inf. Sci. **384**, 263–280 (2017)
10. Rosenstein, M.T., Collins, J.J., De Luca, C.J.: A practical method for calculating largest Lyapunov exponents from small data sets. Phys. D **65**, 117–134 (1993)

11. Stergiou, N., Decker, L.M.: Human movement variability, nonlinear dynamics, and pathology: is there a connection? Hum. Mov. Sci. **30**(5), 869–888 (2011)
12. Takens, F.: Detecting strange attractors in turbulence. In: Rand, D., Young, L.-S. (eds.) Dynamical Systems and Turbulence, Warwick 1980. LNM, vol. 898, pp. 366–381. Springer, Heidelberg (1981). https://doi.org/10.1007/BFb0091924
13. Vlachos, I., Kugiumtzis, D.: State space reconstruction for multivariate time series prediction. Nonlinear Phenom. Complex Syst. **11**(2), 241–249 (2008)
14. Zhang, Ch.-T., Guo, J., Ma, Q.-L., Peng, H., Zhang, X.-D.: Phase space reconstruction and prediction of multivariate chaotic time series. In: Proceedings of the 9th International Conference on Machine Learning and Cybernetics, pp. 2428–2433. IEEE (2010)

Granular Computing (GC) Demonstrates Interactions Between Depression and Symptoms Development in Parkinson's Disease Patients

Andrzej W. Przybyszewski[1(✉)], Jerzy Paweł Nowacki[1],
Aldona Drabik[1], Stanislaw Szlufik[2], Piotr Habela[1],
and Dariusz M. Koziorowski[2]

[1] Polish-Japanese Academy of Information Technology,
02-008 Warsaw, Poland
{przy,jerzy.nowacki,adrabik,
piotr.habela}@pjwstk.edu.pl
[2] Neurology, Faculty of Health Science, Medical University Warsaw,
Warsaw, Poland
stanislaw.szlufik@gmail.com, dkoziorowski@esculap.pl

Abstract. There is high frequency incidence of depressive symptoms in neurodegenerative diseases (ND) but reasons for it is not well understood. Parkinson's disease (PD) is often evoked by strong emotional event and related to reduced level of dopamine (reward hormone). Similarly to PD, in older (over 65 year of age) subjects with late onset Alzheimer's disease (LOAD) have first symptoms related to depression (95%). Present work is devoted to the question if evaluation of depression can help to predict PD symptoms? We have gathered results of: neurological (disease duration, values of Unified Parkinson's Disease Rating Scale (UPDRS)), neuropsychological (depression – Beck test, PDQ39 (life quality), Epworth (sleep problems)) and eye movement (RS – reflexive saccadic) tests. We have tested 24 PD patients only with medical treatment (BMT-group), and 23 PD with medical and recent DBS (deep brain stimulation DBS-group), and 15 older DBS (POP-group) treatments during one and half year with testing every six months (W1, W2, W3). From rules found with help of GC (RST-rough set theory) in BMTW1 (patients BMT during first visit W1) we have predicted UPDRS in BMTW2 and BMTW3 with accuracies (acc.) 0.765 (0.7 without Beck result) and 0.8 (0.7 without Beck result). By using BMTW1 rules we could predict disease progression (UPDRS) of another group of patients – DBSW1 group with accuracy of 0.765 but not DBSW2/W3 patients. By using DBSW2 rules we could predict UPDRS of DBSW3 (acc. = 0.625), POPW1 (acc. = 0.77), POPW2 (acc. = 0.5), POPW3 (acc. = 0.33). By adding depression attribute and by using GC we could make better predictions of disease progressions in many different groups of patients than without it.

Keywords: Neurodegeneration · Rough set theory · KDD ·
Granular computing

© Springer Nature Switzerland AG 2019
N. T. Nguyen et al. (Eds.): ACIIDS 2019, LNAI 11432, pp. 591–601, 2019.
https://doi.org/10.1007/978-3-030-14802-7_51

1   Introduction

It is commonly accepted that clinically important depressive disorders occur in 40–50% of patients with PD and they influence many other clinical aspects of the disease. In addition to triggering innate emotional suffering, depressive disorders harmfully impact quality of life, motor and cognitive deficits, and also functional disability [1].

The significance of depression is mentioned in the Parkinson's Outcomes Project http://parkinson.org/research/Parkinsons-Outcomes-Project is the largest-ever clinical study of Parkinson's disease with over 12,000 participants in five countries. In their outcomes states that Depression and anxiety are the number one factors impacting the overall health of people with Parkinson's.

Depression is also prediagnostic. Depressive symptoms were observed many years before Parkinson's diagnosis in patients in neurological clinics in large studies in different countries [2, 3]. In UK [2] was found that in about 5000 patients 7% was depressed whereas in above 25000 controls only 4% had depression in 5 years before diagnosis. In Rotterdam study [3] they have tested motor and non-motor features over a time period of up 23 years before diagnosis. The early symptoms were related to motor and equilibrium, but about 5 years before diagnosis anxiety and depression was observed.

Depression and anxiety in PD might be physiologically related to a specific loss of dopamine and noradrenaline innervation of cortical and subcortical components of the limbic system [4].

A worldwide study of over 1,000 patients with PD found that more than 50% of the subjects testified clinically substantial depressive symptoms based on Beck depression scores [5]. In majority of PD studies, depressive symptom gravity is mild to moderate. In studies examining the occurrence of PD depression indicate that depressive disorders can advance at any phase in the development of PD [5]. Often, affective disorders precede - 4–6 years before the PD diagnosis - the beginning of motor symptoms [6]. Parkinson's disease (PD) starts from the degeneration of dopamine neurons in the substantia nigra (SN), and later to the neuron death in many other brain's structures.

As SN is one of the main sources of the dopamine (Dopa), its lack causes instabilities in the movement's control, as well as in some patients, depression (Dopa is reward transmitter) in addition to emotional and cognitive problems. As each patient has different neurodegeneration development and compensation in consequence has different disease progression and symptoms that has to be estimated by experienced neurologist in order to find an optimal therapy. This depends on results of tests, neurologist's experience and doctor's time. The knowledge of neurologist is based not only on his/her experience but also intuition to predict results of different therapies for a particular patient.

We have estimated disease progression in different groups of patients that were under different therapies and they were tested during three every half-year visits. We hope that our method will lead to introduce more precise and more automatic follow ups in the perspective possibilities of the remote diagnosis and treatments.

This study is expansion of our previous works by using additional to our granular computing method: rough set theory, a new attribute – the depression that as we will demonstrate plays a significant role in prediction of PD progression in dissimilar groups of patients with various treatments such as medication and/or DBS (deep brain stimulation) procedures.

2 Methods

We have analyzed tests from Parkinson Disease (PD) patients divided into three groups:

- **BMT-group** (the Best Medical Treatment) consists of 23 patients that were only on medication. The major medication in this group was L-Dopa that increases concentration of the transmitter dopamine in the brain as it that is lacking in Parkinson's patients. In most cases PD starts with neurodegeneration in substantia nigra that is response for the release of the dopamine.
- **DBS-group** (Deep Brain Stimulation) consists of 24 patients on medications and with implanted electrodes in the subthalamic nucleus during our study. These patients were more advanced in the disease than patients from BMT-group. Their first visit DBSW1 was before the DBS surgery, and second DBSW2 and third DBSW3 were with implanted stimulating electrodes in the subthalamic nucleus.
- **POP-group** (Post Operative Patients) consists of 15 patients with stimulating electrodes implanted before the beginning of our study. There were the most advanced patients as they have DBS surgery several years before the beginning of our study. Question was how long electrical stimulation is changing brain mechanisms and if we can approximate disease progression in these patients by other less advanced PD?

All together we have four different sessions: #1 to #4 that are all related to combinations of medication (MedOFF/ON) and brain stimulation (DBSOFF/ON). As in BMT-group patient are without stimulating electrodes so they were tested only in two sessions. Similar situation was in DBS-group before surgery (DBSW1 – visit W1 – the first visit).

We have performed the following testing for all patients:

- (a) MedOFF (session #1 without - medication) and MedON (session #3 patients on medications).
- (b) DBSOFF (DBS stimulation switched OFF in: session #1 without – medication; and session #3 with – medication) and DBSON (DBS stimulation switched ON in: session #2 without – medication; and session #4 with – medication). It was possible only in patients with implanted electrodes: DBS-group or POP-groups.

In addition all patients have to keep on with the following procedures: several neuropsychological tests (a new depression – Beck test, and used before PDQ39 (quality of life), and Epworth sleepiness test) and eye movement (RS – reflexive saccadic) and many neurological tests involved in the UPDRS (Unified Parkinson's Disease Rating Scale). All tests were performed in Brodno Hospital, department of Neurology, Faculty of Health Science, Medical University Warsaw, Poland. In the present work, we have tested and measured reflexive fast eye movements (saccades) as described in our previous publications [9]. In summary, every subject was sitting in a stable position without head movements and watching a computer screen before him/her. At the beginning he/she has to fixate in the center of the screen, and to keep on moving light spot. This spot was jumping randomly, ten degrees to the right or ten degrees to the left.

We have recorded simultaneously movements of the light spot and both eyes by means of professionally tested head-mounted saccadometer (Ober Consulting, Poland). On the basis of both signals we have calculated parameters of the saccades: the latency as a delay measured the start of the light spot movement to the beginning of the eye movement; the amplitude of the saccadic, and its max.

All above described procedures were repeated for each session (as mentioned above).

2.1   Theoretical Basis

Our KDD (knowledge discovery database) analysis is based on granular computing implemented in RST (rough set theory proposed by Pawlak [10]).

Our results were converted into the decision table with rows related to different measurements in different or the same subject and columns were related to different attributes. An information system [10] and the *indiscernibility relation,* as well as *lower approximation* and *upper approximation* were described in details before [10, 11].

On the basis of the reduct we have generated rules using four different ML methods (RSES 2.2): exhaustive algorithm, genetic algorithm [12], covering algorithm, or LEM2 algorithm [13].

One can also see the decision table as a triplet: $S = (U, C, D)$ where: C is condition, and D is decision attribute [14]. Each row of the decision table is in a natural way interpreted as a specific rule that links condition and decision attributes for a single measurement of the individual subject. As there are results (rows) related to diverse sessions and patients, they in an automatic way give rules - each one specific for one row. They can be very often contradictory. RS granular computing is approximating human way of thinking. Neurologist is always approximating patient's conditions with certain approximation as patient has some symptoms certainly but other only partly. RS theory implies generalizing all particular rules into general propositions that are always true (lower approximation) and partly true (upper approximation). This is related to discovery of the specific directions in the database (KDD) and determines optimal treatments for different PD patients. The decision attribute D can be interpreted as a single measure of patient's condition estimated by an expert (doctor). One can interpret classification of the data by the information table with the decision attribute submitted by doctor as the *supervised learning (ML)* process with neurologist as the teacher.

It is well recognized that neurodegenerative processes start about 20 years before primary noticeable symptoms in PD and they might be various in diverse patients. It is a famous expression: "no two PDs are the same" so finding optimal treatment is very difficult. Also effects of comparable treatments might give different effects in individual patients. Our algorithms have certain granular properties to cover all individual differences but with certain approximation (RST). The purpose of our computation is to follow interactions: doctor and patients. Significant advantage of our granular computations is abstraction and generalization in various levels that mimics approach of the very experienced doctor. Granular computing follows complex objects classifications as we have found in the visual brain [7, 8]. Our brains make object classification on the basis of inborn mechanisms and individual experience. We want to find enough flexible rules that will determine disease progressions of PD with diverse treatments, and in distinctive disease stages.

We have applied as KDD the RSES 2.2 (based on RST) [15] in order to find RS rules to process different patients. We have verified in our previous publication that the RS method gives better estimations than other classical methods [9].

3 Results

As described in the Methods section our patients were divided into three different groups: 23 BMT patients that were merely on medication, and 24 DBS patients were in addition to medication had electric stimulation of DBS-STN (subthalamic nucleus) with surgery completed during our study, and 15 POP patients with DBS procedure performed earlier.

Comparison of Longitudinal Changes in Tests Results
In BMT-group of patients in visit 1 (W1) had the mean age of 57.8 ± 13 (SD) years. Their confirmed disease duration was 7.1 ± 3.5 years, PDQ39 = 48.3 ± 29 (SD); Epworth 8 ± 5, Beck 14.2 ± 9.7, UPDRS session 1 was 48.3 ± 17.9 statistically ($p < 0.0001$) different than UPDRS equal 23.6 ± 10.3 in the session 3.

PatBMTW2 PDQ39 = 55.6 ± 34.5 (SD); Epworth 8 ± 5, Beck 16.3 ± 12.1; UPDRS in session 1 was 57.3 ± 16.8 ($p < 0.0005$ significantly different than in visit W1); whereas in session 3 it was 27.8 ± 10.8;

PatBMTW3 PDQ39 = 50.6 ± 28 (SD); Epworth 7.3 ± 4, Beck 14.1 ± 9.7; UPDRS in session 1 was 62.2 ± 18.2 ($p < 0.05$ significantly different than in visit W2); in session 3 was 25 ± 11.6.

It was no statistically significant difference in UPDRS between visits for session 3.

In DBS - group, the mean age of patients was 53.7 ± 9.3 years, and disease duration was 10.25 ± 3.9 years. In visit W1 UPDRS was 62.1 ± 16.1 (statistically different $p < 0.0001$ than in BMT-group, visit W1), PDQ39 = 56.5 ± 22.6 (SD); Epworth 9.1 ± 5.4, Beck 14.8 ± 10.0.

In DBS – group, visit W2 that was directly the surgery, in session 1 UPRDS equal 65.3 ± 17.6 became larger than before the surgery (see above) but there were not statistically significant difference; PDQ39 = 44.0 ± 22.1 (SD); Epworth 9.0 ± 4.8, Beck 11.0 ± 8.8.

In DBS – group, visit W3 session 1 UPDRS was 68.7 ± 17.7 and statistically different ($p < 0.03$) than in visit W2; PDQ39 = 46.1 ± 23.0 (SD); Epworth 9.2 ± 4.3, Beck 10.0 ± 8.4.

In POP-group UPDRS in session 1 for visit W1 was: 63.1 ± 18.2; for visit W2 was: 68.9 ± 20.3 to for visit it was W3: $74,2 \pm 18.4$. In session 4 (session with medication and DBS procedures) in visit W1 was 21 ± 11.3, in visit W2 was

23.3 ± 9.5, and in visit W3 was 23,8 ± 10.7. There are some similarities between groups DBS and POP.

3.1 KDD Findings Depression for BMT Group

In the BMT group were patients on only medication treatment with two sessions (no medication - MedOff session #1 and on medication MedOn session #3). They were measured three times every half of the year (W1, W2 and W3).

We have used RSES for the discretization and UPDRS were divided into the following ranges: "(−Inf, 24.0)", "(24.0, 36.0)", "(36.0, 45.0)", "(45.0, Inf)".

We had initially 72 rules for BMTW1 patients, but by generalization and filtering we have reduced them to 7 rules that are presented below without one rule that was specific for only one patient.

Table 1 is a discretized table for three patients: 4, 5, and 7 in two sessions: MedOFF (#1), MedON (#3) with parameters related to Beck depression scale and quality of sleep (Epworth scale), saccades latency values (SccLat), and the decision attribute was UPDRS (last column). Notice that Table 1 above is similar to Table 1 in [14] with an exception that in the previous table we did not use Beck score and PDQ39 has replaced present depression score (now PDQ39 is skipped).

Table 1. Discretized-table extract for BMT patients

P#	tdur	Ses	Beck	Epworth	PDQ39	RSLat	RSDur	RSAmp	RSVel	UPDRS
4	"(-Inf,9.75)"	1	"(-Inf,14.0)"	"(-Inf,3.0)"	*	"(-Inf,181.5)"	*	*	*	"(36.0,45.0)"
4	"(-Inf,9.75)"	3	"(-Inf,14.0)"	"(-Inf,3.0)"	*	"(-Inf,181.5)"	*	*	*	"(-Inf,24.0)"
5	"(9.75,Inf)"	1	"(-Inf,14.0)"	"(3.0,Inf)"	*	"(181.5,395.0)"	*	*	*	"(36.0,45.0)"
5	"(9.75,Inf)"	3	"(-Inf,14.0)"	"(3.0,Inf)"	*	"(181.5,395.0)"	*	*	*	"(24.0,36.0)"
7	"(-Inf,9.75)"	1	"(14.0,Inf)"	"(3.0,Inf)"	*	"(181.5,395.0)"	*	*	*	"(36.0,45.0)"
7	"(-Inf,9.75)"	3	"(14.0,Inf)"	"(3.0,Inf)"	*	"(181.5,395.0)"	*	*	*	"(-Inf,24.0)"

Each row can be written as a rule that will give 23 (number of BMT patients) * 2 (OFF and ON) = 46 very specific rules like that for the first row:

*(P#4)&(dur="(-Inf,9.75)")&(Ses=1)&(Beck="(-Inf,14.0)")&(Epworth="(-Inf,3.0)")
& (RSLat="(-Inf,181.5)") => (UPDRS="(36.0,45.0)")*

$$(1)$$

It means that if the for Pat#4, saccade duration is smaller than 9.7 ms, session is #1, Beck is smaller than 14, Epworth score smaller than 3.0 and saccade latency smaller than 181.5 ms then UPDRS will be between 36 and 45.

By using RST we can generalize rules from above table to the following rules:

$$(dur="(9.75,Inf)")\&(Beck="(14.0,Inf)")\&(Ses=1)\&(Epworth="(3.0,Inf)") => \\ (UPDRS="(45.0,Inf)"[6]) \tag{2}$$

$$(Ses=3)\&(Epworth="(-Inf,3.0)")=>(UPDRS="(-Inf,24.0)"[4]) \tag{3}$$

$$(Beck="(-Inf,14.0)")\&(Ses=3)\&(RSLat="(-Inf,181.5)")=>(UPDRS="(-Inf,24.0)"[4]) \tag{4}$$

$$(dur="(-Inf,9.75)")\&(Ses=3)\&(RSLat="(-Inf,181.5)")=>(UPDRS="(-Inf,24.0)"[3]) \tag{5}$$

$$(dur="(-Inf,9.75)")\&(Ses=1)\&(Epworth="(3.0,Inf)")\&(RSLat="(Inf,181.5)")=> \\ (UPDRS="(24.0,36.0)"[3]) \tag{6}$$

$$(dur="(Inf,9.75)")\&(Beck="(-Inf,14.0)")\&(Ses=1)\&(RSLat="(181.5,395.0)")=> \\ (UPDRS="(45.0,Inf)"[2]) \tag{7}$$

In the second formula (2) states that for the session 1 and saccade duration longer than 9.75 ms and Beck depression score larger than 14 and Epworth larger than 3.0 then UPDRS will be larger than 45.0. Eq. 2 was true in 6 cases.

Table 2. Confusion matrix for UPDRS of BMTW2 group by rules obtained from BMTW1-group

Actual	Predicted				
	"(36.0, 45.0)"	"(−Inf, 24.0)"	"(24.0, 36.0)"	"(45.0, Inf)"	ACC
"(36.0, 45.0)"	0. 0	0.0	0.0	0.0	0.0
"(−Inf, 24.0)"	0.0	8.0	0.0	0.0	1.0
"(24.0, 36.0)"	0.0	1.0	0.0	0.0	0.0
"(45.0, Inf)"	0.0	0.0	3.0	5.0	0.625
TPR	0.0	0.9	0.0	1.0	

TPR: True positive rates for decision classes; ACC: Accuracy for decision classes: the global coverage was 0.37 and the global accuracy was 0.765, the coverage for decision classes was 0.0, 0.73, 0.11, 0.4.

In addition, we have used BMTW1 rules to predict UPDRS in the next two visits: BMTW2 and BMTW3 in patients on medication only, with global accuracies of 0.765 (Table 2) and 0.8, and with the global coverage 0.37 and 0.33. It looks that accuracy are better than without Beck scale (0.7 for both visits), but global coverage with PDQ39 was 1 for both visits W2 and W3. Notice that in Eq. (2) fulfilled for 6 cases, the Beck score is high (above 14) and UPDRS is large (above 45), and in Eq. (4) satisfied in 4 cases the Beck score is low (below 14) and UPDRS is below 24. However, in Eq. 7 in two cases the Beck score is low (below 14) and UPDRS is large (above 45). It means that the depression score is *rough* – is not 100% discriminatory.

3.2   KDD for DBS Group

We have excluded DBSW1 group as these patients do not have implanted electrodes, but we made predictions UPDRS of DBSW3 by rules from DBSW2 (only sessions with DBSON and MedOFF – session 2, MedON – session 4), and we have obtained the global accuracy 0.67 (0.56 without Beck depression score) and global coverage 0.625 (1 without Beck inventory results).

But in DBSW2 decision classes were different than in BMTW1 "(36.5, Inf)" "(28.0, 36.5)" "(19.5, 28.0)" "(−Inf, 19.5)". We have obtained 6 rules with LEM algorithm [10] after filtering one-case rules. There are interesting differences to rules from BMTW1 group e.g.:

$$(RSLat="(-Inf,310.0)")\&(PDQ39="(-Inf,69.5)")\&(Ses=3)\&(Beck="(4.5,23.0)")=> (UPDRS="(-Inf,19.5)"[10])$$

(8)

$$(Epworth="(-Inf,7.5)")\&(RSLat="(-Inf,310.0)")\&(dur="(8.0,12.58)")\&(PDQ39="(-Inf,69.5)")\&(Beck="(-Inf,4.5)")=>(UPDRS="(-Inf,19.5)"[4])$$

(9)

$$(PDQ39="(-Inf,69.5)")\&(Beck="(4.5,23.0)")\&(Ses=1)\&(RSLat="(-Inf,310.0)")\& (dur="(8.0,12.58)")\&(Epworth="(-Inf,7.5)")=>(UPDRS="(28.0,36.5)"[2])$$

(10)

Notice that these rules Eqs. 8–10 have not only Beck scores (depression), but in contrast to BMTW1 rules, also PDQ39 quality of life scale. By adding depression score we have obtained better accuracy (Table 3) than without it [16].

Table 3. Confusion matrix for UPDRS of DBSW3 group by rules obtained from DBSW2-group

Actual	Predicted				
	"(36.0, Inf)"	"(28.0, 36.5)"	"(19.5, 28.0)"	"(−Inf, 19.5)"	ACC
"(36.5, Inf)"	2. 0	1.0	0.0	2.0	0.4
"(28.0, 36.5)"	0.0	3.0	1.0	2.0	0.5
"(19.5, 28.0)"	0.0	0.0	4.0	5.0	0.44
"(−Inf, 19.5)"	0.0	0.0	3.0	11.0	0.92
TPR	1.0	0.75	0.67	0.55	

TPR: True positive rates for decision classes; ACC: Accuracy for decision classes: the global coverage was 0.67 and the global accuracy was 0.625, the coverage for decision classes was 0.5, 0.55, 0.75, 0.75.

3.3   KDD for DBS Group Based on BMT Patients

In the next step, we have applied BMTW1 rules (Eqs. 2–7) to all visits of patients from the DBS group. As before [16] we were successful only for DBSW1 group (patients still before implementation of the stimulating electrodes). Therefore, these patients had only two sessions but with the higher dosage of medication as they were in more advance disease stage in comparison to BMT-group. We have obtained the global accuracy 0.765 (it was 0.64 without Beck [16]) with the global coverage 0.354 (0.5 without Beck [16]), for DBSW2 we got accuracy of 0.85, coverage of 0.3 (0.77 and 0.37 with Beck), for DBSW2 accuracy and coverage were 0.74 and 0.56 (0.8 and 0.33 with Beck depression score).

3.4   KDD for POP Group Based on DBSW2 Patients

Similarly to the previous study [16], we have divided DBSW2 PD into two subgroups: the first one with electric stimulation switched off (DBSOFF), and the second subgroup with electric stimulation switched on – DBSON. The reason was that POP patients were in more advanced stage of the disease and DBS makes POP and DBS patients more similar. Therefore, we made predictions only for UPDRS in POP groups with DBSON and for two sessions: MedOFF and MedON.

We have used rules from the DBSW2 group (see above). Previously [16] using rules without depression inventory (Beck score) we were not able to predict UPDRS of POP patients, as we have found that POP in comparison to DBS rules were contradictory.

Table 4. Confusion matrix for UPDRS of POPW1 group by rules obtained from DBSW2-group

Actual	Predicted				
	"(36.0, Inf)"	"(28.0, 36.5)"	"(19.5,2 8.0)"	"(−Inf, 19.5)"	ACC
"(36.5, Inf)"	1. 0	0.0	1.0	0.0	0.5
"(28.0, 36.5)"	2.0	0.0	1.0	1.0	0.0
"(19.5, 28.0)"	2.0	1.0	3.0	2.0	0.375
"(−Inf, 19.5)"	0.0	1.0	0.0	8.0	0.89
TPR	0.2	0.0	0.6	0.73	

TPR: True positive rates for four decision classes; ACC: Accuracy for decision classes: the global coverage was 0.77 and the global accuracy was 0.52, the coverage for decision classes was 0.4, 0.57, 1.0, 0.9.

Adding depression is important as we could predict disease progression of POPW1 group with accuracy 0.5, coverage 0.77 (Table 4). For POPW2 group we have obtained global accuracy 0.4 and global coverage 0.17. For POPW3 we had global accuracy 0.25 and global coverage 0.33. It means that there are still other long-term effects of brain stimulation that we cannot effectively predict.

4   Discussion

There is permanent problem in handling patients with neurodegenerative diseases: how to find and test if an actual treatment is optimal or even 'near' optimal? It is very important question as the right, optimal therapy may improve patient's quality of life, help caregiver, and prolong patient's activity and his/her life expectancy. Technology made important progress in medical science and introduced new procedures improving patient's handlings. The main problem is the long lasting (about 20 years) neurode-generation processes with the specific for each person compensatory mechanism happening before the first disease symptoms. As plastic mechanisms are influenced by many factors as such as: daily activity – physical and intellectual, profession as cog-nitive training, so-called social brain, diet and physical training. In the consequence, each patient must be handled in an individual, unique way. In order to fulfill it, we have used KDD approach looking for hidden rules with help of data mining and machine learning methods (RST granular computations) that propose universal rules with enough generalization and specificity that determine treatments of individuals from different groups of patients. These general rules are related to the knowledge and experience of the neurologist but are also related to individual patients. Our long-term plans are to expand this granular computing approach not only to study patients with many different treatments, but also to compare many different groups of patients var-ious centers using not exactly the same approach in diagnosis and medications. If we obtain rules that are different for different medical centers we can easy compare them in order to find granules determining more optimal set of treatments for each individual patient.

In this study, we have examined three groups of PD patients in the different disease stages and procedures: BMT, DBS, and POP groups and tried to find common mechanisms between them. Previously we have made effective prediction of the dis-ease progression for BMT and DBS groups of patients. However, we were not suc-cessful to predict disease progression in the patients with long brain electric stimulation (POP group). In this analysis, we have improved our results by adding depression attribute (Beck score). Depression was sufficient in BMT group but for DBS and POP groups the quality of life (PDQ39), with sleepless (Epworth), and eye movement were major attributes that helped to predict UPDRS. **Therefore depression plays a sig-nificant role in the disease progression of PD patients.**

Ethics Statement. This study was carried out in accordance with the recommendations of Bioethics Committee of Warsaw Medical University with written informed consent from all subjects. All subjects gave written informed consent in accordance with the Declaration of Helsinki. The Bioethics Committee of Warsaw Medical University approved the protocol.

References

1. Reijnders, J.S., Ehrt, U., Weber, W.E., et al.: A systematic review of prevalence studies of depression in Parkinson's disease. Mov. Disord. **23**, 183–189 (2008)
2. Schrag, A., Horsfall, L., Walters, K., Noyce, A.: Petersen I prediagnotic presentations of Parkinson's disease in primary care: a case-control study. Lancet Neurol. **14**(1), 57–64 (2015)
3. Darweesh, S.K.L., et al.: Trajectories of prediagnostic functioning in Parkinson's disease. Brain **140**, 429–441 (2017)
4. Remy, P., Doder, M., Lees, A., Turjanski, N., Brook, D.: Depression in Parkinson's disease: loss of dopamine and noradrenaline innervation in the limbic system. Brain **128**(Pt 6), 1314–1322 (2005)
5. GPDS Steering Committee: Factors impacting on quality of life in Parkinson's disease: results from an international survey. Mov. Disord. **17**, 60–67 (2002)
6. Ishihara, L., Brayne, C.: A systematic review of depression and mental illness preceding Parkinson's disease. Acta Neurol. Scand. **113**, 211–220 (2006)
7. Przybyszewski, A.W.: Logical rules of visual brain: from anatomy through neurophysiology to cognition. Cognit. Syst. Res. **11**, 53–66 (2010)
8. Przybyszewski, A.W.: The neurophysiological bases of cognitive computation using rough set theory. In: Peters, J.F., Skowron, A., Rybiński, H. (eds.) Transactions on Rough Sets IX. LNCS, vol. 5390, pp. 287–317. Springer, Heidelberg (2008). https://doi.org/10.1007/978-3-540-89876-4_16
9. Przybyszewski, A.W., Kon, M., Szlufik, S., Szymanski, A., Koziorowski, D.M.: Multimodal learning and intelligent prediction of symptom development in individual parkinson's patients. Sensors **16**(9), 1498 (2016). https://doi.org/10.3390/s16091498
10. Pawlak, Z.: Rough Sets: Theoretical Aspects of Reasoning About Data. Kluwer, Dordrecht (1991)
11. Bazan, J., Nguyen, S.H., Nguyen, T.T., Skowron, A., Stepaniuk, J.: Decision rules synthesis for object classification. In: Orłowska, E. (ed.) Incomplete Information: Rough Set Analysis, pp. 23–57. Physica–Verlag, Heidelberg (1998)
12. Bazan, J., Nguyen, H.S., Nguyen, S.H., Synak, P., Wróblewski, J.: Rough set algorithms in classification problem. In: Polkowski, L., Tsumoto, S., Lin, T. (eds.) Rough Set Methods and Applications, pp. 49–88. Physica-Verlag, Heidelberg (2000)
13. Grzymała-Busse, J.: A new version of the rule induction system LERS. Fundamenta Informaticae **31**(1), 27–39 (1997)
14. Bazan, J.G., Szczuka, M.: The rough set exploration system. In: Peters, J.F., Skowron, A. (eds.) Transactions on Rough Sets III. LNCS, vol. 3400, pp. 37–56. Springer, Heidelberg (2005). https://doi.org/10.1007/11427834_2
15. Bazan, J.G., Szczuka, M.: RSES and RSESlib - a collection of tools for rough set computations. In: Ziarko, W., Yao, Y. (eds.) RSCTC 2000. LNCS (LNAI), vol. 2005, pp. 106–113. Springer, Heidelberg (2001). https://doi.org/10.1007/3-540-45554-X_12
16. Przybyszewski, A.W., Szlufik, S., Habela, P., Koziorowski, D.M.: Rules determine therapy-dependent relationship in symptoms development of Parkinson's disease patients. In: Nguyen, N.T., Hoang, D.H., Hong, T.-P., Pham, H., Trawiński, B. (eds.) ACIIDS 2018. LNCS (LNAI), vol. 10752, pp. 436–445. Springer, Cham (2018). https://doi.org/10.1007/978-3-319-75420-8_42

Measurements of Antisaccades Parameters Can Improve the Prediction of Parkinson's Disease Progression

Albert Sledzianowski[1]([✉]), Artur Szymanski[1], Aldona Drabik[1],
Stanisław Szlufik[2], Dariusz M. Koziorowski[2], and Andrzej W. Przybyszewski[1]

[1] Polish-Japanese Academy of Information Technology, 02-008 Warsaw, Poland
albert.sledzianowski@gmail.com
[2] Neurology, Faculty of Health Science, Medical University of Warsaw,
03-242 Warsaw, Poland

Abstract. In this text we present the results of oculometric experiment consisting the registration of anitsaccades of patients with Parkinson's Disease (PD) in relation to their neurological data. PD is an important and incurable neurodegenerative disease and we are looking for methods optimizing the treatment. In our previous works we used Reflexive Saccades (RS) and Pursuit Ocular Movements (POM) to check what it can tell us about the disease's progression expressed in the Unified Parkinson's Disease Rating Scale (UPDRS). The UPDRS is the most commonly used scale in the clinical studies of Parkinson's disease. In this experiment we examined antisaccades (AS) of 11 PD patients who performed eye movement tests in controlled conditions. We correlated neurological measurements of patient's motoric abilities and data describing their treatment with values of AS parameters. We used RSES and for prediction of the UPDRS scoring groups and Weka methods for presentation of the results. We achieved good results with accuracy of 91% and coverage of 100%. The AS test is a relatively easy and non-invasive method that can be used in the telemedicine in the future.

Keywords: Parkinson's disease · Antisaccades · Eye tracking · Data mining · Machine learning

1 Introduction

Antisaccade is a voluntary eye move in the opposite direction of appearing target [8]. Subject have to suppress a glance towards a suddenly presented peripheral stimulus and look away from it to the mirror location [13]. The eye move schema is presented in Fig. 1. Antisaccades are generally more difficult than eye move towards the stimuli (prosaccades) for some PD patients even impossible to perform. The performance of antisaccades is influenced by parameters interacting with the fixation and/or attention system of oculomotor control [9]. Olk and Kingstone [10] assumed in their research, that prosaccades to new objects are

© Springer Nature Switzerland AG 2019
N. T. Nguyen et al. (Eds.): ACIIDS 2019, LNAI 11432, pp. 602–614, 2019.
https://doi.org/10.1007/978-3-030-14802-7_52

made reflexively and for antisaccades, this reflexive eye movement have to be inhibited thus antisaccades are generated volitionally. This oculomotor inhibition is the main factor leading to long antisaccade latency causing that antisaccades are generally slower than prosaccades. Inhibition is being produced by the reallocation of covert attention from the target location towards the opposite antisaccade location [10]. The prefrontal cortex (PFC) is being found to be crucial for control of reflexive behavior allowing for voluntary reaction [13]. Brain imaging studies showed that cortical and subcortical network is widely active during the generation of antisaccades [15]. The interesting finding has been made by Fischer and Weber [9] who observed that parameters supporting the disengagement of fixation at the time of stimulus onset provoke a reduction of the antisaccadic reaction times and that certain state of disengagement seems to facilitate the occurrence of reflex-like errors.

The ability to suppress reflexive responses in favor of voluntary motor acts is very important for everyday life and variety of neurological diseases result in dysfunctions and errors of this mechanism, what can be observed in the voluntary eye move tasks [15]. In terms of PD disease, various studies have shown that patients have impaired executive function, including deficits in attention, movement initiation, motor planning and decision making leading to impairments in controlling involuntary behavior [12,17]. Such dysfunctions plays important role during execution of voluntary eye movements and resulting in difficulties, as antisaccade requires suppression of an automatic eye movement to a visual stimulus and execute a voluntary eye movement in the opposite direction [17]. Antisaccade deficits in PD have been attributed to fronto-basal ganglia (BG) dysfunction and are similar to those seen in the task switching, whereby one is required to change a response after performing a different behavior [17]. Crevits et al. [11] observed that the degree of advancement of Parkinson's disease significantly increases mean latencies and error rate in the antisaccade tasks. Antisaccades in PD has been described as abnormal, multiple-step and hypometric and associated with a significant decrease in the velocity [14]. PD patients are treated medically or by stimulation of the Subthalamic Nucleus (STN) with an electrode (Deep Brain Stimulation - DBS). In terms of medical treatment, Hood et al. [12] found that Levodopa (L-Dopa) commonly used to improve the symptoms of Parkinson's disease significantly reduces error rate for antisaccades and suggests that L-Dopa improves function of the voluntary frontostriatal system, which is deficient in PD. It has been also observed that PD patients in the medicated state are better able to plan and execute antisaccades [12]. In contrast to L-Dopa, electrical stimulation of the STN, the alternative method to the medical treatment, has been found to have no effect on the antisaccade task. According to Rivaud-Péchoux, et al. [16] STN stimulation improves only the accuracy of the memory guided saccades.

The UPDRS becomes common rating scale of the progression of a Parkinson's disease among neurologists and researchers who want to carry out measurements with objective instruments [5]. It consists of 42 items divided into 4 sections [6]. First 2 sections consist scoring of personal behavior, mood, mental activity and activities of daily living. Next 2 sections consist examination of patient's motor fitness and difficulties during the treatment. In our experiment we decided to compare the results of Sects. 3 and 4 of UPDRS examination with the results of the oculomotor study. The UPDRS III includes clinician-scored monitored motor evaluation and evaluation of complications during the therapy (UPDRS IV) [7]. The UPDRS III refers directly to motor results an UPDRS IV to motor fluctuations, both might have direct correlation with the patient's oculomotor abilities.

The problem in evaluating the scale of Parkinson's disease progression lies in the very individual symptoms of this disease. Every patient diverge substantially in his combinations of symptoms, rates of progression, and reactions to treatment [4]. In the experiment we wanted to check the effectiveness of predictions of neurological evaluations, represented by Sects. 3 and 4. We tried to find out whether there are correlations between the antisaccade parameters collected from oculometric tests and data from the neurological classifications. We used combined results of neurological diagnoses as decision attributes along with oculometric measurements as conditions expressed in the parameters. With such approach we researched correlations between both sources of the data. The aim of this experiment was to test algorithms, allowing for machine-learning evaluation of the UPDRS III and IV, based on type of the treatment and results of the anitsaccade trials. We believe that methods of predictions like presented in this article might extend available data of patient, if patients could perform oculomotor tests using their personal devices in different conditions, not only in the clinical settings. Data evaluation presented in this article could be automated by open-access software running on personal device like PC, tablet or a smart-phone.

2   Methods of the Experiment

We examined 11 patients in the clinical conditions. Patients underwent experimental trials under the supervision of a doctor. Results of patients were collected and divided according to their treatment. Our data distinguished patients who undergo pharmacological (BMT - Best Medical Treatment) treatment basing on the medication of the L-Dopa and the DBS (Deep Brain Stimulation). Patients qualified for DBS surgery are mainly characterized by low sensitivity to L-Dopa [4].

Possible variants of those two parameters described types of different sessions in which the results of patients were considered:

- S1: No treatment - (BMT Off, DBS Off)
- S2: Patients undergo only non-pharmacological treatment (BMT Off, DBS On)
- S3: Patients undergo only pharmacological treatment (BMT On, DBS Off)
- S4: Patients undergo both types of treatment (BMT On, DBS On)

We compared correlations between types of the sessions and UPDRS results. In total our data contained 28 registrations with relevant data from neurological tests including the results of the UPDRS classifications. Not every patients were treated with both pharmacological and surgical treatment, so it was impossible to examine each patient in the four different sessions. The eye moves of the patient were recorded by the eye tracker. We used head-mounted eye-tracker JAZZ-Novo with frequency of 1000 Hz. Patient head was positioned on the chin-rest at a distance of 60–70 cm from the monitor in order to minimize the head movements.

During each experimental trial patients task was to follow horizontal moves of the light spot generated by the eye tracker. At the beginning of each one patients were viewing the fixation point and it was the primary position of the gaze in each antisaccade trial (0°). When trial started fixation point disappeared and at the same moment target of the antisaccade appeared randomly on its left or the right side (10° to the left or right of the fixation point) in arbitrary times between 500–1500 ms. The antisaccade target remained for 100 ms before another trial started. Patient task was to move eyes in opposite direction to the appearing targets with best accuracy and smallest delay. The experiment was conducted in "no-gap" model in contrast to the model introduced by Saslow [21]. Schema on Fig. 1 presents the model of the anisaccade trial.

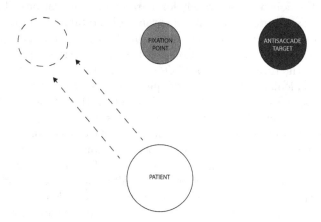

Fig. 1. Model of the antisaccade trial.

All experimental trials were conducted in the same lighting conditions. The data was analyzed by software detecting antisaccades in the eye move signal and calculating its Delay, Duration and Speed parameters. The algorithm searched the oculometric data composed from the time stamps and x-coordinates of the stimuli and patient eyes. The start point of each search window was the moment of appearance of the antisaccade target. The algorithm expected straight-aimed move from fixation point to the opposite direction of the appearing target below the delay threshold of 500 ms. The latency parameter have proved to be a valuable source of detailed and quantitative information in a wide range of neurological conditions [19]. Parameter Latency can also give the information of the impairments of the decision mechanisms described in the section "Introduction". Any record which hadn't passed criteria of move direction (patient performed prosaccade) in the defined window below the maximum acceptable latency were removed from the original data. The record was also rejected if there was non-response or mis-recording (blinks, head movements, etc.) [19]. Calculated statistics of successful antisaccades gave the mean of 6.53 (SD 2.5). This fairly weak indicator may be explained by the fact, that for many PD patients the antisaccade can be a quite difficult task. For particular cases even impossible to perform.

3   Computational Basis

The eye move parameters, mean latency, mean duration and max speed were calculated on the basis of registered appearances of fixation point and the target in correlation with patient's eye moves. Duration (time) and speed (velocity) are base parameters describing the eye move. We choose max speed (peak velocity) because we believe that this parameter is better in showing the oculomotoric capabilities of the patient than the average speed.

We used the following approach for parameters calculations. The latency was measured as a period between appearance of the target (and fixation point disappearance) and start of the eye move in the right direction. Start of the antisaccade (start point of the duration) has been fixed to the moment when gaze speed exceeded the threshold of speed determined for particular trial. The speed threshold was calculated from all subsequent frames of the record by dividing the maximum speed and the average speed. The Duration of the antisaccade was determined as the period when gaze direction has began to follow the opposite direction of the target and when simultaneously the eye speed has started to rise from the starting point of change of the move direction. The means of Delay and Duration parameters were calculated arithmetically for a particular patient. The Max Speed was counted as the maximum from all values collected and calculated for every eye-tracked frame in the period of the antisaccade duration.

After carrying out the oculometric tests we created the dataset from parameters of the anticassade trials (numeric values) and the neurological data. The neurological data contained parameters describing type of the treatment expressed in the symbolic attribute "Session" (S1, S2, S3, S4) and the results of UPDRS classifications (numeric values). In the next phase, the dataset has been used as the input decision table for Rough Set Exploration System (RSES) we used for further analysis. RSES is a data-mining software written at Warsaw University and it has been previously found that RSES deals very good with predicates based on small data [3]. RSES contains a tool set of methods coming from the Rough Set Theory (RST) [1]. RST is founded on the assumption that every object associate some information which are characterized by the same information in view of the available information about them [3]. This approach is related to the granular computing paradigm where every particular granule contains all attributes are related to "and" logic, and where interactions between granules are related to "or" logic [4]. The relation of indiscernibility is the mathematical basis of this theory stating that a set of similar objects forms a basic granule of knowledge and any union of those elementary sets formulates a precise set [1]. In contrast to the precise set, the rough set (RS) cannot be characterized in terms of information about its elements. Each RS has boundary-line of objects which cannot be certainly classified. Each RS contains also associated pair of precise sets (the lower and high approximations). The lower consists of all objects certainty belong to the set and the upper one containing objects which possible belong to the set and difference between the upper and the lower precise sets constitutes the boundary region of the RS [1]. Such approximations are basic operations used in this data-mining methodology [3].

We used decision table as input data for RSES (as proposed by Pawlak [1]) constructed from columns where two last are the condition attributes measured by the neurologist (UPDRS III and IV) and all preceding columns are the decision attributes. Each row of the table describes the rules, by which each patient can be described. Therefore, rules can have many specific conditions, as number of rules equals number of rows. Using this approach, we can describe different UPDRS values in different patients. This approach should also simulate the way in which neurologists might interact with patients, perceiving various patient's symptoms at various levels of granularity. There are large inconsistencies between PD progression, symptoms, between individual patients and also in effects of similar treatments and the task of neurologists is to abstract and consider only those symptoms that are universally significant and serve to determine a specific treatment [4]. For visualization of the results obtained from RSES we used the WEKA data-mining software written at University of Waikato with J48 algorithm generating a pruned or unpruned C4.5 decision tree [2]. C4.5 is an algorithm building decision trees from a set of data using the concept of the information entropy and is probably one of most widely used machine learning tool in the current practice [20].

4 Results

Initial dataset contained 16 attributes and 28 experimental measurements (observations), representing calculated parameters from the antisaccade records mapped to the records from neurological database of particular patients. The example of initial dataset is presented in Table 1. Attributes it the Table 1 were defined as follow:

- Patient ID (ID) - the id of particular patient.
- Session - parameter describing the session type.
- Delay Mean - calculated eye mean delay relative to the movement of the spot.
- Duration Mean - calculated parameter describing duration of particular antisaccade.
- Max Speed - calculated maximum eye speed during particular antisaccade.
- UPDRS III - numeric result of patient's UPDRS III classification.
- UPDRS IV - numeric result of patient's UPDRS IV classification.

The example from combined dataset with mostly numeric data is presented in Table 1. In the first step of analysis we separated the UPDRS III and IV parameters placing it in two different tables. Then we used RSES for reduction of attributes and data discretization using the local method with symbolic attributes, allowing for nominal values analysis. In some cases it give better outputs in terms of sensitivity of discretization, as it is generating much more cuts and it is also slightly faster than the global method [18].

Table 1. Sample of the initial dataset

ID	Session	DelayMean	DurationMean	MaxSpeed	UPDRS-III	UPDRS-IV
13	S4	0.26	1.95	5.05	8	6
14	S1	0.5	4.4	5.05	43	0
14	S2	0.35	4.15	5.08	14	0
14	S3	0.42	4.01	5.24	5	8
14	S4	0.33	2.86	5.29	9	0

The final dataset containing decision important and discretized attributes representing ranges of values after the classified selection is presented in examples of data in Tables 2 and 3 accordingly to different UPDRS scale parameters.

Table 2. Sample of the reduced and discretized dataset with attribute UPDRS III

PatientID	Session	DelayMean	DurationMean	MaxSpeed	UPDRS-III
"13"	"S4"	(−inf–0.335)	(1.85–inf)	(−inf–5.075)	(3–13)
"14"	"S1"	(0.445–inf)	(1.85–inf)	(−inf–5.075)	(25–59)
"14"	"S2"	(0.335–0.445)	(1.85–inf)	(5.075–5.6)	(13–25)
"14"	"S3"	(0.335–0.445)	(1.85–inf)	(5.075–5.6)	(3–13)
"14"	"S4"	(−inf–0.335)	(1.85–inf)	(5.075–5.6)	(3–13)

In the next phase we wanted to find correlations between UPDRS values and the rest of the attributes. We compared different RSES classifiers performing the same cross-validation prediction of attribute UPDRS III and UPDRS IV on the discreatized datasets. The values of the UPDRS has been estimated with various accuracy and coverage depending on used algorithm. The classification has been performed in the method of global 5 Folds cross-validation. We tested different variants of classifications and 5 Folds gave the best predictive results for our dataset.

Table 3. Sample of the reduced and discretized dataset with attribute UPDRS IV

PatientID	Session	DelayMean	DurationMean	MaxSpeed	UPDRS-IV
13	S4	(−inf–0.335)	(1.94–2.725)	(−inf–5.105)	(3–9)
14	S1	(0.375–inf)	(3.885–inf)	(−inf–5.105)	(0–1)
14	S2	(0.335–0.375)	(3.885–inf)	(−inf–5.105)	(0–1)
14	S3	(0.375–inf)	(3.885–inf)	(5.105–inf)	(3–9)
14	S4	(−inf–0.335)	(2.725–3.885)	(5.105–inf)	(0–1)

We compared two different classifiers available in the RSES - the Decision Rules and the Decomposition Tree. The Decision Rules is based on decision table approach where columns are labeled by attributes, rows by objects of interest and entries of the table are attribute values [1]. Rows of a decision table are referred to "if/then" decision rules which give conditions necessary to make decisions specified by the decision attributes [1]. The Decomposition Tree splits dataset into fragments represented as a tree's leafs. Those subsets of data are used for calculation of decision rules and are supposed to be more uniform and easier to cope with decision-wise [17]. The RSES expresses the results of classifications in two main attributes: Total Accuracy (TA) and Total Coverage (TC). The TA represents the ratio of number of correctly classified cases (sum of values on diagonal in confusion matrix) to the number of all tested cases (number of test objects used to obtain this result) [18]. The TC represents ratio of classified objects from the class to the number of all objects in the class (percentage of test objects that were recognized by classifier) [18]. The best results were achieved with the RSES Decomposition Tree. For attribute UPDRS III classification results indicated 0.85 of TA with TC of 0.48. The value of the UPDRS IV has been estimated with TA of 0.91 and TC of 0.39. Other classifiers i.e. Decision Rules gave worse TA of 0.7 but with much better TC of 1. Tables 4 and 5 are showing the results for best classification where columns represent predicted values and rows represent actual values.

In order to better understand and visualize correlations provided by results we derived the decision trees using WEKA J48 classifier [2]. Analysis of visualization of the obtained trees brought interesting observations. When viewing

Table 4. Result of Decomposion Tree classification with 5 Folds Cross Validation for attribute UPDRS III.

UPDRS III	(3, 13)	(13, 25)	(25, 59)	No. of obj.	Accuracy	Coverage
(3, 13)	0.8	0	0	1.6	0.4	0.333
(13, 25)	0.2	0.2	0	1.6	0.1	0.133
(25, 59)	0.2	0	1	1.8	0.7	0.68
Total accuracy: 0.85				Total coverage: 0.48		

Table 5. Result of Decomposion Tree classification with 5 Folds Cross Validation for attribute UPDRS IV.

UPDRS IV	(3, 9)	(0, 1)	(1, 3)	No. of obj.	Accuracy	Coverage
(3, 9)	1.75	0	0.25	2.75	0.938	0.708
(0, 1)	0.25	0.25	0	1.75	0.125	0.25
(1, 3)	0	0	0.25	2.5	0.25	0.083
Total accuracy: 0.91					Total coverage: 0.39	

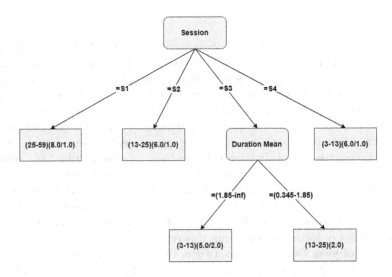

Fig. 2. Decision tree visualization for decision attribute UPDRS III and discretized dataset.

tree describing the UPDRS III attribute (Fig. 2) we can see that it is character-ized by strong correlations with method of treatments represented by attribute Session obscuring the oculometric parameters. However quite strong interplay can be seen between the type of pharmacological treatment of examined patient (S3) and mean duration parameter (Duration Mean). To see direct correlations between UPDRS score and the antisaccade parameters we removed the Session attribute. With unified methods of treatments, the results showed correlations between group of the highest results of the UPDRS III (25–59) and groups of the highest duration (1.85–inf) and delay (0.445–inf) and the group of lowest the speed (−inf–5.075). In Fig. 3 we can see that Duration Mean is the main attribute describing the UPDRS III score group and how values of other oculo-metric parameters are being distributed. A quite different view emerged from the analysis of the decision tree containing attribute UPDRS IV (Fig. 4). Duration Mean applied as the main decision attribute. What seems to be interesting the attribute Session created own branch connected to the group located exactly in middle of Duration Mean values. Additionally attribute Max Speed correlated with the group of highest Mean Duration values.

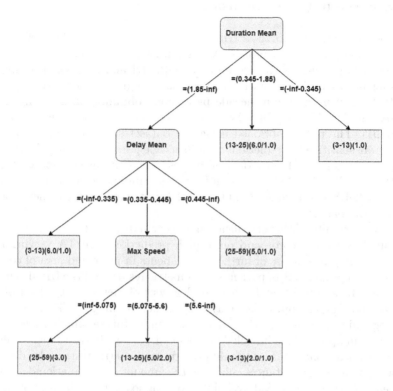

Fig. 3. Decision tree visualization for decision attribute UPDRS III and discretized dataset with attribute Session removed.

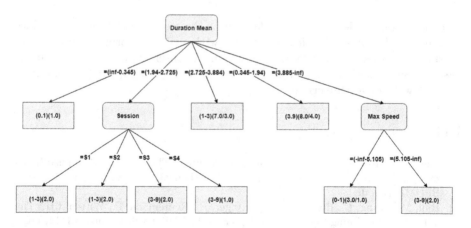

Fig. 4. Decision tree visualization for decision attribute UPDRS IV and discretized dataset.

5   Discussion and Conclusions

As can be seen in the results the attribute Session (describing methods of patients treatment) and attribute Duration Mean (antisaccade parameter) were most sensitive in predicting scoring group of UPDRS III and IV. When comparing results for both UPDRS III an IV, analysis showed greater correlation between UPDRS IV scoring and anitisaccade parameters obtained during oculometric examination. Attribute UPDRS IV also showed better accuracy during predictions (0.91). The results obtained using our dataset suggests that UPDRS IV scale is more sensitive in predictions to anitsaccade parameters than the UPDRS III. In our dataset the Delay Mean also presents as the most important decision attribute. Results of classification for this attribute also shows that we may increase UPDRS predictions by adding antisaccade parameters to neurological dataset of the patient.

We found results of this classification as very indicative taking into account such small group of records and quite high prediction with TA varying from 0.85 to 0.91. It proves sense of further experiments in presented area of correlations, between parameters of patient's oculometric results and the UPDRS motor evaluations. It is hoped that further development of methodology described in this text and similar approaches may help in determining the Parkinson Disease progression. We believe that in the upcoming future, applied algorithms can be used in applications installed on personal devices. Patients then could be free from clinical conditions and could perform oculometric tests under different environments and circumstances enlarging the amount of information describing the disease. Such a widespread availability of diagnostic tools, by increasing the quantity of patients data, should also increase the precision of patient's diagnosis.

6   Ethic Statement

This study was carried out in accordance with the recommendations of Bioethics Committee of Warsaw Medical University with written informed consent from all subjects. All subjects gave written informed consent in accordance with the Declaration of Helsinki. The protocol was approved by the Bioethics Committee of Warsaw Medical University.

References

1. Pawlak, Z.: Rough sets and data mining. Institute of Theoretical and Applied Informatics, Polish Academy of Sciences. http://bcpw.bg.pw.edu.pl/Content/1884/RSDMEAK.pdf
2. Hall, M., Frank, E., Holmes, G., Pfahringer, B., Reutemann, P., Witten, I.H.: The WEKA Data Mining Software: An Update. SIGKDD Explorations, vol. 11, no. 1 (2009)

3. Śledzianowski, A., Szymański, A., Szlufik, S., Koziorowski, D.: Rough set data mining algorithms and pursuit eye movement measurements help to predict symptom development in Parkinson's disease. In: Nguyen, N.T., Hoang, D.H., Hong, T.-P., Pham, H., Trawiński, B. (eds.) ACIIDS 2018. LNCS (LNAI), vol. 10752, pp. 428–435. Springer, Cham (2018). https://doi.org/10.1007/978-3-319-75420-8_41

4. Przybyszewski, A., Szlufik, S., Szymanski, A., Habela, P., Koziorowski, D.: Multimodal learning and intelligent prediction of symptom development in individual Parkinson's patients. Sensors (Basel) 16(9), 1498 (2016)

5. Ramaker, C., Marinus, J., Stiggelbout, A., Van Hilten, M., Bob, J.: Systematic evaluation of rating scales for impairment and disability in Parkinson's disease. Mov. Disord. 17(5), 867–876

6. Bhidayasiri, R., Martinez-Martin, P.: Clinical assessments in Parkinson's disease: scales and monitoring. Int. Rev. Neurobiol. 132, 129–182 (2017)

7. Okai, D., et al.: Parkinson's impulse-control scale for the severity rating of impulse-control behaviors in Parkinson's disease: a semistructured clinical assessment tool. Mov. Disord. Clin. Pract. 3, 494–499 (2016)

8. Crevits, L., Versijpt, J., Hanse, M., De Ridder, K.: Antisaccadic effects of a dopamine agonist as add-on therapy in advanced Parkinson's patients. Neuropsychobiology 42, 202–206 (2000)

9. Fischer, B., Weber, H.: Effects of stimulus conditions on the performance of antisaccades in man. Exp. Brain Res. 116(2), 191–200 (1997)

10. Olk, B., Kingstone, A.: Why are antisaccades slower than prosaccades? A novel finding using a new paradigm. NeuroReport 14(1), 151–155 (2003). Cognitive Neuroscience And Neuropsychology

11. Kitagawa, M., Fukushima, J., Tashiro, K.: Relationship between antisaccades and the clinical symptoms in Parkinson's disease. Neurology 44(12), 2285 (1994)

12. Hood, A.J., Amador, S.C., Cain, A.E., et al.: Levodopa slows prosaccades and improves antisaccades, an eye movement study in Parkinson's disease. J. Neurol. Neurosurg. Psychiatry 78, 565–570 (2007)

13. Everling, S., DeSouza, J.F.X.: Rule-dependent activity for prosaccades and antisaccades in the primate prefrontal cortex. J. Cogn. Neurosci. 17(9), 1483–1496 (2005)

14. Nakamura, T., Funakubo, T., Kanayama, R., Aoyagi, M.: Reflex saccades, remembered saccades and antisaccades in Parkinson's disease. J. Vestib. Res. 4(6), S32 (1996). (Suppl. 1)

15. Everling, S., Fischer, B.: The antisaccade: a review of basic research and clinical studies. Neuropsychologia 36(9), 885–899 (1998)

16. Rivaud-Péchoux, S., Vermersch, A., Gaymard, B., et al.: Improvement of memory guided saccades in Parkinsonian patients by high frequency subthalamic nucleus stimulation. J. Neurol. Neurosurg. Psychiatry 68, 381–384 (2000)

17. Bazan, J.G., Nguyen, H.S., Nguyen, S.H., Synak, P., Wróblewski, J.: Rough set algorithms in classification problem. In: Polkowski, L., Tsumoto, S., Lin, T.Y. (eds.) Rough Set Methods and Applications. Studies in Fuzziness and Soft Computing, vol. 56, pp. 48–88. Physica, Heidelberg (2000). https://doi.org/10.1007/978-3-7908-1840-6_3

18. RSES 2.2 User's Guide Warsaw University. http://logic.mimuw.edu.pl/~rses/RSES_doc_eng.pdf. Accessed 19 Jan 2005

19. Antoniades, C., et al.: An internationally standardised antisaccade protocol. Vis. Res. **84**, 1–5 (2013)
20. Witten, I.H., Frank, E., Hall, M.A.: Data Mining: Practical Machine Learning Tools and Techniques, 3rd edn. Morgan Kaufmann, San Francisco
21. Effects of components of displacement-step stimuli upon latency for saccadic eye movement. J. Opt. Soc. Am. **57**(8), 1024–1029 (1967)

DTI Helps to Predict Parkinson's Patient's Symptoms Using Data Mining Techniques

Artur Chudzik[1]([⊠]) [ID], Artur Szymański[1] [ID], Jerzy Paweł Nowacki[1] [ID], and Andrzej W. Przybyszewski[1,2] [ID]

[1] Polish-Japanese Academy of Information Technology,
Koszykowa 86 St., 02-008 Warsaw, Poland
{artur.chudzik, artur.szymanski,
nowacki, przy}@pjwstk.edu.pl
[2] Department of Neurology, University of Massachusetts Medical School,
65 Lake Avenue, Worcester, MA 01655, USA
andrzej.przybyszewski@umassmed.edu

Abstract. Deep Brain Stimulation (DBS) is commonly used to treat, inter alia, movement disorder symptoms in patients with Parkinson's disease, dystonia or essential tremor. The procedure stimulates a targeted region of the brain through implanted leads that are powered by a device called an implantable pulse generator (IPG). The mentioned targeted region is mainly chosen to be sub-thalamic nucleus (STN) during most of the operations. STN is a nucleus in the midbrain with a size of 3 mm × 5 mm × 9 mm that consist of parts with different physiological functions. The purpose of the study was to predict Parkinson's patient's symptoms defined by Unified Parkinson's Disease Rating Scale (UPDRS) that may occur after the DBS treatment. Parameters had been obtained from 3DSlicer (Harvard Medical School, Boston, MA), which allowed us to track connections between the stimulated part of STN and the cortex based on the DTI (diffusion tensor imaging).

Keywords: Subthalamic nucleus · UPDRS · RSES · MRI · DTI · DBS · Parkinson's disease · Data mining

1 Introduction

Neurodegenerative diseases, in which we could distinguish Parkinson Disease (PD), have their background in neurodegeneration which could be described as progressive loss of structure or function of neurons, including the death of neurons. PD is primarily related to the substantia nigra degeneration which leads to dopamine insufficiency. Standard medication in PD is L-DOPA, which is a precursor of dopamine. However, disease progression affects in L-DOPA efficiency decay which may be revealed in on-off symptom fluctuation.

Thus, the neurologist has often to extend standard medication therapy to DBS (Deep Brain Stimulation) surgery [1]. DBS treatment depends on stimulation of the subthalamic nucleus (STN) which is dorsal to the substantia nigra and medial to the internal capsule. STN is also being known as a "hyper direct pathway" [2] of motor

© Springer Nature Switzerland AG 2019
N. T. Nguyen et al. (Eds.): ACIIDS 2019, LNAI 11432, pp. 615–623, 2019.
https://doi.org/10.1007/978-3-030-14802-7_53

control, contrasting with the direct and indirect pathways implemented elsewhere in the basal ganglia. However, the procedure of application the DBS electrode under the appropriate placement is challenging and may affect in different recovery time and treatment effectiveness.

The searching of localization of the subthalamic nucleus is done mainly by the registrations of neuronal activity via microelectrode recording (MER) [8]. MER is an intraoperative analysis of multi-unit activity (MUA). The commonly used criteria for electrophysiological localization of the STN are qualitative and mainly based on visual and acoustic observations of changes in spike frequencies and background activity. The characteristics of spike trains change during the whole path of brain structures and differ when the electrode passes through the thalamus, zona incerta, lenticular fasciculus, subthalamic nucleus, and the substantia nigra. Bursts in the background activity and sudden increases in the frequency of neuronal spiking are signs that electrode is near to STN. To obtain additional confirmation of the correct electrode placement, supplementary kinesthetic responses measurements aligned with microstimulation are being proceeded. There are two main strategies in searching for STN. First one depends on a single microelectrode which leads to the necessity of multiple passes for correct localization of the motor region. The second one uses 4–5 microelectrode insertions simultaneously. It has to be noticed that any stimulation or manipulation of the non-motor STN region is usually avoided since it can provoke psychiatric and cognitive dysfunctions [7].

To predict the neurological effects related to different electrode-contact stimulations, we have extracted specific parameters acquired from diffusion tensor imaging (DTI). We have demonstrated that with the data mining methods, supported with the rough set theory, it is possible to predict Parkinson's patient's symptoms, according to Unified Parkinson's Disease Rating Scale (UPDRS) [9].

2   Methods

In this research, the subject of study was data acquired from nine patients with advanced PD, which have had DBS electrodes implanted. The primary step was the analysis of the data acquired from the DTI by 3DSlicer software. Those parameters were: two technical values (fiducial region size which determines the tractography radius for selected electrode contact; stopping value - the value of ceasing for the generation of the given tract) and an amount of tract reaching the proximity of given ROI, distinguished between left and right side for every region.

The process of the tractography generation was described in previous works [3, 4]. The generation was carried separately for each contact, and it was on DTI data from the pre-OP DWI (pre-operational diffusion-weighted imaging). The DWI to DTI (diffusion tensor imaging) data was estimated by the use of least squares function approximation. Then, to generate relevant tracts, a proper ROI (region of interest) has been set for each patient, based on electrode contacts. Next, a module called Tractography Interactive Seeding has been applied in order to generate tracts. For every patient, two sets of data have been generated- with a minimal and large (over 30) number of tracts into primary and supplementary motor cortices (Fig. 1). The parameters that were used during the creation of individual tractography were fiducial region size and stopping value, mentioned previously.

The analysis included discovering of the correlation between given attributes aligned with the importance. This operation was performed with the usage of *pandas* library, which is a Python tool for data analysis and statistics [6]. Based on the obtained values, it was possible to select the attributes relevant in the data mining process, which was performed in RSES software.

We have used the RSES 2.1.1 (Rough System Exploration Program) with implementation of RS rules to process our data. An information system [5] can be considered as a pair:

$$S = (U, A)$$

Where:

$U = universe\ of\ objects$
$A = set\ of\ attributes$
$V = set\ of\ values$
$a(u) = unique\ element\ of\ V$
$a \in A$
$u \in U$
A decision table for S is the 3-tuple:

$$S = (U, C, D)$$

Where:

$C = condition\ attributes$
$D = decision\ attributes$

Information table contains rows, where each denotes a particular rule that connects condition and decision attributes for a single measurement of a specific patient.

For results evaluation, we have used a technique called cross-validation, which is a suitable method for estimating the performance of a predictive model, selection of features or parameters adjustment. It is based on the approach of partitioning a data set X into n subsets X_i. Then, given algorithm is performed n times, each time using a different training set $X - X_i$ and validating the results on X_i.

The classifier could be considered to be relevant because of attributes selection that had been done by an algorithm itself. For example, when the left hand tremor is taken as a prediction value, the RSES assumed that relevant attributes are related to, inter alia, hand tracts.

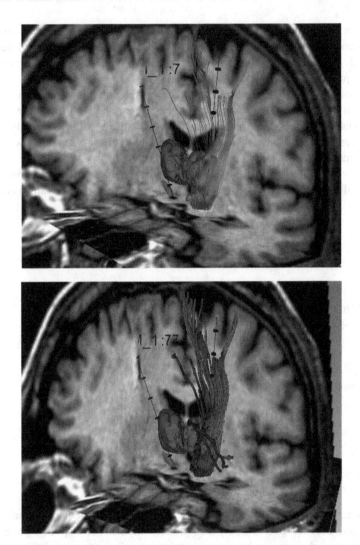

Fig. 1. Screenshots taken from Slicer 3D project of sample patient. We can observe trajectories of both implanted electrodes marked by orange lines, which are aligned with dead tissue visible on MRI slices as a result of surgery. Position of the electrodes is visualized with relevant neighboring structures: STN, Globus Pallidus and Thalamus. DTI tractography is generated from left STN showing connections going to SM and M1 areas of cortex. (Color figure online)

3 Results

The first step was to create a decision table that consists of data attributes obtained from 3DSlicer, based on diffusion tensor imaging, as described previously in the section on Methods. Then, RSES methods were applied, to get decision rules. To achieve that, rows and columns of Table 1 must have been exchanged so that parameters of different patients were in rows, and their results (attributes) were in columns.

Table 1. A fragment of the input dataset. UPDRS <code> - UPDRS value for specific motoric classification of patient' condition; Slicer L/R fiducial region size – Slicer tractography radius for selected electrode contact (in millimeters); Slicer L/R stopping value – Slicer parameter for ceasing generation of the given tract; Slicer L/R tracts lip/hand/foot – number of tract reaching proximity of lip/hand/foot ROI.

Patient ID	10	10	20
UPDRS 21 L Hand action or postural tremor	0	1	2
Slicer R fiducial region size	5	5	5
Slicer R stopping value	0.21	0.21	0.21
Slicer R tracts hand	2	3	3
Slicer R tracts lip	4	15	2
Slicer R tracts foot	15	2	2

In all experiments, we had used the Unified Parkinson Disease Rating Scale (UPDRS) which gave us information about the disease progression in various parts of the body in the context of disease-dependent factor (tremor).

The dataset has been spited into three, smaller subsets, where each was targeted to a different UPDRS marker ("left-hand tremor", "face, lips, chin tremor", "handwriting distortion"). On every subset, we had conducted an experiment based on defined attributes that led to the conclusion of possible UPDRS value after application of DBS treatment on the patient.

3.1 Prediction of the Left-Hand Tremor

Attributes that were relevant during the discretization process were strictly related to Slicer data acquired from the right side of the area, such as fiducial region size, stopping value, tracts for hand, lip and foot (Fig. 2).

The results of the prediction of the left-hand tremor based on given DTI parameters revealed the accuracy of 0.967 with the coverage of 0.425 (Table 2).

3.2 Prediction of the Facial Area Tremor

For this task, chosen attributes were related both to the left and right part of the area. From the left side: fiducial region size, stopping value and lip tracts. From the right side: stopping values, and tracts for lips and foot (Fig. 3).

Prediction result of the "face, chin, lips tremor" was with the accuracy as high as 0.824 with the coverage of 0.7 (Table 3).

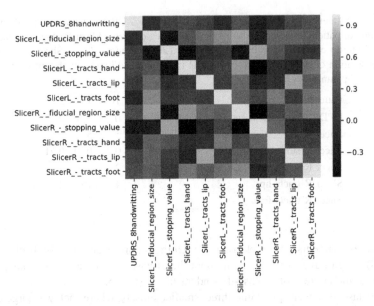

Fig. 2. Computed correlation between Slicer parameters and UPDRS examination: action or postural tremor of Hands – Left hand. The similarity between the two parameters is correlated with the color of the cell in the matrix (lighter shade represents higher similarity and contrariwise). (Color figure online)

Table 2. Confusion matrix of UPDRS #21: action or postural tremor of hands - left hand by rules obtained from 3DSlicer values based on DTI data. Number of tested subjects: 8. Accuracy (total): 0.967. Coverage (total): 0.425. TPR stands for "true positive rate".

		Predicted					
		0	**1**	**2**	**No. of obj.**	**Accuracy**	**Coverage**
Actual	**0**	2.6	0.0	0.0	5.2	**0.8**	0.48
	1	0.2	0.6	0.0	2.0	**0.5**	0.30
	2	0.0	0.0	0.0	0.8	**0.0**	0.00
	TPR	**0.76**	**0.6**	**0.0**			

3.3   Prediction of the Handwriting Disturbances

The last experiment with the detection of UPDRS value change of the "handwriting" test was with the accuracy of 0.878 by the coverage 0.667 (Fig. 4).

The relevant attributes were: DBS and BMT. Furthermore, mainly parameters from the left side Slicer were observed as relevant, such as fiducial region size, stopping value, hand and lip tracts. From the right side, only one technical value (stopping) was taken under consideration (Table 4).

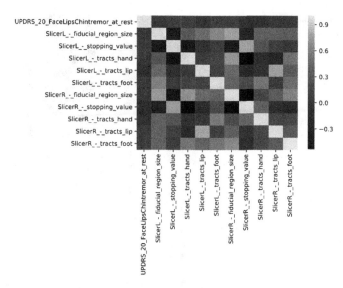

Fig. 3. Computed correlation between Slicer parameters and UPDRS examination: tremor at rest, face, lips, and chin. The similarity between the two parameters is correlated with the color of the cell in the matrix (lighter shade represents higher similarity and contrariwise). (Color figure online)

Table 3. Confusion matrix of UPDRS #20: tremor at rest, face, lips, chin by rules obtained from 3DSlicer values based on DTI data. Total number of tested subjects: 10. Accuracy (total): 0.824. Coverage (total): 0.7.

		Predicted						
		0	**1**	**2**	**3**	**No. of obj.**	**Accuracy**	**Coverage**
Actual	0	4.75	0.50	0.00	0.25	6.5	**0.867**	0.866
	1	0.25	0.75	0.00	0.00	1.5	**0.750**	0.833
	2	0.00	0.00	0.00	0.00	0.5	**0.000**	0.000
	3	0.25	0.00	0.00	0.25	1.5	**0.125**	0.250
	TPR	**0.93**	**0.50**	**0.00**	**0.25**			

4 Discussion

Deep brain stimulation is currently widely applied as a surgical choice of treatment for patients with advanced PD. The benefits of STN stimulation are due to combined mechanisms and involve several adjacent structures. To improve the success of the procedure, more selectivity is needed and both topographical level and stimulation parameters must be enhanced [1].

This article represents the continuation of previous findings presented in [7] that are useful to the surgeon as a tool for confirmation that the subthalamic nucleus is near to the microelectrode path. Furthermore, it extends them even more with data mining techniques to predict the neurological effects related to different electrode-contact stimulations.

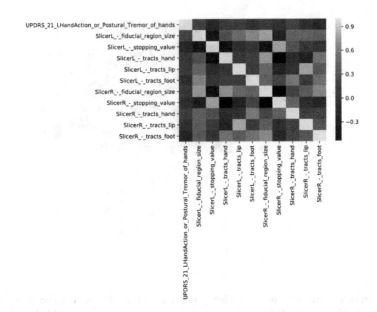

Fig. 4. Computed correlation between Slicer parameters and UPDRS examination: handwriting. The similarity between the two parameters is correlated with the color of the cell in the matrix (lighter shade represents higher similarity and contrariwise). (Color figure online)

Table 4. Confusion matrix of UPDRS #8: handwriting distortion by rules obtained from 3DSlicer values based on DTI data. Total number of tested subjects: 6. Accuracy (total): 0.878. Coverage (total): 0.667.

		Predicted							
		0	1	2	3	4	No. of obj.	Accuracy	Coverage
Actual	0	3.33	0.17	0.00	0.00	0.00	4.00	**0.944**	0.878
	1	0.00	0.00	0.00	0.00	0.00	0.83	**0.000**	0.000
	2	0.00	0.00	0.17	0.00	0.00	0.50	**0.167**	0.083
	3	0.00	0.00	0.33	0.00	0.00	0.33	**0.000**	0.167
	4	0.00	0.00	0.00	0.00	0.00	0.33	**0.000**	0.000
	TPR	**1.00**	**0.00**	**0.17**	**0.00**	0.00			

5 Conclusions

Our recent research described above was meant to determine if data mining can predict possible Parkinson's patient's symptoms based only on the DTI data of patients who go through the DBS surgery. We have applied the rough set theory on the data obtained from DTI after the operation to conclude whether it is possible to create a system that is unbiased of human opinion.

The results have shown that it is possible to introduce a new, autonomous, doctor independent and a highly accurate method of disease course prediction.

What is more, this approach enables a new way for a deduction of an impact of a region-specific stimulation of STN and its effect on patients.

However, since this results have been based on a small data set, further work is required to perform more credible statistics and verification in the sake of elimination of overfitting problem.

References

1. Benabid, A.L., et al.: Deep brain stimulation of the subthalamic nucleus for the treatment of Parkinson's disease. Lancet Neurol. **8**(1), 67–81 (2009)
2. Nambu, A., Tokuno, H., Takada, M.: Functional significance of the cortico–subthalamo–pallidal 'hyperdirect' pathway. Neurosci. Res. **43**(2), 111–117 (2002)
3. Szymański, A., Przybyszewski, A.W.: Rough set rules help to optimize parameters of deep brain stimulation in Parkinson's patients. In: Ślęzak, D., Tan, A.-H., Peters, J.F., Schwabe, L. (eds.) BIH 2014. LNCS (LNAI), vol. 8609, pp. 345–356. Springer, Cham (2014). https://doi.org/10.1007/978-3-319-09891-3_32
4. Szymański, A., Kubis, A., Przybyszewski, A.W.: Data mining and neural network simulations can help to improve deep brain stimulation effects in Parkinson's disease. Comput. Sci. **16**(2), 199 (2015)
5. Pawlak, Z.: Rough set theory and its applications. J. Telecommun. Inf. Technol., 7–10 (2002)
6. McKinney, W.: Data structures for statistical computing in python. In: Proceedings of the 9th Python in Science Conference, vol. 445 (2010)
7. Przybyszewski, A.W., et al.: Multi-parametric analysis assists in STN localization in Parkinson's patients. J. Neurol. Sci. **366**, 37–43 (2016)
8. Benazzouz, A., et al.: Intraoperative microrecordings of the subthalamic nucleus in Parkinson's disease. Mov. Disord. Official J. Mov. Disord. Soc. **17**(S3), S145–S149 (2002)
9. Movement Disorder Society Task Force on Rating Scales for Parkinson's Disease: The unified Parkinson's disease rating scale (UPDRS): status and recommendations. Mov. Disord. **18**(7), 738–750 (2003)

Fractional Calculus in Human Arm Modeling

Tomasz Grzejszczak[1], Piotr Jurgaś[1], Adrian Łęgowski[1],
Michał Niezabitowski[1,2(✉)], and Justyna Orwat[3,4]

[1] Faculty of Automatic Control, Electronics and Computer Science,
Silesian University of Technology, Akademicka 16 Street, 44-100 Gliwice, Poland
{tomasz.grzejszczak,michal.niezabitowski}@polsl.pl
[2] Faculty of Mathematics, Physics and Chemistry, Institute of Mathematics,
University of Silesia, Bankowa 14, 40-007 Katowice, Poland
mniezabitowski@us.edu.pl
[3] Faculty of Mining and Geology, Silesian University of Technology,
Akademicka 2 Street, 44-100 Gliwice, Poland
justyna.orwat@polsl.pl
[4] Faculty of Civil Engineering, Silesian University of Technology,
Akademicka 5 Street, 44-100 Gliwice, Poland

Abstract. Human limbs from kinematic point of view can be considered as simple robots' manipulators. The first part is dedicated to kinematics of human arm modeled as three-link planar manipulation system. For dynamics we propose simple 2-DOF nonlinear model with use of fractional calculus. According to the latest research fractional systems have "natural" damping. This means that even simple model may be able to show some additional properties of the object. Moreover, in presented paper we study the impact of approximation method on solving the inverse kinematics for 3-DOF human limb as well as some parameters of compared methods. This part of research may have some value from visualization point of view. Solving the Inverse Kinematics is the first step in getting full information about the system. The second part of research may be of use in simplifying models. Creating ideologically simple model may let us understand the nature of the world.

Keywords: Human limb · 2-DOF model · Fractional calculus

1 Introduction

The human limbs are dynamical systems that are very complex and hard to describe. However, in our opinion the approach considering only the dynamical

The research presented here was supported by Polish Ministry for Science and Higher Education for Institute of Automatic Control, Silesian University of Technology, Gliwice, Poland under internal grant BKM-508/RAU1/2017 (M.N.). Moreover, the research was done as parts of the project funded by the National Science Centre in Poland granted according to decisions DEC-2015/19/D/ST7/03679 (P.J.), DEC-2017/25/B/ST7/02888 (A.L.).

N. T. Nguyen et al. (Eds.): ACIIDS 2019, LNAI 11432, pp. 624–636, 2019.
https://doi.org/10.1007/978-3-030-14802-7_54

models is not entirely correct. The solution of direct or inverse kinematics may be also of use. With use of both direct and inverse kinematics The visualization of the motion of the arm may be created. Therefore, studying this topic is important. Cruse and Brüwer [8] presented the concept of the human limb kinematics and the manipulation system.

The human arm may be considered as a complex manipulation system with some degree of accuracy. In a great simplification it can be considered as a 3-DOF (degrees of freedom) model. On the other hand, adding additional degrees of freedom provides more complex equations that are more time consuming in terms of calculation, yet it does not change the idea behind the method.

The Inverse Kinematics (IK) solution can be found in various ways, for example by finding the analytical formulas [18]. Usually the following equations are used:

$$\frac{dx}{dt} = J(q)\frac{dq}{dt}, \tag{1}$$

where q is the vector of coordinates in joint space, $J(q)$ is the Jacobian matrix for given q and x is the vector of coordinates in Cartesian space. For given target manipulator position (X_{ref}) the Eq. (1) can be rewritten as:

$$J^{-1}(q_{i-1})\Delta x_i = \Delta q_i, \tag{2}$$

where $\Delta x_i = X_{ref} - x_{i-1}$, $q_i = \Delta q_i + q_{i-1}$, $i = 1, 2, 3, \dots$.

The process of finding solution is iterative and its accuracy depends mainly on the number of iterations. For example, in computer graphics it is necessary to compute the trajectory of joints motion. The realization of the Cartesian requires computing the IK solution for every given X_{ref} (e.g. the center of the hand's palm region). In each step of the algorithm the demanded position (X_{ref}) and previous values of Cartesian and joint space coordinates are taken into account. It changes the previous equations into:

$$q_{i-1} + J^{-1}(q_{i-1})(X_{ref}(j\Delta t) - x_{i-1}) = q_i, \quad j = 1, 2, 3, \dots . \tag{3}$$

By the local nature of the integer-order derivative, the value of x is computed by solving the forward kinematics problem from previous iteration for given $X_{ref}(j\Delta t)$.

It is commonly suggested by researchers to use *Moore-Penrose pseudoinverse* (*MPp*) of matrix in computations. This approach allows to use the expression (3) for even redundant structures. One can write this as:

$$J^{\#}(q_{i-1})\Delta x(i) = \Delta q(i) \tag{4}$$

and rewrite it as:

$$J^{\#}(q_{i-1})(X_{ref} - x_{i-1}) = q_i - q_{i-1}, \quad q_{i-1} + J^{\#}(q_{i-1})(X_{ref} - x_{i-1}) = q_i \tag{5}$$

which gives:

$$q_{i-1} + J^{\#}(q_{i-1})(X_{ref}(j\Delta t) - x_{i-1}) = q_i, \tag{6}$$

where $J^{\#}$ is the *MPp* of matrix J. It is assumed that q_i represents the $i - th$ computed value of joints and x_i is the $i - th$ computed value of Cartesian space coordinates. These values are not time-dependent and hold the computation results.

This simplified method provides an aperiodic joints motion for certain cyclic hand trajectories [11]. It is sometimes called closed loop pseudoinverse (CLP). Due to the fact that human arms are redundant, this problem mostly concerns redundant structures and specific trajectories.

Dynamical models of the human arm can be found in the literature [6, 16, 20, 22]. Due to the important fact that the hand shape is changeable, the dynamics of the described system is not easily describable. Hand shape depends on the muscle contraction state, rotation in joints, etc. Moreover these effects may be caused by various means. The switched linear models [2–4] are proposed to be utilized. Presented work assumes that human arm can be modeled as a two-link planar system. Due to the the simplification, the concept stands and proves to be relatively simple to describe and implement.

Some literature [12, 14, 17] suggest that the human arm has viscoelastic properties. Modeling human limbs is complicated and has to be considered with different effects, e.g. the muscle and skin dumping and elasticity. All these effects are very difficult to describe due to individual properties among people.

Considering human arm as the rigid manipulator can cause the large loss of accuracy. However this approach is simple, standard and can be described in the following way.

The model described in paper [2] consist of two rotational joints. The paper deals with human arm model as the state-space equations. The simplest approach assumes only vertical motion. This simplification makes the model of human arm equals to model of rigid two-link planar manipulator.

However, this [2] study does not consider the damping. Thus, the response, as expected in this kind of structures, will be in the form of undamped oscillations.

The origin of this study is the equation

$$M(q)\frac{d^2q}{dt^2} + C\left(q, \frac{dq}{dt}\right)\frac{dq}{dt} + G(q) = \tau, \tag{7}$$

that is the result of Euler-Lagrange description. Here $M(q)$ - is the inertia matrix, $C(q, \frac{dq}{dt})$ is the Coriolis and centrifugal forces matrix and $G(q)$ is the matrix representing gravity forces. The vector τ represents the forces and moments of the drives. The following case describes only two torques. These matrices are in the forms:

$$M(q) = \begin{bmatrix} d_1 & d_3cos(q_1 - q_2) \\ d_3cos(q_1 - q_2) & d_2 \end{bmatrix},$$

$$C(q, \dot{q}) = \begin{bmatrix} 0 & d_3sin(q_1 - q_2)\dot{q}_2 \\ -d_3sin(q_1 - q_2)\dot{q}_1 & 0 \end{bmatrix}, \quad G(q) = \begin{bmatrix} -d_4gsinq_1 \\ -d_5gsinq_2 \end{bmatrix},$$

where

$$d_1 = m_1a_{c1}{}^2 + m_2a_1{}^2 + I_1, \quad d_2 = m_2a_{c2}{}^2 + I_2,$$

$$d_3 = m_2a_1a_{c2}, \quad d_4 = m_1a_{c1} + m_2a_2, \quad d_5 = m_2a_{c2}$$

and m_i - is the mass of $i-th$ joint, a_i- is the $i-th$ link length, a_{ci} - is the distance from the $i-th$ joint (its coordinate system) to the center of (its) mass, I_i - is the $i-th$ joint moment of inertia.

The solution of the nonlinear differential Eq. (7) is the response for given torques and provides the functions describing the inner coordinates of the system (Fig. 1).

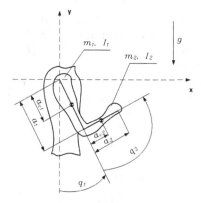

Fig. 1. Simplified model of the human arm (source: [2])

2   Fractional Calculus

The idea of non-integer order derivative and integral is nearly as old as well-known integer-order calculus. It goes back to the 1695 and Leibniz's letter to L'Hospital [23]. There are many definitions of fractional derivative and integral, but here we present the three most popular.

The Riemann-Liouville (RL) defines the derivative of non-integer order $0 < \alpha < 1$ as [25]:

$$_0D_t^\alpha y(t) = \frac{1}{\Gamma(1-\alpha)} \frac{d}{dt} \int_0^t (t-\tau)^{-\alpha} y(\tau) d\tau. \tag{8}$$

The Caputo defines the derivative as [13]:

$$_0D_t^\alpha y(t) = \frac{1}{\Gamma(1-\alpha)} \int_0^t \frac{y'(\tau)}{(t-\tau)^\alpha} d\tau. \tag{9}$$

Finally, Grünwald-Letnikov defines [13,25]:

$$_aD_t^\alpha y(t) = \lim_{\Delta t \to 0} \left[\frac{1}{(\Delta t)^\alpha} \sum_{k=0}^{\frac{t-a}{\Delta t}} y(t - k\Delta t)\gamma(\alpha, k) \right], \tag{10}$$

$$\gamma(\alpha, k) = (-1)^k \frac{\Gamma(\alpha+1)}{\Gamma(k+1)\Gamma(\alpha-k+1)}. \tag{11}$$

The function $\Gamma()$ is the Gamma function.

Usually, the implementation of (10) approximation is done with use of the short-memory principle [21] by formula (12):

$$D^{\alpha}y(t) \approx \frac{1}{(\Delta t)^{\alpha}} \sum_{k=0}^{N} y(t - k\Delta t)\gamma(\alpha, k),$$ (12)

where Δt is sampling time and N is the truncation order [11].

The last definition can be easily used in software implementation, especially when the value of γ can be stored in *lookup table* (LUT) and does not need to be calculated in every iteration.

In this paper we - among others - compare the operator given by (10) with the Al-Alaoui operator [5] after power series expansion (PSE):

$$s^{\alpha} \approx (\frac{8}{7\Delta t}\frac{1 - z^{-1}}{1 + \frac{z^{-1}}{7}})^{\alpha}.$$ (13)

Having the expression (13) we can write that:

$$D^{\alpha}y(t) \approx \left(\frac{8}{7\Delta t}\right)^{\alpha} \sum_{k=0}^{\infty} \left[\sum_{j}^{k} h(\alpha, j, k)\right] y(t - k\Delta t),$$ (14)

where function γ is defined as in (11) and

$$h(\alpha, j, k) = (\frac{1}{7})^{k-j}\gamma(\alpha, j)\frac{\Gamma(-\alpha + 1)}{\Gamma(k - j + 1)\Gamma(-\alpha - k + j + 1)}.$$ (15)

Grünwald-Letnikov operator is simpler then the presented formula, but with use of LUT, proper implementation and truncation the computation is not so time consuming.

Currently researchers are looking for new applications of fractional calculus (FC) in various branches of science. Many researchers prove that the FC can be applied in control theory in order to design new type of controllers [7,19]. In paper [1] presents the study on fractional continuous models. The main requirement in fractional models implementation is the accurate approximation methods. Usually, the approaches are based on the approximation of s^{α} in Laplace domain [5,25].

Some papers proves that using FC for modeling purposes may result in better accuracy with lower number of parameters [24].

3   Inverse Kinematics with Fractional Calculus

Among many applications of FC, one can find a study on the application of FC to the CLP method [11] and a deep insight into the repeatability problem for the redundant manipulators [15]. We believe that similar approach may be used in case of human limbs. In the ideal case, the fractional-order derivative is the global operator with a memory of all past events. In case of desired cyclic hand trajectories, the memory property may force the periodic motion which should prevent the unnatural behavior in computer visualizations. The formula (4) can be rewritten with consideration of the approximation of Grünwald-Letnikov derivative, in $j - th$ time moment, and following the results given in [11,15] as:

$$J^{\#}(q_{i-1})\Delta^{\alpha}x(i) = \Delta^{\alpha}q(i), \tag{16}$$

$$q_i + \sum_{k=1}^{N}\gamma(\alpha,k)q_{i-k} = J^{\#}(q_{i-1})\left[X_{ref}(j\Delta t) + \sum_{k=1}^{N}\gamma(\alpha,k)x_{i-k}\right], \tag{17}$$

where N is the truncation order and

$$\gamma(\alpha,k) = (-1)^k \frac{\Gamma(\alpha+1)}{\Gamma(k+1)\Gamma(\alpha-k+1)}. \tag{18}$$

Having formula (17) we can compute:

$$q_i = J^{\#}(q_{i-1})\left[X_{ref}(j\Delta t) + \sum_{k=1}^{N}\gamma(\alpha,k)x_{i-k}\right] - \sum_{k=1}^{N}\gamma(\alpha,k)q_{i-k}. \tag{19}$$

After substituting Al-Alaoui operator with similar transformations the expression (16) can be rewritten as:

$$q_i = J^{\#}(q_{i-1})\left[X_{ref}(j\Delta t) + \sum_{k=1}^{N}\left[\sum_{j}^{k}h(\alpha,j,k)\right]x_{i-k}\right] - \sum_{k=1}^{N}\left[\sum_{j}^{k}h(\alpha,j,k)\right]q_{i-k}, \tag{20}$$

where $h(\alpha,j,k)$ is given by (15).

The iterative procedure for finding the solution for IK problem can be design with use of Eqs. (19) and (20). The value of q_i is computed for given X_{ref} at the time $j\Delta t$. The reference position as a function of time is the only variable. However, this may cause the large memory usage in computer systems. One should strongly pay attention to requirements.

In paper [11] authors suggest that lowering the differential order α lowers the positioning accuracy. We can confirm that for most studied trajectories. To address this issue we propose a variable order α. It has been proven that integer order derivation maintains high positional accuracy. Having that in mind we can write that derivation order is given by the expression (21):

$$\alpha(c) = \begin{cases} 1 & \text{if } c \geq I - d, \\ \alpha_s & \text{if } c < I - d, \end{cases} \tag{21}$$

where α_s is the initial order of derivative, I is the maximal number of iterations, d defines the number of iterations with integer-order derivation and c is the iteration number between $X_{ref}((j-1)\Delta t)$ and $X_{ref}(j\Delta t)$. Thus, the Eq. (19) can be rewritten as:

$$q_i = J^\#(q_{i-1}) \left[X_{ref}(j\Delta t) + \sum_{k=1}^{N} \gamma(\alpha(c), k) x_{i-k} \right] - \sum_{k=1}^{N} \gamma(\alpha(c), k) q_{i-k} \qquad (22)$$

and the Eq. (20) as:

$$q_i = J^\#(q_{i-1}) \left[X_{ref}(j\Delta t) + \sum_{k=1}^{N} \left[\sum_{j}^{k} h(\alpha(c), j, k) \right] x_{i-k} \right] - \sum_{k=1}^{N} \left[\sum_{j}^{k} h(\alpha(c), j, k) \right] q_{i-k}. \qquad (23)$$

Proposed approach may improve the accuracy (comparing to known from some other papers). This approach has been developed for solving the problem of redundancy for industrial manipulators.

4   Human Arm Fractional Dynamics

In literature it has been stated that the fractional oscillatory models offers very strong damping [10]. The same effect may be noticed in fractional manipulator system (and general pendulum systems) [9]. In the paper there is a prove that simple model based on fractional calculus has very complex properties.

We propose to change the derivatives order in Eq. (7) to non-integer orders. This modification would lead to the following equation:

$$M(q)_0 D_t^{1+\alpha} + C(q, _0 D_t^\alpha)_0 D_t^\alpha + G(q) = \tau, \qquad (24)$$

where α is the order of fractional derivative. Its main influence should be on damping properties of the system.

With use of this method the complex human arm dumping properties can be included. It is assumed that fractional model of human arm will be strongly damped, as it is in other fractional oscillatory systems. Substituting $\alpha = 1$ provides a standard, integer-order approach.

Solving the Euler-Lagrange equation in general form leads to much different equation. Thus, presented approach has to be considered only as the approximation. The accuracy of that approximation needs to be reviewed by proper experimental data.

5   Simulation - Kinematics

The first experiment shows the influence of memory length (N) for two fractional operators (Grünwald-Letnikov and Al-Alaoui). The trajectories for manipulators' joints are presented in Figs. 2, 3, 4, 5, 6, 7. It is obvious that for implementation the smallest passible N would be used. All results are presented for periodical trajectory in Cartesian space.

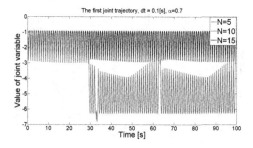

Fig. 2. First joint trajectory with use of Grünwald-Letnikov operator for various N [own source]

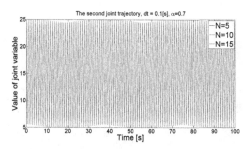

Fig. 3. Second joint trajectory with use of Grünwald-Letnikov operator for various N [own source]

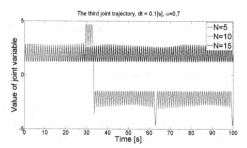

Fig. 4. Third joint trajectory with use of Grünwald-Letnikov operator for various N [own source]

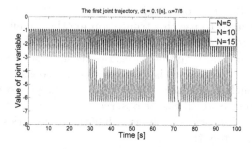

Fig. 5. First joint trajectory with use of Al-Alaoui operator for various N [own source]

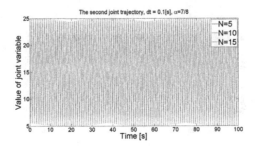

Fig. 6. Second joint trajectory with use of Al-Alaoui operator for various N [own source]

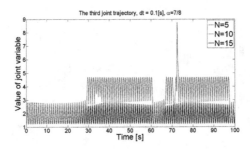

Fig. 7. Third joint trajectory with use of Al-Alaoui operator for various N [own source]

In Figs. 2, 3, 4 one can observe that extending the memory may improve the trajectory shape. For integer operator this path requires rotations of the first joint which is impossible if we consider human arm.

One can observe that there are small differences in (Figs. 2, 3, 4, 5, 6, 7). For $N = 5$ the algorithm does not perform well. The trajectory is stable however, its shape is unacceptable. In this test we require a periodic motion for every joint with extreme values as close to zero as possible. Considering this criterion we conclude that $N = 15$ may be sufficient for our test trajectory. The small number of N meets the other requirement about small memory consumption. A low-cost implementation can be allowed by a proper truncation of the sums in approximation method. With use of the LUT there should not be any performance issues during the computation of γ or $h(\alpha, j, k)$. Since for $N = 15$ we obtain relatively good results, we decided to use this truncation in other simulations.

The joint trajectories for various order α are presented in the Figs. 8, 9, 10. In these figures one can observe that integer-order approach causes the multiple rotations for the first joint. This is obviously wrong considering the nature of human limb. We conclude that for specific cases lowering the order may improve the motion performance which can be observed in the Fig. 10 for $\alpha = 0.4$. Moreover, from Figs. 8, 9, 10, one can observe that the trajectories for $\alpha = 0.4$ and $\alpha = 0.8$ are approximately the same.

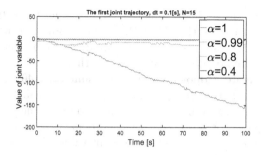

Fig. 8. First joint trajectory with use of Al-Alaoui operator for various α [own source]

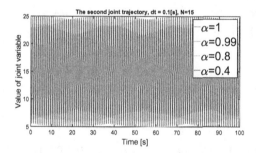

Fig. 9. Second joint trajectory with use of Al-Alaoui operator for various α [own source]

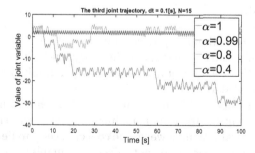

Fig. 10. Third joint trajectory with use of Al-Alaoui operator for various α [own source]

6   Simulation - Dynamics

Solving the Eq. (24) numerically one can try to find the following function $q = f(t, \tau(t))$ that satisfies the Eq. (24). In our studies we used Oustaloup's approximation [25]. As for now, the presented research concerns simulation results only. There are no experimental data at the moment.

In order to evaluate proposed method we simulated 40 s of motion of the human arm. The simulation starts with no torque applied. In 10th second we apply torque to the first joint and in 30th second to the second joint. The responses are compared for various derivation order α in order to expose the influence of this parameter on the trajectory of joint values. We compared responses for $\alpha = 0.1, 0.2, 0.6, 0.8, 0.9$.

All simulation results are presented in figures, that allows for comparison of these two models.

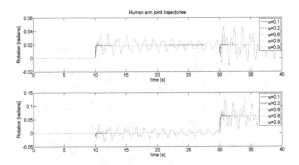

Fig. 11. Comparison for $\alpha = 0.1, 0.2, 0.6, 0.8, 0.9$

According to the literature, the fractional systems are damped by their nature [10]. This can be observed in the Fig. 11.

7   Conclusions

In this paper we presented our approach to fractional order calculus and modeling human arm. This study is only the preliminary study. It is clear that further experimental investigation is required. However, our results are consistent and they fit current knowledge state.

Using fractional calculus for improving CLP algorithm results in very good trajectories. Implementing fractional nonlinear equation as the model of the human arm requires evaluation, however the concept seems to be very interesting. Moreover, presented approach does not affect the singular configurations since the Jacobian matrix is computed without any adjustments (see Figs. 8, 9, 10 for $\alpha = 1$). This is one of the reasons why we use fractional model of human arm.

References

1. Aoun, M., Malti, R., Levron, F., Oustaloup, A.: Numerical simulations of fractional systems: an overview of existing methods and improvements. Nonlinear Dyn. **38**(1), 117–131 (2004)
2. Babiarz, A.: On mathematical modelling of the human arm using switched linear system. In: AIP Conference Proceedings, vol. 1637, pp. 47–54 (2014)
3. Babiarz, A., Czornik, A., Niezabitowski, M., Zawiski, R.: Mathematical model of a human leg: the switched linear system approach. In: 2015 International Conference on Pervasive and Embedded Computing and Communication Systems (PECCS), pp. 1–8. IEEE (2015)
4. Babiarz, A., Klamka, J., Zawiski, R., Niezabitowski, M.: An approach to observability analysis and estimation of human arm model. In: 11th IEEE International Conference on Control & Automation (ICCA), pp. 947–952. IEEE (2014)
5. Barbosa, R.S., Machado, J.A.T.: Implementation of discrete-time fractional-order controllers based on LS approximations. Acta Polytech. Hung. **3**(4), 5–22 (2006)
6. Biess, A., Flash, T., Liebermann, D.G.: Riemannian geometric approach to human arm dynamics, movement optimization, and invariance. Phys. Rev. E **83**(3), 031927 (2011)
7. Cao, J.Y., Cao, B.G.: Design of fractional order controllers based on particle swarm optimization. In: 2006 1ST IEEE Conference on Industrial Electronics and Application, pp. 1–6, May 2006
8. Cruse, H., Brüwer, M.: The human arm as a redundant manipulator: the control of path and joint angles. Biol. Cybern. **57**(1), 137–144 (1987)
9. David, S., Balthazar, J.M., Julio, B., Oliveira, C.: The fractional-nonlinear robotic manipulator: modeling and dynamic simulations. In: AIP Conference Proceedings, pp. 298–305 (2012)
10. David, S.A., Valentim, C.A.: Fractional Euler-Lagrange equations applied to oscillatory systems. Mathematics **3**(2), 258–272 (2015)
11. Duarte, F.B.M., Machado, J.A.T.: Pseudoinverse trajectory control of redundant manipulators: a fractional calculus perspective. In: Proceedings of the 2002 IEEE International Conference on Robotics and Automation, ICRA 2002, Washington, DC, USA, 11–15 May 2002, pp. 2406–2411 (2002)
12. Frolov, A.A., Prokopenko, R., Dufosse, M., Ouezdou, F.B.: Adjustment of the human arm viscoelastic properties to the direction of reaching. Biol. Cybern. **94**(2), 97–109 (2006)
13. Garrappa, R.: A Grünwald-Letnikov scheme for fractional operators of Havriliak-Negami type. Recent Adv. Appl. Model. Simul. **34**, 70–76 (2014)
14. Gomi, H., Osu, R.: Task-dependent viscoelasticity of human multijoint arm and its spatial characteristics for interaction with environments. J. Neurosci. **18**(21), 8965–8978 (1998)
15. da Graça Marcos, M., Duarte, F.B., Machado, J.T.: Fractional dynamics in the trajectory control of redundant manipulators. Commun. Nonlinear Sci. Numer. Simul. **13**(9), 1836–1844 (2008)
16. Van der Helm, F.C., Schouten, A.C., de Vlugt, E., Brouwn, G.G.: Identification of intrinsic and reflexive components of human arm dynamics during postural control. J. Neurosci. Methods **119**(1), 1–14 (2002)
17. Kubo, K., Kanehisa, H., Kawakami, Y., Fukunaga, T.: Influence of static stretching on viscoelastic properties of human tendon structures in vivo. J. Appl. Physiol. **90**(2), 520–527 (2001)

18. Łęgowski, A.: The global inverse kinematics solution in the adept six 300 manipulator with singularities robustness. In: 2015 20th International Conference on Control Systems and Computer Science, pp. 90–97, May 2015

19. Mackowski, M., Grzejszczak, T., Łęgowski, A.: An approach to control of human leg switched dynamics. In: 2015 20th International Conference on Control Systems and Computer Science (CSCS), pp. 133–140, May 2015

20. Mobasser, F., Hashtrudi-Zaad, K.: A method for online estimation of human arm dynamics. In: 28th Annual International Conference of the IEEE 2006 Engineering in Medicine and Biology Society, EMBS 2006, pp. 2412–2416. IEEE (2006)

21. Podlubny, I.: Fractional Differential Equations: An Introduction to Fractional Derivatives, Fractional Differential Equations, to Methods of Their Solution and some of Their Applications, vol. 198. Academic press (1998)

22. Rosen, J., Perry, J.C., Manning, N., Burns, S., Hannaford, B.: The human arm kinematics and dynamics during daily activities-toward a 7 DOF upper limb powered exoskeleton. In: 2005 Proceedings of 12th International Conference on Advanced Robotics, ICAR 2005, pp. 532–539. IEEE (2005)

23. Ross, B.: A brief history and exposition of the fundamental theory of fractional calculus. In: Ross, B. (ed.) Fractional Calculus and Its Applications. LNM, vol. 457, pp. 1–36. Springer, Heidelberg (1975). https://doi.org/10.1007/BFb0067096

24. Sabatier, J., Aoun, M., Oustaloup, A., Grégoire, G., Ragot, F., Roy, P.: Fractional system identification for lead acid battery state of charge estimation. Sign. Process. **86**(10), 2645–2657 (2006)

25. Vinagre, B., Podlubny, I., Hernandez, A., Feliu, V.: Some approximations of fractional order operators used in control theory and applications. Fractional Calc. Appl. Anal. **3**(3), 231–248 (2000)

Computer Vision and Intelligent Systems

Real-Time Multiple Faces Tracking with Moving Camera for Support Service Robot

Muhamad Dwisnanto Putro and Kang-Hyun Jo[(✉)]

School of Electrical Engineering, University of Ulsan, Ulsan, Korea
dputro@islab.ulsan.ac.kr, acejo@ulsan.ac.kr

Abstract. This paper proposes a real-time robot vision system to track multiple faces. This system supports service robots to communicate with consumers simultaneously. The Viola-Jones algorithm to detect faces early in the process, while the Kanade-Lucas-Tomasi algorithm is used to track detected facial features. The system follows the multiple human faces and synchronizes software and hardware to move the webcam in the middle position of the frame. Extraction of the center point from a set of faces as core information for controlling webcam movement. The challenge that can be overcome is that it can maintain multiple faces remains in the middle of the frame with various poses and the use of accessories. This system uses the PID controller which makes the webcam move in a fast to follow the face and maintain the stability and accuracy of actuator movements when speed increases. The experiments were done in the 42.052 frames per second as the maximum speed of system performance.

1 Introduction

The needs in service robots is increasing rapidly in the world. Now developed with extensive and widely applied in industries, offices, and homes. The service robot has the main task of helping people (consumers) in public places by interacting and communicating well. This is the main requirement when this robot does its task. Because of a misunderstanding of communication with consumers, some tasks failed. Frequent problems that occurs when interacting with multiple consumers when the static camera is not able to maintain the position of all consumer's faces to be in the middle of the frame. This causes some information to be lost when multiple consumers communicate in front of the robot. So we need a face tracking system with camera movements to overcome this problem. The process of face tracking begins with face detection using The Viola-Jones Algorithm, then proceed with tracking facial features using the KLT (Kanade-Lucas-Tomasi) algorithm that has been widely used to detect the best features. Multiple faces must be resolved by this algorithm is a major challenge in our paper. The proposed system can maintain all faces in the center of the frame by moving the webcam. Calculation of error values is used to determine deviations

© Springer Nature Switzerland AG 2019
N. T. Nguyen et al. (Eds.): ACIIDS 2019, LNAI 11432, pp. 639–647, 2019.
https://doi.org/10.1007/978-3-030-14802-7_55

from the position of a set of faces that are not in the center of the frame. The proposed system has the advantage of being able to move fast to drive the actuator using a PID control system. The results of this system are supporting service robots to communicate and recognize the expressions of multiple consumer faces simultaneously.

2 Related Work

The robot vision system serves as the senses for service robots. Face detection and tracking are the foundation of the vision system used to recognize humans as consumers [5,7,14]. The vision system with real-time work is mostly used in robots to detect objects quickly and early. Including facial tracking systems, many teams implemented advanced technologies [4,6,9]. Some tasks have been completed by taking the main object as a human face. The Viola-Jones algorithm introduces robust and fast face detection in finding features distinctive on the face. This algorithm uses the training data for the process of image classification based on the value of a feature [11]. Tracking multiple faces in real-time videos is necessary for a service robot to interact with multiple consumers. The famous algorithm is KLT used to track the movements of the best distinctive features [3,10,13]. Future challenges in the face tracking system are about the number of faces, illuminations, accessories, facial expressions, and head orientation are still minimal from their performance [8]. Proposed method in this paper attempts to solve facial tracking problems such as multiple faces, accessories, and pose face with increasing the performance of speed.

3 System Overview

Several approaches that have been presented before are used to build this system. In general, this system uses the Viola-Jones algorithm to perform face detection and KLT algorithms to track features on the face. In this section, we also explain the methods to make the camera movement faster. The whole process of the tracking algorithm shown in Fig. 1. This real-time system input is obtained from a webcam connected to a personal computer via a USB cable transmission and then processes tracking with the help of software. The information from the tracking process is extraction, then sent to Arduino Uno as a controller. The actuators used to consist of two DC servo motors (yaw and pitch axis), each actuator assigned to rotate the webcam.

3.1 Multiple Faces Detection and Tracking

The Viola-Jones algorithm is used to detect faces in this system. This algorithm has a good speed and it is implemented in many real-time systems. Even today many are implemented in various vision technologies [12]. This algorithm has four main components that are, Haar-like features, integral image, AdaBoost

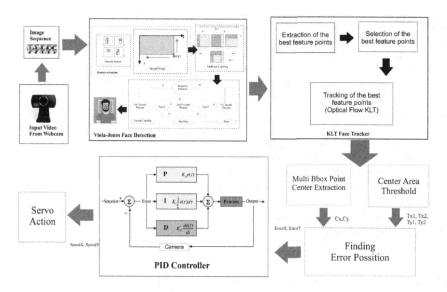

Fig. 1. Flowchart of overall tracking system.

learning, and cascade classifier as a determinant to the face area in the image [15]. The face tracking process on this system uses KLT tracker. The KLT algorithm is generally widely used for object tracking by optical flow estimation. This algorithm extracts detected features from objects in the first frame and performing tracking in subsequent frames.

The process of tracking multiple faces is more challenging than tracking a single face. The new faces detected must be matched with the previous face was detected, by comparing whether the new bounding box overlaps with the previous bounding box. If it is not overlapping, it is considered a new face. This process is explained in Algorithm 1. In this case, the $NextId$ value as the face ID storage will increase. At the end of the algorithm, it is possible to update $BoxScores$ when face already existed before. We set the threshold value for the area intersection between the bounding boxes to be 0.1. If the value of the intersection area is greater than the threshold, then the bounding box in the intersection area that has a larger ID will be removed.

3.2 Area Target on Tracking

The Error value based on the difference from the center point from a set of bounding boxes with the target area. The target location is the surrounding area from the center of the frame (X_c, Y_c). The target area is used to make the target value wider than just one point, so the movement of the actuator is not critical. Small changes from the tracking position often occur, if only using the center point as a target causes the movement of the actuator to be unstable.

Algorithm 1. Multiple Faces Tracking Algorithm

1: Input $NextId, BoxScores$
2: Find match bounding box
3: **if** The new face does not overlap **then**
4:     Features detection and find location
5:     Input new ID, $NextId = NextId + 1$
6:     $BoxScores = 1$
7: **else**
8:     Delete matched bounding box
9:     Replace with new bounding box
10:    Features detection and find location
11:    Update $BoxScores = BoxScores + 1$
12: **end if**
13: Remove faces are no longer tracked
14: Update points tracker

The target area is determined by the following equation:

$$X_c - 20 \leq Tx_c \leq X_c + 20, \quad Y_c - 20 \leq Ty_c \leq y_c + 20, \tag{1}$$

where Tx_c is target area for x-axis coordinates and Ty_c is target area for y-axis coordinates. The ideal position is the center point from a set of faces must be in the target area (errors is zero). This system divides the error value into two parts that are horizontal and vertical error. The resolution of the test performed on this system is 640×480. Figure 2 illustrates the target location of multiple faces tracking. The initial process is to make error corrections to the location of a set of bounding boxes (see Fig. 2(a)). The center point of a set of faces (C_x, C_y) is obtained by considering information of the maximum coordinates $(MaxBBx$ and $MaxBBy)$ and the minimum coordinates $(MinBBx, MinBBy)$ from the center point of the bounding box of each face, showed by the following equation:

$$C_x = min[BB_x] + \frac{max[BB_x] - min[BB_x]}{2}, \ C_y = min[BB_y] + \frac{max[BB_y] - min[BB_y]}{2}, \tag{2}$$

where BBx is the coordinate bounding box for a set of faces on the x-axis, dan BBy on the y-axis. The actuator will move the webcam to make a set of faces covered in the center of the frame. Then maintain the center point of a set of faces to be inside the target area. The expected result is shown in Fig. 2(b) with the center point of the bounding box from a set of faces in the target area. This method attempts to accommodate tracking on all faces and keep them in the center of the frame.

3.3   Speed Control System for Moving Camera

The proposed system uses a PID (Proportional-Integral-Derivative) controller to control the speed of the actuator's motion against input from the computer vision process. This controller makes actuator movement more accurate, acceleration

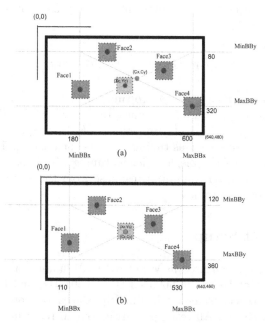

Fig. 2. Error condition when a set of faces is not in the center of the frame (a), Ideal condition when a set of face has been in target area (b).

Table 1. Ziegler-Nichols recommended tunning (proposed by [1]).

The Controller	K_p	K_i	K_d
P	$\frac{T}{L}$	-	-
PI	$0.9\frac{T}{L}$	$\frac{0.27T}{L^2}$	-
PID	$1.2\frac{T}{L}$	$\frac{1.2T}{2L^2}$	$\frac{1.2T}{0.5L^2}$

Table 2. Tracking performance

Video	TPR	FNR	TNR	FPR	Precision	Recall	Accuracy
V1	0.988	0.012	1.000	0.000	1.000	0.990	0.989
V2	0.978	0.022	1.000	0.000	1.000	0.980	0.979
V3	0.973	0.027	0.988	0.012	1.000	0.970	0.974
V4	0.957	0.043	0.956	0.044	1.000	0.960	0.957
V5	0.904	0.096	0.975	0.025	1.000	0.900	0.909
V6	0.895	0.105	1.000	0.000	1.000	0.900	0.902
V7	0.920	0.080	0.549	0.451	1.940	0.920	0.875
V8	0.889	0.101	0.814	0.186	1.980	0.900	0.890

increases and stabilizes when a significant error occurs, mathematically shown in the following equation:

$$u_t = K_p e(t) + K_i \int_0^t e(\tau)d\tau + K_d \frac{de}{dt}, \tag{3}$$

where the output the system respect to time t is $u(t)$, τ is the variable of integration that takes on the values from time 0 to the present t, and $e(.)$ is the error of system. The determination of proportional (K_p), integral (K_i), and derivative (K_d) constants are known as tuning of PID controller [2]. The gain of these parameters obtained by the Ziegler-Nichols tuning with L as time delay and T as time constant based on the initial response from the system. The equation this method is shown in Table 1.

4   Experiment Setup

Experiments were implemented in MATLAB R2017a and conducted on the following hardware: Intel Core I5-6600 CPU @ 3.30 GHz, and 8 GB RAM. The successful system was considered when multiple faces was correctly tracked and followed by the webcam. All tests are carried out in real-time and tested with stable light conditions.

5   Experiment Results

5.1   Multiple Faces Tracking Results

The proposed system achieves good test results for the entire video. We test multiple faces videos with a maximum limit of four human faces. The challenges in this system were successfully resolved by overcoming extreme facial positions and using accessories. The results obtained are the system can maintain a set of faces to always be in the frame by moving the webcam continuously.

System performance testing is carried out to assess the ability of this system to detect and track multiple faces. The video contains a single face and multiple is converted to sequential images. Then true and false tracking analysis is calculated at 1000 sequential images. The results of these calculations are used to determine the rate of system performance such as TPR (true positive rate), FNR (false negative rate), TNR (true negative rate), and FPR (false positive rate). The system proposed to get the results of the accuracy value is good. With a value of 0.934 as an average of accuracy, 0.940 as the average of recall, and 0.99 as the average of precision. Tracking performance results for multiple faces in real-time video are detailed in Table 2.

Our results have decreased in recall and accuracy measurements when tracking more faces. in the V7 video (see Table 2) there are more faces than other videos, face detection errors in areas not faces and more ignore the face on several frames. The more faces there are on the frame, the more features that must be detected and tracked. Because the proposed face tracking system is feature based and requires computation for each frame in detecting and tracking distinctive features. So that some increased errors and losses are obtained when faced with more faces. The decrease in accuracy and recall is also obtained when many accessories are used on the face. There are several re-detection failures when the face is back to the camera or the entire face is closed so that this limits the facial features to be detected. However, the decrease in recall results and accuracy obtained is still within the reasonable limits.

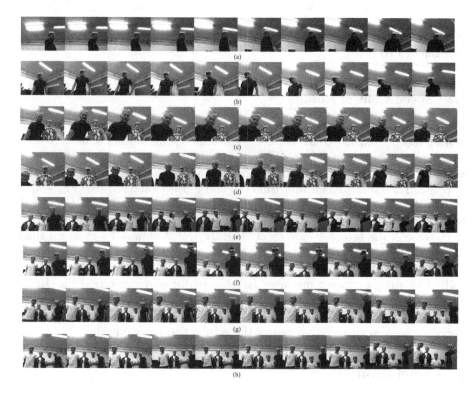

Fig. 3. Results of sequential images from multiple faces tracking with device.

Performance testing of the tracking system is carried out on the following eight videos: V1 is a video with a single face, V2 is a video with a single face using accessories, V3 is a video with two people, V4 is a video with two people using accessories, V5 is a video with three people, V6 is a video with three people using accessories, V7 is a video with four people, and V8 is a video with four people using accessories. The maximum speed of the tracking system process is 42.052 fps for all samples tested. We also display visual results of camera movements to follow and keep faces in the center of the frame (see Fig. 3). These results show ten frames for each video. This proves that our tracking system can maintain a set of faces to be in the center of the frame with various facial positions and using accessories.

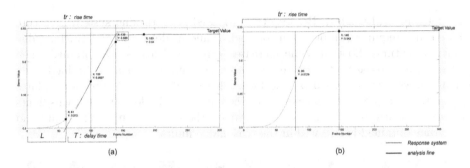

Fig. 4. Response of servo movement system for webcam, initial response (a), response after tuning PID (b).

5.2   Speed Control Results

The PID controller is used to increase the speed of the system to follow multiple moving faces. Even this controller keeps the center point of a set of faces to stay awake in the target area. The beginning of the process, the characteristics of the actuator movement system are determined by analyzing the curve of the initial response. Then found a constant time (L) and a delay time (T) (see Fig. 4(a)).

The conversion process of the number of frames to unit time is calculated based on the information rate of frames per second (fps) obtained from testing. In the case of the actuator movement (Fig. 4(a)), it is obtained a constant time $(L = 61)$ and a delay time $(T = 78)$. And by using the Ziegler-Nichols tuning (Table 1) found the parameter values of gain $K_p = 1.53$, $K_i = 0.41$, and $K_d = 1.41$. This tuning process produces faster actuator movement (see Fig. 4(b)), with a smaller rise time (t_r) than before using the PID Controller. The results obtained also maintain the accuracy and stability of the camera movement when there is a large error rate that causes the actuator to move quickly.

6   Conclusion

This paper presents a real-time multiple faces tracking system applied to service robots, to support the process of the human-robot interaction. The proposed system can follow a set of faces that move around the frame based on the extraction of the center point. Good results were obtained when doing the tracking by being able to maintain a set of faces to remain covered in the middle area of the frame. Various facial positions and the use of accessories are the challenges that has been successfully completed by this system. Based on the speed control result, the PID controller makes this system successful in increasing speed by maintaining the value of the accuracy and stability of the actuator when the webcam moves to catch up with the target quickly.

References

1. Anto, E., Asumadu, J., Okyere, P.: PID control for improving P&O-MPPT performance of a grid-connected solar PV system with ziegler-nichols tuning method. In: IEEE 11th Conference on Industrial Electronics and Applications (ICIEA), pp. 1847–1852 (2016)
2. Basu, A., Mohanty, S., Sharma, R.: Designing of the PID and FOPID controllers using conventional tuning techniques. In: IEEE International Conference on Inventive Computation Technologies (ICICT) (2016)
3. Buddubariki, V., Tulluri, S., Mukherjee, S.: Multiple object tracking by improved KLT tracker over surf features. In: National Conference on Computer Vision, Pattern Recognition, Image Processing and Graphics (NCVPRIPG) (2015)
4. Das, A., Pukhrambam, M., Saha, A.: Real-time robust face detection and tracking using extended haar functions and improved boosting algorithm. In: IEEE International Conference on Green Computing and Internet of Things (ICGCIoT), pp. 981–985 (2015)
5. Harshita, H.: Surveillance robot using raspberry Pi and IoT. In: IEEE International Conference on Design Innovations for 3Cs Compute Communicate Control (2018)
6. Huang, D.-Y., Chen, C.-H., Chen, T.-Y., Wu, J.-H., Ko, C.-C.: Real-time face detection using a moving camera. In: IEEE 32nd International Conference on Advanced Information Networking and Applications Workshops. pp. 609–614 (2018)
7. Jiang, W., Wang, W.: Face detection and recognition for home service robots with end-to-end deep neural networks. In: IEEE International Conference on Acoustics, Speech and Signal Processing (ICASSP), pp. 2232–2236 (2017)
8. Kwon, H., Lee, S.H., Hosseini, S., Moon, J., Koo, H., Cho, N.: Multiple face tracking method in the wild using color histogram features. In: IEEE International Symposium on Signal Processing and Information Technology (ISSPIT), pp. 51–55 (2017)
9. Lee, S., Xiong, Z.: A real-time face tracking system based on a single PTZ camera. In: IEEE China Summit and International Conference on Signal and Information Processing (ChinaSIP), pp. 568–572 (2015)
10. Marques, B., Correia, L., Bezerra, K., Goncalves, L.: Tracking spatially distributed features in KLT algorithms for RGB-D visual odometry. In: IEEE Workshop of Computer Vision (WVC), pp. 67–72 (2017)
11. Pradana, A., Paulus, E., Setiana, D.: Face detection with various angular positions on a bunch of people by comparing viola-jones and kanade-lucas-tomasi methods. Nat. J. Inf. Eng. Educ. (JANAPATI) 5(3), 136–141 (2016)
12. Putro, M., Teguh, A., Winduratna, B.: Adult image classifiers based on face detection using viola-jones method. In: 1st International Conference on Wireless and Telematics (ICWT) (2015)
13. Shi, T.: Good features to track. In: Conference on Computer Vision and Pattern Recognition (CVPR 1994) (1994)
14. Simul, N., Ara, N., Islam, M.: A support vector machine approach for real time vision based human robot interaction. In: IEEE International Conference on Computer and In-formation Technology, pp. 496–500 (2016)
15. Viola, P., Jones, M.: Robust real-time face detection. Int. J. Comput. Vision 57(2), 137–154 (2004)

Modified Stacked Hourglass Networks
for Facial Landmarks Detection

Van-Thanh Hoang⬛ and Kang-Hyun Jo$^{(\boxtimes)}$⬛

School of Electrical Engineering, University of Ulsan, Ulsan, Korea
thanhhv@islab.ulsan.ac.kr, acejo@ulsan.ac.kr

Abstract. Facial landmarks detection is a fundamental research topic in computer vision. This topic has been largely improved recently thanks to the development of convolution neural networks (CNN). This paper proposes a modified version of the Stacked Hourglass Network, which is a state-of-the-art architecture for landmark localization. Instead of using the original residual block, this paper uses the λ-residual-block to get more effective features. The proposed network can achieve better result than other state-of-the-art methods on two very challenging 3D facial landmark datasets, Menpo-3D and 300 W.

Keywords: CNN · Hourglass · Facial landmarks · λ-residual block

1 Introduction

Recently, thanks to the advent of Deep Learning and the development of large datasets, many research works have shown results of fantastic accuracy even on the most challenging computer vision tasks. This paper focuses on the task of landmark localization, in particular, on facial landmark localization, also known as face alignment, arguably one of the most heavily researched topics in computer vision over the last decades.

Face alignment or facial landmark estimation is the task of estimating the position of key-points of faces such as eye-corners, mouth corners in an image. As shown in [3], the accurate face alignment can improve the performance of a face verification system, as well as other application such as 3D face modeling, face animation.

Before the advent of deep neural networks, many different techniques have been used for landmark localization. They almost depend on the task in hand. For example, works in human pose estimation was primarily based on sophisticated extensions [23,24,29,31,37] and pictorial structures [9] due to their ability to accommodate a wide spectrum of human poses and also model large appearance changes.

Recently, the Fully Convolutional Neural Network architectures based on the score-map regression have revolutionized the human pose estimation task [14,20,22,25,32,33,35] to produce results of good accuracy even for the very challenging datasets [1]. Thanks to the similarity between estimation human

© Springer Nature Switzerland AG 2019
N. T. Nguyen et al. (Eds.): ACIIDS 2019, LNAI 11432, pp. 648–657, 2019.
https://doi.org/10.1007/978-3-030-14802-7_56

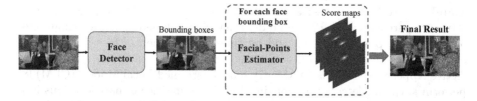

Fig. 1. Overall architecture of the proposed network. From the input image, it uses the faster R-CNN face detector to get the bounding boxes for every face inside the image. Then, for each bounding box, it uses the proposed Facial-Points Estimator to generate the score-maps of all facial points. Finally, it aligns the landmarks of all faces based on the score-maps.

Fig. 2. Example output produced by the proposed Facial Points Estimator. The first left image is the final facial landmarks provided by the max activations across each score map. The other images show sample score maps of some facial key-points (with the original image in behind). From left to right: right jaw, chin, left jaw, left eyebrow, right eye, nose, and mouth.

key-points and facial landmarks tasks, such methods can be freely applied to the problem of facial landmarks alignment.

This paper adopts the top-down approach to align the landmarks for every face in the input image. Firstly, it uses a face detector, which can be MTCNN (Multi-Task Cascade Convolutional Neural Networks) [39] or faster R-CNN (Region-based Convolutional Neural Networks) [16] to get the bounding boxes for every face inside the image. Then, for each bounding box, it crops the input image and uses the proposed Facial-Points Estimator to generate the score-maps of all facial points for the face inside the cropped image. Finally, it aligns the landmarks of all faces based on the score-maps (as can be seen in Fig. 1).

The proposed Facial-Points Estimator is a modified version of the Stacked Hourglass Network [20], which is a state-of-the-art architecture for landmark localization, by replacing the original residual block with the λ-residual block. The examples of output score-maps for some facial key-points are shown in Fig. 2.

2   Related Work

This section reviews related work on face alignment and landmark localization under the two categories: hand-crafted features-based and deep learning-based methods.

Hand-Crafted Features-Based Methods. Zhu et al. [42] proposed the tree structure part model (TSPM) used deformable part-based model for simultaneous detection, pose estimation and landmark localization of face images modeling the face shape in a mixture of trees model. The statistical methods like Active Appearance Models (AAM) [7] and Constrained Local Models (CLM) [8] perform keypoint detection by maximizing the confidence of part locations in a given input image using handcrafted features such as SIFT [21] and HOG. In [2], Asthana et al. proposed a dictionary of the probability response maps followed by linear regression in a CLM framework. Early cascade regression-based methods such as [5,34,41] also used hand-crafted features such as SIFT to capture the appearance of the face image. The major drawback of regression-based methods is their inability to learn models for unconstrained faces in extreme pose.

Deep Learning-Based Methods. Sun et al. [30] used a CNN cascade to regress the facial landmark locations. The work in [39,40] proposed multi-task learning for joint facial landmark localization and attribute classification. The method of [36] and its extended version [34] are within recurrent neural networks. 3DDFA [44] modeled the depth of the face image in a Z-buffer, after which a dense 3D face model was fitted to the image via CNNs. Pose Invariant Face Alignment (PIFA) proposed by Jourabloo et al. [17] predicted the coefficients of 3D to 2D projection matrix via deep cascade regressors.

Landmark Localization. There are many good CNN architectures proposed for key-points estimation task. The stacked hourglass network [20] is the state-of-the-art architecture. Hourglass networks use a stack of 8 very deep hourglass modules to generate the per-pixel labeling task for every key-points. This paper proposes a modified hourglass network to have fewer parameters and achieve better performance.

Face Detection. MTCNN [39] and faster R-CNN [16] are famous in face detection application. Faster R-CNN was proposed by Ren et al. [26] guided by the R-CNN family [10,11] for object detection task. The proposed method uses the Faster R-CNN detector with Resnet-50 backbone for the face detection part.

The proposed method is based on the top-down approach. Firstly, it uses the faster R-CNN face detector to get the bounding boxes for every face inside the image. Then, for each bounding box, Then, for each bounding box, it crops the input image and uses the proposed Facial-Points Estimator to generate the score-maps of all facial points for the face inside the cropped image. Finally, it aligns the landmarks of all faces based on the score-maps.

3   Our Approach

As shown in Fig. 3, the proposed Facial-Points Estimator network has the Residual Networks (ResNet) [13] as the backbone followed by four stacked Modified Hourglass block to generate the score-maps of every facial points.

Deep Residual Network. The ResNet is proposed by He et al. [13] to overcome the degradation problem: "When the network goes deeper and deeper, accuracy

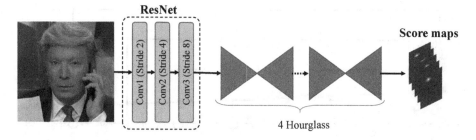

Fig. 3. Architecture of the proposed Facial-Points Estimator. It uses three first blocks of ResNet-50 as backbone. From the input cropped image, it uses ResNet to extract features. Then, these features are fed to four stacked modified hourglasses to generate the score-maps for all facial-points of the face inside the cropped image.

can be saturated and then degrades rapidly". This problem is not caused by overfitting, and adding more layers to a suitably deep model leads to higher error. Their solution is to add the skip connections to make a residual mapping. Figure 4c shows the architecture of an original residual block.

Facial-Points Estimator. The architecture of Facial-Points Estimator is shown in Fig. 3. This proposed network uses three first blocks of ResNet-50 as backbone. From the input cropped image, it extracts features by using ResNet backbone. Then, these features are fed to four stacked modified hourglasses to generate the score-maps for all facial-points of the face inside the cropped image.

Modified Hourglass Block. The architecture of the Modified Hourglass block is shown in Fig. 4a. It uses λ-residual block instead of the original residual block in the main stream. Additionally, it replaces the residual block in the branch stream with a 1×1 Convolution layer to reduce the number of parameters.

The λ-residual block, as shown in Fig. 4b, used by this network is a little bit different from the original residual block in ResNet. Before the addition, the output of the far previous layer is multiplied by a trainable number λ, it also has another branch with a 1×1 convolution layer to be added before going to the next layer.

4 Experiments

4.1 Implementation Details

The proposed network is trained on the 300W-LP Dataset [43], a synthetically expanded version of 300-W [28]. This dataset provides 3D landmarks allowing for training models and conducting experiments. The 3D annotations are actually the 2D projections of the 3D facial landmarks but for simplicity, we will just call them 3D.

Data augmentation is a simple process to increase number of training images. The augmentation used to train the proposed network are randomly: rotation (from $-50°$ to $+50°$), color jittering, flipping, and scale noise (from 0.5 to 1.2).

652   V.-T. Hoang and K.-H. Jo

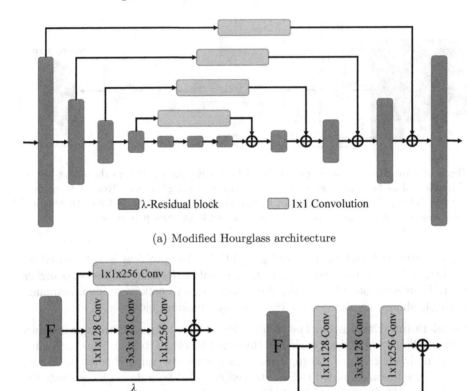

(a) Modified Hourglass architecture

(b) λ-Residual block architecture

(c) Original Residual block architecture

Fig. 4. Architecture of Modified Hourglass block and λ-residual block used in the proposed network.

The proposed method is validated on two kinds of dataset. 300-W test set [27] and Menpo-3D [38]. The 300-W dataset consists of the 600 images used for the evaluation purposes of the 300-W Challenge [27]. They are split into two categories: Indoor and Outdoor. The Menpo dataset is a recently introduced for the Menpo Challenger [38] containing 3D landmark annotations for about 9,000 faces from FDDB [15] and ALFW [19].

Similar to [4], this paper use AUC (Area-Under-the-Curve) score as the metric. This score is calculated based on the threshhold of 7% of Normalized Mean Error (NME). The NME is defined as:

$$\text{MNE} = \frac{1}{N} \sum_{k=1}^{N} \frac{\|x_k - y_y\|_2}{d},$$ (1)

Table 1. AUC scores (at NME 7%) of proposed methods and other state-of-the-art methods on the 300W and Menpo datasets. Scores of compared methods are obtained from [4].

Dataset	**Ours**	[4]	[34]	[40]	[41]
Menpo	**68.2%**	67.5%	67.1%	47.9%	60.5%
300W	**74.6%**	66.9%	58.1%	41.7%	55.9%

where x denotes the ground truth landmarks for a given face, y is the corresponding prediction and d is the square-root of the ground truth bounding box, computed as $d = \sqrt{w_{bbox} * h_{bbox}}$.

Fig. 5. Qualitative results of the proposed network on the Menpo-3D dataset. Red: ground truth. White: proposed network's predictions. The top two rows are the examples of good results when the predicted facial key-points is similar to the ground truth. The last row is the example of highest-error results when predicted facial landmarks is mixed up. The main reasons are very low resolution, bad lighting, and/or face is behind some objects. (Color figure online)

The proposed network is implemented on MXNet open source deep learning framework [6]. The Adam [18] optimizer implemented by MXNet is used for training the proposed network. It is trained for 50 epochs on a computer with AMD Ryzen 7 3.60 GHz CPU, NVIDIA 1080Ti GPU device, and 32-GB RAM. The initialized learning rate is set to 4e−5 and then reduced 10 times at 20 and 35 epochs, respectively. The Xavier's initializer [12] is used to initialize the parameters of weighted layers (e.g. Convolution layers). The other settings are: batch size of 24, weight decay of 0.0001, and momentum of 0.9.

4.2 Experiment Results

Table 1 shows the AUC score at NME 7% of the proposed network and other state-of-the-art methods. The proposed network consistently able to outperform the state-of-the-art methods in both two validation datasets. Specially, it has much higher score than compared methods in 300 W dataset.

Some selected visualization results are shown in Fig. 5. As can be seen, most of predicted facial landmarks are similar to the ground truth. But in case of very low resolution, bad lighting, and/or face is behind some objects, the predicted facial landmarks are mixed up. Additionally, these hard cases do not appear in the training dataset much.

5 Conclusion

This paper proposed a facial-points estimator to align the facial landmarks for faces in an image. The proposed network uses the ResNet-50 as backbone, followed by a modified Hourglass networks. Instead of using the original residual block, it uses the λ-residual block to extract better features.

In the future, because the faster R-CNN for face detector and proposed network use the same ResNet as backbone, it is necessary to combine them into one system to have a smaller system, which can run in real-time.

References

1. Andriluka, M., Pischulin, L., Gehler, P., Schiele, B.: 2D human pose estimation: new benchmark and state of the art analysis. In: Proceedings of the IEEE Conference on Computer Vision and Pattern Recognition, pp. 3686–3693 (2014)
2. Asthana, A., Zafeiriou, S., Cheng, S., Pantic, M.: Robust discriminative response map fitting with constrained local models. In: Proceedings of the IEEE Conference on Computer Vision and Pattern Recognition, pp. 3444–3451 (2013)
3. Bansal, A., Castillo, C.D., Ranjan, R., Chellappa, R.: The Do's and Don'ts for CNN-based face verification. In: Proceedings of the IEEE International Conference on Computer Vision Workshops, pp. 2545–2554 (2017)
4. Bulat, A., Tzimiropoulos, G.: How far are we from solving the 2D & 3D face alignment problem? (and a dataset of 230,000 3D facial landmarks). In: Proceedings of the IEEE International Conference on Computer Vision, p. 4 (2017)
5. Cao, X., Wei, Y., Wen, F., Sun, J.: Face alignment by explicit shape regression. Int. J. Comput. Vision 107(2), 177–190 (2014)

6. Chen, T., et al.: MXNET: a flexible and efficient machine learning library for heterogeneous distributed systems. In: Neural Information Processing Systems, Workshop on Machine Learning Systems (2015)
7. Cootes, T.F., Edwards, G.J., Taylor, C.J.: Active appearance models. IEEE Trans. Pattern Anal. Mach. Intell. **23**, 681–685 (2001)
8. Cristinacce, D., Cootes, T.: Automatic feature localisation with constrained local models. Pattern Recogn. **41**(10), 3054–3067 (2008)
9. Felzenszwalb, P.F., Huttenlocher, D.P.: Pictorial structures for object recognition. Int. J. Comput. Vis. **61**(1), 55–79 (2005)
10. Girshick, R.: Fast R-CNN. In: Proceedings of the IEEE International Conference on Computer Vision, pp. 1440–1448 (2015)
11. Girshick, R., Donahue, J., Darrell, T., Malik, J.: Rich feature hierarchies for accurate object detection and semantic segmentation. In: Proceedings of the IEEE Conference on Computer Vision and Pattern Recognition, pp. 580–587 (2014)
12. Glorot, X., Bengio, Y.: Understanding the difficulty of training deep feedforward neural networks. In: Proceedings of the International Conference on Artificial Intelligence and Statistics, pp. 249–256 (2010)
13. He, K., Zhang, X., Ren, S., Sun, J.: Deep residual learning for image recognition. In: Proceedings of the IEEE Conference on Computer Vision and Pattern Recognition, pp. 770–778 (2016)
14. Insafutdinov, E., Pishchulin, L., Andres, B., Andriluka, M., Schiele, B.: DeeperCut: a deeper, stronger, and faster multi-person pose estimation model. In: Leibe, B., Matas, J., Sebe, N., Welling, M. (eds.) ECCV 2016. LNCS, vol. 9910, pp. 34–50. Springer, Cham (2016). https://doi.org/10.1007/978-3-319-46466-4_3
15. Jain, V., Learned-Miller, E.: FDDB: a benchmark for face detection in unconstrained settings. Technical report UM-CS-2010-009, University of Massachusetts, Amherst (2010)
16. Jiang, H., Learned-Miller, E.: Face detection with the faster R-CNN. In: IEEE International Conference on Automatic Face & Gesture Recognition, pp. 650–657. IEEE (2017)
17. Jourabloo, A., Liu, X.: Pose-invariant 3D face alignment. In: Proceedings of the IEEE International Conference on Computer Vision, pp. 3694–3702 (2015)
18. Kingma, D.P., Ba, J.: Adam: a method for stochastic optimization. In: International Conference on Learning Representations (2015)
19. Koestinger, M., Wohlhart, P., Roth, P.M., Bischof, H.: Annotated facial landmarks in the wild: a large-scale, real-world database for facial landmark localization. In: Proceedings of the IEEE International Conference on Computer Vision Workshops, pp. 2144–2151. IEEE (2011)
20. Newell, A., Yang, K., Deng, J.: Stacked hourglass networks for human pose estimation. In: Leibe, B., Matas, J., Sebe, N., Welling, M. (eds.) ECCV 2016. LNCS, vol. 9912, pp. 483–499. Springer, Cham (2016). https://doi.org/10.1007/978-3-319-46484-8_29
21. Ng, P.C., Henikoff, S.: Sift: predicting amino acid changes that affect protein function. Nucleic Acids Res. **31**(13), 3812–3814 (2003)
22. Pfister, T., Charles, J., Zisserman, A.: Flowing ConvNets for human pose estimation in videos. In: Proceedings of the IEEE International Conference on Computer Vision, pp. 1913–1921 (2015)
23. Pishchulin, L., Andriluka, M., Gehler, P., Schiele, B.: Poselet conditioned pictorial structures. In: Proceedings of the IEEE Conference on Computer Vision and Pattern Recognition, pp. 588–595 (2013)

24. Pishchulin, L., Andriluka, M., Gehler, P., Schiele, B.: Strong appearance and expressive spatial models for human pose estimation. In: Proceedings of the IEEE International Conference on Computer Vision, pp. 3487–3494 (2013)

25. Pishchulin, L., et al.: DeepCut: joint subset partition and labeling for multi person pose estimation. In: Proceedings of the IEEE Conference on Computer Vision and Pattern Recognition, pp. 4929–4937 (2016)

26. Ren, S., He, K., Girshick, R., Sun, J.: Faster R-CNN: towards real-time object detection with region proposal networks. IEEE Trans. Pattern Anal. Mach. Intell. **39**, 1137–1149 (2017)

27. Sagonas, C., Antonakos, E., Tzimiropoulos, G., Zafeiriou, S., Pantic, M.: 300 faces in-the-wild challenge: database and results. Image Vis. Comput. **47**, 3–18 (2016)

28. Sagonas, C., Tzimiropoulos, G., Zafeiriou, S., Pantic, M.: 300 faces in-the-wild challenge: the first facial landmark localization challenge. In: Proceedings of the IEEE International Conference on Computer Vision Workshops, pp. 397–403 (2013)

29. Sapp, B., Taskar, B.: MODEC: multimodal decomposable models for human pose estimation. In: Proceedings of the IEEE Conference on Computer Vision and Pattern Recognition, pp. 3674–3681 (2013)

30. Sun, Y., Wang, X., Tang, X.: Deep convolutional network cascade for facial point detection. In: Proceedings of the IEEE Conference on Computer Vision and Pattern Recognition, pp. 3476–3483 (2013)

31. Tian, Y., Zitnick, C.L., Narasimhan, S.G.: Exploring the spatial hierarchy of mixture models for human pose estimation. In: Fitzgibbon, A., Lazebnik, S., Perona, P., Sato, Y., Schmid, C. (eds.) ECCV 2012. LNCS, vol. 7576, pp. 256–269. Springer, Heidelberg (2012). https://doi.org/10.1007/978-3-642-33715-4_19

32. Tompson, J.J., Jain, A., LeCun, Y., Bregler, C.: Joint training of a convolutional network and a graphical model for human pose estimation. In: Advances in Neural Information Processing Systems, pp. 1799–1807 (2014)

33. Toshev, A., Szegedy, C.: DeepPose: human pose estimation via deep neural networks. In: Proceedings of the IEEE Conference on Computer Vision and Pattern Recognition, pp. 1653–1660 (2014)

34. Trigeorgis, G., Snape, P., Nicolaou, M.A., Antonakos, E., Zafeiriou, S.: Mnemonic descent method: a recurrent process applied for end-to-end face alignment. In: Proceedings of the IEEE Conference on Computer Vision and Pattern Recognition, pp. 4177–4187 (2016)

35. Wei, S.E., Ramakrishna, V., Kanade, T., Sheikh, Y.: Convolutional pose machines. In: Proceedings of the IEEE Conference on Computer Vision and Pattern Recognition, pp. 4724–4732 (2016)

36. Xiong, X., De la Torre, F.: Supervised descent method and its applications to face alignment. In: Proceedings of the IEEE Conference on Computer Vision and Pattern Recognition, pp. 532–539 (2013)

37. Yang, Y., Ramanan, D.: Articulated pose estimation with flexible mixtures-of-parts. In: Proceedings of the IEEE Conference on Computer Vision and Pattern Recognition, pp. 1385–1392. IEEE (2011)

38. Zafeiriou, S., Trigeorgis, G., Chrysos, G., Deng, J., Shen, J.: The menpo facial landmark localisation challenge: a step towards the solution. In: The IEEE Conference on Computer Vision and Pattern Recognition Workshops, p. 2 (2017)

39. Zhang, K., Zhang, Z., Li, Z., Qiao, Y.: Joint face detection and alignment using multitask cascaded convolutional networks. IEEE Signal Process. Lett. **23**(10), 1499–1503 (2016)

40. Zhang, Z., Luo, P., Loy, C.C., Tang, X.: Facial landmark detection by deep multi-task learning. In: Fleet, D., Pajdla, T., Schiele, B., Tuytelaars, T. (eds.) ECCV 2014. LNCS, vol. 8694, pp. 94–108. Springer, Cham (2014). https://doi.org/10.1007/978-3-319-10599-4_7
41. Zhu, S., Li, C., Change Loy, C., Tang, X.: Face alignment by coarse-to-fine shape searching. In: Proceedings of the IEEE Conference on Computer Vision and Pattern Recognition, pp. 4998–5006 (2015)
42. Zhu, X., Ramanan, D.: Face detection, pose estimation, and landmark localization in the wild. In: Proceedings of the IEEE Conference on Computer Vision and Pattern Recognition, pp. 2879–2886. IEEE (2012)
43. Zhu, X., Lei, Z., Li, S.Z., et al.: Face alignment in full pose range: a 3D totalsolution. IEEE Trans. Pattern Anal. Mach. Intell. (2017)
44. Zhu, X., Lei, Z., Liu, X., Shi, H., Li, S.Z.: Face alignment across large poses: a 3D solution. In: Proceedings of the IEEE Conference on Computer Vision and Pattern Recognition, pp. 146–155 (2016)

Ensemble of Predictions from Augmented Input as Adversarial Defense for Face Verification System

Laksono Kurnianggoro and Kang-Hyun Jo[✉]

Graduate School of Electrical Engineering, University of Ulsan, Ulsan, South Korea
laksono@islab.ulsan.ac.kr, acejo@ulsan.ac.kr

Abstract. Face identification is employed in security system. Recently, it is gaining reliable result thanks to the deep learning method. However, the deep learning-based methods are prone against adversarial attack that leads to wrong prediction in presence of simple alteration on the pixel values. Thus the reliability of such system is compromised. This paper proposed a simple defense strategy to improve the reliability of a system in the presence of adversarial attack. By combining the prediction from few samples of altered input image, the effect of adversarial attack can be reduced effectively. The proposed method has been tested using public face dataset in the presence of strong attacks. Experiment results shows that the proposed method is reliable to suppress the adversarial attacks.

Keywords: Face identification · Adversarial defense · Deep learning · Machine learning · Neural network

1 Introduction

Research in deep learning has been creating impact in various research fields, including face verification and identification [15,16,26] which are commonly utilized for security and surveillance system [8,21,22].

Despite of its superior performance, deep learning models are potentially suffers from adversarial attacks [1,4,6,9,13,20]. Therefore, concerns are raised especially in security related applications such as face identification system.

The adversarial attacks are getting more advanced and proven to works on breaking various real life applications via appearance alteration. It was firstly demonstrated in [11] that the adversarial attacks can be realized by printing the perturbed images on papers. Similarly, [7] demonstrated real life attack on traffic signs classifier that able to work on various object scales and distances. Beyond 2D and planar objects, research in [1] shows that attacks can be realized on 3D printed objects which resulting misclassification regardless of its pose.

Attacks on face identity classifier was firstly demonstrated [17]. Successful attack were achieved by printing the perturbation on eye glass frame. An illustration of the scheme is shown in Fig. 1. The deep neural network (DNN) model

© Springer Nature Switzerland AG 2019
N. T. Nguyen et al. (Eds.): ACIIDS 2019, LNAI 11432, pp. 658–669, 2019.
https://doi.org/10.1007/978-3-030-14802-7_57

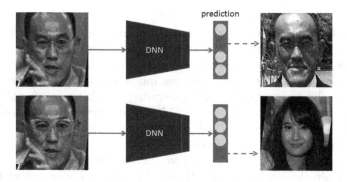

Fig. 1. An illustration of the adversarial attack. Top: a deep neural network model predict correctly the identity of a given input image. Bottom: with the presence maliciously crafted adversarial attack the model prediction is changed.

predicts an identity from a given input image. Without the existence of adversarial perturbation, it could produce correct prediction with high confidence. In the other hand, with presence of adversarial perturbation in a form of eye glass frame, its prediction is changed.

The work in this paper aims to hardening the face verification method against adversarial attack. The original perturbation method from [17] is re-implemented with Carlini-Wagner loss [4] to create strong attacks. Under this attack, a simple strategy which based on input transformation is proposed to defend the deep learning model. Experiments were conducted in several cases to validate its effectiveness in countering the adversarial attacks. There are several contributions in this paper as listed in the following:

- This paper demonstrated a practical attack and defense methods for face classification-based face verification system.
- A simple strategy to counter the adversarial attack is proposed as a patch to an existing face classifier, meaning there is no need to perform re-training of the classifier.
- The work in this paper increase the awareness of adversarial attack existence and the needs to perform precaution against this matter.

The remaining of this paper is organized as follows. Firstly, the summary of related works are discussed. In the following section, the attack method which is based on masked perturbation with Carlini-Wagner loss is presented followed by the proposed defense method which is based on ensemble of classification from transformed input. Experiment setups and results are explained in the subsequently and finally conclusions are provided to wrap up the contents of this paper.

2   Related Works

Research on the defense methods for adversarial attacks has been studied intensively in the past few years due to its importance for the deep learning system.

The most common way to hardening a DNN model is by including the adversarial samples in the training step as demonstrated in [9,19]. In this way the network is expected to not be fooled by the similar adversarial samples used in the training. Albeit simple, this method require to retrain the model which often takes a lot of time. Furthermore, there are infinitively many adversarial sample for a single valid training data while generating an adversarial sample is often time consuming. Therefore, the perturbed samples used in the training should be carefully selected in order to make the training time tractable.

Defensive distillation was proposed in [14]. At a high level, this method trains a teacher model in conventional way. Then, the prediction of this model is used to train the student model. During training and label generation from the teacher model, temperature is set in the softmax layer to make the decision score smoother. Finally, the temperature is set to 1 (i.e. standard softmax layer) at the test time in the student model. This strategy was demonstrated to be effective in defending the classification model.

A more recent study in [2] use thermometer encoding instead of the well known one-hot encoding as the target for training a DNN model. In the conventional case, the label is encoded with a vector of length class number and it has single non-zero value at position correspond to the target label. In the other hand, the thermometer encoding set non zero values from the position that correspond to the target label until the last element of vector. This defend method is built based on the idea that discretization is one way to apply a non-differentiable and non-linear transformation on the input which is expected to increase the non-linearity of model. Thus, allowing a stronger resistance against adversarial attack on the DNN model.

Another remarkable defense method is presented in [12]. An auto-encoder variant model is trained to reduce the perturbation effect before the input image is passed to the final DNN model. This model could be attached to any classification model without the need of classifier re-training. The denoiser model itself is trained in a special way so that the produced output is expected to be predicted correctly by the classifier. Instead of aiming to restore the perturbed image into its original form, this denoiser is trained to produce an image where its corresponding high level feature on the classification model is similar to the one produced by the non-perturbed image.

Unlike the other aforementioned defense methods, this paper aims to explore a method which does not require any retraining so that it is easier to be implemented in the existing application. It is exploring the domain transformation as adversarial defense which can be done easily on the perturbed image. Moreover, it does not focus on denoising-based strategy due to the nature of adversarial attack in the face verification system which often have large magnitude of pixel value alteration.

3   Adversarial Attack on Face Verification System

An adversarial example is a slightly modified image which divert prediction of a classifier. Most of the research on adversarial attacks nowadays focus on visually imperceptible image alteration but produce wrong classification result. Given a classifier C, an adversarial attack tries to find $\hat{x} = x + \delta$ where δ has small magnitude so that it is imperceptible by human eyes but its corresponding prediction $f(\hat{x})$ is incorrect or goes to target label t.

This problem formulation leads to an optimization problem for finding δ according to both constraints (i.e. image similarity and classification result) as shown in (1).

$$minimize \, \|x - \hat{x}\| + L_C\left(f(\hat{x}), t\right) \tag{1}$$

$$L_C = \max\left(f(\hat{x})_k - f(\hat{x})_t\right); where : k \neq t \tag{2}$$

In this case, the classification loss is score difference between the maximum score of any possible outcome labels to the score correspond to a target label as shown in (2).

For real attack on face verification system, small magnitude of perturbation is not absolutely necessary. In this case, physical realization is more important. To achieve it, the pixel value should match the printing constraint while divert the classification result. Thus, the image similarity constraint can be omitted from the optimization problem while adding physical realization constraints as described in (3) where α and β are weights for printability loss and total variation loss, respectively.

$$\underset{\hat{x}}{\mathrm{argmin}}\, L_C\left(f(\hat{x}), t\right) + \alpha L_P + \beta L_V \tag{3}$$

As suggested in [7,17], the value of perturbed pixels should close to any values in set of printable colors P. As formulated in (4), the loss is defined as summation of printability score from all of the pixel values. This score tend to go lower whenever $\hat{x}_{i,j}$ is similar to a printable score p.

$$L_P = \sum_{i,j} \prod_{p \in P} |\hat{x}_{i,j} - p| \tag{4}$$

To enhance the success rate of the printed adversarial attack, it is necessary to have smooth values on the perturbed region due to the fact that fluctuating values might be wiped out by the camera imaging system. Therefore, an additional constraint on pixel value variation is added as shown in (5) which encapsulate the changes of values between current pixel to its two neighbors.

$$L_V = \sum_{i,j} \left(\left(\hat{x}_{i,j} - \hat{x}_{i+1,j}\right)^2 + \left(\hat{x}_{i,j} - \hat{x}_{i,j+1}\right)^2\right)^2 \tag{5}$$

4   The Proposed Defense

As demonstrated in [17], the adversarial attack for face verification system is physically manifested on eye glass frame. The simplest way to defend this particular attack is by covering the region nearby eyes. It can be done by firstly detect the eyes using robust face keypoints detector such as [23] or [24] and then draw a block of uniform pixels around the eye key points.

In the real case, the attacker could utilize any form of adversarial patch not limited to eye glass frame. Therefore, the location and size of the compromised region could not be trivially obtained.

In this work, the defend method is proposed to suppress the effect of adversarial attack on face verification system regardless of the perturbation shape and position.

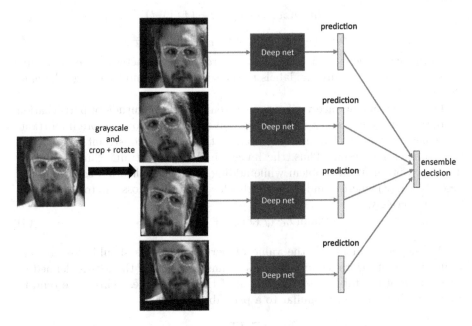

Fig. 2. Illustration of the proposed defense method (best viewed in screen or color). (Color figure online)

The proposed defend method is built based on the fact that data augmentations are commonly utilized in the training phase of a deep learning-based classification model. This defend transforms the input image into grayscale format and then perform cropping and rotation on it as shown in Fig. 2.

Several transformed input images are passed to the classifier and their corresponding predictions are collected to generate the ensemble decision.

There are two types of ensemble decision utilized in this work, aggregation and voting. Aggregate prediction is obtained by summing up all of the predicted scores from each transformed input while the voting result is defined by counting the most voted decision from each prediction result.

The reason behind this strategy is that the adversarial attack is only effective on the data domain where it was crafted. Therefore, this method aims to convert the data domain from RGB color space into grayscale format and perform cropping to reduce the completeness of input data which can be seen as a simple form of dimensionality reduction. In the other hand, the ensemble prediction is utilized with aim to further improve the robustness of defense by combining several protective methods.

5   Experiments and Results

To evaluate the proposed method, 20 identities are selected from VGGface2 dataset [3] and splitted into new training and test data. The statistic of this new dataset is shown in Tables 1 and 2. Three types of evaluations were performed using these dataset to assess the proposed method on man only data, woman only data, and combination of both man and woman (i.e. 20 classes).

The training split is used to train face classifier based on pre-trained InceptionV3 model [18] while test split is used in attack and defense experiments. The classification model was fine tuned for 20 epochs with batch size of 24 in 6 GB GTX 1060 while the gradient descent with momentum of 0.9 was utilized as its optimizer. The learning rate for training was set to 0.001 and decreased with factor 0.1 in epoch 10 and 15.

For both of model training and adversarial attack scenario, the face image was aligned and cropped into a canonical form using method from [24]. Examples of the face image in canonical form are shown in Fig. 1. The images are firstly aligned to make the eyes in horizontal state with 80 pixels distance between them on an image with size 399×399 pixels. The aligned images are then feed to the training algorithm with data augmentation of random cropping and flipping with final image size 299×299 pixels.

Table 1. Dataset statistic for the men faces.

ID (Man)	0	1	2	3	4	5	6	7	8	9	Total
Train	488	264	208	244	302	399	337	362	313	299	3216
Test	50	50	50	50	50	50	50	50	50	50	500

Table 2. Dataset statistic for the women faces.

ID (Woman)	0	1	2	3	4	5	6	7	8	9	Total
Train	501	417	393	299	284	175	182	284	320	329	3184
Test	50	50	50	50	50	50	50	50	50	50	500

The training results for 3 classification models are shown in Table 3. For all of the experiment setups, the trained model are not overfitted and has reliable results.

The adversarial samples are generated using optimization method as formulated in (3) by Adam optimizer [10] with learning rate 0.01 for 300 steps.

Table 3. Classification accuracy (%) for 3 models in this experiments.

Split	Men	Women	All
Train	99.96	100.00	99.10
Test	99.40	99.00	98.20

Using this setup, the attack successfully divert all of model predictions as shown in Table 4. The classification accuracy from perturbed images is zero percent meaning that the model perform poorly against adversarial attack.

A comprehensive experiment result on men dataset for single image defense is shown in Fig. 3. There are 4 types of scenario in which the perturbed image is not defended, random cropping in the RGB is performed, grayscale transformation as defense, and grayscaling followed by random cropping defense. The remaining two cases are classification accuracy on original image without perturbation (99.96%) and accuracy in the presence of initial perturbation mask which is set to one of 30 possible printable color values as suggested in [7]. It is interesting finding that the classification accuracy is dropped even with non optimally crafted perturbation as in the initial mask case.

As shown in Fig. 3, the cropping strategy on color image could reduce the effect of adversarial attack but on average it perform less effective compared to the grayscale transformed input and grayscale+cropping defense. Similar results also happens for women only dataset and 20 classes of men and women dataset. The summary is presented in Table 4. Due to this fact, the method in this paper extend the defense strategy with ensemble of prediction from grayscaling and cropping to increase its robustness.

Table 4. Classification accuracy (%) with single image defense.

Data	Perturbed	Cropped	Grayscale	Grayscale+crop	Initial
Men	0	55.20	65.40	61.00	86.20
Women	0	63.00	69.40	70.20	87.60
All	0	39.30	57.50	57.20	67.80

For the ensemble cases, defense with up to 10 transformations were examined where the results are shown in Figs. 4, 5, and 6. As shown in Fig. 4 for the men dataset case, the ensemble performance tends to go higher as the number of transformation increase. Interestingly, the ensemble of color cropped input image always perform poorly compared to other type of ensemble defense strategy. It confirms the claim that domain transformation is important to prevent the adversarial attack transfer to the target model. This property is further confirmed by the two other experiment results for women dataset and combination dataset as shown in Figs. 5 and 6.

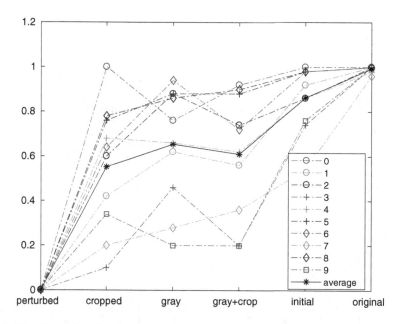

Fig. 3. Details of the classification accuracy under adversarial attack for each identity and defense setup in the men dataset.

There is no clear result that directly conclude which one is better between ensemble of grayscale+cropping and combination of grayscale and RGB cropping. Note that in the former case there are 10 different transformations while for the later case there are 20 augmentations in total (there are 10 crops of RGB input and 10 crops of grayscale input). However, the ensemble of grayscale+crop deliver better performance for men and combination dataset while in fact it uses fewer number of ensemble.

It also shown in the experiment results that the selection of ensemble type does affect the performance. In most cases, the aggregate strategy performs better compared to the voting method.

5.1   Discussion and Future Works

The experiments demonstrated that the proposed method is able to counter the adversarial attack with significant performance thanks to the domain transformation. It should be noted that there are a lot of possible domain transformation that has not been explored in this experiment such as random erasing augmentation [25] or other color space transformations [5] such as opponent colors or color-names space. More comprehensive study of various domain transformation are left for the future works.

Furthermore, there are various possible improvement that will be considered for the future works such as different type of ensemble decision (such as

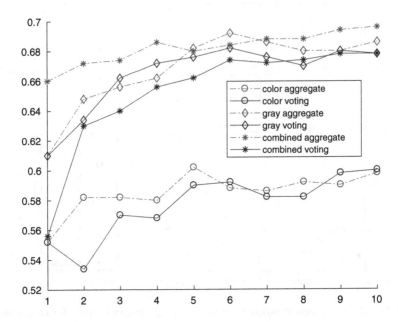

Fig. 4. Accuracy comparison between different number of ensembles for the men dataset.

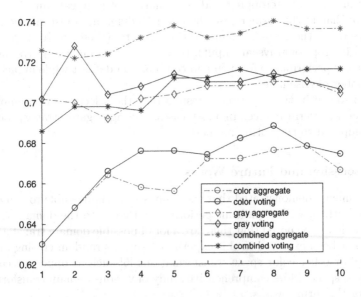

Fig. 5. Accuracy comparison between different number of ensembles for the women dataset.

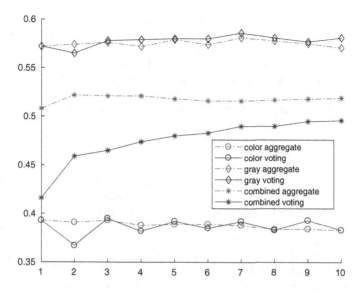

Fig. 6. Accuracy comparison between different number of ensembles for the complete dataset.

weighted voting, decision tree, or multi layer perceptron decision) and types of classification model (architecture) used in the ensemble.

6   Conclusions

An effective defense strategy against adversarial attack on face verification system has been discussed. The proposed method is based on ensemble of classification from domain transformed input data. It has been shown in the experiment that domain transformation is useful to suppress the effect of adversarial attack on a classification model. Furthermore, the proposed ensemble strategy boosts defense performance compared to a single image defense.

References

1. Athalye, A., Sutskever, I.: Synthesizing robust adversarial examples. In: Proceeding of International Conference on Machine Learning (2018)
2. Buckman, J., Roy, A., Raffel, C., Goodfellow, I.: Thermometer encoding: one hot way to resist adversarial examples (2018)
3. Cao, Q., Shen, L., Xie, W., Parkhi, O.M., Zisserman, A.: VGGFace2: a dataset for recognising faces across pose and age. In: 13th IEEE International Conference on Automatic Face & Gesture Recognition (FG 2018), pp. 67–74. IEEE (2018)
4. Carlini, N., Wagner, D.: Towards evaluating the robustness of neural networks. In: IEEE Symposium on Security and Privacy (SP), pp. 39–57. IEEE (2017)

5. Danelljan, M., Shahbaz Khan, F., Felsberg, M., Van de Weijer, J.: Adaptive color attributes for real-time visual tracking. In: Proceedings of the IEEE Conference on Computer Vision and Pattern Recognition, pp. 1090–1097 (2014)
6. Dong, Y., et al.: Boosting adversarial attacks with momentum. In: The IEEE Conference on Computer Vision and Pattern Recognition (CVPR) (2018)
7. Eykholt, K., et al.: Robust physical-world attacks on deep learning visual classification. In: Proceedings of the IEEE Conference on Computer Vision and Pattern Recognition, pp. 1625–1634 (2018)
8. Filonenko, A., Hernández, D.C., Jo, K.H.: Fast smoke detection for video surveillance using CUDA. IEEE Trans. Ind. Inform. **14**(2), 725–733 (2018)
9. Goodfellow, I.J., Shlens, J., Szegedy, C.: Explaining and harnessing adversarial examples. In: International Conference on Learning and Representation (2015)
10. Kingma, D.P., Ba, J.: Adam: a method for stochastic optimization. In: International Conference on Learning and Representation (2015)
11. Kurakin, A., Goodfellow, I., Bengio, S.: Adversarial examples in the physical world. arXiv preprint arXiv:1607.02533 (2016)
12. Liao, F., Liang, M., Dong, Y., Pang, T., Zhu, J., Hu, X.: Defense against adversarial attacks using high-level representation guided denoiser. In: Proceedings of the IEEE Conference on Computer Vision and Pattern Recognition, pp. 1778–1787 (2018)
13. Madry, A., Makelov, A., Schmidt, L., Tsipras, D., Vladu, A.: Towards deep learning models resistant to adversarial attacks. In: International Conference on Learning and Representation (2018)
14. Papernot, N., McDaniel, P., Wu, X., Jha, S., Swami, A.: Distillation as a defense to adversarial perturbations against deep neural networks. In: IEEE Symposium on Security and Privacy (SP), pp. 582–597. IEEE (2016)
15. Parkhi, O.M., Vedaldi, A., Zisserman, A., et al.: Deep face recognition. In: BMVC, vol. 1, p. 6 (2015)
16. Schroff, F., Kalenichenko, D., Philbin, J.: FaceNet: a unified embedding for face recognition and clustering. In: Proceedings of the IEEE Conference on Computer Vision and Pattern Recognition, pp. 815–823 (2015)
17. Sharif, M., Bhagavatula, S., Bauer, L., Reiter, M.K.: Accessorize to a crime: real and stealthy attacks on state-of-the-art face recognition. In: Proceedings of the 2016 ACM SIGSAC Conference on Computer and Communications Security, pp. 1528–1540. ACM (2016)
18. Szegedy, C., Vanhoucke, V., Ioffe, S., Shlens, J., Wojna, Z.: Rethinking the inception architecture for computer vision. In: Proceedings of the IEEE Conference on Computer Vision and Pattern Recognition, pp. 2818–2826 (2016)
19. Szegedy, C., et al.: Intriguing properties of neural networks (2014)
20. Tramèr, F., Kurakin, A., Papernot, N., Goodfellow, I., Boneh, D., McDaniel, P.: Ensemble adversarial training: attacks and defenses. In: International Conference on Learning and Representation (2018)
21. Wahyono, W., Filonenko, A., Jo, K.H.: Unattended object identification for intelligent surveillance systems using sequence of dual background difference. IEEE Trans. Ind. Inform. **12**(6), 2247–2255 (2016)
22. Wahyono, W., Jo, K.H.: Cumulative dual foreground differences for illegally parked vehicles detection. IEEE Trans. Ind. Inform. **13**(5), 2464–2473 (2017)
23. Wei, S.E., Ramakrishna, V., Kanade, T., Sheikh, Y.: Convolutional pose machines. In: Proceedings of the IEEE Conference on Computer Vision and Pattern Recognition, pp. 4724–4732 (2016)

24. Zhang, K., Zhang, Z., Li, Z., Qiao, Y.: Joint face detection and alignment using multitask cascaded convolutional networks. IEEE Signal Process. Lett. **23**(10), 1499–1503 (2016)
25. Zhong, Z., Zheng, L., Kang, G., Li, S., Yang, Y.: Random erasing data augmentation. arXiv preprint arXiv:1708.04896 (2017)
26. Zhu, C., Tao, R., Luu, K., Savvides, M.: Seeing small faces from robust anchors perspective. In: Proceedings of the IEEE Conference on Computer Vision and Pattern Recognition (2018)

Vehicle Categorical Recognition for Traffic Monitoring in Intelligent Transportation Systems

Diem-Phuc Tran[1] and Van-Dung Hoang[2(✉)]

[1] Duy Tan University, Da Nang, Vietnam
phuctd@gmail.com
[2] Quang Binh University, Đồng Hới, Quang Binh, Vietnam
dunghv@qbu.edu.vn

Abstract. Automatic vehicle detection and recognition play a vital role in intelligent transport systems (ITS). However, study results in this field remain certain limitations in terms of accuracy and processing time. This article proposes a solution to improve the accuracy of vehicle recognition in order to support traffic monitoring on vehicle restricted roads. The proposed solution to vehicle recognition consists of two basic stages: (1) Vehicle detection, (2) vehicle recognition. This study focuses on proposing solutions for improving the accuracy of vehicle recognition (stage 2). The vehicle recognition solution is based on the combination of architectural development in deep neural networks, SVM model, and data augmenting solutions. It aims at achieving a greater accuracy than traditional approaches. The proposed solution is experimented, evaluated, and compared with different approaches to the same set of data. Experimental results have shown that the proposed solution brings a higher accuracy than other approaches. Along with an acceptable processing time, this promising solution is able to be applied in practical systems.

Keywords: Deep learning · Vehicle recognition · Feature extraction

1 Introduction

Vehicle detection and recognition are highly applicable to traffic monitoring and classification systems. Along with the development of science and technology, the demand for travel and the number of vehicles are increasing. With the growing number of means of transportation, there are many arising problems in controlling and monitoring the traffic flows, thus worsening traffic circulation management. In order to solve this problem, it is necessary to apply automated systems to effectively control, monitor and manage traffic with high accuracy without human intervention. There are many solutions for a monitoring and decision support system in terms of transportation management (ITS), including using sensors on vehicles to read data from devices mounted on the vehicles, syncing vehicles over an internet connection (internet of things), etc. However, many practical solutions are still theoretical due to limitations in device fabrication, bandwidth, and cost of deployment.

Our solution is to automate the recognition vehicles to support vehicle flow monitoring. This is a system that automatically retrieves vehicle recognition information

© Springer Nature Switzerland AG 2019
N. T. Nguyen et al. (Eds.): ACIIDS 2019, LNAI 11432, pp. 670–679, 2019.
https://doi.org/10.1007/978-3-030-14802-7_58

through the surveillance camera system located in areas where management is required. In addition to applications in controlling moving vehicles, the proposed solution also has a wide range of applications, such as automatic weighing stations, automatic vehicle washers, automatic parking lots, traffic statistics by type of transport, etc.

2 Related works

There are many approaches to detecting and recognizing vehicles in single images or video images extracted from cameras on routes. Two main research directions are using traditional methods only or combining them with deep learning.

Traditional methods include the proposed Gauss mixture model (GMM) and the Kalman filter [2]. GMM is used to recognize vehicles and the Kalman Filter is used to track vehicles under light adaptive conditions. Another suggestion is the Optical Flow estimation method [4], which uses the edge features of images (as determined by the Canny algorithm) to determine the moving vehicles. In feature extraction, there are several methods, such as Scale Invariant Features Transform (SIFT) [5] and Histogram of Oriented Gradients (HOG) [19], followed by using the SVM classifier to determine means of transport [7]. The evaluating results show that the use of the HOG extraction features method and SVM classifier for recognition brings good results. A recent study proposes the use of feature extraction methods to recognize vehicles, vehicle counts, and classification [14]. In this method, GMM is used to segment images, and then Canny edge detector is used to define the boundary and extract features. Generally, traditional methods use the extraction of features associated with the shape, color, and texture of images to represent objects of interest (IO). After that, the classifier architecture is used to recognize the meaning of the transport situation.

Researches of deep learning often use high-performance Convolution Neural Network (CNN) models, such as AlexNet [1], GoogleNet [17], Microsoft ResNet [11], R-CNN [9], Fast R-CNN [8], Faster R-CNN [16], etc. For example, a recent research provides comparisons between the R-CNN and Faster R-CNN models [20] or between the AlexNet and Faster R-CNN models [6], in terms of vehicle recognition. Some proposed approaches bring high accuracy in vehicle recognition, even from satellite images [3, 12, 15] or from the low-resolution video [13]. However, most of the proposals, including the traditional methods and methods that use deep learning, only solve the problem of vehicle detection. There it very little proposals for the recognition of specific vehicles, such as motors, cars, coaches, trucks, etc. Our proposed solution is using the combination of CNN Inception technique [17], data Augmentation solution [16] for vehicle recognition. Based on the assessment of the combination of methods, appropriate models for the detailed recognition of vehicles will be proposed.

3 Overview Architecture

The proposed solution commences with the acquisition of images from the surveillance camera in ITS. Collected images are used to recognize objects of interest and determine the type of transportation. There are many methods for detecting vehicles, yet in this article, we focus on recognition models instead of detecting vehicles. By default, we use the semantic segmentation model based on Segnet's CNN architecture [10, 18]. Vehicles detected will then be extracted to determine regions of interest (ROI). Area of interest is a sample of the vehicle. Depending on the proposed method, it is possible to use the CNN model as well as combine with data augmentation to improve accuracy. Recognition results are used in the ITS system to alert vehicles when they are not allowed to enter the limit line and to handle violations (Fig. 1).

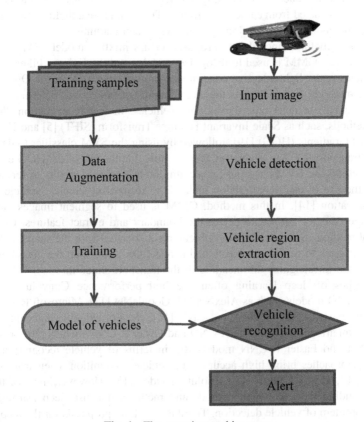

Fig. 1. The overview architecture

4   Proposed Method

4.1   Sequential Deep Learning Architecture

Usually, available pre-trained network models can be used to re-train the vehicle recognition models. However, in our approach, reusing the trained model is inappropriate, as the size of the old models' differs from the actual images obtained simultaneously. Besides, the training parameters do not support accuracy improvement. Some proposed models, such as AlexNet [1], GoogleNet [17],... are only effective for general recognition problems, not for this specific recognition problem. There are many different approaches to building a CNN model in vehicle recognition. In this study, we constructed a 24-layer CNN architecture, shown in Table 1, consisting of the input layer, convolution layer, rectified linear unit layer (ReLu), cross-normalization, max-pooling, and fully connected layer. The network model transforms the input image into a serial hierarchical descriptor. The neural aggregate input is the intensity values of the image applied to the CNN model. Input sample includes $128 \times 128 \times 3$ images. In this model, filters at the first layer concern to three color channels, namely R-G-B. Filters operate independently and jointly among hidden layers, involving three channels of the input image. The final layer handling the feature vector will be extracted into the classification layer. A convolutional layer implements a combination of mapped input images with a filter size $n_x \times n_y$.

Table 1. CNN architecture with 22 hidden layers, 1 input layer, and the final classification layer

TT	Layer type	Parameter
1	Image input	Image size $128 \times 128 \times 3$
2	Convolution	64 $7 \times 7 \times 3$ convolutions with stride [1 1]
3	ReLU	ReLU
4	Normalization	Cross channel normalization
5	Max pooling	3×3 max pooling with stride [1 1]
6	Convolution	64 $7 \times 7 \times 64$ convolutions with stride [1 1]
7	ReLU	ReLU
8	Max pooling	2×2 max pooling with stride [1 1]
9	Convolution	64 $7 \times 7 \times 64$ convolutions with stride [1 1]
10	ReLU	ReLU
11	Normalization	Cross channel normalization
12	Max pooling	2×2 max pooling with stride [1 1]
13	Convolution	64 $7 \times 7 \times 64$ convolutions with stride [1 1]
14	ReLU	ReLU
15	Max pooling	2×2 max pooling with stride [1 1]
16	Convolution	64 $7 \times 7 \times 64$ convolutions with stride [1 1]
17	ReLU	ReLU
18	Normalization	Cross channel normalization
19	Max pooling	2×2 max pooling with stride [1 1]
20	Fully connected	1024 fully connected layer
21	ReLU	ReLU
22	Fully connected	4 fully connected layer
23	Softmax	Softmax
24	Classification output	Crossentropyex with 4 other classes

4.2   Data Augmentation

The training data set classified during the collection is shown in Fig. 2.

In order to improve the accuracy of vehicle recognition, we propose to augment data about 10 times. Images are rotated $[-5^0, 5^0]$, flipped or added noise, yet no changes will be made to the image quality during training. The training data set after augmentation is shown in Table 3.

(a) Motor   (b) Car

(c) Coach   (d) Truck

Fig. 2. Some examples of vehicle categories

5   Experimental Results

5.1   Experimental Data

We have conducted experiments on a real database of vehicles including motors, cars, coaches, trucks taken from actual traffic situations. Camera systems typically receive signals in front of or behind the vehicles in traffic. This dataset is collected from different practical contexts on different traffic routes. The training dataset is divided into 4 different vehicle classes, including motors, cars, coaches, trucks simulated in Fig. 2, with 8,558 vehicle images. Dataset is partitioned into 60% for training and the remaining 40% for evaluation as shown in Table 2.

Table 2. Training data

Categories	Number of samples			Sample size
	Overall	Train	Evaluation	
Motor	2673	1604	1069	128 × 128
Car	2808	1685	1123	128 × 128
Coach	1640	984	656	128 × 128
Truck	1437	862	575	128 × 128

Table 3. Training data after augmentation and balance data

Categories	Number of samples
Motor	16040
Car	16850
Coach	17712
Truck	17240

5.2   Training CNN

Result obtained after CNN model training is shown as follows:

(i) Filter parameters: The first convolution layer uses 64 filters, whose filter's weight
     is shown in Fig. 3.

Fig. 3. The weight values of the filter of the first convolution layer. This layer consists of 64
filters size 7 × 7, each of which is connected to three RGB image input channels.

(ii) Convolution result: The sample images fed into the network through a convolu-
     tion filter and the obtained data show components distinct from the original RGB
     image with various feature result, creating a variety of vehicle features. The output
     value of the convolution set contains a negative value, which should be nor-
     malized by linear adjustment. The output of some layers is shown below, with the
     input pattern of the motor sample (Fig. 4).

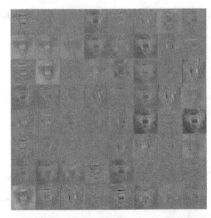

(a) The output of 64 convolutions at the first convolution layer

(b) The linear correction value after the first convolution layer

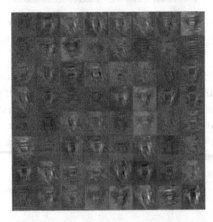

(c) The output of 64 samples at the second Convolution layer

Fig. 4. Some results of linear convolution and linear correction for the input images being motors

5.3   Categorical Vehicle Recognition

Based on the experiment, three different methods have been evaluated on the same set of sample data as shown in Table 2. Methods include: (i) Traditional methods of HOG and SVM; (ii) CNN network; (iii) CNN network in combination with data augmentation.

The accuracy of the HOG and SVM method on the sample data set was 89.31%. Details of the sample size for each type and recognition result are shown in Table 4.

Table 4. Confusion matrix of vehicle recognition using HOG and SVM

	Motor		Car		Coach		Truck	
	1069		1123		656		575	
	#Num	Per(%)	#Num	Per(%)	#Num	Per(%)	#Num	Per(%)
Motor	1029	97.26	16	1.53	15	1.87	9	1.75
Car	25	2.36	989	94.37	77	9.59	32	6.23
Coach	1	0.09	23	2.19	599	74.60	33	6.42
Truck	3	0.28	20	1.91	112	13.95	440	85.60

The evaluated accuracy of the CNN method based on original data was achieved 90.10% on average, as shown in Table 5.

Table 5. Confusion matrix of vehicle recognition using CNN

	Motor		Car		Coach		Truck	
	1069		1123		656		575	
	#Num	Per(%)	#Num	Per(%)	#Num	Per(%)	#Num	Per(%)
Motor	1026	95.98	38	3.38	1	0.15	5	0.87
Car	32	2.99	953	84.86	17	2.59	24	4.17
Coach	6	0.56	104	9.26	617	94.05	58	10.09
Truck	5	0.47	28	2.49	21	3.20	488	84.87

The evaluated accuracy of the CNN method based on data augmentation was achieved 95.59% on average, as shown in Table 6.

Table 6. Confusion matrix of vehicle recognition using CNN and data augmentation

	Motor		Car		Coach		Truck	
	1069		1123		656		575	
	#Num	Per(%)	#Num	Per(%)	#Num	Per(%)	#Num	Per(%)
Motor	1060	99.16	11	0.98	0	0	1	0.17
Car	5	0.47	1057	94.12	8	1.22	13	2.26
Coach	0	0	41	3.65	645	98.32	51	8.87
Truck	4	0.37	14	1.25	3	0.46	510	88.70

In this study, we also evaluated the proposed CNN model to another traditional approach based on HOG feature descriptor and SVM classifier. Results of the comparison are shown in Fig. 5.

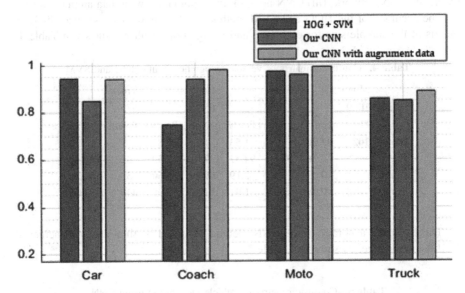

Fig. 5. Comparison of HOG + SVM, CNN model and CNN with augmenting data

6   Conclusions

The experimental results show that traditional models using HOG and SVM still ensure relatively high accuracy in object identification, even in cases of not big enough training data sets. The method of using deep network did not bring high accuracy in cases of small data sets. After data augmentation, CNN model's identification accuracy saw significant changes.

Through our experimental results, our solutions offer the following benefits:

(1) Provide an assessment and selection of effective methods for vehicle recognition.
(2) Provide detailed recognition of vehicles with relatively high accuracy based on experiment and comparison.
(3) Propose practical and cost-effective solutions.

Based on the results of the study, in the near future, we suggest continuing to train and experiment with a variety of vehicles, supporting the actual management of ITS systems.

References

1. Krizhevsky, A., Sutskever, I., Hinton, G.E.: ImageNet classification with deep convolutional neural networks. In: Advances in Neural Information Processing Systems (NIPS 2012), vol. 25, pp. 1106–1114 (2012)
2. Amirullah, I., Bakti, R.Y., Areni, I., Alimuddin, A.A.: Vehicle detection and tracking using Gaussian Mixture Model and Kalman filter, pp. 115–119 (2016)
3. Chen, X., Xiang, S., Liu, C.-L., Pan, C.-H.: Vehicle detection in satellite images by hybrid deep convolutional neural networks. IEEE Geosci. Remote Sens. Lett. **11**, 1797–1801 (2014)
4. Chen, Y., Wu, Q.: Moving vehicle detection based on optical flow estimation of edge, pp. 754–758 (2015)
5. Choi, J.-y., Sung, K.-S., Yang, Y.: Multiple vehicles detection and tracking based on scale-invariant feature transform, pp. 528–533 (2007)
6. Espinosa, J.E., Velastin, S.A., Branch, J.W.: Vehicle detection using alex net and faster R-CNN deep learning models: a comparative study. In: Badioze Zaman, H., et al. (eds.) Advances in Visual Informatics. IVIC 2017. LNCS, vol. 10645, pp. 3–15. Springer, Cham (2017). https://doi.org/10.1007/978-3-319-70010-6_1
7. da Silva Filho, S.G., Freire, R.Z., dos Santos Coelho, L.: Feature extraction for on-road vehicle detection based on support vector machine. In: Conference Proceedings (2017)
8. Girshick, R.: Fast R-CNN (2015)
9. Girshick, R., Donahue, J., Darrell, T., Malik, J.: Rich feature hierarchies for accurate object detection and semantic segmentation (2014)
10. He, K., Zhang, X., Ren, S., Sun, J.: Delving deep into rectifiers: surpassing human-level performance on ImageNet classification, vol. 1502 (2015)
11. He, K., Zhang, X., Ren, S, Sun, J.: Deep residual learning for image recognition, pp. 770–778 (2016)
12. Koga, Y., Miyazaki, H., Shibasaki, R.: Counting vehicles by deep neural network in high resolution satellite images (2017)
13. Bautista, C.M., Dy, C.A., Manalac, M.I., Orbe, R.A., Cordel II, M.: Convolutional neural network for vehicle detection in low resolution traffic videos, pp. 277–281 (2016)
14. Moutakki, Z., Ouloul, M.I., Afdel, K., Amghar, A.: Real-time system based on feature extraction for vehicle detection and classification. Transp. Telecommun. J. **19**, 93–102 (2018)
15. Qu, S., Wang, Y., Meng, G., Pan, C.: Vehicle detection in satellite images by incorporating objectness and convolutional neural network (2016)
16. Ren, S., He, K., Girshick, R., Sun, J.: Faster R-CNN: towards real-time object detection with region proposal networks, pp. 1–10 (2016)
17. Szegedy, C., et al.: Going deeper with convolutions, pp. 1–9 (2015)
18. Badrinarayanan, V., Kendall, A., Cipolla, R.: SegNet: a deep convolutional encoder-decoder architecture for image segmentation. IEEE Trans. Pattern Anal. Mach. Intell. **39**, 2481–2495 (2017)
19. Yan, G., Ming, Y., Yu, Y., Fan, L.: Real-time vehicle detection using histograms of oriented gradients and AdaBoost classification (2016)
20. Yılmaz, A., Guzel, M., Askerbeyli, I., Bostanci, E.: A vehicle detection approach using deep learning methodologies (2018)

Video-Based Vietnamese Sign Language Recognition Using Local Descriptors

Anh H. Vo[1], Nhu T. Q. Nguyen[1], Ngan T. B. Nguyen[1], Van-Huy Pham[1(✉)],
Ta Van Giap[2], and Bao T. Nguyen[3(✉)]

[1] Faculty of Information Technology, Ton Duc Thang University,
Ho Chi Minh City, Vietnam
{vohoanganh,phamvanhuy}@tdtu.edu.vn, nh_babe@yahoo.com.vn,
baonannguyen95@gmail.com
[2] Can Tho Medical College, Can Tho City, Vietnam
giapcmc@gmail.com
[3] University of Education and Technology, Ho Chi Minh, Vietnam
baont@hcmute.edu.vn

Abstract. Sign Language is one of the method for non-verbal communication. It is most commonly used by deaf or dumb people who have hearing or speech problems to communicate among themselves or with normal people. Vietnamese Sign Language (VSL) is a sign language system used in the community of Vietnamese hearing impaired individuals. VSL recognition aims to develop algorithms and methods to correctly identify a sequence of produced signs and to understand their meaning in Vietnamese. However, automatic VSL recognition in video has many challenges due to the orientation of camera, hand position and movement, inter hand relation, etc. In this paper, we present some feature extraction approaches for VSL recognition includes spatial feature, scene-based feature, and especially motion-based feature. Instead of relying on a static image, we specifically capture motion information between frames in a video sequence. We evaluated the proposed framework on our acquired VSL dataset including 23 alphabets, 3 diacritic marks and 5 tones in Vietnamese language with $2D$ camera. Additionally, in order to gain more information of hand movement and hand position, we also used the data augmentation technique. All these helpful information would contribute to an effective VSL recognition system. The experiments achieved the satisfactory results with 86.61%. It indicates that data augmentation technique provides more information about the orientation of hand. Moreover, the combination of spatial, scene and especially motion information could help the system to be able to capture information from both single frame and from multiple frames, and thus the performance of VSL recognition system could be improved.

Keywords: Vietnamese Sign Language (VSL) · VSL recognition ·
Local descriptors · Spatial feature · Scene-based feature ·
Motion-based feature

© Springer Nature Switzerland AG 2019
N. T. Nguyen et al. (Eds.): ACIIDS 2019, LNAI 11432, pp. 680–693, 2019.
https://doi.org/10.1007/978-3-030-14802-7_59

1   Introduction

Sign Language is the natural language of choice for most deaf and dump people. In sign language, each gesture already has assigned meaning, and strong rules of context and grammar may be applied to make recognition tractable. It is also known as grammatically complete and copious as spoken language. Sign language conveys meaning more than only the moving hands. The main four components describe a sign including hand shape, location in relation to the body, movement of hands, and the orientation of palms. However, they are not international language because of the different expressions of signs among other countries. Vietnamese Sign Language (VSL) is based on the established American Sign Language (ASL), but VSL also has some special signs which are not involved in ASL dictionary.

Recognition of sign language has many applications such as intelligent control systems, robotic, human computer interaction, smart home, etc., [4,6,7,12]. Previous researches focused on written and spoken languages while recognition of sign language has not yet studied widely. In this paper, we proposed a framework for Vietnamese Sign Language recognition using local features. Three types of feature in our system are *spatial features, scene-based feature and motion-based feature.* Beside each single features, we also try to combine them together to capture more important information of a sign. Note that both the spatial and scene-based feature are extracted from one frame of video sequence, so they are considered as *single-frame based features.* Conversely, the motion-based feature are called *multi-frame based feature* because it must be calculated from multiple frames from input video sequence. For evaluating our system, we also collect the corpus of continuous VSL datasets consisting of 23 alphabets, 3 diacritic mark and 5 tones only with $2D$ camera.

This paper is organized as follows. The related work is presented in Sect. 2. In Sect. 3, we describe the overview of methodology. Experiments and discussion can be found in Section 5. Finally, we conclude by mentioning our contributions and some future works in Section 6.

2   Related Works

Computer recognition of sign language deals from sign gesture acquisition and continues till alphabet/text/speech generation. Sign gestures can be classified as static and dynamic. However static gesture recognition is simpler than dynamic gesture recognition but both recognition systems are important to the human community. Up to present, there are two main approaches for sign language recognition: sensor and vision-based.

Sensor-Based Approach: Bui et al. [2] used gloves with sensors to identify 23 characters in Vietnamese Sign Language. By using sensors, this approach can remove some outliers from the complex environment, therefor it simplifies the pre-processing stage in sign language recognition system. In addition, the use of gloves makes it able to capture the change of shapes, and also hand movements in video. However, it is not convenient to singers when wearing gloves, and it is not suitable in real world when signers always bring gloves all time.

Vision-Based Approach: There are many works applied this approach because of convenience with users [9]. This approach captures the hand gesture or body part gesture in form of static images or sequence images by camera without gloves or sensor devices, which makes it more suitable for real world. However, it has many challenges like complex background problems, variation in lighting condition, changes of skin color, and the properties and configurations of camera (such as variations of scale, translation, rotation, view and occlusion). In [5, 8,14], Microsoft Kinect camera are used to capture RGB-D image instead of RGB image to improve the results. However, Microsoft Kinect camera is hard to integrated into real world applications. For that reason, in this study we propose to use RGB image in a video sequence to capture motion features of isolated sign for VSL recognition.

3   Approach

We propose a method for Vietnamese Sign Language recognition in video consequence based on local descriptors. In sign language recognition where the motion of the hand and its location in consecutive frames is a key feature in the classification of different signs. Due to that, this framework was implemented with both single-frame features (spatial features, scene-based features) and multi-frame feature (motion feature).

3.1   Preprocessing

Firstly, some key frames are extracted from the input video sequence for each alphabets or words manually. The original images are converted to HSV-color space to segment skin regions. Due to the difference among signers when collecting data, we also applied the data augmentation technique (rotation and noise adding) to create more images with the aim of gaining more information of hand position. More specifically, the original images are rotated $\pm 5°$, $\pm 10°$ and added salt/pepper noise with probability of 0.05. After that, all hand skin regions are normalized to eliminate face region and background region.

3.2   Feature Extraction

In feature extraction step, we combined spatial, scene-based and motion-based features to capture different information among each alphabets of VSL in a video sequence.

Spatial Features. There are three features calculated based on appearance as following:

Local Binary Pattern (LBP): LBP operator was initially proposed by Ojala et al. [10] to express the texture of the image patches. LBP is tolerant to illumination variations, and simply computing. It calculates a code for each pixel by thresholding its value with that of its neighbor and converts the code into a decimal. Given a pixel i_n in the image, i_c is the neighbors of i_n.

$$LBP_P(x_c, y_c) = \sum_{n=0}^{P-1} s(i_n - i_c)2^n s(x) = \begin{cases} 1 & x \geq 0 \\ 0 & x < 0 \end{cases} \quad (1)$$

In this framework, we split hand region into 49 sub-regions and then applied LBP operator in each sub-region. Instead of 256-bin histogram of traditional LBP, we only use 59-bin histogram to reduce dimension for feature vector. Our experiment shows that there is not much difference in recognition result between 256-bin vs 59-bins. As a result, we obtain a 2891 dimensional feature vector for each hand region.

Local Phase Quantization (LPQ): LPQ descriptor has been firstly appointed for use in the classification of textures blur [11]. LPQ is constructed to retain an image in the local invariant information to artifacts generated by different forms of blur. This descriptor uses local phase information extracted from a short-term Fourier transform (STFT) over a rectangular $M \times M$ neighborhoods N_x at each pixel position x of the image $f(x)$ (Fig. 1):

$$F(u, x) = \sum_{y \in N_x} f(x - y)e^{-j2\pi u^T y} \quad (2)$$

where w_u is the basis vector of the 2-D DFT at frequency u, and f_x is the vector containing all M^2 samples from N_x. The local Fourier coefficients are

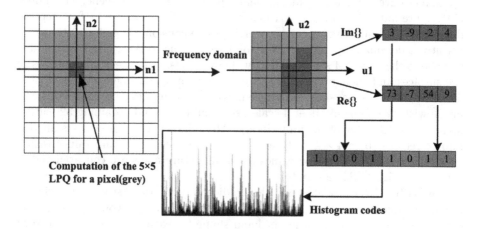

Fig. 1. Local phase quantization.

computed as four frequency points: $u_1 = [a, 0]^T$, $u_2 = [0, a]^T$, $u_3 = [a, a]^T$, $u_4 = [a, -a]^T$, where a is a sufficiently small scalar. It results in a vector $F_x = [F(u_1, x), F(u_2, x), F(u_3, x), F(u_4, x)]$ for each pixel position. The phase information in the Fourier coefficients are decided by examining the signs of the real and imaginary parts of each component in F_x, using a simple scalar quantizer:

$$q_j = \begin{cases} 1 & g_j \geq 0 \\ 0 & g_j < 0 \end{cases} \qquad (3)$$

where $g_j(x)$ is the jth component of the vector $G_x = [Re\{F_x\}, Im\{F_x\}]$. The result of eight bit binary coefficients $g_j(x)$ are represented as integers using binary coding:

$$f_{LPQ}(x) = \sum_{j=1}^{8} q_j 2^{j-1} \qquad (4)$$

Inspired by this idea, we propose the LPQ as an effective method to solve the problem of expressions variations. We applied LPQ operator in 25 sub-regions of hand region, and finally, a histogram of values from all sub-regions are combined into only one feature vector with 6400 dimensions.

Histogram of Oriented Gradients (HOG): HOG descriptor [3] is one of the appearance descriptor which is able to captures edge or gradient structure of local shape. HOG is invariant to local geometric and photo transformations (translation, rotation). The main idea of HOG is to calculate the distribution of local intensity gradient orientation of a detection window.

In this study, we set the size of each block 16×16 pixel and 2×2 for each cell. Finally, we obtain a final feature vector with 8100 dimensions.

Scene-Based Feature. The GIST descriptor was initially proposed in [12] with the aim of developing a low dimensional representation of the scene, which does not require any form of segmentation. The authors propose a set of perceptual dimensions (naturalness, openness, roughness, expansion, ruggedness) that represent the dominant spatial structure of a scene. Due to that, GIST would able to present the shape of scene image by the relationship between the outlines of the surfaces and their properties and perceptual properties: naturalness, openness, roughness, expansion and ruggedness. It is computed by convolving the image with 32 Gabor filters at 4 scales, 8 orientations, and obtaining 32 feature maps of the same size of the input image. After that each feature map is divided into 16 regions (by a 4×4 grid), and then it averages the feature values within each region. The 16 averaged values of all 32 feature maps are concatenated in a $16 \times 32 = 512$ GIST descriptor.

The GIST descriptor presents the gradient information (scales and orientations) for different parts of an image, and due to that, it would able to take over many essential information of hand shape for each alphabet or sign word in our case.

Motion-Based Feature. The extraction of Histograms of Oriented Optical Flow provides a histogram $h_{b,t} = [h_{t,1}, h_{t,2}, \ldots, h_{t,B}]$ at each time instant t, for each block b in the frame, in which each flow vector is binned according to its primary angle from the horizontal axis and weighted according to its magnitude. Every optical flow vector $v = [x, y]^T$, with direction $\theta = tan^{-1}(\frac{y}{x})$ and in the range.

$$\frac{-\pi}{2} + \pi\frac{b-1}{B} \leq \theta < -\frac{\pi}{2} + \pi\frac{b}{B} \tag{5}$$

will contribute by $\sqrt{x^2 + y^2}$ to sum in bin b, $1 \leq b \leq B$, out of a B bins. Finally, the histogram is normalized to sum up to 1.

3.3   Classification

The recognition on sequence images can be enforced into two main approaches: based on static image and based on image sequence approach.

Static Image Based Approach: This approach is presented in [1,13]. Each frame I_i was classified individually and recognition result s was probability vector $\{p_{i,c_1}, p_{i,c_2}, \ldots, p_{i,c_n}\}$ where p_{i,c_j}, $(1 \leq j \leq n)$ is the probability when frame I_i corresponds to one of n classes of Vietnamese signs.

Image Sequence Based Approach: This approach used a model which reflected the importance of modeling temporal motion patterns for signs language recognition as Latents Random Fields (LDCRFs).

In this paper, we inspired by the static image based approach to evaluate one image sequence. Specifically, we used Support Vector Machine to classify alphabets of Vietnamese sign language on each frame individually and the recognition result was the normalized sum of probability of all frames, to predict sign label in one sequence. Given a training set of labeled examples $\{(x_i, y_i), i = 1, \ldots, k\}$ where $x \in R^n$ and $y_i \in \{1, -1\}$, SVM classifies a new test example x basing the following functions:

$$f(x) = sgn(\sum_{i=1}^{l} \alpha_i \gamma_i K(x_i, x) + b) \tag{6}$$

where α_i are Lagrange multipliers of a dual optimization problem that describes the separating hyper-plane; $K(.,.)$ is a kernel function; and b is threshold parameter of hyper-plane. The training sample x_i (with $\alpha_i > 0$) is called support vectors, and SVM would find the hyper-lane that maximizes the distance between the support vectors and the hyper-plane.

4   Vietnamese Sign Language Dataset

Vietnamese alphabet system is more complicated than English one because it is a tone language which consists of six different tones (high level, low rising, high rising glottalized, low falling glottalized, high rising, and low falling), and three diacritic marks $(\breve{A}, \hat{O}, \sigma)$.

In [2], Bui et al. collected VSL dataset only containing 23 alphabets of (A, B, C, D, $Đ$, E, G, H, I, K, L, M, N, O, P, Q, R, S, U, V, X, Y), and signers need to wear gloves with MEMS sensor during the recording process. However, in our project, to increase the challenges of Vietnamese sign language recognition, we only use a $2D$ camera for collecting data, and the signers wear the normal clothes without any sensors. We acquired a dataset of VSL with Alphabet, Diacritic Marks and Tones (VSL-ADT) consisting of 23 alphabets, 3 diacritic marks and 5 tones. Note that in VSL-ADT the high level tone is removed because it does not present more meaning as well as low rising tone, high rising glottalized tone, low falling glottalized tone, high rising tone, and low falling tone. On the other hand, we added 3 diacritic marks of $\breve{A}$, $\hat{O}$, σ. The significant contribution of VSL-ADT is to add 3 diacritic marks ($\breve{A}$, $\hat{O}$, σ) , and 5 tones, which were not included in [2]. Moreover, we only used 2D camera which makes more reasonable and popular in daily life than using gloves with MEMS sensors as in [2].

We split VSL-ADT into two non-intersecting datasets corresponding to when the signers wear short sleeved shirt (VSL-SADT) or long sleeved shirt

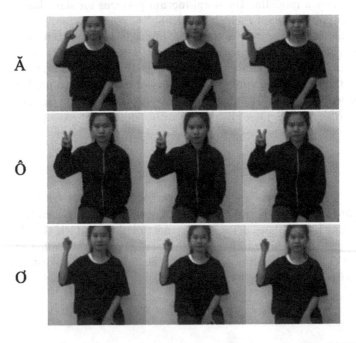

Fig. 2. Diacritic marks with short sleeved shirt wearing signers in VSL-SADT dataset

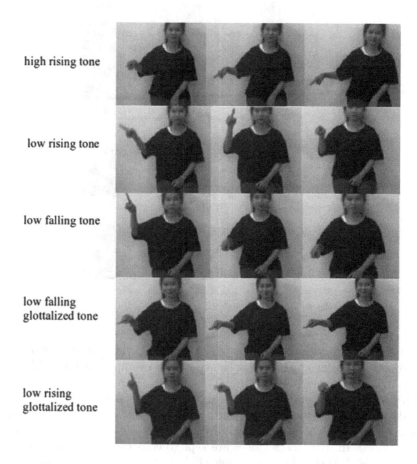

Fig. 3. Five tones of VSL with short sleeved shirt wearing signers

Table 1. Description of VSL-ADT dataset with 23 alphabets, 3 diacritic marks and 5 tones

	Long sleeved shirt	Short sleeved shirt
Resolution	1920 × 1080	1920 × 1080
# video	899	1178
# signer	29	38
# alphabet	31	31
# video for each alphabet	29	38
Time for each video	2–5 s	2–5 s

Ă

Ô

Ơ

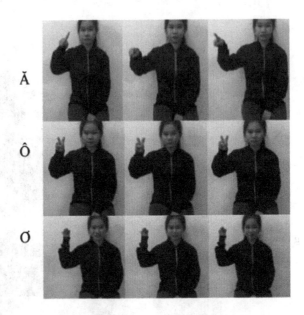

Fig. 4. Diacritic marks with long sleeved shirt wearing signers in VSL-LADT dataset

(VSL-LADT). All videos were collected in laboratory environment with white background and black shirt wearing signers. Some examples of VSL-SADT and VSL-LADT are showed in Figs. 2 and 3.

Five tones on Vietnamese Sign Language Dataset with signer wear long sleeved shirt are the same gesture with when signer wear short sleeved shirt. Human annotators who are teachers of deaf people are also asked to evaluate for each video. In total, 1178 videos are kept to construct VSL-SADT and 899 videos for VSL-LADT (Fig. 4 and Table 1).

5   Experiments

When analyzing the VSL-SADT dataset, we figure out that in some cases there are much changes between frames in a video sequence of one alphabet, while in the other cases, not many differences happen during the whole sequence. Inspiring from that, we decide to separate VSL-SADT into two small datasets which is based on the characteristics of motion gestures. The first dataset (VSL-SADT-01) has a few changes between frames and it totally contains 23 VSL alphabets *A, B, C, D, Đ , E, F, G, H, I, K, L, M, N, O, P, Q, R, S, T, U, V, X, Y* and two diacritic marks of Vietnamese language ($\hat{O}$ and σ). And the second one (VSL-SADT-02) is extended from the first dataset with additional five tones (*low rising tone, high rising tone, low falling tone, high rising glottalized tone, low falling glottalized tone, low falling tone*) and a diacritic mark *(Ă)*.

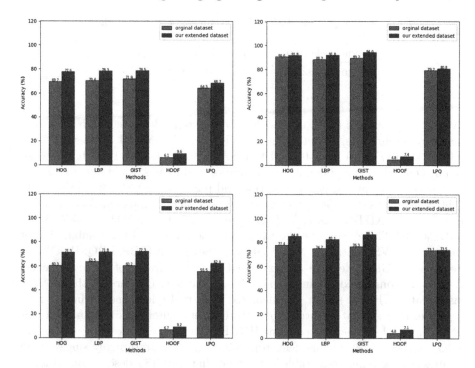

Fig. 5. The accuracy of VSL recognition with different local descriptors on original and extended dataset: VSL-SADT-01 (top-left), VSL-LADT-01 (top-right), VSL-SADT-02 (bottom-left), VSL-LADT-02 (bottom-right)

Similarly, VSL-LADT is also divided into VSL-LADT-01, VSL-LADT-02 with the same conditions. Moreover, as mentioned in Sect. 3, the data augmentation technique (rotation and noise adding) is applied in order to create additional images with the aim of gaining more information of hand position (±5°, ±10° rotation, 0.05 probability of salt/pepper noise). As the result, we have the original dataset and the extended one (-EXT). All videos in each dataset are used for training and testing with ratio of 9 : 1. The system was performed in Matlab environment with CPU Intel Core (TM) i7, 8 GB of RAM.

Firstly, we try to find out the most satisfied kernel for SVM classifier including linear, polynomial, rbf and sigmoid kernel. In this paper, we only used SVM for classification, but some other classifiers such as kNN, Naive Bayes, Random Forest, etc. must be investigated. We will discuss about this matter in the last section of conclusion and future work Sect. 6. Table 2 shows that the best accuracy would be achieved with polynomial kernel (85.48%), followed by rbf (83.87%) and linear 82.26%. And the worst case is sigmoid with only 38.71%. From now then, we only use the polynomial kernel for later experiments.

The next experiment was run by using different local descriptors for VSL recognition task on two datasets: original and our extended one. Figure 5 shows the comparison of the accuracy result between different models on two datasets.

Table 2. Comparison of SVM kernels on the VSL-SADT dataset

Kernel	linear	polynomial	rbf	sigmoid
Accuracy	82.26%	85.48%	83.87%	38.71%

Overall, the accuracy result of models on our extended dataset with data augmentation techniques, is better than using only the orginal dataset on both VSL-SADT-01/VSL-LADT-01 and VSL-SADT-02/VSL-LADT-02. It is a proof for our belief that applying data augmentation techniques would able to capture more useful information of hand motion and position for VSL recognition task. Table 3 illustrates the result experiments of the local descriptors on four extended dataset VSL-SADT-01-EXT, VSL-LADT-01-EXT, VSL-SADT-02-EXT, VSL-LADT-02-EXT. We compare the five local descriptors and the combination of them on four VSL datasets. Note that both the spatial (LBP, LPQ, HOG) and scene-based (GIST) feature are extracted from one frame of video sequence, so they are considered as single-frame based features. Conversely, the motion-based feature (HOOF) must be calculated from multiple frames of input video sequence. Due to that, in order to capture diverse information from both single and multiple frames, we combine HOOF (multiple frame feature) with LBP, HOG, or GIST (static-image feature), and do not integrate many single-frame features as once time. From Table 3, it is clear that the GIST description achieved the highest accuracy in both VSL-SADT-01-EXT (78.85%) and VSL-LADT-01-EXT (94%), because these two datasets only include 23 alphabets and two diacritic marks of VSL, all of which specially have no more motion gesture. In the meanwhile, the combination of GIST and HOOF can capture the motion features between frames in dynamic signs of VSL, and that is reason why this combination would able to get satisfactory accuracy on both VSL-SADT-02-EXT (73.1%) and VSL-LADT-02-EXT (86.61%). In the other view, the results of LADT (93.8%, 86.61%) is better than SADT (78%, 73.1%) in both 01-EXT/02-EXT, it indicate that SADT dataset is more challenges than LADT dataset because it contains

Table 3. VSL recognition accuracy with different types of local feature. The name of local descriptor is in the first column, and the next four columns show the results on 04 extended datasets

Method	VSL-SADT-01-EXT	VSL-SADT-02-EXT	VSL-LADT-01-EXT	VSL-LADT-02-EXT
HOG	78.1%	71.29%	91.8%	84.83%
LBP	78.28%	71.77%	91.8%	82.09%
LPQ	68.6%	62.37%	80.8%	73.54%
GIST	**78.53%**	72.26%	**94%**	86.29%
HOOF	9.6%	9.61%	7.4%	7.1%
HOG + HOOF	78.1%	71.94%	92%	84.35%
LBP + HOOF	77.73%	72.9%	92%	81.61%
GIST + HOOF	78%	**73.1%**	93.8%	**86.61%**

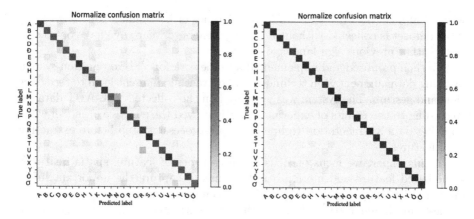

Fig. 6. Confusion matrix with GIST descriptor on VSL-SADT-01-EXT (left) and VSL-LADT-01-EXT (right)

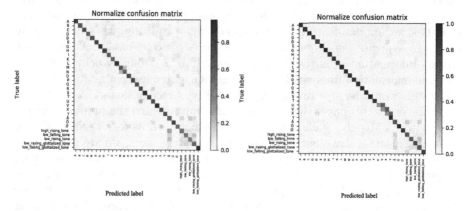

Fig. 7. Confusion matrix with GIST-HOOF combination on VSL-SADT-02-EXT (left) and VSL-LADT-02-EXT (right)

the arm of signer but it is not reflect more necessary information to improve the accuracy of VSL recognition system.

The confusion matrix when applying GIST on VSL-SADT-01-EXT/VSL-LADT-01-EXT is presented in Fig. 6, and the one of combination GIST and HOOF is on Fig. 7.

6 Conclusion and Future Work

Sign language is one of the tool of communication for those with hearing and vocal disabilities. Sign language recognition hence plays very important role in this regard by capturing the sign language video and then recognizing the sign language accurately. This paper deals with the Vietnamese sign language recognition. The first contribution of this work is to acquire the VSL-ADT dataset with 24 alphabets, 3 diacritic marks and 5 tones of Vietnamese language. VSL-ADT dataset is collected without requiring signers to wear any gloves or sensors as the other previous sign language dataset.

We then proposed and implemented a framework for VSL recognition system by using data augmentation technique and extracting different local to recognize the hand gesture. The system was evaluated on the collected VSL-ADT dataset. According to the results of experiments, it is proved the effective of model when applying data augmentation technique in pre-processing step before extracting local features.

In this paper, we focus on ocus on three types of feature: spatial features, scene-based feature and motion-based feature to capture the important information of language signs. The spatial and scene-based feature are considered as single-frame based features because they are extracted from one single frame of video sequence. While the multiple-frame motion-based feature must be calculated from multiple frames from input video sequence. The results of experiments also show that the combination of spatial features or scene-based features together with motion-based features would able to recognize more accurate. Besides, motion-feature has the capability of capturing the changes from frames in sequence images, while spatial features or scene-based contain almost enough information of an each static frame. It is a strong evident that if we combine information from single frame together with one from multiple frames, the system would be able to improve the ability of recognition.

However, in this paper, we have just only implemented the proposed framework with SVM only, while some other classifier such as k-NN, Naive Bayes, Random Forest, ... have not been invested. In future work, we are going to do more experiments on several classifiers to choose the most suitable one. Then, we plan to develop a VSL recognition system in the area of words or sentence instead of alphabet VSL to apply in real-life system. Additionally, we believe that the accuracy of VSL recognition system would be improved when using Convolutional Neural Network.

Acknowledgment. The authors would like to thank the teachers of the deaf people in Binh Duong province, Vietnam. We acknowlegment the support of the students in Ton Duc Thang University.

References

1. Bartlett, M.S., Littlewort, G., Fasel, I., Movellan, J.R.: Real time face detection and facial expression recognition: development and applications to human computer interaction. In: 2003 Conference on Computer Vision and Pattern Recognition Workshop, vol. 5, pp. 53–53, June 2003. https://doi.org/10.1109/CVPRW.2003.10057
2. Bui, T.D., Nguyen, L.T.: Recognizing postures in Vietnamese sign language with mems accelerometers. IEEE Sens. J. **7**(5), 707–712 (2007). https://doi.org/10.1109/JSEN.2007.894132

3. Dalal, N., Triggs, B.: Histograms of oriented gradients for human detection. In: 2005 IEEE Computer Society Conference on Computer Vision and Pattern Recognition (CVPR 2005), vol. 1, pp. 886–893, June 2005. https://doi.org/10.1109/CVPR. 2005.177

4. Donahue, J., et al.: Long-term recurrent convolutional networks for visual recognition and description (2015)

5. Duc, H.V., Huynh, H.H., Phuoc, M.D., Meunier, J.: Dynamic gesture classification for Vietnamese sign language recognition. Int. J. Adv. Comput. Sci. Appl. **8**, 415–420 (2017)

6. Hai, P.T., Thinh, H.C., Phuc, B.V., Kha, H.H.: Automatic feature extraction for Vietnamese sign language recognition using support vector machine. In: 2018 2nd International Conference on Recent Advances in Signal Processing, Telecommunications Computing (SigTelCom), pp. 146–151, January 2018. https://doi.org/10. 1109/SIGTELCOM.2018.8325780

7. Liang, Z.J., Liao, S.B., Hu, B.Z.: 3D convolutional neural networks for dynamic sign language recognition. Comput. J. **61**(11), 1724–1736 (2018). https://doi.org/ 10.1093/comjnl/bxy049

8. Pigou, L., Dieleman, S., Kindermans, P.-J., Schrauwen, B.: Sign language recognition using convolutional neural networks. In: Agapito, L., Bronstein, M.M., Rother, C. (eds.) ECCV 2014. LNCS, vol. 8925, pp. 572–578. Springer, Cham (2015). https://doi.org/10.1007/978-3-319-16178-5_40

9. Ng, J.Y., Hausknecht, M.J., Vijayanarasimhan, S., Vinyals, O., Monga, R., Toderici, G.: Deep networks for video classification (2015)

10. Ojala, T., Pietikäinen, M., Harwood, D.: A comparative study of texture measures with classification based on feature distributions. Pattern Recogn. **29**, 51–59 (1996). https://doi.org/10.1016/0031-3203(95)00067-4

11. Ojansivu, V., Heikkilä, J.: Blur insensitive texture classification using local phase quantization. In: Elmoataz, A., Lezoray, O., Nouboud, F., Mammass, D. (eds.) ICISP 2008. LNCS, vol. 5099, pp. 236–243. Springer, Heidelberg (2008). https:// doi.org/10.1007/978-3-540-69905-7_27

12. Oliva, A., Torralba, A.: Modeling the shape of the scene: a holistic representation of the spatial envelope. Int. J. Comput. Vis. **42**(3), 145–175 (2001). https://doi. org/10.1023/A:1011139631724

13. Shan, C., Gong, S., McOwan, P.W.: Facial expression recognition based on local binary patterns: a comprehensive study. Image Vis. Comput. **27**(6), 803–816 (2009). https://doi.org/10.1016/j.imavis.2008.08.005

14. Vo, D., Nguyen, T., Huynh, H., Meunier, J.: Recognizing Vietnamese sign language based on rank matrix and alphabetic rules. In: 2015 International Conference on Advanced Technologies for Communications (ATC), pp. 279–284, October 2015. https://doi.org/10.1109/ATC.2015.7388335

Mathematical Variable Detection in PDF Scientific Documents

Bui Hai Phong[1(✉)], Thang Manh Hoang[2], Thi-Lan Le[1], and Akiko Aizawa[3]

[1] MICA International Research Institute (HUST - CNRS/UMI2954 - Grenoble INP),
Hanoi University of Science and Technology, Hanoi, Vietnam
{hai-phong.bui,thi-lan.le}@mica.edu.vn
[2] School of Electronics and Telecommunications,
Hanoi University of Science and Technology, Hanoi, Vietnam
thang.hoangmanh@hust.edu.vn
[3] National Institute of Informatics,
2-1-2 Hitotsubashi Chiyoda-ku, Tokyo 101-8430, Japan
aizawa@nii.ac.jp

Abstract. The detection of mathematical expression from PDF documents has been studied and advanced for recent years. In the process, the detection of variables of inline expressions that are represented by alphabetical characters is a challenge. Compared to other components of inline expressions, there are many factors that cause the ambiguities for the detection of variables. In this paper, the error in detecting variables in PDF scientific documents is analytically presented. Novel rules are proposed to improve the accuracy in the detection process. The experimental results on benchmark datasets containing English and Vietnamese documents show the effectiveness of the proposed method. The comparison with existing methods demonstrates the out-performance of the proposed method. Furthermore, pre-trained deep Convolutional Neural Networks are employed and optimized to automatically extract visual features of extracted components from PDF and machine learning algorithms are used to improve the accuracy of the detection.

Keywords: PDF document analysis ·
Mathematical expression extraction · Machine learning ·
Rule-based classification · Deep learning

1 Introduction

In recent years, the detection of mathematical expressions in scientific documents has received many advances. Based on the nature, input documents of the detection can be divided into three categories: stroke (e.g. handwritten), vector (e.g. PDF) and raster (e.g. image) format. The applied techniques for the detection highly depend on the category of input document. Among these document formats, PDF is the format where the detection is considered the most precise one [1]. The attribute and content extracted from PDF document are

© Springer Nature Switzerland AG 2019
N. T. Nguyen et al. (Eds.): ACIIDS 2019, LNAI 11432, pp. 694–706, 2019.
https://doi.org/10.1007/978-3-030-14802-7_60

two top items in $\boxed{S}$ (s_1 and s_2) are popped, and a new item is pushed onto $\boxed{S}$ This new item is a tree formed by making the root s_1 of a dependent of the root of s_2, or the root of s_2 a dependent of the root	keeps a heap $\boxed{H}$ containing multiple parser states T_0... T_m. These states are ordered in the heap according to their probabilities, which are determined by multiplying the probabilities of each of the

Fig. 1. Examples of variables represented by single character (in red) in a sample PDF scientific document (Color figure online)

more proper and there is less noises than those of other formats. These factors are useful for the detection of mathematical expressions. Several researches have investigated to extract inline mathematical expressions from PDF documents. There are many components of inline mathematical expressions: operator, variable or function. An operator is any symbol that indicates an operation to be performed (e.g. $+, -, *, /$). A variable is a quantity that may change or have variety of values in the mathematical problem. Typically, a variable is represented by a single letter (e.g. x, y). A function can be an expression or a law that defines a relationship between variables (e.g. $y = f(x)$) [2]. In general, the accuracy of the detection of inline mathematical expression has been much improved in recent years. However, the improvement of the accuracy of the detection of variable remains a challenge. The work in [3] points out that there are many failure cases in the detection of variable in body text of document. Thus, the work focuses on the detection of variable. The detection of other mathematical components (e.g functions or operators) is not in the scope of the work.

In the paper, we present the method for detecting variables in PDF scientific documents. The content of input documents is assumed to be content stream [4]. The content allows to directly get character information without any Optical Character Recognition (OCR) techniques. The work focuses on the detection of variable that is represented by a single alphabetical character. In Fig. 1, the alphabetical characters "S" and "H" are used as variables of inline expressions. The proposed method attempts to detect this type of variable because its appearance is similar to text and causes more ambiguities in the detection compared to other types of variable (e.g. s_1). In the detection, two key phases are performed. The first phase is the analysis of input PDF document in order to extract information of mathematical variable and textual word. The second phase is the classification of extracted variable and textual words. The contributions of our work are threefold:

(1) Existing methods concentrate on the detection of mathematical specific symbol (e.g. $\sum$, $\int$) or functions (e.g. cos, sin, log). Many failure cases in the detection of variable are remained. Therefore, the frequently encountered errors in the detection of variable are analytically presented in the work. After analyzing the errors, a set of rules are constructed to determine whether a character is a variable. The comparison results with InftyReader [5] that is a commonly used OCR tool demonstrate the effectiveness of the proposed rules.

(2) Taking the advantages of deep Convolutional Neural Networks (CNN) in image classification, two networks that are AlexNet [6] and ResNet-50 [7] are employed to automatically extract visual features of variable and text from documents. These CNNs allow to obtain discriminating features without the usage of domain (expert) specific knowledge. Then, two binary classification algorithms including Support Vector Machine (SVM) [8] and k Nearest Neighbor (kNN) [9] are adopted to discriminate variables from textual words. By using the combination of CNNs and machine learning algorithms, the high accuracy of the classification is achieved.

(3) The experiments of the proposed method are carried out on datasets that contain both English and Vietnamese PDF documents. So far, the existing methods have worked on the detection of mathematical expression from English documents. To our best knowledge, this is the first research that focuses on the detection of variable from Vietnamese documents.

The rest of the paper is organized as follows. Section 2 overviews significant related works. Section 3 presents the detail of the analysis of current errors of the detection of variable in existing methods. In Sect. 4, new approaches are proposed in order to improve the limitations. In Sect. 5, experimental results are shown and discussed. Finally, Sect. 6 gives the conclusion and the future work.

2   Related Work

2.1   PDF Document Analysis

Since developing in the 1990s, PDF (Portable Document Format) format has become a quasi-standard for document presentation. Among many formats such as .doc, XML, images, the majority scientific documents are published in PDF [3]. Therefore, the demand of analyzing PDF document has increased in recent years. For identification mathematical expression, the textual content and mathematical components are necessary to extract. The textual content in PDF files is represented by *content stream*. This is the object that contains a sequence of graphic objects to be painted on a PDF page [10]. Beside, other meta information including coordinates, bounding box, font (font name and size) of each character is extracted precisely from PDF documents. The complex mathematical symbols are represented in a different ways and various strategies are considered to obtain the content of these symbols. The work in [10] attempts to solve the problem by grouping symbols and recalculating the glyph, bounding boxes or labeling symbols based on shapes and context information. The work in [11] renders the symbols to images and after that OCR techniques are applied to these obtained images. However, the method costs extra efforts. The extracted information from PDF files is then input for the detection of mathematical expression.

There are several existing tools that are used for the information extraction from PDF files such as poppler [12], ImageMagic [13]. In our work, Apache PDFBox [14], an open source library, is used to extract information from PDF files. The library fully supports the manipulation, creation, extraction PDF files in Java programming language with high performance. The target for our work is the identification of variable, therefore, the information of characters including content, coordinates, bounding boxes, font (size and name) is extracted. In order to store extracted information from each PDF page, a multi-dimensional vector structure is used. The vector is formed by the elements that are described as follows:

(1) The content (called *glyph*) that is represented by a *string* data type of each character.
(2) The X and Y coordinates that are represented by doubles of the upper left corner the bounding box of each character.
(3) The width and height that are represented by doubles of the bounding box of each character.
(4) The font name of each character that is represented by a *string* data type.

2.2   Mathematical Expression Detection from PDF Document

Most existing methods pay attention to identify mathematical expressions from document images (e.g. [15,16]). The identification of displayed formula has obtained high accuracy, however, the identification of inline formula has faced many difficulties. The existing methods of identification of displayed formula attempt to extract features and apply machine learning algorithms to discriminate formulas from ordinary text lines. Meanwhile, for identification of inline formula, not only feature extraction but also OCR techniques are required [16]. Many efforts are performed to improve the accuracy of the detection of inline expression however, the accuracy highly depends on the quality of input document images. Moreover, processing of a large number of document images is a time-consuming task and requires a lot of computational resources. Fortunately, the extracted information from PDF files is more accurate than that of document images. In addition, the processing of PDF files is generally faster than that of document images. Recently, several works have been investigated for the identification of inline expression in PDF files.

InftyReader [5] is the most commonly used tool for OCR tasks. The software allows to extract and recognize mathematical expression from documents images and PDF documents. The software performs the recognition in four steps: layout analysis, character recognition, structure analysis of mathematical expressions and manual error correction. A large number (about 180000) of character patterns are manually prepared for the detection and recognition. The accuracy highly depends on the quality of input documents and font-types of characters.

The work in [3] applies natural processing language approach to detect inline formula. The method firstly prepares the PDF annotated corpus, then extracts layout features including font, length of words and linguistic features such as context *n-grams* from PDF files. Finally, CRF (Conditional Random Fields) is applied to label whether a word is a part of inline formula. Although the method shows high accuracy (obtained F-measure is from 80.41% to 88.95%) in the detection of math-zone, the detection of variable faces many failure cases.

In recent years, deep Convolutional Neural Networks have shown the outstanding performance in the image classification task. Among the successful networks, AlexNet and ResNet-50 have become popularly for image and document classification. The work in [17] uses deep learning approach to detect formula in PDF documents. A combination of the Convolutional Neural Network and Recurrent Neural Network (RNN) networks is utilized for the detection. The method is reported to performs better than learning-based methods in the detection. However, to obtain the high accuracy, it takes many efforts to train (thousands of mathematical expressions are manually labeled) and optimize the deep networks.

3   Error Analysis of Variable Detection

After investigating and testing the existing methods that are reported in [3] and [5], there are failures cases that often occur in the detection of variable. Each error case is described as follows:

(1) The variable is followed by punctual characters such as comma ","; dot "." or colon ":". These punctual characters cause the ambiguities in the detection of the variable and existing methods detect them as texts. Examples of a variable that is followed by a dot "." (a) and a colon ":" (b) marks are shown in Fig. 2.

(2) The abbreviation in scientific documents can be fail to detect as variable. Font attributes of the abbreviation are similar to those of variables. For instance, a character that is followed by the equals sign "=". This case is shown in Fig. 3.

(3) In some languages, documents contain special single character that can be fail to detect as variable when these characters display in italic style. An example of a character in Czech language that can be fail to detect as a variable is shown in Fig. 4.

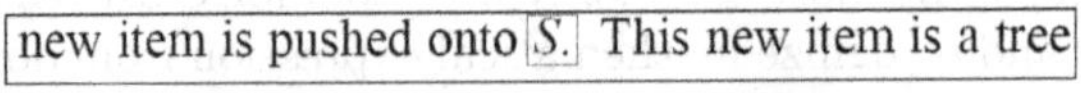

new item is pushed onto $S.$ This new item is a tree

(a) A variable that is followed by a dot "." in an English document

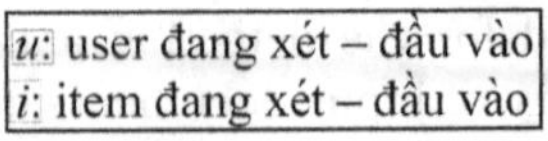

u: user đang xét – đầu vào
i: item đang xét – đầu vào

(b) A variable that is followed by a colon ":" in a Vietnamese document

Fig. 2. Examples of variables

Table 5: Evaluation of ASN antecedent annotation. *P = perfectly,* $\boxed{R = reasonably}$, *I = implicitly, N = not at all*

Fig. 3. An example of abbreviations that are followed by the equals sign

$\boxed{\text{English } such~as, \text{ in Czech } jde\boxed{o}\,(=\text{lit. } It~is~about),}$

Fig. 4. An example of a special character in Czech language that can be fail to detect as a variable

4 Proposed Method of Variable Detection in PDF Document

The workflow of the proposed method for the variable detection is described in Fig. 5. In our work, a word (containing characters that are separated by space character) is a basic unit for the variable detection module. Fig. 6 shows an example of the extracted information from the PDF document by using the Apache PDFBox tool. In order to determine if a word is a variable, two approaches are utilized. The first one is the rule-based classification. In the approach, a set of rules are constructed for the classification. The second one is the combination of pre-trained CNNs (AlexNet and ResNet-50) and machine learning algorithms. The CNNs are used to automatically extract features of variable and word images that are rendered from PDF files. The extracted features are then fed into machine learning classifiers (SVM and kNN) for the classification.

4.1 Rule-Based Classification

In rule-based classification, a rule is a natural way for knowledge representation. A rule for the classification purpose can be described as follows [18]:

Rule: *condition → conclusion*

Where *condition* is a conjunction of observed features or constraints of extracted words. When an observed word satisfies a sequence of conditions, it is categorized into "variable" or "not variable" groups. The rules can be represented by a series of **IF-THEN** pairs. The classification rules are described as follows:

(1) Rule 1: if properties of a word satisfy the following conditions, the word is labeled as a variable.

The font of the word is italic type (the extracted font name from PDF document is checked whether it is an Italic font such as "Times-Italic").

The word contains two characters: the first one is an alphabetical character and the second one is a punctuation mark (e.g. ",", ".", ":").

(2) Rule 2: if properties of a word satisfy the following conditions, the word is labeled as a candidate of a variable.

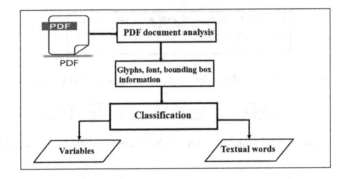

Fig. 5. Workflow of the proposed method

```
Character: s----font: Times-Italic Positions of character:  [(X=167.04071,Y=73.41449) height=6.0719976 width=4.2963257]
Character: 1----font: Times-Italic Positions of character:  [(X=171.36072,Y=74.85449) height=3.8279986 width=3.4799957]
```

(a) Extracted information including font, glyph, bounding box information

two top items in S (s₁ and s₂) are popped, and a
new item is pushed onto S. This new item is a tree
formed by making the root s₁ of a dependent of the

(b) Bounding boxes of characters

Fig. 6. An example of the extracted information from PDF document

The word contains only one alphabetical character. The font of the word is italic type.

(3) Rule 3: if a candidate of a variable is in the list of accent characters (e.g. "ê", "à", "â", "ǎ"), it is classed as "not a variable".

(4) Rule 4: if a candidate of a variable matches one of the following string patterns, it is classed as "not a variable".
The word is followed by an equals sign ("=") and the word after the equals sign contains more than one character.
The word is in the sentence that contains "replaced by" words.
The word is in the sentence that contains "in Czech" words.

(5) Rule 5: if a candidate of a variable that is not mentioned in *Rule 3* and *Rule 4*, it is actually a variable.

The *Rule 1* and *Rule 2* are set by using the features that are used in existing works [3]. The rules including *Rule 3*, *Rule 4* and *Rule 5* are proposed to the detect the specific characters in Vietnamese, Czech languages and abbreviations. The proposed rules are set by the observation of specific cases that can cause the ambiguities of the variable detection in the benchmark datasets.

4.2 Classification Based on the Combination of CNNs and Machine Learning Algorithms

In our work, the pre-trained CNNs, AlexNet and ResNet-50, are used to extract visual features of variable and textual word. The pre-trained CNNs play a role as a powerful feature extractor. The AlexNet and ResNet-50 models have been trained on 1.2 million images that are in the subset of ImageNet database [19]. The pre-trained CNNs using the ImageNet database have shown the effectiveness in the document classification task [21]. Therefore, the models are selected as feature extractors in our work. AlexNet consists of 25 layers. The network architecture consists of 5 convolutional layers and 3 fully connected layers. For AlexNet, each input image is preprocessed and resized to $[227 \times 227 \times 3]$. 4096 features are automatically extracted on the fully connected layer of the CNN for each input image. ResNet-50 consists of 177 layers corresponding to 50 residual blocks. For ResNet-50, each input image is preprocessed and resized to $[224 \times 224 \times 3]$. 1000 features are automatically extracted on the fully connected layer of the CNN for each input image. Input images for CNNs are bounding boxes of glyph extracted from PDF documents and then converted into image format.

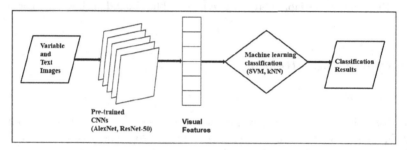

Fig. 7. The classification of variable and textual word by using the combination of CNNs and machine learning algorithms

Table 1. Number of training and testing images in the classification method using CNNs and machine learning algorithms

Datasets	Number of training images	Number of testing images
Vietnamese	660	454
ACL Anthology	560	700

Variable and textual words images are divided into training and testing sets. In the training phase, the extracted features of training images obtained by using the pre-trained CNNs together with labels are used to train SVM and kNN. After that, in the testing phase, the trained SVM and kNN are used to discriminate variable from textual word. In our work, the linear function is used for the SVM and the Euclidean distance is used for the kNN. The classification of variable and textual word by using the combination of CNNs and machine learning algorithms is described in Fig. 7.

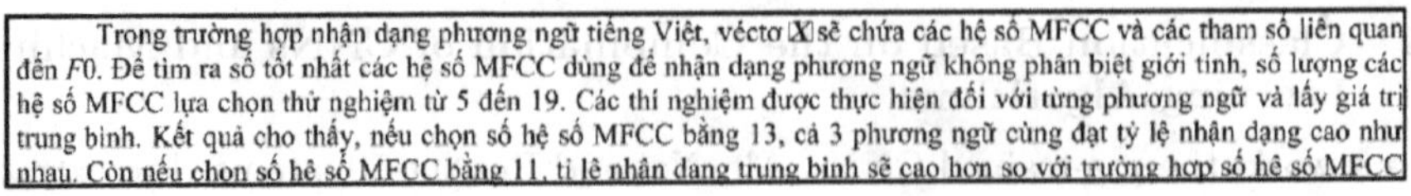

Trong trường hợp nhận dạng phương ngữ tiếng Việt, véctơ *X* sẽ chứa các hệ số MFCC và các tham số liên quan đến *F0*. Để tìm ra số tốt nhất các hệ số MFCC dùng để nhận dạng phương ngữ không phân biệt giới tính, số lượng các hệ số MFCC lựa chọn thử nghiệm từ 5 đến 19. Các thí nghiệm được thực hiện đối với từng phương ngữ và lấy giá trị trung bình. Kết quả cho thấy, nếu chọn số hệ số MFCC bằng 13, cả 3 phương ngữ cùng đạt tỷ lệ nhận dạng cao như nhau. Còn nếu chọn số hệ số MFCC bằng 11, tỉ lệ nhận dạng trung bình sẽ cao hơn so với trường hợp số hệ số MFCC

(a) Detection results in Vietnamese documents

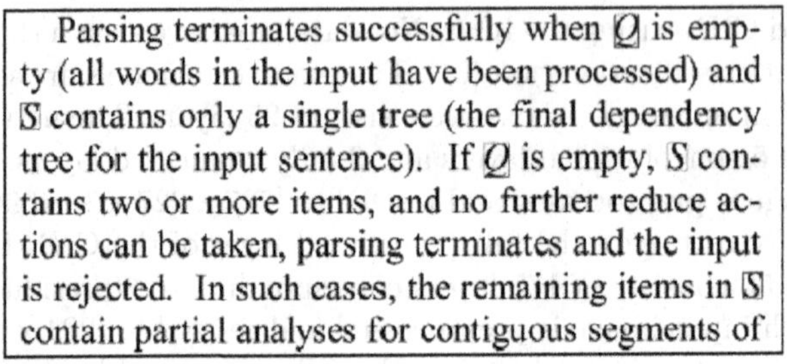

Parsing terminates successfully when *Q* is empty (all words in the input have been processed) and *S* contains only a single tree (the final dependency tree for the input sentence). If *Q* is empty, *S* contains two or more items, and no further reduce actions can be taken, parsing terminates and the input is rejected. In such cases, the remaining items in *S* contain partial analyses for contiguous segments of

(b) Detection results in English documents

Fig. 8. Examples of variable detection results in blue bounding boxes (Color figure online)

5 Experimental Results

5.1 Datasets

In the work, two datasets are used for the experimentation. The first one contains 132 PDF paper pages from the ACL Anthology [20]. The second one contains 122 Vietnamese paper pages. These Vietnamese papers are collected from various scientific journals and conferences on Computer Science topic. Each document contains a minimum of one variable and a maximum of 32 variables and an average of about 7 variables. For the rule-based classification, 805 and 661 variables in the ACL Anthology and Vietnamese datasets are used respectively. For the classification using CNNs and machine learning algorithms, number of training and testing images are described in the Table 1. These images of variables and textual words are converted from PDF documents.

5.2 Performance Evaluation

In our work, the Precision (P), Recall (R) and F1 score are used for the performance evaluation. Precision is the proportion of the true positives against all the positive results; Recall is the proportion of the true positives against all the true results and F1 score is the harmonic mean of precision and recall.

The performance comparison between the proposed method and InftyReader is shown in Table 2. InftyReader version 3.1.4.8 is used for the comparison. The InftyReader doesn't currently support the detection of mathematical expression in Vietnamese language. For English language, the precision and F1 score of

Table 2. Performance comparison between the proposed method and InftyReader OCR tool (Bold value indicates the highest score in each method)

Methods	ACL Anthology			Vietnamese		
	P	R	F1 score	P	R	F1 score
InftyReader	29.45%	20.10%	23.89%	Not totally support		
Rule-based (existing rules)	80.06%	74.98%	77.44%	82.18%	73.45%	77.57%
Rule-based (existing + proposed rules)	**93.30%**	**85.05%**	**89%**	**92.85%**	**89.04%**	**90.91%**
Alexnet+SVM	97.71%	95.18%	96.43%	100%	100%	100%
Alexnet+kNN	96.14%	91.82%	93.93%	100%	100%	100%
Resnet-50+SVM	**99.57%**	**99.07%**	**99.32%**	**100%**	**100%**	**100%**
Resnet-50+kNN	97.14%	93.87%	95.48%	100%	100%	100%

InftyReader are lower than those of our proposed method. InftyReader is powerful in the detection and recognition of specific mathematical expressions. Similar to existing methods, the software focuses on the discrimination between normal text and mathematical expressions based on the variation of geometric layout features and the statistic of mathematical symbols and functions. For the case of variable, the strategy is not really efficient. In contrast, both the feature and context information are extracted in order to overcome the obstacles in our work. These factors allow to obtain better results in the detection.

In general, the combination of CNNs and machine learning algorithms allows to obtain the most accurate results in our testing. The highest performance is achieved by using the ResNet-50 and SVM for both ACL Anthology and Vietnamese datasets. The result shows the efficiency of the feature extraction and classification. In our experimentation, ResNet-50 network outperforms in the feature extraction. In this case, ResNet-50 network obtains better results because it consists more layers in the architecture than Alexnet. The increase in the depth of network architecture should increase the accuracy of the network. However, in order to take the advantages of these deep CNNs, the extra time and computational resources are payed in comparison to those of rule-based method. Examples of detection results of variables that are in blue from English and Vietnamese PDF documents are shown in Fig. 8.

In the rule-based classification method, in order to determine the importance of the proposed rules for the detection, two ablation tests are carried out. Firstly the process of detection is performed by using existing *rules* that are *Rule 1* and *Rule 2*. Then the process of detection is performed by using all the five *Rules*. The performance comparison of applied rules for Vietnamese and ACL Anthology datasets is shown in the Table 2. The comparison shows that the accuracy is much improved for the detection of Vietnamese and ACL Anthology datasets respectively by using the proposed rules.

5.3 Error Analysis and Discussion

Some remaining errors in the rule-based classification method are analyzed as follows:

Some narrative specific symbols in tables are fail to detect as variables. These symbols have the same appearance in the font, size as variables. Besides, there is not much context information to discriminate variables and narrative texts. These factors cause the error for the detection.

The variables in the parentheses are fail to detect as text. The parentheses and italic narrative texts are some times used for highlight items in scientific documents. The similarity of the appearance between variable and text in this case causes the error variable detection.

In Vietnamese scientific documents, variables are typically represented by English alphabetical characters. Meanwhile, narrative texts use Vietnamese characters that often contain accent characters (e.g. "ê", "à", "â", "ă"). In the rule-based method, for English documents, the discrimination of mathematical variable from short words such as "a" or "of" is more difficult than that of long words. The rule-based method discriminates variables from textual words by using layout features and context features. The combination of pre-trained CNNs and machine learning algorithms performs the discrimination by using thousands of visual features. The different features between Vietnamese characters and English Alphabetical characters allow to detect variable accurately by the method using CNNs. Due to the limitation of our linguistic knowledge, the proposed method is performed on the datasets that contain English and Vietnamese PDF papers. Actually, the method can be applied for other languages.

6 Conclusion and Future Work

We have presented the method for detecting variables that are represented by a single alphabetical character in inline expressions in PDF scientific documents. The errors cases of existing methods are analytically presented and novel techniques are proposed in order to improve the accuracy of the detection. After the information extraction from PDF documents, two classification approaches are presented. For the rule-based classification, the obtained F1 score varies from 89% to 90.91% on benchmark datasets containing English and Vietnamese documents. For the classification based on the combination of pre-trained CNNs and machine learning algorithms, the highest F1 score of 99.32% and 100% is achieved for ACL and Vietnamese datasets respectively. The comparison with existing method demonstrates the out-performance of the method.

In the future, more information about the context and linguistic features of variable will be investigated to improve the accuracy of the detection. Moreover, other types of variable such as the variable that is followed by index (e.g. x_i) or variable that appears in tables will be investigated for the detection. The datasets that contain Vietnamese papers will be enlarged and published for researching purposes.

Acknowledgement. This work was supported by JST CREST Grant Number JPMJCR1513, Japan.

References

1. Zanibbi, R., Blostein, D.: Recognition and retrieval of mathematical expressions. Int. J. Doc. Anal. Recogn. **15**, 331–357 (2012)
2. Britannica. https://www.britannica.com. Accessed 4 Sept 2018
3. Iwatsuki, K., et al.: Detecting in-line mathematical expression in scientific documents. In: Proceedings of the 2017 ACM Symposium on Document Engineering, pp. 141–144. ACM (2017)
4. Yu, B., et al.: Extracting mathematical components directly from PDF documents for mathematical expression recognition and retrieval. In: ICSI, pp. 95–104. ACM (2014)
5. Suzuki, M., et al.: INFTY: an integrated OCR system for mathematical documents. In: Proceedings of the 2003 ACM Symposium on Document Engineering. ACM (2003)
6. Krizhevsky, A., et al.: ImageNet classification with deep convolutional neural networks. In: Proceedings of the 25th International Conference on Neural Information Processing Systems, vol. 1. IEEE (2012)
7. He, K., et al.: Deep residual learning for image recognition. In: Proceedings of the 2016 IEEE Conference on Computer Vision and Pattern Recognition. IEEE (2016)
8. SVMlight. http://svmlight.joachims.org/. Accessed 4 Sept 2018
9. Friedman, J., Finke, R.A.: An algorithm for finding best matches in logarithmic expected time. ACM Trans. Math. Softw. **2**, 209–226 (1977)
10. Baker, J., et al.: Extracting precise data on the mathematical content of PDF documents. In: Proceedings of DML-08. Masaryk University Press (2008)
11. Rahman, F., Alam, H.: Conversion of PDF documents into HTML: a case study of document image analysis. In: Proceedings of the Thirty-Seventh Asilomar Conference on Signals, Systems and Computers, pp. 87–91. IEEE (2003)
12. Poppler. https://poppler.freedesktop.org/. Accessed 4 Sept 2018
13. Magick. https://legacy.imagemagick.org/. Accessed 4 Sept 2018
14. PDFBox. https://pdfbox.apache.org/. Accessed 4 Sept 2018
15. Garain, U.: Identification of mathematical expressions in document images. In: Proceedings of the 10th International Conference on Document Analysis and Recognition, pp. 1340–1344. IEEE (2009)
16. Jin, J., Han, X., Wang, Q.: Mathematical formulas detection. In: Proceedings of the Seventh International Conference on Document Analysis and Recognition. IEEE (2003)
17. Gao, L., et al.: A deep learning-based formula detection method for PDF documents. In: Proceedings of the 2017 IEEE Conference on Document Analysis and Recognition. IEEE (2017)
18. Panyr, J.: Information retrieval techniques in rule-based expert systems. In: Bock, H.H., Ihm, P. (eds.) Classification, Data Analysis and Knowledge Organization. Studies in Classification, Data Analysis, and Knowledge Organization, vol. 15, pp. 331–357. Springer, Heidelberg (1991). https://doi.org/10.1007/978-3-642-76307-6_26

19. ImageNet. http://www.image-net.org/. Accessed 4 Sept 2018
20. ACL. http://aclanthology.info/. Accessed 4 Sept 2018
21. Afzal, M.Z., et al.: DeepDocClassifier: document classification with deep convolutional neural network. In: 13th International Conference on Document Analysis and Recognition (ICDAR) (2015)

Predicting Cardiovascular Risk Level Based on Biochemical Risk Factor Indicators Using Machine Learning: A Case Study in Indonesia

Yaya Heryadi[1](✉), Raymond Kosala[2], Raymond Bahana[2],
and Indrajani Suteja[3]

[1] Computer Science Department, BINUS Graduate Program –
Doctor of Computer Science, Bina Nusantara University, Jakarta, Indonesia
yayaheryadi@binus.edu
[2] Computer Science Program, Binus International,
Bina Nusantara University, Jakarta, Indonesia
{rkosala, rbahana}@binus.edu
[3] School of Information System, Bina Nusantara University, Jakarta, Indonesia
indrajani@binus.edu

Abstract. Early detection of cardiovascular risk level remains an important issue in healthcare. It is still considered a very important preventive measure of cardiovascular disease as it gives a significant impact to reducing mortality rates and cardiovascular events. Prior to developing a prediction model of cardiovascular risk, identification of dominant predictor variables is very crucial. Some prominent studies have proposed a vast number of predictor variables. Although some predictor variables might be universal in nature, as the premise of this study, some of the variables might be associated with local lifestyle that governs patient behavior. This paper presents a verificative study on previous studies predicting cardiovascular risk level by using Indonesian adult patients' lab records as the input dataset. In relation to this objective, this study aimed to select dominant biochemical indicators as predictor variables and trained machine learning models as classifier. Finally, this study compared the performance of several prominent classifier models such as: XGBoost, Random Forest, k-NN, Gradient Boosting, Artificial Neural Network (Multilayer Perceptron), Decision Tree, and Ada Boost. The results show that: XGBoost model achieved the best training and testing accuracy (0.965 and 0.964) compared to Random Forest (0.964 and 0.962), 5-NN (0.952 and 0.948), Gradient Boosting (0.948 and 0.940), Artificial Neural Networks (0.945 and 0.933), Decision Tree (0.861 and 0.860) and Ada Boost models (0.748 and 0.718).

Keywords: Cardiovascular risk prediction · Machine learning

1 Introduction

Many prominent studies showed some evidences that exposure to cardiovascular risk factors in early childhood/adolescence may link to cognitive performance of young or adult-age later. Moreover, other evidences also showed that cumulative burden of

© Springer Nature Switzerland AG 2019
N. T. Nguyen et al. (Eds.): ACIIDS 2019, LNAI 11432, pp. 707–717, 2019.
https://doi.org/10.1007/978-3-030-14802-7_61

cardiovascular risk factors from childhood/adolescence links to worse cognitive performance during middle-age regardless exposure during the adult-age [1]. In reducing such effect, early detection of the elevated risk factor is very important. Cardiovascular disease (CVD) which is still considered as a deadly but silent killer is caused by atherosclerosis or deposition of fatty material on the inner artery walls [2]. Untreated CVD may lead to heart attacks and strokes [3] or event premature mortality. Many past studies showed some evidences that early detection of Cardiovascular disease is highly crucial to save patients' lives. One of such measures is predicting cardiovascular risk level of patients based on lab data.

Predicting cardiovascular risk level is an interesting computer vision problem with significant application in healtcare. Many previous studies have explored various predicting variables and the association to cardiovascular risks. These variables include: blood pressure, serum total-cholesterol, high-density lipoprotein cholesterol, triglycerides, low-density lipoprotein (LDL) cholesterol, smoking habit, weight, height and body mass index (BMI) [4–6]. In addition, many statistical models have been employed to predict cardiovascular risk level. For example: linear regression and mixed model regression splines [7] and survival model [8].

The advent of machine learning approach in the past ten years has motivated researchers to develop a robust classifier to recognize the risk of cardiovascular diseases based on various diagnostic features such as biochemical lab data as input features or predictor variables. For example, a recent study reported by [9] explored machine learning model such as: random forest, logistic regression, gradient boosting, and neural networks. Although many studies have been reported, automated recognition of Cardiovascular risk level remains a challenging problem for the following reasons: (1) the blood test matrix data are typically sparse due to incomplete check of the test by many physicians and (2) blood test dataset for research is not widely available.

Although a similar study has been reported by [10], this study is different from the former study in the following aspects. *First*, the former study was based on a cohort analysis in which observation over a period of time was conducted. On the other hand, this study was based on cross-section lab patient records. *Second*, the former and this study used different populations of patients. *Finally*, the former study aimed to explore more predictor variables.

Despite many studies which have explored machine learning for recognizing cardiovascular risk level, to the best of our knowledge, few studies involve patients from Indonesia hospitals as research objects. The premise of this study is that different region where the patient live and different ethnics of the patients might affect the dominant features, which can be used to predict cardiovascular risk level. Hence, it is interesting to explore the question of whether a model for predicting cardiovascular risk trained with a patient dataset from a country can achieve similar accuracy when it is used for predicting patient risk from different countries. Therefore, the objective of this study are: (1) to select dominant biochemical risk factor indicators as predictor variables for predicting cardiovascular risk level from Indonesia hospitals; and (2) to compare several machine learning models as classifiers.

2   Related Works

2.1   Cardiovascular Risk Factors

A vast number of methods have been proposed to predict cardiovascular disease before it actually happens. Some common predictor variables associated with cardiovascular risk involved some basic testing results of blood, urine or tissue such as: systolic and diastolic blood pressure, blood viscosity, hematocrit, plasma viscosity, serum total-cholesterol, low-density lipoprotein cholesterol, high-density lipoprotein cholesterol, triglycerides, glucose, weight, height and body mass index (BMI) [4–6, 9, 11–14]. In addition to blood test results, some variables associated with lifestyles such as drinking and smoking habits, and physical activity [15–17] tend to increase the risk of cardiovascular disease. Considering a vast number of variables associated with increasing cardiovascular risk, there was an increasing interest on which variables are considered as dominant factors for predicting cardiovascular risk level.

2.2   Cardiovascular Risk Level Modelling

Predicting cardiovascular risk level is a task in computer vision field. In particular, the task can be categorized as a classification problem. A plethora of studies addressing this task have been reported, resulted in many proposed models including: decision tree [18]; naïve Bayes model [19]; artificial neural network model [10, 20]; logistic regression [10, 12]; random forest and gradient boosting [10]. Each of the proposed models is different in terms of model interpretability and prediction accuracy. From these proposed machine learning models, tree based models such as: decision tree models, have some advantages such as easier to interpret and invariant to input scale. Some of the prominent studies to address this problem are summarized in the following Table 1.

Decision tree is a prominent machine learning model which can be applied to address classification, regression, and clustering [20]. Classification and Regression Trees (CART) which was proposed by Breiman et al. [21] is one of the decision tree algorithms. The popularity of the algorithm was mainly related to the following aspects: highly interpretable [22], tolerance to missing data [21], representing a non-parametric universal approximators [22], flexible to input data type (numeric or binary data) [23], producing hierarchical features of data representation [22]. The main limitations of the algorithm are: (1) high and unstable variance, (2) not always achieving better accuracy compared to the other machine learning model, and (3) high space complexity [23]. The attempts to overcome this limitation include XGBoost [24–29], Random Forest [30], Ada Boost [31], and Gradient Boosting [32]. In this study, several prominent machine learning models are evaluated namely: kNN [32], Decision Tree [21], and Artificial Neural Networks [33].

Table 1. Some of previous study on predicting cardiovascular risk

Authors	Predictor variables	Model accuracy
Miranda et al. [19]	Age, gender, urea, creatinine, uric acid, cholesterol, triglyceride, HDL, LDL, glucose, glucose 2H, creatine kinase, CK-MB, LDH, TROPK, and TROPT	Naive Bayes model: • Risk level 1: 0.86 • Risk level 2: 0.88 • Risk level 3: 0.86
Heryadi et al. [18]	Glucose, glucose 2H, total cholesterol, triglyceride, HDL, urea, creatinine, uric acid, and LDH	Decision tree 0. 97
Juarez-Orozco et al. [20]	Gender, age, BMI and family history of CAD, chest complaints, diabetes, smoking, Dyslipidemia, hypertension, rest heart rate, resting blood pressure, duke score, abnormal rest ECG, abnormal stress ECG, stress HR, % of max HR, stress BP, rest LVEF and stress LVEF	Artificial neural network 0.75 accuracy
Weng et al. [10]	Age, gender, ethnicity, smoking, HDL, HbA1c, triglycerides, SES Townsend deprivation index, BMI and total cholesterol	Random forest: AUC 0.745
	Ethnicity, age, SES Townsend deprivation index, gender, smoking, atrial fibrillation, chronic kidney disease, rheumatoid arthritis, family history of premature CHD and COPD	Logistic regression: AUC 0.760
	Age, gender, ethnicity, smoking, HDL, triglycerides, total cholesterol, HbA1c, systolic blood pressure and SES Townsend deprivation index	Gradient boosting: AUC 0.761
	Atrial fibrillation, ethnicity, oral corticosteroid prescribed, age, severe mental illness, SES Townsend deprivation index, chronic kidney disease, BMI missing, smoking and gender	MLP: AUC 0.764

3   Research Method

The framework of this study is presented in Fig. 1. The differences between this study and the similar study reported by [10] are that (1) this study addressed imbalanced data using downsampling technique; in contrast, the study reported by [10] did not address imbalanced data; and (2) this study select importance predictor variables from overall data features; in contrast, the study reported by [18, 19] did not implement feature selection.

The dataset for this study was provided by a blood chemical lab of a private hospital in southern part of Jakarta, Indonesia. For confidentiality reason, some variables such as name, address, religion and other identifiers were removed before any

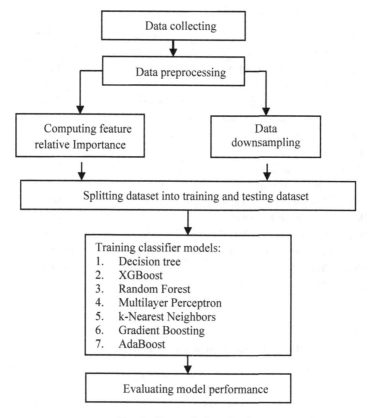

Fig. 1. Research framework

data were analyzed. The initial features of the dataset for this study include 16 features used as representation of the biochemical risk indicators. The features can be categorized into 4 categories as follows:

(1) Lipid level in blood indicators comprise: CHOL (cholesterol), TRIG (triglyceride), HDL (high-density lipoprotein), and LDL (low-density lipoprotein) variables;
(2) Kidney function indicators comprise: UREA (ureum), CREA (creatinine), and UA (uric acid) variables;
(3) Diabetes mellitus indicators comprise: GLU (glucose) and GLU2H (glucose 2-h) variables;
(4) Coronary artery function indicators comprise: CK (creatine kinase), CKMB (creatine kinase-MB), LDH (lactate dehydrogenase), TROPK (troponin) and TROPT (troponin T) variables.

The sample labelling process works in two steps (see Fig. 1). The first step is categorizing each individual feature into initial categorization: risk and non-risk categories. The second is categorizing the sample into four risk levels: risk level 1,

Table 2. Categorical value of each cardiovascular risk factor

Feature	Description	Normal threshold	Risk threshold
GLU	Glucose (mg/dL)	1–110	≥ 111
GLU2J	Glucose 2H (mg/dL)	1–140	≥ 141
CHOL	Cholesterol (mg/dL)	<200	≥ 200
TRIG	Triglyceride (mg/dL)	<200	≥ 200
HDL	High-densitylipoprotein (mg/dL)	>65	<65
LDL	Low-density lipoprotein (mg/dL)	<100	≥ 100
UREA	Urea (mg/dL)	1–39	≥ 40
CREA	Creatinine (mg/dL)	0.1–1.3	≥ 1.4
UA	Uric Acid (mg/dL)	0.1–6.2	≥ 6.3
LDH	Lactate dehydrogenase (U/L)	140–280	>280
CK	Creatinine Kinase (U/L)	21–215	>215
CKMB	Creatine Kinase-MB (U/L)	<25	>25
TROPT	Troponin T (ng/mL)	Negative	Positive
TROPK	Troponin (ng/mL)	Negative	Positive

Source: [19].

Table 3. Definition of cardiovascular risk level

Risk level	Remarks
1	At least one feature is not-normal
2	At least two factors contains two non-normal features, or each factor contains at least a non-normal feature
3	Three factors are not normal, each factor contains at least one non-normal feature, or coronary artery function is not normal
4 (normal)	Each feature is normal

Source: [19].

risk level 2, risk level 3, and normal cardiovascular risk level. Following [19], the threshold values used to categorize each feature into normal and risky categories are summarized in Table 2. Based on the Table 2, doctors' general rules to label each data sample are presented in Table 3.

In this study, 7 models were explored to predict cardiovascular risk level based on the most dominant predictor variables. Following studies reported by [10, 18, 19] the classifier models in this study were: kNN [32], Decision Tree [21], Ada Boost [31], XGBoost [24], Random Forest [30], Gradient Boosting [25] and Artificial Neural Networks (Multilayer Perceptron) [33].

4   Experiment Result and Discussion

The initial number of data in the dataset was 50,518 records of adult patients. Distribution of risk level category in the sample dataset can be grouped into: normal (39,318 or 77.8%), risk level 1 (8,120 or 16.1%), risk level 2 (1,409 or 2.8%), and risk level 3 (1,671 or 3.3%). The number of samples for each category before and after resample can be seen in Fig. 2(a) and (b).

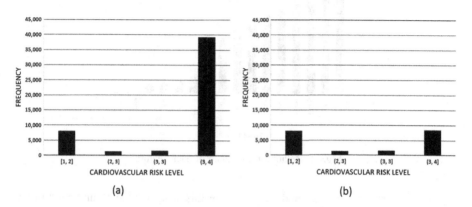

(a)   (b)

Fig. 2. Distribution of cardiovascular risk level from (a) raw data and (b) down-sampled data

Since the dataset was imbalanced, the normal category was downsampled so that the number of samples in the new dataset became: normal (8,500 or 43.1%), risk level 1 (8,120 or 41.2%), risk level 2 (1,409 or 7.2%), and risk level 3 (1,671 or 8.5%). The number of samples from raw dataset and new dataset for each risk level is summarized in Fig. 2. The distribution of the patient age and gender of the new dataset is presented in the following Table 4.

Table 4. Categorical value of age and gender

Feature	Categorical value	N-samples
Age	1: 31–40	4,944 (25.1%)
	2: 41–50	5,661 (28.7%)
	3: >50	9,093 (46.2%)
Gender	1: Male	7,111 (36.1%)
	2: Female	12,587 (63.9%)

This dataset was randomly split into 18,713 patient records (95%) to train the machine-learning algorithms (training dataset) and the remaining sample of 985 patient records (5%) for testing the algorithm (testing dataset).

The relative importance of each biochemical risk factor as predictor variables is computed and summarized in Fig. 3. The figure shows that the predictor variables in this study were dominated by 10 variables namely: UA, GLU, UREA, CHOL, HDL, GLU2 J, CREA, TRIG, AGE and GENDER. Meanwhile, CKMB, LDH, CK, LDL, TROPT and TROPK features were less important to predict the data class than the first 10 variables.

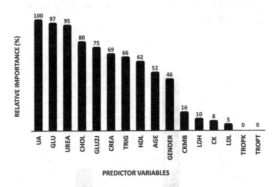

Fig. 3. Relative importance of each feature

The accuracy of classifier model explored in this study can be summarized in the following Table 5.

Table 5. Accuracy comparison of several models

Accuracy	5-NN	Decision Tree	AdaBoost	**XGBoost**	Random forest	Gradient boosting	Multilayer perceptron
Training accuracy	0.952	0.861	0.748	**0.965**	0.964	0.948	0.945
Testing accuracy	0.948	0.860	0.718	**0.964**	0.962	0.940	0.933

The training results are as follows:

(1) XGBoost model predicted 18,053 samples correctly from 18,713 training dataset, resulting in accuracy of 0.965.
(2) Random forest model predicted 18,044 samples correctly from 18,713 training dataset, resulting in accuracy of 0.964.
(3) 5-Nearest Neighbor model predicted 17,819 samples correctly from 18,713 training dataset, resulting in accuracy of 0.952.
(4) Gradient Boosting model predicted 17,733 samples correctly from 18,713 training dataset, resulting in accuracy of 0.948.
(5) Artificial Neural Network (Multilayer Perceptron) model predicted 17,679 samples correctly from 18,713 training dataset, resulting in accuracy of 0.945.

(6) Decision Tree model predicted 16,110 samples correctly from 18,713 training dataset, resulting in accuracy of 0.861.

(7) Ada Boost model predicted 13,999 samples correctly from 18,713 training dataset, resulting in accuracy of 0.748.

The samples of confusion matrix of the trained model can be seen in the following Fig. 4.

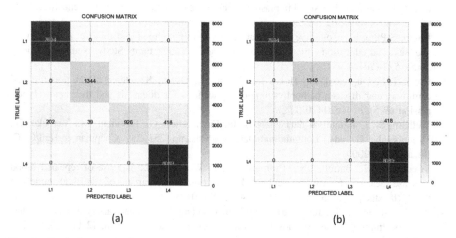

(a)   (b)

Fig. 4. Confusion matrix from model training of (a) XGBoost model, (b) Random forest model

With such performance, we concluded that the XGBoost model achieved high performance as a classifier which maps the selected predictor variables to the cardiovascular risk level.

5   Conclusion

In order to save many patients' lives, early detection of cardiovascular risk is a highly crucial task. This task can be viewed as a classification problem which can be solved using machine learning algorithm. The main issues to address are how the patient data will be represented, what are the main important features, and what machine learning model to be trained as a classifier. The process of predictors variable selection in this study resulted in 10 biochemical risk factors, namely: UA, GLU, UREA, CHOL, HDL, GLUJ, CREA, TRIG, AGE and GENDER. Using these predictor variables as input data representation, the training of 7 machine learning algorithms revealed that XGBoost model achieved the best training and testing accuracy (0.965, 0.964) compared to Random Forest (0.964 and 0.962), 5-NN (0.952 and 0.948), Gradient Boosting (0.948 and 0.940), Artificial Neural Networks (0.945 and 0.933), Decision Tree (0.861 and 0.860) and Ada Boost models (0.748 and 0.718).

References

1. Rovio, S.P., et al.: Cardiovascular risk factors from childhood and midlife cognitive performance: the Young Finns Study. J. Am. Coll. Cardiol. **69**(18), 2279–2289 (2017)
2. Hansson, G.K., Hermansson, A.: The immune system in atherosclerosis. Nat. Immunol. **12**(3), 204–212 (2011)
3. WHO. http://www.who.int/cardiovascular_diseases/en/. Accessed 26 Oct 2017
4. Friedewald, W.T., Levy, R.I., Fredrickson, D.S.: Estimation of the concentration of low-density lipoprotein cholesterol in plasma, without use of the preparative ultracentrifuge. Clin. Chem. **18**(6), 499–502 (1972)
5. Rovio, S.P., et al.: Cognitive performance in young adulthood and midlife: relations with age, sex, and education—the cardiovascular risk in Young Finns Study. Neuropsychology **30**(5), 532 (2016)
6. Cohn, J.N., Duprez, D.A., Hoke, L., Florea, N., Duval, S.: Office blood pressure and cardiovascular disease: pathophysiologic implications for diagnosis and treatment. Hypertension **69**(5), e14–e20 (2017)
7. Welham, S.: Longitudinal data analysis. In: Fitzmaurice, G., Davidian, M., Verbeke, G., Molenberghs, G. (eds.) Longitudinal Data Analysis, pp. 253–289. Chapman & Hall/CRC, Boca Raton (2009)
8. Sweeting, M.J., Barrett, J.K., Thompson, S.G., Wood, A.M.: The use of repeated blood pressure measures for cardiovascular risk prediction: a comparison of statistical models in the ARIC study. Stat. Med. **36**(28), 4514–4528 (2017)
9. Patsch, J.R., et al.: Relation of triglyceride metabolism and coronary artery disease. Studies in the postprandial state. Arteriosclerosis and thrombosis. J. Vasc. Biol. **12**(11), 1336–1345 (1992)
10. Weng, S.F., Reps, J., Kai, J., Garibaldi, J.M., Qureshi, N.: Can machine-learning improve cardiovascular risk prediction using routine clinical data? PLoS ONE **12**(4), e017494 (2017)
11. Kannel, W.B., McGee, D.D., Gordon, T.: A general cardiovascular risk profile: the Framingham study. Am. J. Cardiol. **38**(1), 46–51 (1976)
12. Plekhova, N.G., et al.: Scale of binary variables for predicting cardiovascular risk scale for predicting cardiovascular risk. In: 2018 3rd IEEE Russian-Pacific Conference on Computer Technology and Applications (RPC), pp. 1–4 (2018)
13. Peters, S.A., Woodward, M., Rumley, A., Tunstall-Pedoe, H.D., Lowe, G.D.: Plasma and blood viscosity in the prediction of cardiovascular disease and mortality in the Scottish Heart Health Extended Cohort study. Eur. J. Prevent. Cardiol. **24**(2), 161–167 (2017)
14. Muntner, P., Whelton, P.K.: Using predicted cardiovascular disease risk in conjunction with blood pressure to guide antihypertensive medication treatment. J. Am. Coll. Cardiol. **69**(19), 2446–2456 (2017)
15. Marcovina, S.M., et al.: Biochemical and bioimaging markers for risk assessment and diagnosis in major cardiovascular diseases: a road to integration of complementary diagnostic tools. J. Intern. Med. **261**(3), 214–234 (2007)
16. Miao, C., et al.: Cardiovascular health score and the risk of cardiovascular diseases. PLoS ONE **10**(7), e0131537 (2015)
17. Sun, X., Jia, Z.: A brief review of biomarkers for preventing and treating cardiovascular diseases. J. Cardiovasc. Dis. Res. **3**, 251 (2012)
18. Heryadi, Y., Miranda, E., Warnars, H.L.H.S.: Learning decision rules from incomplete biochemical risk factor indicators to predict cardiovascular risk level for adult patients. In: Proceedings of 2017 IEEE International Conference on Cybernetics and Computational Intelligence (CyberneticsCom), Puket, Thailand (2017)

19. Miranda, E., Irwansyah, E., Amelga, A.Y., Maribondang, M.M., Salim, M.: Detection of cardiovascular disease risk's level for adults using Naive Bayes classifier. Healthc. Inform. Res. **22**(3), 196–205 (2016)

20. Juarez-Orozco, L.E., Knol, R.J.J., Sanchez-Catasus, C.A., Van Der Zant, F.M., Knuuti, J.: Improving the value of clinical variables in the assessment of cardiovascular risk using artificial neural networks. Eur. Heart J. **38**(suppl_1), 227–228 (2017)

21. Breiman, L., Friedman, J., Stone, C.J., Olshen, R.A.: Classification and Regression Trees. CRC Press, Hoboken (1984)

22. Geurts, P., Irrthum, A., Wehenkel, L.: Supervised learning with decision tree-based methods in computational and systems biology. Mol. BioSyst. **5**(12), 1593–1605 (2009)

23. Aertsen, W., Kint, V., van Orshoven, J., Özkan, K., Muys, B.: Comparison and ranking of different modelling techniques for prediction of site index in mediterranean mountain forests. Ecol. Model. **221**, 1119–1130 (2010)

24. Chen, T., Guestrin, C.: XGBoost: a scalable tree boosting system. In: Proceedings of the 22nd ACM SIGKDD International Conference on Knowledge Discovery and Data Mining. ACM (2016)

25. Friedman, J.: Greedy function approximation: a gradient boosting machine. Ann. Stat. **29**(5), 1189–1232 (2001)

26. Bennett, J., Lanning, S.: The Netflix prize. In: Proceedings of the KDD Cup Workshop 2007, New York, pp. 3–6 (2007)

27. Burges, C.: From ranknet to lambdarank to lambdamart: an overview. Learning **11**, 23–581 (2010)

28. He, X., et al.: Practical lessons from predicting clicks on ads at Facebook. In: Proceedings of the Eighth International Workshop on Data Mining for Online Advertising, ADKDD 2014 (2014)

29. Li, P.: Robust Logitboost and adaptive base class (ABC) Logitboost. In: Proceedings of the Twenty-Sixth Conference Annual Conference on Uncertainty in Artificial Intelligence (UAI 2010), pp. 302–311 (2010)

30. Breiman, L.: Random forests. Mach. Learn. **45**(1), 5–32 (2001)

31. Freund, Y., Schapire, R.E.: A decision-theoretic generalization of on-line learning and an application to boosting. J. Comput. Syst. Sci. **55**, 119 (1997)

32. Bishop, C.M.: Pattern Recognition and Machine Learning. Information Science and Statistics. Springer, New York (2006)

33. Bishop, C.M.: Neural Networks for Pattern Recognition. Oxford University Press, Oxford (1995)

Author Index

Author Index

Printed in the United States
By Bookmasters